Subcellular Biochemistry

Volume 83

Series editor
J. Robin Harris
University of Mainz, Mainz, Germany

The book series SUBCELLULAR BIOCHEMISTRY is a renowned and well recognized forum for disseminating advances of emerging topics in Cell Biology and related subjects. All volumes are edited by established scientists and the individual chapters are written by experts on the relevant topic. The individual chapters of each volume are fully citable and indexed in Medline/Pubmed to ensure maximum visibility of the work.

More information about this series at http://www.springer.com/series/6515

J. Robin Harris • Jon Marles-Wright
Editors

Macromolecular Protein Complexes

Structure and Function

Springer

Editors
J. Robin Harris
Institute of Zoology
University of Mainz
Mainz, Germany

Jon Marles-Wright
School of Biology
Newcastle University
Newcastle upon Tyne, UK

ISSN 0306-0225
Subcellular Biochemistry
ISBN 978-3-319-46501-2 ISBN 978-3-319-46503-6 (eBook)
DOI 10.1007/978-3-319-46503-6

Library of Congress Control Number: 2017934081

Printed on acid-free paper

This Springer imprint is published by Springer Nature
The registered company is Springer International Publishing AG
The registered company address is: Gewerbestrasse 11, 6330 Cham, Switzerland

Preface

The last few decades have seen an exceptional expansion of studies on protein molecules and protein complexes by both X-ray crystallography and high-resolution transmission electron microscopy, increasingly incorporating cryo-conditions (cryo-TEM). Advances in cryo-TEM instrumentation and methods now allow the analysis of dynamic structures to resolutions that were once only accessible by X-ray crystallographic analysis. These developments led to cryo-TEM being highlighted as the method of 2015 by Nature Methods[1].

In this book, we have brought together a number of studies of macromolecular complexes from what is now a very large internationally based discipline. These include the bacterial stressosome complex, the Type III CRISPR-Cas prokaryotic intracellular defense system, and the eukaryotic inflammasome complex, to mention but three of the chapters. While numerous topics could have been included in this volume, we have chosen to concentrate on soluble macromolecular complexes in this book. Future volumes in the *Subcellular Biochemistry* series will cover virus protein and nucleoprotein complexes and membrane protein complexes.

The range of material included in the 22 chapters of this book indicates the breadth of knowledge and the recent advancement of structural studies on protein complexes. These studies highlight the importance of closely integrating structural biology with biochemistry and cell biology to understand the functional consequences of the structural arrangements adopted by these complexes. To comment here upon all the interesting chapter topics would duplicate the information given in the Contents list, and certainly it is not in order to select only a few. The strength and diversity of the material reviewed in the chapters here speaks for itself in light of the continual stream of recent and upcoming articles appearing in scientific journals devoted to structural molecular biology.

We would like to thank our chapter authors for their manuscripts and enthusiasm. Electronic publishing and the availability of e-books are making an increasing impact, in particular through the availability of individual chapters and rapid publication of new knowledge. The *Subcellular Biochemistry* series, along with other

[1] Method of the Year 2015. *Nat. Methods* **13,** 1.

established book series, provides a continuing platform for the broad dissemination of knowledge and gives authors an opportunity to reflect on their field and provide a deeper perspective on recent advances in fast-moving fields. We hope that the information presented in this book will be of broad interest and value to molecular structural biologists and the wider scientific community.

Mainz, Germany J. Robin Harris
Newcastle upon Tyne, UK Jon Marles-Wright
July, 2016

Contents

Chapter 1
Structure and Function of the Stressosome Signalling Hub

Jan Pané-Farré, Maureen B. Quin, Richard J. Lewis, and Jon Marles-Wright

Abstract The stressosome is a multi-protein signal integration and transduction hub found in a wide range of bacterial species. The role that the stressosome plays in regulating the transcription of genes involved in the general stress response has been studied most extensively in the Gram-positive model organism *Bacillus subtilis*. The stressosome receives and relays the signal(s) that initiate a complex phosphorylation-dependent partner switching cascade, resulting in the activation of the alternative sigma factor σ^B. This sigma factor controls transcription of more than 150 genes involved in the general stress response. X-ray crystal structures of individual components of the stressosome and single-particle cryo-EM reconstructions of stressosome complexes, coupled with biochemical and single cell analyses, have permitted a detailed understanding of the dynamic signalling behaviour that arises from this multi-protein complex. Furthermore, bioinformatics analyses indicate that genetic modules encoding key stressosome proteins are found in a wide range of bacterial species, indicating an evolutionary advantage afforded by stressosome complexes. Interestingly, the genetic modules are associated with a variety of signalling modules encoding secondary messenger regulation systems, as well as classical two-component signal transduction systems, suggesting a diversification in function. In this chapter we review the current research into stressosome systems and discuss the functional implications of the unique structure of these signalling complexes.

J. Pané-Farré
Division of Microbial Physiology and Molecular Biology, University of Greifswald, Greifswald 17487, Germany
e-mail: janpf@uni-greifswald.de

M.B. Quin
Department of Biochemistry, Molecular Biology and Biophysics, University of Minnesota, St. Paul, MN 55108, USA
e-mail: mbquin@umn.edu

R.J. Lewis
Institute for Cell and Molecular Biosciences, Faculty of Medical Sciences, University of Newcastle, Newcastle upon Tyne NE2 4HH, UK

J. Marles-Wright (✉)
School of Biology, Newcastle University, Devonshire Building, Newcastle upon Tyne NE1 7RU, UK
e-mail: jon.marles-wright1@ncl.ac.uk

J.R. Harris, J. Marles-Wright (eds.), *Macromolecular Protein Complexes*, Subcellular Biochemistry 83, DOI 10.1007/978-3-319-46503-6_1

Keywords Stressosome • Bacillus subtilis • Phosphorylation • Kinase • SigmaB • RsbR • RsbS • YtvA

1.1 Introduction

1.1.1 Environmental Sensing and Signalling in Bacteria

Bacteria have colonized almost every possible environmental niche on earth and are subjected to constant fluctuations in their growth conditions (Aertsen and Michiels 2004). Consequently, they have developed sensing and signalling systems that allow adaptive responses through changes in gene expression and cellular behaviour (Hecker and Völker 2001). These signalling systems can alter some cellular behaviours directly, these include motility (Mitchell and Kogure 2006), the modulation of biochemical pathways by allosteric activation/inhibition (Wang et al. 2008); and affect gene expression through the activation of alternative RNA polymerase sigma factors (Paget 2015), transcription factors, and DNA binding two-component signal transduction systems (Stock et al. 2000; Capra and Laub 2012).

Two-component and hybrid two-component signal transduction systems that regulate gene expression comprise a, usually membrane-embedded but occasionally soluble, sensor histidine kinase that senses environmental signals. The kinase is coupled to a cognate response regulator that mediates a cellular response by affecting the transcription of target genes (Gao and Stock 2009). Upon receipt of the appropriate stimulus, the sensor kinase autophosphorylates on an invariant histidine residue. The phosphoryl group is subsequently transferred to a conserved aspartic acid on the cognate response regulator, triggering a conformational change that activates the effector domain of this protein. Hybrid two-component systems combine the function of both the sensory kinase and the response regulator in a single polypeptide, but function in essentially the same way as the classical two-component systems (Capra and Laub 2012). A number of recent reviews of two component signalling are available, but given the dynamic nature of the field, an exhaustive reference list is impossible because of space limitations (Gao and Stock 2009; Krell et al. 2010; Lowe et al. 2012; Bhate et al. 2015).

RNA polymerase sigma factors are essential for the correct initiation of gene transcription through the recognition of promoters (Paget 2015). In addition to the major sigma factors that are required for the transcription of essential housekeeping genes, many bacteria possess alternative sigma factors that control the expression of distinct subsets of genes. The first identification of an alternative sigma factor was in 1979 by Haldenwang and Losick (Haldenwang and Losick 1979), which was at that time erroneously linked to the sporulation process of *Bacillus subtilis*. Subsequently, it became clear that this particular sigma factor was actually induced by general stress (Moran et al. 1981) and the interest in σ^B waned somewhat. Today, σ^B is known as the sigma factor required for the general stress response in *B. subtilis* and related bacteria, controlling a large regulon of nearly two hundred genes involved in cellular stress (Price et al. 2001; Petersohn et al. 2001). Owing to the potentially large number of genes that alternative sigma factors can regulate, and the resulting heavy metabolic burden on the cell, the activity of sigma factors are regulated by complex signalling cascades, which include phosphorylation, proteolysis, modulation by the alarmone ppGpp, and partner-switching, to ensure their correct and timely function (Österberg et al. 2011).

1.1.2 **B. subtilis** *σ^B Partner Switching Cascade*

The σ^B-controlled general stress response of *B. subtilis* is regulated by an intricate, phosphorylation-dependent partner switching cascade comprising over a dozen currently-known components (Fig. 1.1) (Hecker et al. 2007). In unstressed cells, σ^B is prevented from interacting productively with RNA polymerase because σ^B is sequestered in a transcriptionally-inactive complex with its cognate anti-sigma factor, RsbW (where Rsb stands for Regulator of sigma B) (Benson and Haldenwang 1993). Upon the imposition of stress, RsbW releases σ^B, allowing it to bind to RNA polymerase to initiate transcription of the general stress regulon (Alper et al. 1996; Kim et al. 2004a). This effect is mediated by the anti-anti sigma factor RsbV. In its dephosphorylated state, RsbV has a greater affinity for RsbW than RsbW does for σ^B. The switch in binding partner of RsbW from σ^B to RsbV causes the release of σ^B, allowing σ^B to interact with RNA polymerase (Yang et al. 1996). Under non-stressful conditions, RsbV is maintained in a phosphorylated state by the kinase function of RsbW, and when phosphorylated, RsbV has a much lower affinity for RsbW than RsbW has for σ^B and consequently σ^B is preferentially bound by RsbW (Delumeau et al. 2002).

RsbV sits at a branch-point in the signalling cascade, where both energy and environmental stresses are integrated (Voelker et al. 1996). The activator/phosphatase pair, RsbQ/P, responds to the energy status of the cell and controls the energy stress branch of the σ^B pathway. The enzymatic activity of RsbQ is required for the correct response to energy stress (Brody et al. 2001), presumably because the reaction product of the RsbQ α/β hydrolase is needed to stimulate the phosphatase activity of RsbP, towards phosphorylated RsbV (RsbV-P), by binding to its Per-Arndt-Sim (PAS) regulatory domain. When the ATP:ADP ratio falls, the RsbQ/P couple dephosphorylates RsbV, inducing the partner-switching of RsbW away from σ^B and towards RsbV to liberate σ^B (Vijay et al. 2000; Kaneko et al. 2005; Nadezhdin et al. 2011).

In the pathway that responds to environmental stresses, such as high salt and ethanol, the RsbV phosphorylation status is controlled by the phosphatase RsbU, which also dephosphorylates RsbV-P (Yang et al. 1996). In this pathway RsbU is activated by the RsbT kinase. When the cell is unstressed, RsbT is bound by structural proteins, including RsbR and RsbS, which self-assemble into a ~1.8 MDa macromolecular complex known as the stressosome. Upon receipt of an environmental stress

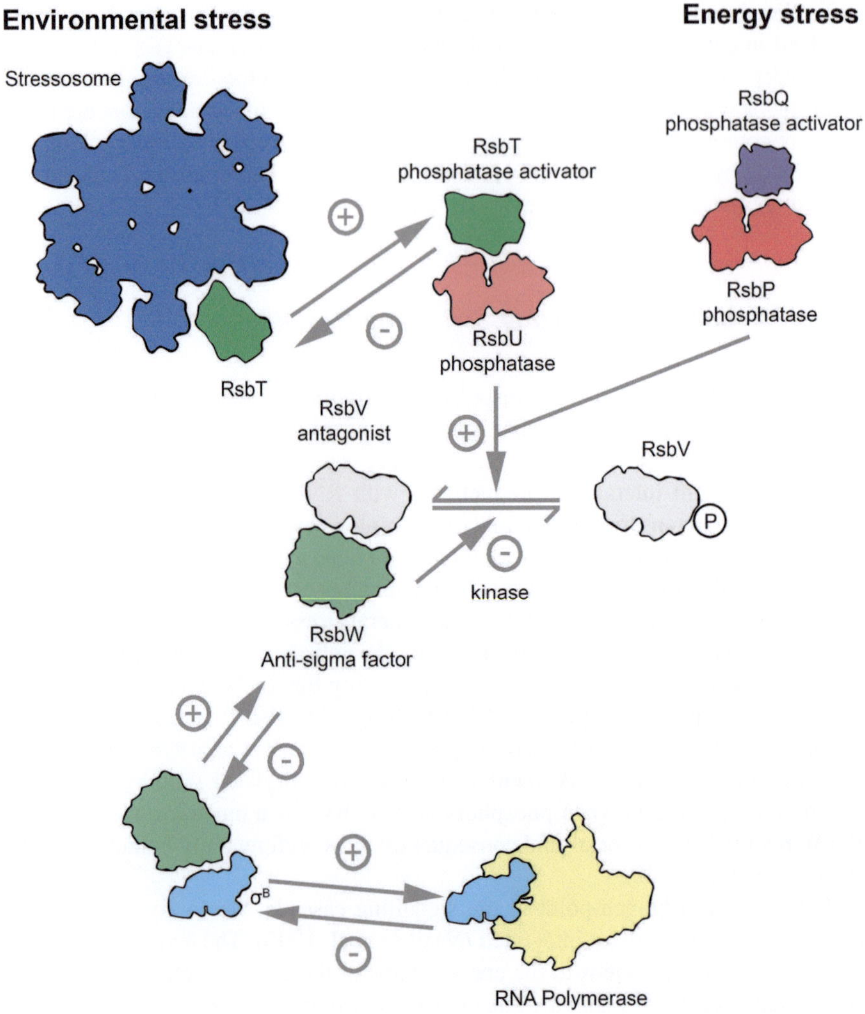

Fig. 1.1 σ^B cascade. The *Bacillus subtilis* σ^B cascade is illustrated to show how partner-switching induces the activity of RNA polymerase and acts to regulate gene expression. Pre-stress, the anti-sigma factor RsbW sequesters σ^B and prevents it from directing RNA polymerase to σ^B-controlled promoters. In this state, RsbV is phosphorylated (RsbV-P) by the kinase activity of RsbW and hence RsbV is inactivated. Under stress conditions, RsbV-P becomes dephosphorylated by one of two phosphatases and attacks the RsbW:σ^B complex and liberates σ^B to direct transcription of its regulon to provide the cell with stress-resistance. RsbV is also the point at which the environmental and energy stress responses converge. Under energetic stress, the phosphatase RsbP is activated by RsbQ and dephosphorylates RsbV-P to allow it to form complexes with RsbW. Environmental stresses are integrated by the stressosome, which sequester the RsbU phosphatase-activator, RsbT, in the absence of stress. Under environmentally stressful conditions, RsbT phosphorylates the STAS domains of the stressosome proteins and disassociates, because of a presumed reduced affinity for the phosphorylated proteins, and RsbT switches its binding partner to the phosphatase RsbU. The RsbT:RsbU complex activates RsbV by its dephosphorylation. The phosphatase RsbX acts to remove phosphoryl groups from the stressosome and to mediate the duration of the stress response by 'resetting' the system. *Ringed plus signs* indicate positive regulatory events affecting σ^B activity, while *ringed minus signs* indicate those that are negative regulatory events

signal, RsbT phosphorylates the structural proteins, initiating its release from the stressosome, and leading to the subsequent activation of RsbU (Kang et al. 1998; Chen et al. 2003; Kim et al. 2004a). The structural proteins that form the stressosome include RsbS; RsbR and the paralogues of RsbR, YkoB, YojH and YqaH (Delumeau et al. 2006), which have been collectively re-named RsbRA, RsbRB, RsbRC and RsbRD, respectively (Kim et al. 2004b); and the blue-light sensor, YtvA (Akbar et al. 2001; Losi et al. 2002; Jurk et al. 2013). The relative stoichiometries of the RsbR paralogues in the stressosome, assuming a single entity is formed in the cell, and the exact mechanism by which environmental stress signals are perceived and transduced by the stressosome are not known at this time. For simplicity, most structural studies have concentrated on the RsbR/RsbS/RsbT triumvirate as a surrogate of the likely more complex stressosome assemblies found in the cell.

By use of the σ^B-dependent promoter of the *ctc* gene (which encodes a component of the large subunit of the ribosome (Truitt et al. 1988; Schmalisch et al. 2002)) fused to a *lacZ* reporter, σ^B activity has been monitored in real time by several groups. From a combination of studies, it is clear that the σ^B signalling cascade is temporally limited. Upon the imposition of environmental stress, σ^B activity rises to a maximum after 20–30 min, after which time σ^B activity begins to decline back to base-line levels. The resetting of the system takes place in two independent steps:

1. RsbW phosphorylates RsbV, resulting in the partner switching of RsbW to resequester and inactivate σ^B, switching off gene transcription
2. The phosphatase RsbX dephosphorylates RsbR-P and RsbS-P (Chen et al. 2004), which resets the system through the re-sequestration of RsbT by the stressosome (Eymann et al. 2011)

The system is also regulated transcriptionally, due to the presence of a σ^B promoter upstream of RsbV, which leads to increased amounts of the RsbW anti-sigma factor and RsbX phosphatase, which are located downstream of RsbV in the *B. subtilis* genome (See 1.7.1) (Dufour et al. 1996).

1.1.3 RsbRST Module Distribution

RsbR, RsbS and RsbT form the core components of the stressosome and the genes encoding these proteins are co-located in the genome at the start of the *rsb* operon. These three proteins form the RsbRST module, which is distributed widely across bacteria and is found in representatives of the Firmicutes, Actinobacteria, proteobacteria, Bacteroides, cyanobacteria and Deinococcus groups. Whilst the core of the module is well conserved, the N-terminal domain of RsbR is less well maintained between species. Furthermore, the downstream components in the *rsb* operon vary considerably, including secondary messenger signalling and two-component regulators, as well as alternative sigma factors, indicating that the stressosome complex may have evolved as a signalling hub controlling a diversity of cellular functions (Pané-Farré et al. 2005).

1.1.4 Chapter Outline

In this chapter we will discuss how the structure of the stressosome relates to its function as a signal integration and transduction hub, and the complex signalling behaviours seen in *B. subtilis*. We will assess the current structural knowledge of the individual components of the stressosome as determined by X-ray crystallography and the structure of the stressosome determined by single particle cryo-EM. We will highlight the modularity of this complex and show that it has been widely adopted among bacterial species as a signalling hub with distinct inputs and outputs. Furthermore, we will discuss the structural basis for its signalling mechanism and the role of different inputs in modulating the output level of the complex. The challenges of structural studies on flexible and heterogeneous complexes will be discussed with reference to the cryo-EM structure. The chapter will close with a discussion of unanswered questions on stressosome signalling.

1.2 Stressosome Components

1.2.1 Stressosome Composition

The simplest stressosome complex comprises the RsbS scaffold protein, which consists of a single STAS (Sulphate Transporter and Anti-Sigma factor antagonist (Sharma et al. 2011)) domain and RsbR, which has a variable N-terminal domain and a conserved C-terminal STAS domain (Chen et al. 2003). The STAS domain scaffold is well-conserved between RsbS and the RsbR paralogues (Fig. 1.2) and the RbsR paralogues can form heterogeneous stressosome complexes both *in vivo* and *in vitro* (Delumeau et al. 2006; Jurk et al. 2013). On the imposition of stress, RsbT phosphorylates conserved serine and threonine residues on the STAS domains of RsbS and the RsbR paralogues (Kim et al. 2004a). In the RsbR/RsbS stressosome surrogate, the rate of phosphorylation of RsbS on Ser59 is quicker *in vitro* than on either RsbR phosphorylation site (Thr171, Thr205) (Chen et al. 2003), strongly suggesting that RsbT interacts predominantly with RsbS in the stressosome. How the kinase is held in an inactive state in the absence of stress signals is unknown. Moreover, the effect and purpose of phosphorylation in the stressosome is also unclear (Kim et al. 2004a; Chen et al. 2004; Liebal et al. 2013; Gaidenko and Price 2014). It is generally agreed, however, that RsbT must dissociate from the stressosome to activate RsbU and the downstream components of the σ^B cascade in order to respond to the imposition of stress, (Kang et al. 1998). In the absence of RsbT, the RsbX phosphatase is able to dephosphorylate RsbS-P and RsbR-P (Chen et al. 2004). RsbX has a differential activity against the different phosphorylation sites on these proteins, which has been suggested to relate to an adaptive response of the stressosome to sustained, or repeated, stresses (Eymann et al. 2011).

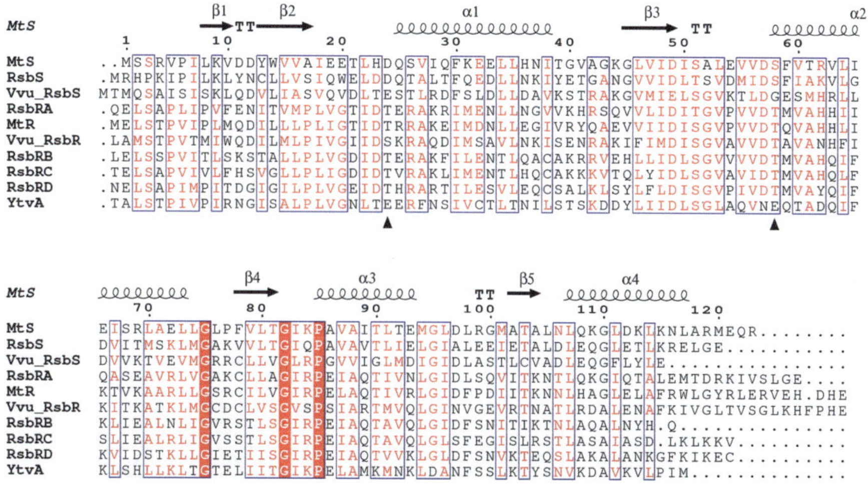

Fig. 1.2 Alignment of STAS domain sequences. Multiple sequence alignment generated for the STAS domains of stressosome proteins from *Bacillus subtilis* (RsbS, RsbRA-D, YtvA), *Moorella thermoacetica* (MtS, MtR), and *Vibrio vulnificus* (Vvu_RsbS, Vvu_RsbR) to highlight sequence conservation in these proteins. Regions outlined in *blue* show areas of sequence conservation, residues *coloured red* indicate partial conservation and those outlined in *red* are completely conserved across the aligned sequences. The secondary structure elements from the X-ray crystal structure of *Mt*S (PDBID: 2VY9) are shown above the alignment with residues numbered for the *Mt*S sequence. The positions of conserved serine and threonine residues phosphorylated by RsbT kinase are shown with *black triangles* below the alignments. Note that RsbS and its homologues are only phosphorylated on the conserved serine at position 58 in *Mt*S and 59 in RsbS. VVu_RsbS has a glycine at this position, although there is a serine at position 62 in this protein; there is currently no published experimental evidence that the Vvu_RsbS protein is phosphorylated. Figure 1.2 was prepared using EsPript (Gouet et al. 2003)

1.2.2 Structure of RsbS

RsbS is a single STAS domain protein, and STAS domains are also found in anion transporters, the SpoIIAA anti-anti-sigma factor antagonist in *B. subtilis* (Kovacs et al. 1998), and as essential components of the stressosome. STAS domains appear to function primarily as a scaffold for the recruitment of other proteins, particularly in partner-switching networks (Aravind and Koonin 2000). The *B. subtilis* RsbS protein forms stable stressosome complexes with the RsbR paralogues both *in vivo* and *in vitro* (Chen et al. 2003; Kim et al. 2004b; Delumeau et al. 2006; Reeves et al. 2010). The RsbS STAS domain has a five-stranded beta sheet core with three alpha helices on one face and a single C-terminal helix on the other face (Fig. 1.3). Phosphorylation takes place on a conserved serine/threonine at the N-terminus of helix 2, the central of the three helices of the STAS domain. Structures of the STAS domains of SpoIIAA (Seavers et al. 2001) and RsbS (Marles-Wright et al. 2008; Quin et al. 2012) in both phosphorylated and non-phosphorylated forms do not reveal any significant conformational changes in protein structure, suggesting that

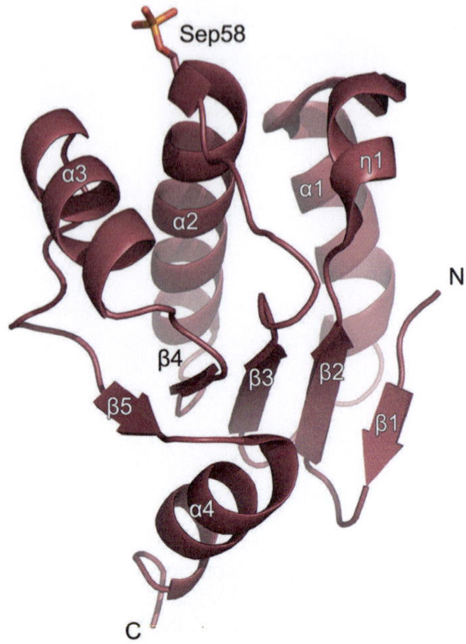

Fig. 1.3 X-ray crystal structure of MtS. The X-ray crystal structure of the phosphorylated form of MtS (PDBID: 3TZB) is shown as a cartoon with secondary structure elements labelled from the N- to C-terminus. Serine 58 is phosphorylated in this structure and the phosphoryl group is depicted as *orange and red sticks*

release of RsbT from the stressosome as a function of phosphorylation on RsbS-Ser59, and/or its equivalent residue in RsbR, Thr205 (and/or the non-equivalent Thr171), is likely to occur solely by electrostatic and physical repulsion. Similarly, the anti-anti-sigma factor SpoIIAA is released from its complex with cognate dual-function kinase and anti-sigma factor SpoIIAB by the same mechanism (Seavers et al. 2001; Masuda et al. 2004). However, in the absence of a high resolution structure of a phosphorylated stressosome complex, rather than isolated RsbS and RsbS-P structures, significant conformational changes in STAS domain architecture during stressosome activation cannot be excluded (Kumar et al. 2010).

1.2.3 Structure of RsbR

The *B. subtilis* RsbR protein has a C-terminal STAS domain and an N-terminal non-heme globin domain (Murray et al. 2005). While there is no crystal structure of the RsbR STAS domain, it has 30% sequence identity to that of RsbS and will presumably adopt the same fold. Unlike RsbS and other characterised single domain STAS proteins, RsbR is phosphorylated on two conserved threonine residues, Thr171 and Thr205 (Gaidenko et al. 1999; Eymann et al. 2011). Thr205 is equivalent to Ser59 of RsbS and is likely to be located at the N-terminus of helix 2, whereas the unique phosphorylation site of RsbR, Thr171, is positioned at the N-terminus of helix 1. Thr171 tends to be occupied by serine or threonine in the two-domain RsbR paralogues, but glutamate or aspartate in the single domain RsbS paralogues and YtvA, suggesting that the Asp/Glu residues mimic the effect of phosphorylation on RsbR in the latter group of proteins, which apparently increases the activity of RsbT towards RsbS (Chen et al. 2004). Whilst phospho-ablative mutations have been made in RsbR and the effects on phosphorylation patterns, rates and signalling have been measured (Chen et al. 2004; Gaidenko and Price 2014), similar experiments have not been performed to the best of our knowledge on Asp25Thr (or alanine) mutations in RsbS in order to decipher the role of phosphorylation on Thr171 in RsbR.

The structure of the N-terminal domain of RsbR (N-RsbR) displays an all alpha-helical non-heme globin fold (Murray et al. 2005) (Fig. 1.4a, b). A comparison of N-RsbR to other globin-like proteins, such as the HemAT aerotaxis sensor (Hou et al. 2000; Zhang and Phillips 2003) shows that RsbR does not possess a heme binding pocket and also lacks the heme coordinating residues found in the hemoglobins (Murray et al. 2005). In the absence of a heme binding pocket, it is thought that RsbR may interact with an as yet unidentified small molecule ligand in an analogous way to the amino acid-binding globin sensor proteins (Kitanishi et al. 2011), or to the ligand-binding sites in the non-heme globin sensors that regulate the entry of *Bacillus anthracis* into sporulation (Stranzl et al. 2011).

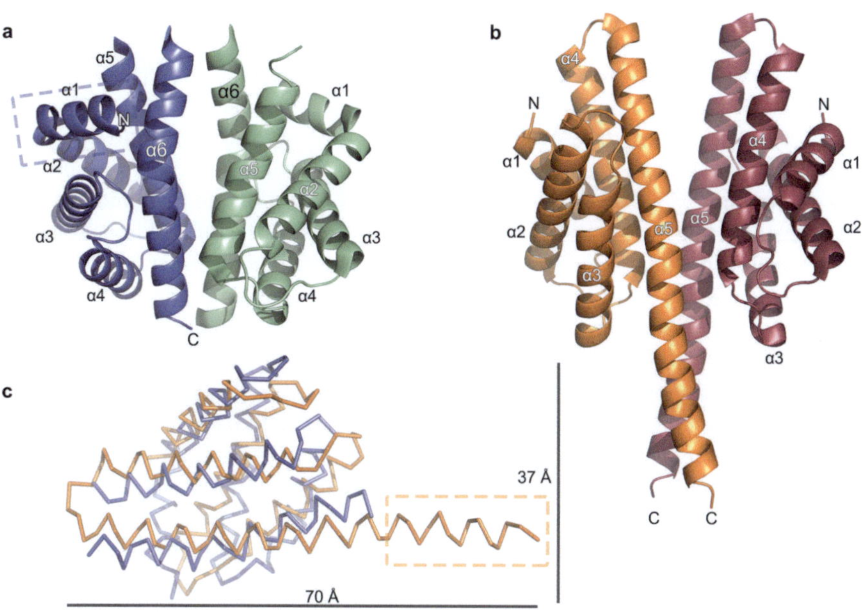

Fig. 1.4 X-ray crystal structures of RsbR and MtR. The X-ray crystal structures of the *B. subtilis* RsbR (**a**) (PDBID: 2BNL) and *Moorella thermoacetica MtR* (**b**) (PDBID: 3TZA) are shown as cartoons with secondary structure elements labelled. Both proteins are depicted as the physiologically relevant dimer forms. (**c**) Secondary structure comparison of RsbR and *Mt*R monomers; RsbR is shown in *blue*, *Mt*R in *orange*. While the two structures have the same non-heme globin fold, RsbR has an additional N-terminal α-helix (*blue-dashed box*) and in the X-ray crystal structure lacks the C-terminal portion of the α-helix that links to the STAS domain (*orange-dashed box*)

The N-terminal domain of the *Moorella thermoacetica* RsbR orthologue, *Mt*R (N-*Mt*R), has only 12% sequence identity to that of *B. subtilis* RsbR, yet an almost identical fold (Fig. 1.4c). The N-*Mt*R structure has an extended alpha helix at the C-terminus of the globin domain that links this with the C-terminal STAS domain (Quin et al. 2012). This extended, terminal helix is equivalent to the 'J' helix of the LOV (light-oxygen-voltage) domain of the RsbR paralogue YtvA, which has been suggested to convey blue light-dependent signalling (Jurk et al. 2010).

The N-terminal domains of the RsbR paralogues encoded in the *B. subtilis* genome share limited sequence identity, although they are each found in stressosomes *in vivo* and can associate with RsbS to form stressosome-like structures using purified recombinant proteins *in vitro* (Delumeau et al. 2006). While the structures of their N-terminal domains are yet to be determined, they are likely to form sensory domains given the fact that a number of them belong to the PAS (Per-Arnt-Sim) domain superfamily, which are usually found in signalling systems as sensors (Ponting and Aravind 1997).

1.2.4 Structure of YtvA

While the exact role of the RsbR paralogues is not known, it appears that the detection of blue light by YtvA is modulated by the other RsbR paralogues (van der Steen et al. 2012). Indeed, YtvA is the only RsbR related stressosome component in *B. subtilis*, and potentially *Listeria monocytogenes*, for which the activating signal has been confirmed (Möglich and Moffat 2007; Ondrusch and Kreft 2011). This protein has a C-terminal STAS domain and an N-terminal LOV domain that binds a flavin mononucleotide cofactor and undergoes a light-dependent conformational change in the nucleotide binding site (Losi et al. 2002; Möglich and Moffat 2007; Herrou and Crosson 2011) (Fig. 1.5a). This structural rearrangement is transmitted to the 'J' helix, which links the sensory domain to the STAS domain, resulting in the movement of this helix away from the axis of the dimer interface (Möglich and Moffat 2007). These structural changes hint at the signal transduction mechanism from the sensory domains of RsbR proteins to the STAS domain, although solution small-angle X-ray scattering experiments on the full length YtvA failed to show any significant structural rearrangements upon illumination with blue light (Jurk et al. 2010).

The solution structure of the full length *B. subtilis* YtvA protein was published recently in the PDB (PDBID: 5MWG) (Fig. 1.5b). While the dimer interface in the crystal structure of the YtvA LOV domain is formed primarily by interactions between the beta-sheets of the two monomers, the solution structure has a distinct dimer arrangement with an additional N-terminal α-helix modelled to form the primary point of contact between the monomers. This particular arrangement is consistent with the dimer arrangement seen in the crystal structures of both the *B. subtilis* and *M. thermoacetica* RsbR proteins (Fig. 1.4). These intriguing structural differences may be biologically significant, or represent artefacts of the different

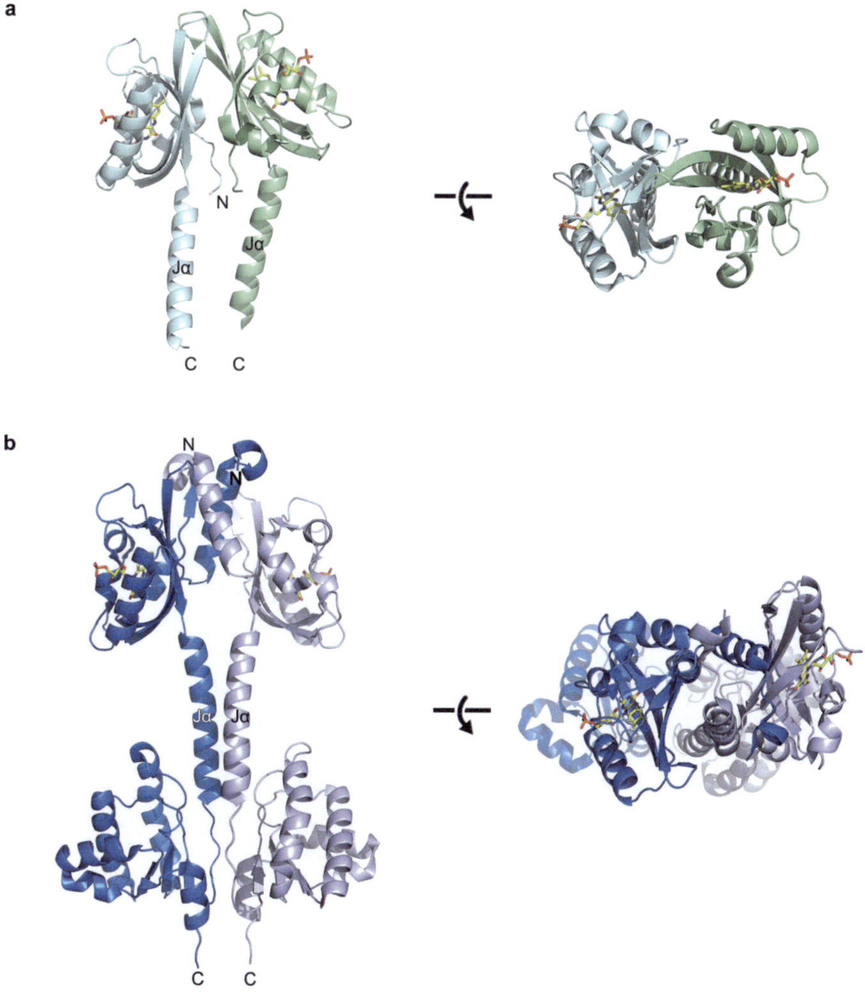

Fig. 1.5 **Structures of YtvA**. (**a**) Crystallographic model of the LOV domain of *B. subtilis* YtvA in the dark state (PDBID: 2PR5). The model comprises a dimer in the asymmetric unit, which are shown as cartoons with the bound FMN cofactor as sticks coloured by atom. In the light state structure (PDBID: 2PR6) the Jα helices are displaced by up to 2 Å away from the dimer axis. (**b**) Solution structure of the full length *B. subtilis* YtbA protein (PDBID: 2MWG). The LOV domain is shown at the *top* and STAS domain to the *bottom* in the *left* panel. The full length structure displays a distinct dimer organisation to the isolated LOV domain with an N-terminal helix, which is not modelled in the crystal structure, forming substantial contacts in the dimer interface

constructs and methods used to produce the two structures. The orientation of the LOV domain in relation to the STAS domains in the full length dimer structure hints at a mechanism of signal transduction through the Jα helix to alter the orientation of the LOV domain relative to the STAS domain, either through a movement of the LOV domain itself, or of the STAS domains in the stressosome core.

YtvA is the only RsbR paralogue that is not phosphorylatable, encoding glutamate at both positions equivalent to Thr171 and Thr205 in RsbR. The ready light-to-dark reversibility of YtvA is indicated by a half-life of the photo-excited state of some 40 min (Losi et al. 2003). The effect of mutation of Glu142 and Glu202 in YtvA to serine/threonine, resulting in the potential for phosphorylation of these residues by RsbT, and the potential for phosphoregulation of YtvA and its impact on σ^B activity, are as-of-yet unconducted experiments that might explain the regulatory role of YtvA in stressosome signalling.

1.2.5 Structure of RsbT

To date there is no X-ray crystal structure of the RsbT protein kinase. However, there is a structure of a *Bacillus stearothermophilus* homologue, the SpoIIAB anti-sigma factor, with which it shares 28% sequence identity (Campbell et al. 2002). This protein binds to the sporulation sigma factor, σ^F, in an ATP dependent manner to suppress the activity of the sigma factor. SpoIIAB also binds in an ADP-dependent fashion to the anti-anti-sigma factor protein SpoIIAA, which is a STAS domain protein that disrupts the binding between SpoIIAB and σ^F to release the sigma factor (Alper et al. 1994). The nucleotide-dependencies of these interactions are probably a reflection of the rather poor stability of recombinant SpoIIAB in the absence of ADP or ATP (Lord et al. 1996). The homologous RsbW can bind to both its cognate sigma factor (σ^B) and anti-anti-sigma factor *in vitro* in the absence of additional nucleotide (Delumeau et al. 2002). Both SpoIIAB and RsbT are members of the GHKL superfamily of protein kinases and the structure of SpoIIAB displays an α/β sandwich with an antiparallel β sheet flanked by alpha helices (Dutta and Inouye 2000). The ATP binding pocket is found in a deep crevice between the helices and the β sheet. The pocket has a highly flexible lid that is thought to communicate the nucleotide binding status of the protein through changes in its structure. Unlike SpoIIAB, which acts as a dimer, RsbT appears to act as a monomer in solution, because it lacks a C-terminal helix found in SpoIIAB that mediates dimerisation (Campbell et al. 2002; Masuda et al. 2004; Delumeau et al. 2006).

1.2.6 Structure of RsbX

RsbX is the protein phosphatase responsible for resetting the stressosome to the resting state after the imposition of stress and its phosphorylation by RsbT. RsbX acts on both RsbS-P and RsbR-P, although it displays differential activity against the different proteins *in vitro* and *in vivo* (Chen et al. 2003; Eymann et al. 2011). RsbX is a member of the PP2C family of protein phosphatases (Das et al. 1996) and the structures of both the *B. subtilis* and *M. thermoacetica* homologues (*Mt*X) have been determined by X-ray crystallography (Quin et al. 2012; Teh et al. 2015). These

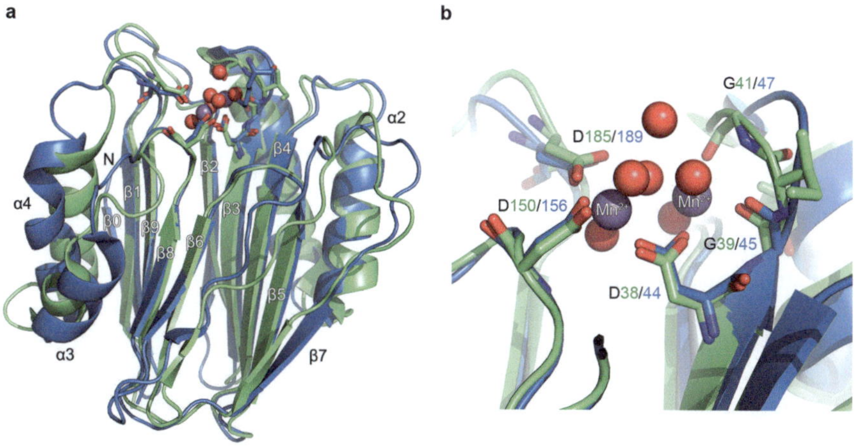

Fig. 1.6 X-ray crystal structure of RsbX and *Mt* X. (**a**) Crystallographic models of both the *B. subtilis* RsbX (PDBID: 3ZT9) (*green*) and *Moorella thermoacetica MtX* (PDBID: 3W40) (*blue*) protein phosphatases have been published. The models are depicted as cartoons with secondary structure features labelled from the N- to the C-terminus. Manganese ions required for the catalytic activity of the protein bound in the active site of the protein are depicted as *purple spheres*, with water molecules shown as *red spheres*. (**b**) Metal binding site of RbsX and *Mt*X showing residues coordinating the bound manganese ions. The two manganese ions (*purple spheres*) are coordinated by the carboxylic acid groups of aspartic acid residues and the backbone carbonyl groups of two glycine residues (shown as stick representations). A number of ordered solvent residues are also present in the coordination shell of the manganese ions (*red spheres*)

proteins share only 25% sequence identity yet superimpose with an RMSD Cα of 1.9 Å over 182 Cα. The two proteins display the same phosphatase fold with an αββα architecture, with the central beta-sandwich flanked by two pairs of alpha helices on the outer faces of the protein (Fig. 1.6). The catalytic centre of the protein is located at one edge of the beta-sandwich and centred around a cluster of acidic residues, which are highly conserved in the PPM phosphatase family (Quin et al. 2012). These residues coordinate two divalent metal ions that are required for the function of the protein. RsbX and *Mt*X show a strong preference *in vitro* for the presence of Mn^{2+} in this metal binding site for their activity against both their native and synthetic phosphatase substrates (Quin et al. 2012; Teh et al. 2015). A depression on the surface of these proteins above the active site provides a potentially ideal site for interaction with their targets (Quin et al. 2012; Teh et al. 2015). The dephosphorylation reaction is likely to proceed in a manner analogous to other members of the PP2C family, where a metal-ion bridged water molecule acts as a nucleophile against the phosphate group on the target protein and a second water molecule protonates the dephosphorylated serine/threonine residue (Das et al. 1996; Barford et al. 1998).

1.3 Stressosome Complexes

The RsbR paralogues and RsbS form a stable 1.5 MDa stressosome complex in *B. subtilis* that sequesters RsbT in the absence of stress. These complexes can be isolated directly from *B. subtilis* cells and are present throughout the life of individual bacterial cells (Kim et al. 2004b; Gaidenko and Price 2014). Minimal stressosome complexes can be reconstituted *in vitro* by mixing RsbS with the RsbR paralogues to simplify the structural biology, although RsbR has a tendency to self-associate into stressosome-like particles in the absence of RsbS. In the absence of RsbS, stressosome complexes are unable to sequester RsbT, and deletion of the *rsbS* gene in *B. subtilis* leads to constitutive activation of RsbT and a resulting small-colony phenotype. This is presumably caused by deleterious effects of the products of the σ^B regulon, or a negative effect of competition for cellular resources with σ^A regulated housekeeping genes (Kang et al. 1996).

The minimal stressosome complexes of individual RsbR paralogues and RsbS are competent to bind RsbT (Delumeau et al. 2006) and can be phosphorylated by the kinase on conserved residues in their C-terminal STAS domains (Reeves et al. 2010). RsbR paralogues in stressosomes can also be exchanged *in vitro*, implying that these complexes have some dynamic properties in solution (Delumeau et al. 2006). However, immunofluorescence experiments using antibodies specific for the N-terminal domain of RsbR show that RsbR – and hence the stressosome - forms punctate foci that persist in cells throughout the stress response and its recovery (Marles-Wright et al. 2008). These results suggest that in wildtype *B. subtilis* the stressosome always contains RsbR and that the RsbR paralogues may exchange into stressosomes through the life of the cell. It is not known if distinct populations of stressosomes with different RsbR paralogues exist at the same time within the cell. *B. subtilis* stressosomes purified by anti-N-RsbR affinity purification contain at least three paralogues of RsbR (Delumeau et al. 2006); thus, single, double, and triple knockouts are effectively complemented by the remaining paralogues in normally growing *B. subtilis* (Kim et al. 2004b). A quadruple knockout of the RsbR paralogues leads to constitutive σ^B activation in much the same manner as the RsbS knockout, as no competent stressosome complexes are able to form in the absence of the RsbR paralogues, leading to the presence of free RsbT in the cell, up-regulated RsbU activity and σ^B liberated to interact with RNA polymerase.

1.4 Production of Recombinant *B. subtilis* Stressosomes for Structural Analysis

1.4.1 RsbR/RsbS Binary Complex

Recombinant *B. subtilis* stressosome complexes for structural analysis were produced by plasmid-based co-expression of RsbR and RsbS from a bi-cistronic operon in *Escherichia coli* and purified to homogeneity in a multi-step protocol

(Marles-Wright et al. 2008). As with any high molecular weight protein complex destined for structural analysis, the purification protocol was extensively refined to ensure the final sample was as homogeneous as possible. The optimised protocol included anion-exchange, size-exclusion gel-filtration, hydrophobic interaction chromatography, and a second gel-filtration step. Using this protocol, stressosome complexes were separated from ribosomes, and other high-molecular weight proteins and macromolecular complexes from the *E. coli* host strain. Despite the relatively high levels of protein expression and the multi-step purification protocol, it was still possible to identify contamination by the cubic core of the pyruvate dehydrogenase complex in some micrographs (Marles-Wright and Lewis, unpublished observations).

1.4.2 RsbR/S/T Ternary Complex

Production of the RsbR/S/T ternary complex was achieved by expression of RsbT as an N-terminal GST-fusion to enhance protein production levels and solubility. The RsbT protein was purified by glutathione affinity chromatography in the presence of 1 mM ADP to ensure the nucleotide-binding site of the kinase was occupied, and subsequent on-column cleavage of the GST-tag was achieved using the HRV-3C protease. Purified ADP-loaded RsbT was mixed in excess with purified RsbR/S minimal stressosomes and subjected to an additional round of size-exclusion gel-filtration chromatography to remove unbound RsbT (Marles-Wright et al. 2008).

1.5 Cryo-EM Structure of the *Bacillus subtilis* Stressosome

Initial negative stain electron microscopy studies of recombinant *B. subtilis* stressosomes identified a 20 nm ring-shaped structure (Fig. 1.7) (Chen et al. 2003). Due to the absence of distinct views other than the characteristic ring, the complex was initially thought to form a doughnut-like oligomer that exhibited a preferential orientation on the carbon-film of the EM grid (Delumeau et al. 2006). Collection of data on unstained samples by cryo-EM and single-particle analysis allowed the calculation of 3D–reconstructions for a stressosome core structure comprising the full length RsbS protein and an N-terminally truncated RsbR; the full length RsbR/RsbS stressome; and a ternary complex between RsbR/RsbS and the kinase RsbT. We will discuss each of these in turn below.

1.5.1 RsbR/RsbS Core Structure

The single-particle cryo-EM reconstruction of the $RsbR_{(146–274)}$:RsbS core STAS-domain complex was determined to a resolution of 6.5 Å by imposing icosahedral symmetry restraints on the calculation of the molecular envelope as indicated by

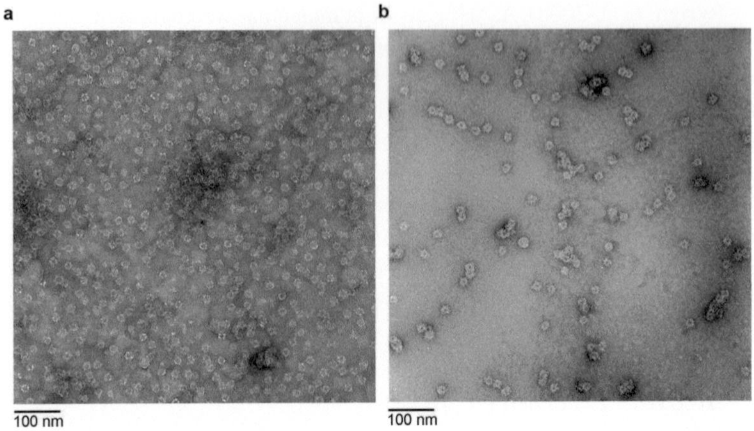

Fig. 1.7 Negatively-stained transmission electron micrograph of the *B. subtilis* stressosome. Purified recombinant *B. subtilis* stressosome complexes were stained with 1% uranyl acetate and imaged by transmission electron microscopy. (**a**) RsbR/RsbS stressosome complexes have a ring-shaped appearance in projection, with rough edges. They are approximately 25–30 nm in size when imaged by TEM. (**b**) RsbR/RsbS/RsbT ternary complexes have a similar appearance to the RsbR/RsbS complexes with more pronounced features at their edges. The ternary complex has a similar diameter to the binary complex. These images were taken with the assistance of Professor J. Robin Harris

analysis of the initial Eigenimages from the first round of reference-free class-averaging. The reconstructed density was a hollow shell with an outer radius of 90 Å and an inner radius of 45 Å. The final reconstruction displayed clear tubes and flat sheets of electron density consistent with the alpha helices and beta sheet seen in the crystal structure of *Mt*S (Fig. 1.8). The position of the STAS domains in the EM-derived molecular envelope places the N-terminus of the STAS domain at the external surface. This distribution is consistent with their covalent attachment to the C-terminus of the N-terminal domains of RsbR, to place these domains on the outside of the stressosome complex so as to most easily interact with other proteins, ligands and stimuli. Indeed, it has already been suggested that the groove in the dimer interface of N-RsbR could be utilised for binding upstream signalling partners, based on its structural similarity to HemAT and the location of the unique 'Z' helix of HemAT in the dimer interface groove; the binding of KaiC peptides in the circadian clock complex KaiA/KaiC; and the interaction of *Vitreoscilla* haemoglobin with a partner dioxygenase (Murray et al. 2005).

The arrangement of the STAS domains across the icosahedral dimer interface of the core stressosome reconstruction was distinct to the dimeric arrangement seen within the *Mt*S crystal structure; the primary interactions in the stressosome core are between the first β-strand and the C-terminal helix of the STAS domain, whereas the crystallographic dimer of *Mt*S is formed between α3 and β5 across the two chains in the dimer, and probably does not represent a stable assembly as determined by the PISA server (Krissinel and Henrick 2007) .

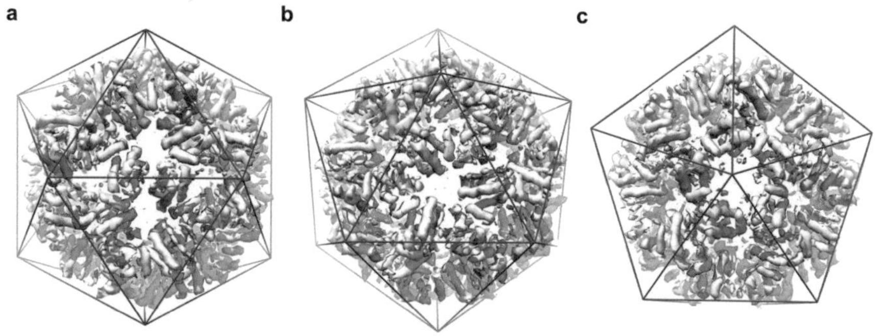

Fig. 1.8 Single particle reconstruction of the $RsbR_{(146–274)}$/RsbS stressosome core complex. The final experimental EM-derived icosahedral reconstruction of the $RsbR_{146–274}$/RsbS stressosome core complex is shown as a *blue* surface contoured at 3σ. Views down the icosahedral two (**a**), three (**b**) and fivefold (**c**) axes are shown with an icosahedral net for reference. The stressosome core has radius of 90 Å

1.5.2 RsbR/RsbS Structure

The cryo-EM reconstruction of the full-length RsbR/RsbS minimal stressosome was determined to a nominal resolution of 8.0 Å. Inspection of experimental Eigenimages generated from the initial particle set showed that the complex possessed a mixed-symmetry, with an icosahedral core and projections from this core that obeyed a lower, D_2 symmetry (Fig. 1.9). While unusual, this type of symmetry mismatch has been found in in the capsids of bacteriophage, where nucleic-acid portal translocases inhabit unique vertices (Dube et al. 1993), and also in multi-component proteasome complexes (Beuron et al. 1998). Analysis of this symmetry mismatch and manual model building with a child's magnetic toy led to a single plausible model for the RsbR/RsbS stressosome containing 40 copies of RsbR and 20 copies of RsbS (Fig. 1.10). In this model, the core of the stressosome is comprised solely of 60 STAS domains, but with an additional 20 projections that obey D_2 symmetry to yield 20 peripheral 'turrets' made up of dimers of N-RsbR (Fig. 1.11). This model was validated experimentally by the production of Eigenimages from back-projections of the final stressosome model with D_2 symmetry imposed (Fig. 1.9c). The final reconstruction had a core with a radius of 90 Å and total radius with the projections of 150 Å. The density for the peripheral projections was less well defined than the core density, implying a level of heterogeneity in the position of the N-termini of RsbR in the complex relative to the core.

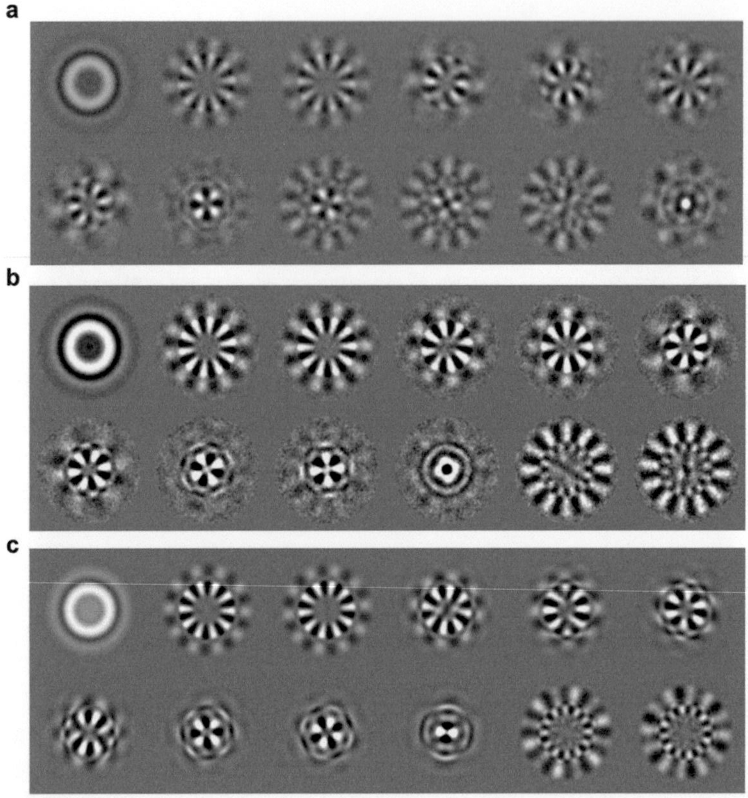

Fig. 1.9 Experimental Eigenimages of the RsbR/RsbS stressosome complex. (**a**) Eigenimages showing symmetry elements associated RsbR/RsbS stressosome showing the mixed symmetry that appears due to the N-terminal projections of RsbR protein. (**b**) Eigenimages from the full length RsbR/RsbS reconstruction, which appear to show a clear tenfold symmetry (**c**) Eigenimages from re-projections of a model RsbR/RsbS with imposed D_2 symmetry, which also exhibit a clear tenfold symmetry, suggesting the tenfold feature is a consequence of the centring and not a structural feature

1.5.3 RsbR/RsbS/RsbT Ternary Complex

The ternary complex of RsbR/RsbS and RsbT was determined to 8.3 Å resolution and displayed the same symmetry mismatch as the binary RsbR/RsbS complex. This reconstruction had identical core and total dimensions to the RsbR/RsbS complex, while the presence of additional density features was apparent above the regions of the core not occupied by the N-RsbR turrets (Fig. 1.12). These additional density features were attributed to the presence of the RsbT protein. In this reconstruction the turrets appeared less well distinguished than in the RsbR/RsbS complex implying a greater degree of positional heterogeneity than in the binary complex.

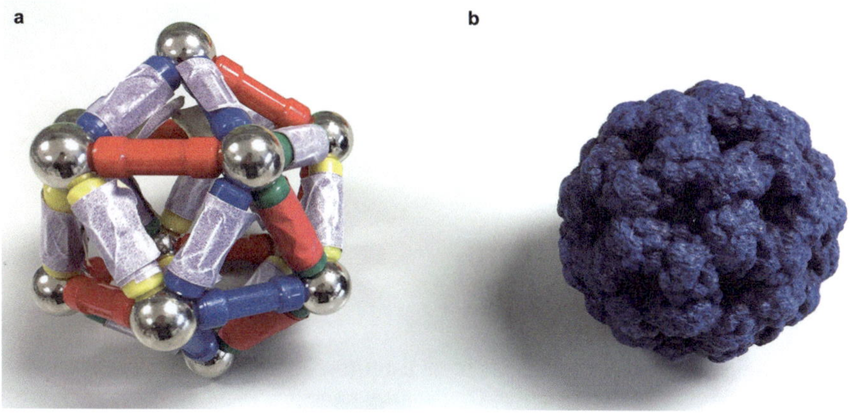

Fig. 1.10 Stressosome model building. To aid with the determination of the subunit arrangement within the stressosome reconstructions a model of the stressosome core was built using a magnetic modelling set (**a**). In this model the magnetic bars represent dimers of RsbR (*blue*) and RsbS (*red*), and the steel balls serve as anchor points between the bars, but represent holes at the fivefold axes of the stressosome. The model was built using a set of rules for protein:protein interactions determined from solution studies of the individual components, namely: (1) RsbS is a dimer in solution, but doesn't interact to form higher-order structures. (2) RsbR is a dimer in solution and can interact with itself to form higher-order structures; however, higher-order interactions with RsbS are stronger than self-interactions. (3) The final model would take the form of an icosahedron. These rules lead the model that was ultimately shown to represent the cryo-EM data. A 3D–printed model of the $RsbR_{(146–274)}$:RsbS stressosome core is shown in (**b**), where all STAS domains are coloured *blue*

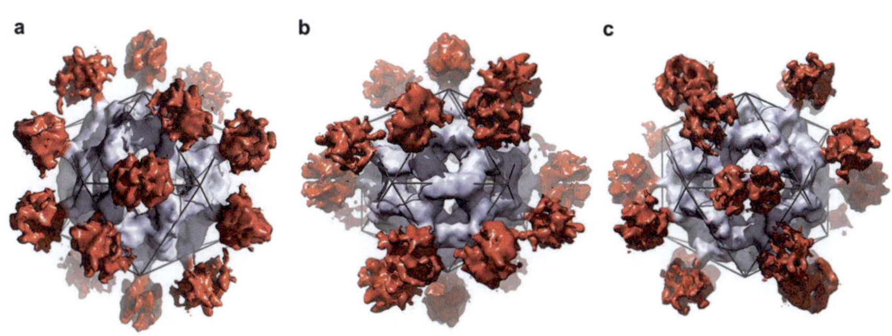

Fig. 1.11 Single particle reconstruction of the RsbR/RsbS stressosome complex. The final experimental EM-derived D_2 symmetry reconstruction of the RsbR/RsbS stressosome complex is shown as a surface contoured at 2σ. The RsbR and RsbS core STAS domains are coloured *blue*, while the N-terminal RsbR signalling domains are coloured *red*. Views down the three unique twofold symmetry axes are shown in panels **a**–**c**, with an icosahedral net for reference. The total radius of the stressosome is 150 Å

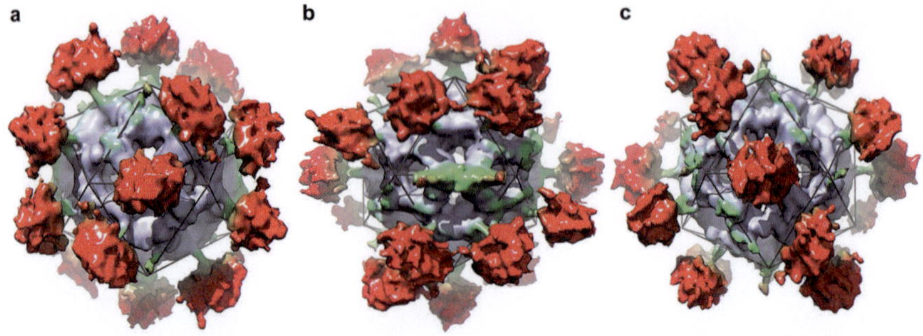

Fig. 1.12 Single particle reconstruction of the RsbR/RsbS/RsbT ternary stressosome complex. The final experimental EM-derived D_2 symmetry reconstruction of the RsbR/RsbS: RsbT ternary stressosome complex is shown as a surface contoured at 1.5σ. The RsbR and RsbS core STAS domains are coloured *blue*, the N-terminal RsbR signalling domains are coloured *red*, the additional density present when this reconstruction is compared to the RsbR/RsbS reconstruction is attributed to the presence of RsbT and is shown in *green*. Views down the three unique twofold symmetry axes are shown in panels **a–c**, with an icosahedral net for reference. The total radius of the stressosome is 150 Å as before

1.5.4 Pseudo-atomic Model of the B. subtilis *Stressosome*

With cryo-EM reconstructions available for the core stressosome, RsbR/RsbS binary complex, and RsbR/RsbS/RsbT ternary complex it was possible to dock the crystal structures of the individual stressosome components into the reconstructed density maps to produce a pseudo-atomic model of the stressosome complexes.

Because the turret-like peripheral projections from the RsbR/RsbS and RsbR/RsbS/RsbT reconstructions were absent from the $RsbR_{(146–274)}$/RsbS complex, these regions were attributed to the N-terminal domains of RsbR. The initial model was based on the structure of the *B. subtilis* N-RsbR protein (Murray et al. 2005; Marles-Wright et al. 2008). The crystallographic dimer fitted the density well; however, the model lacked the linker between the N-terminal globin domain and the C-terminal STAS domain; therefore, the density visible between the turrets and core stressosome was not occupied by this model. The subsequent structure of the *M. thermoacetica* RsbR homologue, *Mt*R, had fifteen additional C-terminal residues that formed four turns of an α-helix that extended from the bottom of the globin domain to the N-terminus of the STAS domain (Quin et al. 2012). Docking the *Mt*R model into the stressosome reconstructions accounted completely for the turret density and the neck region between the peripheral turrets and the inner core (Fig. 1.13).

A monomer of the *Mt*S structure was used as a model for the STAS domain and was fitted to the clear secondary structure elements seen in the stressosome core reconstruction (Fig. 1.14). With the position of RsbR determined by the position of the turrets in the RsbR/RsbS reconstructions, it was possible to assign the positions of the RsbR and RsbS STAS domains with reference to the turrets. The three recon-

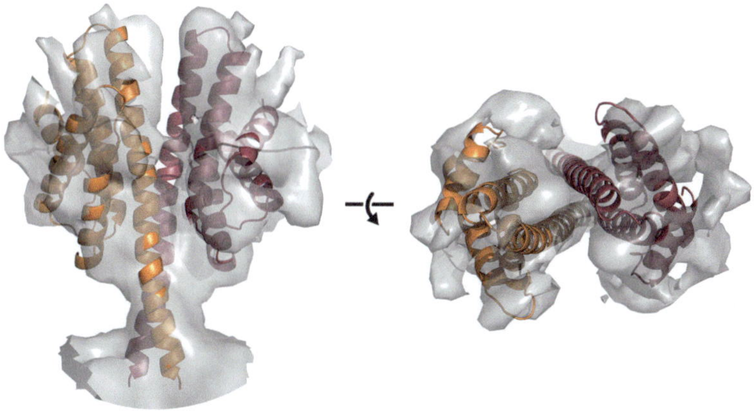

Fig. 1.13 Docking of MtR to the stressosome reconstruction envelope. Orthogonal views of the secondary structure fit of the X-ray crystal structure of MtR in the RsbR/RsbS stressosome reconstruction; the density map is contoured at 2σ. The model of MtR is shown as a secondary structure cartoon coloured *orange* and *red*

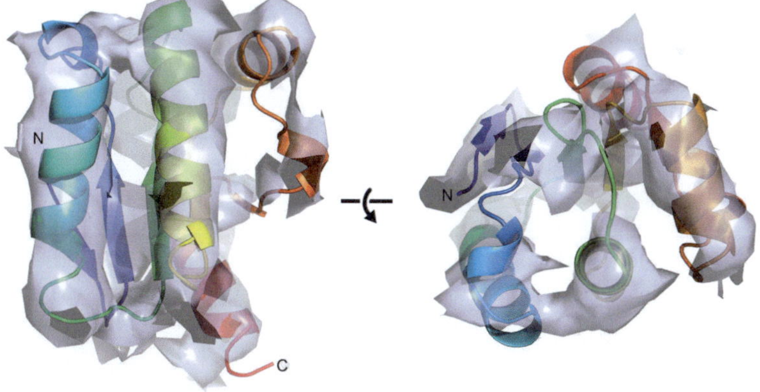

Fig. 1.14 Docking of MtS model to the stressosome core reconstruction envelope. Orthogonal views of the secondary structure fit of the X-ray crystal structure of MtS (PDBID: 3TZA) in the icosahedral stressosome core ($RsbR_{146–274}$/RsbS) reconstruction. The density map is contoured at 2.5σ to emphasise α-helices. The model of MtS is shown as a secondary structure cartoon and colour ramped from *blue* at the N-terminus, to *red* at the C-terminus

structions were aligned to the same coordinate frame and STAS domains below the turrets were assigned as RsbR and those with no turret as RsbS.

The RsbR/RsbS/RsbT ternary complex reconstruction showed clear additional density above the positions assigned as RsbS in the RsbR/RsbS reconstruction; this density was therefore attributed to the presence of the RsbT protein. The co-crystal structure of the SpoIIAA/SpoIIAB complex was used to guide fitting of a SpoIIAB-based homology model of the RsbT kinase in the RsbR/RsbS/RsbT reconstruction

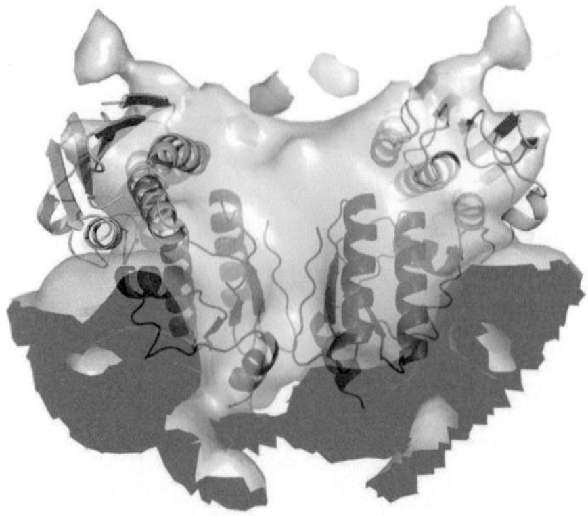

Fig. 1.15 Docking of RsbS and RsbT models to the ternary stressosome complex reconstruction envelope. Secondary structure fit of a homology model of RsbS produced from MtS (PDBID: 2VY9) and a homology model of of RsbT produced from the ***Bacillus stearothermophilus*** SpoIIAB structure (PDBID: 1TH8) in the RsbR/RsbS/RsbT ternary stressosome complex reconstruction. The density map is contoured at 2σ; the model of RsbT is shown as a secondary structure cartoon coloured *green*, while RsbS is shown in *red*. In this model RsbT is poised above the RsbS protein in a pre-phosphorylation state

(Fig. 1.15) (Masuda et al. 2004). This placed RsbT above RsbS and in proximity to RsbR to form a phosphorylation-competent complex in which RsbS was closest to the ATP-binding site in RsbT, in accordance with the faster rate of phosphorylation of RsbS than RsbR in minimal stressosomes *in vitro* (Chen et al. 2003).

Using the three reconstructions it was possible to produce a complete pseudo-atomic model of the stressosome to gain insight into the architecture of the complex and how its structure influences its signalling mechanism (Fig. 1.16). These insights are discussed in Sect. 1.6 below, with reference to the validating experimental evidence

1.5.5 Other Stressosomes

Since the discovery of the stressosome as an ordered macromolecular complex (Chen et al. 2003) a number of studies have presented negatively stained electron micrographs of stressosome complexes from *B. subtilis* and *M. thermoacetica* (Delumeau et al. 2006; Marles-Wright et al. 2008; Quin et al. 2012). The gross appearance of these complexes in two-dimensions are essentially the same, with a 90 Å core radius and turrets giving a total radius of 150 Å. To date the cryo-EM reconstructions of the *B. subtilis* stressosome are the only published examples for this class of protein complex (Marles-Wright et al. 2008). Further examples of reconstructions from other species will determine whether the structural paradigm seen for the *B. subtilis* stressosome holds in different bacteria and will help shed

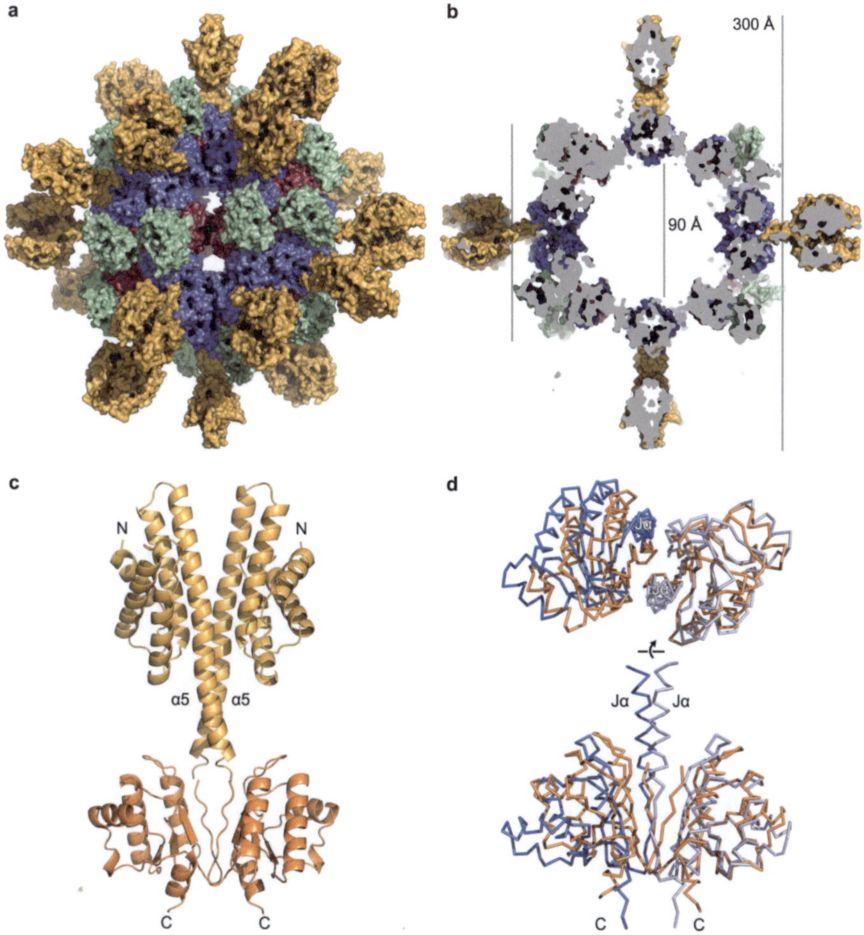

Fig. 1.16 Pseudo-atomic model of the RsbR/RsbS/RsbT stressosome ternary complex. (**a**) A pseudo-atomic model of the RsbR/RsbS/RsbT stressosome was produced using the fitted coordinates of homology models of each component of the complex. The N-terminus of RsbR was modelled using the MtR crystal structure (PDBID: 3TZA) and is shown in orange; the STAS domains of RsbR and RsbS were modelled using the structure of MtS (PDBID: 2VY9) and are shown in *blue* and *red* respectively; and RsbT was modelled using the structure of SpoIIAB (PDBID: 1TH8) and is shown in *green*. (**b**) Cross section of the RsbR/RsbS/RsbT pseudo-atomic model showing the empty central cavity and dimensions of the complex. The cavity has a diameter of 90 Å, while the external extent of the core of the stressosome is 180 Å. The RsbR signalling domains extend from the complex to give an overall diameter of 300 Å. (**c**) Cartoon model of the orientation of the two domains of RsbR/MtR from the pseudo-atomic stressosome model, showing the arrangement of the STAS domain dimer in the core. Signalling domain at top and STAS domain at the bottom. (**d**) Ribon models of a structural alignment of the STAS domain dimer from YtvA (*blue*) and RsbR in the stressome core. The two models show essentially the same monomer orientation in the dimer, with a small 10° shift in the relative position of the monomers around the twofold axis

further light onto the structure and mechanism of this widely distributed bacterial signalling complex.

1.6 Structure/Function Relationships in Stressosomes

The complex structure of the *B. subtilis* stressosome underlies its central role in integrating the diverse environmental signals that activate the σ^B pathway. The knowledge gained from the cryo-EM reconstructions and pseudo-atomic models permitted a number of insights into the relationship between the structure and function of the stressosome.

1.6.1 Activation

Inspection of the initial stressosome reconstructions highlights the potential mobility of the sensory turrets, given the fact that these regions appear less well defined than the core of the complex. These reconstructions were produced before the introduction of the gold-standard FSC resolution estimation methods, or the introduction of more sensitive detector technologies and modern algorithms that are driving the current resolution revolution in electron microscopy. Nonetheless, the structures may suggest a model of signal perception by the N-terminal domains and subsequent signal transduction through movement of the linking helixes that is transmitted to the C-terminal STAS domains, with subsequent activation of the RsbT kinase. The pseudo-atomic model produced from the stressosome reconstructions illustrates the relationships between the individual proteins that make up the complex (Fig. 1.16a, b). There is a large cavity at the centre of the stressosome; however, it is not thought that any proteins reside within this cavity and its presence is likely to be a secondary consequence of the architecture of the complex, where the STAS domains of the core are spaced to allow the recruitment of the RsbT kinase and to allow space for the RsbR N-terminal domains. Having a central cavity may allow structural rearrangements in the complex upon receipt of activating signals that would not be possible with a solid core.

The exact nature of the primary activating signals for the *B. subtilis* stressosome is not known. While each of the RsbR paralogues have a distinct N-terminal domain, only the activating signal for the YtvA blue-light receptor has been identified, though the details of how this signal modulates stressosome activation is still an area of active debate (Jurk et al. 2010; van der Steen and Hellingwerf 2015). Illumination of crystals of the LOV domain of YtvA with blue light leads to a shift in the Jα-helix region of the protein that links the N-terminal LOV domain to the C-terminal STAS domain (Möglich and Moffat 2007). This movement was suggested to propagate signal perception to changes in the relative orientation of the STAS domain (Möglich and Moffat 2007) and thus induce gross structural changes in the stressosome core to allow signal transduction to activate RsbT. Solution scattering studies of full length YtvA show minimal structural rearrangements upon illumination with blue-light, leading to the observation that this particular switch 'does not flip' (Jurk et al. 2010). Further work by this group, using NMR, showed that the surface accessibility of some residues in the STAS domain change in response to blue light and that these structural changes propagated by light absorption are not substantial in magnitude and thus may not be observed in low resolution SAXS studies. The subtle structural changes in YtvA may rather affect the network of protein-protein interactions with the partner proteins in the stressosome (Jurk et al. 2011). It is clear from this work on YtvA that the magnitude of changes in individual stressosome components upon activation is likely to be small on the scale of individual proteins, but it tells little about the changes in the structure, or thermodynamics of the whole system.

It is notable that the modelled organisation of the full length RsbR/MtR proteins in the stressosome is consistent with the dimer organisation of the full length structure of YtvA in solution (Figs. 1.5b and 1.16c). Furthermore, the STAS domain dimers in the stressosome core are consistent with the STAS domains seen in the YtvA structure, with only minor differences in their relative orientation (Fig. 1.16d). The differences seen in the position of the N-terminal portion of the STAS domain in relation to the Jα helix indicates a level of conformational flexibility in this region of the stressosome core that further reinforces the hypothesis that any signals perceived by the N-terminal RsbR/YtvA paralogues could be transmitted to the stressosome core to permit STAS domain phosphorylation by RsbT and its release from the complex.

It is also possible that the stressosome core provides a rigid scaffold that allows movement only in the N-terminal domains of the RsbR paralogues and YtvA. In this model, signalling by the stressosome would be a result of the differential positioning of the N-terminal domains in the activated/resting state and their influence on the probability of activating the kinase function of RsbT. The effect of mutations in RsbR seen in the work by Gaidenko and colleagues may be a consequence of constraints on the movement of the N-terminal domain of RsbR about the STAS domain scaffold in the stressosome core with resultant changes in the activation of RsbT, rather than direct interactions with RsbT itself (Gaidenko et al. 2011). The lack of significant conformational changes seen in YtvA low resolution solution studies (Jurk et al. 2010) could be explained in this model by a requirement for the core stressosome STAS scaffold to provide a rigid body to push against when activated.

In the context of a native stressosome with twenty peripheral turrets, small changes induced by external signals could be propagated and amplified allosterically to activate stressosome phosphorylation by RsbT and its subsequent release. Indeed, the σ^B response to environmental stimuli in live *B. subtilis* cells obeys a sigmoidal curve, indicating that co-operativity plays an important role in the signal transduction cascade that most likely stems from the stressosome (Marles-Wright et al. 2008).

Single cell fluorescence experiments and mathematical models of stressosome activation show that the response to different environmental stresses obeys the same kinetics and magnitude of response, indicating a key role for the stressosome in mediating and normalising the environmental stress response and highlighting the role of RsbR in the activation of RsbT (Locke et al. 2011; Liebal et al. 2013; Young et al. 2013). The hypothesis that structural changes in the sensory domains of the RsbR paralogues affect RsbT binding to the stressosome through direct interactions with the kinase, or through signalling via the STAS domains, has been tested both *in vitro* (Murray et al. 2005) and *in vivo* through the mutagenesis of both the N-terminal domain of RsbR and its linker helix (Gaidenko et al. 2011; Gaidenko et al. 2012). Mutagenesis of residues that line a surface groove in the dimer interface of the RsbR sensory domain abrogated the binding of RsbT to reconstituted stressosomes *in vitro*. However, the introduction of these mutants into a strain of *B. subtilis* lacking the other RsbR paralogues, failed to show corresponding changes to σ^B activation in vivo (Gaidenko et al. 2011). Strains bearing other mutations in the dimer interface showed increased basal activity of σ^B, of particular note was the effect seen in the glutamic acid 136 to lysine mutation, located in the linking helix that connects the two domains of RsbR. While a number of mutants displayed an enhanced basal σ^B activity, there was little effect on signalling outputs upon the imposition of stress. Mutagenesis of the linker helix in RsbR has shown that changes to residues in the helix-interface enhance basal stressosome output, whilst those on the surface of the linker helix diminish it (Gaidenko et al. 2012). The enhanced basal stressosome activity seen does not appear to diminish its capacity for responding to specific stresses, this may be a consequence of the presence of a pool of RsbT that is still bound to stressosomes that is not completely exhausted by the high basal activity, or an as yet unidentified stressosome signalling mechanism that is independent of the N-terminal RsbR domains and/or phosphorylation (Gaidenko and Price 2014). Taken together, these observations suggest that the perception of stresses by the stressosome and modulation of the output signal is more complex than can be explained by a mechanism where there is a direct link between stress perception by the N-terminal RsbR domains and transduction to the C-terminal STAS domains, with subsequent activation of the RsbT kinase.

The identification of putative heme-containing RsbR paralogues in *Vibrio vulnificus, Chromobacterium violaceum* and *Ruegeria species* (Pané-Farré et al. 2005) provides a potential means by which to directly probe stressosome activation, given the known haem-containing signalling systems and their responsiveness to diatomic gases (Zhang and Phillips 2003). Future structural analyses of stressosome complexes will hopefully cast light on the changes that take place upon signal perception and give insight into how this is propagated to activate RsbT.

1.6.2 Stressosome Signalling Response

While the exact nature of the activating signals for the stressosome-mediated branch of the σ^B pathway are not known, and the mechanism of signal transduction is still a matter of some debate, the nature and dynamics of this response have been the subject of a number of single-cell systems biology studies (Locke et al. 2011; Young et al. 2013). Experiments on population-level activation of the σ^B pathway showed that the energy stress branch of the *B. subtilis* general stress response obeys a simple hyperbolic activation model, while the stressosome mediated environmental stress branch displays a cooperative Hill response (Marles-Wright et al. 2008). The cooperative nature of the environmental stress response was attributed to the stressosome, which carries multiple copies of RsbT, and would be ideally placed to rapidly release a bolus of the kinase in response to stresses perceived by the RsbR sensory domains.

A number of elegant single-cell studies were performed by the Elowitz group to investigate the dynamics of the σ^B response in *B. subtilis* (Locke et al. 2011; Young et al. 2013). In their first study they show that the energy stress response displays stochastic frequency-modulated pulsing behaviour on a single-cell level (Locke et al. 2011). This behaviour is proposed to allow the bacteria to respond to unpredictable environments, while stochastic differences across populations present a bet-hedging strategy to mitigate the high cost of activating the σ^B regulon. In contrast to the frequency-modulated behaviour exhibited by the energy stress pathway, subsequent studies showed that the stressosome-mediated environmental stress response displayed amplitude-modulated pulsing (Young et al. 2013). The biological consequences of this response to stress are a rapid activation of the σ^B regulon in high-stress conditions and a change in the magnitude of response as levels of stress increase. The structure of the stressosome can be used to provide a molecular explanation for this response: through the release of a bolus of RsbT in response to environmental stresses perceived by the stressosome, the σ^B pathway can be activated rapidly to provide a timely response to high-stress conditions. The tuneable magnitude of the stressosome response is thought to be a consequence of the release of variable amounts of RsbT from the complex as a function of the magnitude of the perceived stress. The presence of multiple RsbR paralogues in *B. subtilis* and other species provide for the possibility for the modulation of the stressosome in response to different stresses and cellular contexts, where different paralogues may be present in the stressosome in different growth conditions (Pané-Farré et al. 2005).

1.6.3 Phosphorylation Dynamics of Stressosome

Activation of the stressosome is accompanied by the phosphorylation of RsbS and RsbR paralogues on conserved serine and threonine residues by the RsbT protein kinase (Kim et al. 2004a). The stressosome appears to be primed through phosphorylation of Thr171 in RsbR by RsbT in unstressed cells; in strains where this residue is mutated, the stressosome-mediated response is much diminished (Kim et al. 2004a). Phosphorylation of RsbS at Ser59, and at the equivalent position in RsbR, Thr205, follows the imposition of stress and the subsequent release of RsbT. The activity of the RsbX phosphatase against both RsbS-P and RsbR-P is required to reset the stressosome to a pre-stress state for the recruitment of RsbT back to the resting state (Chen et al. 2004). In high-stress conditions, RsbR is more likely to be phosphorylated on both Thr171 and Thr205 and this correlates with a diminished stress response (Eymann et al. 2011).

Mathematical models of the dephosphorylation of RsbS-P and RsbR-P by RsbX are not entirely consistent with all the published experimental data (Liebal et al. 2013). However, the mathematical models do highlight the dynamic behaviour of the RsbX phosphatase against stressosome components. The rapid dephosphorylation of RsbS-P is in marked contrast to the slow dephosphorylation of RsbR-P, and the models for these reactions explain the experimental observation that in high-stress conditions RsbR is not fully dephosphorylated (Eymann et al. 2011). The presence of two regulatory loops for stressosome signalling, the first mediated by the RsbX phosphatase to reset the signal, and the slow kinetics of dephosphorylation of Thr205-P on RsbR-P, allow for the rapid resetting of the stressosome in the first case, and modulation of the response to high stress in the latter case. These insights into stressosome phosphorylation dynamics correlate with the behaviours modelled by Elowitz and colleagues as described above in Sect. 1.6.2.

1.7 Evolution of the Stressosome Signalling Pathway

1.7.1 Structure and Conservation of the Stressosome Gene Cluster

In *B. subtilis*, the stressosome genes *rsbR*, *rsbS* and *rsbT* form a gene cluster that is co-transcribed with the σ^B structural gene and four additional genes encoding key regulators of σ^B activity as part of an eight-gene operon (Fig. 1.17). Transcription of this operon is initiated from two different promoters: a σ^A-dependent promoter upstream of of *rsbR* leads to synthesis of an mRNA comprising the entire operon (*rsbR-rsbS-rsbT-rsbU-rsbV-rsbW-sigB-rsbX*), whilst a σ^B-dependent promoter located between the *rsbU* and *rsbV* genes increases transcription of the downstream half of the operon (*rsbV-rsbW-sigB-rsbX*) following stress (Wise and Price 1995). Identical operons are present in many low-GC Gram-positive bacteria including several *Bacillus* species (e.g. *B. licheniformis*, *B. halodurans*, *B. pumilus*, *B. coagulas* and *Oceanobacillus iheyensis*) and close relatives like the *Listeria* (e.g. *L. monocytogenes* and *L. innocua*). However, stressosome-related genes are absent from the genomes of members of the *Bacillus cereus* group, although these organisms encode a σ^B protein, its principal regulators RsbV, RsbW and a PP2C-type phosphatase, RsbY (Fig. 1.17). Like RsbU, RsbY acts as a positive regulator of σ^B activity in *B. cereus* (Schaik et al. 2005). Genome comparison and experimental data suggest that in the absence of a stressosome, the phosphatase activity of RsbY is regulated by a hybrid sensor histidine kinase, termed RsbK, encoded in close proximity of the *sigB* operon. According to the current model RsbK controls the phosphorylation status of the N-terminal REC domain (CheY-homologous <u>rec</u>eiver domain) of RsbY and

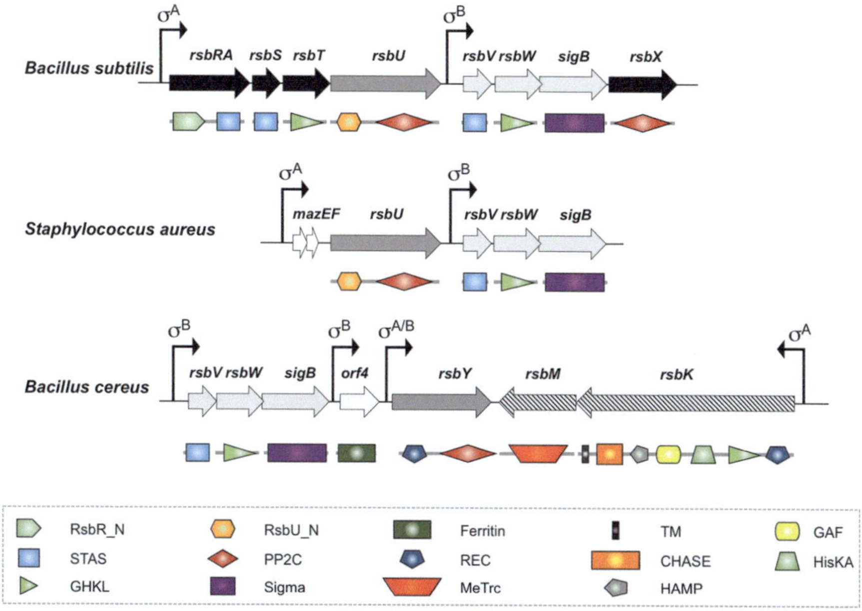

Fig. 1.17 σ^B-operon structure and domain organization of the encoded proteins. In *B. subtilis*, genes encoding the stressosome proteins RsbR, RsbS, and RsbT are organized with the σ^B structural gene and additional regulators of σ^B activity in an eight-gene operon. While the core of the σ^B-operon (*light grey*), is conserved in close relatives of *B. subtilis* including *B. cereus* and *S. aureus*, different solutions – including the RsbRSTX module (shaded *black*) or a hybrid kinase (*black stripes*) – merging at the control of a PP2C-type phosphatase (*dark grey*) have evolved to control the phosphorylation dependent interaction within the σ^B partner-switching module. Domain organization of the encoded proteins is shown below open reading frames (*arrows*). Transcription start sites and the initiating sigma factor are indicated. Abbreviations of domains detected with SMART (http://smart.embl-heidelberg.de) are: RsbR_N (RsbR N-terminus), STAS (sulphate transporter and anti-sigma factor antagonist), GHKL: (Gyrase, Hsp90, Histidine Kinase, MutL), RsbU_N (RsbU N-terminus), PP2C (Protein phosphatase 2C), Sigma (sigma factor), Ferritin (Ferritin family), REC (cheY-homologous receiver domain), MeTrc (Methyltransferase, chemotaxis proteins), TM (transmembrane domain), CHASE (extracellular sensory domain), HAMP (Histidine kinases, Adenylyl cyclases, Methyl binding proteins, Phosphatases domain), GAF (domain present in phytochromes and cGMP-specific phosphodiesterases), HisKA (His Kinase A (phosphoacceptor) domain)

thereby the activity of the C-terminally located PP2C domain (de Been et al. 2010, 2011). In addition, the activity of the hybrid kinase RsbY is modulated by methylation via RsbM, which is encoded down-stream of RsbY (Chen et al. 2012).

A similar situation is found in the genus *Staphylococcus*, which does not encode components of the stressosome, but has retained a truncated version of the *B. subtilis sigB* operon with the structure *rsbU-rsbV-rsbW-sigB* (Fig. 1.17). In contrast to *B. cereus* RsbY, the staphylococcal RsbU protein shows high similarity over the entire length of its sequence to the *B. subtilis* RsbU protein. In this context it is noteworthy that residues in the N-terminus of *B. subtilis* RsbU that are important for the interac-

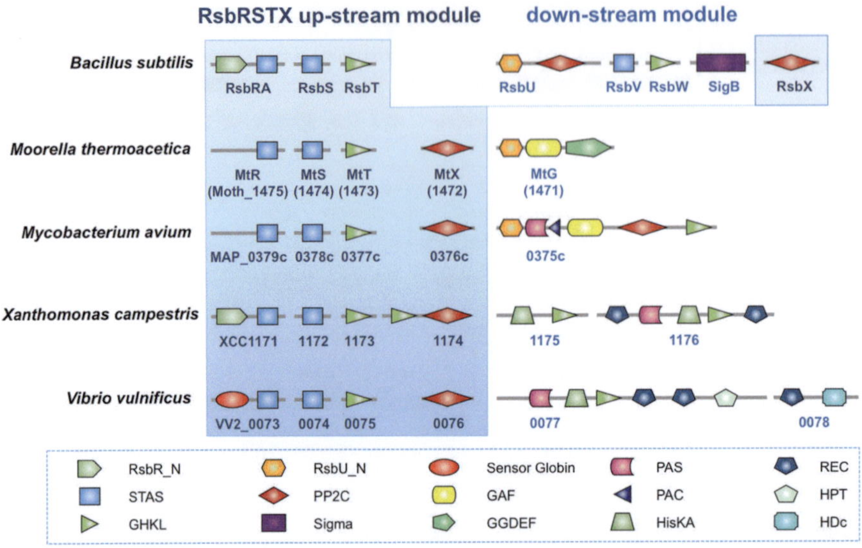

Fig. 1.18 Architecture of different output modules associated with the RsbRSTX stressosome module. Examples shown represent species from which experimental data on stressosome function or expression is available. Locus-tags and, if available, protein names are indicated. Abbreviations of domains detected with SMART (http://smart.embl-heidelberg.de) are: RsbR_N (RsbR N-terminus), STAS (sulphate transporter and anti-sigma factor antagonist), GHKL: (Gyrase, Hsp90, Histidine Kinase, MutL), RsbU_N (RsbU N-terminus), PP2C (Protein phosphatase 2C), Sigma (sigma factor), REC (cheY-homologous receiver domain), Sensor globin (heme coupled globin sensor), GGDEF (diguanylate cyclase), GAF (domain present in phytochromes and cGMP-specific phosphodiesterases), HisKA (His Kinase A (phosphoacceptor) domain), PAS (Per-Arnt-Sim protein), PAC (Motif C-terminal to PAS motifs), HTP (Histidine Phosphotransfer domain), HDc (Metal dependent phosphohydrolases with conserved 'HD' motif)

tion with RsbT are not conserved in *Staphylococcus aureus* RsbU (Hardwick et al. 2007). Furthermore, expression of *S. aureus* RsbU in *B. subtilis* leads to unrestricted σ^B activation and thus it remains an open question as to how RsbU, and hence σ^B activity, is controlled in the staphylococci (Pané-Farré et al. 2009).

Mirroring the situation observed in the *B. cereus* group and the staphylococci, many species that do not encode σ^B still encode the stressosome gene cluster (see Sect. 1.1.3). Contrary to *B. subtilis* and the Listeria, however, the gene encoding the negative feedback phosphatase RsbX is always located immediately down-stream of the stressosome genes, thus forming a highly conserved RsbRSTX module (Fig. 1.18). This module, according to the *B. subtilis* model, would provide the minimal set of proteins required to form a functional stressosome complex (RsbR, RsbS, RsbT) and a negative regulator (RsbX) to reset activated stressosomes. Indeed, the RsbRSTX module is almost always associated with genes encoding various proteins with signal transduction related functions including transcription factors, hybrid sensor kinases or proteins involved in the turnover of secondary messengers (Pané-Farré et al. 2005). Since these down-stream modules are usually separated by only a few base pairs from the stressosome module it is likely that they are functionally linked to the stressosome, thereby providing distinct regulatory outputs.

1.7.2 Evolution of the Stressosome

Almost nothing is known at present about the origin and evolution of the stressosome. The observation that the stressosome gene cluster can be found in a huge variety of distantly related bacteria and even some archaea (*Methanosarcina horonobensis*, *Halorhabdus utahensis* and *Candidatus Methanoregula boonei*) suggests that the stressosome genes were distributed across the microbial world by horizontal gene transfer. Indeed, the strong association of stressosome genes in a highly conserved RsbRSTX module may favour transfer between species and would thus provide an excellent building block that could be easily combined with various output modules to control different cellular processes. In agreement with this hypothesis, a study by Marri and colleagues suggested that the stressosome genes entered the genome of *Mycobacterium avium* subsp. *paratuberculosis* by horizontal gene transfer (Marri et al. 2006). The intriguing possibility that the stressosome genes are transferred between species as a unit raises the interesting question: how is this sensing complex integrated into the regulatory circuits of the recipient organism? In *B. subtilis*, signals perceived by the stressosome are conveyed to σ^B by the physical interaction of the serine threonine kinase RsbT with the N-terminal four-helical bundle of RsbU that acts as a recruitment domain for RsbT (Delumeau et al. 2004). This interaction does not require RsbT kinase activity (Kang et al. 1998). It is tempting to speculate that the *B. subtilis sigB*-operon could have arisen by an insertion of a group of genes including *rsbUVW* and *sigB* between the *rsbT* gene and the gene encoding an RsbX-like phosphatase. Domains showing sequence similarity to the N-terminal domain of *B. subtilis* RsbU can be identified in a number of down-stream proteins associated with the stressosome module in phylogenetically diverse species (Fig. 1.19); in such cases the answer as to how the stressosome communicates with the associated down-stream modules is evident. Consistent with that notion, a surface area identified to be crucial for the interaction of RsbU with RsbT in *B. subtilis* (Hardwick et al. 2007) shows a high degree of amino acid conservation in a comparison of RsbRSTX associated RsbU N-terminus-like domains (Fig. 1.19). Intriguingly, an isolated RsbU N-terminal domain located down-stream of an RsbRSTX gene cluster and up-stream of a GAF domain protein can be found in the actinobacterium *Modestobacter marinus* (Ponting and Aravind 1997). This observation could indicate the formation of a new stressosome signalling cascade, or alternatively, document the decay of the same.

In the majority of species, however, no RsbU N-terminus-like domain can be identified in the down-stream encoded regulators. If these proteins indeed receive regulatory input from the nearby-encoded stressosome, how is signal transfer realised? A possible answer may lay in the evolutionary origin of the RsbT protein, a member of the widely distributed GHKL protein family. GHKL domains are an

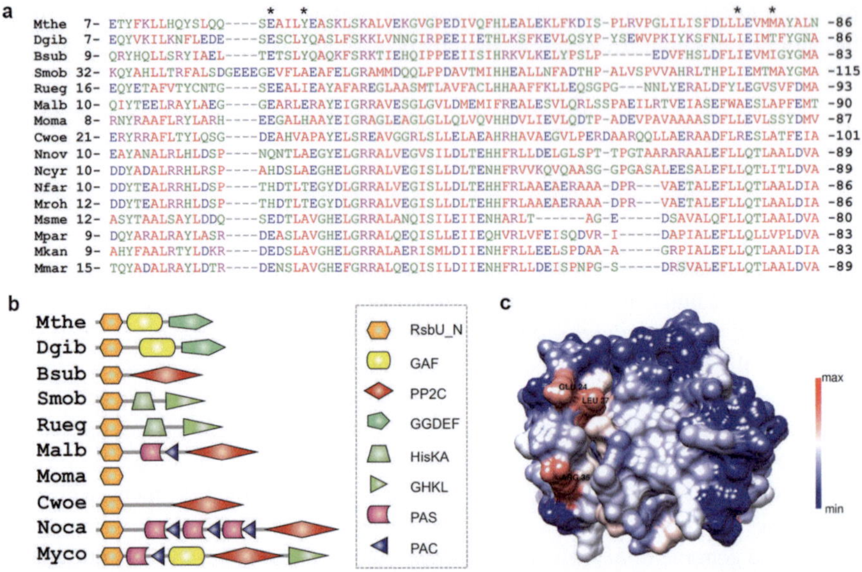

Fig. 1.19 The RsbU N-terminal domain. Sequences of proteins with an RsbU N-terminal domain were retrieved via the Pfam web-page (http://pfam.xfam.org). (**a**) Sequence alignment of RsbU N-terminal domains present in proteins encoded down-stream of RsbRSTX modules. For clarity, only the *B. subtilis* RsbU N-terminus was included as representative for species with an *rsbUVW*σ[B]*rsbX* down-stream gene cluster. Protein residues within the RsbU N-terminus identified as important for RsbT:RsbU interaction in *B. subtilis* (Hardwick et al. 2007) are labelled with an asterisk. (**b**) Domain architectures of proteins associated with RsbRSTX modules displaying an RsbU N-terminal domain. (**c**) Structure of the *B. subtilis* RsbU N-terminus showing the amino acid conservation across RsbU N-termini presented in the above alignment. Abbreviations are as follows: Bsub (*Bacillus subtilis*), Cwoe (*Conexibacter woesei*), Dgib (*Desulfotomaculum gibsoniae*), Malb (Methylomicrobium album), Mkan (*Mycobacterium kansasii*), Mmar (*Modestobacter marinus*), Mmar (*Mycobacterium marinum*), Mpar (*Mycobacterium paratuberculosis*), Mroh (*Mycobacterium rhodesiae*), Msme (*Mycobacterium smegmatis*), Mthe (*Moorella thermoacetica*), Myco (genus *Mycobacterium*), Nocar (genus *Nocardia*), Ncyr (*Nocardia cyriacigeorgica*), Nfar (*Nocardia farcinica*), Nnov (*Nocardia nova*), Rueg (*Ruegeria* sp.), Smob (*Symbiobacter mobilis*)

integral part of the sensor kinases of bacterial two component signal systems, in which the GHKL domain phosphorylates a conserved histidine residue within an N-terminally associated phospho-accepting histidine/dimerisation domain (HisKA). Assuming a conserved mode of communication between the stressosome and the putative downstream module via the RsbT orthologue, it is interesting to note that HisKA domains represent the least common denominator in the majority of downstream modules devoid of RsbU N-terminus-like domains.

Further complexity is added to the stressosome by the observation that the RsbX-like phosphatase of many species often has an additional N-terminal GHKL domain related to RsbT. The presence of an additional domain with potential kinase activity in these RsbRSTX modules underscores the importance of phosphorylation events during stressosome signalling. However, a recent study by Gaidenko and Price showed that stressosome dependent signalling is also supported in *B. subtilis* in the absence of a phosphoryltable RsbS protein, leading to the suggestion that RsbS phosphorylation is a recent evolutionary addition that overlays a primordial signalling mechanism to increase the sensitivity of the stressosome by increasing the rate of RsbT dissociation (Gaidenko and Price 2014). Presumably, there are some genetic or growth differences in the strains used in this study in comparison to previous studies that pointed towards the inviability of *rbsS* null and point mutants (Kang et al. 1996).

Whilst orthologues of RsbT and the STAS domains of RsbS and RsbR orthologues are highly conserved within the RsbRST module across species, significant variation occurs in the N-termini of RsbR orthologues. The STAS domain of RsbR orthologues has been found to be linked to non-heme globin as well as heme globin sensory domains and members of the PAS domain family, which occur in tandem within a single RsbR protein (e.g. in the genus *Aeromonas* (Pané-Farré et al. 2005; Sharma et al. 2011)). In some species, in addition to the promoter-proximal RsbR protein, further genes encoding RsbR paralogues can be identified. The occurrence of RsbR paralogues is particularly prominent in the Gram-positive bacteria. Usually, the N-termini of the RsbR paralogues appear to be related to the N-terminus of the RsbR protein encoded within the RsbRSTX modules, but a diversity of N-terminal domains may be seen across the RsbR paralogues of one organism (see Sect. 1.2.3).

The observation that RsbR proteins are equipped with different N-termini and some species even encode multiple RsbR paralogues provides further evidence for the notion that the RsbR N-terminus plays a crucial role in stressosome signal perception. However, experiments testing this hypothesis in *B. subtilis* do not unequivocally support this conclusion for the family of RsbR co-antagonists displaying non-heme globin N-terminal domains (Gaidenko et al. 2011, 2012) and the role of this domain in signalling remains unclear.

Finally, proteins similar to RsbR but not encoded within a stressosome gene cluster can be identified in several species of green non-sulphur bacteria. For instance, the photosynthetic bacterium *Chloroflexus aurantiacus* encodes 17 full-length RsbR paralogues some of which display multiple N-terminal PAS domains. Given the early evolutionary origin of green non-sulphur bacteria, it will be interesting to determine if these proteins also interact to form complex protein assemblies like the stressosome of *B. subtilis*, which could point to an evolutionary origin of the stressosome and would imply that stressosome-like protein assemblies may be a much more widely distributed paradigm than currently anticipated.

In summary, the available structural and sequence data indicate that stressosome dependent signalling may show substantial differences between species regarding signal perception, integration and transmission to the various output modules. Comparative genomics of stressosome proteins and down-stream regulators in combination with targeted experiments will be required to understand shared and specific mechanisms of signal transduction involving the stressosome.

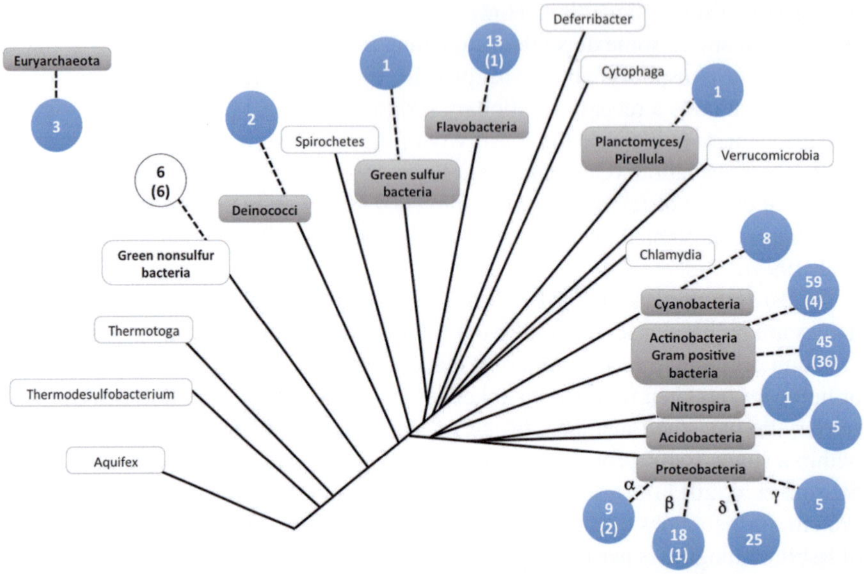

Fig. 1.20 Distribution of the RST stressosome module. To estimate the occurrence of RSTX modules within the microbial world, taxonomic information for the RsbR protein family (K17763) was retrieved from the KEGG webpage (http://www.kegg.jp). Phyla in which the genes for RsbR proteins are organized in a *rsbRSTX* gene cluster are highlighted in *grey*. Numbers in *blue circles* indicate the total number of species with RsbRSTX modules within one phylum. *Numbers* in *parentheses* indicate the number of species that in addition to the promoter-proximal RsbR in the RsbRSTX cluster also encode at least one RsbR paraloge. No RsbRSTX module was identified in the *green* nonsulfur bacteria, but multiple RsbR-like proteins were detected in six members (*white circle*) of this phylum

1.7.3 Stressosome Function in Other Organisms

While the *B. subtilis* stressosome and the σ^B pathway is well characterised, there is little data available on stressosome function in other species. The stressosome module can be identified in microorganisms that thrive in virtually every ecological niche and there seems to be no preference for a specific habitat or lifestyle. This includes bacteria living in soil, deep-sea and estuarine marine environments, as well as plant and animal symbionts and pathogens (Pané-Farré et al. 2005) (Fig. 1.20).

In the Gram-positive pathogen *L. monocytogenes* the stressosome controls σ^B-activity in response to both environmental and nutritional stress (Chaturongakul and Boor 2004; Chaturongakul and Boor 2006). Furthermore, by replacing the *B. subtilis rsbR* gene with its *L. monocytogenes* counterpart, the response to environmental and nutritional stress of the latter was transferred into the *B. subtilis* host. However, the presence of RsbR paralogues in *B. subtilis* abrogated the nutritional stress sensing properties of the *L. monocytogenes* RsbR protein (Martinez et al. 2010). These results provide evidence that strengthens the hypothesis that the RsbR proteins directly sense stress signals within the cell.

There is also evidence that the stressosome has been adapted to control secondary messenger levels, rather than the activity of a transcription factor. The *M. thermoacetica* orthologue RsbT, termed *Mt*T, inhibits the activity of a downstream-encoded GG(D/E)EF-type diguanylate cyclase (Schirmer and Jenal 2009), *Mt*G (Quin et al. 2012). Activation of the stressosome and hence inhibition of the *Mt*G diguanylate cyclase activity is thus assumed to control the cellular level of the second messenger cyclic di-GMP (c-di-GMP) in *M. thermoacetica*. C-di-GMP is one of the most common and important bacterial secondary messengers and has been shown to be involved in the regulation of many process often involving lifestyle changes that occur during development, virulence gene expression, biofilm formation, the cell cycle, or the switch between a sessile and a motile state (Römling et al. 2013; Ryan 2013). However, the *in vivo* role of the *M. thermoacetica* stressosome has not yet been investigated.

For the actinobacterium and opportunistic human pathogen *Mycobacterium marinum* transcription of stressosome genes was observed to be increased during cold stress, suggesting a role for the stressosome in cold adaptation in this species (Pettersson et al. 2013). For the γ-proteobacterium *Xanthomonas campestris*, a plant pathogen infecting cabbage, phosphorylation of the RsbR orthologue has been reported during transition from exponential growth to stationary phase (Musa et al. 2013). Since phosphorylation of RsbR correlates with stress activation, stressosome function may be required during stationary phase adaptation in *X. campestris*. The *M. marinum* downstream protein associated with the RsbRSTX cluster has a domain similar to the N-terminus of RsbU, and which is linked to a C-terminal PAS-PP2C-GAF-GHKL multi-domain polypeptide and may be involved in sigma-factor regulation via the PP2C domain. A functional prediction is, at present, not possible for the two potential *X. campestris* downstream proteins, which display a combination of REC, GHKL and HisKA domains, but do not possess an RsbU N-terminus-like domain (Fig. 1.18).

A stressosome gene cluster has also been identified in a number of *Vibrio* species including *V. vulnificus* and *V. mimicus*, two species that cause seafood associated gastroenteritis; *V. coralliilyticus*, a coral pathogen; and *V. orientalis*, a luminescent bacterium. Interestingly, whole genome sequencing has shown that the stressosome gene cluster is particularly well conserved in clinical *V. vulnificus* isolates and less frequently found in the genome of environmental strains (Morrison et al. 2012; Williams et al. 2014), suggesting that the stressosome may be advantageous during the establishment of an infection. The presence of an N-terminal heme-binding sensor globin domain in the *V. vulnificus* RsbR orthologue strongly suggests a function in the sensing of oxygen or other diatomic gases for the *V. vulnificus* stressosome. The stressosome genes are associated in *V. vulnificus* with two open reading frames encoding a hybrid histidine kinase and an HDc domain that likely functions in the degradation of c-di-GMP (Fig. 1.17). Hence, as in *M. thermoacetica*, control of c-di-GMP levels may represent the regulatory output of the *V. vulnificus* stressosome.

1.8 Conclusions

The stressosome has been the subject of extensive research into its role in the general stress response of the model Gram-positive bacterium *B. subtilis* since its discovery and isolation over 20 years ago. Our structural studies of the stressosome complex showed how it is able to sequester a pool of the RsbT activator kinase that can be released upon the receipt of stress signals by the sensory domains of RsbR and related proteins. The mechanism of activation and signal transduction by the stressosome remains to be determined and is a subject of great interest and debate within the field. With the explosion in the number of sequenced bacterial genomes seen in the last 10 years, the distribution of stressosome signalling systems across different phyla points to their utility. Furthermore, the diversity of downstream signalling modules seen in different species hints at a variety of different physiological roles for the stressosome. Future work will hopefully reveal the mechanism of stressosome signalling and shed light onto the nature of stressosome signalling in diverse bacterial species.

References

Aertsen A, Michiels CW (2004) Stress and how bacteria cope with death and survival. Crit Rev Microbiol 30:263–273. doi:10.1080/10408410490884757

Akbar S, Gaidenko TA, Kang CM et al (2001) New family of regulators in the environmental signaling pathway which activates the general stress transcription factor σ^B of Bacillus subtilis. J Bacteriol 183:1329–1338. doi:10.1128/JB.183.4.1329-1338.2001

Alper S, Duncan L, Losick R (1994) An adenosine nucleotide switch controlling the activity of a cell type-specific transcription factor in *B. subtilis*. Cell 77:195–205

Alper S, Dufour A, Garsin DA et al (1996) Role of adenosine nucleotides in the regulation of a stress-response transcription factor in Bacillus subtilis. J Mol Biol 260:165–177. doi:10.1006/jmbi.1996.0390

Aravind L, Koonin EV (2000) The STAS domain – a link between anion transporters and antisigma-factor antagonists. Curr Biol 10:R53–R55

Barford D, Das AK, Egloff MP (1998) The structure and mechanism of protein phosphatases: insights into catalysis and regulation. Annu Rev Biophys Biomol Struct 27:133–164. doi:10.1146/annurev.biophys.27.1.133

Benson AK, Haldenwang WG (1993) Bacillus subtilis σ^B is regulated by a binding protein (RsbW) that blocks its association with core RNA polymerase. Proc Natl Acad Sci U S A 90:2330–2334

Beuron F, Maurizi MR, Belnap DM et al (1998) At sixes and sevens: characterization of the symmetry mismatch of the ClpAP chaperone-assisted protease. J Struct Biol 123:248–259. doi:10.1006/jsbi.1998.4039

Bhate MP, Molnar KS, Goulian M, DeGrado WF (2015) Signal transduction in histidine kinases: insights from new structures. Structure 23:981–994. doi:10.1016/j.str.2015.04.002

Brody MS, Vijay K, Price CW (2001) Catalytic function of an alpha/beta hydrolase is required for energy stress activation of the σ^B transcription factor in Bacillus subtilis. J Bacteriol 183:6422–6428. doi:10.1128/JB.183.21.6422-6428.2001

Campbell EA, Masuda S, Sun JL et al (2002) Crystal structure of the *Bacillus stearothermophilus* anti-sigma factor SpoIIAB with the sporulation sigma factor sigmaF. Cell 108:795–807. doi:10.1016/S0092-8674(02)00662-1

Capra EJ, Laub MT (2012) Evolution of two-component signal transduction systems. Annu Rev Microbiol 66:325–347. doi:10.1146/annurev-micro-092611-150039

Chaturongakul S, Boor KJ (2004) RsbT and RsbV contribute to σ^B-dependent survival under environmental, energy, and intracellular stress conditions in *Listeria monocytogenes*. Appl Environ Microbiol 70:5349–5356. doi:10.1128/AEM.70.9.5349-5356.2004

Chaturongakul S, Boor KJ (2006) σ^B activation under environmental and energy stress conditions in *Listeria monocytogenes*. Appl Environ Microbiol 72:5197–5203. doi:10.1128/AEM.03058-05

Chen C-CCC, Lewis RJRJ, Harris R et al (2003) A supramolecular complex in the environmental stress signalling pathway of Bacillus subtilis. Mol Microbiol 49:1657–1669. doi:10.1046/j.1365-2958.2003.03663.x

Chen C-C, Yudkin MD, Delumeau O (2004) Phosphorylation and RsbX-dependent dephosphorylation of RsbR in the RsbR-RsbS complex of Bacillus subtilis. J Bacteriol 186:6830–6836. doi:10.1128/JB.186.20.6830-6836.2004

Chen LC, Chen JC, Shu JC et al (2012) Interplay of RsbM and RsbK controls the σB activity of Bacillus cereus. Environ Microbiol 14:2788–2799. doi:10.1111/j.1462-2920.2012.02788.x

Das AK, Helps NR, Cohen PT, Barford D (1996) Crystal structure of the protein serine/threonine phosphatase 2C at 2.0 A resolution. EMBO J 15:6798–6809

de Been M, Tempelaars MH, van Schaik W et al (2010) A novel hybrid kinase is essential for regulating the σ^B-mediated stress response of Bacillus cereus. Environ Microbiol 12:730–745. doi:10.1111/j.1462-2920.2009.02116.x

de Been M, Francke C, Siezen RJ, Abee T (2011) Novel σ^B regulation modules of Gram-positive bacteria involve the use of complex hybrid histidine kinases. Microbiology 157:3–12. doi:10.1099/mic.0.045740-0

Delumeau O, Lewis RJ, Yudkin MD (2002) Protein-protein interactions that regulate the energy stress activation of σ^B in Bacillus subtilis. J Bacteriol 184:5583–5589. doi:10.1128/JB.184.20.5583-5589.2002

Delumeau O, Dutta S, Brigulla M et al (2004) Functional and structural characterization of RsbU, a stress signaling protein phosphatase 2C. J Biol Chem 279:40927–40937. doi:10.1074/jbc.M405464200

Delumeau O, Chen C-C, Murray JW et al (2006) High-molecular-weight complexes of RsbR and paralogues in the environmental signaling pathway of Bacillus subtilis. J Bacteriol 188:7885–7892. doi:10.1128/JB.00892-06

Dube P, Tavares P, Lurz R, van Heel M (1993) The portal protein of bacteriophage SPP1: a DNA pump with 13-fold symmetry. EMBO J 12:1303–1309

Dufour A, Voelker U, Voelker A, Haldenwang WG (1996) Relative levels and fractionation properties of Bacillus subtilis σ^B and its regulators during balanced growth and stress. J Bacteriol 178:3701–3709

Dutta R, Inouye M (2000) GHKL, an emergent ATPase/kinase superfamily. Trends Biochem Sci 25:24–28

Eymann C, Schulz S, Gronau K et al (2011) In vivo phosphorylation patterns of key stressosome proteins define a second feedback loop that limits activation of Bacillus subtilis σ(B). Mol Microbiol 80:798–810. doi:10.1111/j.1365-2958.2011.07609.x

Gaidenko TA, Price CW (2014) Genetic evidence for a phosphorylation-independent signal transduction mechanism within the Bacillus subtilis stressosome. PLoS One 9:e90741. doi:10.1371/journal.pone.0090741

Gaidenko TA, Yang X, Lee YM, Price CW (1999) Threonine phosphorylation of modulator protein RsbR governs its ability to regulate a serine kinase in the environmental stress signaling pathway of Bacillus subtilis. J Mol Biol 288:29–39. doi:10.1006/jmbi.1999.2665

Gaidenko TA, Bie X, Baldwin EP, Price CW (2011) Substitutions in the presumed sensing domain of the Bacillus subtilis stressosome affect its basal output but not response to environmental signals. J Bacteriol 193:3588–3597. doi:10.1128/JB.00060-11

Gaidenko TA, Bie X, Baldwin EP, Price CW (2012) Two surfaces of a conserved interdomain linker differentially affect output from the RST sensing module of the Bacillus subtilis stressosome. J Bacteriol 194:3913–3921. doi:10.1128/JB.00583-12

Gao R, Stock AM (2009) Biological insights from structures of two-component proteins. Annu Rev Microbiol 63:133–154. doi:10.1146/annurev.micro.091208.073214

Gouet P, Robert X, Courcelle E (2003) ESPript/ENDscript: extracting and rendering sequence and 3D information from atomic structures of proteins. Nucleic Acids Res 31:3320–3323. doi:10.1093/nar/gkg556

Haldenwang WG, Losick R (1979) A modified RNA polymerase transcribes a cloned gene under sporulation control in Bacillus subtilis. Nature 282:256–260

Hardwick SW, Pané-Farré J, Delumeau O et al (2007) Structural and functional characterization of partner switching regulating the environmental stress response in Bacillus subtilis. J Biol Chem 282:11562–11572. doi:10.1074/jbc.M609733200

Hecker M, Völker U (2001) General stress response of Bacillus subtilis and other bacteria. Adv Microb Physiol 44:35–91. doi:10.1016/S0065-2911(01)44011-2

Hecker M, Pané-Farré J, Völker U (2007) SigB-dependent general stress response in Bacillus subtilis and related gram-positive bacteria. Annu Rev Microbiol 61:215–236. doi:10.1146/annurev.micro.61.080706.093445

Herrou J, Crosson S (2011) Function, structure and mechanism of bacterial photosensory LOV proteins. Nat Rev Microbiol 9:713–723. doi:10.1038/nrmicro2622

Hou S, Larsen RW, Boudko D et al (2000) Myoglobin-like aerotaxis transducers in Archaea and Bacteria. Nature 403:540–544. doi:10.1038/35000570

Jurk M, Dorn M, Kikhney A et al (2010) The switch that does not flip: the blue-light receptor YtvA from Bacillus subtilis adopts an elongated dimer conformation independent of the activation state as revealed by a combined AUC and SAXS study. J Mol Biol 403:78–87. doi:10.1016/j.jmb.2010.08.036

Jurk M, Dorn M, Schmieder P (2011) Blue flickers of hope: secondary structure, dynamics, and putative dimerization interface of the blue-light receptor YtvA from Bacillus subtilis. Biochemistry 50:8163–8171. doi:10.1021/bi200782j

Jurk M, Schramm P, Schmieder P (2013) The blue-light receptor YtvA from Bacillus subtilis is permanently incorporated into the stressosome independent of the illumination state. Biochem Biophys Res Commun 432:499–503. doi:10.1016/j.bbrc.2013.02.025

Kaneko T, Tanaka N, Kumasaka T (2005) Crystal structures of RsbQ, a stress-response regulator in Bacillus subtilis. Protein Sci 14:558–565. doi:10.1110/ps.041170005

Kang CM, Brody MS, Akbar S et al (1996) Homologous pairs of regulatory proteins control activity of Bacillus subtilis transcription factor σ^B in response to environmental stress. J Bacteriol 178:3846–3853

Kang CM, Vijay K, Price CW (1998) Serine kinase activity of a Bacillus subtilis switch protein is required to transduce environmental stress signals but not to activate its target PP2C phosphatase. Mol Microbiol 30:189–196

Kim T-J, Gaidenko TA, Price CW (2004a) In vivo phosphorylation of partner switching regulators correlates with stress transmission in the environmental signaling pathway of Bacillus subtilis. J Bacteriol 186:6124–6132. doi:10.1128/JB.186.18.6124-6132.2004

Kim T-J, Gaidenko TA, Price CW (2004b) A multicomponent protein complex mediates environmental stress signaling in Bacillus subtilis. J Mol Biol 341:135–150. doi:10.1016/j.jmb.2004.05.043

Kitanishi K, Kobayashi K, Uchida T et al (2011) Identification and functional and spectral characterization of a globin-coupled histidine kinase from Anaeromyxobacter sp. Fw109-5. J Biol Chem 286:35522–35534. doi:10.1074/jbc.M111.274811

Kovacs H, Comfort D, Lord M et al (1998) Solution structure of SpoIIAA, a phosphorylatable component of the system that regulates transcription factor sigmaF of Bacillus subtilis. Proc Natl Acad Sci U S A 95:5067–5071

Krell T, Lacal J, Busch A et al (2010) Bacterial sensor kinases: diversity in the recognition of environmental signals. Annu Rev Microbiol 64:539–559. doi:10.1146/annurev.micro.112408.134054

Krissinel E, Henrick K (2007) Inference of macromolecular assemblies from crystalline state. J Mol Biol 372:774–797. doi:10.1016/j.jmb.2007.05.022

Kumar A, Lomize A, Jin KK et al (2010) Open and closed conformations of two SpoIIAA-like proteins (YP_749275.1 and YP_001095227.1) provide insights into membrane association and ligand binding. Acta Crystallogr Sect F Struct Biol Cryst Commun 66:1245–1253. doi:10.1107/S1744309109042481

Liebal UW, Millat T, Marles-wright J et al (2013) Simulations of stressosome activation emphasize allosteric interactions between RsbR and RsbT. BMC Syst Biol 7:3. doi:10.1186/1752-0509-7-3

Locke JCW, Young JW, Fontes M et al (2011) Stochastic pulse regulation in bacterial stress response. Science 334:366–369. doi:10.1126/science.1208144

Lord M, Magnin T, Yudkin MD (1996) Protein conformational change and nucleotide binding involved in regulation of sigmaF in Bacillus subtilis. J Bacteriol 178:6730–6735

Losi A, Polverini E, Quest B, Gärtner W (2002) First evidence for phototropin-related blue-light receptors in prokaryotes. Biophys J 82:2627–2634. doi:10.1016/S0006-3495(02)75604-X

Losi A, Quest B, Gärtner W (2003) Listening to the blue: the time-resolved thermodynamics of the bacterial blue-light receptor YtvA and its isolated LOV domain. Photochem Photobiol Sci 2:759–766

Lowe EC, Baslé A, Czjzek M et al (2012) A scissor blade-like closing mechanism implicated in transmembrane signaling in a Bacteroides hybrid two-component system. Proc Natl Acad Sci U S A 109:1–6. doi:10.1073/pnas.1200479109

Marles-Wright J, Grant T, Delumeau O et al (2008) Molecular architecture of the "stressosome," a signal integration and transduction hub. Science 322:92–96. doi:10.1126/science.1159572

Marri PR, Bannantine JP, Paustian ML, Golding GB (2006) Lateral gene transfer in Mycobacterium avium subspecies paratuberculosis. Can J Microbiol 52:560–569. doi:10.1139/w06-001

Martinez L, Reeves A, Haldenwang W (2010) Stressosomes formed in Bacillus subtilis from the RsbR protein of *Listeria monocytogenes* allow **σ**B activation following exposure to either physical or nutritional stress. J Bacteriol 192:6279–6286. doi:10.1128/JB.00467-10

Masuda S, Murakami KS, Wang S et al (2004) Crystal structures of the ADP and ATP bound forms of the Bacillus anti-sigma factor SpoIIAB in complex with the anti-anti-sigma SpoIIAA. J Mol Biol 340:941–956. doi:10.1016/j.jmb.2004.05.040

Mitchell JG, Kogure K (2006) Bacterial motility: links to the environment and a driving force for microbial physics. FEMS Microbiol Ecol 55:3–16

Möglich A, Moffat K (2007) Structural basis for light-dependent signaling in the dimeric LOV domain of the photosensor YtvA. J Mol Biol 373:112–126. doi:10.1016/j.jmb.2007.07.039

Moran CP, Lang N, Banner CD et al (1981) Promoter for a developmentally regulated gene in Bacillus subtilis. Cell 25:783–791

Morrison SS, Williams T, Cain A et al (2012) Pyrosequencing-based comparative genome analysis of Vibrio vulnificus environmental isolates. PLoS One 7:e37553. doi:10.1371/journal.pone.0037553

Murray JW, Delumeau O, Lewis RJ (2005) Structure of a nonheme globin in environmental stress signaling. Proc Natl Acad Sci U S A 102:17320–17325. doi:10.1073/pnas.0506599102

Musa YR, Bäsell K, Schatschneider S, Vorhölter FJ, Becher D, Niehaus K (2013) Dynamic protein phosphorylation during the growth of Xanthomonas campestris pv. campestris B100 revealed by a gel-based proteomics approach. J Biotechnol 167(2):111–122. doi:10.1016/j.jbiotec.2013.06.009. Epub 2013 Jun 20

Nadezhdin EV, Brody MS, Price CW (2011) An α/β hydrolase and associated Per-ARNT-Sim domain comprise a bipartite sensing module coupled with diverse output domains. PLoS One 6:e25418. doi:10.1371/journal.pone.0025418
Ondrusch N, Kreft J (2011) Blue and red light modulates SigB-dependent gene transcription, swimming motility and invasiveness in *Listeria monocytogenes*. PLoS One 6:e16151. doi:10.1371/journal.pone.0016151
Österberg S, del Peso-Santos T, Shingler V (2011) Regulation of alternative sigma factor use. Annu Rev Microbiol 65:37–55. doi:10.1146/annurev.micro.112408.134219
Paget MS (2015) Bacterial sigma factors and anti-sigma factors: structure, function and distribution. Biomolecules 5:1245–1265. doi:10.3390/biom5031245
Pané-Farré J, Lewis RJ, Stülke J et al (2005) The RsbRST stress module in bacteria: a signalling system that may interact with different output modules. J Mol Microbiol Biotechnol 9:65–76. doi:10.1159/000088837
Pané-Farré J, Jonas B, Hardwick SW et al (2009) Role of RsbU in controlling SigB activity in *Staphylococcus aureus* following alkaline stress. J Bacteriol 191:2561–2573. doi:10.1128/JB.01514-08
Petersohn A, Brigulla M, Haas S et al (2001) Global analysis of the general stress response of Bacillus subtilis. J Bacteriol 183:5617–5631. doi:10.1128/JB.183.19.5617-5631.2001
Pettersson BMF, Nitharwal RG, Das S et al (2013) Identification and expression of stressosomal proteins in *Mycobacterium marinum* under various growth and stress conditions. FEMS Microbiol Lett 342:98–105. doi:10.1111/1574-6968.12118
Ponting CP, Aravind L (1997) PAS: a multifunctional domain family comes to light. Curr Biol 7:R674–R677
Price CW, Fawcett P, Cérémonie H et al (2001) Genome-wide analysis of the general stress response in Bacillus subtilis. Mol Microbiol 41:757–774
Quin MB, Berrisford JM, Newman JA et al (2012) The bacterial stressosome: a modular system that has been adapted to control secondary messenger signaling. Structure 20:350–363. doi:10.1016/j.str.2012.01.003
Reeves A, Martinez L, Haldenwang W (2010) Expression of, and in vivo stressosome formation by, single members of the RsbR protein family in Bacillus subtilis. Microbiology 156:990–998. doi:10.1099/mic.0.036095-0
Römling U, Galperin MY, Gomelsky M (2013) Cyclic di-GMP: the first 25 years of a universal bacterial second messenger. Microbiol Mol Biol Rev 77:1–52. doi:10.1128/MMBR.00043-12
Ryan RP (2013) Cyclic di-GMP signalling and the regulation of bacterial virulence. Microbiology 159:1286–1297. doi:10.1099/mic.0.068189-0
Schaik Van W, Tempelaars MHMH, Zwietering MHMH et al (2005) Analysis of the role of RsbV, RsbW, and RsbY in regulating σ^B activity in Bacillus cereus. J Bacteriol 187:5846–5851. doi:10.1128/JB.187.16.5846
Schirmer T, Jenal U (2009) Structural and mechanistic determinants of c-di-GMP signalling. Nat Rev Microbiol 7:724–735. doi:10.1038/nrmicro2203
Schmalisch M, Langbein I, Stülke J (2002) The general stress protein Ctc of Bacillus subtilis is a ribosomal protein. J Mol Microbiol Biotechnol 4:495–501
Seavers PR, Lewis RJ, Brannigan JA et al (2001) Structure of the Bacillus cell fate determinant SpoIIAA in phosphorylated and unphosphorylated forms. Structure 9:605–614
Sharma AK, Rigby AC, Alper SL (2011) STAS domain structure and function. Cell Physiol Biochem 28:407–422. doi:10.1159/000335104
Stock AM, Robinson VL, Goudreau PN (2000) Two-component signal transduction. Annu Rev Biochem 69:183–215. doi:10.1146/annurev.biochem.69.1.183
Stranzl GR, Santelli E, Bankston LA et al (2011) Structural insights into inhibition of Bacillus anthracis sporulation by a novel class of non-heme globin sensor domains. J Biol Chem 286:8448–8458. doi:10.1074/jbc.M110.207126
Teh AH, Makino M, Hoshino T et al (2015) Structure of the RsbX phosphatase involved in the general stress response of Bacillus subtilis. Acta Crystallogr D Biol Crystallogr 71:1392–1399. doi:10.1107/S1399004715007166

Truitt CL, Weaver EA, Haldenwang WG (1988) Effects on growth and sporulation of inactivation of a Bacillus subtilis gene (ctc) transcribed in vitro by minor vegetative cell RNA polymerases (E-sigma 37, E-sigma 32). Mol Gen Genet 212:166–171

van der Steen JB, Hellingwerf KJ (2015) Activation of the general stress response of Bacillus subtilis by visible light. Photochem Photobiol 91:1032–1045. doi:10.1111/php.12499

van der Steen JB, Avila-Pérez M, Knippert D et al (2012) Differentiation of function among the RsbR paralogs in the general stress response of Bacillus subtilis with regard to light perception. J Bacteriol 194:1708–1716. doi:10.1128/JB.06705-11

Vijay K, Brody MS, Fredlund E, Price CW (2000) A PP2C phosphatase containing a PAS domain is required to convey signals of energy stress to the σ^B transcription factor of Bacillus subtilis. Mol Microbiol 35:180–188

Voelker U, Voelker A, Haldenwang WG (1996) Reactivation of the Bacillus subtilis anti-sigma B antagonist, RsbV, by stress- or starvation-induced phosphatase activities. J Bacteriol 178:5456–5463

Wang E, Bauer MC, Rogstam A et al (2008) Structure and functional properties of the Bacillus subtilis transcriptional repressor Rex. Mol Microbiol 69:466–478. doi:10.1111/j.1365-2958.2008.06295.x

Williams TC, Blackman ER, Morrison SS et al (2014) Transcriptome sequencing reveals the virulence and environmental genetic programs of Vibrio vulnificus exposed to host and estuarine conditions. PLoS One 9:e114376. doi:10.1371/journal.pone.0114376

Wise AA, Price CW (1995) Four additional genes in the sigB operon of Bacillus subtilis that control activity of the general stress factor σ^B in response to environmental signals. J Bacteriol 177:123–133

Yang X, Kang CM, Brody MS, Price CW (1996) Opposing pairs of serine protein kinases and phosphatases transmit signals of environmental stress to activate a bacterial transcription factor. Genes Dev 10:2265–2275

Young JW, Locke JCW, Elowitz MB (2013) Rate of environmental change determines stress response specificity. Proc Natl Acad Sci U S A 110:4140–4145. doi:10.1073/pnas.1213060110

Zhang W, Phillips GN (2003) Structure of the oxygen sensor in Bacillus subtilis: signal transduction of chemotaxis by control of symmetry. Structure 11:1097–1110

Chapter 2
The Canonical Inflammasome: A Macromolecular Complex Driving Inflammation

Tom P. Monie

Abstract The inflammasome is a multi-molecular platform crucial to the induction of an inflammatory response to cellular danger. Recognition in the cytoplasm of endogenously and exogenously derived ligands initiates conformational change in sensor proteins, such as NLRP3, that permits the subsequent rapid recruitment of adaptor proteins, like ASC, and the resulting assembly of a large-scale inflammatory signalling platform. The assembly process is driven by sensor-sensor interactions as well as sensor-adaptor and adaptor-adaptor interactions. The resulting complex, which can reach diameters of around 1 micron, has a variable composition and stoichiometry. The inflammasome complex functions as a platform for the proximity induced activation of effector caspases, such as caspase-1 and caspase-8. This ultimately leads to the processing of the inflammatory cytokines pro-IL1β and pro-IL18 into their active forms, along with the cleavage of Gasdermin D, a key activator of cell death via pyroptosis.

Keywords Innate immunity • Death domain • Inflammation • Inflammasome • NLR • AIM2 • ASC • Caspase • Interleukin • IL-1beta • Cell death • Pyroptosis • Supramolecular-organising centre

---

T.P. Monie (✉)
Elsie Widdowson Laboratory, Medical Research Council, Cambridge, UK

Institute of Continuing Education, University of Cambridge, Cambridge, UK
e-mail: tpm22@cam.ac.uk

J.R. Harris, J. Marles-Wright (eds.), *Macromolecular Protein Complexes*, Subcellular Biochemistry 83, DOI 10.1007/978-3-319-46503-6_2

2.1 Introduction

The inflammasome is a large supramolecular signalling complex that forms in the cell cytoplasm in response to a wide range of exogenous and endogenous danger signals. These can include bacterial and viral proteins, toxins, ATP, uric acid, mitochondrial damage, potassium efflux and crystalline material.

The primary function of the inflammasome is to activate caspase-1, which leads to the processing of pro-interleukin (IL)-1β and pro-IL-18 to their mature forms (Fig. 2.1) (Schroder and Tschopp 2010; Davis et al. 2011; Lamkanfi and Dixit 2014; Bryant et al. 2015). The subsequent secretion of these cytokines from the cell leads to the induction of a pro-inflammatory immune response. Caspase-1 activation can also result in cleavage of the protein Gasdermin D and the subsequent induction of cell death via a process known as pyroptosis (Kayagaki et al. 2015; Shi et al. 2015; He et al. 2015).

The composition and stoichiometry of the inflammasome is variable and influenced by the nature of the activating stimulus. For example, bacterial flagellin leads to inflammasome formation via activation of NAIP (neuronal apoptosis inhibitory protein); uric acid promotes NLRP3 (nucleotide-binding, leucine-rich repeat containing receptor (NLR) containing a pyrin 3)-dependent inflammasome formation; and AIM2 (absent in melanoma 2) drives inflammasome formation in response to dsDNA. However, broadly speaking, the inflammasome consists of a receptor or sensor that recognises and/or responds to a ligand, an adaptor molecule that helps drive complex formation and an effector caspase recruited via interactions with the adaptor protein. Upon activation, the effector caspase drives protein processing to mediate an effective cellular response.

Within the literature, inflammasomes have been classified in a variety of different ways. Here I will focus on the classical, or canonical, inflammasomes. These represent the molecular structures upon which most of our understanding of inflammasome biology, function and assembly is based. Specifically, this will include inflammasomes formed by the activation of NLRP1, NLRP3, the NAIP proteins, and AIM2. As a note of clarity, the NAIP-driven inflammasomes have historically been referred to as the NLRC4 (NLR containing a CARD (caspase recruitment domain) 4)-inflammasome, as it was originally believed that NLRC4 was the protein responsible for detecting the activating bacterial stimuli in these complexes. It is now clear that this role is in fact provided by the NAIP family of proteins.

In this chapter I will focus on four separate areas relevant to the assembly of a functional inflammasome complex. First I will describe the major protein constituents of the inflammasome. This will be followed by details of the molecular processes involved in initiating inflammasome assembly. I will then discuss the nature of the protein interactions required for actual assembly of an inflammasome, and finish with consideration of how inflammasome assembly and function can be regulated at the molecular level.

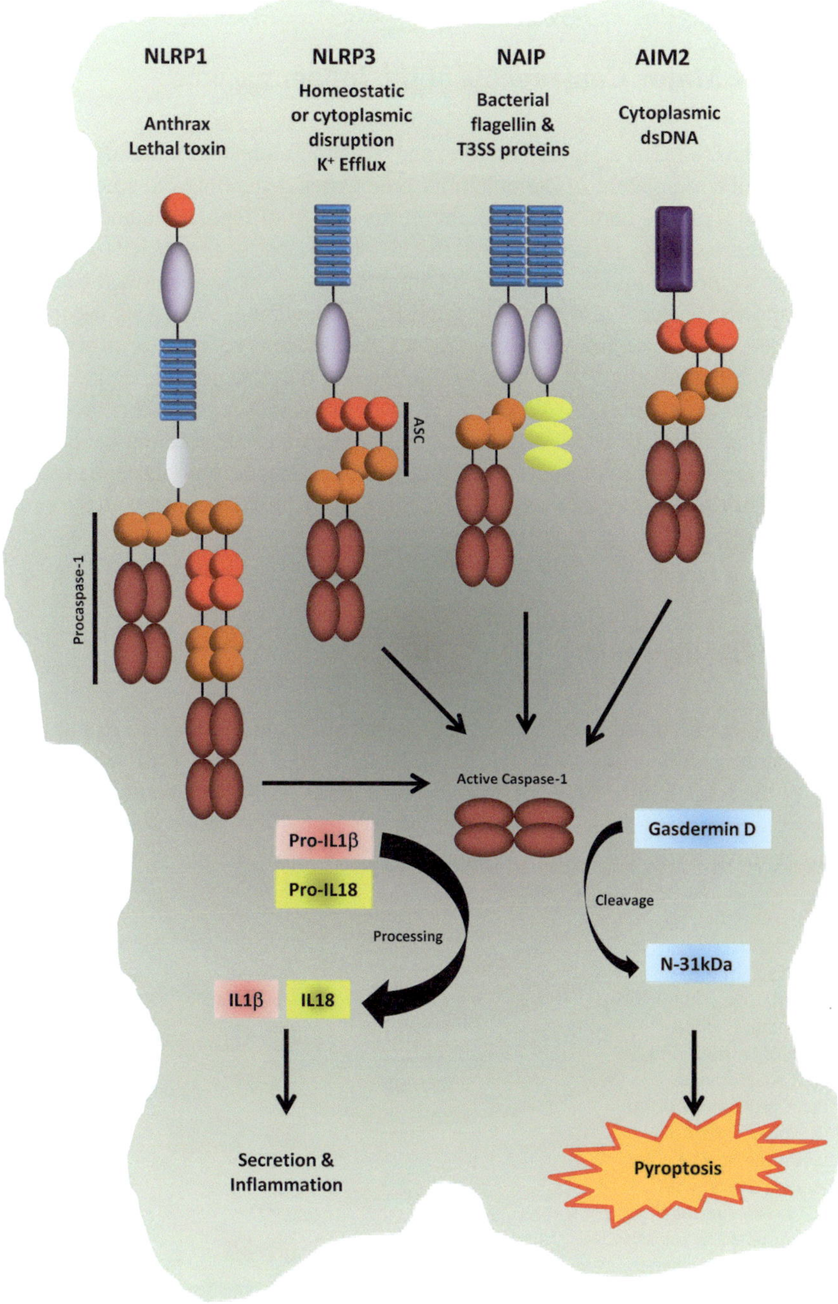

Fig. 2.1 The canonical inflammasome signalling pathway. A simplified schematic representation of canonical inflammasome signalling from the point of sensor activation, through inflammasome assembly, to the activation of caspase-1 and subsequent generation of active forms of IL-1β, IL-18, and Gasdermin D. The colour scheme is shared with Fig. 2.2. Protein domains are represented as follows: CARD – *orange spheres*; Pyrin domain – *red spheres*; NACHT - *light purple ovals*; LRRs – *blue rectangles*; FIIND – *grey oval*; HIN200 – *purple rectangle*; BIR – *yellow ovals*; caspase domains – *deep pink ovals*. In the case of NLRP1 signalling both ASC and procaspase-1 can be recruited to the NLRP1 CARD

2.2 The Major Constituents of the Inflammasome

The inflammasome is a highly complex and dynamic molecular structure. The precise protein composition of the inflammasome varies depending on, amongst other things, cell type, the nature of the stimulus, the degree of regulation, and the actual stage of the assembly process. As was first demonstrated for the NLRP1 inflammasome, at its most simplistic level the inflammasome can be reconstituted using just three purified proteins: a receptor (e.g. NLRP1), an adaptor (e.g. ASC (apoptosis-associated speck-like protein containing a CARD)), and an effector protein (e.g. procaspase-1) (Faustin et al. 2007). Within the cell the process can be, and usually is, much more complex. Aside from the potentially vast repertoire of associated proteins that may help activate and/or regulate inflammasome assembly, it has been shown using immunofluoresence that a single inflammasome complex can in fact be formed from, and simultaneously contain, multiple different receptors, adaptors and effector proteins (Man et al. 2014).

2.2.1 The Receptors

The receptors involved in activating the canonical inflammasome are examples of cytoplasmic pattern recognition receptors. NLRP1, NLRP3 and NAIP are all members of the NLR family, whilst AIM2 is part of the PYHIN (Pyrin and HIN (hemopoietic expression, interferon-inducibility, nuclear localization)) family. NLR proteins have a tripartite domain structure (Fig. 2.2), consisting of an N-terminal

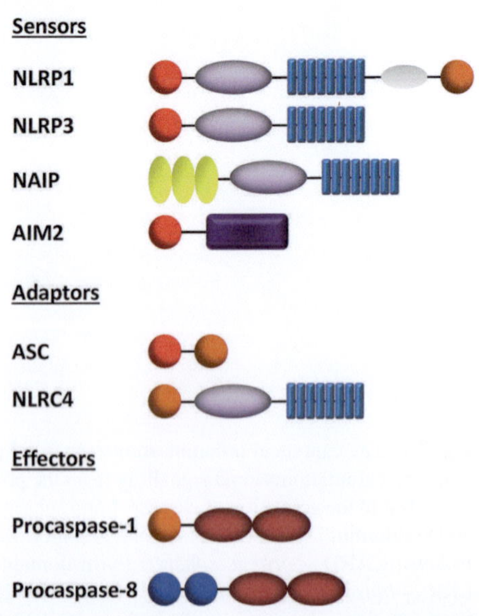

Fig. 2.2 The protein components of the inflammasome. Schematic representation of the domain organisation of the sensor, adaptor, and effector proteins involved in the formation of the inflammasome. The colour scheme is shared with Fig. 2.1. Protein domains are represented as follows: CARD – *orange spheres*; Pyrin domain – *red spheres*; NACHT – *light purple ovals*; LRRs – *blue rectangles*; FIIND – *grey oval*; HIN200 – *purple rectangle*; BIR – *yellow ovals*; caspase domains – *deep pink ovals*

signalling domain, a central NACHT (NAIP, CIITA (MHC class II transcription activator), HET-E (incompatibility locus protein from Podospora anserina) and TP1 (telomerase-associated protein)) domain, which contains the nucleotide-binding domain and a C-terminal leucine rich repeat (LRR) domain. Contrastingly, AIM2 is formed from an N-terminal Pyrin domain and a C-terminal HIN domain (Fig. 2.2). The HIN domain contains two OB (oligonucleotide/oligosaccharide binding) folds that directly bind dsDNA (Jin et al. 2012).

2.2.1.1 NLRP Family Members

Of the 14 human NLRP family members, there is clear, widespread and conclusive evidence that NLRP1 and NLRP3 form functional inflammasomes. It has also been suggested that NLRP6, NLRP7 and NLRP12 may form inflammasomes (Wang et al. 2002; Elinav et al. 2011; Khare et al. 2012; Vladimer et al. 2012). However, the physical and mechanistic evidence for this is less clear and requires further validation. Here I will concentrate on NLRP1 and NLRP3-driven inflammasomes.

NLRP1 is highly expressed in a range of cell types, including lymphocytes, myeloid cells, and epithelial cells in the respiratory system. It is widely accepted as being the first protein described as being able to form an inflammasome (Martinon et al. 2002). Reconstitution of the human NLRP1 inflammasome has been demonstrated using recombinant and purified human NLRP1, ASC and procaspase-1. In the presence of muramyl dipeptide (MDP), a putative NLRP1 ligand, procaspase-1 was processed to its active form (Faustin et al. 2007). The biological relevance of MDP as a ligand for NLRP1 remains to be verified, however, Lethal Toxin (LeTx) from *Bacillus anthracis* (Anthrax) has been shown to be a clear and bona-fide activator of murine NLRP1b (Cordoba-Rodriguez et al. 2004; Fink et al. 2008).

NLRP1 has a unique domain structure in that it possesses both a FIIND (Function to Find) domain and a CARD at its C-terminus (Fig. 2.2). Human NLRP1 has an N-terminal Pyrin domain, however, this appears to be absent in all of the protein products of the *Nlrp1a, Nlrp1b* and *Nlrp1c* genes in mice that encode orthologs of NLRP1. The CARD of NLRP1 can interact directly with the CARD of procaspase-1, suggesting that this interaction could serve to recruit procaspase-1 to the inflammasome complex for activation (Damiano et al. 2004). There is currently no evidence that the Pyrin domain of human NLRP1 is involved in recruiting ASC and indeed absence of the Pyrin domain does not stop IL-1β release following stimulation of human NLRP1 (Finger et al. 2012). However, ASC has been reported to be essential for full NLRP1 functionality and is recruited to the C-terminal CARD of NLRP1 (Finger et al. 2012). Interestingly, cleavage of NLRP1 is required for activation and subsequent assembly of the NLRP1 inflammasome (Finger et al. 2012; Chavarría-Smith and Vance 2013) (see Sect. 2.3.1).

NLRP3 has a tripartite structure of an N-terminal Pyrin domain, a central NACHT, and a C-terminal set of LRRs (Fig. 2.2) and is the most heavily studied inflammasome forming protein. A wide variety of exogenous and endogenous danger signals and events have been connected with activation of the NLRP3 inflammasome. These include potassium efflux, lysosomal damage, pore-forming toxins, mitochondrial damage, high ATP concentrations, cell swelling, and crystalline materials, to name but a few (Schroder and Tschopp 2010; Davis et al. 2011; Latz et al. 2013; Lamkanfi and Dixit 2014; Bryant et al. 2015). The molecular diversity of these stimuli makes it unlikely that NLRP3 directly interacts with these ligands. It does however appear to play a crucial role in the response to cellular homeostatic disruption and indeed it may be pertinent to view NLRP3 as a sensor of cellular disruption and imbalance independent of cause. Activation of NLRP3 results in the recruitment of ASC via Pyrin: Pyrin interactions and the formation of a large speck-like structure in the cytoplasm (Schroder and Tschopp 2010) (see Sects. 2.3.2 and 2.4). Procaspase-1 is recruited to the CARD of ASC leading to its activation and the subsequent processing and secretion of pro-IL1β (Fig. 2.1). Mutations in NLRP3 are associated with a series of auto-inflammatory conditions known as the cryopyrin-associated periodic syndromes (CAPS) (Hoffman et al. 2001; Broderick et al. 2015).

2.2.1.2 NAIP

The NAIP proteins function as direct sensors of bacterial proteins that include flagellin and components of the type III secretion system, such as PrgJ in *Salmonella* spp. (Kofoed and Vance 2011; Zhao et al. 2011). The NAIP proteins are members of the NLRB sub-family, as they have baculovirus inhibitor repeat (BIR) domains at their N-terminus (Fig. 2.2). Humans have one and mice have four functional NAIP proteins. In mice NAIP5 and NAIP6 respond to flagellin, whilst NAIP1 and NAIP2 are specific for the needle and rod proteins from bacterial type III secretion systems, respectively. Human NAIP was originally reported to recognise needle proteins from bacterial secretion systems and not to respond to flagellin (Yang et al. 2013). However, more recently, it has been demonstrated, that in macrophages at least, human NAIP responds to both stimulus (Kortmann et al. 2015) and therefore seems provide functional coverage for the specificity of the individual murine NAIPs.

2.2.1.3 AIM2

AIM2 is the prototypical member of the PYHIN family of proteins. It utilises two OB-folds in its C-terminal domain to bind to dsDNA in the cytoplasm (Jin et al. 2012). Interactions between the Pyrin domain of AIM2 and the Pyrin domain of ASC facilitate formation of the inflammasome complex and the subsequent recruitment of procaspase-1 (Roberts et al. 2009; Bürckstümmer et al. 2009; Fernandes-Alnemri et al. 2009; Hornung et al. 2009). As with other canonical inflammasomes, this results in the activation of caspase-1, which leads to processing of pro-IL-1β and pro-IL18. Another member of the PYHIN family, interferon gamma-inducible protein 16, has also been reported to form an inflammasome in response to infection with Kaposi's Sarcoma-associated herpes virus, which is also a dsDNA virus (Kerur et al. 2011).

2.2.2 *The Adaptors*

The principle role of the adaptor molecules is to recruit the effector caspases to the growing signalling complex. Until very recently, it was thought that ASC was the only adaptor associated with the inflammasome and that some inflammasomes, most notably those involving NLRC4, could assemble and function in both ASC-independent and ASC-dependent manners. However, as highlighted in Sect. 2.2.1.2, as our understanding of the molecular processes underlying inflammasome function and assembly have improved, it has become apparent that NLRC4 is not a ligand-specific receptor, but instead functions as an integral component in the formation of the NAIP-dependent inflammasome. The ability of NLRC4 to directly interact with and recruit procaspase-1 means it is more biologically accurate to now think of NLRC4 as an adaptor protein.

2.2.2.1 ASC

ASC is a relatively small protein (22 kDa) that consists of two discrete functional domains, an N-terminal Pyrin Domain and a C-terminal CARD (Fig. 2.2). Structural analysis of full-length ASC has shown that these domains function independently of one another and are connected by a flexible linker (de Alba 2009). Homotypic, and occasionally heterotypic, interactions with other Pyrin and CARD-containing proteins (see Sect. 2.4.2) allow ASC to act as a critical component in facilitating caspase recruitment and activation in response to receptor activation. The detection of discrete aggregates, or 'specks', of ASC by immunohistochemistry and immunofluoresence is routinely used as a marker of inflammasome activation. Overexpression studies of ASC have shown that both the CARD and Pyrin domain have roles in ASC aggregation (Masumoto et al. 1999, 2001; Moriya et al. 2005; Proell et al. 2013; Lu et al. 2014b; Schmidt et al. 2016) and that this process can result in the formation of filamentous structures (Masumoto et al. 2001; Lu et al. 2014b; Schmidt et al. 2016).

2.2.2.2 NLRC4

NLRC4 was for a long-time viewed as the sensor for bacterial flagellin. However, the discovery that the NAIP proteins actually provide the ligand sensing function in the NAIP/NLRC4 inflammasome (Kofoed and Vance 2011; Zhao et al. 2011) (see Sects. 2.2.1.2. and 2.3.3) has led to the reclassification of NLRC4 as an adaptor protein in the formation of the inflammasome. However, NLRC4 is a member of the NLR family and consequently possesses a tripartite domain arrangement broadly similar to that of NLRP1, NLRP3 and NAIP (Fig. 2.2). As such its biochemical behaviour, functional role, and mechanisms of action display characteristics similar to both the inflammasome receptors and the adaptor ASC. NLRC4 interacts with the

NAIP proteins directly to initiate inflammasome formation and is also capable of binding and recruiting, via its CARD, procaspase-1 (Vance 2015). NLRC4 can also interact with the CARD of ASC, an interaction that appears to enhance, but not be essential for, the activity of the NAIP/NLRC4 inflammasome.

2.2.3 *The Effectors*

The effectiveness of highly sensitive and rapidly activated systems to detect danger signals requires an ability to initiate appropriate effector mechanisms within the cell. In the case of canonical inflammasome signalling, the key effector molecule is caspase-1 (Schroder and Tschopp 2010). Caspase-8 has however also been shown to be recruited to the inflammasome complex and helps to drive appropriate inflammatory responses (Man et al. 2013; Monie and Bryant 2015).

2.2.3.1 Caspase-1

The caspase-1 protein exists as an inactive precursor, or zymogen, in the cell known as procaspase-1 (Fig. 2.2). Procaspase-1 possesses an N-terminal CARD attached to the catalytic p20 and p10 caspase domains. This precursor is activated as a result of proximity-induced dimerisation leading to autoprocessing of the molecule (Salvesen and Dixit 1999; Shi 2004). Autoprocessing results in the release of the caspase domains from the CARD and the cleavage of the linker between the caspase domains themselves in order to permit conformational rearrangement into the correct spatial organisation for caspase-1 function. Once activated, caspase-1 processes proIL-1β and proIL-18 to their active forms (Dinarello 1998). It also, along with caspase-11, cleaves Gasdermin D to release a 31 kDa N-terminal fragment that drives pyroptotic cell death (Kayagaki et al. 2015; Shi et al. 2015; He et al. 2015).

2.2.3.2 Caspase-8

Traditionally caspase-8 has been associated with apoptotic and necroptotic cell death pathways, particularly those involving receptor-interacting serine/threonine kinase (RIPK) 3 (Vandenabeele et al. 2010; Khan et al. 2014). More recently, the importance of caspase-8 in inflammatory pathways has become apparent. Like caspase-1, caspase-8 is translated as an inactive zymogen, procaspase-8. The N-terminal prodomain of procaspase-8 is formed from two Death Effector Domains (DEDs), which facilitate the interaction with protein partners and bring procaspase-8 molecules into a conformation permissive for autocatalytic processing to create active caspase-8.

In the inflammatory response, caspase-8 has been connected to both the canonical and non-canonical inflammasome pathways (Monie and Bryant 2015). The

DEDs of procaspase-8 interact with the Pyrin domain of ASC (Masumoto et al. 2003, 2011) to facilitate the recruitment of procaspase-8 to the inflammasome, a process that occurs simultaneously with procaspase-1 recruitment. Active caspase-8 is able to process proIL-1β (Maelfait et al. 2008). In response to infection of macrophages with *Salmonella Typhimurium*, or stimulation of AIM2, caspase-8 is recruited to and activated in inflammasome complexes (Pierini et al. 2012; Sagulenko et al. 2013). Interestingly, in dendritic cells, caspase-8 appears to negatively regulate the NLRP3 inflammasome (Kang et al. 2013).

2.3 The Initiation of Inflammasome Activation and Assembly

The active inflammasome complex is formed in the cytoplasm in response to direct stimulation. Control of the activation process is important to avoid unwanted and potentially harmful inflammatory side-effects (see also Sect. 2.5). Various mechanisms exist to help ensure that the actual process of inflammasome activation is appropriate and occurs in response to specific stimuli. These include the proteolytic processing of receptors, receptor auto-inhibition, transcriptional control and co-factor mediated repression. In this section I will focus predominantly on the mechanisms that facilitate the initial steps required to precipitate the formation of the actual inflammasome complex. Further consideration of some of the processes regulating inflammasome signalling is provided in Sect. 2.5.

2.3.1 The NLRP1 Inflammasome

Activation of the NLRP1 inflammasome requires direct cleavage of the NLRP1 protein. This occurs in two distinct locations, in the C-terminal FIIND domain via an autocatalytic process and near to the N-terminus through the action of LeTx. Autolytic cleavage of the FIIND domain is dependent on His1186 (human NLRP1 numbering) de-protonating Ser1213, which subsequently initiates cleavage via nucleophilic attack (D'Osualdo et al. 2011; Finger et al. 2012). The process of N-terminal cleavage has been clearly demonstrated in rodent NLRP1b as resulting from LeTx activity, but may also potentially be initiated by other extrinsic proteases, raising the possibility that NLRP1b might be a broad sensor of pathogen-derived proteases (Chavarría-Smith and Vance 2015).

NLRP1b cleavage in the N-terminus occurs in an unstructured region of the protein. The actual site of cleavage somewhat surprisingly varies with mouse strain. NLRP1b from NOD mice is cut between residues Lys44 and Leu45, and NLRP1b from BALB mice is preferentially cleaved between Lys38 and Leu39 (Hellmich et al. 2012; Chavarría-Smith and Vance 2013). Mutation of residues in and around the cleavage site disrupts NLRP1b activation, and the introduction of an artificial cleavage site was used to confirm that it is the actual cleavage of NLRP1b, and not some other cellular protein, that is the crucial factor in initiating NLRP1b driven inflammasome formation (Chavarría-Smith and Vance 2013). Not all of the NLRP1b present in the cytoplasm needs to be cleaved in order to initiate inflammasome activation (Hellmich et al. 2012; Levinsohn et al. 2012; Chavarría-Smith and Vance 2013). This is consistent with the concept of an activation threshold that regulates an "all-or-nothing" cellular response. The biochemical mechanisms underpinning NLRP1b cleavage and activation require further elucidation and it remains to be seen if the role of cleavage of the N-terminus is simply to release the protein from an auto-inhibited conformation. It is also unclear whether the N-terminus of human NLRP1 also requires cleaving in order to initiate activation and assembly of the inflammasome. If cleavage is necessary, then logic would dictate that the most likely sites would either be in the linker region between the Pyrin domain and the NACHT, or in the disordered loop that replaces the third helix of the Pyrin domain (Hiller et al. 2003).

2.3.2 The NLRP3 Inflammasome

Switching on the NLRP3 inflammasome requires two steps. The first of these is a 'priming' stage, often referred to as 'signal 1', and causes transcriptional upregulation of the genes encoding NLRP3, proIL-1β, and proIL-18 (Latz et al. 2013). This leads to a related increase in protein levels, although not for pro-IL18 (Schroder et al. 2012), and ensures the cell will be ready to mediate an inflammatory response to NLRP3 stimulation. Signal 1 is often provided through stimulation of Nuclear Factor kappa B (NFκB) signalling pathways, such as via the detection of bacterial lipopolysaccharide by Toll-like receptor 4. Although the NAIP/NLRC4 and AIM2 inflammasomes do not need priming in the same manner as the NLRP3 inflammasome, it does appears that the efficiency of their activation can be improved by high levels of signal 1 molecules (Schroder et al. 2012).

The second stage of NLRP3 inflammasome activation is that of 'sensing', also known as 'signal 2', in which a danger signal is recognised by NLRP3 and initiates the cascade of protein-protein interactions that will ultimately produce a functional NLRP3-driven inflammasome complex (Latz et al. 2013; Lamkanfi and Dixit 2014; Bryant et al. 2015). Since the first report of the NLRP3 inflammasome, a diverse range of molecules and cellular changes have been described as activators of the NLRP3 inflammasome (Rock et al. 2010; Davis et al. 2011; Latz et al. 2013; Wen et al. 2013; Lamkanfi and Dixit 2014). The chemical and structural diversity of these activators makes it implausible that NLRP3 directly interacts with and senses each of them individually. Consequently it is probably more accurate to think of NLRP3 as a sensor of homeostatic disruption, rather than a specific danger signal. This homeostatic disruption can occur in a number of ways, including lysosomal disruption, damage to mitochondria, generation of reactive oxygen species (ROS), and the alteration of ionic and osmotic balance.

Damaged lysosomes release their contents into the cytoplasm. This can occur following phagocytosis of crystalline or particulate material (Hornung et al. 2008; Dostert et al. 2008; Eisenbarth et al. 2008; Cassel et al. 2008; Halle et al. 2008; Duewell et al. 2010) and results in stimulation of the NLRP3 inflammasome. However, many NLRP3 activators, such as ATP and the ionophore nigericin, do not cause lysosomal damage. Consequently, whilst clearly relevant for specific stimuli and inflammatory conditions, like gout (Martinon et al. 2006) and atherosclerosis (Duewell et al. 2010), lysosomal damage is not the all encompassing mechanism underlying NLRP3 inflammasome activation.

The generation of ROS is a common innate effector mechanism in the response to cellular danger, particularly that from pathogens and following mitochondrial perturbation. ROS production has therefore often been associated with explanations of NLRP3 activation (Allen et al. 2009; Schroder et al. 2010; Zhou et al. 2010, 2011). However, the mechanism of action of ROS remains unclear. Conflicting evidence has been presented for a role of ROS-induced proteins, such as the redox related protein Thioredoxin-interacting protein (Zhou et al. 2010; Masters et al. 2010), whilst the use of ROS inhibitors suggests that ROS may be more important for the 'priming' stage and not the 'sensing' element of NLRP3 activation (Bauernfeind et al. 2011). As well as producing ROS, other elements of mitochondrial damage and dysfunction have been associated with NLRP3 activation. These include mitochondrial DNA and the phospholipid cardiolipin (Nakahira et al. 2011; Shimada et al. 2012; Iyer et al. 2013). Once again the mechanistic details are unclear and the actual role of mitochondria in NLRP3 activation remains an area of uncertainty for the field (Lawlor and Vince 2014).

Changes in ionic concentration, such as increased levels of calcium, both within and outside the cell, have been reported to initiate NLRP3-inflammasome activation (Murakami et al. 2012; Lee et al. 2012; Rossol et al. 2012). However, it is potassium efflux that has been most strongly linked to the NLRP3 inflammasome. The action of purigenic P2X7 receptors activated via ATP results in potassium efflux from the cell, which in turn activates NLRP3 (Kanneganti et al. 2007; Pétrilli et al. 2007; Muñoz-Planillo et al. 2013). Potassium efflux has now been proposed to be the unifying process underpinning all NLRP3 activating stimuli (Muñoz-Planillo et al. 2013). However, this assertion requires extensive validation and does not address whether reduced potassium concentration could exert conformational changes compatible with NLRP3 activation. Of course, it remains possible that an as yet unidentified cellular metabolite produced in response to cellular disruption may in fact serve as a directly sensed activatory ligand.

2.3.3 The NAIP/NLRC4 Inflammasome

Although multiple NLR proteins have been visualised in a single inflammasome structure (Man et al. 2014) the NAIP/NLRC4 inflammasome is currently unique in being entirely dependent on two distinct NLR family members. This functional dependence has parallels with some of the genetic pairings seen in plant-based responses to pathogen threat. The functional importance of the NAIP proteins as the ligand sensing partner was brought to forefront of the field by the combined work of the Vance and Shao labs. Both research groups independently demonstrated the importance of NAIP2 and NAIP5 in the detection of PrgJ, the rod protein of the type III secretion system in *Salmonella* spp., and flagellin, respectively (Kofoed and Vance 2011; Zhao et al. 2011). Further work by these groups, and others, has subsequently extended our understanding of the ligand repertoire and specificity conferred by the NAIP proteins (Vance 2015). The recent generation of mice defective in specific NAIP proteins has now provided genetic support to the biochemical evidence regarding ligand specificity (Rauch et al. 2016; Zhao et al. 2016). It is now clear that, at least in the case of murine NAIP proteins, NAIP1 and NAIP2 respond to needle and rod components of bacterial type III secretion systems, whilst NAIP5 and NAIP6 detect bacterial flagellin. The human NAIP protein, despite initial reservations, appears able to detect both classes of ligand (Kortmann et al. 2015). Interestingly, it has been suggested that the ligand specificity of these proteins is conferred by alpha helices in the NACHT region of the protein, and not via the LRRs (Tenthorey et al. 2014). Following ligand detection, the NAIP proteins interact with NLRC4 and initiate inflammasome assembly through the formation of wheel-like structures that serve as nucleation points for the recruitment of effector caspases (see Sect. 2.4.1).

2.3.4 The AIM2 Inflammasome

The initiation of signalling via the AIM2 inflammasome occurs as a result of the OB-folds in the AIM2 HIN200 region recognising and binding to the sugar-phosphate backbone of dsDNA within the cytoplasm. This interaction is based upon electrostatic attraction between positively-charged residues in the HIN200 domains and the negatively-charged phosphate groups of the DNA. Consequently, there is no sequence-specific element to the recognition process, although the dsDNA fragments must be at least 80 nucleotides in length (Jin et al. 2012; Jin et al. 2013). Fragments shorter than this do not permit sufficient side-by-side accumulation of AIM2 to enable self-oligomerisation and the propagation of downstream signalling and the ultimate formation of an AIM2-inflammasome (Lu et al. 2014a; Hou and Niu 2015). The interaction between the HIN200 domains and the dsDNA serves to separate the Pyrin domain from the HIN200 domains thereby releasing the Pyrin domain from its auto-inhibited state and freeing it to interact with the Pyrin domain of ASC (Jin et al. 2012; Jin et al. 2013).

2.4 The Mechanism of Inflammasome Assembly

Assembly of a functional inflammasome complex capable of activating procaspase-1 is dependent on an extensive network of protein-protein interactions. Intra-molecular interactions, post-translational modifications and interactions with accessory and regulatory proteins are all important for the overall activation, regulation and function of the inflammasome. However, the mechanism by which the inflammasome 'speck' actually assembles can be broadly summarised through two separate categories of protein-protein interaction: those that stimulate the formation of wheel-like sensor driven complexes early in the process of inflammasome activation; and those that facilitate the formation of protein filaments and aggregates that ultimately recruit effector caspases.

2.4.1 Formation of Helical Wheels

The similarities in domain and functional organisation between NLR family members and the apoptosome forming proteins Apaf-1 (apoptotic protease-activating factor 1) and pro-caspase-9 originally led to the understandable assumption that they would share similar mechanisms of complex assembly (Davis et al. 2011). This framework suggests that detection of ligand and nucleotide binding leads to a conformational change in the sensor molecule that then permits NACHT-mediated oligomerisation of the NLR to create a wheel-like structure. Recruitment of the effector caspase is then facilitated by CARD-CARD interactions. As with many biological systems it has become apparent that, at least in the case of the inflammasome, the process is somewhat more complex.

In 2007 the first visualization of an 'inflammasome' was performed by workers in the group of John Reed who used negative-staining electron microscopy (EM) to observe the structures formed by preparations of recombinant NLRP1. Similar to the apoptosome, wheel-like structures with an outer diameter of roughly 30 nm were observed. However, unlike the apoptosome which maintains a strict stoichiometrical relationship of seven Apaf-1 subunits the NLRP1 wheels appeared to contain either 5 or 7 protomers. More recent work on the NAIP/NLRC4 inflammasome has confirmed that the inflammasome forming NLR proteins are indeed capable of forming wheel-like structures.

The wheel-like structures formed by NAIP5 and NLRC4 and visualized by negative-stain EM are almost twice the diameter of those formed by NLRP1. Consistent with this they are formed from either 11 or 12 constituent units (Halff et al. 2012). Higher resolution studies using cryo-EM suggest that PrgJ stimulated formation of the NAIP2-NLRC4 inflammasome creates wheels with 10–12 spokes (Hu et al. 2015; Zhang et al. 2015). Interestingly, both these studies demonstrated that these complexes only contain a single NAIP2 protein implying that the sensing of ligand by the NAIP component results in conformational changes that drive the

NACHT-mediated oligomerisation of NLRC4. Differences in the surfaces of NLRC4 and the NAIPs suggests that a specific oligomerisation surface may exist on NLRC4, whereas the NAIPs instead possess a catalytic surface to initiate interaction with NLRC4, but cannot then be incorporated further into the growing structure. Ligand detection by the NAIPs therefore acts as a molecular switch controlling an "all or nothing" approach to activation of the NAIP/NLRC4 inflammasome. In addition to forming discrete wheel-like structures there have also been reports that assembly of the NAIP/NLRC4 inflammasome can lead to the construction of a multimeric helical filament like structure. It is yet to be seen if this structure has physiological relevance (Diebolder et al. 2015).

In light of the ability of both NLRP1 and NAIP to drive the formation of wheel-like structures it would therefore seem likely that NLRP3 will also be capable of forming similar complexes, particularly given that in response to infection with *Salmonella enterica* serovar *Typhimurium* (S. Typhi), or stimulation with flagellin, NLRP3 and NLRC4 have been visualized in the same inflammasome complex (Man et al. 2014) and shown to interact with one another in a NACHT-dependent manner (Qu et al. 2016). However, helical wheel formation does not appear to be a pre-requisite for the formation of an inflammasome. AIM2, which is of course not an NLR protein, has not currently been reported to form wheel-like structures. Given its lack of a NACHT domain this is not necessarily surprising. Instead it appears to form chain-like structures with AIM2 molecules sequentially bound along the activating dsDNA ligand (Jin et al. 2012).

2.4.2 Assembly of Filament Fragments

One might ask what the physiological purpose of the structures described above is in relation to inflammasome assembly? It would appear that their role is to act as a nucleation point in the cell for the recruitment of ASC and effector caspase proteins. This results in the formation of a large maromolecular speck and helps to explain how formation of 30 nm wheel structures could lead to the production of cytoplasmic complexes of around 1 micron in size. It also serves to amplify the signal initiated by inflammasome activation. The recruitment of ASC and the effector caspases is dependent on the interactions between members of the death domain superfamily (Fig. 2.2).

The Death Domain superfamily refers to a collection of protein folds that form helical bundles and consists of CARD, Pyrin, Death domain (DD), and DED modules (Fig. 2.2). Homotypic interactions, i.e. CARD with CARD, form the basis of many macromolecular signalling complexes involved in inflammatory and cell death pathways such as the death-inducing signaling complex, the piddosome, the apoptosome, the ripoptosome, and the myddosome, as well as the inflammasome (Kersse et al. 2011b; Dickens et al. 2012). These homotypic interactions are mediated by three specific protein-protein interfaces known as Type I, II and III (Fig. 2.3). Each interface consist of two unique non-overlapping 'a' and 'b' surfaces.

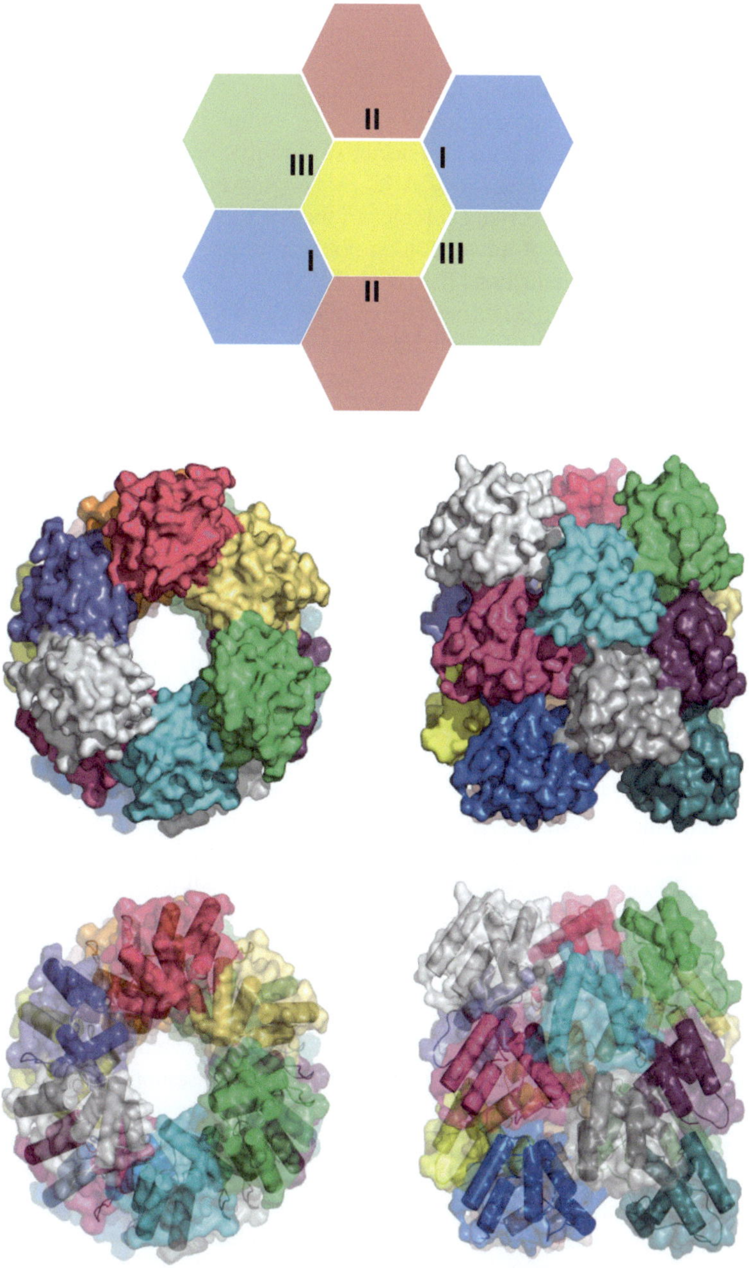

Fig. 2.3 Homotypic interactions drive inflammasome growth. (*Top*) Schematic representation of the three types of interface that occur in homotypic interactions between members of the death domain protein superfamily. Each hexagon representations a death domain fold protomer and the types of interface are labelled I, II and III. The central (*yellow*) domain can therefore possess six different interface contacts. Surface (*middle*) and surface and cartoon (*bottom*) representations of the filament structure formed by polymerisation of the ASC Pyrin domain. Each individual Pyrin is coloured differently. The left hand image represents the view from above; the right hand image represents the view from the side. Images were generated using PyMol (Schrodinger Inc) from PDB file 3 J63

Consequently, any death domain family member can in theory form up to 6 discrete interactions during the course of macromolecular complex formation (Kersse et al. 2011b). Heterotypic interactions also occur, such as that between the DEDs of pro-caspase-8 and the Pyrin domain of ASC (Masumoto et al. 2003; Vajjhala et al. 2015); and the recently described complex of the RIPK2 CARD and the DD of p75. In this instance the sites of interaction do not map to any of the typical interfaces seen with homotypic interactions (Lin et al. 2015).

Much work has been performed to identify key functional residues in these domains that are important for inflammasome assembly. These residues are often charged in nature and include for example Arg10, Asp27, Glu41, Lys42, Arg55 and Asp59 of the caspase-1 CARD (Kersse et al. 2011a; Proell et al. 2013); and Arg125, Glu130, Asp134, Tyr137, Glu144, Arg160, Asp191 of the ASC CARD and Glu13, Lys21, Arg41, Asp48, Asp51 of the ASC Pyrin domain (Vajjhala et al. 2012; Proell et al. 2013). Hydrophobic residues do however also play a role as demonstrated by the type II interface between the ASC and caspase-1 CARDs, in which Tyr187 in the ASC CARD is crucial to the interaction (Schmidt et al. 2016). The complexity of these interactions and the range of residues involved in forming and stabilizing the interfaces has made it hard to design therapeutics to disrupt these interfaces and reiterates the fact that inflammasome assembly is a complex multi-faceted procedure (Vajjhala et al. 2012, 2014).

A number of recent studies have demonstrated that in the context of inflammasome assembly the caspase-1 CARD (Lu et al. 2016) and the ASC Pyrin (Cai et al. 2014; Lu et al. 2014b) domain are able to form filamentous assemblies as a result of these homotypic interactions (Fig. 2.3). Similar structures are also seen in other innate immune pathways including Retinoic acid inducible gene I signalling (Hou et al. 2011; Wu et al. 2014) and the formation of the B-cell lymphoma 10 signalosome (Qiao et al. 2013). Gold-labelling studies have shown that the NLRP3 and AIM2 Pyrin domains are both capable of serving as initiators of ASC filament formation (Lu et al. 2014b). Somewhat similarly to the relationship between NAIP and NLRC4 neither the AIM2 nor NLRP3 Pyrin domain are incorporated into the growing filament, but are only located at the site of initiation. It seems therefore that the process of nucleation is a key concept in inflammasome assembly.

Until recently filament formation by inflammasome-forming proteins had only been demonstrated using overexpression systems, *in vitro* approaches, and with procedures containing harsh purification and preparation steps. Death domain family members have a natural propensity to form aggregates and as such the potential physiological relevance of these filaments was less clear. However, using domain specific antibodies Schmidt and colleagues have now observed the endogenous presence of ASC filaments (Schmidt et al. 2016). It still remains to be determined how exactly these filaments, and indeed the nucleating wheel-like complexes, combine to form the endogenous ring-like inflammasome complex observed upon infection with *Salmonella* (Man et al. 2014).It may well be that under normal physiological conditions the filaments interact more closely with one another, or form a branched network of interactions that lead to speck formation. Improvements in super-resolution microscopy and single-molecule based studies may soon provide understanding.

2.4.3 *When and Where Does the Inflammasome Assemble?*

Inflammasome assembly is a rapid process and is driven by the aggregation of ASC. It can occur within moments of appropriate cellular stimulation and once initiated will continue to completion. The dynamics of ASC:ASC interactions mean that normally only a single inflammasome speck is formed in the cell (Bryant et al. 2015), although this is not an absolute rule. Mutations in NLRC4 associated with auto-inflammatory disorders result in an increased tendency for cells to form multiple specks suggesting that the dynamics of inflammasome nucleation may play an important role (Canna et al. 2014; Romberg et al. 2014; Kitamura et al. 2014). Under normal conditions the first nucleation event that passes a detection threshold and begins ASC recruitment is likely to lead to a single speck (Cheng et al. 2010), whilst with the mutant proteins there are likely to be numerous pre-activated complexes ready to act as nucleation points upon appropriate stimulation.

The inflammasome complex forms within the cytoplasm of the cell. It is unclear whether this is associated with any particular cytoplasmic structure of organelle. Suggestions have been made that the NLRP3 inflammasome associates with mitochondria as a result of interaction with the mitochondrial anti-viral signalling protein (Subramanian et al. 2013). However, a more comprehensive analysis of the location of the NLRP3 inflammasome failed to show any association with seven different organelles, including mitochondria (Wang et al. 2013).

2.5 How Is the Inflammasome Regulated?

The activation of the inflammasome needs to be tightly controlled in order to not only avoid inappropriate, or excessive, inflammatory responses, but to also ensure that danger signals can be quickly detected at a threshold below that which would be overly dangerous to the host. There are various approaches used to ensure the appropriate activation of the inflammasome. These might include regulation at the transcriptional level reflected in cell-specific expression patterns of the signalling components; could involve translational regulation; may be reliant on other protein factors; or could be the result of post-translational modifications. What is certain is that as our understanding of inflammasome biology continues to develop we will become aware of more and more regulatory controls that are used to maintain the delicate balance between beneficial and potentially harmful processes.

2.5.1 *Receptor Auto-inhibition*

The presence of disease-specific mutations in both NLRP3 and NLRC4 that lead to the clinical development of inflammatory disorders such as CAPS and recurrent macrophage activation syndrome have already been mentioned (Hoffman et al.

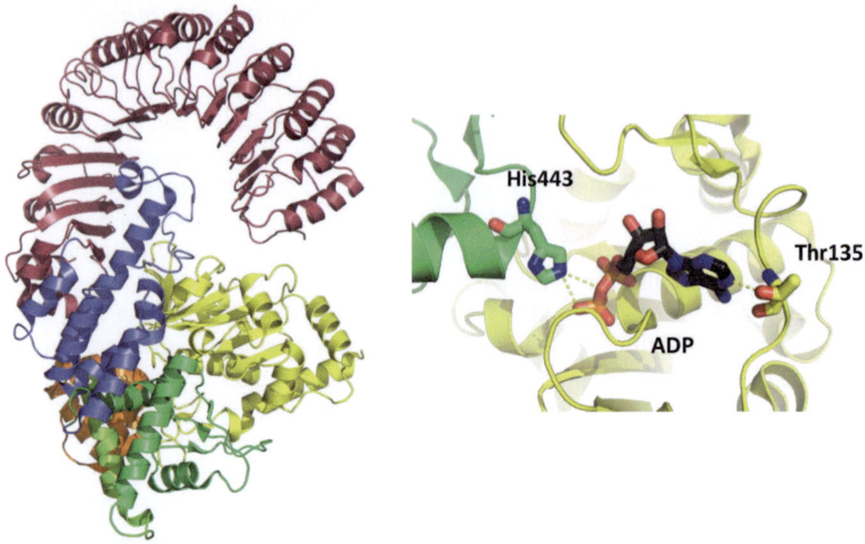

Fig. 2.4 Autoinhibition of NLRC4. Cartoon representation (*top*) of the autoinhibited conformation of the NLRC4 adaptor protein. Domains are coloured as follows – NACHT – *yellow*; Winged Helix Domain – *green*; Helical Domain 1 – *orange*; Helical Domain 2 – *blue*; LRRs – *purple*. The CARD was not present in the crystal structure. The *bottom* image shows the hydrogen bonds (*dotted lines*) formed between the bound ADP molecule and His443 and Thr135 that help maintain the autoinhibited state

2001; Canna et al. 2014; Romberg et al. 2014; Kitamura et al. 2014; Broderick et al. 2015). The functional impact of these mutations are believed to result from a disruption of receptor auto-inhibition that permits the ligand-independent formation of signalling platforms and the subsequent processing and release of IL-1β.

Models outlining mechanisms by which inflammasome forming proteins can mediate auto-inhibition have generally focussed upon intra-domain interactions that limit the process of ligand recognition and nucleotide-binding and/or the availability of the domains required for complex formation and downstream signal transduction. These models have often been developed based upon the current state of knowledge regarding the regulation of apoptosome activation and plant R protein signalling. A significant improvement in our understanding of the process of auto-regulation was provided by the solving of the crystal structure of the auto-inhibited form of the NACHT and LRR domains of NLRC4 (Hu et al. 2013). The auto-inhibited conformation was stabilised by hydrogen bonding between a bound adenosine diphosphate molecule within the nucleotide binding domain and residues in the Walker A site of the NLRC4. Further hydrogen bonding with Thr135 and His443 were crucial for maintaining the inactive conformation. Indeed, mutation of His 443 to leucine abolished the hydrogen bonding with the β-phosphate moiety, destabilised the structure and resulted in ligand-independent activation of NLRC4 signalling (Fig. 2.4). The LRR domain also helps maintain an inactivate conformation by folding over and occupying the space that would be taken by an additional protomer during complex formation. This is driven by both hydrophobic stacking interactions and hydrogen bonding and is consistent with the autoactivation of NLR family members upon removal of their LRRs. Further important protein-protein interactions occur between Helical Domain 2 and the NACHT (Hu et al. 2013). In particular Helical Domain 2 interacts with a helix known to be important for inter-molecular complex formation in other ATPases (Fig. 2.4). Removal of Helical Domain 2, as well the LRRs, from NLRC4 resulted in a higher level of auto-activation supportive of an important regulatory role for Helical Domain 2. The ability of mutations in the NBD to lead to NLR autoactivation makes it appealing to ascribe wide applicability of the mechanisms displayed by NLRC4 to other NLR proteins. However, we currently lack the structural evidence to be certain that this is the case, although mechanisms are likely to share common themes, and it remains to be determined how binding of adenosine triphosphates, rather than adenosine diphosphate alters the dynamics of the regulatory interactions. Indeed this could be a key step in the processes required to release the receptors from an auto-inhibited state.

2.5.2 Accessory Proteins

2.5.2.1 Pyrin Only Proteins

The importance of homotypic interactions between DD family members in the formation of an active inflammasome complex has been discussed earlier (Sect. 2.4.2). The crucial nature of both Pyrin:Pyrin and CARD:CARD interactions for the recruitment of adaptor proteins, complex assembly and caspase activation make them a highly attractive target for inflammasome regulation. Unsurprisingly therefore, there exist a range of proteins that through adoption of the same structural folds are able to disrupt the activation, assembly and signalling processes of the inflammasome. Here I will consider how Pyrin-only proteins (POPs) and CARD-only proteins (COPs; Sect. 2.5.2.2), both of which appear to be restricted to higher primates (Kersse et al. 2007), serve to regulate inflammasome activity.

Four POPs, POP1–4, have so far been identified in the human genome. POP4, which is encoded from the *NLRP2P* pseudogene appears to be truncated to the extent that it only possesses sufficient sequence to form the first two helices of the PYD fold. POP1 is likely to be the result of gene duplication of the ASC PYD. POP1 shows high (64%) sequence identity with the PYD of ASC (Stehlik et al. 2003) and the two domains are capable of interacting directly with one another in an interaction driven by negatively charged acidic residues on ASC and positively charged basic residues on POP1 (Srimathi et al. 2008; Vajjhala et al. 2012). POP1 has been clearly shown to interfere with NFκB signalling, but somewhat surprisingly, given its similarity to and ability to interact with ASC, does not appear to regulate inflammasome activation (Le and Harton 2013). Rather the opposite, as there is evidence from in vitro assays suggesting that POP1 is in fact able to enhance inflammasome activation (Stehlik et al. 2003). Clearly, there is a need for further research investigating the precise functional role of POP1, but it may be that it can substitute for the ASC PYD in the formation of large maromolecular assemblies or filaments.

POP2 clearly interferes with inflammasome function. At the sequence level POP2 is closest to the PYDs of NLRP2 and NLRP7 and has been shown using yeast-2-hybrid studies to be capable of interacting with ASC, NLRP1, NLRP2, NLRP4 and NLRP12 (Dorfleutner et al. 2007a). It is less clear than with POP1 as to how these interactions come about, but they may potentially involve negatively charged residues (Glu6, Asp8, Glu16) located on the first helix of POP2 (Atianand and Harton 2011; Dorfleutner et al. 2015). Truncated forms of POP2 that only contain the first alpha helix are still able to inhibit inflammasome signalling (Atianand and Harton 2011). In the case of POP2 its ability to interact with ASC, coupled with the broadly divergent sequences of the two proteins suggests that it has the potential to be a broad inhibitor of inflammasome signalling. To date there is no clear evidence of an interaction between NLRP3 and POP2.

It has recently been shown that POP3 functions to interfere with signalling from the AIM2 and IFI16 inflammasomes (Khare et al. 2014). Somewhat analogously to POP1, POP3 is thought to derive from a duplication of the AIM2 PYD exon - there is 61% identity at the amino acid level – and is capable of interacting directly with the AIM2 PYD. POP3 also interacts with the IFI16 PYD, but does not appear to be capable of interacting with the ASC PYD. This differential binding underpins the selectively with which POP3 disrupts AIM2 and IFI16 inflammasome activation, but does not interfere with other inflammasomes.

The importance of POPs as immunoregulators of inflammasome activation is highlighted by the presence of viral Pyrin-only proteins (vPOPs) in members of the *Poxviridae*. For example, the Shope fibroma virus and myxoma virus 13 L proteins both adopt a Pyrin fold, bind to the PYD of ASC and disrupt the NLRP3 inflammasome mediated activation of IL-1β (Johnston et al. 2005; Dorfleutner et al. 2007b). As with human POP2 this interaction appears reliant on residues that form the first alpha helix of the PYD (Johnston et al. 2005). vPOPs have been identified in other poxviruses, but have yet to be reported in other virus families.

2.5.2.2 CARD Containing Proteins

Three COPs have been identified in the human genome – CARD16 (pseudo-ICE), CARD17 (INCA) and CARD18 (ICEBERG) – which arose as a result of gene duplication. In comparison with the POPs less is known about how these proteins function in an immunoregulatory capacity. However, all three COPs are capable of binding to the CARD of caspase-1 and inhibiting its activation (Humke et al. 2000; Druilhe et al. 2001; Lee et al. 2001; Lamkanfi et al. 2004). However, whereas CARD17 is without doubt able to inhibit inflammasome-driven caspase-1 activation, it is currently unclear as to whether CARD16 and CARD18 can also interfere with this aspect of caspase-1 activation. CARD17 regulates the inflammasome by stopping the formation of caspase-1 CARD-driven filaments. It binds to the caspase-1 CARD and serves as a capping structure that terminates filament expansion. In contrast CARD18, and given its sequence similarities most likely CARD16 as well, are able to be incorporated into and form part of the caspase-1 filament (Lu et al. 2016). Consequently, they do not inhibit inflammasome-driven processing of caspase-1. CARD16 and CARD18 do however interfere with the interaction between the CARDs of caspase-1 and RIPK2 and stop caspase-1 activation via this pathway (Le and Harton 2013; Dorfleutner et al. 2015), suggesting that the interfaces used to co-ordinate caspase-1 autoactivation differ between the RIPK2-dependent and inflammasome dependent pathways. It is unknown how, or indeed if, these COPs influence ASC-independent NAIP/NLRC4-dependent caspase-1 activation.

2.5.3 Post-translational Modifications

Post-translational modification is a commonly used process by which eukaryotes regulate the function of their proteome. The cellular effects of post-translational modifications are wide ranging and can include changes in protein localisation, activation or inhibition of enzymatic activity, creation or disruption of binding surfaces, and even protein degradation. An extensive repertoire of modifications can be employed and include phosphorylation, carbonylation, glycosylation, lipidation, myristylation, and ubiquitination amongst others. Here I will consider just two of these processes, phosphorylation and ubiquitination, as these have the clearest associations with inflammasome signalling.

2.5.3.1 Phosphorylation

As our understanding of the phenotypic and cell-wide impact of inflammasome-mediated signalling has improved, researchers have begun to focus upon the molecular events that regulate and control this process. This has resulted in the identification of both broad-based kinase-dependent regulatory mechanisms as well as residue-specific phosphorylation events. There remains however some controversy over the precise requirements for some of the proposed phosphorylation events.

Depletion of protein kinase R (PKR), a sensor of viral dsDNA and key driver of the antiviral response, results in a reduction in IL-1β secretion and pyroptotic death following exposure to a wide-range of inflammasome activating stimuli (Lu et al. 2012). This impact appears to be mediated via direct interaction between PKR, which undergoes extensive autophosphorylation upon inflammasome activation, and NLRP1, NLRP3, AIM2 and NLRC4. This lack of binding specificity raises the possibility that the process may also involve other, as yet unidentified, protein partners.

Phosphorylation of ASC, in either a Syk-dependent or Jnk-dependent manner, is an important factor in the formation of ASC specks in macrophages following activation of NLRP3 or AIM2 (Hara et al. 2013). The activation of the NAIP/NLRC4 inflammasome was not affected by alterations in ASC phosphorylation, consistent with its ability to signal independently of ASC. ASC most likely experiences numerous phosphorylation events during inflammasome activation. Of these, phosphorylation of Ser144 in murine ASC has been shown to have critical importance. Mutation of Ser144, in the CARD, resulted in complete loss of ASC phosphorylation suggesting that it acts as a key regulator in ASC activation. It also stopped the formation of ASC specks (Hara et al. 2013). In addition to driving inflammasome activation phosphorylation of ASC is also used as a negative regulator of inflammasome function. It has recently been shown that IKKα directly binds ASC, phosphorylates Ser293, and results in nuclear retention of ASC (Martin et al. 2014). This ASC:IKKα complex translocates to the cytoplasm upon macrophage activation, but ASC is only released once specific phosphatases, such as PP2A, are activated following cellular receipt of a second signal. It is also possible, but experimentally unvalidated, that the activating Syk and Jnk mediated phosphorylation of ASC (Hara et al. 2013) induces dissociation of the ASC:IKKα complex and thereby permits inflammasome activation.

2.5.3.2 Ubiquitination

In order for the NLRP3 inflammasome to be activated it first needs priming (see Sect. 2.3.1). This process does not just involve upregulation of NLRP3 gene and protein expression, but is also reliant on deubiquitination. In 2012 Juliana and colleagues reported that stimulation of human and murine cells with either signal 1 (LPS) or signal 2 (ATP or nigericin) resulted in a significant reduction in the level of ubiquitination observed on immunoprecipitated NLRP3 (Juliana et al. 2012). Pharmacological inhibition of deubiquitinases stopped this change from happening. The mechanisms by which these stimuli encourage deubiquitination are likely to vary as changes in ubiquitination mediated by signal 1, but not signal 2, are inhibited by the scavenging of mitochondrial-derived ROS (Juliana et al. 2012). More recently it has been suggested that one of the deubiquitinases involved in this process is likely to be BRCC3 (Py et al. 2013).

It is not just NLRP3 that displays functional regulation mediated by ubiquitin. The chemical inhibition of deubiquitinases impairs the processing and secretion of IL-1β by stopping the oligomerisation of ASC and hence abrogating inflammasome assembly (Lopez-Castejon et al. 2013). This may be related to degradation of inflammasome components as it has been shown that K63-linked ubiquitination signals for autophagosome-mediated destruction of the AIM2 inflammasome (Shi et al. 2012). In contrast, linear ubiquitination of ASC, mediated by the HOIL-1 L component of the linear ubiquitination assembly complex (LUBAC), does not stimulate protein degradation, but instead appears to be a crucial component for the assembly of a functional inflammasome complex. Indeed HOIL-$1^{-/-}$ mice were unable to assemble active NLRP3 inflammasomes and displayed resistance to lethal infection models (Rodgers et al. 2014).

2.5.4 NEK7

The use of unbiased CRISPR/Cas9 screening, forward genetics, and liquid chromatography mass spectroscopy approaches have all recently identified the NIMA ('never in mitosis gene a')–related serine-threonine kinase family member NEK7 as an upstream regulator of the NLRP3 inflammasome. Activation of the NLRP3 inflammasome, but not the AIM2 or NAIP/NLRC4 inflammasomes, is abrogated in *Nek7*$^{-/-}$ murine macrophages, in cells lacking *Nek7* following the use of gene specific disruption by CRISPR/Cas9, and in mice with N-ethyl-N-nitrosourea–induced mutations in *Nek7* (He et al. 2016; Schmid-Burgk et al. 2016; Shi et al. 2016). The precise molecular details of how NEK7 regulates NLRP3 activation are still being determined. However, the catalytic domain of NEK7 interacts directly with NLRP3 at a point in the signalling cascade between potassium efflux, induced by nigericin or crystalline stimulation, and the formation of an active, signalling competent NLRP3 inflammasome (He et al. 2016; Shi et al. 2016). This interaction requires both the NLRP3 NBD and LRR domains suggesting that the interaction with NEK7 may play an important role in the release of NLRP3 from an autoinhibited state, but is independent on the catalytic activity of NEK7 (He et al. 2016; Shi et al. 2016). NEK7 plays an important role in the progression of mitosis and it has been suggested that the levels of cellular NEK7 are insufficient to allow both mitosis and NLRP3 activation to proceed simultaneously (Shi et al. 2016).

2.5.5 Chemical Regulation

Production of IL-1β following activation of the inflammasome is a key driver of clinical inflammation in a wide range of acute, and particularly chronic, diseases. There exist highly successful IL-1 neutralising biologics such as anakinra and canakinumab that can be used to effectively treat a range of IL-1 associated myopathies. However, the effectiveness of these treatments can be short-lived in some patients and may also lead to undesirable systemic effects such as immunosuppression. A major goal therefore for both academic and industrial researchers is to identify therapeutic approaches that can specifically target inflammasome activation in a disease-relevant manner. Initial progress has been somewhat slow, but early in 2015 two research groups led by Professors O'Neill and Dixit reported on the successful small-molecule mediated inhibition of NLRP3 inflammasome function by the sulfonylurea MCC950 and the ketone metabolite β-hydroxybutyrate respectively (Coll et al. 2015; Youm et al. 2015). Neither molecule was active against other canonical inflammasomes, although MCC950 was also effective against the non-canonical caspase-11 inflammasome.

The identification of these small molecule inhibitors, and their success in alleviating inflammation in animal models of disease, opens up a range of possibilities for the further development of specific inflammasome inhibitors. With an IC_{50} of less than 10 nM MCC950 is highly potent and may be a suitable starting point in chemical space for the rational design of additional inhibitors. The activity of β-hydroxybutyrate at physiologically relevant concentrations (5–10 mM) raises the attractive proposition that there may be a whole range of endogenous metabolites with immunomodulatory molecules awaiting discovery.

2.6 Concluding Remarks

Our understanding of the molecular processes that drive the activation and early stages of inflammasome formation has dramatically improved over recent years. Along with improvements in the visualisation of the seemingly widespread process of filament formation it would appear that we are getting closer to understanding how this supramolecular complex is switched on, and how it assembles, following stimulation. As we become more certain of the molecular contacts involved the possibility of designing specific inhibitors of the inflammasome becomes a more realistic option. This could reduce the blanket reliance on the use of anti-IL-1 biologics for the treatment of chronic inflammatory conditions and lead to the creation of targeted therapeutics designed to disrupt inflammasome-based signalling in a disease-specific manner.

References

Allen IC, Scull MA, Moore CB et al (2009) The NLRP3 inflammasome mediates in vivo innate immunity to influenza A virus through recognition of viral RNA. Immunity 30:556–565. doi:10.1016/j.immuni.2009.02.005

Atianand MK, Harton JA (2011) Uncoupling of Pyrin-only protein 2 (POP2)-mediated dual regulation of NF-κB and the inflammasome. J Biol Chem 286:40536–40547. doi:10.1074/jbc.M111.274290

Bauernfeind F, Bartok E, Rieger A et al (2011) Cutting edge: reactive oxygen species inhibitors block priming, but not activation, of the NLRP3 inflammasome. J Immunol 187:613–617. doi:10.4049/jimmunol.1100613

Broderick L, De Nardo D, Franklin BS et al (2015) The inflammasomes and autoinflammatory syndromes. Annu Rev Pathol 10:395–424. doi:10.1146/annurev-pathol-012414-040431

Bryant CE, Orr S, Ferguson B et al (2015) International Union of Basic and Clinical Pharmacology. XCVI. Pattern recognition receptors in health and disease. Pharmacol Rev 67:462–504. doi:10.1124/pr.114.009928

Bürckstümmer T, Baumann C, Blüml S et al (2009) An orthogonal proteomic-genomic screen identifies AIM2 as a cytoplasmic DNA sensor for the inflammasome. Nat Immunol 10:266–272. doi:10.1038/ni.1702

Cai X, Chen J, Xu H et al (2014) Prion-like polymerization underlies signal transduction in antiviral immune defense and inflammasome activation. Cell 156:1207–1222. doi:10.1016/j.cell.2014.01.063

Canna SW, de Jesus AA, Gouni S et al (2014) An activating NLRC4 inflammasome mutation causes autoinflammation with recurrent macrophage activation syndrome. Nat Genet 46:1140–1146. doi:10.1038/ng.3089

Cassel SL, Eisenbarth SC, Iyer SS et al (2008) The Nalp3 inflammasome is essential for the development of silicosis. Proc Natl Acad Sci U S A 105:9035–9040. doi:10.1073/pnas.0803933105
Chavarría-Smith J, Vance RE (2013) Direct proteolytic cleavage of NLRP1B is necessary and sufficient for inflammasome activation by anthrax lethal factor. PLoS Pathog 9:e1003452. doi:10.1371/journal.ppat.1003452
Chavarría-Smith J, Vance RE (2015) The NLRP1 inflammasomes. Immunol Rev 265:22–34. doi:10.1111/imr.12283
Cheng J, Waite AL, Tkaczyk ER et al (2010) Kinetic properties of ASC protein aggregation in epithelial cells. J Cell Physiol 222:738–747. doi:10.1002/jcp.22005
Coll RC, Robertson AAB, Chae JJ et al (2015) A small-molecule inhibitor of the NLRP3 inflammasome for the treatment of inflammatory diseases. Nat Med 21:248–255. doi:10.1038/nm.3806
Cordoba-Rodriguez R, Fang H, Lankford CSR, Frucht DM (2004) Anthrax lethal toxin rapidly activates caspase-1/ICE and induces extracellular release of interleukin (IL)-1beta and IL-18. J Biol Chem 279:20563–20566. doi:10.1074/jbc.C300539200
D'Osualdo A, Weichenberger CX, Wagner RN et al (2011) CARD8 and NLRP1 undergo autoproteolytic processing through a ZU5-like domain. PLoS One 6:e27396. doi:10.1371/journal.pone.0027396
Damiano JS, Oliveira V, Welsh K, Reed JC (2004) Heterotypic interactions among NACHT domains: implications for regulation of innate immune responses. Biochem J 381:213–219. doi:10.1042/BJ20031506
Davis BK, Wen H, Ting JP-Y (2011) The inflammasome NLRs in immunity, inflammation, and associated diseases. Annu Rev Immunol 29:707–735. doi:10.1146/annurev-immunol-031210-101405
de Alba E (2009) Structure and interdomain dynamics of apoptosis-associated speck-like protein containing a CARD (ASC). J Biol Chem 284:32932–32941. doi:10.1074/jbc.M109.024273
Dickens LS, Powley IR, Hughes MA, Macfarlane M (2012) The "complexities" of life and death: death receptor signalling platforms. Exp Cell Res 318:1269–1277. doi:10.1016/j.yexcr.2012.04.005
Diebolder CA, Halff EF, Koster AJ et al (2015) Cryoelectron tomography of the NAIP5/NLRC4 inflammasome: implications for NLR activation. Structure 23:2349–2357. doi:10.1016/j.str.2015.10.001
Dinarello CA (1998) Interleukin-1 beta, interleukin-18, and the interleukin-1 beta converting enzyme. Ann N Y Acad Sci 856:1–11
Dorfleutner A, Bryan NB, Talbott SJ et al (2007a) Cellular pyrin domain-only protein 2 is a candidate regulator of inflammasome activation. Infect Immun 75:1484–1492. doi:10.1128/IAI.01315-06
Dorfleutner A, Talbott SJ, Bryan NB et al (2007b) A shope fibroma virus PYRIN-only protein modulates the host immune response. Virus Genes 35:685–694. doi:10.1007/s11262-007-0141-9
Dorfleutner A, Chu L, Stehlik C (2015) Inhibiting the inflammasome: one domain at a time. Immunol Rev 265:205–216. doi:10.1111/imr.12290
Dostert C, Pétrilli V, Van Bruggen R et al (2008) Innate immune activation through Nalp3 inflammasome sensing of asbestos and silica. Science 320:674–677. doi:10.1126/science.1156995
Druilhe A, Srinivasula SM, Razmara M et al (2001) Regulation of IL-1beta generation by Pseudo-ICE and ICEBERG, two dominant negative caspase recruitment domain proteins. Cell Death Differ 8:649–657. doi:10.1038/sj.cdd.4400881
Duewell P, Kono H, Rayner KJ et al (2010) NLRP3 inflammasomes are required for atherogenesis and activated by cholesterol crystals. Nature 464:1357–1361. doi:10.1038/nature08938
Eisenbarth SC, Colegio OR, O'Connor W et al (2008) Crucial role for the Nalp3 inflammasome in the immunostimulatory properties of aluminium adjuvants. Nature 453:1122–1126. doi:10.1038/nature06939
Elinav E, Strowig T, Kau AL et al (2011) NLRP6 inflammasome regulates colonic microbial ecology and risk for colitis. Cell 145:745–757. doi:10.1016/j.cell.2011.04.022

Faustin B, Lartigue L, Bruey J et al (2007) Reconstituted NALP1 inflammasome reveals two-step mechanism of caspase-1 activation. Mol Cell 25:713–724. doi:10.1016/j.molcel.2007.01.032

Fernandes-Alnemri T, Yu J-W, Datta P et al (2009) AIM2 activates the inflammasome and cell death in response to cytoplasmic DNA. Nature 458:509–513. doi:10.1038/nature07710

Finger JN, Lich JD, Dare LC et al (2012) Autolytic proteolysis within the function to find domain (FIIND) is required for NLRP1 inflammasome activity. J Biol Chem 287:25030–25037. doi:10.1074/jbc.M112.378323

Fink SL, Bergsbaken T, Cookson BT (2008) Anthrax lethal toxin and Salmonella elicit the common cell death pathway of caspase-1-dependent pyroptosis via distinct mechanisms. Proc Natl Acad Sci U S A 105:4312–4317. doi:10.1073/pnas.0707370105

Halff EF, Diebolder CA, Versteeg M et al (2012) Formation and structure of a NAIP5-NLRC4 inflammasome induced by direct interactions with conserved N- and C-terminal regions of flagellin. J Biol Chem 287:38460–38472. doi:10.1074/jbc.M112.393512

Halle A, Hornung V, Petzold GC et al (2008) The NALP3 inflammasome is involved in the innate immune response to amyloid-beta. Nat Immunol 9:857–865. doi:10.1038/ni.1636

Hara H, Tsuchiya K, Kawamura I et al (2013) Phosphorylation of the adaptor ASC acts as a molecular switch that controls the formation of speck-like aggregates and inflammasome activity. Nat Immunol 14:1247–1255. doi:10.1038/ni.2749

He W, Wan H, Hu L et al (2015) Gasdermin D is an executor of pyroptosis and required for interleukin-1β secretion. Cell Res 25:1285–1298. doi:10.1038/cr.2015.139

He Y, Zeng M, Yang D et al (2016) NEK7 is an essential mediator of NLRP3 activation downstream of potassium efflux. Nature 530:354–357. doi:10.1038/nature16959

Hellmich KA, Levinsohn JL, Fattah R et al (2012) Anthrax lethal factor cleaves mouse nlrp1b in both toxin-sensitive and toxin-resistant macrophages. PLoS One 7:e49741. doi:10.1371/journal.pone.0049741

Hiller S, Kohl A, Fiorito F et al (2003) NMR structure of the apoptosis- and inflammation-related NALP1 pyrin domain. Structure 11:1199–1205. doi:10.1016/j.str.2003.08.009

Hoffman HM, Mueller JL, Broide DH et al (2001) Mutation of a new gene encoding a putative pyrin-like protein causes familial cold autoinflammatory syndrome and Muckle-Wells syndrome. Nat Genet 29:301–305. doi:10.1038/ng756

Hornung V, Bauernfeind F, Halle A et al (2008) Silica crystals and aluminum salts activate the NALP3 inflammasome through phagosomal destabilization. Nat Immunol 9:847–856

Hornung V, Ablasser A, Charrel-Dennis M et al (2009) AIM2 recognizes cytosolic dsDNA and forms a caspase-1-activating inflammasome with ASC. Nature 458:514–518. doi:10.1038/nature07725

Hou X, Niu X (2015) The NMR solution structure of AIM2 PYD domain from *Mus musculus* reveals a distinct α2-α3 helix conformation from its human homologues. Biochem Biophys Res Commun:2–6. doi:10.1016/j.bbrc. 2015.04.046

Hou F, Sun L, Zheng H et al (2011) MAVS forms functional prion-like aggregates to activate and propagate antiviral innate immune response. Cell 146:448–461. doi:10.1016/j.cell.2011.06.041

Hu Z, Yan C, Liu P et al (2013) Crystal structure of NLRC4 reveals its autoinhibition mechanism. Science 341:172–175. doi:10.1126/science.1236381

Hu Z, Zhou Q, Zhang C et al (2015) Structural and biochemical basis for induced self-propagation of NLRC4. Science 350:399–404. doi:10.1126/science.aac5489

Humke EW, Shriver SK, Starovasnik MA et al (2000) ICEBERG: a novel inhibitor of interleukin-1beta generation. Cell 103:99–111

Iyer SS, He Q, Janczy JR et al (2013) Mitochondrial cardiolipin is required for Nlrp3 inflammasome activation. Immunity 39:311–323. doi:10.1016/j.immuni.2013.08.001

Jin T, Perry A, Jiang J et al (2012) Structures of the HIN domain: DNA complexes reveal ligand binding and activation mechanisms of the AIM2 inflammasome and IFI16 receptor. Immunity 36:561–571. doi:10.1016/j.immuni.2012.02.014

Jin T, Perry A, Smith P et al (2013) Structure of the absent in melanoma 2 (AIM2) pyrin domain provides insights into the mechanisms of AIM2 autoinhibition and inflammasome assembly. J Biol Chem 288:13225–13235. doi:10.1074/jbc.M113.468033

Johnston JB, Barrett JW, Nazarian SH et al (2005) A poxvirus-encoded pyrin domain protein interacts with ASC-1 to inhibit host inflammatory and apoptotic responses to infection. Immunity 23:587–598. doi:10.1016/j.immuni. 2005.10.003

Juliana C, Fernandes-Alnemri T, Kang S et al (2012) Non-transcriptional priming and deubiquitination regulate NLRP3 inflammasome activation. J Biol Chem 287:36617–36622. doi:10.1074/jbc.M112.407130

Kang T, Yang S, Toth B et al (2013) Caspase-8 blocks kinase RIPK3-mediated activation of the NLRP3 inflammasome. Immunity 38:27–40. doi:10.1016/j.immuni.2012.09.015

Kanneganti T-D, Lamkanfi M, Kim Y-G et al (2007) Pannexin-1-mediated recognition of bacterial molecules activates the cryopyrin inflammasome independent of toll-like receptor signaling. Immunity 26:433–443. doi:10.1016/j.immuni.2007.03.008

Kayagaki N, Stowe IB, Lee BL et al (2015) Caspase-11 cleaves gasdermin D for non-canonical inflammasome signalling. Nature 526:666–671. doi:10.1038/nature15541

Kersse K, Vanden Berghe T, Lamkanfi M, Vandenabeele P (2007) A phylogenetic and functional overview of inflammatory caspases and caspase-1-related CARD-only proteins. Biochem Soc Trans 35:1508–1511. doi:10.1042/BST0351508

Kersse K, Lamkanfi M, Bertrand MJM et al (2011a) Interaction patches of procaspase-1 caspase recruitment domains (CARDs) are differently involved in procaspase-1 activation and receptor-interacting protein 2 (RIP2)-dependent nuclear factor κB signaling. J Biol Chem 286:35874–35882. doi:10.1074/jbc.M111.242321

Kersse K, Verspurten J, Vanden Berghe T, Vandenabeele P (2011b) The death-fold superfamily of homotypic interaction motifs. Trends Biochem Sci 36:541–552. doi:10.1016/j.tibs.2011.06.006

Kerur N, Veettil MV, Sharma-Walia N et al (2011) IFI16 acts as a nuclear pathogen sensor to induce the inflammasome in response to Kaposi sarcoma-associated herpesvirus infection. Cell Host Microbe 9:363–375. doi:10.1016/j.chom.2011.04.008

Khan N, Lawlor KE, Murphy JM, Vince JE (2014) More to life than death: molecular determinants of necroptotic and non-necroptotic RIP3 kinase signaling. Curr Opin Immunol 26:76–89. doi:10.1016/j.coi.2013.10.017

Khare S, Dorfleutner A, Bryan NB et al (2012) An NLRP7-containing inflammasome mediates recognition of microbial lipopeptides in human macrophages. Immunity:1–13. doi:10.1016/j.immuni.2012.02.001

Khare S, Ratsimandresy RA, de Almeida L et al (2014) The PYRIN domain-only protein POP3 inhibits ALR inflammasomes and regulates responses to infection with DNA viruses. Nat Immunol 15:343–353. doi:10.1038/ni.2829

Kitamura A, Sasaki Y, Abe T et al (2014) An inherited mutation in NLRC4 causes autoinflammation in human and mice. J Exp Med 211:2385–2396. doi:10.1084/jem.20141091

Kofoed EM, Vance RE (2011) Innate immune recognition of bacterial ligands by NAIPs determines inflammasome specificity. Nature 477:592–595. doi:10.1038/nature10394

Kortmann J, Brubaker SW, Monack DM (2015) Cutting edge: inflammasome activation in primary human macrophages is dependent on flagellin. J Immunol 195:815–819. doi:10.4049/jimmunol.1403100

Lamkanfi M, Dixit VM (2014) Mechanisms and functions of inflammasomes. Cell 157:1013–1022. doi:10.1016/j.cell.2014.04.007

Lamkanfi M, Denecker G, Kalai M et al (2004) INCA, a novel human caspase recruitment domain protein that inhibits interleukin-1beta generation. J Biol Chem 279:51729–51738. doi:10.1074/jbc.M407891200

Latz E, Xiao TS, Stutz A (2013) Activation and regulation of the inflammasomes. Nat Rev Immunol 13:397–411. doi:10.1038/nri3452

Lawlor KE, Vince JE (2014) Ambiguities in NLRP3 inflammasome regulation: is there a role for mitochondria? Biochim Biophys Acta 1840:1433–1440. doi:10.1016/j.bbagen.2013.08.014
Le HT, Harton JA (2013) Pyrin- and CARD-only proteins as regulators of NLR functions. Front Immunol 4:275. doi:10.3389/fimmu.2013.00275
Lee SH, Stehlik C, Reed JC (2001) Cop, a caspase recruitment domain-containing protein and inhibitor of caspase-1 activation processing. J Biol Chem 276:34495–34500. doi:10.1074/jbc.M101415200
Lee G-S, Subramanian N, Kim AI et al (2012) The calcium-sensing receptor regulates the NLRP3 inflammasome through Ca2+ and cAMP. Nature 492:123–127. doi:10.1038/nature11588
Levinsohn JL, Newman ZL, Hellmich KA et al (2012) Anthrax lethal factor cleavage of Nlrp1 is required for activation of the inflammasome. PLoS Pathog 8:e1002638. doi:10.1371/journal.ppat.1002638
Lin Z, Tann JY, Goh ETH et al (2015) Structural basis of death domain signaling in the p 75 neurotrophin receptor. Elizab Theatr 4:e11692. doi:10.7554/eLife.11692
Lopez-Castejon G, Luheshi NM, Compan V et al (2013) Deubiquitinases regulate the activity of caspase-1 and interleukin-1β secretion via assembly of the inflammasome. J Biol Chem 288:2721–2733. doi:10.1074/jbc.M112.422238
Lu B, Nakamura T, Inouye K et al (2012) Novel role of PKR in inflammasome activation and HMGB1 release. Nature 488:670–674. doi:10.1038/nature11290
Lu A, Kabaleeswaran V, Fu T et al (2014a) Crystal structure of the F27G AIM2 PYD mutant and similarities of its self-association to DED/DED interactions. J Mol Biol 426:1420–1427. doi:10.1016/j.jmb.2013.12.029
Lu A, Magupalli VG, Ruan J et al (2014b) Unified polymerization mechanism for the assembly of ASC-dependent inflammasomes. Cell 156:1193–1206. doi:10.1016/j.cell.2014.02.008
Lu A, Li Y, Schmidt FI et al (2016) Molecular basis of caspase-1 polymerization and its inhibition by a new capping mechanism. Nat Struct Mol Biol 23:416–425. doi:10.1038/nsmb.3199
Maelfait J, Vercammen E, Janssens S et al (2008) Stimulation of toll-like receptor 3 and 4 induces interleukin-1beta maturation by caspase-8. J Exp Med 205:1967–1973. doi:10.1084/jem.20071632
Man SM, Tourlomousis P, Hopkins L et al (2013) Salmonella infection induces recruitment of Caspase-8 to the inflammasome to modulate IL-1β production. J Immunol 191:5239–5246. doi:10.4049/jimmunol.1301581
Man SM, Hopkins LJ, Nugent E et al (2014) Inflammasome activation causes dual recruitment of NLRC4 and NLRP3 to the same macromolecular complex. Proc Natl Acad Sci U S A 111:7403–7408. doi:10.1073/pnas.1402911111
Martin BN, Wang C, Willette-Brown J et al (2014) IKKα negatively regulates ASC-dependent inflammasome activation. Nat Commun 5:4977. doi:10.1038/ncomms5977
Martinon F, Burns K, Tschopp J (2002) The inflammasome: a molecular platform triggering activation of inflammatory caspases and processing of proIL-beta. Mol Cell 10:417–426
Martinon F, Pétrilli V, Mayor A et al (2006) Gout-associated uric acid crystals activate the NALP3 inflammasome. Nature 440:237–241. doi:10.1038/nature04516
Masters SL, Dunne A, Subramanian SL et al (2010) Activation of the NLRP3 inflammasome by islet amyloid polypeptide provides a mechanism for enhanced IL-1β in type 2 diabetes. Nat Immunol 11:897–904. doi:10.1038/ni.1935
Masumoto J, Taniguchi S, Ayukawa K et al (1999) ASC, a Novel 22-kDa protein, aggregates during apoptosis of human promyelocytic leukemia HL-60 cells. J Biol Chem 274:33835–33838. doi:10.1074/jbc.274.48.33835
Masumoto J, Taniguchi S, Sagara J (2001) Pyrin N-terminal homology domain- and caspase recruitment domain-dependent oligomerization of ASC. Biochem Biophys Res Commun 280:652–655. doi:10.1006/bbrc.2000.4190
Masumoto J, Dowds TA, Schaner P, et al. (2003) ASC is an activating adaptor for NF-B and caspase-8-dependent apoptosis. Biochem Biophys Res Commun 303:69–73. doi:10.1016/S0006-291X(03)00309-7

Monie TP, Bryant CE (2015) Caspase-8 functions as a key mediator of inflammation and pro-IL-1β processing via both canonical and non-canonical pathways. Immunol Rev 265:181–193
Moriya M, Taniguchi S, Wu P et al (2005) Role of charged and hydrophobic residues in the oligomerization of the PYRIN domain of ASC. Biochemistry 44:575–583. doi:10.1021/bi048374i
Motani K, Kushiyama H, Imamura R et al (2011) Caspase-1 induces apoptosis-associated speck-like protein containing a caspase-recruitment domain (ASC)-mediated necrosis independently of its catalytic activity. J Biol Chem 286:33963–33972. doi:10.1074/jbc.M111.286823
Muñoz-Planillo R, Kuffa P, Martínez-Colón G et al (2013) K^+ efflux is the common trigger of NLRP3 inflammasome activation by bacterial toxins and particulate matter. Immunity 38:1142–1153. doi:10.1016/j.immuni.2013.05.016
Murakami T, Ockinger J, Yu J et al (2012) Critical role for calcium mobilization in activation of the NLRP3 inflammasome. Proc Natl Acad Sci U S A 109:11282–11287. doi:10.1073/pnas.1117765109
Nakahira K, Haspel JA, Rathinam VAK et al (2011) Autophagy proteins regulate innate immune responses by inhibiting the release of mitochondrial DNA mediated by the NALP3 inflammasome. Nat Immunol 12:222–230. doi:10.1038/ni.1980
Pétrilli V, Papin S, Dostert C et al (2007) Activation of the NALP3 inflammasome is triggered by low intracellular potassium concentration. Cell Death Differ 14:1583–1589. doi:10.1038/sj.cdd.4402195
Pierini R, Juruj C, Perret M et al (2012) AIM2/ASC triggers caspase-8-dependent apoptosis in Francisella-infected caspase-1-deficient macrophages. Cell Death Differ 19:1709–1721. doi:10.1038/cdd.2012.51
Proell M, Gerlic M, Mace PD et al (2013) The CARD plays a critical role in ASC foci formation and inflammasome signalling. Biochem J 449:613–621. doi:10.1042/BJ20121198
Py BF, Kim M-S, Vakifahmetoglu-Norberg H, Yuan J (2013) Deubiquitination of NLRP3 by BRCC3 critically regulates inflammasome activity. Mol Cell 49:331–338. doi:10.1016/j.molcel.2012.11.009
Qiao Q, Yang C, Zheng C et al (2013) Structural architecture of the CARMA1/Bcl10/MALT1 signalosome: nucleation-induced filamentous assembly. Mol Cell 51:766–779. doi:10.1016/j.molcel.2013.08.032
Qu Y, Misaghi S, Newton K et al (2016) NLRP3 recruitment by NLRC4 during salmonella infection. J Exp Med 37:782–789. doi:10.1084/jem.20132234
Rauch I, Tenthorey JL, Nichols RD et al (2016) NAIP proteins are required for cytosolic detection of specific bacterial ligands in vivo. J Exp Med 213:657–665. doi:10.1084/jem.20151809
Roberts TL, Idris A, Dunn JA et al (2009) HIN-200 proteins regulate caspase activation in response to foreign cytoplasmic DNA. Science 323:1057–1060
Rock KL, Latz E, Ontiveros F, Kono H (2010) The sterile inflammatory response. Annu Rev Immunol 28:321–342. doi:10.1146/annurev-immunol-030409-101311
Rodgers MA, Bowman JW, Fujita H et al (2014) The linear ubiquitin assembly complex (LUBAC) is essential for NLRP3 inflammasome activation. J Exp Med 211:1333–1347. doi:10.1084/jem.20132486
Romberg N, Al Moussawi K, Nelson-Williams C et al (2014) Mutation of NLRC4 causes a syndrome of enterocolitis and autoinflammation. Nat Genet 46:1135–1139. doi:10.1038/ng.3066
Rossol M, Pierer M, Raulien N et al (2012) Extracellular Ca(2+) is a danger signal activating the NLRP3 inflammasome through G protein-coupled calcium sensing receptors. Nat Commun 3:1329. doi:10.1038/ncomms2339
Sagulenko V, Thygesen SJ, Sester DP et al (2013) AIM2 and NLRP3 inflammasomes activate both apoptotic and pyroptotic death pathways via ASC. Cell Death Differ 20:1149–1160. doi:10.1038/cdd.2013.37
Salvesen GS, Dixit VM (1999) Caspase activation: the induced-proximity model. Proc Natl Acad Sci U S A 96:10964–10967

Schmid-Burgk JL, Chauhan D, Schmidt T et al (2016) A genome-wide CRISPR screen identifies NEK7 as an essential component of NLRP3 inflammasome activation. J Biol Chem 291:103–109. doi:10.1074/jbc.C115.700492
Schmidt FI, Lu A, Chen JW et al (2016) A single domain antibody fragment that recognizes the adaptor ASC defines the role of ASC domains in inflammasome assembly. J Exp Med 213:771–790. doi:10.1084/jem.20151790
Schroder K, Tschopp J (2010) The inflammasomes. Cell 140:821–832. doi:10.1016/j.cell.2010.01.040
Schroder K, Zhou R, Tschopp J (2010) The NLRP3 inflammasome: a sensor for metabolic danger? Science 327:296–300. doi:10.1126/science.1184003
Schroder K, Sagulenko V, Zamoshnikova A et al (2012) Acute lipopolysaccharide priming boosts inflammasome activation independently of inflammasome sensor induction. Immunobiology 217:1325–1329. doi:10.1016/j.imbio.2012.07.020
Shi Y (2004) Caspase activation: revisiting the induced proximity model. Cell 117:855–858. doi:10.1016/j.cell.2004.06.007
Shi C-S, Shenderov K, Huang N-N et al (2012) Activation of autophagy by inflammatory signals limits IL-1β production by targeting ubiquitinated inflammasomes for destruction. Nat Immunol 13:255–263. doi:10.1038/ni.2215
Shi J, Zhao Y, Wang K et al (2015) Cleavage of GSDMD by inflammatory caspases determines pyroptotic cell death. Nature 526:660–665. doi:10.1038/nature15514
Shi H, Wang Y, Li X et al (2016) NLRP3 activation and mitosis are mutually exclusive events coordinated by NEK7, a new inflammasome component. Nat Immunol 17:250–258. doi:10.1038/ni.3333
Shimada K, Crother TR, Karlin J et al (2012) Oxidized mitochondrial DNA activates the NLRP3 inflammasome during apoptosis. Immunity 36:401–414. doi:10.1016/j.immuni.2012.01.009
Srimathi T, Robbins SL, Dubas RL et al (2008) Mapping of POP1-binding site on pyrin domain of ASC. J Biol Chem 283:15390–15398. doi:10.1074/jbc.M801589200
Stehlik C, Krajewska M, Welsh K et al (2003) The PAAD/PYRIN-only protein POP1/ASC2 is a modulator of ASC-mediated nuclear-factor-kappa B and pro-caspase-1 regulation. Biochem J 373:101–113. doi:10.1042/BJ20030304
Subramanian N, Natarajan K, Clatworthy MR et al (2013) The adaptor MAVS promotes NLRP3 mitochondrial localization and inflammasome activation. Cell 153:348–361. doi:10.1016/j.cell.2013.02.054
Tenthorey JL, Kofoed EM, Daugherty MD et al (2014) Molecular basis for specific recognition of bacterial ligands by NAIP/NLRC4 inflammasomes. Mol Cell 54:17–29. doi:10.1016/j.molcel.2014.02.018
Vajjhala PR, Mirams RE, Hill JM (2012) Multiple binding sites on the pyrin domain of ASC protein allow self-association and interaction with NLRP3 protein. J Biol Chem 287:41732–41743. doi:10.1074/jbc.M112.381228
Vajjhala PR, Kaiser S, Smith SJ et al (2014) Identification of multifaceted binding modes for pyrin and ASC pyrin domains gives insights into pyrin inflammasome assembly. J Biol Chem 289:23504–23519. doi:10.1074/jbc.M114.553305
Vajjhala PR, Lu A, Brown DL et al (2015) The inflammasome adaptor ASC induces procaspase-8 death effector domain filaments. J Biol Chem 290:29217–29230. doi:10.1074/jbc.M115.687731
Vance RE (2015) The NAIP/NLRC4 inflammasomes. Curr Opin Immunol 32:84–89. doi:10.1016/j.coi. 2015.01.010
Vandenabeele P, Galluzzi L, Vanden Berghe T, Kroemer G (2010) Molecular mechanisms of necroptosis: an ordered cellular explosion. Nat Rev Mol Cell Biol 11:700–714. doi:10.1038/nrm2970
Vladimer GI, Weng D, Paquette SWM et al (2012) The NLRP12 inflammasome recognizes Yersinia pestis. Immunity 37:96–107. doi:10.1016/j.immuni.2012.07.006
Wang L, Manji GA, Grenier JM et al (2002) PYPAF7, a novel PYRIN-containing Apaf1-like protein that regulates activation of NF-kappa B and caspase-1-dependent cytokine processing. J Biol Chem 277:29874–29880. doi:10.1074/jbc.M203915200

Wang Y, Yang C, Mao K et al (2013) Cellular localization of NLRP3 inflammasome. Protein Cell 4:1–7. doi:10.1007/s13238-013-2113-2

Wen H, Miao EA, Ting JP-Y (2013) Mechanisms of NOD-like receptor-associated inflammasome activation. Immunity 39:432–441. doi:10.1016/j.immuni.2013.08.037

Wu B, Peisley A, Tetrault D et al (2014) Molecular imprinting as a signal-activation mechanism of the viral RNA sensor RIG-I. Mol Cell 55:511–523. doi:10.1016/j.molcel.2014.06.010

Yang J, Zhao Y, Shi J, Shao F (2013) Human NAIP and mouse NAIP1 recognize bacterial type III secretion needle protein for inflammasome activation. Proc Natl Acad Sci U S A 110:14408–14413. doi:10.1073/pnas.1306376110

Youm Y-H, Nguyen KY, Grant RW et al (2015) The ketone metabolite β-hydroxybutyrate blocks NLRP3 inflammasome–mediated inflammatory disease. Nat Med 21:263–269. doi:10.1038/nm.3804

Zhang L, Chen S, Ruan J et al (2015) Cryo-EM structure of the activated NAIP2-NLRC4 inflammasome reveals nucleated polymerization. Science 350:404–409. doi:10.1126/science.aac5789

Zhao Y, Yang J, Shi J et al (2011) The NLRC4 inflammasome receptors for bacterial flagellin and type III secretion apparatus. Nature 477:596–600. doi:10.1038/nature10510

Zhao Y, Shi J, Shi X et al (2016) Genetic functions of the NAIP family of inflammasome receptors for bacterial ligands in mice. J Exp Med 213:647–656. doi:10.1084/jem.20160006

Zhou R, Tardivel A, Thorens B et al (2010) Thioredoxin-interacting protein links oxidative stress to inflammasome activation. Nat Immunol 11:136–140. doi:10.1038/ni.1831

Zhou R, Yazdi AS, Menu P, Tschopp J (2011) A role for mitochondria in NLRP3 inflammasome activation. Nature 469:221–225. doi:10.1038/nature09663

Chapter 3
The Ferritin Superfamily

Alejandro Yévenes

Abstract Iron is very important in many biological processes and the ferritin protein family has evolved to store iron and to maintain cellular iron homeostasis. The deletion of the coding gene for the H subunit of ferritin leads to early embryonic death in mice and mutations in the gene for the L subunits in humans has been observed in neurodegenerative diseases, such as neuroferritinopathy. Thus, understanding how ferritin works is imperative and many studies have been conducted to delineate the molecular mechanism of ferritins and bacterioferritins. In the ferritin protein family, it is clear that a catalytic center for iron oxidation, the routes for iron to reach this center and the ability to nucleate an iron core, are common requirements for all ferritins. However, there are differences in the structural and mechanistic details of iron oxidation and mineralization. Although a common mechanism has been proposed for all ferritins, this mechanism needs to be further explored. There is a mechanistic diversity related to structural variation in the ferritin protein family. It is clear that other factors appear to affect the mechanism of iron oxidation and mineralization. This review focusses on the structural features of the ferritin protein family and its role in the mechanism of iron mineralization.

Keywords Ferritin • Bacterial ferritin • Bacterioferritin • Iron biomineral • Ferroxidase activity • Ion channel • Nanomaterial • Diiron oxygenase

Abbreviations

AfFtn	*Archaeoglobus fulgidus* ferritin
AvBFR	*Azotobacter vinelandii* bacterioferritin
BFR	Bacterioferritin
BfMF	Bull frog M ferritin
CjFtn	*Campylobacter jejuni* ferritin
CtFtn	*Chlorobium tepidum* ferritin

A. Yévenes (✉)
Departamento de Química Física, Facultad de Química, Pontificia Universidad Católica de Chile, Santiago, Chile
e-mail: ayevenes@uc.cl

J.R. Harris, J. Marles-Wright (eds.), *Macromolecular Protein Complexes*, Subcellular Biochemistry 83, DOI 10.1007/978-3-319-46503-6_3

DdBFR	*Desulfovibrio desulfuricans* bacterioferritin
DFP	Diferric peroxo
Dps	DNA binding protein from starved cells
EcBFR	*Escherichia coli* bacterioferritin
EcFtnA	*Escherichia coli* ferritin A
FC	ferroxidase center
HpFtn	*Helicobacter pylori* ferritin
HoSF	Horse spleen ferritin
HuHF	Human H ferritin
HuLF	Human L ferritin
HMFt	human mitochondrial ferritin
ITC	Isothermal titration calorimetry
MRI	Magnetic resonance imaging
MD	Molecular dynamics
MtbBFrB	*Mycobacterium tuberculosis* bacterioferritin B
NMR	Nuclear magnetic resonance
PaFtn	*Pseudomonas aeruginosa* ferritin
PaBFR	*Pseudomonas aeruginosa* bacteriferritin
PfFtn	*Pyrococcus furiosus* ferritin
PmFtn	*Pseudo-nitzschia* ferritin
TmFtn	*Thermotoga maritima* ferritin
VcFtn	*Vibrio cholerae* ferritin

3.1 Introduction

Ferritin has evolved to store iron as a ferric oxyhydroxide mineral. Particulate or bound iron is encapsulated within members of the ferritin family of proteins. For living organisms the maintenance of iron homeostasis is very important, if we consider that Fe(II) is readily oxidized to Fe(III) by dioxygen and other reactive oxygen species, such as superoxide, hydrogen peroxide and water. Besides, Fe(II) reacts with hydrogen peroxide to generate the hydroxyl radical through the Fenton reaction (Boukhalfa and Crumbliss 2012), and Fe(III) is involved in the destructive catalytic cycle of Haber-Weiss (Kehrer 2000). Furthermore, the propensity of Fe(II) to react with dioxygen to generate the highly insoluble $Fe(OH)_3(H2O)_3$ (Ksp, of ~10−38) (Boukhalfa and Crumbliss 2012) limits the bio-availability of Fe(II).

The chemical properties of iron show the importance of ferritin in cellular iron homeostasis. Besides, deletion of the coding gene for the H subunit of ferritin leads to early embryonic death in mice (Ferreira et al. 2000) and mutations in the gene of the L subunits in humans has been observed in neurodegenerative diseases such as neuroferritinopathy (Curtis et al. 2001). Furthermore, in Parkinson's or Alzheimer's disease (Jellinger et al. 1990) and in acquired immunodeficiency syndrome (Drakesmith et al. 2005) an increase of the ferritin expression level and the concentration of Fe(III) has been reported. Also, other roles for ferritin have been suggested (Watt 2011), including a role in immunity and autoimmunity (Recalcati et al. 2008) and in lipid metabolism (Bu et al. 2012). Thus, understanding how ferritin works is imperative, and many studies have been conducted to delineate the molecular mechanism of ferritins and bacterioferritins.

3.2 Structural Overview of Ferritins

Ferritin is a member of the four α-helix bundle structure superfamily (Andrews 2010), which also includes soluble methane monooxygenase, ribonucleotide reductase, rubrerythrin, bacterioferritin (BFR), DNA binding protein from starved cells (Dps), Dps-like proteins, and the recently identified archaeoferritin (Ebrahimi et al. 2012). Furthermore, ferritin is a member of the ferritin-superfamily, possessing iron-storage capacity, this family also includes bacterial ferritin and bacterioferritin (Carrondo 2003), Dps and Dps-like proteins (Haikarainen and Papageorgiou 2010; Calhoun and Kwon 2011; Wiedenheft et al. 2005) and archaeoferritin.

The quaternary structure of ferritin consists of 24 subunits and each subunit is characterized by four tightly packed α-helices (A–D), a loop sequence connecting the B–C helices, and a short fifth α-helix (E) that adopts an acute angle with respect to the bundle near the C-terminus tail (Le Brun et al. 2010) (*see* Fig. 3.1a). However, helix E adopts a different orientation in *C. tepidum ferritin* (CtFtn) where it forms an angle of almost 90° with respect to the four-helical bundle (Arenas-Salinas et al. 2014) (Fig. 3.1a). Despite significant differences in the primary sequences, the tertiary structures of the different ferritins are remarkably similar, however, significant differences in quaternary structure have been observed. Where Dps and Dps-like proteins form 12-mer assemblies,where ferritin, bacterial ferritin and BFR are made of 24 identical subunits (24-mer). Interestingly, BFR has a heme group between pairs of subunits (Fig. 3.1b). The primary function of the Dps and Dps-like proteins appears to be the protection of DNA against oxidative damage (Haikarainen and Papageorgiou 2010; Grant et al. 1998). Archaeoferritin is a monomeric protein that assembles upon aerobic addition of Fe(II) and its physiological function is unknown (Ebrahimi et al. 2012). This review will focus on ferritins, bacterial ferritins and bacterioferritin, as they share similar structures and properties.

3.2.1 Mechanism of Ferritin Self-Assembly

Recent studies have provided evidence that dimers are the first intermediate in the ferritin self-assembly pathway. Alanine scanning mutagenesis of *E. coli* bacterioferritin (EcBFR) shows that mutations at the threefold and twofold symmetry axis abolished formation of the 24-mer protein shell completely and only dimers were formed (Zhang et al. 2010). The self-assembly of human H ferritin (HuHF) was

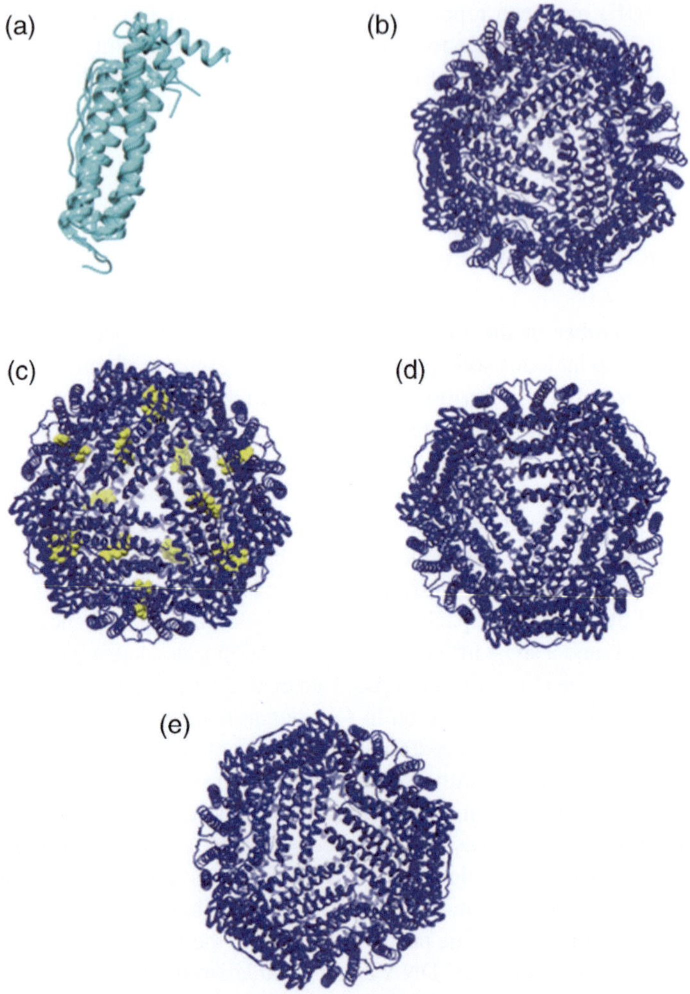

Fig 3.1 Structural alignment of ferritins subunits and cartoon representation of the macromolecular assembly of ferritins. (**a**) Structural alignment of the subunits of EcFtnA (1EUM.pdb), HuHF (2FHA.pdb), EcBFR (3E1M.pdb) and CtFtn (4CMY.pdb) where the helix E adopts a different orientation with respect to the rest of ferritin protein family. Cartoon representation of the macromolecular assembly of HuHF (**b**) (2FHA.pdb), EcBFR (**c**) showing the HEM group in *yellow* (2E1M.pdb), EcFtnA (**d**) (1EUM.pdb) and CtFtn (**e**) (4CMY.pdb). The figure and the structural alignment was made with Chimera (Pettersen et al. 2004). The figure was prepared using the crystal structures available in the Protein Data Bank

studied using a recent engineering strategy, which was termed reverse *metal-templated interface redesign* (Huard et al. 2013; Salgado et al. 2007, 2008). The formation of a 24-mer shell was only observed after the formation of a stable dimer. Thus, this interesting study defined a path for self-assembly of the 24-mer shell through the formation of dimers. The loop between helix B and helix C of one subunit interacts with this same loop of another subunit, generating a twofold symmetry. It has been shown that deletion of two residues in this loop in HuHF abolished formation of the 24-mer complex (Levi et al. 1989b) and a Asp80Lys mutation in this loop reduced the solubility and the stability of bull frog M ferritin (BfMF) (Bernacchioni et al. 2014). Therefore, a dimer appears to be the first intermediate during the self-assembly of the ferritin protein family and as soon as dimmers are formed, the formation of the 24-mer protein shell spontaneously occurs. Furthermore, a role of salt concentration in the mechanism of self-assembly has been described for bacterial ferritin. For example, it has been described that for *Archaeoglobus fulgidus* ferritin (AfFtn) self-assembly only occurs in the presence of a high salt concentration (Johnson et al. 2005; Sana et al. 2013), and it has been shown that the Apo *Thermotoga maritia* ferritin (TmFtn) exists largely as dimers at the cytoplasmic ionic strength of moderately halophilic bacteria (Ceci et al. 2011). In a similar way, EcBFR dissociates into dimers to the same extent as TmFtn at pH values around neutrality and 0.15 M NaCl (Andrews et al. 1995). Ceci et al. (2011) suggesting that TmFtn assembly is stabilized by iron micelles, as was also described for AfFtn after incorporation of 500 Fe/24-mer (Johnson et al. 2005), however, this characteristic has not been described in general for members of the ferritin protein family.

3.2.2 Structural Variety Amongst the Catalytic Ferroxidase Centers of Ferritins

Animal ferritins consist of two related subunits, the heavy and the light chain (H and L chain), that assemble in different proportions to form the 24-subunit protein cage (Cho et al. 2009; Toussaint et al. 2007a, b; Ha et al. 1999; Trikha et al. 1995; Granier et al. 1997). The HuHF, bacterial ferritins and BFR contain a highly conserved catalytic center called the ferroxidase center (FC), where iron is oxidized as the initial step of iron mineralization. This center possesses two iron binding sites (A and B), however, a third binding site (site C) has been found near the ferroxidase center in some Archaeal and bacterial ferritins and in BFR, as well as in mammalian ferritin (Le Brun et al. 2010; Toussaint et al. 2007a, b; Ha et al. 1999; Trikha et al. 1995; Granier et al. 1997; Masuda et al. 2010a; Hamburger et al. 2005; Stillman et al. 2001; Tatur et al. 2007; Johnson et al. 2005; Cho et al. 2009; Ceci et al. 2011; Ilari et al. 2000; Macedo et al. 2003; Ren et al. 2003; Khare et al. 2011; Pfaffen et al. 2013; Hitchings et al. 2014; Ebrahimi et al. 2012).

The 3-dimensional structure of representatives members from all three ferritin subfamilies have been determined by X-ray crystallography, from a range of species, including human (Toussaint et al. 2007a, b), frog (Ha et al. 1999; Trikha et al. 1995), horse (Granier et al. 1997), soybean (Masuda et al. 2010a), insect (Hamburger et al. 2005), *E. coli* (Stillman et al. 2001), *Campylobacter jejuni* (PDB code: 1krq; unpublished work), *Pyrococcus furiosus* (Tatur et al. 2007), *A. fulgidus*

(Johnson et al. 2005), *Helicobacter pylori* (Cho et al. 2009), *T. maritima* (Ceci et al. 2011), *Listeria innocua* (Ilari et al. 2000), *Desulfovibrio desulfuricans* (Macedo et al. 2003), *Bacillus brevis* (Ren et al. 2003), *Mycobacterium tuberculosis* (Khare et al. 2011), *Vibrio cholerae* (PDB code:3QZ3; unpublished work), pinnate diatom (Pfaffen et al. 2013), *Streptomyces coelicolor* (Hitchings et al. 2014) and others. Fundamental to the available structures of the ferritin protein family, the FC has been classified as belonging to one of three distinct types: H-chain-type, Ftn-type and bacterioferritin type.

3.2.2.1 The H-Chain-Type Ferroxidase Centers

The BfMF was the first eukaryotic ferritin ferroxidase center observed in an Fe(III) bound state (Bertini et al. 2012). The Fe(III) at site A is coordinated to Glu23, Glu58, which bridges the two iron ions and His61. At site B, Fe(III) is coordinated by the bridges Glu58 and Glu103 (Bertini et al. 2012)(*see* Fig. 3.2a). Close to the FC, Tyr34/Tyr30 and Gln41/137 are strictly conserved amongst eukaryotic ferritins. Tyrosine radical formation has been detected in the catalytic reaction of the FC, however, the function of this tyrosine is unclear because it does not appear to be essential in the oxidation reaction (Chen-Barrett et al. 1995) (*see* Sect. 3.3). A similar coordination environment is observed in the FC of HuHF, derived from its crystal structure containing Zn(II) (Toussaint et al. 2007a, b). In BfMF the di-iron distance between site A and B is 3.1 Å, consistent with a bridged diferric center (Bertini et al. 2012). Recently, a third iron binding site in mammalian ferritins was identified. The analysis of the crystal structure of HuHF in presence of Tb(III), Zn(II) and Mg(II) showed the presence of three metal ion binding sites (Lawson et al. 1991; Toussaint et al. 2007a, b). Depending on the metal ion, the conformation of site C in HuHF varied and the amino acid ligands of the metal ion in site C are different: for Tb(III) Glu61 and Glu64 (Lawson et al. 1991); for Zn(II) His57 and Glu61(Toussaint et al. 2007a, b); and for Mg(II) Glu140 (Masuda et al. 2010a). Mutation of the coordinating residues of site C in HuHF decreased the rate of catalysis of Fe(II) oxidation in the FC (Masuda et al. 2010a, b). Therefore, site C also appears to be important for the catalysis of Fe(II) oxidation (*see* Sect. 3.3). In human L-chain ferritin (HuLF), which apparently oxidizes Fe(II) with a rate just above the rate of background oxidation of free Fe(II) by molecular oxygen (Levi et al. 1988, 1989a, b), the coordination environment of sites A and B is absent (Lawson et al. 1991; Boyd et al. 1985). This observation suggests that sites A and B are essential for fast catalysis of Fe(II) oxidation by HuHF.

In addition, three metal ion binding sites have been described in BfMF. Specifically, three Co(II) ion binding sites were observed in the X-ray structure of BfMF (Tosha et al. 2010), two sites in the middle of each subunit in the same place as the FC in HuHF and a third site with two slightly different conformations (each conformation <50% occupancy), in a position that corresponds to that of site C in HuHF. As was described to HuHF, the site directed mutation of the residues of site A, B, or C in BfMF reduced the rate of Fe(II) oxidation (Tosha et al. 2008; Behera and Theil

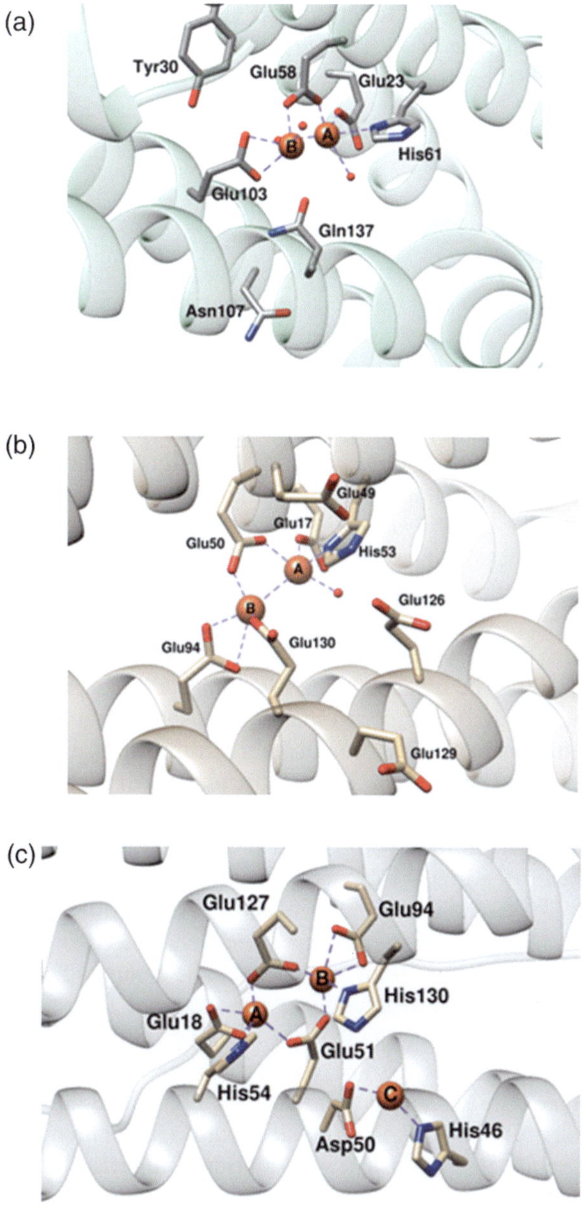

Fig. 3.2 Structural variety amongst the catalytic Ferroxidase centers of Ferritins. In (**a**) is shown the FC of BfFM (3RBC.pdb); the iron at site (**a**) is coordinated to Glu23 and Glu58, which bridges the two iron ions and His61. In the site (**b**), iron is coordinated by the bridges Glu58 and Glu103. In (**b**) is shown the FC of CtFtn (4CMY.pdb); the iron at site (**a**) is coordinated by Glu17 and His53, and bridging to Glu50. In site (**b**), iron is coordinated to the bridging Glu50 and terminal Glu94 and Glu130. The additional Glu130 ligand is actually a bridging ligand to a third iron-binding site (**c**), which is also coordinated by three further glutamates (Glu49, Glu126 and Glu129). In (**c**) is shown the FC of EcBFR (3E1M.pdb); iron at site (**a** and **b**) are coordinated by Glu18/Glu94 and histidine (His54/130), respectively. Additionally, iron at site (**a** and **b**) is coordinated to two bridging glutamate (Glu51 and Glu127). The figures were made with Chimera (Pettersen et al. 2004). The figure was prepared using the crystal structures available in the Protein Data Bank

2014). The X-ray structure of BfMF loaded with two Fe(II) per subunit (Bertini et al. 2012) shows that the two Fe(II) are located in site A and B of the FC. In this way, the amino acids that form sites A, B and C in BfMF align with those observed in HuHF.

3.2.2.2 Bacterial Ferritin Ferroxidase Centers

The bacterial ferritin ferroxidase center closely resembles that found in H-chain ferritin. In CtFtn, the amino acid residue side chain that coordinates the iron in site A is the same as in vertebrate ferritins, involving terminal Glu17 and His53, and bridging Glu50. In site B, iron is coordinated to the bridging Glu50 and terminal Glu94 and Glu130 residues. The additional Glu130 ligand (compared to H-chain-type centers), is actually a bridging ligand to a third iron-binding site (site C), which is also coordinated by three further glutamates (Glu49, Glu126 and Glu129) (Arenas-Salinas et al. 2014) (*see* Fig. 3.2b). This coordination environment is similar to the one that is found in *Escherichia coli* (EcFtnA) (Stillman et al. 2001). Sequence comparisons have revealed that the FC ligands are highly conserved in other prokaryotic ferritins (Tatur et al. 2007). The same coordination environment that is found in EcFtnA can also be found in two Archaeal ferritins, from *A. fulgidus* and *P. furiosus*, which have been characterized in an iron loaded form (Tatur et al. 2007; Johnson et al. 2005). The iron found in *P. furiosus* FC (Tatur et al. 2007) shows a lower occupancy at site B compare to site A, and also shows a di-iron distance that is shorter than the distance that is seen in other ferritin structures, but which is consistent with a di-Fe(III) center. The crystal structure of AfFtn showed that the Fe-A to Fe-B distance was 3.18 Å, where an electron density was clearly observed between the irons, which would be consistent with a oxo-bridge and occupancy of Fe(III). In this case, the iron at site C is 6 Å away from Fe-B (Johnson et al. 2005), and site C is a 6.3 Å from this center. In eukaryotes, the ferritins of frog and pinnate diatom *Pseudo-nitzschia* multiseries have been characterized in an iron-bound form. The *P. nitzschia* ferritin (PmFtn) appears to be composed of a single subunit type, with a sequence that is weakly related to prokaryotic ferritins (Marchetti et al. 2009). Its FC contains a Fe-A site that has a similar coordinating environment as those of the other eukaryotic ferritins. However, its Fe-B is distinct, because it has another glutamate (Glu130) besides the bridging Glu48 and Glu94. The iron-iron distance after an overnight soak of its crystal with Fe(II) solution was 3.6 Å, longer that in other ferritins, but consistent with a bridged diferric center (Pfaffen et al. 2013). The Glu30 in EcFtnA also acts as a bridging ligand to the third iron-binding site. Therefore, PnFtn is an Ftn-type ferritin with coordination at site C, somewhat similar to what is seen in prokaryotic ferritins (Pfaffen et al. 2013).

The role of the FC in bacterial ferritin has been studied by site-directed mutagenesis, where amino acid replacement in sites A, B and C reduced the rate of Fe(II) oxidation in *Pyrococcus furiosus* ferritin (PfFtn) (Ebrahimi et al. 2012) and EcFtnA (Treffry et al. 1998a, b). In the crystal structure of bacterial ferritin, site A has always been found to be fully occupied with metal ions. However, in PmFtn, sites

A, B or C are not occupied with metal ions (Ebrahimi et al. 2015). A full occupancy of sites A and B of the FC has been observed for BfMF and *Pseudomonas aeruginosa* ferritin (PaFtn) at neutral pH and for CtFtn at pH 9.0 (Arenas-Salinas et al. 2014). Taking in account the available data and the comparison of the metal ion coordination sites of the FC of different ferritins, it is possible to consider that all ferritins possess two iron binding sites in their FC (Site A and B) and a third iron binding site in the vicinity of the FC (site C). Site C of the FC in some ferritins shows a different conformation and in all structures, it is either partially occupied with metal ions or it is vacant. This suggests that this site can act as a transient iron binding site. More studies are needed to define the role of site C in the mechanism of action of ferritin. It has however been suggested that site C acts as a gateway of the FC, which is responsible for the translocation of Fe(II)/Fe(III) into or out of the FC (Ebrahimi et al. 2012)(see Sect. 3.3).

3.2.2.3 Bacterioferritin Ferroxidase Centers

The analysis of the ferroxidase center of BFR has been possible due to several iron loaded crystal structures that have been obtained; these include those from *Desulfovibrio desulfuricans*, *E. coli*, *Azotobacter vinelandii* and *P. aeruginosa* (Macedo et al. 2003; Crow et al. 2009; Swartz et al. 2006; Weeratunga et al. 2010). BFR contains an intra-subunit di-iron site and it is different from the H-chain type and bacterial ferritin centers (*see* Fig. 3.2c). In EcBFR, the iron at sites A and B are coordinated by Glu18/Glu94 and His54/130, respectively. Besides, the iron at sites A and B is coordinated by two bridging glutamates (Glu51 and Glu127) (Crow et al. 2009). This symmetrical FC is similar to the dinuclear iron centers that are seen in ribonucleotide reductase and methane monooxygenase (Nordlund et al. 1990; Rosenzweig et al. 1993). Due to this characteristic, BFR belongs to class II of the dinuclear iron centers. The inter Fe(III)/Fe(III) distance is 3.6 Å in the EcBFR structure (Crow et al. 2009), longer than other prokaryotic ferritins, but similar to inter Fe(III)/Fe(III) distance of *D. desulfuricans* and *A. vinelandii* BFR (3.7 and 3.5 Å, respectively) (Macedo et al. 2003; Swartz et al. 2006). This distance increases to 4 Å in the Fe(II)/Fe(II) forms (Macedo et al. 2003; Crow et al. 2009; Swartz et al. 2006). The His130 shows flexibility and it is not always coordinated to Fe(II) at site B (Swartz et al. 2006). There is no site C in the BFR FC, which is similar to the bacterial ferritin FC center, but an additional iron binding site has been characterized in EcBFR, facing the inner cavity. In this site, the Fe(II) is coordinated by Asp50, His 46 and three water molecules (Crow et al. 2009)(Fig. 3.2c). Despite the fact that this site could be similar to site C found in bacterial ferritin, in BFR this site is located at 9.2 Å from the FC. However, the Asp50 in BFR is equivalent but not structurally identical to the site C residue Glu49 in EcFtnA. This inner iron binding site is important in the mechanism of iron mineralization in BFR (*see* Sect. 3.3). Another important structural feature of BFR is that it can contain up to 12 hemes per 24-mer protein, the heme group participates in iron release, as is also described in Sect. 3.3 (Andrews et al. 1995; Frolow et al. 1994).

Table 3.1 Amino acids composition of the threefold channels in ferritins and bacterial ferritins

PDB code	Organisms	Outer entrance		Middle		Inner entrance	
2JD6	PfFtn	Ala106	Glu109	Tyr114	Arg117	Glu121	
4CMY	CtFtn	Leu109		Arg110	Phe117	Gln118	
1SQ3	AfFtn	Glu108	Met111	Phe116	Tyr119	Asn120	
3EGM	HpFtn	Lys110		Ile109	Phe117	Asn118	
1EUM	EcFtnA	His106	Met109	Tyr114	Phe117	Asn118	Gln121
3QZ3	VcFtn	Thr116		Phe115	Phe123	Asn124	
1VLG	TmFtn	Ser111		Val119		Ser120	Lys123
3QD8	MtFtn	Arg114		Glu122		Gln123	
1KRQ	CjFtn	Thr110		Leu109	Phe117	Asn118	
3R2K	PaFtn	Glu108		Arg116		Asp117	Lys120
3E6R	PmFtn	Ser109	Thr110	Met117		Asn121	
2FG8	HuLF	Ser122	Cy130	Asp131	Glu134	Thr135	
3AJO	HuHF	Thr122	Cys130	Asp131	Glu134	Thr135	

Table 3.2 Amino acids composition of the fourfold channels in ferritins and bacterial ferritins

PDB code	Organisms	Outer entrance		Middle		Inner entrance	
2JD6	PfFtn	Gln149		Met153		Lys156	Glu157
4CMY	CtFtn	Asn145	Met147	Arg150	Arg154	Asp156	Glu157
1SQ3	AfFtn			Open pore			
3EGM	HpFtn	His149		Leu153		Gln156	
1EUM	EcFtnA	Glu149		Phe153		Lys156	Glu157
3QZ3	VcFtn	Lys155		Phe159		Lys162	
1VLG	TmFtn	Gln149		Ser151		Gln155	
3QD8	MtFtn	Glu155		Asn158		Arg162	
1KRQ	CjFtn	Gly148		Asn149		Leu153	
3R2K	PaFtn	Glu147		Asn148		Gln151	
3E6R	PmFtn	Leu150		Ser153		Thr157	
2FG8	HuLF	Leu165		Leu169		Leu173	
3AJO	HuHF	Leu165		Leu169		His173	

3.2.3 *Structural Variety in the Three and Fourfold Channel Amongst Ferritins (Possible Role of the B-Channel)*

In the ferritin protein family, the formation of the 4-3-2 symmetrical structure of the protein shell creates channels that allow the communication of their inner cavity with the outside world. However, the analysis of the amino acid residues that form the three and fourfold channels in ferritins and bacterial ferritins show that these amino acid are not conserved among the ferritin protein family (*see* Tables 3.1 and 3.2). Early studies in horse spleen ferritin (HoSF) shows that Cd(II) binds to its threefold channel (Stefanini et al. 1987) and the chemical modification of Cys126 present in its threefold channel decreases the Fe(II) oxidation rate (Douglas and

Ripoll 1998). Using a theoretical approach, it has been determined that the residues forming the HuHF threefold channel possess an electrostatic gradient, which allows iron to enter into its inner cavity (Douglas and Ripoll 1998; Laghaei et al. 2013). It has been shown that Zn(II) binds to the threefold channel, inhibiting the ferroxidase activity of HuHF (Bou-Abdallah et al. 2003; Toussaint et al. 2007a, b). Studies in BfMF show that Co(II) or Mg(II) are present in the threefold channel (Tosha et al. 2010). Furthermore, the flash frozen crystal of BfMF, which was aerobically soaked with Fe(II), shows the presence of this cation in the threefold channel (Bertini et al. 2012). However, similar studies show that iron was not found in the threefold or fourfold channel of PmFtn (Pfaffen et al. 2013). It has been proposed that Fe(II) reaches the FC by a diffusion controlled rate after entry into the inner cavity by the threefold channel (Bou-Abdallah et al. 2008). When residues belonging to site C are changed, a decrease Fe(II) oxidation is observed in Soybean ferritin and HuHF (Masuda et al. 2010a, b; Ensign et al. 2004). These results are in agreement with molecular dynamics (MD) simulation studies that predict that site C is present in an Fe(II) pathway to the FC on HuHF (Laghaei et al. 2013) and by the presence of Zn(II) in the crystal structure of human mitochondrial ferritin (HMFt) (Langlois et al. 2004).

Although bacterial ferritins present similar channels to those present in eukaryotic ferritin, there is no experimental evidence regarding the involvement of these channels as the iron entry route. There is only the crystal structure of HpFtn, which shows iron in its fourfold channel, which was suggest as the iron entry route to the FC and from there to the inner cavity (Cho et al. 2009). From multiple sequence alignment, it was concluded that, unlike the threefold channels in eukaryotic ferritin, the threefold channels in bacterial ferritin are hydrophobic, while the fourfold channels in some microorganism are hydrophilic. Tables 3.1 and 3.2 show that there is more variety in the amino acid composition of the threefold channel in bacterial ferritins than there is in the eukaryotic ferritins. The fourfold channel of BfMF, HuLf and HuHF, HMFt and plant ferritin are hydrophobic and all of them present a hydrophilic and negatively charged threefold channel. In contrast, prokaryotic ferritins have both hydrophilic and hydrophobic three and fourfold channel (*see* Tables 3.1 and 3.2). PfFtn has a hydrophilic threefold channel with a mix of positive and negative residues and a hydrophobic fourfold channel. The threefold channel of bacterioferritin B (MtBfrB) of *Mycobacterium tuberculosis* is similar to the PfFtn threefold channel, but MtBfrB has a hydrophobic fourfold channel. On another hand, the CtFtn has hydrophilic threefold and fourfold channels with a mix of positively and negatively charged residues. PaFtn has a hydrophilic positively charged threefold channel, but a hydrophobic fourfold channel. TmFtn and *Helicobacter pylori* ferritin (HpFtn) have a hydrophobic threefold channel, but a hydrophilic fourfold channel. EcFtnA, *Vibrio cholerae* ferritin (VcFtn) and *Campylobacter jejuni* ferritin (CjFtn) have hydrophobic three- and fourfold channels. An exceptional case is AfFtn, which has a hydrophobic threefold channel that forms an open pore (Johnson et al. 2005).

In general term, the quaternary structure of bacterioferritin is similar to that of ferritin and similar channels, as have been described for ferritin, are observed in bacterioferritin. However, in bacterioferritin, the threefold channels are mainly hydrophilic and the amino acids that form the fourfold channel are mainly hydrophobic. In general, three channels have been proposed as a possible pathway of Fe(II) to the ferroxidase center of bacterioferritin: the fourfold channels, B-channel and the ferroxidase channel. The ferroxidase channel was proposed based on the conformational changes of a histidine of the ferroxidase center that were observed in the crystal structures of *Azotobacter vinelandii* bacterioferritin (AvBFR), *Pseudomonas aeruginosa* bacteriferritin (PaBFR) and *Desulfovibrio desulfuricans* bacterioferritin (DdBFR) (Swartz et al. 2006; Weeratunga et al. 2010; Macedo et al. 2003). In this pathway, the iron enters directly to the FC, where it is oxidized and then transferred to the inner cavity. The observation of Ba(II) or Fe(III) ions at the fourfold channel in the crystal structure of AvBFR (Liu et al. 2004; Swartz et al. 2006); and K(I) in the crystal structure of PaBFR (Weeratunga et al. 2010) has been used to support the fourfold channel as the iron entry route. However, Wong et al. (2015) proposed that the B-channel is a major route for iron entry into both the ferroxidase center and the iron storage cavity of bacterioferritin. The site-directed mutagenesis of the residues belonging to the B-channel of EcBFR produced a significant decrease in the rates of initial oxidation of Fe(II) at the FC. In fact, the crystal structure of the Asp132Phe variant showed that this substitution caused a steric blockage of the B-channel (Wong et al. 2015). Studies by MD simulation predict a route where iron can reach the FC from the B-channel through site C of the FC (Weeratunga et al. 2010). However, the exact role of this site in bactrioferritin is not known. Also, Wong et al. (2015) reported that the examination of ferritins from *E. coli*, *C. jejuni* (PDB ID 1KRQ), M. tuberculosis (Nordlund et al. 1990), *Vibrio cholera* (PBD ID 3QZ3), T. maritime (PDB ID 1Z4A) (Salgado et al. 2008), and *P. furiosus* (Salgado et al. 2007) revealed that B-type channels in these structures are larger than those observed in the structure of EcBFR and should readily permit iron passage. Considering the difference in the composition of the residues belonging to the three and fourfold channels in bacterial ferritins, site C of the FC acquires an important role as the probable iron entry route to the FC from the B channels. In a similar way to what is found in eukaryotic ferritin, the mutation of site C in PfFtn decreases the Fe(II) oxidation rate (Tatur et al. 2007). In addition, the rearrangement of the side chains of two of the residues of site C (Glu51 and Glu131) observed in AfFtn is in agreement with a role for site C in the iron entry to the FC (Johnson et al. 2005). Similar conclusions were obtained when the crystal structure of BfMF in presence of Co(II), Cu(II) or Fe(II) were analyzed (Behera and Theil 2014).

3.3 Mechanistic Variations in Mineralization Amongst Ferritins

Initial studies show that in anaerobic conditions and at an iron concentration below or equal to that which is required to fill the FC of HuHF, two Fe(II) bind at sites A and B of the FC (Tosha et al. 2010). The two Fe(II) are subsequently oxidized by O_2,

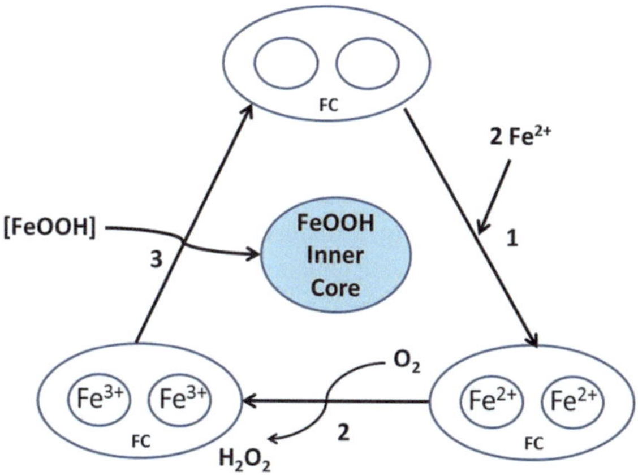

Fig. 3.3 Mechanism for H-chain-type ferritins. Two Fe(II) bind at the site A and B of the FC (*1*). Then, the two Fe(II) are oxidized by O_2 and this is reduced to hydrogen peroxide by a two-electron process where each Fe(II) contributes one electron, resulting in an iron/O_2 ratio of 2:1 (*2*). After the initial oxidation of the Fe(II), the μ-1,2-oxodiferric species is not stable at the FC and undergoes a hydrolysis reaction, giving [2FeOOH] which forms part of the nascent inner mineral core (*3*)

which is reduced to hydrogen peroxide by a two-electron process where each Fe(II) contributes one electron, resulting in an iron/O_2 ratio of 2:1 (Jameson et al. 2002; Zhao et al. 2003; Bou-Abdallah et al. 2002a, b) (see Fig. 3.3):

$$[\mathrm{P}: FC(Fe(II)_2)]^{z+4} + \mathrm{O}_2 + \mathrm{H}_2\mathrm{O} \rightarrow [\mathrm{P}: FC(Fe(III)_2)\mathrm{O}]^{z+4} + \mathrm{H}_2\mathrm{O}_2 \quad (3.1)$$

In Eq. 3.1, P represents the iron bound to ferritin, FC represents the iron at the ferroxidase center, and z is the net charge on the protein. This reaction proceeds through a blue diferric peroxo (DFP) intermediate (Bou-Abdallah et al. 2002a, b, 2005a, b; Pereira et al. 1998; Moenne-Loccoz et al. 1999; Zhao et al. 2005). Although this has not been detected in all H-chain ferritins, a DFP species has been proposed as a general feature of the H-chain FC. During the FC reaction, a tyrosil radical close to the FC has also been detected (Chen-Barrett et al. 1995), however, the role of this radical is unclear as several H-chain variants exhibit wild-type ferroxidase activity without the detectable formation of this radical. After the initial oxidation of the Fe(II), the μ-1,2-oxodiferric species is not stable at the FC and undergoes a hydrolysis reaction, leading to proton release (Yang et al. 1998) and giving [2FeOOH] which forms part of the nascent inner mineral core (*see* Eq. 3.2, Fig. 3.3):

$$[\mathrm{P}: FC(Fe(II)_2)]^{z+4} + \mathrm{O}_2 + 4\mathrm{H}_2\mathrm{O} \rightarrow 2[(Fe(III)OOH(\mathrm{core})]^{z} + \mathrm{H}_2\mathrm{O}_2 + 4\mathrm{H}^{+} \quad (3.2)$$

Studies with transferrin suggest that Fe(III) remains stable at the FC after its oxidation (Ebrahimi et al. 2012). However, this is in disagreement with other studies in which migration of Fe(III) out the FC and formation of small iron clusters have been demonstrated directly by time-resolved Mössbauer spectroscopy in HuHF and frog ferritin (Bou-Abdallah et al. 2002a, b; Bauminger et al. 1993; Pereira et al. 1997). The rate of Fe(III) migration from the FC into the inner cavity is enhanced by the addition of Fe(II), implying a displacement mechanism (Ebrahimi et al. 2012; Bou-Abdallah et al. 2005a, b). The latter suggests that the FC is a gated iron site, facilitating transfer of iron into the cavity (Bou-Abdallah et al. 2005a, b; Yang et al. 1998; Waldo and Theil 1993). NMR experiments indicate that Fe(III) takes a route along the long axis of the subunit, emerging at the fourfold axis, where it enters the cavity. Once a mineral core has formed, oxidation of Fe(II) is catalyzed by the growing core surface (Theil 2011; Yang et al. 1998; Xu and Chasteen 1991) (*see* Eq. 3.3).

$$4[\text{Cavity}: Fe(II)]^{z+2} + O_2 + 6H_2O \rightarrow 4[Fe(III)OOH(\text{core})]^{z} + 8H^{+} \quad (3.3)$$

Cavity indicates that the iron is located in the protein cavity but that it is not yet part of the iron mineral. In Eq. 3.3, the iron to oxygen ratio is 4:1, suggesting that the two reaction pathways (Eqs. 3.1 and 3.3) can be distinguished. Besides, the H_2O_2 at the FC can also act as an oxidant of Fe(II) at the core surface (Zhao et al. 2003) suggestion that the Fe(II):O_2 ratio will tend toward 4:1. Therefore, once a significant core has formed, further mineralization in H-chain-type ferritin occurs via a complex of parallel oxidation pathways. However, recently Ebrahami et al. (2012) analyzed the properties of HuHF with the PfFtn, and these results were used to suggest a new mechanism of iron mineralization in ferritin. The main result of this study was the identification of a third iron binding site in HuHF, in this way both ferritins present three iron sites. In fact, these sites share similar properties, one site of high affinity and two sites of low affinity. The high affinity site was assigned to site A, and the low affinity to site B and C. The authors also proposed that the Fe(III) formed remains at the FC, because it is accessible to transferrin and through Mossbauer spectroscopy the stepwise movement of the iron was determined through the FC to the nucleation site and from there to the inner cavity. When Fe(III) reaches the site C it is less accessible and essentially unavailable in the nucleation center. Electron paramagnetic resonance experiments showed that a mixed-valent Fe(II)-Fe(III) it is produced in the FC, where addition of excess Fe(II) decreases the formation of this intermediate, suggesting that there is electron transfer from the external Fe(II) to the Fe(III) at the nucleation center. In summary, Ebrahimi et al. (2012) proposed a common mechanism for the HuHF and PfFtn. The mechanism proposed that both system possess a third Fe(II) binding site and that at least some of the oxidized Fe(III) remains at the FC upon oxidation of iron. Then, additional Fe(II) forces the Fe(III) to move from the FC to the inner cavity of ferritin. Furthermore, it was proposed as a common mechanism for all ferritins, based on their study of PfFtn and HuHF (Ebrahimi et al. 2012, 2013, 2015). In this mechanism, iron can be

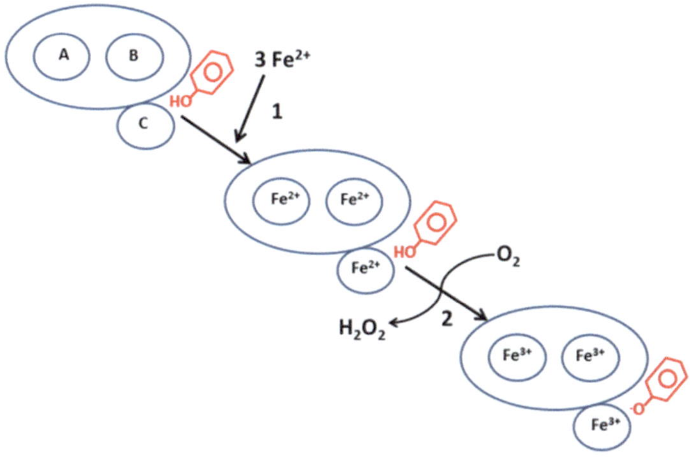

Fig. 3.4 Mechanism of Fe(II) oxidation by ferritin. In this mechanism iron can distribute among of three sites, if iron bind to site *A* and *B*, the mechanism of iron oxidation follow the Fig. 3.3. If three iron bind to site *A*, *B* and *C* (*1*), two Fe(II) in the ferroxidase center are simultaneously oxidized to form a blue intermediate. The third Fe(II) either reacts with this intermediate or it is oxidized by the peroxide that is released as the *blue* intermediate decays. The fourth electron for complete reduction of molecular oxygen to water is proposed to be provided by the conserved tyrosine in the vicinity of the ferroxidase center (*2*)

oxidized by two pathways (see Fig. 3.4). The first pathway is generally accepted in the literature for HuHF (Chasteen and Harrison 1999; Bou-Abdallah et al. 2002a, b, 2005a, b; Zhao et al. 2001, 2003; Yang et al. 1998) and proposes that iron is oxidized at the A and B sites of the FC and hydrogen peroxide is the product of the dioxygen reduction, giving an Fe(II)/O_2 stoichiometry of 2:1, as was described above in Fig. 3.3 and Eq. 3.1. In the second pathway, iron is simultaneously oxidized at the A, B and C site of the FC, this mechanism suggests an important role for Tyr24, which form a Tyr cation radical that is neutralized by oxidation of additional Fe(II) at an unspecified site to give a Fe(II)/O_2 stoichiometry of 4:1 (Fig. 3.4).

The third iron site has also been proposed to function as a transit site from the FC in a number of ferritins, including EcFtnA, soybean and frog H-chain ferritin (Masuda et al. 2010a, b; Ebrahimi et al. 2012). In general, EcFtnA displays similar properties to those seen in the HuHF, both have a similar Fe(II) oxidation stoichiometry of 48 Fe(II)/shell (Zhao et al. 2001, 2003; Yang et al. 1998;Treffry et al. 1997). Iron oxidation is dependent of the functionality of sites A and B (Chasteen and Harrison 1999; Chasteen 1998; Harrison and Arosio, 1996), producing two related colored reaction intermediates (Bou-Abdallah et al. 2005a, b) and generating a tyrosil radical (Chen-Barrett et al. 1995). However, when 48 Fe(II) per EcFtnA and HuHF 24-mer are added, both proteins differ in the stoichiometry of Fe(II)/O2 oxidation, 3 and 2, respectively (Bou-Abdallah et al. 2014). Site C is highly conserved in bacterial ferritins (Stillman et al. 2003; Tatur et al. 2007; Cho et al. 2009; Ebrahimi et al. 2009; Johnson et al. 2005; Le Brun et al. 2010; Treffry et al. 1998a, b; Pereira et al. 2012). In particular, site C of EcFtnA is very important and modulates the stoichimetry and kinetics of iron oxidation. The elimination of Site C in EcFtnA produces a decrease in the Fe(II)/O_2 stoichiometry from 3 to 2 for the first 48 Fe(II) added to the protein (Treffry et al. 1998b), which is similar to HuFtn. Also, the site in Eco FtnA controls the rate of regeneration of its FC, elimination of site C allows for the regeneration of their ferroxidase activity within a few hours, instead of a day that it is needed in WT E.coli FtnA (Bauminger et al. 1999).

Significantly, the iron coordination in site C in HuFH and EcFtnA are different. The C site residues Glu129 and Glu139 in EcFtnA are replaced by Lys and Ala in HuFH. However, both proteins maintain a Glu residue in site C (Glu140 and Glu106 in HuFH and EcFtnA, respectively). Another important difference is that mutations Glu140Ala and Glu140Gln in HuHF reduce the oxidation rate just to 50% (Masuda et al. 2010b; Bauminger et al. 1999), similar result can be found when the same mutations are introduced in soybean ferritin (Masuda et al. 2010a; Bauminger et al. 2000; Bou-Abdallah et al. 2014). However, the Glu129Arg, Glu129Cys and Glu129Qln mutations in PfFtn significantly reduce the oxidation rate (Ebrahimi et al. 2012; Bauminger et al. 1999; Bou-Abdallah et al. 2014). Therefore, site C can achieve different functions in these two ferritins.

In the unified mechanism described above, site C and the tyrosine radical perform essential roles (Masuda et al. 2010b; Ebrahimi et al. 2012, 2013), which are not required for rapid iron oxidation in EcFtnA (Pereira et al. 1998; Bou-Abdallah et al. 2005a, b; Treffry et al. 1998b; Bauminger et al. 1999, 2000; Stillman et al. 2003). Also, from X-ray crystallography and Mossbauer spectroscopy it has been shown that site C is not a transit site, iron does not appreciably turnover at the ferroxidase center (Bauminger et al. 1999, 2000). Bou-Abdalla and colleagues analyzed the binding of iron to the apo EcFtnA by ITC and reported that the binding of iron produced two strong, slightly endergonic binding sites and one undefined weak binding per subunit (Bou-Abdallah et al. 2014). In contrast, PfFtn shows one strong highly exergonic and two weak highly endergonic binding sites per subunit (Ebrahimi et al. 2012). Bou-Abdallah et al. (2014) proposed that, in the case of EcFtnA, once the ferroxidase center is saturated with iron, iron at site C inhibits the turnover of Fe(III) at sites A and B, they therefore serve as a redox cofactors, as was proposed for EcBFR (Baaghil et al. 2003; Crow et al. 2009) (*see* below).

In EcBFR, two Fe(II) ions bind to each FC, most likely *via* a pore through the subunit that connects the center with the surrounding solution (Macedo et al. 2003; Crow et al. 2009; Weeratunga et al. 2010). The oxidation of iron in the FC occurs rapidly in the presence of O_2 to generate a μ-oxo/hydroxo bridged di-Fe(III) form, with the reduction of O_2 to H_2O_2 (Yang et al. 2000; Bou-Abdallah et al. 2002a, b). As was described for bacterial ferritin, H_2O_2 is a better oxidant than O_2 (Yang et al. 2000; Bou-Abdallah et al. 2002a, b). In this way, the H_2O_2 generated at one FC is subsequently use to oxidize Fe(II) at another FC. This means that the ratio of Fe(II):O_2 is 4:1, with an overall reduction of O_2 to H_2O. A DPS intermediate has not been detected in BFRs to date, possibly due to rapid decay. In BFRs, the bridged di-Fe(III) center is stable, and several data support a mechanism in which the FC center functions as a true enzyme cofactor site, where it is continually cycling its

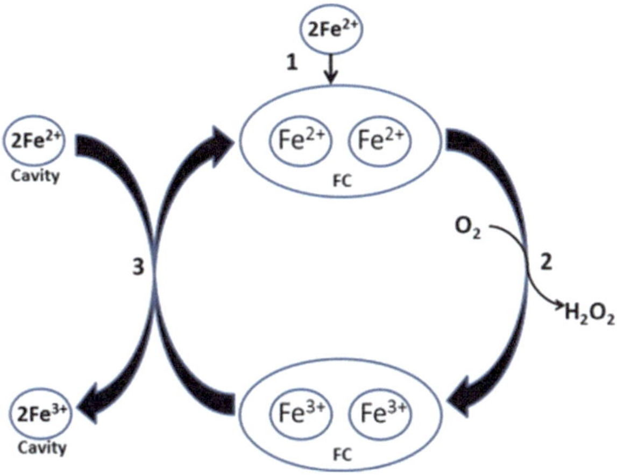

Fig. 3.5 Mechanism of Fe(II) oxidation by bacterioferritin. Iron binds to site A and B (*1*) and is oxidize by with O_2 or H_2O_2 (*2*). Once the FC is saturated, the excess of iron enters into the central cavity and binds on the inner face of the protein shell at internal nucleation sites, where oxidation occur (*3*). Then, the resulting electrons are channelled to the FC, which iron is reduced back to its Fe(II) state, and the resulting di-Fe(II) FC reacts again with O_2 or H_2O_2 in continuous cycles until all the iron has been incorporated to the inner mineral core

redox state, *see* Fig. 3.5 (Le Brun et al. 1993; Crow et al. 2009; Yang et al. 2000; Baaghil et al. 2003; Lawson et al. 2009).

This is completely different to the role of FC in ferritin and bacterial ferritin. The properties of the BFRs FC are consistent with the close similarity between these center and the di-iron sites of other well characterized di-iron enzymes, such as methane monooxygenase (Le Brun et al. 1995). The proposed mechanism for BFR, once the FC is saturated, is that the excess of iron enters into the central cavity and binds on the inner face of the protein shell at internal nucleation sites where oxidation occurs (Crow et al. 2009). Then, the resulting electrons are channeled to the FC, which is reduced back to its Fe(II) state, and the resulting di-Fe(II) FC reacts again with O_2 or H_2O_2 in continuous cycles, until all the iron has been incorporated into the inner mineral core (Watt et al. 1992). The substitution of the surface site residue does not affect the FC activity, but does produce an 80% decrease in the rate of mineralization (Crow et al. 2009), demonstrating the importance of these surface sites in the mineralization mechanism. It has also been determined that Asp126 is important for core formation, but not for the FC reaction (Wong et al. 2009). This mechanism agrees with the observation showing that the BFR FC is essential for mineralization and that the rate of FC catalyzed Fe(II) oxidation is dependent on the mineral core (Baaghil et al. 2003). However, this mechanism cannot be common to all BFRs, as it has been determined that iron in the PaBFR FC is not stable (Weeratunga et al. 2010), suggesting that in some BFRs, FC appears to function as a gated iron site.

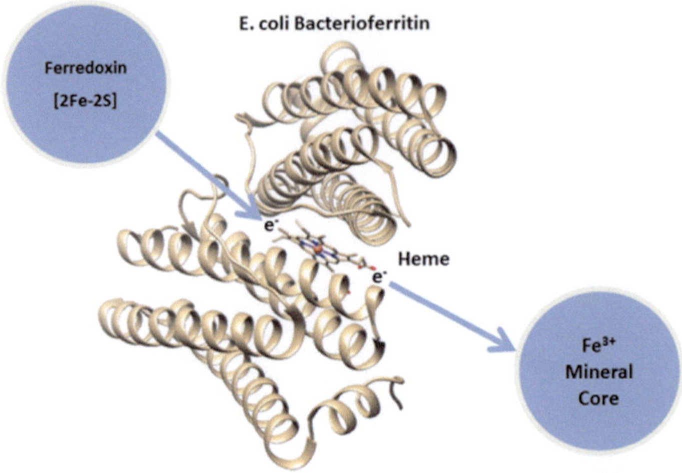

Fig. 3.6 Mechanism of Fe(III) release from bacterioferritin. Ferredoxin binds at the twofold symmetry axis of EcBFR, the Fe(II) is release after that the [2Fe-2S] cluster of ferrodoxin allow the reduction of the iron inner mineral core via an electron transfer that involves the hem group of EcBFR. The Fig. was prepared using the crystal structure of EcBFR (PDB 3E1M)

3.3.1 Mechanism of Iron Release in Ferritin

A mechanism for iron release in BFR has been described in which ferrodoxin can be the electron donor that reaches the mineral inner core through the heme group of EcBFR (Quail et al., 1996) (see Fig. 3.6). One the mechanisms described for iron release from ferritin includes a role for transferrin, a Fe(III) binding protein. Transferrin can scavenge iron from the FC or from the mineral core, but the rate varies between minutes and days (Ebrahimi et al. 2012). However, if this mechanism has a physiological role, it still requires further investigation. It has been suggested that this mechanism may have an important role for scavenging the Fe(III) from serum ferritin in humans (Kell and Pretorius 2014). The iron can be released from ferritin by reduction of Fe(III). *In vitro*, iron can be released from ferritin after chemical reduction, followed by the formation of a complex with a chelating agent for Fe(II), such as bipyridil (Funk et al. 1985; Sirivech et al. 1974) or flavorproteins (Watt et al. 1988). An interesting electrochemical study suggested a role of the PfFtn FC Fe(III) as a mediator to pass electrons from an external reducing agent to Fe(III) in the mineral core (Tatur et al. 2009). The only biologically reductants that have been found to mediate the reduction of Fe(III) under physiological conditions are the dihydroflavins (Jones et al. 1978) or flavoproteins (Watt et al. 1988). Also, it is has been proposed that mild denaturing conditions (1–10 mM Urea or 0.1 mM guanidine hydrochloride) can promote the iron release from ferritin (Liu et al. 2003). Mutations in the threefold channel amino acid residues can also increase the rate of the iron dissolution after the addition of a reducing agent (Tosha et al. 2012). However, is required further investigation to determine their physiological role.

3.4 Application of the 24-mer Ferritin

Protein assemblies represent a class of highly self-organized molecules, which have the potential to be used as novel templates for nanotechnology. In this context, the ferritin proteins offer great future potential, because they form natural nanocages, where inorganic-organic hybrid materials can be synthesized (Theil and Behera 2013) (see Fig. 3.7). Ferritins present several interesting properties, as the different size of their inner cavity, the sequence variations between the ferritins protein cages from different organisms, and their different thermal and chemical stabilities. All are important when considering the use ferritin as a nanoreactor for biotechnological applications (Campan et al. 2011; Yamashita et al. 2010; Uchida et al. 2010; Maham et al. 2009). Another useful property relating to the synthesis of nano-particles is that ferritins are stable up to 80 °C, 6 M urea or guanidine hydrochloride or 1% w/v SDS (Liu and Theil 2005). Further, ferritin stability in organic solvents can be improved by derivatizing carboxylates on the external surface of the protein with long chain hydrocarbons (Liu et al. 2006; Li et al. 2007; Meldrum et al. 1992; Wade et al. 1991, 1993; Wong et al. 1999).

Another useful property of ferritin is its ability to disassemble at low pH and reassemble at high pH, allowing production of different sized nanoparticles inside

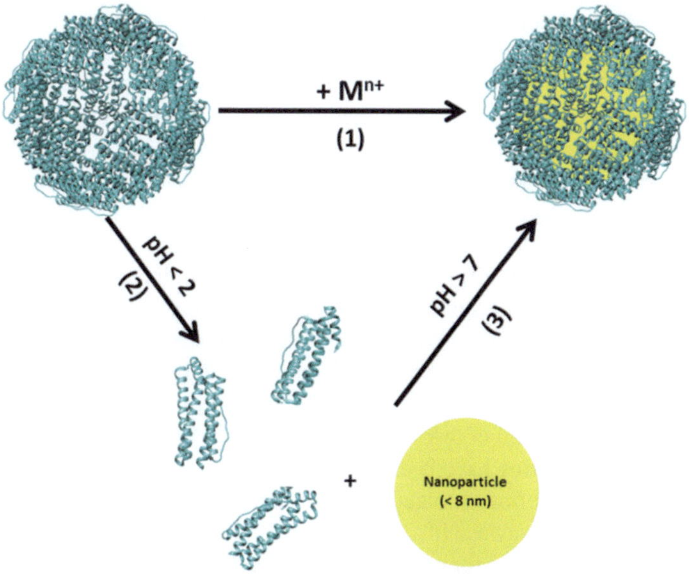

Fig. 3.7 General methods fot the synthesis and incorporation of nanoparticles inside the ferritin inner cavity. (*1*) A positive charged metal diffuse into the ferritin and the nanoparticles is synthesized inside its inner cavity. (*2*) Intact ferritin can disassembly by incubation a pH < 2 and incubated with previously synthesized nanoparticles, then (*3*) the nanoparticles is stored inside ferritin inner cavity when the pH increase to 7, allowing the reassembling of the ferritin protein shell

the protein cages (Lambet et al. 2010)(see Fig. 3.7). These properties have been used in the synthesis and delivery of magnetic resonance imaging (MRI) contrast agents (Sana et al. 2012; Li et al. 2012; Wood 2011; Terashima et al. 2011; Ziv et al. 2010) and for drug delivery and catalysis (Maham et al. 2009; Zheng et al. 2010; Xing et al. 2009; Mann 2009; Abe et al. 2009). *In vitro*, purified ferritin cages can incorporate a variety of metal ions (Fe, Au, Pd, Rh, Pt, Ni, Cr, Cd, Ti, Tb, Co, Cu and Zn) (Tosha et al. 2010; Uchida et al. 2010; Iwahori and Yamashita 2008). Also, it has been possible to synthesize nano-particles of gold, silver, lead, copper and nickel after incubation of apo-ferritins with ions, followed by a reducing agent (Uchida et al. 2010). Usefully, the engineering of additional metal binding sites has allowed the synthesis of silver and gold nanoparticles (Butts et al. 2008). Taking advantage of the thermal resistance of PfFtn, it has been possible to synthesize magnetite (Uchida et al. 2010). The apo-form of ferritin has an inner cavity that could be used to entrap not just metallic nanoparticle derivatives, but also organic molecules. Thus, ferritin has been reported while containing hydrophilic drugs, imaging agents (MRI and fluorophores) (Terashima et al. 2011; Kitagawa et al. 2012; Flenniken et al. 2009), radionuclides or nuclear medicines (Maham et al. 2009), anti-tumor drugs (Xing et al. 2009; Yang et al. 2007a, b) and DNA probes (Maham et al. 2009). Furthermore, genetic or chemical modification of the surface of ferritin could help to target these complexes to specific cell types, thereby improving their future use in nano-medicine. An example of the use of ferritin in nano-medicine is the development of ferritin with the peptide RGD-4 attached to its surface. This allows the specific binding of this complex to integrins v3 and v5 (Terashima et al. 2011; Kitagawa et al. 2012), which can be used for the imaging of vascular inflammation and angiogenesis and for affecting the metabolism of C32 human melanoma cells.

In an indirect method, the superparamagnetism of the inner core of ferritin can be used as an endogenous enhancer of the transverse nuclear magnetic relaxation of water. Thus it has been used to determine iron levels associated with hypertransfusion treatments of sickle cell anemia, thalassemia, and in hemochromatosis and Alzheimer disease (Wood 2011). On another hand, a similar approach, which was described earlier, has been used to check transgene expression (Ono et al. 2009). It has been reported that improvement of the MRI contrasting ability of ferritin complexes occurs upon an increase of the iron biomineral or by entrapment of synthetic magnetite or other nanoparticles (Li et al. 2012; Terashima et al. 2011; Uchida et al. 2008). Another application of ferritin as nano-reactor is in nano-catalysis and nano-electronics. Thus, ferritins have been used as sites for organometallic catalysis, to produce polymers of defined molecular weight and with a narrow size distribution (Wang et al. 2011; Takezawa et al. 2011; Abe et al. 2009). In this way, ferritin cages have been used as templates for metal oxide nanoparticles and metal oxide semiconductors and also to produce building blocks for the fabrication of electronic devices, such as the floating nanodot memory device or thin film transistor flash memory (Yamashita et al. 2010; Uchida et al. 2010).

3.5 Concluding Remarks

In the ferritin protein family, it is clear that a catalytic center for iron oxidation, the routes for iron to reach this center and the ability to nucleate an iron core are common requirements for all ferritins. However, there are differences in the structural and mechanistic details of iron oxidation and mineralization. Therefore, ferritins have been classified in three distinct types, H-chain-type, Ftn-type and BFR-type. Although a common mechanism has been proposed for all ferritins, this mechanism needs to be further explored. There is a mechanistic diversity that is related to structural variation in the ferritin protein family. For example, the highly related BFR from *E. coli* and *P. aeruginosa* have different mechanisms of mineralization, suggesting that just the first coordination sphere ligands do not allow the prediction of a mechanism. It is clear that other factors appear to affect the mechanism of iron oxidation and mineralization. A significant challenge for the future is to correlate the high quality structural data on ferritins with the biochemical diversity observed in the ferritin protein family. Further studies should contribute to our understanding of the role of this protein family in iron homeostasis and be used to develop treatments for iron metabolism disorders and new biotechnological applications.

References

Abe S, Hirata K, Ueno T, Morino K, Shimizu N, Yamamoto M, Takata M, Yashima E, Watanabe Y (2009) Polymerization of phenylacetylene by rhodium complexes within a discrete space of apo-ferritin. J Am Chem Soc 131:6958–6960

Andrews SC (2010) The Ferritin-like superfamily: evolution of the biological iron storeman from a rubrerythrin-like ancestor. Biochim Biophys Acta 1800:691–705

Andrews SC, Le Brun NE, Barynin V, Thomson AJ, Moore GR, Guest JR, Harrison PM (1995) Site-directed replacement of the coaxial heme ligands of bacterioferritin generates heme-free variants. J Biol Chem 270:23268–23274

Arenas-Salinas M, Townsend PD, Brito C, Márquez V, Marabolli V, Gonzalez-Nilo F, Matias C, Watt RK, López-Castro JD, Domínguez-Vera J, Pohl E, Yévenes A (2014) The crystal structure of ferritin from Chlorobium tepidum reveals a new conformation of the 4-fold channel for this protein family. Biochimie 106:39–47

Baaghil S, Lewin A, Moore GR, Le Brun NE (2003) Core formation in Escherichia coli bacterioferritin requires a functional ferroxidase center. Biochemistry 42:14047–14056

Bauminger ER, Harrison PM, Hechel D, Nowik I, Treffry A (1993) The function of H and L chains in iron sequestration in ferritin. Nucl Inst Methods Phys Res B 76:403–404

Bauminger ER, Treffry A, Quail MA, Zhao ZW, Nowik I, Harrison PM (1999) Stages in iron storage in the ferritin of Escherichia coli (EcFtnA): analysis of Mössbauer spectra reveals a new intermediate. Biochemistry 38:7791–7802

Bauminger ER, Treffry A, Quail MA, Zhao Z, Nowik I, Harrison PM (2000) Metal binding at the active centre of the ferritin of Escherichia coli (EcFtnA). A Mössbauer spectroscopic study. Inorg Chim Acta 297:171–180

Behera RK, Theil EC (2014) Moving Fe2+ from ferritin ion channels to catalytic OH centers depends on conserved protein cage carboxylates. Proc Natl Acad Sci U S A 111:7925–7930

Bernacchioni C, Ghini V, Pozzi C, Di Pisa F, Theil EC, Turano P (2014) Loop electrostatics modulates the intersubunit interactions in ferritin. ACS Chem Biol 9(11):2517–2525. doi:10.1021/cb500431r

Bertini I, Lalli D, Mangani S, Pozzi C, Rosa C, Theil EC, Turano P (2012) Structural insights into the ferroxidase site of ferritins from higher eukaryotes. J Am Chem Soc 134:6169–6176

Bou-Abdallah F, Lewin AC, Le Brun NE, Moore GR, Chasteen ND (2002a) Iron detoxification properties of Escherichia coli bacterioferritin. Attenuation of oxyradical chemistry. J Biol Chem 277:37064–37069

Bou-Abdallah F, Papaefthymiou GC, Scheswohl DM, Stanga SD, Arosio P, Chasteen ND (2002b) mu-1,2-Peroxobridged di-iron(III) dimer formation in human H-chain ferritin. Biochem J 364:57–63

Bou-Abdallah F, Arosio P, Levi S, Janus-Chandler C, Chasteen ND (2003) Defining metal ion inhibitor interactions with recombinant human H- and L-chain ferritins and site-directed variants: an isothermal titration calorimetry study. J Biol Inorg Chem 8:489–497

Bou-Abdallah F, Zhao GH, Mayne HR, Arosio P, Chasteen ND (2005a) Origin of the unusual kinetics of iron deposition in human H-chain ferritin. J Am Chem Soc 127:3885–3893

Bou-Abdallah F, Woodhall MR, Velásquez-Campoy A, Andrews SC, Chasteen ND (2005b) Thermodynamic analysis of ferrous ion binding to Escherichia coli ferritin EcFtnA. Biochemistry 44:13837–13846

Bou-Abdallah F, Zhao G, Biasiotto G, Poli M, Arosio P, Chasteen ND (2008) Facilitated diffusion of iron(II) and dioxygen substrates into human H-chain ferritin. A fluorescence and absorbance study employing the ferroxidase center substitution Y34W. J Am Chem Soc 130:17801–17811

Bou-Abdallah F, Yang H, Awomolo A, Cooper B, Woodhall MR, Andrews SC, Chasteen ND (2014) Functionality of the three-site ferroxidase center of Escherichia coli bacterial ferritin (EcFtnA). Biochemistry 53:483–495

Boukhalfa H, Crumbliss AL (2012) Chemical aspects of siderophore mediated iron transport. BioMetals 15:325–339

Boyd D, Vecoli C, Belcher DM, Jain SK, Drysdale JW (1985) Structural and functional relationships of human ferritin H and L chains deduced from cDNA clones. J Biol Chem 260:11755–11761

Bu W, Liu R, Cheung-Lau JC, Dmochowski IJ, Loll PJ, Eckenhoff RG (2012) Ferritin couples iron and fatty acid metabolism. FASEB J 26:2394–2400

Butts CA, Swift J, Kang SG, Di Costanzo L, Christianson DW, Saven JG, Dmochowski IJ (2008) Directing noble metal ion chemistry within a designed ferritin protein. Biochemistry 47:12729–12739

Calhoun LN, Kwon YM (2011) Structure, function and regulation of the DNA-binding protein Dps and its role in acid and oxidative stress resistance in Escherichia coli: a review. J Appl Microbiol 110:375–386

Campan M, Lionetti V, Aquaro GD, Forini F, Matteucci M, Vannucci L, Chiuppesi F, Di Cristofano C, Faggioni M, Maioli M, Barile L, Messina E, Lombardi M, Pucci A, Pistello M, Recchia FA (2011) Ferritin as a reporter gene for in vivo tracking of stem cells by 1.5-T cardiac MRI in a rat model of myocardial infarction. Am J Physiol Heart Circ Physiol 300:H2238–H2250

Carrondo MA (2003) Ferritins, iron uptake and storage from the bacterioferritin viewpoint. EMBO J 22:1959–1968

Ceci P, Forte E, Di Cecca F, Fornara M, Chienacone E (2011) The characterization of Thermotoga marítima ferritin reveals an unusual subunit dissociation behavior and efficient DNA protection form iron-mediated oxidative stress. Extremophiles 15:431–439

Chasteen ND (1998) Ferritin. Uptake, storage, and release of iron. In: Sigel H, Sigel A (eds) Metal ions in biological systems, vol 35. Marcel Dekker Inc., New York, pp 479–514

Chasteen ND, Harrison PM (1999) Mineralization in ferritin: an efficient means of iron storage. J Struct Biol 126:182–194

Chen-Barrett Y, Harrison PM, Treffry A, Quail MA, Arosio P, Santambrogio P, Chasteen ND (1995) Tyrosyl radical formation during the oxidative deposition of iron in human apoferritin. Biochemistry 34:7847–7853

Cho KJ, Shin HJ, Lee JH, Kim KJ, Park SS, Lee Y, Lee C, Park SS, Kim KH (2009) The crystal structure of ferritin from Helicobacter pylori reveals unusual conformational changes for iron uptake. J Mol Biol 390:83–98

Crow A, Lawson TL, Lewin A, Moore GR, Le Brun NE (2009) Structural basis for iron mineralization by bacterioferritin. J Am Chem Soc 131:6808–6813

Curtis AR, Fey C, Morris CM, Bindoff LA, Ince PG, Chinnery PF, Coulthard A, Jackson MJ, Jackson AP, McHale DP (2001) Mutation in the gene encoding ferritin light polypeptide causes dominant adult-onset basal ganglia disease. Nat Genet 28:350–354

Douglas T, Ripoll DR (1998) Calculated electrostatic gradients in recombinant human H-chain ferritin. Protein Sci 7:1083–1091

Drakesmith H, Chen N, Ledermann H, Screaton G, Townsend A, Xu XN (2005) HIV-1 Nef down-regulates the hemochromatosis protein HFE, manipulating cellular iron homeostasis. Proc Natl Acad Sci U S A 102:11017–11022

Ebrahimi KH, Hagedoorn PL, Jongejan JA, Hagen WR (2009) Catalysis of iron core formation in Pyrococcus furiosus ferritin. J Biol Inorg Chem 14:1265–1274

Ebrahimi KH, Bill E, Hagedoorn PL, Hagen WR (2012) The catalytic center of ferritin regulates iron storage via Fe(II)eFe(III) displacement. Nat Chem Biol 8:941–948

Ebrahimi KH, Hagedoorn PL, Hagen WR (2013) A conserved tyrosine in ferritin is a molecular capacitor. Chem Bio Chem 14:1123–1133

Ebrahimi KH, Hagedoom PL, Hagne W (2015) Unity in the biochemistry of the iron-storage proteins ferritin and bacterioferritin. Chem Rev 115:295–326

Ensign D, Young M, Douglas T (2004) Photocatalytic synthesis of copper colloids from CuII by the ferrihydrite core of ferritin. Inorg Chem 43:3441–3446

Ferreira C, Bucchini D, Martin ME, Levi S, Arosio P, Grandchamp B, Beaumont C (2000) Early embryonic lethality of H ferritin gene deletion in mice. J Biol Chem 275:3021

Flenniken ML, Uchida M, Liepold LO, Kang S, Young MJ, Douglas T (2009) A library of protein cage architectures as nanomaterials. Curr Top Microbiol Immunol 327:71–93

Frolow F, Kalb AJ, Yariv J (1994) Structure of a unique twofold symmetric haem-binding site. Nat Struct Biol 1:453–460

Funk F, Lenders JP, Crichton RR, Schneider W (1985) Reductive mobilisation of ferritin iron. Eur J Biochem 152:167–172

Granier T, Gallois B, Dautant A, Langlois d'Estaintot B, Precigoux G (1997) Comparison of the structures of the cubic and tetragonal forms of horse-spleen apoferritin. Acta Crystallogr D Biol Crystallogr 53:580-587

Grant R, Filman D, Finkel S, Kolter R, Hogle J (1998) The crystal structure of Dps, a ferritin homolog that binds and protects DNA. J Nat Struct Mol Biol 5:294–303

Ha Y, Shi D, Small GW, Theil EC, Allewell NM (1999) Crystal structure of bullfrog M ferritin at 2.8 A resolution: analysis of subunit interactions and the binuclear metal center. J Biol Inorg Chem 4:243–256

Haikarainen T, Papageorgiou A (2010) Dps-like proteins: structural and functional insights into a versatile protein family. Cell Mol Life Sci 67:341–351

Hamburger AE, West AP Jr, Hamburger ZA, Hamburger P, Bjorkman PJ (2005) Crystal structure of a secreted insect ferritin reveals a symmetrical arrangement of heavy and light chains. J Mol Biol 349:558–569

Harrison PM, Arosio P (1996) The ferritins: molecular properties, iron storage function and cellular regulation. Biochim Biophys Acta 1275:161–203

Hitchings MD, Townsend P, Pohl E, Facey PD, Jones DH, Dyson PJ, Del Sol R (2014) Tale of tails: deciphering the contribution of terminal tails to the biochemical properties of two Dps proteins from Streptomyces coelicolor. Cell Mol Life Sci 24:4911–4926

Huard DJ, Kane KM, Tezcan FA (2013) Re-engineering protein interfaces yields copper-inducible ferritin cage assembly. Nat Chem Biol 9:169–176

Ilari A, Stefanini S, Chiancone E, Tsernoglou D (2000) The dodecameric ferritin from Listeria innocua contains a novel intersubunit iron-binding site. Nat Struct Biol 7:38–43

Iwahori K, Yamashita I (2008) Size-controlled one-pot synthesis of fluorescent cadmium sulfide semiconductor nanoparticles in an apoferritin cavity. Nanotechnology 19:495601

Jameson GNL, Jin W, Krebs C, Perreira AS, Tavares P, Liu XF, Theil EC, Huynh BH (2002) Stoichiometric production of hydrogen peroxide and parallel formation of ferric multimers through decay of the diferric-peroxo complex, the first detectable intermediate in ferritin mineralization. Biochemistry 41:13435–13443

Jellinger K, Paulus W, Grundke-Iqbal I, Riederer P, Youdim MJ (1990) Brain iron and ferritin in Parkinson's and Alzheimer's diseases. J Neural Transm Park Dis Dement Sect 2:327–340

Johnson E, Cascio D, Sawaya MR, Gingery M, Schröder I (2005) Crystal structures of a tetrahedral open pore ferritin from the hyperthermophilic archaeon Archaeoglobus fulgidus. Structure 13:637–648

Jones T, Spencer R, Walsh C (1978) Mechanism and kinetics of iron release from ferritin by dihydroflavins and dihydroflavin analogues. Biochemistry 17:4011–4017

Kehrer P (2000) The Haber–Weiss reaction and mechanisms of toxicity. Toxicology 149:43–50

Kell DB, Pretorius E (2014) Serum ferritin is an important inflammatory disease marker, as it is mainly a leakage product from damaged cells. Metallomics 6:748–773

Khare G, Gupta V, Nangpal P, Gupta RK, Sauter NK, Tyagi AK (2011) Ferritin structure from Mycobacterium tuberculosis: comparative study with homologues identifies extended C-terminus involved in ferroxidase activity. PLoS One 6:e18570

Kitagawa T, Kosuge H, Uchida M, Dua MM, Iida Y, Dalman RL, Douglas T, McConnell MV (2012) RGD-conjugated human ferritin nanoparticles for imaging vascular inflammation and angiogenesis in experimental carotid and aortic disease. Mol Imaging Biol 14:315–324

Laghaei R, Evans DG, Coalson RD (2013) Metal binding sites of human H-chain ferritin and iron transport mechanism to the ferroxidase sites: a molecular dynamics simulation study. Proteins 81:1042–1050

Langlois d'Estaintot B, Santambrogio P, Granier T, Gallois B, Chevalier JM, Précigoux G, Levi S, Arosio P (2004) Crystal structure and biochemical properties of the human mitochondrial ferritin and its mutant Ser144Ala. J Mol Biol 340:277–293

Lawson DM, Artymiuk PJ, Yewdall SJ, Smith JMA, Livingstone JC, Treffry A, Luzzago A, Levi S, Arosio P, Cesareni G, Thomas CD, Shaw WV, Harrison PM (1991) Solving the structure of human H ferritin by genetically engineering intermolecular crystal contacts. Nature 349:541–544

Lawson TL, Crow A, Lewin A, Yasmin S, Moore GR, Le Brun NE (2009) Monitoring the iron status of the ferroxidase center of Escherichia coli bacterioferritin using fluorescence spectroscopy. Biochemistry 48:9031–9039

Le Brun NE, Wilson MT, Andrews SC, Guest JR, Harrison PM, Thomson AJ, Moore GR (1993) Kinetic and structural characterization of an intermediate in the biomineralization of bacterioferritin. FEBS Lett 333:197–202

Le Brun NE, Andrews SC, Guest JR, Harrison PM, Moore GR, Thomson AJ (1995) Identification of the ferroxidase centre of Escherichia coli bacterioferritin. Biochem J 312:385–392

Le Brun NE, Crow A, Murphy ME, Mauk AG, Moore GR (2010) Iron core mineralization in prokaryotic ferritins. Biochim Biophys Acta 1800:732–744

Levi S, Luzzago A, Cesareni G, Cozzi A, Franceschinelli F, Albertini A, Arosio PJ (1988) Mechanism of ferritin iron uptake: activity of the H-chain and deletion mapping of the ferro-oxidase site. A study of iron uptake and ferro-oxidase activity of human liver, recombinant H-chain ferritins, and of two H-chain deletion mutants. Biol Chem 263:18086–18092

Levi S, Salfeld J, Franceschinelli F, Cozzi A, Dorner MH, Arosio P (1989a) Expression and structural and functional properties of human ferritin L-chain from Escherichia coli. Biochemistry 28:5179–5184

Levi S, Luzzago A, Franceschinelli F, Santambrogio P, Cesareni G, Arosio P (1989b) Mutational analysis of the channel and loop sequences of human ferritin H-chain. Biochem J 264:381–388

Li M, Viravaidya C, Mann S (2007) Polymer-mediated synthesis of ferritin-encapsulated inorganic nanoparticles. Small 3:1477–1481

Li K, Zhang ZP, Luo M, Yu X, Han Y, Wei HP, Cui ZQ, Zhang XE (2012) Multifunctional ferritin cage nanostructures for fluorescence and MR imaging of tumor cells. Nanoscale 4:188–193

Liu X, Theil EC (2005) Ferritins: dynamic management of biological iron and oxygen chemistry. Acc Chem Res 38:167–175

Liu X, Jin W, Theil EC (2003) Opening protein pores with chaotropes enhances Fe reduction and chelation of Fe from the ferritin biomineral. Proc Natl Acad Sci U S A 100:3653–3658

Liu HL, Zhou HN, Xing WM, Zhao JF, Li SX, Huang JF, Bi RC (2004) 2.6 A resolution crystal structure of the bacterioferritin from Azotobacter vinelandii. FEBS Lett 573:93–98

Liu G, Wang J, Wu H, Lin Y (2006) Versatile apoferritin nanoparticle labels for assay of protein. Anal Chem 78:7417–7423

Macedo S, Romao CV, Mitchell E, Matias PM, Liu MY, Xavier AV, LeGall J, Teixeir M, Lindley P, Carrondo MA (2003) The nature of the di-iron site in the bacterioferritin from Desulfovibrio desulfuricans. Nat Struct Biol 10:285–290

Maham A, Tang Z, Wu H, Wang J, Lin Y (2009) Protein-based nanomedicine platforms for drug delivery. Small 5:1706–1721

Mann S (2009) Self-assembly and transformation of hybrid nano-objects and nanostructures under equilibrium and non-equilibrium conditions. Nat Mater 8:781–792

Marchetti A, Parker MS, Moccia LP, Lin EO, Arrieta AL, Ribalet F, Murphy MEP, Maldonado MT, Armbrust EV (2009) Ferritin is used for iron storage in bloom-forming marine pennate diatoms. Nature 457:467–470

Masuda T, Goto F, Yoshihara T, Mikami B (2010a) Crystal structure of plant ferritin reveals a novel metal binding site that functions as a transit site for metal transfer in ferritin. J Biol Chem 285:4049–4059

Masuda T, Goto F, Yoshihara T, Mikami B (2010b) The universal mechanism for iron translocation to the ferroxidase site in ferritin, which is mediated by the well conserved transit site. Biochem Biophys Res Commun 400:94–99

Meldrum FC, Heywood BR, Mann S (1992) Magnetoferritin: in vitro synthesis of a novel magnetic protein. Science 257(5069):522–523

Moenne-Loccoz P, Krebs C, Herlihy K, Edmondson DE, Theil EC, Huynh BH, Loehr TM (1999) The ferroxidase reaction of ferritin reveals a diferric mu-1,2 bridging peroxide intermediate in common with other O2-activating non-heme diiron proteins. Biochemistry 38:5290–5295

Nordlund P, Sjoberg BM, Eklund H (1990) Three-dimensional structure of the free radical protein of ribonucleotide reductase. Nature 345:593–598

Ono K, Fuma K, Tabata K, Sawada M (2009) Ferritin reporter used for gene expression imaging by magnetic resonance. Biochem Biophys Res Commun 388:589–594

Pereira AS, Tavares P, Lloyd SG, Danger D, Edmondson DE, Theil EC, Huynh BH (1997) Rapid and parallel formation of Fe3+ multimers, including a trimer, during H-type subunit ferritin mineralization. Biochemistry 36:7917–7927

Pereira AS, Small W, Krebs C, Tavares P, Edmondson DE, Theil EC, Huynh BH (1998) Direct spectroscopic and kinetic evidence for the involvement of a peroxodiferric intermediate during the ferroxidase reaction in fast ferritin mineralization. Biochemistry 37:9871–9876

Pereira AS, Timóteo CG, Guilherme M, Folgosa F, Naik SG, Duarte AG, Huynh BH, Tavares P (2012) Spectroscopic evidence for and characterization of a trinuclear ferroxidase center in bacterial ferritin from Desulfovibrio vulgaris Hildenborough. J Am Chem Soc 134:10822–10832

Pettersen EF, Goddard TD, Huang CC, Couch GS, Greenblatt DM, Meng EC, Ferrin TE (2004) UCSF Chimera--a visualization system for exploratory research and analysis. J Comput Chem 25:1605–1612

Pfaffen S, Abdulqadir R, Le Brun NE, Murphy MEP (2013) Mechanism of ferrous iron and oxidation by ferritin from a pennate diatom. J Biol Chem 288:14917–14925

Quail MA, Jordan P, Grogan JM, Butt JN, Lutz M, Thomson AJ, Andrews SC, Guest JR (1996) Spectroscopic and voltammetric characterisation of the bacterioferritin-associated ferredoxin of Escherichia coli. Biochem Biophys Res Commun 229:635

Recalcati S, Invernizzi P, Arosio P, Cairo GJ (2008) New functions for an iron storage protein: the role of ferritin in immunity and autoimmunity. J Autoimmun 30:84–89

Ren B, Tibbelin G, Kajino T, Asami O, Ladenstein R (2003) The multi-layered structure of Dps with a novel di-nuclear ferroxidase center. J Mol Biol 329:467–477

Rosenzweig AC, Frederick CA, Lippard SJ, Nordlund P (1993) Crystal structure of a bacterial non-haem iron hydroxylase that catalyses the biological oxidation of methane. Nature 366:537–543

Salgado EN, Faraone-Mennella J, Tezcan FA (2007) Controlling protein-protein interactions through metal coordination: assembly of a 16-helix bundle protein. J Am Chem Soc 129:13374–13375

Salgado EN, Lewis RA, Faraone-Mennella J, Tezcan FA (2008) Metal-mediated self-assembly of protein superstructures: influence of secondary interactions on protein oligomerization and aggregation. J Am Chem Soc 130:6082–6084

Sana B, Poh CL, Lim S (2012) A manganese-ferritin nanocomposite as an ultrasensitive T2 contrast agent. Chem Commun (Camb) 48:862–864

Sana B, Johnson E, Le Magueres P, Criswell A, Cascio D, Lim SJ (2013) The role of nonconserved residues of Archaeoglobus fulgidus ferritin on its unique structure and biophysical properties. Biol Chem 288:32663–32672

Sirivech S, Frieden E, Osaki S (1974) The release of iron from horse spleen ferritin by reduced flavins. Biochem J 143:311–315

Stefanini S, Vecchini P, Chiancone E (1987) On the mechanism of horse spleen apoferritin assembly: a sedimentation velocity and circular dichroism study. Biochemistry 26:1831–1837

Stillman TJ, Hempstead PD, Artymiuk PJ, Andrews SC, Hudson AJ, Treffry A, Guest JR, Harrison PM (2001) The high-resolution X-ray crystallographic structure of the ferritin (EcFtnA) of Escherichia coli; comparison with human H ferritin (HuHF) and the structures of the Fe(3þ) and Zn(2þ) derivatives. J Mol Biol 307:587–603

Stillman TJ, Connolly PP, Latimer CL, Morland AF, Quail MA, Andrews SC, Treffry A, Guest JR, Artymiuk PJ, Harrison PM (2003) Insight into the effects on metal binding of the systematic substitution of five key glutamate ligands in the ferritin of Escherichia coli. J Biol Chem 278:26275–26286

Swartz L, Kuchinskas M, Li HY, Poulos TL, Lanzilotta WN (2006) Redox-dependent structural changes in the Azotobacter vinelandii bacterioferritin: new insights into the ferroxidase and iron transport mechanism. Biochemistry 45:4421–4428

Takezawa Y, Bockmann P, Sugi N, Wang Z, Abe S, Murakami T, Hikage T, Erker G, Watanabe Y, Kitagawa S, Ueno T (2011) Incorporation of organometallic Ru complexes into apo-ferritin cage. Dalton Trans 40:2190–2195

Tatur J, Hagen WR, Matias PM (2007) Crystal structure of the ferritin from the hyperthermophilic archaeal anaerobe Pyrococcus furiosus. J Biol Inorg Chem 12:615–630

Tatur J, Hagen WR, Heering HA (2009) Voltammetry of Pyrococcus furiosus ferritin: dependence of iron release rate on mediator potential. Dalton Trans 15:2837–2841

Terashima M, Uchida M, Kosuge H, Tsao PS, Young MJ, Conolly SM, Douglas T, McConnell MV (2011) Human ferritin cages for imaging vascular macrophages. Biomaterials 32:1430–1437

Theil EC (2011) Ferritin protein nanocages use ion channels, catalytic sites, and nucleation channels to manage iron/oxygen chemistry. Curr Opin Chem Biol 15:304–311

Theil EC, Behera EK (2013) The chemistry of nature's iron biominerals in ferritin protein nanocages. In: Ueno T, Watanabe Y (eds) Coordination chemistry in protein cages: principles, design, and applications. Wiley, Hoboken. doi:10.1002/9781118571811

Tosha T, Hasan MR, Theil EC (2008) The ferritin Fe2 site at the diiron catalytic center controls the reaction with O2 in the rapid mineralization pathway. Proc Natl Acad Sci U S A 105:18182–18187

Tosha T, Ng HL, Bhattasali O, Alber T, Theil EC (2010) Moving metal ions through ferritin-protein nanocages from three-fold pores to catalytic sites. J Am Chem Soc 132:14562–14569
Tosha T, Behera RK, Ng HL, Bhattasali O, Alber T, Theil EC (2012) Ferritin protein nanocage ion channels: gating by N-terminal extensions. J Biol Chem 287:13016–13025
Toussaint L, Bertrand L, Hue L, Crichton RR, Declercq JP (2007a) High-resolution X-ray structures of human apoferritin H-chain mutants correlated with their activity and metal-binding sites. J Mol Biol 365:440–452
Toussaint L, Bertrand L, Hue L, Crichton RR, Declercq JP (2007b) High resolution Xray structures of human apoferritin H-chain mutants correlated with their activity and metal-binding sites. J Mol Biol 365:440–452
Treffry A, Zhao Z, Quail MA, Guest JR, Harrison PM (1997) Dinuclear center of ferritins: studies of iron binding and oxidation show differences in the two iron sites. Biochemistry 36:432–441
Treffry A, Zhao Z, Quail MA, Guest JR, Harrison PM (1998a) The use of zinc(II) to probe iron binding and oxidation by the ferritin (EcFtnA) of Escherichia coli. J Biol Inorg Chem 3:682–688
Treffry A, Zhao Z, Quail MA, Guest JR, Harrison PM (1998b) How the presence of three iron binding sites affects the iron storage function of the ferritin (EcFtnA) of Escherichia coli. FEBS Lett 432:213–218
Trikha J, Theil EC, Allewell NM (1995) High resolution crystal structures of amphibian red-cell L ferritin: potential roles for structural plasticity and salvation in function. J Mol Biol 248:949–967
Uchida M, Terashima M, Cunningham CH, Suzuki Y, Willits DA, Willis AF, Yang PC, Tsao PS, McConnell MV, Young MJ, Douglas T (2008) A human ferritin iron oxide nano-composite magnetic resonance contrast agent. Magn Reson Med 60:1073–1081
Uchida M, Kang S, Reichhardt C, Harlen K, Douglas T (2010) The ferritin superfamily: supramolecular templates for materials synthesis. Biochim Biophys Acta 1800:834–845
Wade VJ, Levi S, Arosio P, Treffry A, Harrison PM, Mann S (1991) Influence of site-directed modifications on the formation of iron cores in ferritin. J Mol Biol 221:1443–1452
Wade VJ, Treffry A, Laulhere JP, Bauminger ER, Cleton MI, Mann S, Briat JF, Harrison PM (1993) Structure and composition of ferritin cores from pea seed (Pisum sativum). Biochim Biophys Acta 1161:91–96
Waldo GS, Theil EC (1993) Formation of iron(III)-tyrosinate is the fastest reaction observed in ferritin. Biochemistry 32:13262–13269
Wang Z, Takezawa Y, Aoyagi H, Abe S, Hikage T, Watanabe Y, Kitagawa S, Ueno T (2011) Definite coordination arrangement of organometallic palladium complexes accumulated on the designed interior surface of apo-ferritin. Chem Commun (Camb) 47:170–172
Watt RK (2011) The many faces of the octahedral ferritin protein. BioMetals 24:489–500
Watt GD, Jacobs D, Frankel RB (1988) Redox reactivity of bacterial and mammalian ferritin: is reductant entry into the ferritin interior a necessary step for iron release? Proc Natl Acad Sci U S A 85:7457–7461
Watt GD, Frankel RB, Jacobs D, Huang HQ (1992) Fe2+ and phosphate interactions in bacterial ferritin from Azotobacter vinelandii. Biochemistry 31:5672–5679
Weeratunga SK, Lovell S, Yao H, Battaile KP, Fischer CJ, Gee CE, Rivera M (2010) Structural studies of bacterioferritin B from Pseudomonas aeruginosa suggest a gating mechanism for iron uptake via the ferroxidase center. Biochemistry 49:1160–1175
Wiedenheft B, Mosolf J, Willits D, Yeager M, Dryden KA, Young M, Douglas T (2005) An archaeal antioxidant: characterization of a Dps-like protein from Sulfolobus solfataricus. Proc Natl Acad Sci U S A 102:10551–10556
Wong KK, Colfen H, Whilton NT, Douglas T, Mann S (1999) Synthesis and characterization of hydrophobic ferritin proteins. J Inorg Biochem 76:187–195

Wong SG, Tom-Yew SAL, Lewin A, Le Brun NE, Moore GR, Murphy MEP, Mauk AG (2009) Structural and mechanistic studies of a stabilized subunit dimer variant of Escherichia coli bacterioferritin identify residues required for core formation. J Biol Chem 284:18873–18881
Wong SG, Grigg JC, Le Brun NE, Moore GR, Murphy ME, Mauk AG (2015) The B-type channel is a major route for iron entry into the ferroxidase center and central cavity of bacterioferritin. J Biol Chem 290:3732–3739
Wood JC (2011) Impact of iron assessment by MRI. Hematology Am Soc Hematol Educ Program 2011:443–450
Xing R, Wang X, Zhang C, Zhang Y, Wang Q, Yang Z, Guo Z (2009) Characterization and cellular uptake of platinum anticancer drugs encapsulated in apoferritin. J Inorg Biochem 103:1039–1044
Xu B, Chasteen ND (1991) Iron oxidation chemistry in ferritin. Increasing Fe/O2 stoichiometry during core formation. J Biol Chem 266:19965–19970
Yamashita I, Iwahori K, Kumagai S (2010) Ferritin in the field of nanodevices. Biochim Biophys Acta 1800:846–857
Yang XK, Chen-Barrett Y, Arosio P, Chasteen ND (1998) Reaction paths of iron oxidation and hydrolysis in horse spleen and recombinant human ferritins. Biochemistry 37:9743–9750
Yang X, Le Brun NE, Thomson AJ, Moore CR, Chasteen ND (2000) The iron oxidation and hydrolysis chemistry of Escherichia coli bacterioferritin. Biochemistry 39:4915–4923
Yang Z, Wang X, Diao H, Zhang J, Li H, Sun H, Guo Z (2007a) Encapsulation of platinum anticancer drugs by apoferritin. Chem Commun (Camb) 33:3453–3455
Yang Z, Wang X, Diao H, Zhang J, Li H, Sun H, Guo Z (2007b) Encapsulation of platinum anticancer drugs by apoferritin. Chem Commun (Camb) 33:3453–3455
Zhang Y, Raudah S, Teo H, Teo GW, Fan R, Sun X, Orner BP (2010) Alanine-shaving mutagenesis to determine key interfacial residues governing the assembly of a nano-cage maxi-ferritin. J Biol Chem 285:12078–12086
Zhao G, Bou-Abdallah F, Yang X, Arosio P, Chasteen ND (2001) Is hydrogen peroxide produced during iron(II) oxidation in mammalian apoferritins? Biochemistry 40:10832–10838
Zhao GH, Bou-Abdallah F, Arosio P, Levi S, Janus-Chandler C, Chasteen ND (2003) Multiple pathways for mineral core formation in mammalian apoferritin. The role of hydrogen peroxide. Biochemistry 42:3142–3150
Zhao GH, Su MH, Chasteen ND (2005) mu-1,2-peroxo diferric complex formation in horse spleen ferritin. A mixed H/L-subunit heteropolymer. J Mol Biol 352:467–477
Zheng B, Yamashita I, Uenuma M, Iwahori K, Kobayashi M, Uraoka Y (2010) Site-directed delivery of ferritin-encapsulated gold nanoparticles. Nanotechnology 21:045305
Ziv K, Meir G, Harmelin A, Shimoni E, Klein E, Neeman M (2010) Ferritin as a reporter gene for MRI: chronic liver over expression of H-ferritin during dietary iron supplementation and aging. NMR Biomed 23:523–531

Chapter 4
Antibody Recognition of Immunodominant Vaccinia Virus Envelope Proteins

Dirk M. Zajonc

Abstract Vaccinia Virus (VACV) is an enveloped double stranded DNA virus and the active ingredient of the smallpox vaccine. The systematic administration of this vaccine led to the eradication of circulating smallpox (variola virus, VARV) from the human population. As a tribute to its success, global immunization was ended in the late 1970s. The efficacy of the vaccine is attributed to a robust production of protective antibodies against several envelope proteins of VACV, which cross-protect against infection with pathogenic VARV. Since global vaccination was ended, most children and young adults do not possess immunity against smallpox. This is a concern, since smallpox is considered a potential bioweapon. Although the smallpox vaccine is considered the gold standard of all vaccines and the targeted antigens have been widely studied, the epitopes that are targeted by the protective antibodies and their mechanism of binding had been, until recently, poorly characterized. Understanding the precise interaction between the antibodies and their epitopes will be helpful in the design of better vaccines against other diseases. In this review we will discuss the structural basis of recognition of the immunodominant VACV antigens A27, A33, D8, and L1 by protective antibodies and discuss potential implications regarding their protective capacity.

Keywords Antibodies • Vaccination • Vaccinia virus • X-ray crystallography • Antibody-antigen complexes

---

D.M. Zajonc (✉)
Division of Cell Biology, La Jolla Institute for Allergy and Immunology (LJI), La Jolla, CA 92037, USA

Department of Internal Medicine, Faculty of Medicine and Health Sciences, Ghent University, Ghent 9000, Belgium
e-mail: dzajonc@liai.org

J.R. Harris, J. Marles-Wright (eds.), *Macromolecular Protein Complexes*, Subcellular Biochemistry 83, DOI 10.1007/978-3-319-46503-6_4

Abbreviations

VACV	Vaccinia virus
VARV	variola virus
MPXV	monkeypox virus
ECTV	ectromelia virus
MAb	monoclonal antibody
IMV	intracellular mature virus
EEV	extracellular enveloped virus
GAG	glycosaminoglycan
HS	heparan sulfate
CS	chondroitin sulfate
GlcA	glucuronic acid
GalNAc	N-acetyl galactosamine
HC	heavy chain
LC	light chain
CDR	complementarity determining region
H	hydrogen (bond)
vdW	van der Waals
BSA	buried surface area
sc	shape correlation
Fab	fragment of antigen binding
K_D	equilibrium binding constant
EFC	entry fusion complex
SAM	single alanine scanning mutagenesis

4.1 Vaccinia Virus

VACV is a 190 kbp double-stranded DNA virus that encodes more than 200 ORFs. Together with cowpox (CPXV), monkeypox (MPXV), smallpox (VARV) and ectromelia (ECTV), among others, VACV belongs to the genus of orthopox viruses. VACV exists as two different infectious particles, the intracellular mature virus (IMV) and the extracellular enveloped virus (EEV). The IMV expresses and utilizes the adhesion molecules A27, D8, and H3, to bind to cell surface molecular like glycosaminoglycans (GAGs), heparan sulfate (A27 and H3), chondroitin sulfate (D8) (Lin et al. 2000; Hsiao et al. 1998, 1999; Matho et al. 2014). The IMV also expresses A26, which binds to laminin (Chiu et al. 2007). The entry fusion complex of the IMV allows VACV to fuse with the host membrane and ultimately to infect the host cell. Twelve proteins that are conserved across the poxvirus family are required for this process (Ojeda et al. 2006b; Townsley et al. 2005b; Senkevich et al. 2004; Brown et al. 2006; Izmailyan et al. 2006; Ojeda et al. 2006a; Senkevich and Moss 2005; Nichols et al. 2008; Senkevich et al. 2005; Bisht et al. 2008b;

Satheshkumar and Moss 2009; Townsley et al. 2005a). Following viral replication within the host cell, newly synthesized VACV particles exit the cell as EEV particles. In addition to IMV, these particles also have a second host cell-derived lipid bilayer that contains specific EEV envelope proteins, including A33. A33 is an adhesion molecule that binds a receptor of unknown identity and is required for long-range dissemination within the host (Roper et al. 1998; Wolffe et al. 1998, 2001).

Inoculation with VACV elicits neutralizing antibodies against major envelope proteins, including A27, A33, B5, D8, H3, and L1, conferring protection against smallpox (Davies et al. 2005; Davies et al. 2007; Lawrence et al. 2007; Benhnia et al. 2008; Putz et al. 2006). Maximal protection is obtained with vaccines combining membrane proteins from both IMV and EEV. Although the vaccine efficacy has never been precisely measured in controlled experiments, neutralizing antibodies are found up to 30 years after standard immunization. Smallpox was the first disease that had been eradicated through global vaccination efforts, resulting in a termination of the wide-spread vaccination of the general human population (Henderson 2011). Yet, the precise interaction between the antibodies and their epitopes remained poorly understood, until recently.

In this review will focus on the interaction between several murine antibodies and the VACV antigens: A27, A33, D8 and L1, the only antigens for which structural data of the antibody-antigen complex has been obtained to date (Fig. 4.1).

4.2 IMV Antigen A27

A27 is one of the three GAG adhesion molecules (A27, D8, and H3) and binds to heparan sulfate. A27 is a homo-trimeric protein that is attached to the viral membrane by binding to the transmembrane protein A17 through its C-terminal leucine zipper domain (residues 80–101). The GAG binding site is located at the N-terminus of the mature protein (residues 21–30) (Wang et al. 2014; Howard et al. 2008). The central region of A27 consists of a coiled coil domain (residues 43–84), which is used to interact with the membrane fusion suppressor protein, A26, through intermolecular disulfide bond formation (Cys71, Cys72). The crystal structure of an N-terminal fragment of A27, containing the heparan sulfate binding site and coiled coil domain (residues 21–84), has recently been determined, however only the central fragment (residues 47–84) is ordered, suggesting flexibility of the N-terminal GAG binding domain (Chang et al. 2013). The A27 structure illustrates the complexity and antiparallel nature of the A27 homo-trimer, yet structural information about the N-terminal and C-terminal extremities is missing.

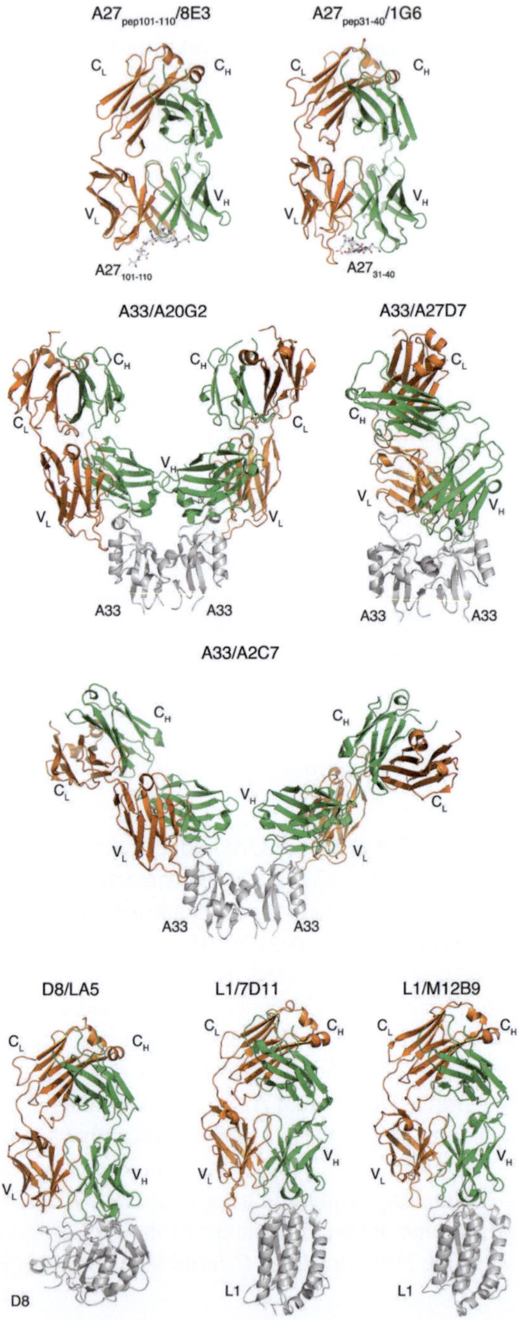

Fig. 4.1 Overview of VACV antigen-antibody (Fab) complexes. The antigens are colored in *grey*, the Fab light (L) chain in *orange* and the Fab heavy (H) chain in *green*. Variable (V) and constant (C) antibody domains are indicated

4.2.1 Anti A27 MAbs

To understand the function of anti-A27 antibodies, especially their protective capacity and their interaction with A27, several murine monoclonal antibodies (MAbs) have been generated and further separated into four distinct specificity groups (Kaever et al. 2016). Three groups (I, III and IV) bind to linear peptides, while group II only binds to VACV lysate and recombinant A27, suggesting it recognized a conformational and discontinuous epitope. Only group I antibodies neutralized MV in a complement-dependent manner and protected against VACV challenge, while a group II mAb protected partially, but did not neutralize (Kaever et al. 2016). The epitope for group I mAbs was mapped to a region adjacent to the GAG binding site and suggested that group I mAbs could potentially interfere with cellular adhesion of A27, while the group IV antibodies bound to the remote C-terminal extremity of A27 and were not neutralizing (Kaever et al. 2016). For details on antibody composition and structure see (Sundberg and Mariuzza 2002; Kumagai and Tsumoto 2001).

4.2.2 Crystal Structure of the $A27_{peptide31\text{–}40}$-1G6 Complex

The crystal structure of the neutralizing group I MAb, 1G6, bound to the A27 peptide "KREAIVKADE" illustrating that both light and heavy chains of the antibody are equally important in binding the peptide (Table 4.1 and Fig. 4.2). The peptide is bound in a crevice formed between the complementarity determining region (CDR) L1 loop and the H2 loop and was supported mainly by the underlying H3 loop. The H chain buried 385 Å^2 on the peptide, while the L chain buried 366 Å^2. 1G6 formed polar contacts (hydrogen bonds and salt bridges) with seven amino acids of the A27 peptide, with five residues of the L chain interacting with four amino acids of the antigen (Arg32, Glu33, Ala34, and Asp39), while five residues of the H chain interacted with four amino acids (Lys37, Ala38, Asp39, Glu40) of the antigen (Table 4.1, Fig. 4.2). L1 appears to form the majority of polar interactions with A27, while the H chain uses H2 and H3 to bind the antigen. H1 is the only CDR that does not contact the peptide.

The structural data corroborated the alanine scanning mutagenesis, which identified several important side chain interactions of the peptide antigen with the antibody. A27 residues Glu33, Ile35, Val36, Lys37, and Asp39 were sensitive to alanine substitution. Peptide residues Glu33 and Asp39 formed a hydrogen (H) bond with L1 residues Ser32 and Tyr37, respectively, while Asp39 also forms a salt bridge with H2 residue Lys55. Peptide residues Lys37 formed a salt bridge with both H2 residues Asp59 and Asp61, while Glu40 forms salt bridge with H3 residue Arg104 and an H bond with Tyr106. Disruption of any of these contacts led to loss of binding, suggesting that these interactions were equally important for binding (Kaever et al. 2016). I35 and V36 are hydrophobic amino acids and their side chains do not

Table 4.1 Structural analysis of antibody-peptide binding

Antibody	**8E3**	**1G6**	**A20G2**	**A27D7**	**A2C7**	**LA5**	**7D11**	**M12B9**
Antigen	A27 peptide 101–110	A27 peptide 31–40	A33 monomer	A33 dimer	A33 monomer	D8	L1	L1
V-Gene and allele, J-Gene and allele	IGHV2–4-1*01, IGHJ3*01	IGHV8–8*01, IGHJ4*01	IGHV5–12*02, IGHJ4*01	IGHV2–6-5*01, IGHJ2*01	IGHV5–12*02, IGHJ4*01	IGHV1S127*01, IGHJ1*01	IGHV1S26*01, IGHJ3*01	IGHV1S26*01, IGHJ3*01
	IGLV3*01, IGLJ2*01	IGKV1–110*01, IGKJ1*01	IGKV1–133*01, IGKJ2*01	IGKV4–91*01, IGKJ4*01	IGKV1–110*01, IGKJ1*01	IGKV4–55*01, IGKJ5*01	IGKV8–21*01, IGKJ1*01	IGKV1–110*01, IGKJ1*01
Isotype	IgG2a	IgG2a	IgG2a	IgG2a	IgG2b	IgG2a	IgG2a	IgG2a
Total BSA Å^2 (antibody/antigen)	HC: 355	HC: 385	HC: 430	HC: 630	HC: 334	HC: 1872	HC: 1340	HC: 1366
	LC: 480	LC: 366	LC: 435	LC: 525	LC: 365	LC: 562	LC: 225	LC: 219
Shape complementarity (0–1)	n.d.	n.d.	HC: 0.68	HC: 0.71	HC: 0.67	HC: 0.62	HC: 0.65	HC: 0.70
			LC: 0.69	LC: 0.74	LC: 0.68	LC: n.d.	LC: 0.26	LC: 0.62
Contact CDRs[a]	**L1, L2, L3, H1, H2**, H3	**L1, L2**, L3, **H2, H3**	**L1, L3, H1, H2**, H3	**L1, L2, L3, H1, H2, H3**	**L1**, L2, **L3, H1, H2, H3**	L1, **L3, H1, H2, H3**	L1, **H1, H2, H3**	L1, L3, **H1, H2, H3**
Number of Antigen contact residues	6	7	16	24	11	23	12	16
Contact overview[b]	HC: 6 HB, (vdW n.d.)	HC: 3 HB, 5 SB, (vdW n.d.)	HC: 8 HB, 2 SB, 66 vdW	HC: 14 HB, 1 SB, 80 vdW	HC: 4 HB, 47 vdW	HC: 16 HB/ 11 SB/ 115 vdW	HC: 8 HB/ 2 SB/ (vdW n.c.)	HC: 11 HB/ 5 SB/ 99 vdW
	LC: 5 HB, 4 SB, (vdW n.d)	LC: 7 HB, 4 SB, (vdW n.d.)	LC: 8 HB, 52 vdW	LC: 10 HB, 66 vdW	LC: 7 HB, 1 SB, 53 vdW	LC: 1 HB/ 15vdW	LC: (vdW n.c.)	LC: 1 HB/ 1 SB

Neutralizing in vitro	No	Yes	Yes	Yes	Yes	Yes	Yes	Yes
Protect against[c] VACV / ECTV	No/n.t.	Yes/n.t.	Yes/ No	Yes/Yes	Yes/No	n.t./n.t.	Yes/n.t.	Yes/No
KD (SPR/BLI)	n.t.	n.t.	66 pM (BLI)	14 nM (BLI)	65 pM (BLI)	0.18 nM (SPR)	n.t.	91 nM (BLI)

[a]CDRs/ residues in bold form multiple contacts
[b]*HB* Hydrogen bond, *SB* salt bridge, *vdW* van der Waals, *n.d.* not determined, *n.c.* not counted
[c]*n.t.*= not tested

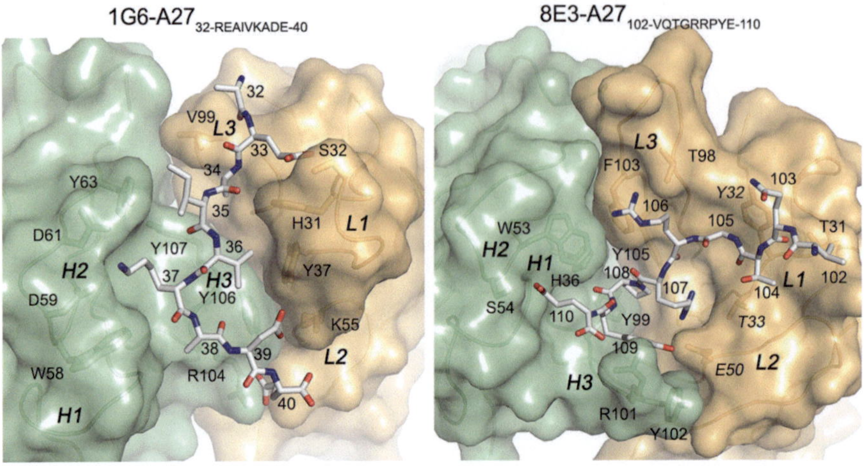

Fig. 4.2 Antibody binding to A27 peptides. The molecular surfaces of the antigen binding sites illustrate intimate binding to A27 peptides. Several contact residues of the Fab are labeled in italics, while peptide residues are numbered according to amino acid position in the A27 protein. Heavy chain in *green* and light chain in *orange*. Peptide as *grey sticks*. CDR loops in *italics*

engage in polar interactions, however Ile33 packed against H2 residue Tyr63, while Val36 packs against L1 residues His31 and Tyr37. These hydrophobic interactions appear to be critical to the overall stability of the complex (Fig. 4.2).

4.2.3 Crystal Structure of the $A27_{peptide101–110}$-8E3 Complex

The group IV MAb 8E3 bound to the A27 peptide DVQTGRRPYE surprisingly with a higher contribution of the L chain than the H chain. However, in contrast to the MAb 1G6, 8E3 forms the binding side perpendicular to the H/L chain interface, while 1G6 bound the peptide in a crevice formed between the H and L chain. As a result, 8E3 binds the N-terminal residues exclusively with the L chain and the C-terminal residues exclusively with the H chain, while for 1G6 certain residues can interact with both antibody chains simultaneously (Fig. 4.2). The H chain of 8E3 interacted less broadly with the antigen and buried 355 $Å^2$, while the L chain buried 480 $Å^2$ on the peptide (Table 4.1).

The antibody 8E3 formed polar contacts with seven amino acids of the antigen. Six residues of the L chain interacted with five amino acids of the peptide (Gln103, Thr104, Arg106, Arg107, and Tyr109), while four residues of the H chain interacted with two amino acids (Pro108 and Glu110) of the peptide (Table 4.1, Fig. 4.2). While the L chain appeared to make more contacts with the peptide overall, alanine scanning of the peptide suggested that the N-terminal half of the peptide was not forming specific interactions with the antibody. H bond interactions with A27 residues Gln103 and Gly105 are formed with backbone oxygen groups, indicating that they are independent of the amino acid sequence. Gly105 to alanine substitution likely induced a steric clash with the antibody surface. As a result the amino acids that were forming specific interactions with the antibody were located within the C-terminal half and include Arg107, Tyr109 and possibly Glu110. Pro108 to alanine exchange likely also induces a steric clash with 8E3. Arg107 and Tyr109 are involved in both a salt bridge and H bond with L2 residue Glu50, which therefore seems to be a major hot spot for binding the antigen, while Glu110 of the peptide formed a H bond with H2 residues Ser54 and H3 residue Arg101, which appeared to be less crucial for binding based on the alanine scanning data (Kaever et al. 2016).

A27 is one of the immunodominant antigens of VACV and antibodies targeting A27 have been extensively studied (Meyer et al. 1994; Rodriguez and Esteban 1987; Rodriguez et al. 1985). Especially group I antibodies had been previously identified and the epitope mapped to the heparan binding domain (Meyer et al. 1994). Therefore, the group I epitope appears to constitute a major binding site for eliciting protective antibodies such as 1G6.

4.3 EEV Antigen A33

A33 is expressed on the EEV membrane as a 23 kDa, homodimeric type II transmembrane protein that undergoes both O- and N-glycosylation (Isaacs et al. 1992; Roper et al. 1996; Payne 1992). A33 controls the incorporation of A36 into the EEV particle and subsequently the production of actin tails. Therefore, A33 plays an important role in effective cell-to-cell spread within the host (Roper et al. 1998; Wolffe et al. 1998, 2001). A33 is also required for proper trafficking of the VACV antigen B5 to the EEV-specific membrane and proper formation of infectious EEV (Wolffe et al. 2001; Chan and Ward 2010). A33 contains a membrane-proximal cysteine on the A33 ectodomain that forms an intermolecular disulfide bridge. However, this cysteine is not required for the production of infectious extracellular virus (Chan et al. 2010). The crystal structure of the ectodomain of A33 revealed an unusual C-type lectin-fold domain, similar in overall architecture to several NK cell ligands (Su et al. 2010).

4.3.1 Anti A33 MAbs

Recent characterization of a panel of anti A33 antibodies using a peptide ELISA indicated that all the anti-A33 MAbs obtained by immunizing mice with live VACV bound only to recombinant A33 but not to any linear peptide, in contrast to most antibodies targeting A27 (Matho et al. 2015). Interestingly, the three antibodies: A2C7, A27D7, and A20G2, which had been characterized in more detail, bound to A33 with different stoichiometries. A single A27D7 Fab bound to one A33 dimer, while A2C7 and A20G2 each bound with one Fab to one A33 subunit (2 Fab's per

A33 dimer) (Matho et al. 2015). While all three MAbs were neutralizing *in vitro*, protected against intranasal VACV challenge *in vivo*, A27D7 also protected mice against ectromelia challenge, identifying A27D7 as a potent cross-species orthopox virus neutralizing antibody (Matho et al. 2015).

4.3.2 Crystal Structure of the A33-A2C7 Complex

The A33/A2C7 complex shows two Fab molecules symmetrically bound to the A33 dimer (Fig. 4.1). Each A2C7- Fab molecule elicits, therefore, identical contacts with a discontinuous and conformational epitope at the membrane-distal extremity of each A33 subunit (Fig. 4.3). The A33 monomer is bound in a deep cavity formed by both L and H chain, with L1 and H2 forming the rim of a binding cup. The L chain buried 365 Å^2 on A33, while the H chain buried a total of 334 Å^2 between A33, leading to a total BSA of 699 Å^2 between Fab and A33, which corresponds to only around 5% of total protein surfaces. The size of the binding interface is comparable to the peptide binding anti A27 MAbs but considerably smaller than the typical range found in antibody-protein antigen complexes (1400–1700 Å^2 BSA), (Davies et al. 1990). Despite recognizing a rather small epitope, A2C7 used sixteen residues (eight from each L and H chain) to interact with eleven A33 residues. Similar to the peptide-specific anti A27 MAbs, both L and H chain of A2C7 appeared to be equally important in antigen binding.

Shape correlation (Sc) measures the geometric surface complementarity of protein-protein interfaces and reflects their specificity (Lawrence and Colman 1993). Both heavy and light chains appeared to bind with high specificity to the antigen (Sc = 0.67 and 068, respectively). For antigen-antibody interfaces, Sc values of 0.64–0.68 (Jones and Thornton 1996) have been reported, illustrating that the A2C7/A33 interaction is very specific. This interface is held together by an extensive network of hydrogen bonds and salt bridges involving every CDR loop (Matho et al. 2015). Gln173 is a major contact residue of the A2C7 epitope, as it reaches into the base of the A2C7 binding cup and intersects with of almost every CDR loop (Fig. 4.3). It forms multiple hydrogen bonds via its backbone and side chain (H1, H2, H3, and L3). Replacing Gln173 with arginine led to a complete loss of A2C7 binding. Since the Gln173Arg substitution is found in ECTV the binding data suggested that A2C7 would be unable to protect against ectromelia challenge and no protection was indeed found using an ectromelia challenge model (Matho et al. 2015). A33 residue Asp170 is located on a flexible loop and is contacted by both H and L chain. It forms a salt bridge with L2 residue Lys55 and a hydrogen bond with L1 residue Tyr32. However, the Asp170Ala mutation did not affect the binding affinity of A2C7-MAb, suggesting it is not a critical binding residue (Matho et al. 2015).

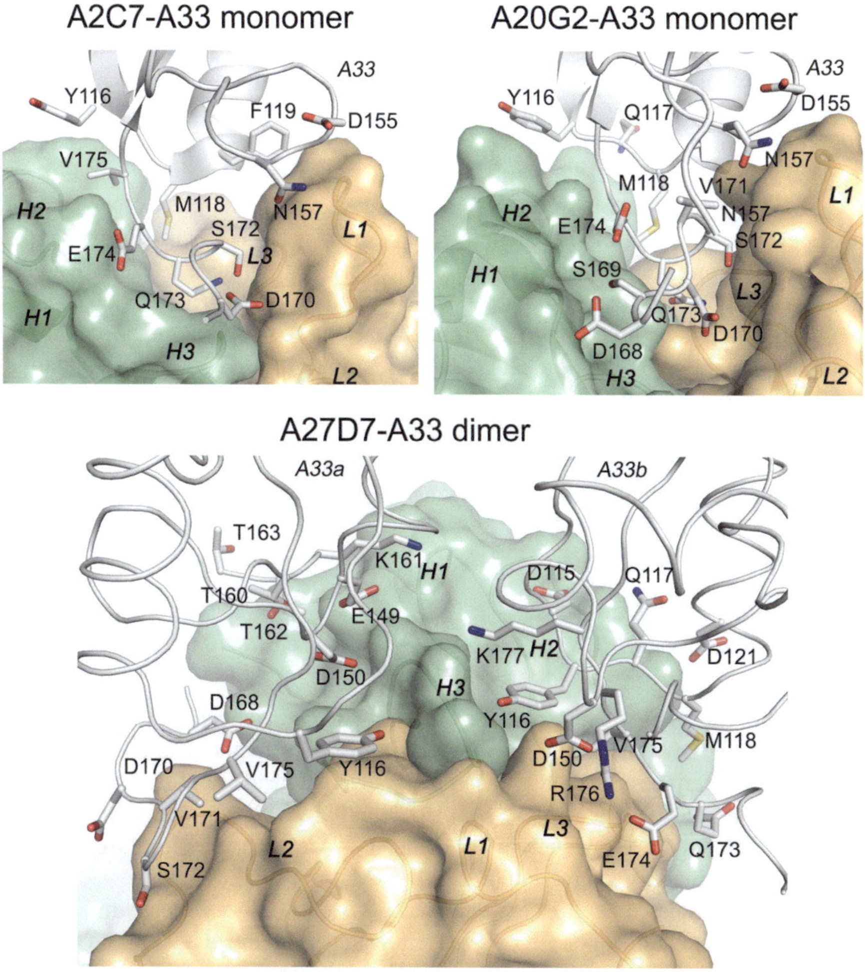

Fig. 4.3 A33-antibody interfaces. Fab A2C7 and A20G2 bind the A33 monomer in a crevice formed between L (*orange*) and H (*green*) chain, while A27D7 binds the A33 dimer. Heavy chain in *green* and light chain in *orange*. A33 in *grey*. CDR loops in *italics*

4.3.3 Crystal Structure of the A33-A20G2 Complex

A20G2 also bound with 1 Fab molecule to one A33 monomer (two Fab's per A33 dimer, Fig. 4.1). A20G2 targets a similar epitope compared to A2C7. Sixteen residues of each A33 subunit were in contact with sixteen residues of the A20G2 Fab (Table 4.1). Compared to the binging cup formed by A2C7, A20G2 is open ended at both ends, explaining the larger binding surface. Both H and L chain grasp the tip of each A33 subunit similar to a tweezer, and each chain buried roughly 430 Å^2 on A33. The Sc value is 0.68 and 0.69 for both antibody/antigen binding interfaces

(Table 4.1). Despite numerous amino acid differences in the CDRs, A20G2 approaches the antigen with an overall similar topology compared to A2C7, with the exception of CDRs L2 and H3. This is reflected in an overall change in binding angle (Fig. 4.1). L1 plays a major role in A33 binding for both A2C7 and A20G2, as it clamps onto one side of each A33 subunit (A33 residues Phe119, Asp155, Gly156, Asn157, Asp170, and Ser172, while Ser154 and Asn157 were exclusive to A20G2) (Fig. 4.3). L3 sits at the base of the binding site, and H1 and H2 target a similar epitope at the top edges of A33. In both complexes, Gln173 is inserted deep into the center of the binding site. Similar to A2C7, the Gln173Arg mutation also fully abrogated A20G2 binding, explaining the inability of this MAb to protect against ectromelia challenge (Matho et al. 2015). In addition, the Asp170Ala mutation did not have any impact on binding of A20G2 to A33, similar to A2C7. The A33 mutation Leu118Arg that is located on both epitopes fully impaired A2C7 binding, while not affecting binding of A20G2. This was likely due to a clash with L3 of A2C7, which approached A33 from a different angle. In conclusion, A2C7 and A20G2 target similar but not identical epitopes mostly due to sequence differences within H3 and L3 (Matho et al. 2015).

4.3.4 Crystal Structure of the A33-A27D7 Complex

A27D7 is unique in its binding to A33. A single A27D7-Fab bound at the A33 dimer interface, with the L and H chains contacted both A33 subunits (Figs. 4.1 and 4.3). The binding interfaces for both L and H chains were considerably larger compared to A2C7 and A20G2 (LC = 525.3 Å^2, HC = 630.2 Å^2) for a total of 1155.5 Å^2 (A27D7-A33 dimer). Sc values of 0.71 and 0.74 were very high and indicate a highly specific binding interaction (Table 4.1) (Matho et al. 2015).

Eighteen residues of the H chain contacted sixteen A33 residues, while fifteen residues of the L chain contacted twelve A33 residues (Matho et al. 2015). CDR H1 interacted only with one A33 subunit, while CDR H2 interacted only with the other A33 subunit of the dimer. Among the many hydrogen bonds and salt bridges formed between A27D7 and A33, a few residues that appeared to make important contacts were targeted for alanine scanning mutagenesis. Lys161 formed a salt bridge with H1 residue Asp31. Asp115 is a central residue of the H2 interface, as it formed six potential hydrogen bonds with H2 residues Gly53, Gly54, Gly55, and Thr56. Surprisingly, the Lys161Ala mutation did not impair A27D7 binding, while the Asp115Ala mutation only slightly affected A33 binding. CDR H3 is centrally positioned in the A33 dimer groove, and contacted residues of both A33 subunits. H3 formed three distinct hydrogen bonds between Ala97, Ser98 and Lys177 of A33. The third hydrogen bond is formed between A33 residue Tyr116 and H3 residue Tyr99 (Matho et al. 2015).

The A27D7 light chain does not elicit any salt bridge with A33, and its interaction with the antigen is mostly driven by van der Waals contacts. In particular L2 and L3 each interact specifically with only one of the A33 subunits (Fig. 4.3). L1 is relatively short compared to group II MAbs A20G2 and A2C7 (12 vs. 16 residues). A single L1 residue, Tyr32, formed one hydrogen bond with Asp150 and also accounted for all vdW contacts with a single A33 subunit. L2 interacts solely with the other A33 subunit, forming a total of two hydrogen bonds and 30 vdW contacts. CDR L3 (Ser92 and Leu94) formed an extensive hydrogen bond network with A33 residues Gln173, Glu174, Val175 and Arg176. However, A27D7 bound the A33 mutant Gln173Arg still with high affinity, suggesting that the many contacts formed between A27D7 and A33 could compensate for the various single amino acid substitutions on A33, in contrast to the antibodies A2C7 and A20G2, which both failed to bind the Gln173Ala mutant (Matho et al. 2015). Since A27D7 binding appeared resistant to single alanine-scanning mutagenesis, A33 residues that were bound by the antibody on both A33 subunits were targeted by mutagenesis, since these mutations would disrupt interactions on both A33 monomers. Tyr116, Asp150, and Val175 form contacts with the Fab in both A33 subunits, and among those, Tyr116 was chosen for mutagenesis. The Tyr116Ala mutation led to a significant decrease in binding affinity, however, still retained sub nanomolar affinity (Matho et al. 2015). Even double mutants that had significantly reduced binding affinity compared to wildtype still bound with low nanomolar affinity, suggesting that a large number of energetically favorable interactions still remained intact and enabled the specific binding interaction. Based on the binding data, A33 mutants were created that mimicked the epitope of the orthopox strains such as cowpox Brighton (CPXV-Br), MPXV, and ECTV. The A33 variant Gln117Lys/Leu118Ser both mimicked the epitope of CPXV and MPXV, while the A33 variant Gln117Lys/Leu118Ser/Gln173Arg was specific for ECTV. Both antibodies A2C7 and A20G2 failed to bind to the ECTV variant, while A20G2 still bound with a ~9-fold lower affinity to the CPXV/MPXV variant. As expected however, A27D7 was able to bind both A33 variants with high affinity and as predicted from the binding data was able to protect mice against ectromelia challenge, suggesting that this antibody is a potent cross-species orthopox virus neutralizing antibody (Matho et al. 2015).

4.4 IMV Antigen D8

D8 is a 32 kDa type 1 membrane protein of the IMV that binds to cell surface chondroitin sulfate (CS) but not to HS (Hsiao et al. 1999). The large N-terminal domain of D8 shares significant homology to human carbonic anhydrases and adopts a typical CAH fold. This N-terminal domain is characterized by a centrally located positively charged crevice that is lined by several lysine, histidine and arginine residues, suitable for the binding of CS. Using a combination of site directed mutagenesis and antibody cross-blocking, the positively charged crevice was identified as the GAG binding site (Matho et al. 2014; Matho et al. 2012). The last thirty residues of the ectodomain were disordered in the crystal structure and contain an extracellular membrane-proximal cysteine that is used for disulfide-linked dimerization. Recombinant full-length D8 forms a hexamer in solution, in which 3 disulfide-linked dimers non-covalently associate, while the N-terminal GAG binding domain only forms monomers (Matho et al. 2014, 2012). Using a glycan microarray, CS-E [(GlcA-4S,6S-GalNAc)n] was identified as the preferred molecular species of chondroitin sulfate that is bound by D8 (Matho et al. 2014). While the biological significance of hexamerization is not clear, it was suggested that it would increase binding affinity to CS for viral adhesion the host cell (Matho et al. 2014).

4.4.1 Anti D8 Antibodies

Recently a panel of murine monoclonal anti D8 antibodies were characterized and further separated into four different specificity groups, based on their cross-blocking ability (Sela-Culang et al. 2014). The grouping suggested that D8 provided a large surface area that was targeted differently by these antibodies. Peptide ELISA further suggested that group II antibodies targeted a linear epitope called peptide 58 (residues 91–110), while groups I, III and IV targeted a conformational epitope (Sela-Culang et al. 2014). Using EM, the binding sites for group I, II, and III were identified, while a crystal structure was obtained for the group IV antibody LA5 bound to D8. Since only the group IV antibody fully blocked binding of D8 to CS-E (Matho et al. 2014), this antibody will be discussed in more detail.

4.4.2 Crystal Structure of the D8-LA5 Complex

LA5 is an IgG2a isotype (Meng et al. 2011), and its Fab bound recombinant D8 with high affinity (KD = 0.18 nM, Table 4.1). The binding interface between D8 and LA5 was very large, with a total buried surface area of 2434 Å^2. The H chain dominated the binding footprint, with a total of 1872 Å^2 buried surface, which is in contrast to the 562 Å^2 covered by the L chain (Table 4.1). Since the shape complementarity was not very high, water molecules were found especially between D8 and the L chain of LA5 to increase the binding specificity (Matho et al. 2012). The LA5 D8 epitope was formed by twenty-three residues covering the entire sequence of D8 (Gln3, Leu5, Thr39, Gly40, Lys41, Arg44, Lys108, Asn145, Ile174, Asn175, His176, Ser177, Ser204, Leu205, Ile215, Glu217, Tyr219, Arg220, Asn221, Tyr223, Lys224, plus N218 and N226 if including water-mediated contacts), with seven residues sharing contacts with at least two CDR's of LA5 (Thr39, Lys41, Asn145, Asn175, His176, Glu217 and Arg220). Interestingly, the D8 epitope residues Arg44, Lys48 and Arg220 were also involved in CS-E binding, since replacement with alanine either reduced (Arg220Ala) or fully abrogated (Arg44Ala/Arg220Ala and Lys48Ala/Arg220Ala) D8 binding to CS-E. Together with the CS-E microarray data, this demonstrated that LA5 binding overlapped with the binding site for CS-E and that several D8 contact for both CS-E and LA5 were shared. Eighteen residues of LA5 contacted the antigen D8, from five of the six CDRs of LA5. While the L chain only formed one polar interaction (H-bond) and 15 vdW contacts, the H chain bound through an intricate network of polar interactions (16 hydrogen bonds and 11 salt bridges), with an additional 115 vdW interactions contributing to the vast majority of the Fab-Ag interactions (Table 4.1). CDR H2 and H3 seemed to contribute equally in terms of the number of interactions with D8, while H1 contributes the least to the binding interface, with only three H-bonds and 43 vdW interactions.

4.4.3 Possible Function of LA5

Most noticeable, LA5 bound with CDR H1, H2, and H3 directly above the GAG binding crevice of D8. Previous CS-E docking experiments placed the carboxylate of glucuronic acid (GlcA) pointing down into the binding crevice, while the 4′ and 6′ sulfates of the N-acetyl galactosamine (GalNAc) were mostly placed at the top of the binding crevice, where its negative charge would be neutralized by the various electrostatic interactions with the positively-charged groove lining residues of D8 (Fig. 4.4). The residues that when replaced with alanine impacted D8 binding to CS-E (Lys41, Arg44 and Arg220), all formed salt-bridges in the D8-CS-E model, corroborating their importance in CS-E binding (Matho et al. 2014). Based on the combined structural and biochemical data Lys48 and Lys98 likely form the entrance portal to the crevice, while Arg220 and Lys41 form the exit portal. In the computational model, CS-E binding extended beyond these portals and the CS-E ends sat flat above the surface of D8, forming less specific polar contacts. However, the exact length of the CS-E molecule that interacts with D8 had not been experimentally determined (Fig. 4.4). Computational docking further suggested the exit portal to form multiple contacts with the sulfates of CS-E, especially both 4′ and 6′ sulfates bound to Arg220, while at the exit portal, only a single salt-bridge was formed with GlcA, while the 6′ sulfate sat deep inside the crevice. Obviously structural data is necessary to really determine the precise interactions between D8 and CS-E. However, the current model suggested that the exit portal of the D8 crevice is structurally optimized for the binding of CS-E, while the exit portal is structurally slightly more promiscuous in its binding specificity (Fig. 4.4). Strikingly in the LA5-D8 structure, a triad of negatively charged residues of CDR H2 (Asp53, Glu56 and Glu58) bound inside the positively charge crevice of D8 (Fig. 4.4), likely mimicking the negatively charged 4′-, and 6′ sulfate groups. Strategies of true receptor-mimicry have been found especially in antibodies targeting the sialic acid binding site of influenza hemagglutinin, where an aspartate residue of the antibody is in an identical position compared to the carboxylate of sialic acid (Lee et al. 2014).

In conclusion, the binding site of LA5 on D8 overlapped directly with the predicted binding site of CS-E and would likely outcompete CS-E binding to D8 through its high, sub-nanomolar binding affinity or block initial binding to CS-E. As a result, LA5 likely prevents D8 binding to CS on host cells, thus leading to a

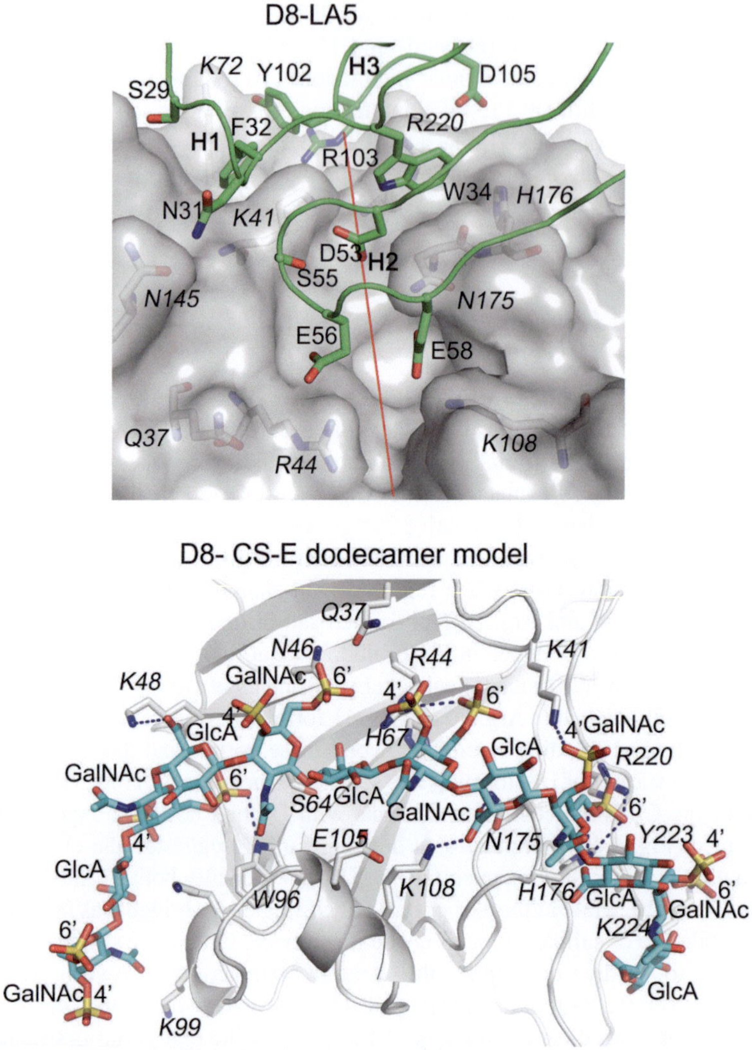

Fig. 4.4 LA5 binding to D8 and D8/CS-E model. The Fab of LA5 binds with the three CDRs of the H chain (H1, H2, and H3 in *green*) directly above the CS-E binding crevice (*red line*) of D8 (*grey surface*). Computational model of a CS-E dodecamer (*cyan sticks*) binding to D8 (*grey cartoon*). D8 residues in *italics*, LA5 residues in regular font

reduced infectivity through the D8/CS axis. This is exemplified by a reduced VACV virulence in BALB/c mice using D8 deletion strains (Rodriguez et al. 1992). However, loss of D8 activity, can be compensated for by VACV binding to HS through the adhesion molecules A27 and H3, since D8-negative virus replicates efficiently in cultured cells (Niles and Seto 1988).

4.5 IMV Antigen L1

L1 is a 250-amino-acid myristoylated protein with a C-terminal transmembrane domain that spans residues 186–204 (Franke et al. 1990; Su et al. 2005). L1 forms a bundle of α-helices that pack against a pair of two-stranded β-sheets (Su et al. 2005). L1 associates with the virus-encoded multiprotein entry-fusion complex (EFC) and plays an essential role in viral entry (Bisht et al. 2008a). It is an immunodominant neutralizing antibody target in mice, though it is a less common target in humans (Benhnia et al. 2008). A VACV L1 mutant had been described that escaped neutralization by the antibody 2D5 and where residue Asp35 had been replaced with asparagine (Ichihashi and Oie 1996). Interestingly, the residue Asp35 in question, is located at the tip of L1, where various loops connect the β-strands and α-helices, and a common target of neutralizing antibody responses (Kaever et al. 2014; Su et al. 2005, 2007). However, not all MAbs that target Asp35 can bind the Asp35Asn escape mutant (Kaever et al. 2014; Su et al. 2007).

4.5.1 Anti L1 Antibodies

Murine anti L1 MAbs had been previously characterized by several labs and identified at least three different specificity groups (Aldaz-Carroll et al. 2005; Ichihashi and Oie 1996; Kaever et al. 2014; Su et al. 2007; Wolffe et al. 1995). In one study, the three antibody groups had been characterized side by side (Kaever et al. 2014). While the majority of antibodies obtained in this study belonged to group I, which targeted a conformational and discontinuous epitope containing the above mentioned Asp35, group II and III MAbs recognized a unique epitope. The group II MAb 8C8 bound a conformational epitope that is separate from the group I MAbs but otherwise unknown, while the group III MAb 39D4 targeted a linear epitope (residues 121–140) that overlapped with a previously identified epitope of L1 (residues 118–128) (Aldaz-Carroll et al. 2005). The group I MAbs neutralized IMV *in vitro* and *in vivo* in the absence of complement. The group II MAb 8C8 failed to neutralize MV, likely because its epitope was not well exposed on the virion (Kaever et al. 2014). The group III MAb 39D4 only neutralized less than 20% and neutralization efficacy could not be increased by the addition of complement, which this antibody can fix. In addition, both group II and III MAbs bound recombinant L1 with greatly reduced binding affinity compared to a the group I MAb M12B9 (K_D = 17 and 32 nM vs. 0.091) (Kaever et al. 2014). The two group I antibodies 7D11 and M12B9 that were produced by different labs had both been characterized structurally and, surprisingly, recognize a highly conserved epitope centered around Asp35 of L1 and have very similar H chain sequences (Kaever et al. 2014; Su et al. 2007).

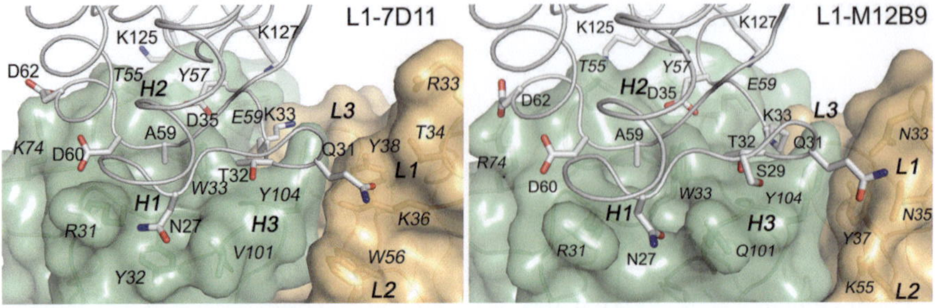

Fig. 4.5 M12B9 and 7D11 binding to the L1 antigen. Antibodies 7D11 and M12B9 bind to VACV-L1 with highly similar binding chemistries, predominantly using the H chain (*green*). Light chain in *orange*, VACV-L1 in *grey*. CDR loops and Fab residues in *italics*. VACV-L1 residues in regular font

4.5.2 Crystal Structure of the L1-7D11 Complex

L1 was the first VACV antigen for which a crystal structure in complex with a Fab was obtained (Su et al. 2007). The antibody 7D11 is a group I antibody and the crystal structure revealed a very H chain centric footprint on top of the L1 molecule (Fig. 4.1). 7D11 binds to four long loops that connect the bundle of helices at the tip of L1, presumably away from the viral membrane. The H chain buried a total of 1340 Å^2, while the L chain participated little in overall binding and buried 225 Å^2 on L1. Shape complementarity is 0.65 for the H chain and 0.26 for the L chain, correlating with an H chain centric footprint. L1 residues Asn27, Gln31, Thr32, Lys33, Asp35, Ser58, Ala59, Asp60, Ala61, Asp62, and Lys125 are bound by the M12B9 heavy chain using vdW interactions, while L1 residues Lys33, Asp35, Ala59, Asp60 also form eight hydrogen bonds and two salt bridges with CDR1–3 residues of the H chain (Table 4.1, Fig. 4.5). Interestingly 7D11 binds to the VCV escape mutant Asp35Asn (Su et al. 2007), which is also found in ECTV. However, whether 7D11 protects against ectromelia challenge is currently unknown.

4.5.3 Crystal Structure of the L1-M12B9 Complex

M12B9 also belongs to the group I antibodies and not surprisingly has a highly similar H chain sequence compare to 7D11, while the L chain is very different (Table 4.1). VACV-L1 residues Glu25, Asn27, Ala28, Ser29, Gln31, Thr32, Lys33, Asp35, Ser58, Ala59, Asp60, Ala61, Asp62, Lys125, Lys127, and Ser153 are bound directly by the M12B9 heavy chain (Fig. 4.5) (Kaever et al. 2014). The heavy chain forms a total of eleven H-bonds, five salt bridges, and ninety-nine van der Waals (vdW) interactions. The H chain buried 1366 Å^2 surface between VACV-L1 and has a shape complementarity of 0.70, while the LC buried a total 219 Å^2 with a shape complementarity of 0.62. In total, the M12B9 light chain formed one potential salt bridge and one hydrogen bond involving L chain residue Asp98 (CDR L3), Asn34 (CDR L1) and VACV-L1 residue Lys33 and Gln31, respectively, while in the 7D11/L1 structure, there were no interactions between the light chain and VACV-L1 within 4 Å distance (Su et al. 2007). Surprisingly, however, the escape mutant Asp35Asn was resistant to neutralization by any group I MAbs characterized in this particular study, suggesting an inability of these MAbs to bind to Asn35 instead of Asp35 (Kaever et al. 2014). While binding of M12B9 to the recombinant VACV-L1 containing the Aps35Asn was not tested, no binding of M12B9 to the VACV-L1 mutant Asp35Ala was observed, demonstrating the importance of Asp35 for group I MAb binding. Since ECTV contains the Asp35Asn variation, and M12B9 did not protect against ECTV challenge it could be speculated that M12B9 does not bind the Asp35Asn variant found in the ECTV L1 ortholog (EVM072). However, that would contrast 7D11 binding, which was not affected by the same mutation.

4.5.4 *7D11 Versus M12B9 Comparison*

Both L1-7D11 and L1-M12B9 complex structures bound a highly conserved epitope on L1, correlating well with the high sequence similarity of the H chains (Kaever et al. 2014). The overall H chain sequence identity was 83% and differences are found in the framework residues and CDR H3, while H1 and H2 are very conserved (only 1 amino acid difference). As a result, the structures superimposed well with a low root mean square deviation of 1.00 Å. The only gross structural differences are within the H3 loop (only 30% conserved sequence) (Kaever et al. 2014). While M12B9 formed a hydrogen bond between CDR H3 residue Gln101 and Asn27, 7D11 formed a salt-bridge between CDR H3 residue Asp102 and L1 residue Lys33. Since the Ans27Ala mutant retained full binding to M12B9, this contact did not appear important for M12B9 binding, while on the other hand the additional salt bridge (adjacent to Asp35 of L1), formed by 7D11 could compensate for Asp35Asn mutation to retain binding (Fig. 4.5). Both MAbs, however, share the majority of polar interactions that are formed by CDRs H1 and H2, while CDR H3 adopted a similar orientation in both complexes but almost exclusively contacted L1 through non-directed vdW interactions (Kaever et al. 2014).

4.6 Concluding Remarks

Antibody responses against VACV had been structurally characterized for the antigens A27, A33, D8 and L1. Antibodies were either obtained by immunizing mice with VACV with or without following boost with recombinant protein or by immunizing mice only with recombinant protein. Immunizing mice with VACV resembles closely the natural antibody response following infection. During that response the obtained antibodies exhibit certain features that appear to be driven by the nature of the antigen. The majority of studied antibodies appeared to target a conformational and discontinuous, rather than a liner peptide, with the exception of A27. For A27 only few antibodies had been obtained that require folded A27 protein for binding. A27 is a rather small protein and contains α-helical segments and disordered domains (Chang et al. 2013). Except for the antiparallel trimerization that creates a composite binding surface for antibodies, the A27 structure itself is linear and it might very well be that several of the linear peptides recapitulate the tertiary structure found in the A27 protein. Both L and H chain contributed equally to the binding, a feature that has been described for antibodies binding small molecule, peptides and haptens (Kumagai and Tsumoto 2001).

For A33, two groups of antibodies have been characterized. Those that bind with one Fab to one A33 monomer, essentially having identical epitopes on each A33 monomer, and those antibodies that recognize a composite and unique binding surface in the A33 dimer, to which only a single Fab can bind. The latter binding mode is characterized by a larger binding foot print (buried surface area), which also appears to be more resistant to single alanine scanning mutagenesis (SAM). SAM has been a technique to identify epitope residues but depending on the antibody, will only identify hot spot residues and not residues that are not crucial for antibody binding. Since the A33 dimer presented a rather small surface area and unique protein shape, most antibodies that target A33 cross-blocked each other and, hence, antibody cross-blocking is not a powerful tool for epitope identification for small protein antigens. For A33, both L and H chain contribute equally to the binding and employ a "tweezer-like" binding mode, essentially grabbing onto A33 from both sides. This was largely due to longer CDR loops (specially CDR L1) that formed a central cavity in the antibody paratope (the antibody surface that binds the epitope of the antigen).

D8 is unique in that it presented a large surface for antibodies to bind. Not surprisingly the most antibody specificity groups are found for anti D8 MAbs. While some MAbs target the CS-E receptor binding site on D8, more or less mimicking the negatively charged ligand with an array of aspartate and glutamate residues, other MAbs bind to different D8 surfaces. The study of anti D8 antibodies revealed that there is no preferred or immunogenic site on D8. Antibodies are found targeting any accessible site on D8 and all anti D8 antibodies bound the N-terminal GAG binding domain. No antibody has been identified to date that binds to the C-terminal stalk region that is necessary for oligomerization. This suggests that the C-terminal domain is not accessible in VACV and a weak target for antibody responses. Interestingly, antibodies that target D8 rely excessively on the H chain in recognizing the antigen, with little contribution from the L chain. This is surprising given the fact, that the antibody LA5 has the largest binding footprint of all anti VACV antibodies for which structural information has been acquired. Similarly, anti-L1 MAbs bind their antigen predominantly with the H chain, at the tip of the molecule, where L1 itself forms 4 different loops that are contacted by the antibodies. Since L1 is part of the EFC, neutralizing anti-L1 MAbs do not require the presence of complement for efficient neutralization, while MAbs recognizing A27, A33, and D8 all require the presence of complement for neutralization, which increases both viral coating with complement and is required for opsonization.

Acknowledgements I would like to thank Bjoern Peters, Yan Xiang, Shane Crotty and Mark Buller for a terrific collaboration and all lab members, especially Michael H. Matho, Tom Kaever, Xiangzhi Meng and Andrew Schlossman without whom the studies would have been nearly impossible. Many thanks also to Yan Xiang for critically reading the manuscript. This project has been funded in whole or in part with federal funds from the National Institute of Allergy and Infectious Diseases, National Institutes of Health, Department of Health and Human Services, under contract no. HHSN272200900048C

References

Aldaz-Carroll L, Whitbeck JC, Ponce de Leon M, Lou H, Pannell LK, Lebowitz J, Fogg C, White CL, Moss B, Cohen GH, Eisenberg RJ (2005) Physical and immunological characterization of a recombinant secreted form of the membrane protein encoded by the vaccinia virus L1R gene. Virology 341(1):59–71. doi:S0042-6822(05)00416-2 [pii]10.1016/j.virol.2005.07.006

Benhnia MR, McCausland MM, Su HP, Singh K, Hoffmann J, Davies DH, Felgner PL, Head S, Sette A, Garboczi DN, Crotty S (2008) Redundancy and plasticity of neutralizing antibody responses are cornerstone attributes of the human immune response to the smallpox vaccine. J Virol 82(7):3751–3768. doi:10.1128/JVI.02244-07

Bisht H, Weisberg AS, Moss B (2008a) Vaccinia virus l1 protein is required for cell entry and membrane fusion. J Virol 82(17):8687–8694. doi:10.1128/JVI.00852-08

Bisht H, Weisberg AS, Moss B (2008b) Vaccinia virus l1 protein is required for cell entry and membrane fusion. J Virol 82(17):8687–8694. doi:10.1128/JVI.00852-08

Brown E, Senkevich TG, Moss B (2006) Vaccinia virus F9 virion membrane protein is required for entry but not virus assembly, in contrast to the related L1 protein. J Virol 80(19):9455–9464. doi:10.1128/JVI.01149-06

Chan WM, Ward BM (2010) There is an A33-dependent mechanism for the incorporation of B5-GFP into vaccinia virus extracellular enveloped virions. Virology 402(1):83–93. doi:10.1016/j.virol.2010.03.017

Chan WM, Kalkanoglu AE, Ward BM (2010) The inability of vaccinia virus A33R protein to form intermolecular disulfide-bonded homodimers does not affect the production of infectious extracellular virus. Virology 408(1):109–118. doi:10.1016/j.virol.2010.09.021

Chang TH, Chang SJ, Hsieh FL, Ko TP, Lin CT, Ho MR, Wang I, Hsu ST, Guo RT, Chang W, Wang AH (2013) Crystal structure of vaccinia viral A27 protein reveals a novel structure critical for its function and complex formation with A26 protein. PLoS Pathog 9(8):e1003563. doi:10.1371/journal.ppat.1003563

Chiu WL, Lin CL, Yang MH, Tzou DL, Chang W (2007) Vaccinia virus 4c (A26L) protein on intracellular mature virus binds to the extracellular cellular matrix laminin. J Virol 81(5):2149–2157. doi:10.1128/JVI.02302-06

Davies DR, Padlan EA, Sheriff S (1990) Antibody-antigen complexes. Annu Rev Biochem 59:439–473. doi:10.1146/annurev.bi.59.070190.002255

Davies DH, Liang X, Hernandez JE, Randall A, Hirst S, Mu Y, Romero KM, Nguyen TT, Kalantari-Dehaghi M, Crotty S, Baldi P, Villarreal LP, Felgner PL (2005) Profiling the humoral immune response to infection by using proteome microarrays: high-throughput vaccine and diagnostic antigen discovery. Proc Natl Acad Sci U S A 102(3):547–552. doi:10.1073/pnas.0408782102

Davies DH, Molina DM, Wrammert J, Miller J, Hirst S, Mu Y, Pablo J, Unal B, Nakajima-Sasaki R, Liang X, Crotty S, Karem KL, Damon IK, Ahmed R, Villarreal L, Felgner PL (2007) Proteome-wide analysis of the serological response to vaccinia and smallpox. Proteomics 7(10):1678–1686. doi:10.1002/pmic.200600926

Franke CA, Wilson EM, Hruby DE (1990) Use of a cell-free system to identify the vaccinia virus L1R gene product as the major late myristylated virion protein M25. J Virol 64(12):5988–5996

Henderson DA (2011) The eradication of smallpox--an overview of the past, present, and future. Vaccine 29(Suppl 4):D7–D9. doi:10.1016/j.vaccine.2011.06.080

Howard AR, Senkevich TG, Moss B (2008) Vaccinia virus A26 and A27 proteins form a stable complex tethered to mature virions by association with the A17 transmembrane protein. J Virol 82(24):12384–12391. doi:10.1128/JVI.01524-08

Hsiao JC, Chung CS, Chang W (1998) Cell surface proteoglycans are necessary for A27L protein-mediated cell fusion: identification of the N-terminal region of A27L protein as the glycosaminoglycan-binding domain. J Virol 72(10):8374–8379

Hsiao JC, Chung CS, Chang W (1999) Vaccinia virus envelope D8L protein binds to cell surface chondroitin sulfate and mediates the adsorption of intracellular mature virions to cells. J Virol 73(10):8750–8761

Ichihashi Y, Oie M (1996) Neutralizing epitope on penetration protein of vaccinia virus. Virology 220(2):491–494. doi:10.1006/viro.1996.0337

Isaacs SN, Wolffe EJ, Payne LG, Moss B (1992) Characterization of a vaccinia virus-encoded 42-kilodalton class I membrane glycoprotein component of the extracellular virus envelope. J Virol 66(12):7217–7224

Izmailyan RA, Huang CY, Mohammad S, Isaacs SN, Chang W (2006) The envelope G3 L protein is essential for entry of vaccinia virus into host cells. J Virol 80(17):8402–8410. doi:10.1128/JVI.00624-06

Jones S, Thornton JM (1996) Principles of protein-protein interactions. Proc Natl Acad Sci U S A 93(1):13–20

Kaever T, Meng X, Matho MH, Schlossman A, Li S, Sela-Culang I, Ofran Y, Buller M, Crump RW, Parker S, Frazier A, Crotty S, Zajonc DM, Peters B, Xiang Y (2014) Potent neutralization of vaccinia virus by divergent murine antibodies targeting a common site of vulnerability in L1 protein. J Virol 88(19):11339–11355. doi:10.1128/JVI.01491-14

Kaever T, Matho MH, Meng X, Crickard L, Schlossman A, Xiang Y, Crotty S, Peters B, Zajonc DM (2016) Linear epitopes in A27 are targets of protective antibodies induced by vaccination against smallpox. J Virol. doi:10.1128/JVI.02878-15

Kumagai I, Tsumoto K (2001) Antigen–Antibody Binding. In: eLS. John Wiley & Sons, Ltd. doi:10.1002/9780470015902.a0001117.pub2

Lawrence MC, Colman PM (1993) Shape complementarity at protein/protein interfaces. J Mol Biol 234(4):946–950. doi:S0022-2836(83)71648-7[pii]10.1006/jmbi.1993.1648

Lawrence SJ, Lottenbach KR, Newman FK, Buller RM, Bellone CJ, Chen JJ, Cohen GH, Eisenberg RJ, Belshe RB, Stanley SL Jr, Frey SE (2007) Antibody responses to vaccinia membrane proteins after smallpox vaccination. J Inf Dis 196(2):220–229. doi:10.1086/518793

Lee PS, Ohshima N, Stanfield RL, Yu W, Iba Y, Okuno Y, Kurosawa Y, Wilson IA (2014) Receptor mimicry by antibody F045-092 facilitates universal binding to the H3 subtype of influenza virus. Nat Commun 5:3614. doi:10.1038/ncomms4614

Lin CL, Chung CS, Heine HG, Chang W (2000) Vaccinia virus envelope H3L protein binds to cell surface heparan sulfate and is important for intracellular mature virion morphogenesis and virus infection in vitro and in vivo. J Virol 74(7):3353–3365

Matho MH, Maybeno M, Benhnia MR, Becker D, Meng X, Xiang Y, Crotty S, Peters B, Zajonc DM (2012) Structural and biochemical characterization of the vaccinia virus envelope protein D8 and its recognition by the antibody LA5. J Virol 86(15):8050–8058. doi:10.1128/JVI.00836-12

Matho MH, de Val N, Miller GM, Brown J, Schlossman A, Meng X, Crotty S, Peters B, Xiang Y, Hsieh-Wilson LC, Ward AB, Zajonc DM (2014) Murine anti-vaccinia virus D8 antibodies target different epitopes and differ in their ability to block D8 binding to CS-E. PLoS Pathog 10(12):e1004495. doi:10.1371/journal.ppat.1004495

Matho MH, Schlossman A, Meng X, Benhnia MR, Kaever T, Buller M, Doronin K, Parker S, Peters B, Crotty S, Xiang Y, Zajonc DM (2015) Structural and functional characterization of anti-a33 antibodies reveal a potent cross-species orthopoxviruses neutralizer. PLoS Pathog 11(9):e1005148. doi:10.1371/journal.ppat.1005148

Meng X, Zhong Y, Embry A, Yan B, Lu S, Zhong G, Xiang Y (2011) Generation and characterization of a large panel of murine monoclonal antibodies against vaccinia virus. Virology 409(2):271–279. doi:10.1016/j.virol.2010.10.019

Meyer H, Osterrieder N, Czerny CP (1994) Identification of binding sites for neutralizing monoclonal antibodies on the 14-kDa fusion protein of orthopox viruses. Virology 200(2):778–783. doi:10.1006/viro.1994.1241

Nichols RJ, Stanitsa E, Unger B, Traktman P (2008) The vaccinia virus gene I2L encodes a membrane protein with an essential role in virion entry. J Virol 82(20):10247–10261. doi:10.1128/JVI.01035-08

Niles EG, Seto J (1988) Vaccinia virus gene D8 encodes a virion transmembrane protein. J Virol 62(10):3772–3778

Ojeda S, Domi A, Moss B (2006a) Vaccinia virus G9 protein is an essential component of the poxvirus entry-fusion complex. J Virol 80(19):9822–9830. doi:10.1128/JVI.00987-06

Ojeda S, Senkevich TG, Moss B (2006b) Entry of vaccinia virus and cell-cell fusion require a highly conserved cysteine-rich membrane protein encoded by the A16L gene. J Virol 80(1):51–61. doi:10.1128/JVI.80.1.51-61.2006

Payne LG (1992) Characterization of vaccinia virus glycoproteins by monoclonal antibody precipitation. Virology 187(1):251–260

Putz MM, Midgley CM, Law M, Smith GL (2006) Quantification of antibody responses against multiple antigens of the two infectious forms of Vaccinia virus provides a benchmark for smallpox vaccination. Nat Med 12(11):1310–1315. doi:10.1038/nm1457

Rodriguez JF, Esteban M (1987) Mapping and nucleotide sequence of the vaccinia virus gene that encodes a 14-kilodalton fusion protein. J Virol 61(11):3550–3554

Rodriguez JF, Janeczko R, Esteban M (1985) Isolation and characterization of neutralizing monoclonal antibodies to vaccinia virus. J Virol 56(2):482–488

Rodriguez JR, Rodriguez D, Esteban M (1992) Insertional inactivation of the vaccinia virus 32-kilodalton gene is associated with attenuation in mice and reduction of viral gene expression in polarized epithelial cells. J Virol 66(1):183–189

Roper RL, Payne LG, Moss B (1996) Extracellular vaccinia virus envelope glycoprotein encoded by the A33R gene. J Virol 70(6):3753–3762

Roper RL, Wolffe EJ, Weisberg A, Moss B (1998) The envelope protein encoded by the A33R gene is required for formation of actin-containing microvilli and efficient cell-to-cell spread of vaccinia virus. J Virol 72(5):4192–4204

Satheshkumar PS, Moss B (2009) Characterization of a newly identified 35-amino-acid component of the vaccinia virus entry/fusion complex conserved in all chordopoxviruses. J Virol 83(24):12822–12832. doi:10.1128/JVI.01744-09

Sela-Culang I, Benhnia MR, Matho MH, Kaever T, Maybeno M, Schlossman A, Nimrod G, Li S, Xiang Y, Zajonc D, Crotty S, Ofran Y, Peters B (2014) Using a combined computational-experimental approach to predict antibody-specific B cell epitopes. Structure. doi:10.1016/j.str.2014.02.003

Senkevich TG, Moss B (2005) Vaccinia virus H2 protein is an essential component of a complex involved in virus entry and cell-cell fusion. J Virol 79(8):4744–4754. doi:10.1128/JVI.79.8.4744-4754.2005

Senkevich TG, Ward BM, Moss B (2004) Vaccinia virus A28L gene encodes an essential protein component of the virion membrane with intramolecular disulfide bonds formed by the viral cytoplasmic redox pathway. J Virol 78(5):2348–2356
Senkevich TG, Ojeda S, Townsley A, Nelson GE, Moss B (2005) Poxvirus multiprotein entry-fusion complex. Pro Natl Acad Sci U S A 102(51):18572–18577. doi:10.1073/pnas.0509239102
Su HP, Garman SC, Allison TJ, Fogg C, Moss B, Garboczi DN (2005) The 1.51-Angstrom structure of the poxvirus L1 protein, a target of potent neutralizing antibodies. Proc Natl Acad Sci U S A 102(12):4240–4245. doi:10.1073/pnas.0501103102
Su HP, Golden JW, Gittis AG, Hooper JW, Garboczi DN (2007) Structural basis for the binding of the neutralizing antibody, 7D11, to the poxvirus L1 protein. Virology 368 (2):331–341. doi:S0042-6822(07)00439-4[pii]10.1016/j.virol.2007.06.042 [doi]
Su HP, Singh K, Gittis AG, Garboczi DN (2010) The structure of the poxvirus A33 protein reveals a dimer of unique C-type lectin-like domains. J Virol 84, 2009/12/25 edn. doi:10.1128/JVI.02247–09
Sundberg EJ, Mariuzza RA (2002) Molecular recognition in antibody-antigen complexes. Adv Protein Chem 61:119–160
Townsley AC, Senkevich TG, Moss B (2005a) The product of the vaccinia virus L5R gene is a fourth membrane protein encoded by all poxviruses that is required for cell entry and cell-cell fusion. J Virol 79(17):10988–10998. doi:10.1128/JVI.79.17.10988-10998.2005
Townsley AC, Senkevich TG, Moss B (2005b) Vaccinia virus A21 virion membrane protein is required for cell entry and fusion. J Virol 79(15):9458–9469. doi:10.1128/JVI.79.15.9458-9469.2005
Wang DR, Hsiao JC, Wong CH, Li GC, Lin SC, Yu SS, Chen W, Chang W, Tzou DL (2014) Vaccinia viral protein A27 is anchored to the viral membrane via a cooperative interaction with viral membrane protein A17. J Biol Chem 289(10):6639–6655. doi:10.1074/jbc.M114.547372
Wolffe EJ, Vijaya S, Moss B (1995) A myristylated membrane protein encoded by the vaccinia virus L1R open reading frame is the target of potent neutralizing monoclonal antibodies. Virology 211(1):53–63. doi:10.1006/viro.1995.1378
Wolffe EJ, Weisberg AS, Moss B (1998) Role for the vaccinia virus A36R outer envelope protein in the formation of virus-tipped actin-containing microvilli and cell-to-cell virus spread. Virology 244(1):20–26. doi:10.1006/viro.1998.9103
Wolffe EJ, Weisberg AS, Moss B (2001) The vaccinia virus A33R protein provides a chaperone function for viral membrane localization and tyrosine phosphorylation of the A36R protein. J Virol 75(1):303–310. doi:10.1128/JVI.75.1.303-310.2001

Chapter 5
The Peroxiredoxin Family: An Unfolding Story

Zhenbo Cao and John Gordon Lindsay

Abstract Peroxiredoxins (Prxs) are a large and conserved family of peroxidases that are considered to be the primary cellular guardians against oxidative stress in all living organisms. Prxs share a thioredoxin fold and contain a highly-reactive peroxidatic cysteine in a specialised active-site environment that is able to reduce their peroxide substrates. The minimal functional unit for Prxs are either monomers or dimers, but many dimers assemble into decameric rings. Ring structures can further form a variety of high molecular weight complexes. Many eukaryotic Prxs contain a conserved GGLG and C-terminal YF motif that confer sensitivity to elevated levels of peroxide, leading to hyperoxidation and inactivation. Inactive forms of Prxs can be re-reduced by the enzyme sulfiredoxin, in an ATP-dependent reaction. Cycles of hyperoxidation and reactivation are considered to play an integral role in a variety of H_2O_2-mediated cell signalling pathways in both stress and non-stress conditions. Prxs are also considered to exhibit chaperone-like properties when cells are under oxidative or thermal stress. The roles of various types of covalent modifications, e.g. acetylation and phosphorylation are also discussed. The ability of Prxs to assemble into ordered arrays such as nanotubes is currently being exploited in nanotechnology.

Keywords Peroxiredoxin • Redox • Peroxide • Hyperoxidation • Sulfiredoxin • Oxidative stress • Antioxidation • Signalling pathway • Chaperone • Protein complex

---

Z. Cao • J.G. Lindsay (✉)
Institute of Molecular, Cell and Systems Biology, Davidson Building, College of Medicine, Veterinary and Life Sciences, University of Glasgow, Glasgow G12 8QQ, UK
e-mail: Gordon.Lindsay@glasgow.ac.uk

J.R. Harris, J. Marles-Wright (eds.), *Macromolecular Protein Complexes*, Subcellular Biochemistry 83, DOI 10.1007/978-3-319-46503-6_5

5.1 Peroxiredoxins and Their Classification

Peroxiredoxins (Prxs) were first identified as a new class of peroxidases in 1994 (Chae et al. 1994). Prior to this landmark discovery, catalase and glutathione peroxidase were thought to provide the main line of defence against oxidative stress in living organisms in addition to non-enzymatic reductants and free radical scavengers, such as glutathione, vitamin E and ascorbate. A novel aspect of Prxs was that they did not rely on a heme prosthetic group or an active selenocysteine moiety for activity, instead employing one or two standard cysteine residues in catalysis. In the past 20 years, Prxs have become one of the most intensively studied protein families, owing to their ubiquitous distribution, catalytic efficiency, capacity for reversible and irreversible hyperoxidation and unusual ability to assume multiple oligomeric states with differing properties, e.g. chaperone activity (Wood et al. 2003b; Perkins et al. 2015). More recently, attention has focused on elucidating their emerging roles in vital cell signalling pathways concerned with governing cellular responses to their immediate environment under both stress and non-stress conditions (Latimer and Veal 2016; Netto and Antunes 2016).

Prxs are now considered to be the primary cellular guardians against oxidative stress in all living organisms. As a group, they are responsible for the removal of more than 90% of cellular peroxides based on their relative abundance, catalytic power and presence in multiple intracellular compartments in eukaryotes (Winterbourn 2008). Classification has been problematic, because of the diverse nature of this large family of enzymes. Nelson and co-workers used the Deacon Active Site Profiler (DASP) tool to categorise Prxs in six evolutionary subfamilies, based on sequence similarity, namely: Prx1/AhpC, Prx5, Prx6, Tpx, PrxQ/BCP and AhpE (Nelson et al. 2011). The Prx1 group is commonly referred to as the "typical 2-Cys" Prx subfamily, since its members almost exclusively contain two active-site cysteines and function as a basic homodimeric unit in which a peroxidatic cysteine (C_P) situated in the N-terminal region of one monomer forms an inter-chain disulfide bond with a resolving cysteine (C_R) located in the C-terminal domain of its partner. In mammalian cells, there are four distinct enzymes in this group: PrxI and PrxII are located in the cytosol and nucleus, PrxIII is found in the mitochondrial matrix and PrxIV is confined to the endoplasmic reticulum. PrxV is an "atypical 2-Cys" Prx in which C_P and C_R are on the same polypeptide chain. PrxV is present in mitochondria, peroxisomes and the cytosol, while the "1-Cys" enzyme, PrxVI is also located in the cytosol.

One of the most interesting features of 2-Cys Prxs is their ability to form oligomeric rings (toroids) from their minimal functional dimeric units. Thus, PrxsI, II, IV and VI can exist as stable decamers whereas PrxIII is a dodecamer. Octameric toroids have also been observed in the case of the 1-Cys Prx, AhpE from *Mycobacterium tuberculosis* (Li et al. 2005).

In 1998, sequence alignment and secondary structure predictions revealed that Prxs share a common thioredoxin (Trx) fold, which consists of an N-terminal βαβ motif and a C-terminal ββα motif, linked by a central loop region containing another

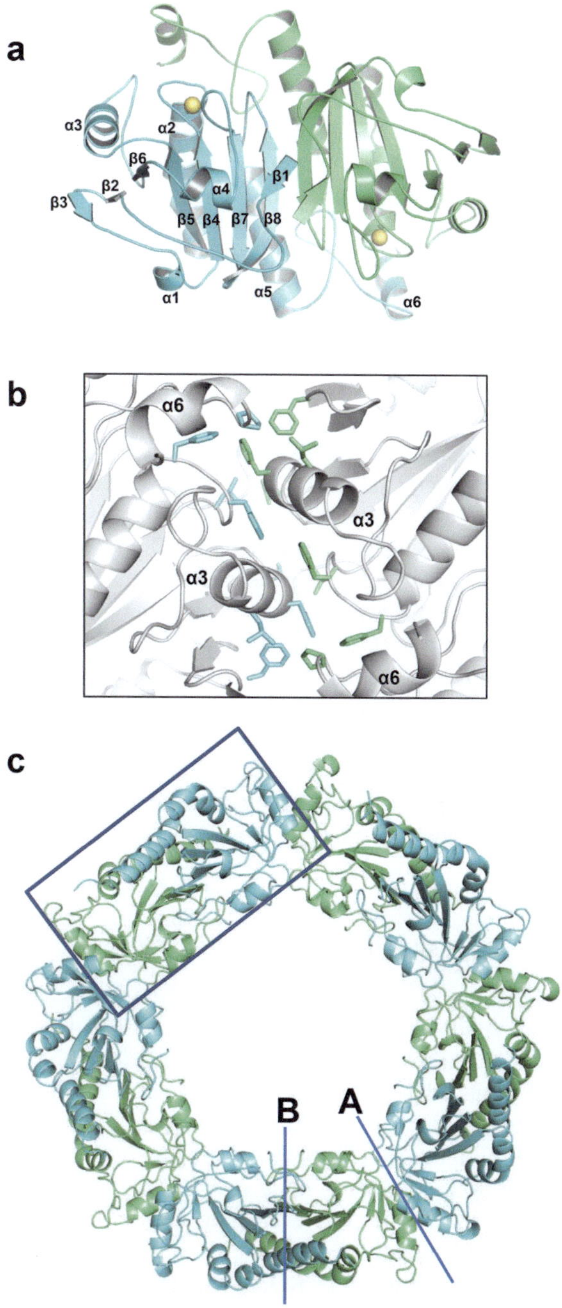

Fig. 5.1 Decamerisation of human PrxIV. (**a**) B-type dimer: The anti-parallel β-sheet in the core of the structure between monomers creates an extended 14-stranded sheet (β2–β8). The C_P is shown as *yellow* ball. (**b**) A-type dimer: The hydrophobic side chains are highlighted. *Cyan* and *green* represent different monomers. (**c**) Toroidal decamer of human PrxIV: Five B-type dimers (marked by a *box*) form a decamer via A-type interfaces. A and B-type interfaces are marked as *solid lines*

α-helix (Schroder and Ponting 1998). Prxs contain several additions to the basic Trx fold, namely an N-terminal extension, an insertion between the second β-strand and α-helix and in some cases, a C-terminal extension, e.g. in the ubiquitous Prx1/AhpC subfamily. Structural data from all Prx subfamilies have revealed that they share a common core organisation comprising seven β-strands and five α-helices (Fig. 5.1a).

With a few exceptions (e.g. monomeric BCP family members), all Prxs function as dimers or higher-order oligomers. Two different types of subunit interfaces (termed A and B) are employed to form the basic dimeric subunit and then mediate dimer-dimer assembly into toroidal structures (Sarma et al. 2005). Crystal structures of dimeric PrxIV revealed that B-type dimers are organised in a 'head-to-tail' fashion in which the β8-strands of adjacent subunits interact in an anti-parallel manner to generate a stable 14-stranded β-sheet involving the 7 central β-strands of each subunit (Cao et al. 2011). The C-terminal region also extends across the two-fold axis, further stabilising the dimeric interface. In contrast, PrxV forms a dimer via an A-type (alternative) interface, utilising hydrogen bonding and hydrophobic interactions (Declercq et al. 2001).

For PrxsI-IV and some members of the PrxVI family, five or six B-type dimers can further associate through their A-type interfaces to produce decameric or dodecameric toroids with a large central cavity. Higher-order complexes have also been observed. These involve ring stacking, ring catenation or even larger oligomeric assemblies (Cao et al. 2005; Gourlay et al. 2003; Jang et al. 2004). The A-type interface is less extensive and mainly involves residues from the α3 and α6-helices (Fig. 5.1b), the C_P loop, and the loops surrounding the α4-helix. In this group of Prxs, decamers and dodecamers are stabilised in the reduced state and exhibit a pronounced tendency to revert to dimers upon oxidation. This redox dependence in the oligomeric state is accounted for by the associated conformational changes in the C_P loop, which forms an integral part of the dimer-dimer interface.

5.2 The Multifaceted Roles of Prxs

The importance of individual Prxs to normal growth and development has been evaluated in several knockout mouse models. Mice lacking PrxI were viable and fertile but had a shortened lifespan, owing to the development, at about 9 months, of severe hemolytic anemia and several types of malignant tumor. These data suggest that PrxI functions as a tumor suppressor, with an important role in cellular defence against ROS in ageing mice (Neumann et al. 2003). PrxII −/− mice appeared healthy and fertile. However, they also suffered from hemolytic anemia with a decreased hematocrit count and the presence of numerous morphologically abnormal cells. High levels of ROS were detected in the blood and many red blood cell (RBC) proteins were subject to oxidative damage, suggesting that PrxII plays a major role in protecting RBCs from oxidative stress (Lee et al. 2003). Thus, although PrxI and PrxII are both located in the cytosol and share similar anti-oxidant activities, current evidence indicates that they have overlapping, but distinctive roles in cellular protection in mouse models. In contrast, the main phenotype in PrxIII-deficient mice is increased sensitivity to lipopolysaccharide-induced oxidative stress, whereas PrxIV null mice display reduced sperm viability (Li et al. 2009; Iuchi et al. 2009).

A host of altered cellular responses have now been described to be associated with the deficiency of a specific Prx family member or linked to their overproduction. These range from roles in breast cancer progression and apoptosis, to alterations in inflammatory responses and protection against oxidative, UV or radiation damage (Cha et al. 2009; Klichko et al. 2016; Ito et al. 2014). A plethora of recent publications have also implicated Prxs in many fundamental biological processes, including the control of the cell cycle, the protection of DNA from oxidative damage and an involvement in circadian rhythmicity (Phalen et al. 2006; Lee et al. 2016; Edgar et al. 2012). Thus, while the pleiotropic effects on cells induced by periods of oxidative stress can be linked to specific roles for individual Prx family members in aiding cell survival, Prxs are also increasingly recognised as integral components in major cell signalling pathways, with roles in monitoring and adjusting cell responses to their immediate aerobic environment under normal conditions (Chang et al. 2007; Tavender et al. 2008; Mukhopadhyay et al. 2006; Stresing et al. 2013).

5.3 The Prx Active Site and Reactivity of the Peroxidatic Cysteine

In both 1-Cys and 2-Cys Prxs, the initiation of the reaction cycle begins with an attack by the peroxidatic cysteine (C_P) thiolate (C_P-S^-) on its peroxide substrate, leading to production of a sulfenic acid intermediate (C_P-SOH). At this stage Prxs are in the fully-folded (FF) conformation (see Fig. 5.2 for details) with C_P located towards the end of the α2-helix. In addition, the side chain of C_P is favourably positioned in the active-site pocket, inducing C_P-S^- formation that is required for the nucleophilic displacement (S_N2) reaction, while also providing a suitable environment for optimal orientation and positioning of the substrate (Hall et al. 2010).

Many Prxs display extremely rapid reaction rates, in the range 10^5–10^8 M^{-1} s^{-1} for H_2O_2 for bacterial and mammalian Prxs (Parsonage et al. 2008; Peskin et al. 2007). Three highly-conserved residues, including a proline and threonine/serine in the C_P loop motif, PXXXT(S)C_P and a more-distantly located arginine, are implicated in stabilising/activating the C_P-S^- thiolate and hydrogen bonding with the bound peroxide substrate, ensuring its optimal positioning to achieve these high catalytic rates.

In the FF conformation, the active-site C_P thiolate is located approximately 14 Å from the resolving cysteine (C_R) in the C-terminal extension of the other chain. A major conformational change, involving partial unwinding of the α2-helix in the C_P active-site loop, occurs upon oxidation with the substrate bringing the C_P-SOH

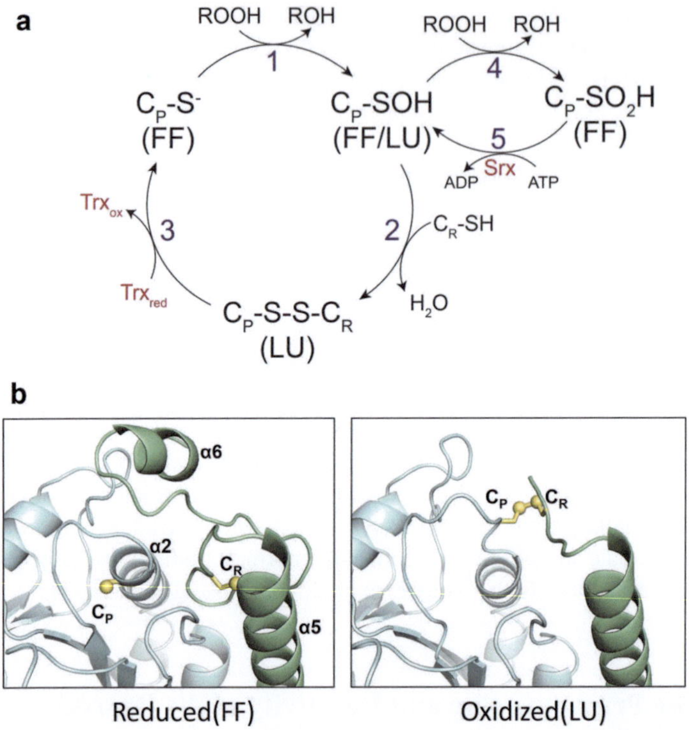

Fig. 5.2 The Prx catalytic cycle. (**a**) Schematic representation of Prx catalytic cycle. (*1*) *peroxidation*, the peroxidatic cysteine (C_P-S^-) reduces peroxide. (*2*) *resolution*, the resolving cysteine (C_R-SH) reacts with sulfenylated C_P to form an inter-chain disulfide bond. (*3*) *recycling*, reduced thioredoxin (Trx_{red}) is oxidised resulting in the regeneration of C_P. (*4*) *hyperoxidation*, occasionally C_P in the sulfenic acid state can react with a second peroxide and become hyperoxidised to sulfinic acid. This form of Prx is inactive. (*5*) *reactivation*, the inactivated form can be rescued through an ATP-dependent reaction catalysed by Srx. The C_P loop conformation is indicated as fully-folded (FF) or locally-unfolded (LU). (**b**) Structure of active-site in reduced and oxidised PrxIV. Different chains colored *cyan* and *green*. The cysteine side chains are shown as yellow sticks and sulfur atoms are shown as *yellow balls*. Note the C-terminal α6-helix needs to be unwound to permit disulfide formation

intermediate into close proximity (2–3 Å) with C_R (located in the C-terminal domain of its partner subunit) prior to disulfide bond formation.

The C_P-SOH intermediate is proposed to equilibrate between the FF and LU conformations until disulfide bond generation locks the enzyme into the LU state (Cao et al. 2011; Perkins et al. 2013). By using rapid reaction and competitive kinetic approaches, Peskin and co-workers have established a rate constant for disulfide formation at 2/second for PrxII (Peskin et al. 2013). The key conserved Arg residue in the active-site pocket (and a backbone amide group) appears to stabilise and activate the C_P thiolate for nucleophilic attack on its peroxide substrate in the FF conformation (Hall et al. 2010). In a recent crystal structure of the bovine PrxIII F190L mutant, this key Arg has been detected in a novel conformation that extended away from the C_P thiolate, indicating the dynamic nature of these interactions (Cao et al. 2015).

5.4 Hyperoxidation of Prxs and Sulfiredoxin Mediated Reactivation

Shortly after the discovery of the Prx family in 1994, it became evident that several of its members could be inactivated by their common substrate H_2O_2, even at physiological (submicromolar) concentrations (Yang et al. 2002; Wagner et al. 2002; Konig et al. 2002). Initially, these hyperoxidised states in which the catalytic, peroxidatic cysteine C_P was further oxidised beyond its normal transition state of the sulfenic acid (Cys-OH) intermediate, were thought to be dead-end products with hyperoxidation being seen as a limitation in the catalytic properties of Prxs. However, Wood and co-workers noted that hyperoxidation was an exclusive feature of eukaryotic peroxiredoxins, whereas their prokaryotic counterparts were not subject to inactivation under similar conditions (Wood et al. 2003a). It was further suggested that the susceptibility to hyperoxidation to their sulfinylated (C_P-O_2H) or sulfonylated (C_P-O_3H) states had been selected for during evolution and that inactivated (hyperoxidised) peroxiredoxins may act as key mediators in evoking cell signalling responses to oxidative stress. The extent of hyperoxidation *in vivo* or *in vitro* can be monitored as these inactivated species carry a higher net negative charge, allowing for their separation on 2-D gels (Mitsumoto et al. 2001) or 1-D isoelectric focusing gels (Cao et al. 2014). Antibodies specific for the hyperoxidised states of all typical 2-Cys Prxs are now widely employed to quantitate their susceptibility to inactivation or the extent of hyperoxidation occurring in living cells subjected to various treatment regimens (Woo et al. 2003).

Two specific motifs that were exclusively present in the sensitive members of eukaryotic 2-cys Prxs were also identified by Wood and co-workers, namely an invariant GGLG sequence residing in the loop region following the α3-helix and a YF motif in the C-terminal α-helical domain (Wood et al. 2003a). As seen in Fig. 5.3, these two motifs are predicted to stabilise the FF state by packing against the C_P loop. Indeed, the YF motif would interfere with the locally-unfolded conformation of the C_P loop if it did not also move away during the reaction cycle (Wang et al. 2012). Recent evidence from a crystal structure of the TSAII C48S mutant indicates that an unfolded C-terminus is needed to permit unfolding of the C_P loop due to steric hindrance effects (Nielsen et al. 2016). Thus, the conformational stability of the C-terminus is likely to be a sensitive regulator of Prx function.

These initial studies indicated that the balance between FF and LU states is crucial to determining the rate of reaction with H_2O_2. Thus, stabilisation of the FF was anticipated to slow the transition to the LU state, thereby favoring hyperoxidation by allowing time for a reaction with a second or third molecule of substrate. As expected, no similar stabilisation occurs in robust members of the same Prx family,

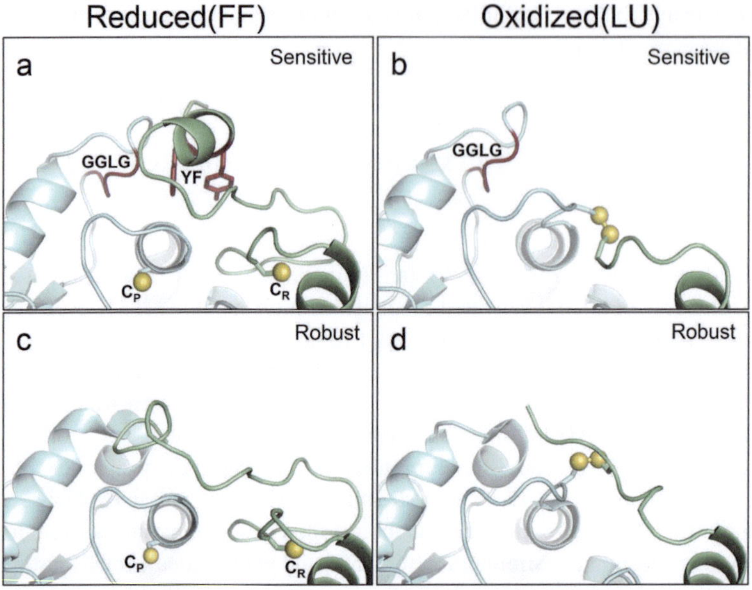

Fig. 5.3 Structural differences between sensitive (eukaryotic) and robust (prokaryotic) Prx1s. Comparison of active-site region of human PrxIV (**a** and **b**) and AhpC from *S. typhimurium* (S.t.) (**c** and **d**). Different chains in the homodimer are colored in *cyan* and *green*. C_P and C_R are shown as *yellow balls*. The conserved GGLG and YF motifs in sensitive Prx1s are colored in *red*. The reduced (FF) conformation is on the *left* side of the panel and the oxidised (LU) conformation is on the *right*. Comparison of **a** and **c** clearly show that the GGLG and YF motifs are blocking the unwinding of C_P from its FF conformation

which lack these motifs (Fig. 5.3). Subsequently, interchanging C-terminals between different Prxs has also established that other C-terminal residues (apart from YF) are instrumental in fine-tuning the precise sensitivity of individual Prxs to the prevailing conditions in the appropriate intracellular compartment. For example, the mitochondrial enzyme, PrxIII is more resistant than PrxII located in the cytosol, although both possess identical GGLG and YF motifs. Exchanging their C-terminal domains, however, reverses this sensitivity (Haynes et al. 2013). The reaction of the C_P-SOH intermediate with C_R in PrxII is 11-times slower than in PrxIII, which is in agreement with the hypothesis that the rate of transition between FF and LU conformations is the major determinant of sensitivity to hyperoxidation (Peskin et al. 2013).

In this context, as discussed previously, the highly-conserved YF motif is a major determinant in ensuring the susceptibility to hyperoxidation in eukaryotic Prxs. Interestingly, a similar YL motif is found in many prokaryotes; however, a mutant bovine PrxIII containing the YL motif is fully active, but also insensitive to H_2O_2-induced inactivation (Cao et al. 2015). Moreover, crystal structures of both the oxidised and the reduced forms of this PrxIII mutant confirm the presence of a disordered C-terminal region, which is consistent with a role for YF in stabilising this domain (Cao et al. 2015). These observations are in agreement with the current hypothesis that the rate of transition from the FF state to the LU state, which is necessary for disulfide bond formation, determines the extent of the resistance to hyperoxidation. Partial unwinding of the α2-helix and the C-terminal helix are required to bring the active C_P and C_R cysteine closer together, from 14 Å apart in the reduced state to within 2–3 Å when facilitating disulfide bond formation. Thus, the slower this transition, the more opportunity for the Cys-OH intermediate to react with a second or third molecule of hydrogen peroxide, i.e. to attain its hyperoxidised sulfinylated or sulfonylated states.

In general, it now appears that eukaryotic Prxs are more sensitive to hyperoxidation than their prokaryotic counterparts as a result of the slower transition from FF to LU states during the reaction cycle (Wood et al. 2003a), which is governed by the stability of the C-terminal α-helix and its conserved YF motif in particular. Surprisingly, although PrxIV is as sensitive to hyperoxidation as PrxII *in vitro*, it is protected from hyperoxidation in the oxidising environment of the endoplasmic reticulum (ER), even upon exposure to high levels of H_2O_2 (Cao et al. 2014). This appears to reflect the lack of an efficient ER recycling system, resulting in the low turnover of PrxIV. Variation in H_2O_2 concentration *in vivo*, for example by localised H_2O_2 accumulation, could also be a factor influencing the extent of Prx hyperoxidation (Woo et al. 2010).

Hyperoxidised forms of Prxs have also been detected in cells that are not subjected to oxidative stress (Schroder et al. 2000; Musicco et al. 2009). Interestingly, regular cycles of reduced and hyperoxidised Prx accumulation have been observed in several cell types, indicating that they are intimately linked to the control of circadian rhythms (Edgar et al. 2012).

Recognition of the biological significance of hyperoxidation in eukaryotic systems was quickly reinforced by the discovery of sulfiredoxin (Srx), a cytoplasmic enzyme that was initially shown to re-reduce PrxI in a Mg^{2+} and ATP-dependent manner (Biteau et al. 2003). *In vitro* studies of the mechanism of Prx reactivation have demonstrated that its action is confined to the 2-Cys Prxs and that Srx has been reported to enter the mitochondrial compartment where it promotes reactivation of hyperoxidised PrxIII, in addition to its cytoplasmic role in the maintenance and restoration of PrxI and II function (Noh et al. 2009). Detailed descriptions of the enzymatic mechanism of the re-reduction of sulfinylated (Cys-SO_2H) active-site cysteines have been published, although it appears that further oxidation to the sulfonylated state (Cys-SO_3H) represents an irreversible step in the process (Jonsson et al. 2008).

5.5 Role of Hyperoxidised Prxs in H_2O_2-Mediated Signalling

The inherent susceptibility of eukaryotic Prxs to inactivation in non-stress conditions initially gave rise to the 'floodgate' model, in which transient local increases in H_2O_2 were proposed to result in temporary Prx inhibition and the resultant oxidation of downstream targets. To date the best documented example of a 'floodgate-type' mechanism is represented by the regulation of corticosteroid production from cholesterol in adrenal mitochondria (Kil et al. 2012). Steroid synthesis is subject to negative feedback in which H_2O_2 is formed indirectly from the inadvertent leakage of electrons onto molecular oxygen during the NADPH-dependent cytochrome P450-catalysed step in their production. Thus, the local elevation of H_2O_2 is potentiated by transient PrxIII inactivation during times of increased steroid production, enabling its increased access to the cytoplasm, where it stimulates p38 mitogen-activated protein kinase and thereby inhibites cholesterol import into mitochondria and the suppression of glucocorticoid biosynthesis. A variation on the floodgate theme has been reported in which growth factor signalling leads to the stimulation of an NADPH oxidase (NOX1) and hence increased H_2O_2 production. Src kinase is also activated under these conditions, catalysing the phosphorylation/inactivation of PrxI on Tyr194 and allowing local increases in H_2O_2 (Woo et al. 2010). Elevated peroxide levels also mediate the inactivation of Protein Tyrosine Phosphatases and influence a host of downstream signalling events (Chiarugi and Cirri 2003). Importantly, phosphorylated PrxI is confined exclusively to the vicinity of NOX1-associated lipid rafts, ensuring local control of regulation.

The interplay between phosphorylation state, hydrogen peroxide production and growth factor regulated signalling in this case indicates similar control mechanisms may exist for a variety of cell signalling pathways. In this respect, it is also of interest that a number of covalent modifications of Prxs have been reported that modulate their activity and susceptibility to inactivation. These include Ser and Tyr phosphorylation, Tyr nitration, S-nitrosylation, glutathionylation of active-site and non-catalytic cysteines, N- or C-terminal acetylation and protein truncation. Many of these post-translational modifications have been shown to enhance or inhibit Prx function. In several cases, these are reported to influence Prx activity by affecting the balance and rate of transition between the fully-folded (FF) and locally-unfolded (LU) conformations during oxidation-reduction. For example, phosphorylation of Thr90, which is located close to the conserved GGLG motif in human PrxI, causes inactivation, as does phosphorylation of Tyr194 (Jang et al. 2006; Woo et al. 2010). In both cases, introduction of a negatively-charged phosphate in the vicinity of the active-site is predicted to destabilise the FF conformation, causing a loss of activity. Similarly nitration of the conserved Tyr in the C-terminal YF motif of human PrxII stimulated activity and protected against hyperoxidation, possibly by increasing the rate of disulfide formation via destabilisation of the C-terminal α-helix (Romero-Puertas et al. 2007). In addition, N-terminal acetylation of PrxII has been reported to prevent its irreversible hyperoxidation, without altering its affinity for H_2O_2 (Seo et al. 2009). In contrast, acetylation of Lys197 or 196 in the C-terminal domains of PrxI and PrxII, respectively, leads to enhanced peroxidase activity and protection against H_2O_2-induced inactivation (Parmigiani et al. 2008). Finally, glutathionylation of the active-site cysteine (Cys83) in human PrxI also appears to influence the decamer-dimer transition, stabilising the dimeric state and abolishing chaperone activity (Park et al. 2011).

In addition to the extensive range of modifications described above, the ability of PrxI to interact with signalling molecules, such as PTEN phosphatase and mammalian ste-20-like kinase-1 (MST1), which are implicated in cell cycle regulation and apoptosis, provide new insights into the complex network of regulatory pathways controlling cell and tissue responses in normal and oxidative stress conditions (Morinaka et al. 2011). Hyperoxidation of 2-Cys Prxs, linked to alterations in their oligomeric state, have also been associated with alterations in the cell cycle, an indication that the oxidation status and quaternary structure of Prxs may act as reporters, monitoring local peroxide levels and generating the appropriate cellular responses (Phalen et al. 2006). In similar fashion, Prxs have also been shown to act as intermediates in so-called redox relays, by transferring oxidising equivalents directly or indirectly to key regulatory components, e.g. transcription factors or signalling kinases/phosphatases. In one example, PrxII was shown to form a mixed disulfide with STAT3, giving rise to disulfide-bonded dimers/tetramers of this transcription factor that could be re-reduced by thioredoxin (Sobotta et al. 2015). Importantly, depletion of PrxII enhanced the rate of STAT3-mediated transcription, suggesting that PrxII-induced oligomerisation of STAT3 inhibits transcription.

Another well-documented case is found in *Schizosaccharomyces pombe,* where the thiol peroxidase (Tpx1) oxidises the AP1-like transcription factor (Pap1) by disulfide exchange, concealing a nuclear export signal for Pap1 (Calvo et al. 2013). The resultant nuclear accumulation of Pap1 leads to upregulation of genes involved in antioxidant defence. Interestingly, at higher H_2O_2 levels, Tpx1 becomes hyperoxidised, rendering it unable to activate Pap1. At this stage, a mitogen activated signalling pathway, employing the protein kinase sty-1, is recruited to perform a similar role. This secondary pathway is also dependent on Tpx1, but operates independently of its state of hyperoxidation.

Recent studies by Day and co-workers have revealed that Tpx1 is reduced by thioredoxin at low H_2O_2 concentrations, however, following Tpx1 hyperoxidation, Trx is available to reduce Pap1 and the disulfides of antioxidant defence proteins (Day et al. 2012). In this regard, the redox status of Tpx1 determines whether it acts in an adaptive mode at low H_2O_2, ensuring upregulation of antioxidant defence genes. After its inactivation, at elevated H_2O_2 levels, Trx can focus on the repair of cellular damage when cell survival is the priority.

5.6 Alterations in Oligomeric State and Chaperone-Like Functions

As mentioned previously, 2-cys Prxs contain a basic functional dimeric unit in which the N-terminal active site, C_P, of one subunit forms a transient disulfide bond with the C-terminal resolving cysteine, C_R, of its partner. Although functional as dimers, 2-cys Prxs generally exist as higher-order oligomers, specifically as octameric, decameric and dodecameric toroids. As seen in Fig. 5.4, small stacks of two

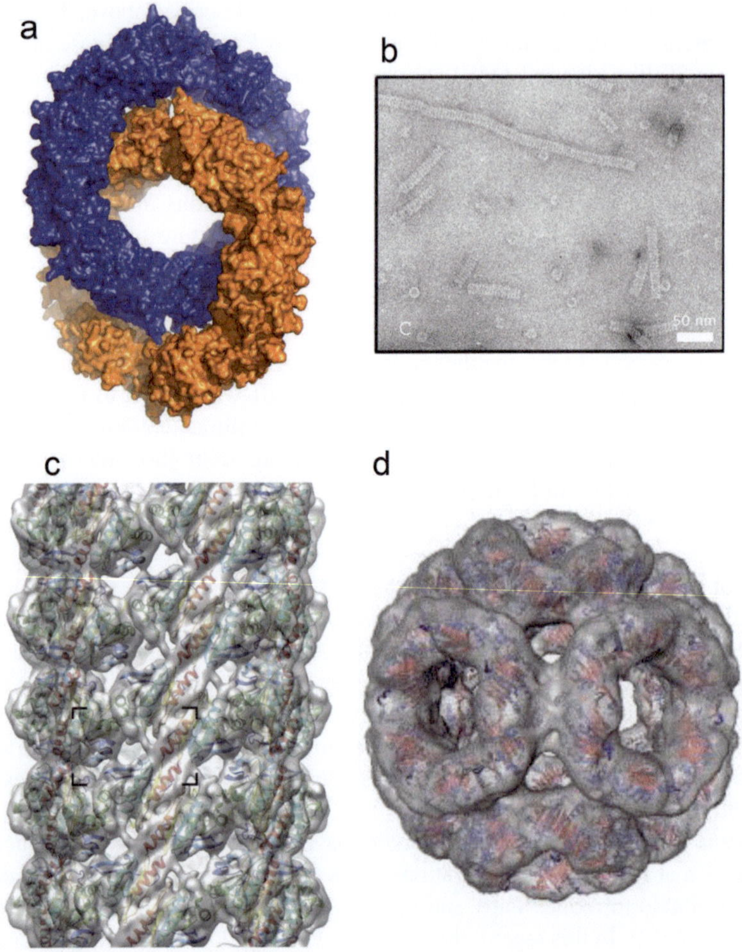

Fig. 5.4 Oligomerisation of the Prx toroid. (**a**) Crystal structure of bovine PrxIII F190L showing two interlocking dodecameric rings forming a catenane (Adapted from Cao et al. 2015). (**b**) Transmission electron microscopy showing stacked dodecameric rings of bovine PrxIII C47S (Adapted from Gourlay et al. 2003). (**c**) Cryo-EM 3D-reconstruction of human PrxIII. Bovine PrxIII crystal structure was docked into the electron density (Adapted from Radjainia et al. 2015). (**d**) The 16 Å 3D-reconstruction of human PrxII dodecahedron (Adapted from Meissner et al. 2007)

to three rings are also readily observed in negatively-stained EM images, while extensive stacking has been reported on occasions, e.g. with PrxI and III in which 50 or more rings coalesce to produce long nanotubes (Gourlay et al. 2003; Phillips et al. 2014). Human PrxII has also been reported to assemble into a novel dodecahedral structure after incubation in 1% PEG at pH 6.5 and similar 'spherical' structures have been observed for yeast TSAI and II (Meissner et al. 2007; Jang et al. 2004).

While ATP-dependent molecular chaperones, such as Hsp-70 and Hsp-40 classes of heat shock proteins, are necessary to direct the ordered folding of nascent polypeptide chains as they emerge from the ribosome, ATP-independent chaperones, generally referred to as 'holdases', are crucial for cellular protection under stress conditions, which favor protein unfolding and the formation of harmful aggregates. The *in vitro* switch from peroxidase to chaperone function amongst the Prxs is usually induced by thermal, oxidative stress or low pH conditions (Jang et al. 2004; Teixeira et al. 2015; Saccoccia et al. 2012). This can also be achieved by phosphorylation or hyperoxidation in some cases (Jang et al. 2004, 2006). For example, for yeast TSAI and human PrxII, treatment with thioredoxin and H_2O_2 to induce the rapid recycling that is required for extensive hyperoxidation, promotes the formation of high molecular weight complexes with chaperone activity (Jang et al. 2004; Moon et al. 2005). The chaperone action of Prxs appears to require the presence of two or more stacked rings, although in several instances, including PrxI and PrxIII, long nanotubes are formed that are composed of 50 or more stacked toroids. The novel dodecahedral assemblies that have been observed in 3D reconstructions of negative stain EM images of yeast TsaI and human PrxII, are also reported to exert chaperone activity (Meissner et al. 2007). Hanzen and co-workers have reported an interaction between Tsa I and Hsp 70, showing that hyperoxidation of Tsa I acts as a redox switch to recruit Hsp70 and Hsp104 during aging. This provides an alternative explanation to account for the chaperone function triggered by peroxiredoxin hyperoxidation (Hanzen et al. 2016).

A considerable amount of data exists on the fine structure of nanotubes and in particular on how mutations, truncations, post-translation modifications or hyperoxidation can affect their appearance, physical properties and putative chaperone activity (Angelucci et al. 2013; Gourlay et al. 2003; Jang et al. 2004). In many cases, the degree of exposure of hydrophobic surfaces in these assemblies correlates well with their ability to act as molecular chaperones. In our studies, the C47S active-site mutant of PrxIII has a pronounced tendency to form long nanotubes devoid of chaperone activity (Gourlay et al. 2003). Thus, it appears that hyperoxidation, or general perturbation of the active-site region, can induce the formation of large high molecular weight complexes. In one case involving the mitochondrial enzyme tryparedoxin peroxidase (mTXNPx) from a Leishmania parasite, the reduced decameric form was shown to exhibit chaperone-like activity above 35 °C (Teixeira et al. 2015). Thermal stress promoted conformational changes and exposure of hydrophobic surfaces as assessed by circular dichroism and enhanced bis-ANS fluorescence. Negative-stain EM images of mTXNPx with an unfolded client protein, luciferase, showed that luciferase was located in the central cavity of the decamer, implicating the inwardly-facing N-terminal regions of the peroxidase in this interaction. These authors also demonstrated that mTXNPx maintained unfolded polypeptides in an assembly-competent state, which required mediation of ATP-dependent chaperones, e.g. members of the Hsp 70/40 family, to complete the renaturation process. Such a model is consistent with reports that the unfolded proteins also bind internally to hollow nanotubular forms of Prxs at their extremities and that

N-terminal modifications, e.g. His tags, can interfere with chaperone function (Teixeira et al. 2015).

The toroidal structure of many 2-Cys Prxs, which enclose a large central cavity, is reminiscent of the Hsp60/Hsp10 chaperone family, where unfolded polypeptides can undergo multiple rounds of ATP-dependent refolding in an isolated environment. In addition, their ability to switch between an array of oligomeric states with differing physical and functional properties, at least *in vitro*, has prompted the idea that most Prxs can block protein aggregation and, in concert with ATP-requiring chaperones, initiate the ordered refolding of aggregated/denatured polypeptide chains. However, while there is increasing evidence for a key *in vivo* role for Prxs in protein renaturation as described above, the physiological relevance of many of the widely-reported chaperone-like actions of high molecular weight Prx complexes is still an open question in our view.

It is well documented that for most 2-Cys peroxiredoxins, the switch between dimers and decamers/ dodecamers is redox-dependent with the higher oligomeric state stabilised in the reduced FF conformation. Since unfolding of the α_2-helix in the active-site region is essential for disulfide bond formation during oxidation by H_2O_2 and as this region also forms part of the A-type interface involved in dimer-dimer interactions, it is unsurprising that dimer-dimer affinity is altered during the redox cycle that leads to dimer-decamer/dodecamer conversion *in vitro* (Cao et al. 2007). However, in the crowded environment of the cell, where protein concentrations are in the range 100–200 mg/ml, although subunit affinities will undoubtedly be altered, the overall oligomeric state is unlikely to be affected by the redox conditions. A cryo-EM and crystallography study of human PrxIII found that both FF and LU states show stacked ring formation, also suggesting that this formation could be favored in the crowded environment inside the mitochondrion (Yewdall et al. 2016).

Similarly, any exposure of hydrophobic surfaces in recombinant, wild-type or mutant Prxs will contribute to apparent chaperone activity in *in vitro* assays involving protection against thermal aggregation of citrate synthase or insulin. BSA, for example, can act as an effective chaperone or refolding agent in such *in vitro* assays thanks to its ability to bind hydrophobic patches on the surfaces of unfolded proteins (Lindsay et al. 2000).

Interestingly, in the case of the yeast cytoplasmic 2-Cys Prxs TSAI and TSAII, there is evidence to suggest that their reported chaperone functions are of physiological relevance (MacDiarmid et al. 2013). TSAI *null* mutants are sensitive to Zn^{2+} deficiency, a condition that leads to accumulation of intracellular protein aggregates. However, expression of an inactive TSAI mutant lacking the resolving cys, is able to rescue this phenotype and prevent protein aggregation, as is the overexpression of known ATP-independent chaperones, e.g. Hsp26 and Hsp42. TSAII can also substitute for its cytoplasmic partner, provided it is upregulated as it is normally present in much lower levels. TSAII has also been reported to protect cells against ribosomal protein aggregate induced by exposure to oxidative stress (Rand and Grant 2006).

One of the most fascinating aspects of chaperone structure and function is their unique ability to form a variety of structures and oligomeric states *in vitro* and the possible implications for function. However, in many instances, the *in vivo* significance and indeed existence of these high molecular weight complexes in an intracellular environment remains to be established, as they are usually generated *in vitro* under stress conditions, such as thermal/oxidative shock or low pH, often using mutant Prxs. For example, the ability of PrxIII to form long nanotubes was confined to an active-site C47S mutant, whereas the wild-type and hyperoxidation resistant (YL) mutants are present as dodecameric toroids (Gourlay et al. 2003; Cao et al. 2015). Interestingly, crystal structures of oxidised and reduced PrxIII at 2.4 and 2.2 Å respectively, show that that both are present as two physically-interlinked dodecamers inclined at an angle of 55° to each other in a so-called 2-ring catenane (Cao et al. 2015). Inter-ring contacts are maintained by an extensive network of hydrogen bonds. At present, it is unclear whether the physically-interlinked 2-ring toroidal form exists *in vivo* or if it represents an artefact of crystal packing, made possible by the presence of the increased diameter of the central cavity (70 Å) in this unusual dodecameric toroid. This novel structure is also favored because inter-ring contacts are maintained by multiple hydrogen bonds involving amino acid residues that are poorly conserved in other members of the Prx family. EM analysis of PrxIII oligomers after crystal dissolution indicates that they have largely reverted to their dodecameric state in dilute solution, indicating that catenane assembly/disassembly is a dynamic process (Cao et al. 2007). However, as catenane formation is concentration-dependent, it is probable that PrxIII exists largely in this conformation in the crowded environment of the mitochondrial matrix. Interestingly, the novel properties of a newly-discovered enzyme catalysing CS_2 conversion to H_2S and CO_2 in the acidothermophile Acidianus A1-3, have been attributed to its unusual structure, namely a catenated, hexadecameric oligomer (Smeulders et al. 2011; van Eldijk et al. 2013).

5.7 Concluding Remarks

The dramatic increase in research on peroxiredoxins in the past 20 years was driven initially by the realisation that these ubiquitous and versatile enzymes constitute the main line of defence against ROS and Reactive Nitrogen Species (RNS) in the vast majority of living systems. ROS and RNS are generated intracellularly as natural by-products of metabolism and from exogenous sources including xenobiotics, UV and ionising radiation. Since the majority of ROS are rapidly converted to H_2O_2, its main route of removal in all intracellular compartments is via the various Prx family members that operate as efficient catalytic machines and are normally present in high abundance.

Subsequently, the landmark discovery that eukaryotic Prxs were amenable to temporary inactivation via hyperoxidation of their conserved active-site cysteines during the catalytic cycle under both non-stress and stress conditions coupled with the realisation that these modified Prxs could be re-reduced/reactivated in an ATP-dependent reaction by Srx, added impetus to this already vibrant research field.

Further recognition that Prxs could exist in a range of oligomeric states with differing enzymatic and chaperone-like properties that could be fine-tuned by a series of post-translation modifications confirmed their central status in eliciting appropriate cellular responses to the fluctuations in the aerobic environment under normal or extreme conditions. In this context, a huge volume of current data is focused on the interaction of Prxs with key signalling molecules, e.g. transcription factors, members of the Tyr/Ser kinase and phosphatase families that are implicated in fundamental cell processes, such as cell cycle regulation, circadian rhythmicity, DNA repair and induction of apoptosis. Research in this emerging field continues at an impressive rate. Undoubtedly, research over the next few years will provide detailed insights into the exact nature of the molecular mechanisms by which Prxs interact with and modulate a host of cellular activities concerned with adaption to changes in their immediate environment.

While Prxs are currently the subject of intensive investigation by cell and molecular biologists, their ability to form large supramolecular assemblies, such as nanotubes, is now being exploited in nanofabrication. Recently a PrxI mutant was employed as a molecular template to promote metal-induced self-assembly of 1-D nanoscopic structures housing linear arrays of Ni^{2+}-functionalised gold nanoparticles in their central cavities (Ardini et al. 2014). Formation of protein-metal complexes is increasingly being exploited in the assembly of electronic nano-devices for a variety of technological applications.

References

Angelucci F, Saccoccia F, Ardini M, Boumis G, Brunori M, Di Leandro L, Ippoliti R, Miele AE, Natoli G, Scotti S, Bellelli A (2013) Switching between the alternative structures and functions of a 2-Cys peroxiredoxin, by site-directed mutagenesis. J Mol Biol 425(22):4556–4568. doi:10.1016/j.jmb.2013.09.002

Ardini M, Giansanti F, Di Leandro L, Pitari G, Cimini A, Ottaviano L, Donarelli M, Santucci S, Angelucci F, Ippoliti R (2014) Metal-induced self-assembly of peroxiredoxin as a tool for sorting ultrasmall gold nanoparticles into one-dimensional clusters. Nanoscale 6(14):8052–8061. doi:10.1039/c4nr01526f

Biteau B, Labarre J, Toledano MB (2003) ATP-dependent reduction of cysteine-sulphinic acid by *S. cerevisiae* sulphiredoxin. Nature 425(6961):980–984. doi:10.1038/nature02075

Calvo IA, Boronat S, Domenech A, Garcia-Santamarina S, Ayte J, Hidalgo E (2013) Dissection of a redox relay: H2O2-dependent activation of the transcription factor Pap1 through the peroxidatic Tpx1-thioredoxin cycle. Cell Rep 5(5):1413–1424. doi:10.1016/j.celrep.2013.11.027

Cao Z, Roszak AW, Gourlay LJ, Lindsay JG, Isaacs NW (2005) Bovine mitochondrial peroxiredoxin III forms a two-ring catenane. Structure 13(11):1661–1664. doi:S0969–2126(05)00312–6 [pii], 10.1016/j.str.2005.07.021

Cao Z, Bhella D, Lindsay JG (2007) Reconstitution of the mitochondrial PrxIII antioxidant defence pathway: general properties and factors affecting PrxIII activity and oligomeric state. J Mol Biol 372(4):1022–1033. doi:10.1016/j.jmb.2007.07.018

Cao Z, Tavender TJ, Roszak AW, Cogdell RJ, Bulleid NJ (2011) Crystal structure of reduced and of oxidized peroxiredoxin IV enzyme reveals a stable oxidized decamer and a non-disulfide-bonded intermediate in the catalytic cycle. J Biol Chem 286(49):42257–42266. doi:10.1074/jbc.M111.298810

Cao Z, Subramaniam S, Bulleid NJ (2014) Lack of an efficient endoplasmic reticulum-localized recycling system protects peroxiredoxin IV from hyperoxidation. J Biol Chem 289(9):5490–5498. doi:10.1074/jbc.M113.529305

Cao Z, McGow DP, Shepherd C, Lindsay JG (2015) Improved catenated structures of bovine peroxiredoxin III F190L reveal details of ring-ring interactions and a novel conformational state. PLoS One 10(4):e0123303. doi:10.1371/journal.pone.0123303

Cha MK, Suh KH, Kim IH (2009) Overexpression of peroxiredoxin I and thioredoxin1 in human breast carcinoma. J Exp Clin Cancer Res 28:93. doi:10.1186/1756-9966-28-93

Chae HZ, Robison K, Poole LB, Church G, Storz G, Rhee SG (1994) Cloning and sequencing of thiol-specific antioxidant from mammalian brain: alkyl hydroperoxide reductase and thiol-specific antioxidant define a large family of antioxidant enzymes. Proc Natl Acad Sci U S A 91(15):7017–7021

Chang XZ, Li DQ, Hou YF, Wu J, Lu JS, Di GH, Jin W, Ou ZL, Shen ZZ, Shao ZM (2007) Identification of the functional role of peroxiredoxin 6 in the progression of breast cancer. Breast Cancer Res 9(6):R76. doi:10.1186/bcr1789

Chiarugi P, Cirri P (2003) Redox regulation of protein tyrosine phosphatases during receptor tyrosine kinase signal transduction. Trends Biochem Sci 28(9):509–514. doi:10.1016/S0968-0004(03)00174-9

Day AM, Brown JD, Taylor SR, Rand JD, Morgan BA, Veal EA (2012) Inactivation of a peroxiredoxin by hydrogen peroxide is critical for thioredoxin-mediated repair of oxidized proteins and cell survival. Mol Cell 45(3):398–408. doi:10.1016/j.molcel.2011.11.027

Declercq JP, Evrard C, Clippe A, Stricht DV, Bernard A, Knoops B (2001) Crystal structure of human peroxiredoxin 5, a novel type of mammalian peroxiredoxin at 1.5 Å resolution. J Mol Biol 311(4):751–759. doi:10.1006/jmbi.2001.4853

Edgar RS, Green EW, Zhao Y, van Ooijen G, Olmedo M, Qin X, Xu Y, Pan M, Valekunja UK, Feeney KA, Maywood ES, Hastings MH, Baliga NS, Merrow M, Millar AJ, Johnson CH, Kyriacou CP, O'Neill JS, Reddy AB (2012) Peroxiredoxins are conserved markers of circadian rhythms. Nature 485(7399):459–464. doi:10.1038/nature11088

Gourlay LJ, Bhella D, Kelly SM, Price NC, Lindsay JG (2003) Structure-function analysis of recombinant substrate protein 22 kDa (SP-22). A mitochondrial 2-CYS peroxiredoxin organized as a decameric toroid. J Biol Chem 278(35):32631–32637. doi:10.1074/jbc.M303862200, M303862200 [pii]

Hall A, Parsonage D, Poole LB, Karplus PA (2010) Structural evidence that peroxiredoxin catalytic power is based on transition-state stabilization. J Mol Biol 402(1):194–209. doi:10.1016/j.jmb.2010.07.022

Hanzen S, Vielfort K, Yang J, Roger F, Andersson V, Zamarbide-Fores S, Andersson R, Malm L, Palais G, Biteau B, Liu B, Toledano MB, Molin M, Nystrom T (2016) Lifespan control by redox-dependent recruitment of chaperones to misfolded proteins. Cell. doi:10.1016/j.cell.2016.05.006

Haynes AC, Qian J, Reisz JA, Furdui CM, Lowther WT (2013) Molecular basis for the resistance of human mitochondrial 2-Cys peroxiredoxin 3 to hyperoxidation. J Biol Chem 288(41):29714–29723. doi:10.1074/jbc.M113.473470

Ito T, Kimura S, Seto K, Warabi E, Kawachi Y, Shoda J, Tabuchi K, Yamagata K, Hasegawa S, Bukawa H, Ishii T, Yanagawa T (2014) Peroxiredoxin I plays a protective role against UVA irradiation through reduction of oxidative stress. J Dermatol Sci 74(1):9–17. doi:10.1016/j.jdermsci.2013.12.002

Iuchi Y, Okada F, Tsunoda S, Kibe N, Shirasawa N, Ikawa M, Okabe M, Ikeda Y, Fujii J (2009) Peroxiredoxin 4 knockout results in elevated spermatogenic cell death via oxidative stress. Biochem J 419(1):149–158. doi:10.1042/BJ20081526

Jang HH, Lee KO, Chi YH, Jung BG, Park SK, Park JH, Lee JR, Lee SS, Moon JC, Yun JW, Choi YO, Kim WY, Kang JS, Cheong GW, Yun DJ, Rhee SG, Cho MJ, Lee SY (2004) Two enzymes in one; two yeast peroxiredoxins display oxidative stress-dependent switching from a peroxidase to a molecular chaperone function. Cell 117(5):625–635. doi:10.1016/j.cell.2004.05.002, S0092867404004878 [pii]

Jang HH, Kim SY, Park SK, Jeon HS, Lee YM, Jung JH, Lee SY, Chae HB, Jung YJ, Lee KO, Lim CO, Chung WS, Bahk JD, Yun DJ, Cho MJ, Lee SY (2006) Phosphorylation and concomitant structural changes in human 2-Cys peroxiredoxin isotype I differentially regulate its peroxidase and molecular chaperone functions. FEBS Lett 580(1):351–355. doi:10.1016/j.febslet.2005.12.030

Jonsson TJ, Johnson LC, Lowther WT (2008) Structure of the sulphiredoxin-peroxiredoxin complex reveals an essential repair embrace. Nature 451(7174):98–101. doi:10.1038/nature06415

Kil IS, Lee SK, Ryu KW, Woo HA, Hu MC, Bae SH, Rhee SG (2012) Feedback control of adrenal steroidogenesis via H2O2-dependent, reversible inactivation of peroxiredoxin III in mitochondria. Mol Cell 46(5):584–594. doi:10.1016/j.molcel.2012.05.030

Klichko VI, Orr WC, Radyuk SN (2016) The role of peroxiredoxin 4 in inflammatory response and aging. Biochim Biophys Acta 1862(2):265–273. doi:10.1016/j.bbadis.2015.12.008

Konig J, Baier M, Horling F, Kahmann U, Harris G, Schurmann P, Dietz KJ (2002) The plant-specific function of 2-Cys peroxiredoxin-mediated detoxification of peroxides in the redox-hierarchy of photosynthetic electron flux. Proc Natl Acad Sci U S A 99(8):5738–5743. doi:10.1073/pnas.072644999

Latimer HR, Veal EA (2016) Peroxiredoxins in regulation of MAPK signalling pathways; sensors and barriers to signal transduction. Mol Cells 39(1):40–45. doi:10.14348/molcells.2016.2327

Lee TH, Kim SU, Yu SL, Kim SH, Park DS, Moon HB, Dho SH, Kwon KS, Kwon HJ, Han YH, Jeong S, Kang SW, Shin HS, Lee KK, Rhee SG, Yu DY (2003) Peroxiredoxin II is essential for sustaining life span of erythrocytes in mice. Blood 101(12):5033–5038. doi:10.1182/blood-2002-08-2548

Lee S, Chung JM, Yun HJ, Won J, Jung HS (2016) New insight into multifunctional role of peroxiredoxin family protein: determination of DNA protection properties of bacterioferritin comigratory protein under hyperthermal and oxidative stresses. Biochem Biophys Res Commun 469(4):1028–1033. doi:10.1016/j.bbrc.2015.12.099

Li S, Peterson NA, Kim MY, Kim CY, Hung LW, Yu M, Lekin T, Segelke BW, Lott JS, Baker EN (2005) Crystal structure of AhpE from mycobacterium tuberculosis, a 1-Cys peroxiredoxin. J Mol Biol 346(4):1035–1046. doi:10.1016/j.jmb.2004.12.046

Li L, Kaifu T, Obinata M, Takai T (2009) Peroxiredoxin III-deficiency sensitizes macrophages to oxidative stress. J Biochem 145(4):425–427. doi:10.1093/jb/mvp011

Lindsay H, Beaumont E, Richards SD, Kelly SM, Sanderson SJ, Price NC, Lindsay JG (2000) FAD insertion is essential for attaining the assembly competence of the dihydrolipoamide dehydrogenase (E3) monomer from *Escherichia coli*. J Biol Chem 275(47):36665–36670. doi:10.1074/jbc.M004777200

MacDiarmid CW, Taggart J, Kerdsomboon K, Kubisiak M, Panascharoen S, Schelble K, Eide DJ (2013) Peroxiredoxin chaperone activity is critical for protein homeostasis in zinc-deficient yeast. J Biol Chem 288(43):31313–31327. doi:10.1074/jbc.M113.512384

Meissner U, Schroder E, Scheffler D, Martin AG, Harris JR (2007) Formation, TEM study and 3D reconstruction of the human erythrocyte peroxiredoxin-2 dodecahedral higher-order assembly. Micron 38(1):29–39. doi:10.1016/j.micron.2006.04.010

Mitsumoto A, Nakagawa Y, Takeuchi A, Okawa K, Iwamatsu A, Takanezawa Y (2001) Oxidized forms of peroxiredoxins and DJ-1 on two-dimensional gels increased in response to sublethal levels of paraquat. Free Radic Res 35(3):301–310

Moon JC, Hah YS, Kim WY, Jung BG, Jang HH, Lee JR, Kim SY, Lee YM, Jeon MG, Kim CW, Cho MJ, Lee SY (2005) Oxidative stress-dependent structural and functional switching of a human 2-Cys peroxiredoxin isotype II that enhances HeLa cell resistance to H2O2-induced cell death. J Biol Chem 280(31):28775–28784. doi:10.1074/jbc.M505362200

Morinaka A, Funato Y, Uesugi K, Miki H (2011) Oligomeric peroxiredoxin-I is an essential intermediate for p53 to activate MST1 kinase and apoptosis. Oncogene 30(40):4208–4218. doi:10.1038/onc.2011.139

Mukhopadhyay SS, Leung KS, Hicks MJ, Hastings PJ, Youssoufian H, Plon SE (2006) Defective mitochondrial peroxiredoxin-3 results in sensitivity to oxidative stress in Fanconi anemia. J Cell Biol 175(2):225–235. doi:10.1083/jcb.200607061

Musicco C, Capelli V, Pesce V, Timperio AM, Calvani M, Mosconi L, Zolla L, Cantatore P, Gadaleta MN (2009) Accumulation of overoxidized Peroxiredoxin III in aged rat liver mitochondria. Biochim Biophys Acta 1787(7):890–896. doi:10.1016/j.bbabio.2009.03.002

Nelson KJ, Knutson ST, Soito L, Klomsiri C, Poole LB, Fetrow JS (2011) Analysis of the peroxiredoxin family: using active-site structure and sequence information for global classification and residue analysis. Proteins 79(3):947–964. doi:10.1002/prot.22936

Netto LE, Antunes F (2016) The roles of peroxiredoxin and thioredoxin in hydrogen peroxide sensing and in signal transduction. Mol Cells 39(1):65–71. doi:10.14348/molcells.2016.2349

Neumann CA, Krause DS, Carman CV, Das S, Dubey DP, Abraham JL, Bronson RT, Fujiwara Y, Orkin SH, Van Etten RA (2003) Essential role for the peroxiredoxin Prdx1 in erythrocyte antioxidant defence and tumour suppression. Nature 424(6948):561–565. doi:10.1038/nature01819

Nielsen MH, Kidmose RT, Jenner LB (2016) Structure of TSA2 reveals novel features of the active-site loop of peroxiredoxins. Acta Crystallogr Sect D, Struct Biol 72(Pt 1):158–167. doi:10.1107/S2059798315023815

Noh YH, Baek JY, Jeong W, Rhee SG, Chang TS (2009) Sulfiredoxin translocation into mitochondria plays a crucial role in reducing hyperoxidized peroxiredoxin III. J Biol Chem 284(13):8470–8477. doi:10.1074/jbc.M808981200

Park JW, Piszczek G, Rhee SG, Chock PB (2011) Glutathionylation of peroxiredoxin I induces decamer to dimers dissociation with concomitant loss of chaperone activity. Biochemistry 50(15):3204–3210. doi:10.1021/bi101373h

Parmigiani RB, Xu WS, Venta-Perez G, Erdjument-Bromage H, Yaneva M, Tempst P, Marks PA (2008) HDAC6 is a specific deacetylase of peroxiredoxins and is involved in redox regulation. Proc Natl Acad Sci U S A 105(28):9633–9638. doi:10.1073/pnas.0803749105

Parsonage D, Karplus PA, Poole LB (2008) Substrate specificity and redox potential of AhpC, a bacterial peroxiredoxin. Proc Natl Acad Sci U S A 105(24):8209–8214. doi:10.1073/pnas.0708308105

Perkins A, Nelson KJ, Williams JR, Parsonage D, Poole LB, Karplus PA (2013) The sensitive balance between the fully folded and locally unfolded conformations of a model peroxiredoxin. Biochemistry 52(48):8708–8721. doi:10.1021/bi4011573

Perkins A, Nelson KJ, Parsonage D, Poole LB, Karplus PA (2015) Peroxiredoxins: guardians against oxidative stress and modulators of peroxide signaling. Trends Biochem Sci 40(8):435–445. doi:10.1016/j.tibs.2015.05.001

Peskin AV, Low FM, Paton LN, Maghzal GJ, Hampton MB, Winterbourn CC (2007) The high reactivity of peroxiredoxin 2 with H(2)O(2) is not reflected in its reaction with other oxidants and thiol reagents. J Biol Chem 282(16):11885–11892. doi:10.1074/jbc.M700339200

Peskin AV, Dickerhof N, Poynton RA, Paton LN, Pace PE, Hampton MB, Winterbourn CC (2013) Hyperoxidation of peroxiredoxins 2 and 3: rate constants for the reactions of the sulfenic acid of the peroxidatic cysteine. J Biol Chem 288(20):14170–14177. doi:10.1074/jbc.M113.460881

Phalen TJ, Weirather K, Deming PB, Anathy V, Howe AK, van der Vliet A, Jonsson TJ, Poole LB, Heintz NH (2006) Oxidation state governs structural transitions in peroxiredoxin II that correlate with cell cycle arrest and recovery. J Cell Biol 175(5):779–789. doi:10.1083/jcb.200606005

Phillips AJ, Littlejohn J, Yewdall NA, Zhu T, Valery C, Pearce FG, Mitra AK, Radjainia M, Gerrard JA (2014) Peroxiredoxin is a versatile self-assembling tecton for protein nanotechnology. Biomacromolecules 15(5):1871–1881. doi:10.1021/bm500261u

Radjainia M, Venugopal H, Desfosses A, Phillips AJ, Yewdall NA, Hampton MB, Gerrard JA, Mitra AK (2015) Cryo-electron microscopy structure of human peroxiredoxin-3 filament reveals the assembly of a putative chaperone. Structure 23(5):912–920. doi:10.1016/j.str.2015.03.019

Rand JD, Grant CM (2006) The thioredoxin system protects ribosomes against stress-induced aggregation. Mol Biol Cell 17(1):387–401. doi:10.1091/mbc.E05-06-0520

Romero-Puertas MC, Laxa M, Matte A, Zaninotto F, Finkemeier I, Jones AM, Perazzolli M, Vandelle E, Dietz KJ, Delledonne M (2007) S-nitrosylation of peroxiredoxin II E promotes peroxynitrite-mediated tyrosine nitration. Plant Cell 19(12):4120–4130. doi:10.1105/tpc.107.055061

Saccoccia F, Di Micco P, Boumis G, Brunori M, Koutris I, Miele AE, Morea V, Sriratana P, Williams DL, Bellelli A, Angelucci F (2012) Moonlighting by different stressors: crystal structure of the chaperone species of a 2-Cys peroxiredoxin. Structure 20(3):429–439. doi:10.1016/j.str.2012.01.004

Sarma GN, Nickel C, Rahlfs S, Fischer M, Becker K, Karplus PA (2005) Crystal structure of a novel Plasmodium falciparum 1-Cys peroxiredoxin. J Mol Biol 346(4):1021–1034. doi:10.1016/j.jmb.2004.12.022

Schroder E, Ponting CP (1998) Evidence that peroxiredoxins are novel members of the thioredoxin fold superfamily. Protein Sci 7(11):2465–2468. doi:10.1002/pro.5560071125

Schroder E, Littlechild JA, Lebedev AA, Errington N, Vagin AA, Isupov MN (2000) Crystal structure of decameric 2-Cys peroxiredoxin from human erythrocytes at 1.7 A resolution. Structure 8(6):605–615

Seo JH, Lim JC, Lee DY, Kim KS, Piszczek G, Nam HW, Kim YS, Ahn T, Yun CH, Kim K, Chock PB, Chae HZ (2009) Novel protective mechanism against irreversible hyperoxidation of peroxiredoxin: Nalpha-terminal acetylation of human peroxiredoxin II. J Biol Chem 284(20):13455–13465. doi:10.1074/jbc.M900641200

Smeulders MJ, Barends TR, Pol A, Scherer A, Zandvoort MH, Udvarhelyi A, Khadem AF, Menzel A, Hermans J, Shoeman RL, Wessels HJ, van den Heuvel LP, Russ L, Schlichting I, Jetten MS, Op den Camp HJ (2011) Evolution of a new enzyme for carbon disulphide conversion by an acidothermophilic archaeon. Nature 478(7369):412–416. doi:10.1038/nature10464

Sobotta MC, Liou W, Stocker S, Talwar D, Oehler M, Ruppert T, Scharf AN, Dick TP (2015) Peroxiredoxin-2 and STAT3 form a redox relay for H2O2 signaling. Nat Chem Biol 11(1):64–70. doi:10.1038/nchembio.1695

Stresing V, Baltziskueta E, Rubio N, Blanco J, Arriba MC, Valls J, Janier M, Clezardin P, Sanz-Pamplona R, Nieva C, Marro M, Petrov D, Sierra A (2013) Peroxiredoxin 2 specifically regulates the oxidative and metabolic stress response of human metastatic breast cancer cells in lungs. Oncogene 32(6):724–735. doi:10.1038/onc.2012.93

Tavender TJ, Sheppard AM, Bulleid NJ (2008) Peroxiredoxin IV is an endoplasmic reticulum-localized enzyme forming oligomeric complexes in human cells. Biochem J 411(1):191–199. doi:10.1042/BJ20071428

Teixeira F, Castro H, Cruz T, Tse E, Koldewey P, Southworth DR, Tomas AM, Jakob U (2015) Mitochondrial peroxiredoxin functions as crucial chaperone reservoir in Leishmania infantum. Proc Natl Acad Sci U S A 112(7):E616–E624. doi:10.1073/pnas.1419682112

van Eldijk MB, van Leeuwen I, Mikhailov VA, Neijenhuis L, Harhangi HR, van Hest JC, Jetten MS, Op den Camp HJ, Robinson CV, Mecinovic J (2013) Evidence that the catenane form of CS2 hydrolase is not an artefact. Chem Commun 49(71):7770–7772. doi:10.1039/c3cc43219j

Wagner E, Luche S, Penna L, Chevallet M, Van Dorsselaer A, Leize-Wagner E, Rabilloud T (2002) A method for detection of overoxidation of cysteines: peroxiredoxins are oxidized in vivo at the active-site cysteine during oxidative stress. Biochem J 366(Pt 3):777–785. doi:10.1042/BJ20020525

Wang X, Wang L, Wang X, Sun F, Wang CC (2012) Structural insights into the peroxidase activity and inactivation of human peroxiredoxin 4. Biochem J 441(1):113–118. doi:10.1042/BJ20110380

Winterbourn CC (2008) Reconciling the chemistry and biology of reactive oxygen species. Nat Chem Biol 4(5):278–286. doi:nchembio.85 [pii], 10.1038/nchembio.85

Woo HA, Kang SW, Kim HK, Yang KS, Chae HZ, Rhee SG (2003) Reversible oxidation of the active site cysteine of peroxiredoxins to cysteine sulfinic acid. Immunoblot detection with antibodies specific for the hyperoxidized cysteine-containing sequence. J Biol Chem 278(48):47361–47364. doi:10.1074/jbc.C300428200

Woo HA, Yim SH, Shin DH, Kang D, Yu DY, Rhee SG (2010) Inactivation of peroxiredoxin I by phosphorylation allows localized H(2)O(2) accumulation for cell signaling. Cell 140(4):517–528. doi:10.1016/j.cell.2010.01.009

Wood ZA, Poole LB, Karplus PA (2003a) Peroxiredoxin evolution and the regulation of hydrogen peroxide signaling. Science 300(5619):650–653. doi:10.1126/science.1080405

Wood ZA, Schroder E, Robin Harris J, Poole LB (2003b) Structure, mechanism and regulation of peroxiredoxins. Trends Biochem Sci 28(1):32–40

Yang KS, Kang SW, Woo HA, Hwang SC, Chae HZ, Kim K, Rhee SG (2002) Inactivation of human peroxiredoxin I during catalysis as the result of the oxidation of the catalytic site cysteine to cysteine-sulfinic acid. J Biol Chem 277(41):38029–38036. doi:10.1074/jbc.M206626200

Yewdall NA, Venugopal H, Desfosses A, Abrishami V, Yosaatmadja Y, Hampton MB, Gerrard JA, Goldstone DC, Mitra AK, Radjainia M (2016) Structures of human peroxiredoxin 3 suggest self-chaperoning assembly that maintains catalytic state. Structure. doi:10.1016/j.str.2016.04.013

Chapter 6
α_2-Macroglobulins: Structure and Function

Irene Garcia-Ferrer, Aniebrys Marrero, F. Xavier Gomis-Rüth, and Theodoros Goulas

Abstract α_2-macroglobulins are broad-spectrum endopeptidase inhibitors, which have to date been characterised from metazoans (vertebrates and invertebrates) and Gram-negative bacteria. Their structural and biochemical properties reveal two related modes of action: the "Venus flytrap" and the "snap-trap" mechanisms. In both cases, peptidases trigger a massive conformational rearrangement of α_2-macroglobulin after cutting in a highly flexible bait region, which results in their entrapment. In some homologs, a second action takes place that involves a highly reactive β-cysteinyl-γ-glutamyl thioester bond, which covalently binds cleaving peptidases and thus contributes to the further stabilization of the enzyme:inhibitor complex. Trapped peptidases are still active, but have restricted access to their substrates due to steric hindrance. In this way, the human α_2-macroglobulin homolog regulates proteolysis in complex biological processes, such as nutrition, signalling, and tissue remodelling, but also defends the host organism against attacks by external toxins and other virulence factors during infection and envenomation. In parallel, it participates in several other biological functions by modifying the activity of

---

I. Garcia-Ferrer
Proteolysis Lab, Structural Biology Unit, "María de Maeztu" Unit of Excellence, Molecular Biology Institute of Barcelona (CSIC), Barcelona Science Park; c/Baldiri Reixac, 15-21, 08028 Barcelona, Spain

Present address: EMBL Grenoble, 71 Avenue des Martyrs; 38042 CS 90181, Grenoble Cedex 9, France
e-mail: irgarcia@embl.fr.

A. Marrero
Proteolysis Lab, Structural Biology Unit, "María de Maeztu" Unit of Excellence, Molecular Biology Institute of Barcelona (CSIC), Barcelona Science Park; c/Baldiri Reixac, 15-21, 08028 Barcelona, Spain

Present address: Department of Chemistry, University of Zürich, Winterthurerstrasse 190, 8057 Zurich, Switzerland
e-mail: aniebrys.marrero@chem.uzh.ch.

F.X. Gomis-Rüth • T. Goulas (✉)
Proteolysis Lab, Structural Biology Unit, "María de Maeztu" Unit of Excellence, Molecular Biology Institute of Barcelona (CSIC), Barcelona Science Park; c/Baldiri Reixac, 15-21, 08028 Barcelona, Spain
e-mail: thgcri@ibmb.csic.es

J.R. Harris, J. Marles-Wright (eds.), *Macromolecular Protein Complexes*, Subcellular Biochemistry 83, DOI 10.1007/978-3-319-46503-6_6

cytokines and regulating hormones, growth factors, lipid factors and other proteins, which has a great impact on physiology. Likewise, bacterial α_2-macroglobulins may participate in defence by protecting cell wall components from attacking peptidases, or in host-pathogen interactions through recognition of host peptidases and/or antimicrobial peptides. α_2-macroglobulins are more widespread than initially thought and exert multifunctional roles in both eukaryotes and prokaryotes, therefore, their on-going study is essential.

Keywords Macroglobulin • Proteinase inhibitor • Conformational change • Growth factor and cytokine regulator

6.1 Introduction

The ability of organisms to tightly control internal processes and protect themselves against antagonism or parasitism determines their success for well-being and reproduction. From single-cellular archaea to more complicated multicellular eukaryotes, and from invading parasites to defending hosts, a consortium of peptidases and their inhibitors are involved in the regulation of many physiological processes and in the protection against invasion and virulence (Armstrong 2006). Many peptidases and inhibitors have been characterised to date, forming a tightly balanced equilibrium.

α_2-Macroglobulins (α_2Ms) are the largest and among the most relevant inhibitors, due to their universal ability to counteract a broad spectrum of endopeptidases. These inhibitors are multi-domain proteins that use two related mechanisms to act as molecular traps instead of catalytic-centre blocks (Sottrup-Jensen 1989). Rather than simple inhibitors, they should be envisaged as sophisticated binding proteins that are activated by limited proteolysis and augment the clearance of binding proteins from the circulation. The liver is the principal site of synthesis in mammals and α_2Ms are found in blood, presumably to have better access to all parts of the body, where they contribute to the bulk of the innate immune system (Andus et al. 1983). Besides peptidase regulation, α_2Ms play a major role in controlling the action of many other effector molecules and modulating the physiology of the host, with many implications in health and disease (Borth 1992).

α_2Ms are anciently diverging molecules, which were initially described in eukaryotes, together with their homologs from the complement component family (Sottrup-Jensen et al. 1985). Human α_2M was first described back in 1946 (Cohn et al. 1946) as a major protein of blood serum, and bacterial orthologs were serendipitously discovered six decades later (Budd et al. 2004). Many years of intensive research has revealed several of their characteristics, which has resulted in a vast number of publications. Here, the unique structural and mechanistic features that enable α_2Ms to interact with a wide variety of peptidases and other physiologically relevant proteins are discussed.

6.2 Evolutionary Origin

α_2Ms belong to the thioester-containing proteins (TEPs) with members sharing a common evolutionary origin and conserved structural and functional features. Other TEPs are the peptidase inhibitors pregnancy zone protein (PZP) and α_1-inhibitor-3 (α_1I3), the complement components C3, C4 and C5, the cell surface antigen CD109, the complement C3 and PZP-like α_2M domain-containing 8 (CPAMD8), and insect and nematode TEPs (Nonaka 2011; Sottrup-Jensen et al. 1985). All these proteins undergo proteolytic processing and structural rearrangements as part of their role in the immune system of host defence. In most cases, they are characterized by the presence of an active thioester bond and have essentially the same domain organisation as the complement C3. However, in some members, the anaphylatoxin (ANA) domain is replaced with a bait region domain (BRD) or a hyper-variable region, and the C345C domain at their C-terminus is missing (Fig. 6.1) (Marrero et al. 2012; Janssen et al. 2005; Baxter et al. 2007; Garcia-Ferrer et al. 2015). Insect TEPs show sequence relatedness halfway between complement components and α_2M. However, like C3, they have a conserved histidine to convert the reactive glutamine/glutamate residue of the activated thioester to an intermediate form that favours its subsequent covalent linkage to hydroxyl groups rather than amines, as found in α_2Ms (Dodds et al. 1996; Law and Dodds 1997; Janssen et al. 2005; Baxter et al. 2007). In accordance with their structural differences, C3 and insect TEPs are functionally different from α_2Ms and are involved in bacterial opsonization, lysis or phagocytosis, rather than in peptidase trapping, inhibition and clearance (Sahu and Lambris 2001; Armstrong 2010). CD109 is a glycosyl-phosphatidyl-inositol-linked glycoprotein that was originally found in endothelial cells, platelets and activated T-cells with complement-like activity (Lin et al. 2002). CPAMD8 has a C-terminal Kazal-type serine peptidase inhibitor-like domain and is mainly expressed in kidney, brain and testis, with a poorly characterized function (Li et al. 2004).

Currently, all members of the TEP superfamily are sub-classified to the α_2M family, which includes α_2M, PZP, α_1I3, CD109 and CPAMD8, and the family of structural homologs of C3, which includes complement components C3, C4 and C5 (Sottrup-Jensen et al. 1985; Fujito et al. 2010; Baxter et al. 2007; Nonaka 2011). Insect TEPs could form a distinct family; however, phylogenetic analysis clearly indicates that they belong to the α_2Ms, which suggests that their functional similarity to mammalian C3 was caused by convergent molecular evolution (Nonaka 2011). The prevalence hypothesis, based on sequence similarity and exon-intron organization studies, states that all members of the TEP superfamily emerged from a common α_2M ancestor and underwent several gene duplication events. One of the duplicated α_2M genes was modified to form a C3/C4/C5 ancestral gene, which was then further duplicated to form the C4 and C3/C5 genes. Finally, the C3/C5 gene was duplicated to produce the C3 and C5 genes (Sahu and Lambris 2001; Nonaka 2011).

All α_2Ms have considerable importance in the innate immune response in metazoans, and emerged at least 600–700 million years ago, predating the emergence of

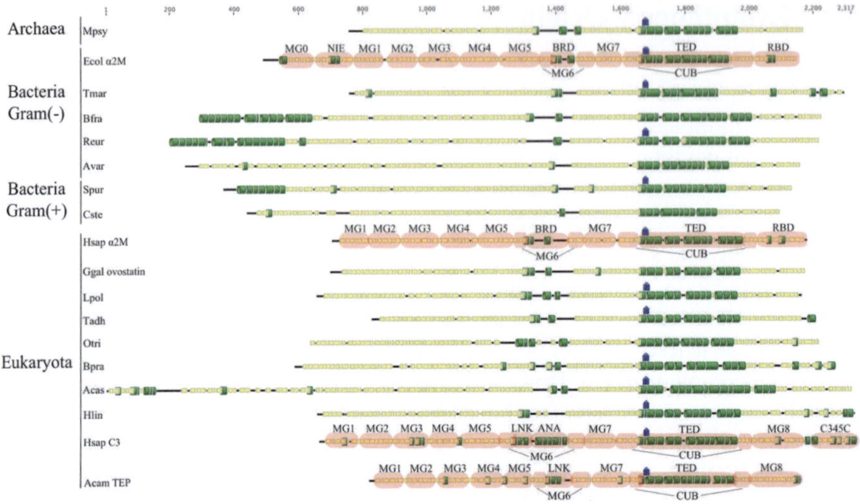

Fig. 6.1 Selected thioester-containing proteins (TEPs) from archaea, bacteria and eukaryota. The sequences are aligned based on the thioester bond motif (CXEQ) found in TED domains. No further alignment was performed due to the excessive phylogenetic distances. The secondary structure is based on existing structural models or on predictions by JPred (Drozdetskiy et al. 2015). β-strands are shown by *yellow bars*, α-helixes by *green barrels*, and the thioester bond by a *blue arrow*. The domains of homologs with known three-dimensional structures are further annotated with *red highlights*. MG, macroglobulin-like domains; BRD, bait region domain; CUB, C1r/C1s, Uegf, and Bmp1 found domain; TED, thioester domain; RBD, receptor-binding domain; LNK, linker domain; ANA, anaphylatoxin domain. Sequence names are shown with a four letter abbreviation, and further defined herein with their UniProt access number and a short description: Mpsy, *Methanolobus psychrophilus* (K4MDH4; Psychrophilic methanogen); Ecol α_2M, *Escherichia coli* α_2M (P76578; Gut habitant); Tmar, *Thermotoga maritima* (Q9X079; Extremophile); Bfra, *Bacteroides fragilis* (A0A0K6BRA4; Gut habitant); Reur, *Rhodopirellula europaea* (M2A507; Marine bacterium); Avar, *Anabaena variabilis* (Q3MAL1; Cyanobacterium); Spur, *Streptomyces purpurogeneiscleroticus* (A0A0M8ZE73; Soil habitant); Cste, *Clostridium stercorarium* (L7VLL5; Cellulolytic thermophile); Hsap α_2M, *Homo sapiens* α_2M (P01023); Ggal, *Gallus gallus* ovostatin (P20740; Egg white protein); Lpol, *Limulus polyphemus* (O01717; Ancient crab); Tadh, *Trichoplax adhaerens* (B3RVT5; Placozoa); Otri, *Oxytricha trifallax* (J9J1G5; Protozoa); Bpra, *Bathycoccus prasinos* (K8F745; Alga); Acas, *Acanthamoeba castellanii* (L8GY08; Amoebozoa); Hlin, *Haliplanella lineata* (D4QA01; Sea anemone); Hsap C3, *Homo sapiens* complement C3 (P01024); Agam TEP, *Anopheles gambiae* TEP (Q9GYW4; Mosquito)

immunoglobulins (Igs) (Sahu and Lambris 2001). Specifically, the most ancient member of the α_2M family described so far is from the hemolymph of the horseshoe crab, *Limulus polyphemus*, one of the oldest surviving multicellular organisms, which dates back over 500 million years (Enghild et al. 1990; Quigley et al. 1991; Rudloe 1979). This protein incorporates both glycerol and methylamine into the thioester, and could be considered a link between peptidase inhibitors and the complement components. Several other α_2Ms were biochemically characterized (Table 6.1 and references therein), and thousands have been identified in genome sequencing projects (Starkey and Barrett 1982; Budd et al. 2004). Accordingly, they are found in several eumetazoan phyla, including Cnidaria (e.g. sea anemone

Table 6.1 List of characterized α_2Ms

Name	Organism/UniProt no.	Localization	Mass (kDa)/Oligom./Thioester			References
α_2M	*Homo sapiens*/P01023	Blood serum	720	Tetramer	+	Barrett et al. (1979), Marrero et al. (2012), and Sottrup-Jensen (1989)
α_2ML1	*Homo sapiens*/A8K2U0	Epidermis	180	Monomer	+	Galliano et al. (2006)
PZP	*Homo sapiens*/P20742	Pregnancy blood serum	720	Tetramer	+	Sand et al. (1985) and Sottrup-Jensen et al. (1984a)
Ovostatin	*Gallus gallus*/P20740	Egg white	780	Tetramer	–	Nagase and Harris (1983)
α_1M	Rabbit	Blood serum	~	Tetramer	+	Banbula et al. (2005)
α_2M	*Biomphalaria glabrata* (gastropod mollusc)	Hemolymph	800	Tetramer	+	Bender and Bayne (1996)
α_2M	*Ornithodoros moubata* (tick)	Plasma	420	Dimer	+	Kopacek et al. (2000)
α_1I3	*Rattus norvegicus*/P14046	Blood serum	174	Monomer	+	Enghild et al. (1989)
α_1M	*Rattus norvegicus*/Q63041	Blood serum	~	Tetramer	~	Xiao et al. (2000)
α_2M	*Limulus polyphemus* (horseshoe crab)	Hemolymph and blood cells	354	Dimer	+	Enghild et al. (1990), Armstrong et al. (1991), and Husted et al. (2002)
α_2M	*Erinaceus europaeus* (hedgehog)	Plasma	800	Tetramer	~	De Wit and Weström (1987)
$\alpha_2\beta$M	*Erinaceus europaeus* (hedgehog)	Plasma	450–550	Dimer	~	De Wit and Weström (1987)
α_2M	*Penaeus vannamei* (white shrimp)	Hemolymph	360	Dimer	+	Gollas-Galvan et al. (2003)
IrA2M	*Ixodes ricinus* (tick)	Hemolymph	440	Dimer	+	Buresova et al. (2009)
α_2M	*Chelonia mydas japonica* (turtle)	Blood serum	~	~	+	Ikai et al. (1988)
Ovostatin	*Chelonia mydas japonica* (turtle)	Egg white	~	~	–	Ikai et al. (1988)
αM	*Rana catesbiana* (frog)	Blood serum	180	Monomer	+	Rubenstein et al. (1993)
α_2M	*Astacus astacus* (crayfish)	Hemolymph	390	Dimer	+	Stöcker et al. (1991)

(continued)

Table 6.1 (continued)

Name	Organism/UniProt no.	Localization	Mass (kDa)/Oligom./ Thioester			References
α_2M	*Octopus vulgaris* (mollusc)	Hemolymph	360	Dimer	+	Thøgersen et al. (1992)
α_2M	*Struthio camelus* (ostrich)	Blood serum	779	Tetramer	+	Van Jaarsveld et al. (1994)
α_2M	*Helix pomatia* (gastropod mollusc)	Hemolymph	697	Tetramer	+	Yigzaw et al. (2001)
α_2M	*Pacifastacus leniusculus* (crayfish)	Hemolymph	380	Dimer	+	Hergenhahn et al. (1988)
α_2M	*Libinia emarginata* and *Cancer borealis* (crab)	Hemolymph	480–460	Dimer	+/−	Armstrong et al. (1985)
α_2M	*Homarus americanus* (lobster)	Hemolymph	342	Dimer	+	Spycher et al. (1987)
α_2M	*Cyprinus carpio* (bony fish carp)	Blood serum	380	Dimer	+	Mutsuro et al. (2000)
α_2M	*Farfantepenaeus paulensis* (shrimp)	Blood serum	389	Dimer	+	Perazzolo et al. (2011)
ECAM	*Escherichia coli*/P76578	Inner membrane lipoprotein	183	Monomer	+	Doan and Gettins (2008), Neves et al. (2012), Garcia-Ferrer et al. (2015), and Fyfe et al. (2015)
Sa-α_2M	*Salmonella enterica* ser. Typhimurium/Q8ZN46	Inner membrane lipoprotein	179	Monomer	+	Wong and Dessen (2014)
MagD (YfaS)	*Pseudomonas aeruginosa*/PA4489	Inner membrane lipoprotein	165	Monomer	−	Robert-Genthon et al. (2013)

Where: (+) presence, (−) absence, and (~) unknown

Haliplanella lineata), Protostomia (e.g. Ecdysozoa: Arthropoda and Nematoda; Lophotrochozoa: Mollusca and Annelida), Deuterostomia (Echinozoa and Chordata) and Platyhelminthes (e.g. *Schistosoma haematobium*) (Fujito et al. 2010; Nonaka 2011). In addition, some more recent initial studies have demonstrated that bacterial α_2Ms are found almost exclusively in free-living, symbiotic and pathogenic Gram-negative bacteria, including proteobacteria, fusobacteria, spirochetes, bacteroidetes, deinococcus, cyanobacteria, planctomycetes and thermotogae (Budd et al. 2004). They were probably acquired one or more times from metazoan hosts and were then spread widely through other colonizing bacterial species by more than ten independent horizontal gene transfer events. However, this is not definite, particularly bearing in mind the putative presence of α_2Ms in free-living *Thermotoga maritima* and archaea species (e.g. *Methanolobus psychrophilus*) and the evidence of lateral gene transfer events through their symbiosis (Nelson et al. 1999; Budd et al. 2004; Wong and Dessen 2014). Considerable evidence of putative TEPs can also be identified in protozoa, such as Amoebozoa (e.g. *Acanthamoeba castellanii*), Alveolata (e.g. *Oxytricha trifallax*), green algae Viridiplantae (e.g. *Bathycoccus prasinos*) and in some Gram-positive bacteria (e.g. *Clostridium stercorarium*, *Halothermothrix orenii* and *Streptomyces purpurogeneiscleroticus*) (Fig. 6.1). All these proteins share an overall similar secondary structure and carry a consensus sequence for a thioester bond; however, more detailed investigation are required to reveal their extent and evolutionary relationship, and to assign them a biological role.

6.3 Structural Aspects of α_2Ms

6.3.1 Primary Structure, Sequence Motifs and Post-translational Modifications

α_2Ms are large macromolecules with 1474 amino acid residues in the human homolog, 1653 in *Escherichia coli*, and over 2000 in other putative sequences (Fig. 6.1) (Doan and Gettins 2008; Sottrup-Jensen et al. 1984b). Close inspection of the sequences reveals several motifs that are important for protein functionality. Among the most important and highly conserved is the motif that encodes the formation of the hyperactive thioester bond (C-X-E-Q: X for G or L) (Sottrup-Jensen et al. 1985). It is present in almost all the TEPs that have been characterized to date and will be further discussed below.

All primary structures start with a 17- (*E. coli* α_2M; hereafter ECAM) to 40-residue (chicken egg-white α_2M; hereafter ovostatin) peptide that is required for signalling protein translocation, either to the blood serum and hemolymph in vertebrates and invertebrates, or to the periplasm in Gram-negative bacteria (Nielsen et al. 1994; Doan and Gettins 2007; Sottrup-Jensen 1989). In some bacterial homologs, a "lipobox" consensus motif succeeds the signal peptide (in ECAM: L-A-G|C-D), which contains a cysteine that will become the N-terminus of the mature protein after post-translational addition of a diacylglycerol moiety to the side chain of the cysteine, and of a palmitoyl group through amide linkage to the N-terminus. In addition, the presence of an aspartic acid at position two, following the cysteine, indicates that the mature lipoprotein should be retained by the inner membrane, and is therefore localized in the periplasmic space (Doan and Gettins 2008; Pugsley 1993; Seydel et al. 1999).

In contrast to the bacterial homologs, eukaryotic α_2Ms are stabilized by an extensive network of disulphide bonds (Doan and Gettins 2008; Robert-Genthon et al. 2013; Wong and Dessen 2014). Tetrameric human α_2M and ovostatin contain twelve intra-chain disulphide bonds and two inter-chain bonds between adjacent monomers (Jensen and Sottrup-Jensen 1986; Marrero et al. 2012; Sottrup-Jensen et al. 1984b; Nielsen et al. 1994). Intra-chain disulphides are also present in monomeric

α_2ML1 from epidermis, and dimeric α_2Ms from *L. polyphemus*, which is further stabilized by an inter-chain bond between dimers (Galliano et al. 2006; Iwaki et al. 1996; Husted et al. 2002).

Extensive glycosylation is also found in eukaryotic homologs, with ovostatin binding 56 glucosamine molecules per subunit, with 12 of its glycosylation motifs (N-X-S/T) likely to be glycan-linked (Nielsen et al. 1994). Human α_2M has a total of 32 glycosylation sites per tetramer, with an average of 31 glycan-chains attached per molecule. These consist of mannose, galactose, N-acetyl-glucosamine, sialic acid and fucose residues (Dunn and Spiro 1967b). In *L. polyphemus*, out of seven potential glycosylation sites, six carry common glucosamine-based carbohydrate groups, whereas one carries a carbohydrate chain containing both glucosamine and galactosamine (Husted et al. 2002). Glycosides generally protrude from the surface of the structures, where they contribute to protein solubility and stability, and participate in interactions with other molecules (Marrero et al. 2012; Sottrup-Jensen et al. 1984b; Dunn and Spiro 1967a, b; Arnold et al. 2006; Goulas et al. 2014; Paiva et al. 2010). In bacterial α_2Ms, glycosylation has not been reported so far and is not expected, since this modification is almost exclusively restricted to eukaryotic organisms, with the exception of some Gram-negative bacteria, such as *Campylobacter jejuni* (Nothaft and Szymanski 2010; Wacker et al. 2002).

6.3.2 Secondary Structure Conservation and Domain Organization

The first structural details at atomic resolution were derived from individual α_2M domains of human and bovine origin, and from full-length models of their close homologs found in complement components (Doan and Gettins 2007; Jenner et al. 1998; Janssen et al. 2005; Huang et al. 2000). Human α_2M consists of 11 domains, corresponding to five structurally different moieties (Fig. 6.1) (domain terminology based on (Marrero et al. 2012; Garcia-Ferrer et al. 2015). The first seven domains, termed macroglobulin-like (MG) domains, MG1-MG7, are seven-stranded antiparallel β-sandwiches comprising a three- and a four-stranded sheet, and are arranged as a central scaffold, into which additional elements are inserted. These insertions result in domains that vary in length, spanning approximately 70 to 128 residues. Exceptionally, in bacterial homologs, there are two extra domains at the N-terminal, an MG0 and a NIE domain (from N-terminal domain of induced ECAM), with the former being a variant of MG domains with an extra inserted short β-strand (Garcia-Ferrer et al. 2015; Wong and Dessen 2014; Fyfe et al. 2015). In domain MG6, an irregularly folded bait-region domain (BRD) is inserted (66 to 126 residues). The BRD is stabilized by three helices, as well as through interactions with adjacent MG domains and the C-terminal receptor-binding domain (RBD). BRD is in an extended conformation and freely accessible to endopeptidases. Following MG7, an all-β CUB domain (C1r/C1s, Uegf, and Bmp1 found domain) is found adjacent to MG6 and it consists of two four-stranded antiparallel β-sheets unrelated to the MG fold.

Inserted within and placed below the CUB domain, the thioester domain (TED) features a ~310-residue helical domain with an α/α-toroid topology. It consists of six concentric α-hairpins, arranged as a sixfold α-propeller around a central axis. This architecture gives rise to a thick disc with two parallel flat sides, the entry and the exit side, which are shaped by the N- and C-termini of the ring of inner helices. In the native state of α_2M, a buried thioester bond is formed between the side chains of the cysteine and the glutamine/glutamate residues within the TED segment (C-G-E-Q). After TED, the polypeptide chain rejoins CUB, which leads to the C-terminal domain, RBD, also known as MG8. RBD features the third type of β-sandwich architecture found within α_2Ms, and forms a variant of the MG fold into which a β-α-β motif is inserted, thus giving rise to a four-stranded and a five-stranded twisted sheet. In the isolated structures of human and bovine RBD, a calcium-binding site is present in the loop region at one end of the β-sandwich (Huang et al. 2000; Jenner et al. 1998). Calcium binding principally affects this loop region and does not significantly perturb the stable core structure of the domain.

6.3.3 Tertiary Structure and Conformational Changes During Induction

6.3.3.1 Early Studies

Determination of the α_2M structure proved a long process, due to difficulties arising from sample heterogeneity owing to glycosylation, but also from the large size and intrinsic flexibility of the molecule (Andersen et al. 1991, 1994; Dolmer et al. 1995; Goulas et al. 2014). The first low resolution models appeared in 1968 and were obtained by negative-staining and cryo-electron microscopy (cryo-EM) (Andersen et al. 1995; Bloth et al. 1968). These models described human α_2M as graceful monograms of the letters "H", "I" and the Russian letter "Ж", which succeeded the preceding denominations "lip", "padlock", "cross", "four-petaled flower", and "eye" (Larquet et al. 1994). Technological advances and the combination of techniques, such as cryo-EM, combined with image processing using three-dimensional reconstructions and low-resolution X-ray models, provided a more accurate view of the markedly different shapes of the molecule, both in the native and induced forms. This contributed to rational and conclusive explanations of the mechanism of action (Kolodziej et al. 2002; Marrero et al. 2012; Garcia-Ferrer et al. 2015).

6.3.3.2 From Tetramers to Monomers and Vice-Versa

The α_2M orthologs studied to date have a molecular mass under denaturing conditions of 150–200 kDa, which gives rise in the native state to monomeric, dimeric or tetrameric arrangements of 165–800 kDa (Table 6.1) (Armstrong et al. 1991; Barrett et al. 1979; Doan and Gettins 2008; Enghild et al. 1989). Like most eukaryotic homologs, human α_2M is a homotetramer (Fig. 6.2C) made up of the non-covalent association of two dimers, a top and a bottom dimer, in an antiparallel or head-to-tail orientation. Each dimer consists of two monomers covalently linked by two symmetric inter-chain disulphide bonds. This structure was initially proposed from sequencing studies of human α_2M showing that the two subunits are linked in an antiparallel orientation by two disulphide bonds close to their N-termini, and subsequently validated by antibody-labelled cryo-EM models and X-ray structures (Kolodziej et al. 1996; Marrero et al. 2012; Qazi et al. 2000; Jensen and Sottrup-Jensen 1986). The tetramers may dissociate into dimers, either after limited reduction of the intra-chain disulphide bonds or by pH or salt-induced dissociation, but they maintain their overall architecture and the labile thioester bond unreacted. The dimers or even monomers are active towards peptidases and resemble the characteristics of the monomeric and dimeric homologs (Armstrong et al. 1991; Doan and Gettins 2007; Gonias and Pizzo 1983; Pochon et al. 1989; Shanbhag et al. 1996). However, reaction with peptidases or chemical induction renders the monomers or dimers prone to di- or tetramerisation, which relates with the observation that induced tetramers are more difficult to dissociate and tightly engaged (Barrett et al. 1979). Monomeric α_2Ms, including the bacterial homologs and rat α_1I3, act similarly by forming stable elongated dimers after induction, but only by specific peptidases (Fig. 6.2F, G) (Enghild et al. 1989; Garcia-Ferrer et al. 2015; Gonias and Pizzo 1983). Besides bait cleavage, in the bacterial homolog ECAM a second proteolytic cleavage at the interface between the MG0 and NIE domain is required before dimer formation (Garcia-Ferrer et al. 2015).

6.3.3.3 Conserved Multi-domain Structure

Superposition of induced monomers from both bacterial and human α_2Ms reveals that they conserve the overall shape (Fig. 6.2A, B). A detailed inspection, however, indicates that only a core consisting of domains MG2, MG3, MG6 and MG7 can be reasonably well-matched (Garcia-Ferrer et al. 2015). The first six MG domains (MG1-MG6) are arranged as a compact ellipsoidal one and a half-turn right-handed superhelix resembling a distorted “key ring”. Preceding this arrangement, two structurally flexible domains (MG0 and NIE) protrude in bacterial homologs, and possibly act as a membrane anchor for cell wall attachment. Inserted within MG6, BRD is framed by the key ring scaffold (Fig. 6.2D) and lines most of the vertical dimension of the formed cavity. It is stabilized by several interactions with MG, RBD and CUB domains, and partly defined in the Fourier density map, which indicates conserved intrinsic flexibility (Marrero et al. 2012; Wong and Dessen 2014; Garcia-Ferrer et al. 2015).

In induced α_2Ms, perpendicularly attached to MG3 and MG6, MG7 closes the MG-superhelix like a plug and forms a protruding structure in the upper limit of the molecule, which was initially attributed to the RBD domains (Qazi et al. 1999; Kolodziej et al. 2002; Marrero et al. 2012). MG7 leads to the hook of the molecule that includes the CUB, TED and RBD domains. The TED domain, inserted into

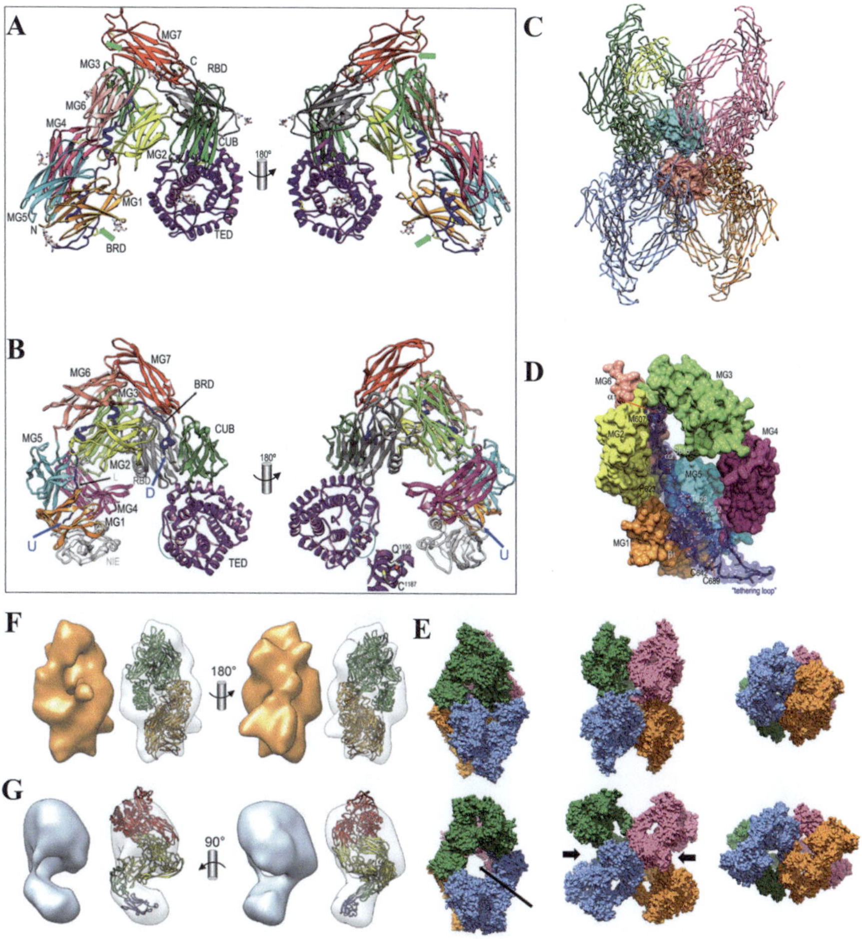

Fig. 6.2 Three-dimensional experimental structures of human and *E. coli* α_2-macroglobulin (hα_2M and ECAM, respectively). (**A**) Ribbon plot of a methylamine-induced hα_2M monomer in front view (convex face; left) and back view (concave face; right). *Green arrows* pinpoint the anchor points of the flexible bait region. (**B**) Ribbon plot of monomeric-induced ECAM in front and back views. The domain colours are as in A, and the visible polypeptide chain ends upstream (U) and downstream (D) of the cleaved bait region, as well as the thioester (*cyan ellipse*). (**C**) Possible representation of the hα_2M in H-view as a coil with two trapped molecules of HIV-1 peptidase tentatively modelled as *cyan* and *salmon pink solid surfaces*. Each monomer is represented in a separate colour. The RBD, which is only defined for monomer D, is further shown in *yellow*. (**D**) Detail of the methylamine-induced hα_2M structure depicting domains MG1-MG6 as rendered surfaces coloured according to A, and BRD as a *dark-blue ribbon* with *grey* side chains – some of which were truncated after C – superposed with its semi-transparent Connolly surface. Secondary structure elements of BRD are labelled, as are the tethering loop and selected residues for reference. (**E**) X-view, H-view and End-view of the experimentally induced hα_2M (*upper row*) and the native tentatively modelled tetramer (*lower row*). *Black arrows* pinpoint the large entrance to the central cavity. (**F**, **G**) Dimeric-induced ECAM and monomeric native ECAM single-particle cryo-EM reconstructions. The cryo-EM maps (*solid surface*) and the fitting of the respective atomic models (*semi-transparent surfaces*) are shown in two views (Figures are reproduced with permission from Garcia-Ferrer et al. 2015; Marrero et al. 2012)

CUB, in its native conformation, maintains its thioester bond buried against RBD, but also interacts with MG2 and CUB domains. The C-terminal domain, RBD, occupies a key position in the human α_2M structure and based on cryo-EM models of antibody-labelled native protein, it is buried between the ends of two dimers and forms a chisel-like feature (Qazi et al. 1999; Kolodziej et al. 2002). In the induced form, RBD interacts with TED and CUB domains, but is rather flexible and detected only in one of the four monomers of the tetramer (Fig. 6.2C). In bacterial homologs, it further interacts with the MG7 domain and after activation, it stabilizes the hook structure protruding from the MG-superhelix through contacts with domains MG2 and MG3 (Garcia-Ferrer et al. 2015; Wong and Dessen 2014; Marrero et al. 2012).

6.3.3.4 Apertures and Prey-Trapping Chambers

Access of peptidases to the bait region varies in monomeric and tetrameric forms. In monomers, the bait protrudes from the "key ring" scaffold, and is freely accessible to the bulk solvent and to peptidases (Garcia-Ferrer et al. 2015; Wong and Dessen 2014). In tetramers, initial hypotheses suggested that it is also externally oriented due to the limited access of the peptidases through the narrow openings of the central cavities (Kolodziej et al. 2002). However, biochemical and more detailed structural information showed that the bait resides within the tetramer (Marrero et al. 2012; Bowen and Gettins 1998). Modelling studies suggested that peptidases possibly enter native α_2M through two rather circular apertures of ~40 Å, at the interface between vicinal monomers on opposite sides of the tetramer, which may permit access to molecules of at least 20–25 kDa (Fig. 6.2E). These apertures would be framed symmetrically by MG1, MG2, and TED of each vicinal monomer and would be independently accessible, which explains how two peptidases can be caught and why small proteins like insulin or other nucleophiles can be bound during the process of complex formation (Marrero et al. 2012). Moreover, the propensity of native α_2M to dissociate into dimers at low protein concentration (10 μg/ml) or under particular biochemical conditions suggests that connectivity between the monomers is minimal, with the apertures probably being more relaxed and able to enlarge to permit better access to molecules (Boisset et al. 1996; Kolodziej et al. 2002).

After activation, several apertures are still present but with restricted size. In human α_2M, twelve entrances are found with some of them modulating the entrance from the bulk solvent by glycan chains acting like plugs (Schroeter et al. 1992; Marrero et al. 2012). The maximum dimensions of the apertures match the estimated size that would prevent non-covalently bound prey from escaping rapidly, whilst allowing small protein substrates or inhibitors to enter the central cavity (25–30 Å). In a similar manner, when monomeric α_2Ms dimerize after activation, a number of much larger entrances are created, which in turn explains why the ECAM protects only very high-molecular-mass proteins from digestion (Garcia-Ferrer et al. 2015; Doan and Gettins 2008).

In tetrameric α_2Ms, internalized peptidases are hosted in a central part of around 60 Å in diameter, termed the "prey chamber", which is delimited at the top and bottom by the segments of MG3 and MG4 that participate in disulphide-mediated dimerization (Marrero et al. 2012). The prey chamber is restricted at its centre, at half particle height, to a diameter of approximately 30 Å by loops provided by the four TEDs (the "narrowing belt"), and complemented by additional elongated volumes provided by the concave face of each of the four monomers ("substrate antechambers"). Overall, the prey chamber has space for up to two peptidases of 20–30 kDa, one above and one below the narrowing belt, that is, one for each disulphide-linked dimer (Fig. 6.2C). This is consistent with the maximal 2:1 stoichiometry of inhibition determined for peptidase binding by tetrameric α_2Ms (Barrett and Starkey 1973). Based on its size, the chamber could also accommodate a single, larger molecule of up to around 80–90 kDa, if the loops of the cavity-narrowing belt were folded back towards the inner wall of the cavity. In a similar manner, monomeric ECAM forms a central prey chamber of ~40 Å in diameter during dimerization, which may accommodate a single small-to-medium sized endopeptidase at best (Garcia-Ferrer et al. 2015).

6.3.3.5 Mechanisms of Peptidase Entrapment

Monomeric and tetrameric α_2Ms employ two approaches for trapping peptidases, named respectively as "the snap trap" and "Venus flytrap" mechanisms (Garcia-Ferrer et al. 2015; Marrero et al. 2012). In both cases, the principal trigger of activation is the peptidase attack on the bait region, which results in massive conformational change from the native to the induced form. This change can also be triggered by small amine molecules after direct reaction with the thioester (Barrett et al. 1979; Doan and Gettins 2008), but are restricted only in homologs with an active thioester bond and structural characteristics that permit the transition without bait cleavage (Garcia-Ferrer et al. 2015; Wong and Dessen 2014). Several structural studies of native (Kolodziej et al. 1996; Wong and Dessen 2014), methylamine- (Marrero et al. 2012; Qazi et al. 1999), trypsin-, chymotrypsin-, and plasmin-transformed α_2M (Kolodziej et al. 1998; Garcia-Ferrer et al. 2015) show how variations in the shape of the native and induced molecules are related to peptidase entrapment (Kolodziej et al. 2002).

Overall, tetrameric native α_2Ms are larger and broader than the induced molecules, with the conformational change causing strong lateral compression and vertical stretching of the native particle (Fig. 6.2E). This extraordinary change in the architecture functions to completely engulf one to two molecules of peptidases within the central cavity and irreversibly encapsulate them (Kolodziej et al. 2002). At the same time, the thioester bond is exposed to the solvent and rapidly reacts with exposed lysines extended from the surface of the peptidase. Trapped peptidases are still active, but unable to escape due to steric hindrance formed by the chambers or the covalent crosslinks with α_2M. The mechanism is so efficient that lack of thioester, as in ovostatin, does not compromise peptidase entrapment (Nagase and Harris 1983).

In the snap-trap mechanism, the conformational rearrangement in monomeric α_2M serves mainly to expose the thioester bond to the solvent and react with the attacking peptidase (Garcia-Ferrer et al. 2015; Enghild et al. 1989; Doan and Gettins 2008). In this way, α_2M nails peptidases, which thereafter are trapped and can no longer act freely. Also, the overall structure is more compact, and in some cases leads to dimerization, but without any obvious influence on inhibitory function. The efficiency of this mechanism relies only on the thioester's ability to act fast and crosslink attacking peptidases, and therefore, is much less efficient inhibitor than the tetrameric homologs.

6.3.4 Bait Region and Bait-Region Domain

The function of the BRD is to orient and expose the bait region in an extended conformation, so that attacking peptidases can easily approach and cut the chain (Marrero et al. 2012; Garcia-Ferrer et al. 2015). It is also the trigger for a massive conformational change, which leads to the entrapment of the attacking peptidase by α_2M. Changes in amino acid residues in the bait region can render molecules functional, but these changes might alter the specificity against non-cutting peptidases, such as tobacco-etch-virus peptidase (Van Rompaey et al. 1995; Garcia-Ferrer et al. 2015; Ikai et al. 1999). More radical modifications, such as removal of the entire BRD, give rise to α_2M with lack of functionality and greater propensity to digestion (Garcia-Ferrer et al. 2015; Gettins et al. 1995).

Typically, the bait region includes target peptidic bonds that are susceptible to attack by several endopeptidases, thus making α_2Ms ideal scavengers for the diverse array of enzymes produced by parasites or host that might attack during the lifetime of any given organism (Sottrup-Jensen et al. 1989; Sottrup-Jensen 1989). However, close inspection of the amino acid sequence shows limited conservation even in phylogenetically close species. Also, the eukaryotic bait domains and regions are much larger than the bacterial ones suggesting that this stretch may be under evolutionary pressure to provide reactivity to destructive peptidases of endogenous and exogenous origin, but restricts reactivity to essential endogenous peptidases of the tissue fluids (Garcia-Ferrer et al. 2015; Armstrong 2006).

6.3.5 The Reactive β-Cysteinyl γ-Glutamyl Thioester Bond

The thioester segment is a 15-atom thio-lactone ring composed of four residues (C-X-E-Q), located at the beginning of the first toroid helix of the TED on the domain entry face (Budd et al. 2004; Sottrup-Jensen 1989, 1994). The thiol-esterified glutamine/ate residue is encoded as glutamine, and the bond may be formed in reactions reminiscent of those of transglutaminases (Hall and Söderhall 1994). The intervening X (glycine or leucine) and glutamate residues are widely conserved across thioester-containing proteins (Armstrong and Quigley 1999; Doan and Gettins 2008).

Similar to the structures of complement and insect TEPs, in α_2M the thioester is stable against hydrolysis, even though it is near to the surface of the molecule. The surrounding aromatic and hydrophobic residues arising from the TED and RBD domains form a hydrophobic pocket that is conserved in bacterial and eukaryotic variants (Baxter et al. 2007; Janssen et al. 2005; Garcia-Ferrer et al. 2015; Marrero et al. 2012). Removal of the RBD domain, or even single mutation of those residues, results in a properly folded α_2M, but with an open thioester bond as was also observed with complement C3 (Janssen et al. 2005; Wong and Dessen 2014). In contrast, the proceeding domains from MG7 do not play any role in the formation and stability of the thioester bond at least in bacterial species, so expression of α_2M fragments, from MG7 to the RBD domain, results in a well-folded protein with a formed thioester bond (Garcia-Ferrer et al. 2015).

An asparagine residue in human α_2M (N1088) was also found to be important. It plays a multifunctional role from the formation of the thioester bond to its subsequent reaction with nucleophiles and the conformational change induced by hydrolysis (Suda et al. 1997). Hydrolysis most likely proceeds *via* direct attack of a nucleophile on the thioester. Nucleophiles can be lysines protruding from the surface of attacking peptidases, or small amines like methylamine, ethylamine and ammonia (Barrett et al. 1979). In contrast, in complement and insect TEPs, this residue is a histidine that participates in the hydrolysis of the thioester bond *via* an acyl-imidazole intermediate and leads to covalent crosslinks with hydroxyls (Fredslund et al. 2006; Janssen et al. 2005; Suda et al. 1997). In some cases, as in *L. polyphemus*, instead of reacting with peptidases, the thioester crosslinks the opposite chains of α_2M and does not encage the entrapped peptidase molecule, so binding is again entirely non-covalent (Dolmer et al. 1996; Quigley et al. 1991).

6.3.6 Receptor Recognition and Endocytosis

After induction by peptidases, human α_2M exposes the C-terminal RBD domain, which is recognized by the ligand-binding domains of low-density lipoprotein receptor-related protein (LRP1 or CD91). This is a 600 kDa endocytic membrane-bound receptor, an essential member of the low density lipoprotein receptor family. It is expressed in a broad spectrum of cell types as a single-chain precursor, which is processed into a 85 kDa transmembrane β-chain and an approximately 515 kDa α-chain, non-covalently associated with the extracellular part of the β-chain (Lillis et al. 2008). The α-chain contains four clusters of 2, 8, 10, and 11 ligand-binding domains, known as complement-like repeats (CR), flanked by epidermal-growth-factor-type repeats, and β-propeller modules (Andersen et al. 2000). A functional CR consists of approximately 40 residues, with three conserved disulphide linkages, and it coordinates one calcium ion (Daly et al. 1995). The rest of the amino acids are variable, yielding CR domains that can bind more than 30 ligands of different sizes and structures in addition to α_2M. They can all be replaced by receptor-associated protein (RAP), thus making LRP a multifunctional protein (Lillis et al. 2008).

α_2M binds with high-affinity to cluster 2 of LRP1, which contains eight CRs, and especially with CR3-4-5 domains rather than CR5-6-7 or the single repeats individually (Dolmer and Gettins 2006). Structural models of RBD in complex with CRs are not available, but biochemical studies evince that they are formed through interactions accomplished *via* a combination of charge-charge or polar contacts, together with significant hydrophobic interactionism in a similar manner as for RAP/CR complexes (Jensen et al. 2006). RBD exposes lysines (K1393 and K1397) as well as upstream hydrophobic residues, yielding a conserved motif (ΨKXΨK; Ψ is hydrophobic and X any amino acid residue) (Jensen et al. 2006; Nielsen et al. 1996). Mutation of the lysines or lack of them, as in chicken homolog ovostatin, is deleterious and results in complete failure of complex formation (Nielsen et al. 1996). In bacterial homologs, besides the overall structural similarity of RBD with the human ones, the α-helices are distorted and do not expose any lysine, thus lacking any possibility for interaction with LRP1 (Garcia-Ferrer et al. 2015).

Ligands bound extracellularly by LRP1 at neutral pH are rapidly internalized (the half-life of the complexes is 2–5 min) and then released in the endosomes (pH 6), leading to their subsequent lysosomal degradation. Dissociation of ligands is crucial for receptor recycling and proper function (Lillis et al. 2008). Complexes that are formed *in vivo* are not influenced by pH variations, therefore they would remain uninfluenced during the process of endocytosis. There is a mechanism of dissociation at acidic pH that involves the protonation of a histidine residue, which serves as a switch, allowing tighter binding of the neighbouring β-propeller domain than the binding ligands, which consequently leads to LRP1 discharging from bound molecules (Dolmer and Gettins 2006; Rudenko et al. 2002).

6.4 Role of α_2Ms in the Control of Proteolytic Activity

The best characterized function of the α_2M homologs is peptidase neutralization by a three-step process involving binding, inhibition and marking for endocytosis and intracellular degradation (Armstrong 2010). This process resembles the opsonization found in the homologous complement components and insect TEPs, which are covalently bound to the surfaces of foreign cells and marked for immune destruction (Le et al. 2012; Janssen et al. 2005). α_2M is an abundant protein in human plasma at concentrations ranging from 2 to 4 mg/ml, and is the second most abundant protein in the hemolymph of the cephalopod sepia, and the third most abundant in *L. polyphemus* (Enghild et al. 1990; Vanhoorelbeke et al. 1993; Sottrup-Jensen 1989). The concentration may vary with age and acute phase responses such as trauma, infection and inflammation, with more than one homolog potentially present in each organism and in the same or different organs. In humans, different α_2Ms can be found in blood serum, epidermis and pregnancy serum (Sand et al. 1985; Galliano et al. 2006; Barrett et al. 1979; Tunstall et al. 1975; Cray et al. 2009). In bacteria, and more precisely in *E. coli*, threefold overexpression of ECAM has been observed in the membranes under anaerobic conditions, but there is no further information available (Brokx et al. 2004).

In contrast to most inhibitors that selectively inhibit peptidases of a specific catalytic mechanism, α_2M targets all peptidases, regardless of the catalytic class or family (serine, cysteine, metallo, aspartic) (Kantyka et al. 2010; Hibbetts et al. 1999). In this case, specificity is characterized mainly by the accesses that α_2Ms assign to the peptidases to approach and cut the bait region, and to initiate the process of entrapment. As mentioned above (in the section Apertures and prey-trapping chambers), peptidases access the bait region of tetrameric α_2M through entrances of limited size. In some instances, they are too large to fit and to enter the cavity where the bait is positioned, and they fail to react with α_2M, as the case of the collagenase of *Clostridium perfringens* that has a molecular mass larger than 100 kDa (Abe et al. 1989). Moreover, the amino acid sequence of the bait region plays a central role in the universality of α_2Ms, and several peptidases fail to cut within it due to lack of the necessary sequences (e.g. the lysyl-specific endopeptidase from *Porphyromonas gingivalis* Kgp and *Achromobacter lyticus*) (Ikai et al. 1999; Gron et al. 1997). Non-reacting peptidases confer a type of immunity towards α_2M that can be of vital importance to the physiology of the organisms. For example, the clotting enzymes of *L. polyphemus* are not inhibited by endogenous α_2M. This is of physiological importance during blood clotting (Armstrong et al. 1984). However, the more exposed bait region of the monomeric and dimeric homologs broadens their inhibitory range towards peptidases of variable size. In this case, peptidase immunity is based on the presence of a well-positioned lysine residue on the surface that can be easily crosslinked by the thioester bond. In addition, the more open character of these α_2Ms permits the peptidases to have better access to larger peptides than in the case of tetramers, where they have to pass through narrow entrances to be trapped (Garcia-Ferrer et al. 2015; Marrero et al. 2012; Galliano et al. 2006; Enghild et al. 1989).

α_2Ms control several essential proteolytic events that take place in all living organisms ranging from digestion of complex molecules for the assimilation and physiological turnover of proteins, to activation of peptide hormones and effector proteins secreted as zymogens, blood clotting and clot resolution, inflammation and remodelling of the extracellular matrix (Lopez-Otin and Bond 2008). However, under some circumstances, endogenous peptidases have the potential for undesirable destructive action, and they should not survive to act beyond the limits of their intended function. The primary aim of human α_2M is rapid inhibition of excess peptidases released during tissue injury, such as those liberated by neutrophils at the site of inflammation. During phagocytosis and neutrophil turnover, neutrophil elastase, proteinase 3 and cathepsin G are released into the extracellular space as active peptidases, but tightly regulated in the extracellular and pericellular space to avoid degradation of connective tissue proteins including elastin, collagen, and proteoglycans (Korkmaz et al. 2010). Human chymase, a chymotrypsin-like peptidase, activates many important biological mediators, such as angiotensin, interleukin-1-β (IL-1-β), big endothelin, and interstitial collagenase with a potential role in hypertension and atherosclerosis. It is stored in mast cell secretory granules and its activity is potentially controlled shortly after exocytosis by several inhibitors, including α_2M (Raymond et al. 2009; Walter et al. 1999). In turn, ADAMTSs (a disintegrin and metalloproteinase with thrombospondin motifs) are peptidases with 19 paralogs in humans, which participate in many functions including processing of pro-collagen, remodelling of extracellular matrix, and organization of connective tissue, coagulation and inflammation (Kelwick et al. 2015). ADAMTSs are controlled by α_2Ms; uncontrolled activity results in pathologies including arthritis and cancer (Luan et al. 2008; Tortorella et al. 2004). α_2Ms also play an important role in hemostasis, by controlling peptidases that participate in blood-coagulation and fibrinolysis. These include factor VIIa, thrombin and plasmin (Petersen et al. 2009; Kremers et al. 2013; Schaller and Gerber 2011). α_2Ms can also function as master inhibitors of cartilage-degrading factors (Wang et al. 2014), and as inhibitors of both the esterolytic and kinin-producing activities of plasma kallikrein (Harpel 1970).

α_2M is also involved in attenuating infection by pathogens and parasite growth in various species by counteracting their peptidases. Secreted peptidases must be neutralized to restrict the steps of invasion, nutrition of parasites, and parasite-mediated inactivation of elements of host immunity (Armstrong 2010). α_2M inhibits several peptidases from bacteria, such as neutral peptidase from *Fusiformis nodosus* (causing ovine foot-root) (Merritt et al. 1971; Werb et al. 1973), collagenase from *Clostridium histolyticus* (causing gangrene) (Giroux and Vargaftig 1978), alkaline peptidases from *Bacillus subtilis* (Dolovich and Wicher 1971), peptidases from *Pseudomonas aeruginosa* and *Serratia marcescens* (causing agents of corneal lesion ulceration and septic shock) (Horvat et al. 1989; Molla et al. 1987; Khan et al. 1995), immune inhibitor A metallopeptidase from *Bacillus anthracis* (causing agent of anthrax) (Arolas et al. 2016), several gingipains from *Porphyromonas gingivalis* involved in periodontitis and arthritis (Gron et al. 1997), the enterotoxin fragilysin from *Bacteroides fragilis* (agent of diarrhoea) (Moncrief et al. 1995), and many other peptidases from important parasites (Borth 1992; Scharfstein 2006). It also protects from injected peptidases, such as the hemorrhagic and collagenolytic snake-venom metallopeptidase from *Bothrops jararaca* and *Crotalus atrox* venom (Anai et al. 1998; Werb et al. 1973). Indirectly, it may also function by delivering foreign proteins to macrophages and enhancing the ability of mammals to produce anti-peptidase antibodies, for example, against the major cysteine peptidase of *Trypanosoma cruzi* (agent of Chagas disease) (Morrot et al. 1997).

The role of α_2Ms in bacteria is still largely unknown. They are found in both pathogenically invasive and saprophytically colonizing species, which suggests that they are involved in colonization rather than virulence (Budd et al. 2004). *In vitro* studies with *E. coli* do not discriminate between ECAM knockout mutants and wild-type strains. Only after exposure to peptidases at physiological concentrations, a decrease in the mutant survivability can be observed. It seems that ECAM protects membrane-attached proteins and other cell components from host peptidases (Garcia-Ferrer et al. 2015). Moreover, in the same operon with ECAM, a penicillin binding protein (Pbp1C) co-occurs that functions as peptidoglycan transglycosylase for cell wall biosynthesis (Budd et al. 2004). Therefore, these two proteins may be coupled as a periplasmic defence and repair system, acting in a concerted fashion. In addition, in *P. aeruginosa*, *E. coli* and several other strains, a second α_2M homolog (MagD) is present that forms multi-molecular complexes with proteins from co-transcribed genes. Together, they may be involved in bacterial pathogenicity and/or defence against the host immune system, but this remains undefined to date (Robert-Genthon et al. 2013).

6.5 Other Physiological Functions

The role of α_2M is not restricted to peptidase binding and inhibition but is extended to interactions with several other endogenous and exogenous proteins, with important physiological implications (Table 6.2 and references therein). To date, the nature of these interactions were studied biochemically, due to difficulties in the crystallization of α_2M to obtain high-resolution X-ray models. Consequently, the exact interacting surfaces are still unknown. They can be observed in native (e.g. interleukin-6; IL-6) and induced α_2M (e.g. nerve-growth factor-β; NGF-β) after the exposure of cryptic sites by reactions with peptidases (e.g. plasmin or highly reactive amine), and can be located on the surface (e.g. β-amyloid peptide-binding plasma protein; Aβ) or inside the prey chamber (e.g. transforming growth factor-β; TGF-β) (Mettenburg et al. 2002; Webb et al. 1998; Liebl and Koo 1993; Matsuda et al. 1989). So far, hydrophobic, ionic and covalent interactions are reported. TGF-β and other growth factors, cytokines and hormones may form complexes with α_2M *via* a common mechanism involving interactions between exposed tryptophan and/or other hydrophobic residues, and a hydrophobic region in induced α_2M (Webb et al. 1998). Also, after induction, the highly reactive cysteine and glutamine/ate residues of the thioester can form disulphide and/or thioester-based covalent bonds with molecules in close proximity. The presence of the molecules inside the α_2M chamber during induction may be required for trapping, as in the case of insulin that is only crosslinked by the thioester when it concurs with trypsin or elastase (Chu et al. 1991). However, disulphide-based interactions can also be formed at a later stage, after induction, if the molecules are small enough to pass through the narrowed entrances of the induced α_2M, and to reach free cysteines arising from thioester hydrolysis. In many cases, molecules from the same or even different class compete for the same binding site on α_2M, which interferes with complex formation. For example, neurotrophins can be displaced by NGF-β, and heparin interacts directly with TGF-β, thus liberating it from the complex with α_2M and restoring its activity (McCaffrey et al. 1989; Wolf and Gonias 1994). All these binding partners probably share a common mechanism that is dependent on local topological structures rather than amino acid sequences, since most of these ligands lack sequence homology. Besides the amino acid sequence, posttranslational modifications such as glycosylation participate in direct interactions with lectin-like binding proteins in both native and induced α_2M (Arnold et al. 2006).

Table 6.2 α_2M interactions with other molecules and implications

Type	Name	α_2M form/Interaction	Implications of the complex	References
Interleukin and Cytokines	IL-1-β	Induced. Disulphide-based. Promoted by Zn^{++}.	Biological activity is retained, and α_2M may act as a spatial and temporal regulator of IL-1β.	Borth et al. (1990) and Lindroos et al. (1995)
	IL-6	Native. No competence from IL-l-α, IL-10 and IL-2.	Biological activity is retained, and α_2M may act as a carrier protein in the circulation. Protection against peptidases at inflammatory sites.	Matsuda et al. (1989)
	IL-8	Induced. Non-covalent.	α_2M does not inhibit IL-8-induced neutrophil degranulation or chemotaxis and protects it from peptidase degradation.	Kurdowska et al. (1997)
	Leptin	Induced. Hydrophobic-based.	Binding of leptin to induced α_2M and its rapid clearance by the α_2M receptor may significantly influence the bioavailability of leptin in human plasma.	Birkenmeier et al. (1998)
	Tumour necrosis factor-α (TNF-α)	Induced by plasmin or methylamine. Non-covalent.	TNF-α, preincubated with either α_2M-plasmin or α_2M-methylamine, remained a potent necrogen for cultured L929 cells. The complex can be removed from the circulation by the α_2M-receptor pathway.	Wollenberg et al. (1991)
Hormones	Hepcidin	Both	The α_2M-hepcidin complex decreased ferroportin expression in J774 cells more effectively than hepcidin alone.	Peslova et al. (2009)
	Inhibin, activin, follistatin	Mainly induced, except activin that binds efficiently in native form. Disulphide-based.	All the complexes may be removed from the circulation by the α_2M-receptor pathway, but activin can be maintained in the circulation by binding native α_2M.	Phillips et al. (1997) and Niemuller et al. (1995)
	Insulin	Induced. Thioester-based. Only if insulin is present during peptidase induction.	Implications in the regulation of carbohydrate and fat metabolism.	Chu et al. (1991)

Growth factors	Basic fibroblast growth factor (βFGF)	Induced. Disulphide-based. αFGF and TGF-β compete for binding to α_2M, whereas PDGF does not.	The complex does not bind to low affinity FGF binding sites and binds poorly to high affinity βFGF binding sites on BHK-21 cells. In addition, it has decreased ability to stimulate plasminogen activator production in bovine capillary epithelial cells.	Dennis et al. (1989)
	Growth hormone (GH)	Induced. Non-covalent.	GH plays an important role in the control of growth.	Kratzsch et al. (1995)
	Transforming growth factor (TGF-$\beta_{1,2}$)	Mainly induced, except TGF-β_2 that binds both. Interactions between topologically exposed tryptophans and/or other hydrophobic residues and a hydrophobic region in α_2M.	Complexes neutralize the activity in fetal bovine heart endothelial cell proliferation assays and can be cleared from the circulation. The complex can be dissociated in the presence of heparin and recover TGF-β activity.	Feige et al. (1996), LaMarre et al. (1991a), Liu et al. (2001), Webb et al. (1998), and McCaffrey et al. (1989)
	Neurotrophins (NT-3, NT-4), nerve growth factor-β (NGF-β), brain-derived neurotrophic factor (BDNF), and ciliary neurotrophic factor (CNTF)	Mainly induced. Non-covalent. NGF-β exhibits the strongest interaction.	Protection of NGF-β from proteolysis. Binding of α_2M to NGF-β and its precursor pro-NGF causes neurodegeneration in a p75(NTR)- and pro-NGF-dependent manner.	Wolf and Gonias (1994), Liebl and Koo (1993), Barcelona and Saragovi (2015), and Ronne et al. (1979)
	Vascular endothelial growth factor (VEGF)	In the interior of induced and the exterior of native α_2M. Binds in a different site than TGF-β.	α_2M does not impact the ability of VEGF to induce cell proliferation or upregulate Ca^{++}.	Bhattacharjee et al. (2000) and Soker et al. (1993)
	Epidermal growth factor (EGF)	Induced. Thioester-based.	~	Gettins and Crews (1993)
	Platelet-derived growth factors (PDGF)	Both. Disulphide-based. All the isoforms bind to the same position.	The reversible nature of the PDGF/α_2M complex could allow for targeted PDGF release near mesenchymal cells that possess PDGF receptors.	Bonner et al. (1991)
	Insulin-like growth factor biding proteins (IGFBP1/2)	~	The decreased concentration of the complexes in the circulation provides a greater amount of free IGFBP-2.	Sunderic et al. (2015)

(continued)

Table 6.2 (continued)

Type	Name	α_2M form/Interaction	Implications of the complex	References
Other	Carboxypeptidase B (CPB)	Native. Non-covalent.	Plasma CPB activity is not significantly affected by binding to α_2M. The complex may function as a "shuttle" to modulate the clearance of CPB from circulation.	Valnickova et al. (1996)
	Mannan binding lectin (MBL)	Both. Direct binding of oligomannose glycans of α_2M to the lectin domains of MBL.	The lectin interaction may represent a mechanism for the localization of α_2M to the MBL/MBL-associated serine proteases complex for the inhibition of spontaneously or inappropriately activated serine peptidases in the serum.	Arnold et al. (2006)
	Apolipoprotein E (ApoE)	Mainly induced. Non-covalent. Fats, sphingomyelin, very low-density lipoprotein, PDGF and Aβ inhibit complex formation.	ApoE plays a central role in the transport of lipids among different organs and cell types.	Krimbou et al. (1998)
	Eosinophil cationic protein (ECP)	Induced. Non-covalent.	The interaction with α_2M may reflect a mechanism by which the organism protects itself against the deleterious effects of the highly cytotoxic protein ECP.	Petersen and Venge (1987) and Bystrom et al. (2011)
	Tartarate-resistant acid phosphatase (TRAP)	Both. Ionic interactions	Complex affects both enzyme activity and immunoreactivity. α_2M may modulate clearance of TRAP from the circulation and possibly from areas of inflammation.	Brehme et al. (1999)
	β_2-Microglobulin	Native	Putative participation of α_2M in modulating β_2-microlgobulin metabolism.	Gouin-Charnet et al. (2000)
	Defensin HNP-1	Induced. Disulphide-based.	α_2M may function as a scavenger of defensins and other peptide mediators in inflamed tissues, and may act as a mechanism for the regulation of inflammation.	Panyutich and Ganz (1991)
	Myelin basic protein (MBP)	Both	Complex protects myelin from proteolysis and may protect extracellular compartments from immunogenic myelin fragments. α_2M mediates clearance form circulation.	Gunnarsson and Jensen (1998)
	β-amyloid peptide-binding plasma protein (Aβ)	Induced. Binding to amino acid sequence (1314–1365) corresponding to α_2M receptor domain.	α_2M and its receptor, low density lipoprotein receptor-related protein, function together to facilitate the cellular uptake and degradation of Aβ.	Mettenburg et al. (2002) and Du et al. (1997)

	Osteogenic growth peptide (OGP)	Both	Native α_2M enhances the immediate availability of OGP to its target cells. Induced α_2M may participate in the removal of OGP from the system.	Gavish et al. (1997)
	Lipoprotein lipase (LPL)	Native. Non-covalent.	~	Vilella et al. (1994)
	CpG oligodeoxy-nucleotides	Induced	Complex enhances the immunostimulatory properties significantly, and protects it from degradation by nucleases.	Anderson et al. (2008)
	Chaperonine activity	Both forms and hypochloride-induced.	α_2M is a specialized chaperone that prevents the extracellular accumulation of mis-folded and pathogenic proteins, particularly during innate immune activity.	Wyatt et al. (2013), (2014)
	Interacts with unfolded proteins			
	Limulin	Induced	α_2M inhibits limulin, a sialic acid-binding lectin that is the mediator of hemolysis in the plasma of the horseshoe crab *Limulus polyphemus*.	Swarnakar et al. (2000)
	Protein G related α_2M binding (GRAB)	Native	It is a virulence factor that traps α_2M and regulates proteolytic activity at the bacterial surface and the host-microbe relation during *S. pyogenes* infections.	Rasmussen et al. (1999)
	Protein G	Native	Protein involved in binding of Group G streptococcal strains with α_2M. It is a virulence factor.	Müller and Rantamäki (1995)
	Ricin A-chain	Induced. Disulphide-based.	Ricin is a highly toxic, naturally occurring lectin produced in the seeds of the oil plant, *Ricinus communis*. Complex formation with α_2M decreases the activity of ricin.	Ghetie et al. (1991)
	α_2M from *Pseudomonas aeruginosa* (MagD)	Native	It forms a complex with three other molecular partners encoded by the same operon, MagA, MagB and MagF. It may play a role in virulence and/or bacterial defence.	Robert-Genthon et al. (2013)

By regulating the distribution and activity of many cytokines, hormones, growth factors and other proteins, α_2M modulates cell proliferation and cell survival pathways. In complex, molecules may retain or lose their function, and α_2M may acquire new characteristics with beneficial (e.g. more effective hepcidin or inhibition of eosinophil cationic protein) or detrimental (e.g. inhibition of NGF-β) implications for physiology (Peslova et al. 2009; Barcelona and Saragovi 2015; Bystrom et al. 2011). Non-covalent bound cytokines can be displaced by heparin or specific cytokine receptors and released during tissue injury and inflammation (LaMarre et al. 1991b). Moreover, induced α_2M serves as a system for balancing or preventing the harmful systemic or local effects of excess cytokines, such as TGF-β and tumour necrosis factor-α (TNF-α), and of toxins such as ricin-A, by removing them from the circulation in a concerted system with LRP1 (LaMarre et al. 1991a; Ghetie et al. 1991). In many cases, α_2M protects the bound molecule from lysis by peptidases or nucleases due to steric hindrances (Anderson et al. 2008; Matsuda et al. 1989). Also, it may serve as a chaperone, binding misfolded potential pathogenic proteins and preventing their extracellular accumulation during stress conditions, and particularly during innate immune system activity (Wyatt et al. 2013, 2014). α_2M influences protein function not only through complex formation, but also by direct modifications through enzymatic reactions. Uniquely, human α_2M exhibits esterase activity that acts specifically on ghrelin, a major hormone secreted from endocrine cells in the stomach that is implicated in feelings of satiety and in the body's reward system (Eubanks et al. 2011).

Besides its physiological role in the human body, α_2M is exploited by bacteria for their own wealth. Through two mechanisms, involving protein G and protein G-related α_2M-binding protein (GRAB), bacteria bind α_2M on their membrane surface, thereby protecting important virulence determinants from host-attacking peptidases and proteolytic degradation. This regulation of proteolytic activity on the bacterial surface can affect the host-microbe relation during *Streptomyces pyogenes* infections (Rasmussen et al. 1999; Müller and Rantamäki 1995). Early observations suggest that α_2Ms of bacterial origin interact with several other proteins that are potentially involved in virulence or defence, however, the exact function is still unknown (Robert-Genthon et al. 2013; Budd et al. 2004).

6.6 Conclusions

α_2Ms represent an evolutionary conserved arm of the innate immune system that regulates the distribution and activity of many proteins, including peptidases, cytokines, hormones and many other physiological effectors (Armstrong and Quigley 1999). It is a sophisticated means to spatially and temporally restrict and regulate key physiological processes. Due to its importance in physiology, several efforts were made to understand the mechanism of action *in vivo* and *in vitro* at biochemical and structural levels. However, further information is required to obtain a detailed picture of how it interacts with proteins. Clear knowledge of the molecule and its interactions at molecular level may allow the rational design of adjuvants, and may also lead to the development of new therapeutic agents against infectious diseases.

Through simple sequence similarity searches, thousands of putative sequences can be identified from both α_2M-type inhibitors and complement proteins. The massive information available, however, still needs to be filtered and verified and more detailed studies at protein level must to be completed to reveal the real extent and functional role of α_2Ms in several organisms. It would not be surprising to identify roles and activities other than peptidase inhibition that involve participation in more complex systems and networks for either housekeeping or protective functions against external attacks. In addition to mammalian homologs, the more recently identified bacterial α_2Ms seem to interact with many other proteins forming macromolecular complexes, which are released after proteolytic attack (Robert-Genthon et al. 2013; Budd et al. 2004). They could serve, therefore, as sensors and signal transmitters for the presence of potentially damaging endogenous or exogenous peptidases, leading to the initiation or pausing of important functions. This could be beneficial to the organism as a fast response to environment changes during parasitism or mutualism.

Acknowledgements This study was funded in part by grants from European, Spanish, and Catalan agencies (FP7-PEOPLE-2011-ITN-290246 "RAPID"; FP7-HEALTH-2012-306029-2 "TRIGGER"; BFU2015-64487-R; BIO2013-49320-EXP; MDM-2014-0435; 2014SGR9). TG acknowledges "Juan de la Cierva" research contracts (JCI-2012-13573) from the Spanish Ministry for Economy and Competitiveness. The Structural Biology Unit of IBMB is a "María de Maeztu" Unit of Excellence of the Ministry of Economy and Competitiveness.

References

Abe K, Yamamoto K, Sinohara H (1989) Proteinase inhibitory spectrum of mouse murinoglobulin and alpha-macroglobulin. J Biochem 106:564–568

Anai K, Sugiki M, Yoshida E, Maruyama M (1998) Inhibition of a snake venom hemorrhagic metalloproteinase by human and rat alpha-macroglobulins. Toxicon 36(8):1127–1139

Andersen GR, Jacobsen L, Thirup S, Nyborg J, Sottrup-Jensen L (1991) Crystallization and preliminary X-ray analysis of methylamine-treated alpha-2-macroglobulin and 3 alpha-2-macroglobulin-proteinase complexes. FEBS Lett 292:267–270

Andersen GR, Koch T, Sørensen AH, Thirup S, Nyborg J, Dolmer K, Linda J, Sottrup-Jensen L (1994) Crystallisation of proteins of the alpha-2-macroglobulin superfamily. Ann N Y Acad Sci 737:444–446

Andersen GR, Koch T, Dolmer K, Sottrup-Jensen L, Nyborg J (1995) Low resolution X-ray structure of human methylamine-treated alpha-2-macroglobulin. J Biol Chem 270(42):25133–25141

Andersen OM, Christensen LL, Christensen PA, Sorensen ES, Jacobsen C, Moestrup SK, Etzerodt M, Thogersen HC (2000) Identification of the minimal functional unit in the low density lipoprotein receptor-related protein for binding the receptor-associated protein (RAP). A conserved acidic residue in the complement-type repeats is important for recognition of RAP. J Biol Chem 275(28):21017–21024

Anderson RB, Cianciolo GJ, Kennedy MN, Pizzo SV (2008) Alpha-2-macroglobulin binds CpG oligodeoxynucleotides and enhances their immunostimulatory properties by a receptor-dependent mechanism. J Leukoc Biol 83(2):381–392

Andus T, Gross V, Tran-Thi TA, Schreiber G, Nagashima M, Heinrich PC (1983) The biosynthesis of acute-phase proteins in primary cultures of rat hepatocytes. Eur J Biochem 133:561–571

Armstrong PB (2006) Proteases and protease inhibitors: a balance of activities in host-pathogen interaction. Immunobiology 211(4):263–281

Armstrong PB (2010) Role of apha-2-macroglobulin in the immune responses of invertebrates. Invertebr Surviv J 7:165–180

Armstrong PB, Quigley JP (1999) alpha-2-macroglobulin: an evolutionarily conserved arm of the innate immune system. Dev Comp Immunol 23:375–390

Armstrong PB, Levin J, Quigley JP (1984) Role of endogenous proteinase inhibitors in the regulation of the blood clotting system of the horseshoe crab, Limulus polyphemus. Thromb Haemost 52(2):117–120

Armstrong PB, Rossner MT, Quigley JP (1985) An alpha-2-macroglobulinlike activity in the blood of chelicerate and mandibulate arthropods. J Exp Zool 236(1):1–9

Armstrong PB, Mangel WF, Wall JS, Hainfield JF, Van Holde KE, Ikai A, Quigley JP (1991) Structure of alpha-2-macrolgobulin from the arthropod Limulus polyphemus. J Biol Chem 266(4):2526–2530

Arnold JN, Wallis R, Willis AC, Harvey DJ, Royle L, Dwek RA, Rudd PM, Sim RB (2006) Interaction of mannan binding lectin with alpha-2-macroglobulin via exposed oligomannose glycans: a conserved feature of the thiol ester protein family? J Biol Chem 281(11):6955–6963

Arolas JL, Goulas T, Pomerantsev AP, Leppla SH, Gomis-Ruth FX (2016) Structural basis for latency and function of immune inhibitor A metallopeptidase, a modulator of the Bacillus anthracis secretome. Structure 24(1):25–36

Banbula A, Chang LS, Beyer WF, Bohra CL, Cianciolo GJ, Pizzo SV (2005) The properties of rabbit alpha-1-macroglobulin upon activation are distinct from those of rabbit and human alpha-2-macroglobulin. J Biochem 138(5):527–537

Barcelona PF, Saragovi HU (2015) A pro-nerve growth factor (proNGF) and NGF binding protein, alpha-2-macroglobulin, differentially regulates p75 and TrkA receptors and is relevant to neurodegeneration ex vivo and in vivo. Mol Cell Biol 35(19):3396–3408

Barrett AJ, Starkey PM (1973) The interaction of alpha-2-macrolgobulin with proteinases. Biochem J 133:709–724

Barrett AJ, Brown MA, Sayers CA (1979) The electrophoretically 'slow' and 'fast' forms of the alpha-2-macroglobulin molecule. Biochem J 181:401–418

Baxter RHG, Chang C-I, Chelliah Y, Blandin S, Levashina EA, Deisenhofer J (2007) Structural basis for conserved complement factor-like function in the antimalarial protein TEP1. Proc Natl Acad Sci U S A 104(28):11615–11620

Bender RC, Bayne CJ (1996) Purification and characterisation of the tetrameric alpha-macroglobulin proteinase inhibitor from the gastropod mollusc Biomphalaria glabrata. Biochem J 316:893–900

Bhattacharjee G, Aspilin IR, Wu SM, Gawdi G, Pizzo SV (2000) The conformational-dependent interaction of alpha-2-macroglobulin with vascular endothelial growth factor: a novel mechanism of alpha-2-macroglobulin/growth factor binding. J Biol Chem 275(35):26806–26811

Birkenmeier G, Kämpfer I, Kratzsch J, Schellenberger W (1998) Human leptin forms complexes with alpha-2-macroglobulin which are recognized by the alpha-2-macroglobulin receptor/low density lipoprotein receptor-related protein. Eur J Endocrinol 139:224–230

Bloth B, Chesebro B, Svehag SE (1968) Ultrastructural studies of human and rabbit alpha-M-globulins. J Exp Med 127:749–762

Boisset N, Taveau J-C, Pochon F, Lamy J (1996) Similar architectures of native and transformed human alpha-2-macroglobulin suggest the transformation mechanism. J Biol Chem 271(42):25762–25769

Bonner JC, Goodell AL, Lasky JA, Hoffman (1991) Reversible binding of platelet-derived growth factor-AA, AB, and BB isoforms to a similar site on the "slow" and "fast" conformations of alpha-2-macroglobulin. J Biol Chem 267(18):12837–12844
Borth W (1992) alpha-2-Macroglobulin, a multifunctional bidning protein with targeting characteristics. FASEB J 6:3345–3353
Borth W, Scheer B, Urbansky A, Luger TA, Sottrup-Jensen L (1990) Binding of IL-1beta to alpha-2-macroglobulin and release by thioredoxin. J Immunol 145(11):3747–3754
Bowen ME, Gettins PG (1998) Bait region involvment in the dimer-dimer interface of human alpha-2-macroglobulin and in mediating gross conformational change. J Biol Chem 273(3):1825–1831
Brehme CS, Roman S, Shaffer J, Wolfert R (1999) Tartrate-resistant acid phosphatase forms complexes with alpha-2-macrolgobulin in serum. J Bone Miner Res 14:311–318
Brokx SJ, Ellison M, Locke T, Bottorff D, Frost L, Weiner JH (2004) Genome-wide analysis of lipoprotein expression in Escherichia coli MG1655. J Bacteriol 186(10):3254–3258
Budd A, Blandin S, Levashina E, Gibson JT (2004) Bacterial alpha-2-macroglobulins: colonization factors acquired by horizontal gene transfer from the metazoan genome? Genome Biol 5:R38
Buresova V, Hajdusek O, Franta Z, Sojka D, Kopacek P (2009) IrAM-An alpha-2-macroglobulin from the hard tick Ixodes ricinus: characterization and function in phagocytosis of a potential pathogen Chryseobacterium indologenes. Dev Comp Immunol 33(4):489–498
Bystrom J, Amin K, Bishop-Bailey D (2011) Analysing the eosinophil cationic protein-a clue to the function of the eosinophil granulocyte. Respir Res 12:10
Chu CT, Rubenstein DS, Enghild JJ, Pizzo SV (1991) Mechanism of insulin incorporation into alpha 2-macroglobulin: implications for the study of peptide and growth factor binding. Biochemistry 30(6):1551–1560
Cohn E, Strong L, Hughes W, Mulford D, Ashworth J, Melin M, Taylor H (1946) Preparation and properties of serum and plasma proteins. IV. A system for the separation into fractions of the protein and lipoprotein components of biological tissues and fluids. J Am Chem Soc 68(3):459–475
Cray C, Zaias J, Altman NH (2009) Acute phase response in animals: a review. Comp Med 59(6):517–526
Daly NL, Scanlon MJ, Djordjevic JT, Kroon PA, Smith R (1995) Three-dimensional structure of a cysteine-rich repeat from the low-density lipoprotein receptor. Proc Natl Acad Sci U S A 92:6334–6338
De Wit CA, Weström BR (1987) Purification and characterisation of alpha-2-, alpha-2-beta-, and beta-macroglobulin inhibitors in the hedghog, Erinaceus europaeus: beta-Macroglobulin identified as the plasma antihemorrhagic factor. Toxicon 25(11):1209–1219
Dennis PA, Saksela O, Harpel P, Rifkin DB (1989) alpha-2-macroglobulin is a bidning protein for basic fibroblast growth factor. J Biol Chem 264(13):7210–7216
Doan N, Gettins PG (2007) Human alpha-2-macroglobulin is composed of multiple domains, as predicted by homology with complement component C3. Biochem J 407(1):23–30
Doan N, Gettins PG (2008) alpha-Macroglobulins are present in some gram-negative bacteria: characterization of the alpha-2-macroglobulin from Escherichia coli. J Biol Chem 283(42):28747–28756
Dodds AW, Ren XD, Willis AC, ASK L (1996) The reaction mechanism of the internal thioester in the human complement component C4. Nature 379:177–179
Dolmer K, Gettins PG (2006) Three complement-like repeats compose the complete alpha-2-macroglobulin binding site in the second ligand binding cluster of the low density lipoprotein receptor-related protein. J Biol Chem 281(45):34189–34196
Dolmer K, Jenner LB, Jacobsen L, Andersen GR, Koch TJ, Thirup S, Sottrup-Jensen L, Nyborg J (1995) Crystallisation and preliminary X-ray analysis of the receptor-binding domain of human and bovine alpha-2-macroglobulin. FEBS Lett 372:93–95

Dolmer K, Husted LB, Armstrong PB, Sottrup-Jensen L (1996) Localisation of the major reactive lysine residue involved in the self-crosslinking of proteinase-activated Limulus alpha-2-macroglobulin. FEBS Lett 393:37–40

Dolovich J, Wicher V (1971) The binding of Bacillus subtilis alkaline proteinases to alpha-2-macroglobulin. J Lab Clin Med 77:951–957

Drozdetskiy A, Cole C, Procter J, Barton GJ (2015) JPred4: a protein secondary structure prediction server. Nucleic Acids Res 43(W1):W389–W394

Du Y, Ni B, Glinn M, Dodel RC, Bales KR, Zhang Z, Hyslop PA, Paul SM (1997) alpha-2-Macroglobulin as a beta-amyloid peptide-binding plasma protein. J Neurochem 69:299–305

Dunn JT, Spiro RG (1967a) The alpha-2 macroglobulin of human plasma. Isolation and composition. J Biol Chem 242(23):5549–5555

Dunn JT, Spiro RG (1967b) The alpha-2-macroglobulin of human plasma. Studies on the carbohydrate units. J Biol Chem 242(23):5556–5563

Enghild JJ, Salvesen G, Thøgersen IB, Pizzo SV (1989) Proteinase binding and inhibition by the monomeric alpha-macroglobulin rat alpha-1-inhibitor-3. J Biol Chem 264(19):11428–11435

Enghild JJ, Thøgersen IB, Salvesen G, Fey GH, Figler NL, Gonias SL, Pizzo S (1990) Alpha-macroglobulin from Limulus polyphemus exhibits proteinase inhibitory activity and participates. Biochemistry 29(43):10070–10080

Eubanks LM, Stowe GN, De Lamo MS, Mayorov AV, Hixon MS, Janda KD (2011) Identification of alpha-2-macroglobulin as a major serum ghrelin esterase. Angew Chem Int Ed 50(45):10699–10702

Feige JJ, Negoescu A, Keramidas M, Souchelnitskiy S, Chambaz EM (1996) alpha-2-Macroglobulin: a binding protein for transforming growth factor-beta and various cytokines. Horm Res 43(3–5):227–232

Fredslund F, Jenner L, Husted LB, Nyborg J, Andersen GR, Sottrup-Jensen L (2006) The structure of bovine complement component 3 reveals the basis for thioester function. J Mol Biol 361(1):115–127

Fujito NT, Sugimoto S, Nonaka M (2010) Evolution of thioester-containing proteins revealed by cloning and characterization of their genes from a cnidarian sea anemone, Haliplanella lineate. Dev Comp Immunol 34(7):775–784

Fyfe CD, Grinter R, Josts I, Mosbahi K, Roszak AW, Cogdell RJ, Wall DM, Burchmore RJ, Byron O, Walker D (2015) Structure of protease-cleaved Escherichia coli alpha-2-macroglobulin reveals a putative mechanism of conformational activation for protease entrapment. Acta Crystallogr D Biol Crystallogr 71(Pt 7):1478–1486

Galliano MF, Toulza E, Gallinaro H, Jonca N, Ishida-Yamamoto A, Serre G, Guerrin M (2006) A novel protease inhibitor of the alpha-2-macroglobulin family expressed in the human epidermis. J Biol Chem 281(9):5780–5789

Garcia-Ferrer I, Arede P, Gomez-Blanco J, Luque D, Duquerroy S, Caston JR, Goulas T, Gomis-Ruth FX (2015) Structural and functional insights into Escherichia coli alpha-2-macroglobulin endopeptidase snap-trap inhibition. Proc Natl Acad Sci U S A 112(27):8290–8295

Gavish H, Bab I, Tartakovsky A, Chorev M, Mansur N, Greenberg Z, Namdar-Attar M, Muhlrad A (1997) Human alpha-2-macroglobulin is an osteogenic growth peptide-binding protein. Biochemistry 36(48):14883–14888

Gettins PG, Crews BC (1993) Epidermal growth factor binding to human alpha-2-macroglobulin. Implications for alpha-2-macroglobulin-growth factor interactions. Biochemistry 32(31):7916–7921

Gettins PG, K-h H, Crews BC (1995) alpha-2-Macroglobulin bait region variants. J Biol Chem 270(23):14160–14167

Ghetie MA, Uhr JW, Vitetta ES (1991) Covalent binding of human alpha-2-macroglobulin to deglycosylated ricin A chain and its immunotoxins. Cancer Res 51:1482–1487

Giroux E, Vargaftig BB (1978) Clostridio peptidase B inhibition by plasma macroglobulins and microbial antiproteases. Biochim et Biophys Acta – Enzymol 525(2):429–437

Gollas-Galvan T, Sotelo-Mundo RR, Yepiz-Plascencia G, Vargas-Requena C, Vargas-Albores F (2003) Purification and characterisation of alpha-2-macroglobulin from the white shrimp (Penaeus vannamei). Comp Biochem Physiol C Toxicol Pharmacol 134(4):431–438

Gonias SL, Pizzo SV (1983) Characterisation of functional human alpha-2-macroglobulin half-molecules isolated by limited reduction with dithiothreitol. Biochemistry 22:536–546

Gouin-Charnet A, Laune D, Granier C, Mani JC, Pau B, Mourad G, Argiles A (2000) alpha-2-Macroglobulin, the main serum antiprotease binds beta-2-microglobulin, the light chain of the class I major histocompatibility complex, which is involved in human disease. Clin Sci 98:427–433

Goulas T, Garcia-Ferrer I, Garcia-Pique S, Sottrup-Jensen L, Gomis-Ruth FX (2014) Crystallization and preliminary X-ray diffraction analysis of eukaryotic alpha-2-macroglobulin family members modified by methylamine, proteases and glycosidases. Mol Oral Microbiol 29(6):354–364

Gron H, Oike R, Potempa J, Travis J, Thogersen B, Enghild JJ, Pizzo SV (1997) The potential role of alpha-2-macroglobulin in the control of cysteine proteinases (gingipains) from Porphyromonas gingivalis. J Periodontal Res 32:61–68

Gunnarsson M, Jensen PE (1998) Binding of soluble myelin basic protein to various conformational forms of alpha-2-macrolgobulin. Arch Biochem Biophys 359:192–198

Hall M, Söderhall K (1994) Crayfish alpha-macroglobulin as a substrate for transglutaminases. Comp Biochem Physiol 108(1):65–72

Harpel PC (1970) Human plasma alpha-2-macroglobulin. An inhibitor of plasma kallikrein. J Exp Med 132(2):329–352

Hergenhahn H-G, Hall M, Söderhall K (1988) Purification and characterisation of an alpha-2-macroglobulin-like proteinase inhibitor from plasma of the crayfish Pacifastacus leniusculus. Biochem J 255:801–806

Hibbetts K, Hines B, Williams D (1999) An overview of proteinase inhibitors. J Vet Intern Med 13:302–308

Horvat RT, Clabaugh M, Duval-Jobe C, Parmely MJ (1989) Inactivation of human gamma interferon by Pseudomonas aeruginosa proteases: elastase augments the effects of alkaline protease despite the prsence of alpha-2-macroglobulin. Infect Immun 57(6):1668–1674

Huang W, Dolmer K, Liao X, Gettins PG (2000) NMR solution structure of the receptor binding domain of human alpha-2-macroglobulin. J Biol Chem 275(2):1089–1094

Husted LB, Sorensen ES, Armstrong PB, Quigley JP, Kristensen L, Sottrup-Jensen L (2002) Localization of carbohydrate attachment sites and disulfide bridges in Limulus alpha-2-macroglobulin. Evidence for two forms differing primarily in their bait region sequences. J Biol Chem 277(46):43698–43706

Ikai A, Osada T, Nishigai M (1988) Conformational changes of alpha-macroglobulin and ovomacroglobulin from the green turtle (Chelonia mydas japonica). J Biochem 103:218–224

Ikai A, Ookata K, Shimizu M, Nakamichi N, Ito M, Matsumura T (1999) A recombinant bait region mutant of human alpha-2-macroglobulin exhibiting an altered proteinase-inhibiting spectrum. Cytotechnology 31:53–60

Iwaki D, Kawabata S, Miura Y, Kato A, Armstrong PB, Quigley JP, Nielsen KL, Dolmer K, Sottrup-Jensen L, Iwanaga S (1996) Molecular cloning of Limulus alpha-2-macrolgobulin. Eur J Biochem 242:822–831

Janssen BJ, Huizinga EG, Raaijmakers HC, Roos A, Daha MR, Nilsson-Ekdahl K, Nilsson B, Gros P (2005) Structures of complement component C3 provide insights into the function and evolution of immunity. Nature 437(7058):505–511

Jenner LB, Husted L, Thirup S, Sottrup-Jensen L, Nyborg J (1998) Crystal structure of the receptor-binding domain of alpha-2-macroglobulin. Structure 6(5):595–604

Jensen PE, Sottrup-Jensen L (1986) Primary structure of human alpha-2-macroglobulin. J Biol Chem 261(34):15863–15869

Jensen GA, Andersen OM, Bonvin AM, Bjerrum-Bohr I, Etzerodt M, Thogersen HC, O'Shea C, Poulsen FM, Kragelund BB (2006) Binding site structure of one LRP-RAP complex: implications for a common ligand-receptor binding motif. J Mol Biol 362(4):700–716

Kantyka T, Rawlings ND, Potempa J (2010) Prokaryote-derived protein inhibitors of peptidases: a sketchy occurrence and mostly unknown function. Biochimie 92(11):1644–1656

Kelwick R, Desanlis I, Wheeler GN, Edwards DR (2015) The ADAMTS (A Disintegrin and Metalloproteinase with Thrombospondin motifs) family. Genome Biol 16:113

Khan MH, Shibuya Y, Kambara T, Yamamoto T (1995) Role of alpha-2-macroglobulin and bacterial elastase in guinea-pig pseudomonal septic shock. Int J Exp Pathol 76:21–28

Kolodziej SJ, Schroeter JP, Strickland DK, Stoops JK (1996) The novel three-dimensional structure of native humna alpha-2-macroglobulin and comparisons with the structure of the methylamine derivative. J Struct Biol 116:366–376

Kolodziej SJ, Klueppelberg HU, Nolasco N, Ehses W, Strickland DK, Stoops JK (1998) Three-dimensional structure of the human plasmin alpha-2-macroglobulin complex. J Struct Biol 123(2):124–133

Kolodziej SJ, Wagenknecht T, Strickland DK, Stoops JK (2002) The three-dimensional structure of the human alpha-2-macroglobulin dimer reveals its structural organization in the tetrameric native and chymotrypsin alpha-2-macroglobulin complexes. J Biol Chem 277(31):28031–28037

Kopacek P, Weise C, Saravanan T, Vitova K, Grubhoffer L (2000) Characterization of an—alpha-macroglobulin-like glycoprotein isolated from the plasma of the soft tick Ornithodoros moubata. Eur J Biochem 267:465–467

Korkmaz B, Horwitz MS, Jenne DE, Gauthier F (2010) Neutrophil elastase, proteinase 3, and cathepsin G as therapeutic targets in human diseases. Pharmacol Rev 62(4):726–759

Kratzsch J, Selisko T, Birkenmeier G (1995) Identification of transformed alpha-2-macroglobulin as a growth hormone-binding protein in human blood. J Clin Endocrinol Metab 80(2):585–590

Kremers R, Bloemen S, Al Dieri R, Hemker CH, Karlaftis V, Attard C, de Laat B, Monagle P, Ignjatovic V (2013) alpha-2-Macroglobulin is a major determinant of a lower thrombin generation in infants and children compared to adults. Blood 122(21):2344–2344

Krimbou L, Tremblay M, Davignon J, Cohn JS (1998) Association of apolipoprotein E with alpha-2-macroglobulin in human plasma. J Lipid Res 39:2372–2386

Kurdowska A, Carr FK, Stevens MD, Baughman RP, Martin TR (1997) Studies on the interaction of IL-8 with human plasma alpha-2-macroglobulin. J Immunol 158(4):1930–1939

LaMarre J, Hayes AM, Wollenberg GK, Hussaini I, Hall SW, Gonias SL (1991a) An apha-2-macrolgobulin receptor-dependent mechanism for the plasma clearance of transforming growth factor-beta-1 in mice. J Clin Investig 87:39–44

LaMarre J, Wollenberg G, Gonias SL, Hayes AM (1991b) Cytokine binding and clearance properties of proteinase-activated alpha 2-macroglobulins. Lab Investig 65(1):3–14

Larquet E, Boisset N, Pochon F, Lamy J (1994) Architecture of native human alpha-2-macroglobulin studied by cryoelectron microscopy and three-dimensional reconstruction. J Struct Biol 113:87–98

Law ASK, Dodds AW (1997) The internal thioester and the covalent binding properties of the complement proteins C3 and C4. Protein Sci 6:263–274

Le BV, Williams M, Logarajah S, Baxter RH (2012) Molecular basis for genetic resistance of Anopheles gambiae to Plasmodium: structural analysis of TEP1 susceptible and resistant alleles. PLoS Pathog 8(10):e1002958

Li ZF, Wu XH, Engvall E (2004) Identification and characterization of CPAMD8, a novel member of the complement 3/alpha-2-macroglobulin family with a C-terminal Kazal domain. Genomics 83(6):1083–1093

Liebl DJ, Koo PH (1993) Comparative binding of neurotrophins (NT-3, CNTF, and NGF) and various cytokines to alpha-2-macroglobulin. Biochem Biophys Res Commun 193(3):1255–1261

Lillis AP, Van Duyn LB, Murphy-Ullrich JE, Strickland DK (2008) LDL receptor-related protein 1: unique tissue-specific functions revealed by selective gene knockout studies. Physiol Rev 88(3):887–918

Lin M, Sutherland DR, Horsfall W, Totty N, Yeo E, Nayar R, Wu XF, Schuh AC (2002) Cell surface antigen CD109 is a novel member of the alpha-2-macroglobulin/C3, C4, C5 family of thioester-containing proteins. Blood 99(5):1683–1691

Lindroos PM, Coin PG, Osornio-Vargas AR, Bonner JC (1995) Interleukin 1 beta (IL-1 beta) and the IL-1 beta-alpha 2-macroglobulin complex upregulate the platelet-derived growth factor alpha-receptor on rat pulmonary fibroblasts. Am J Respir Cell Mol Biol 13(4):455–465

Liu Q, Ling TY, Shieh HS, Johnson FE, Huang JS, Huang SS (2001) Identification of the high affinity binding site in transforming growth factor-beta involved in complex formation with alpha 2-macroglobulin. Implications regarding the molecular mechanisms of complex formation between alpha 2-macroglobulin and growth factors, cytokines, and hormones. J Biol Chem 276(49):46212–46218

Lopez-Otin C, Bond JS (2008) Proteases: multifunctional enzymes in life and disease. J Biol Chem 283(45):30433–30437

Luan Y, Kong L, Howell DR, Ilalov K, Fajardo M, Bai XH, Di Cesare PE, Goldring MB, Abramson SB, Liu CJ (2008) Inhibition of ADAMTS-7 and ADAMTS-12 degradation of cartilage oligomeric matrix protein by alpha-2-macroglobulin. Osteoarthr Cartil 16(11):1413–1420

Marrero A, Duquerroy S, Trapani S, Goulas T, Guevara T, Andersen GR, Navaza J, Sottrup-Jensen L, Gomis-Rüth XF (2012) The crystal structure of human alpha-2-macroglobulin reveals a unique molecular cage. Angew Chem Int Ed 51:3340–3344

Matsuda T, Hirano T, Nagasawa S, Kishimoto T (1989) Identification of alpha-2-macroglobulin as a carrier protein for IL6. J Immunol 142(1):148–152

McCaffrey TA, Falcone DJ, Brayton CF, Agarwal LA, Welt FGP, Weksler BB (1989) Transforming growth factor-beta activity is potentiated by heparin via dissociation of the transforming growth factor-beta/alpha-2-macroglobulin inactive complex. J Cell Biol 109:441–448

Merritt GC, Egerton JR, Loi JS (1971) Inhibition of Fusiformis nodosus protease and bovine trypsin by serum alpha-macroglobulin. J Comp Pathol 81(3):353

Mettenburg JM, Webb DJ, Gonias SL (2002) Distinct binding sites in the structure of alpha-2-macroglobulin mediate the interaction with beta-amyloid peptide and growth factors. J Biol Chem 277(15):13338–13345

Molla A, Matsumura Y, Yamamoto T, Okamura R, Maeda H (1987) Pathogenic capacity of proteases from Serratia marcescens and Pseudomonas aerginosa and their suppression by chicken egg white ovomacroglobulin. Infect Immun 55(10):2509–2517

Moncrief JS, Obiso R, Barroso L, Kling JJ, Wright RL, van Tassell RL, Lyerly DM, Wilkins TD (1995) The enterotoxin of Bacteroides fragilis is a metelloprotease. Infect Immun 63(1):175–181

Morrot A, Strickland DK, Higuchi M, Reis M, Pedrosa R, Scharfstein J (1997) Human T cell responses against the major cysteine proteinase (cruzipain) of Trypanosoma cruzi: role of the mutlifucntional alpha-2-macroglobulin receptor in antigen presentation by monocytes. Int Immunol 9(6):825–834

Müller H-P, Rantamäki LK (1995) Binding of native alpha-2-macroglobulin to human group G Streptococci. Infect Immun 63(8):2833–2839

Mutsuro J, Nakao M, Fujiki K, Yano T (2000) Multiple forms of alpha-2-macroglobulin from a bony fish, the common carp (Cyprinus carpio): striking sequence diversity in functional sites. Immunogenetics 51(10):847–855

Nagase H, Harris ED (1983) Ovostatin: a novel proteinase inhibitor from chicken egg white. J Exp Med 258(12):7481–7489

Nelson KE, Clayton RA, Gill SR, Gwinn ML, Dodson RJ, Haft DH, Hickey EK, Peterson JD, Nelson WC, Ketchum KA, McDonald L, Utterback TR, Malek JA, Linher KD, Garrett MM, Stewart AM, Cotton MD, Pratt MS, Phillips CA, Richardson D, Heidelberg J, Sutton GG, Fleischmann RD, Eisen JA, White O, Salzberg SL, Smith HO, Venter JC, Fraser CM (1999)

Evidence for lateral gene transfer between archaea and bacteria from genome sequence of Thermotoga maritima. Nature 399:323–329

Neves D, Estrozi LF, Job V, Gabel F, Schoehn G, Dessen A (2012) Conformational states of a bacterial alpha-2-macroglobulin resemble those of human complement C3. PLoS One 7(4):e35384

Nielsen KL, Sottrup-Jensen L, Nagase H, Thøgersen HC, Etzerodt M (1994) Amino acid sequence of hen ovomacroglobulin (ovostatin) deduced from cloned cDNA. DNA Seq 5(2):111–119

Nielsen KL, Holtet TL, Etzerodt M, Moestrup SK, Gliemann J, Sottrup-Jensen L, Thogersen HC (1996) Identification of residues in alpha-macroglobulin important for binding to the alpha-2-macrolgobulin receptor/low density lipoprotein receptor relatef protein. J Biol Chem 271(22):12909–12912

Niemuller CA, Randall KJ, Webb DJ, Gonias SL, LaMarre J (1995) Alpha-2-macroglobulin conformation determines binding affinity for activin A and plasma clearance of activin A/alpha-2-macroglobulin complex. Endocrinology 136(12):5343–5349

Nonaka M (2011) The complement C3 protein family in invertebrates. Invertebr Surviv J 8:21–32

Nothaft H, Szymanski CM (2010) Protein glycosylation in bacteria: sweeter than ever. Nat Rev Microbiol 8(11):765–778

Paiva MM, Soeiro MN, Barbosa HS, Meirelles MN, Delain E, Araujo-Jorge TC (2010) Glycosylation patterns of human alpha-2-macroglobulin: analysis of lectin binding by electron microscopy. Micron 41(6):666–673

Panyutich A, Ganz T (1991) Activated alpha-2-macroglobulin is a principal defensin-binding protein. Am J Respir Cell Mol Biol 5:101–106

Perazzolo LM, Bachere E, Rosa RD, Goncalves P, Andreatta ER, Daffre S, Barracco MA (2011) alpha-2-Macroglobulin from an Atlantic shrimp: biochemical characterization, sub-cellular localization and gene expression upon fungal challenge. Fish Shellfish Immunol 31(6):938–943

Peslova G, Petrak J, Kuzelova K, Hrdy I, Halada P, Kuchel PW, Soe-Lin S, Ponka P, Sutak R, Becker E, Huang ML, Suryo Rahmanto Y, Richardson DR, Vyoral D (2009) Hepcidin, the hormone of iron metabolism, is bound specifically to alpha-2-macroglobulin in blood. Blood 113(24):6225–6236

Petersen CG, Venge P (1987) Interaction and complex-formation between the eosinophil cationic protein and alpha-2-macroglobulin. Biochem J 245:781–787

Petersen LC, Elm T, Ezban M, Krogh TN, Karpf DM, Steino A, Olsen EH, Sorensen BB (2009) Plasma elimination kinetics for factor VII are independent of its activation to factor VIIa and complex formation with plasma inhibitors. Thromb Haemost 101(5):795–990

Phillips DJ, McFarlane JR, Hearn MTW, Krester DM (1997) Inhibin, activin and follistatin bind preferentially to the transformed species of alpha-2-macroglobulin. J Endocrinol 155:65–71

Pochon F, Barrey M, Delain E (1989) Dissociation of alpha-2-macroglobulin into functional half-molecules by mild acid treatment. Biochim Biophys Acta 996:132–138

Pugsley A (1993) The complete general secretory pathway in Gram-negative bacteria. Microbiol Rev 57(1):50–108

Qazi U, Gettins PG, Strickland DK, Stoops JK (1999) Structural details of proteinase entrapment by human alpha-2-macroglobulin emerge from three-dimensional reconstructions of Fab labeled native, half-transformed, and transformed molecules. J Biol Chem 274(12):8137–8142

Qazi U, Kolodziej SJ, Gettins PG, Stoops JK (2000) The structure of the C949S mutant human alpha-2-macroglobulin demonstrates the critical role of the internal thiol esters in its proteinase-entrapping structural transformation. J Struct Biol 131(1):19–26

Quigley JP, Ikai A, Arakawa H, Osada T, Armstrong PB (1991) Reaction of proteinases with alpha-2-macroglobulin from the american horseshoe crab, Limulus. J Biol Chem 266(26):19426–19431

Rasmussen M, Müller H-P, Björck L (1999) Protein GRAB of Streptococcus pyogenes regulates proteolysis at the bacterial surface by bnding alpha-2-macroglobulin. J Biol Chem 274(22):15336–15344
Raymond WW, Su S, Makarova A, Wilson TM, Carter MC, Metcalfe DD, Caughey GH (2009) alpha-2-Macroglobulin capture allows detection of mast cell chymase in serum and creates a reservoir of angiotensin II-generating activity. J Immunol 182(9):5770–5777
Robert-Genthon M, Casabona MG, Neves D, Coute Y, Ciceron F, Elsen S, Dessen A, Attree I (2013) Unique features of a Pseudomonas aeruginosa alpha-2-macroglobulin homolog. MBio 4(4):e00309–e00313
Ronne H, Anundi H, Rask L, Peterson PA (1979) Nerve growth factor binds to serum alpha-2-macroglobulin. Biochem Biophys Res Commun 87(1):330–336
Rubenstein DS, Thogersen IB, Pizzo SV, Enghild JJ (1993) Identification of monomeric alpha-macroglobulin proteinase inhibitors in birds, reptiles, amphibians and mammals, and purification and characterisation of a monomeric alpha-macroglobulin proteinase inhibitor from the American bullfrog Rana catesbiana. Biochem J 290:85–89
Rudenko G, Henderson K, Ichtchenko K, Brown MS, Goldstein JL, Deisenhofer J (2002) Structure of the LDL receptor extreacellualar doamin at endosomal pH. Science 298:2353–2358
Rudloe A (1979) Limulus polyphemus: a review of the ecologically significant literature. In: Cohen E (ed) Biomedical applications of the horse shoe crab. Alan R. Liss, New York, pp 27–35
Sahu A, Lambris JD (2001) Structure and biology of complement protein C3, a connecting link between innate and acquired immunity. Immunol Rev 180:35–48
Sand O, Folkersen J, Westergaard JG, Sottrup-Jensen L (1985) Characterisation in human pregnancy zone protein. Comparison with human alpha-2-macrolgobulin. J Biol Chem 260(29):15723–15735
Schaller J, Gerber SS (2011) The plasmin-antiplasmin system: structural and functional aspects. Cell Mol Life Sci 68(5):785–801
Scharfstein J (2006) Parasite cysteine proteinase interactions with alpha-2-macroglobulin or kininogens: differential pathways modulating inflammation and innate immunity in infection by pathogenic trypanosomatids. Immunobiology 211(1–2):117–125
Schroeter JP, Kolodziej SJ, Wagenknecht T, Bretaudiere J-P, Tapon-Bretaudiere J, Strickland DK, Stoops JK (1992) Three-dimensional structures of the human alpha-2-macroglobulin-methylamine and chymotrypsin complexes. J Struct Biol 109:235–247
Seydel A, Gounon P, Pugsley A (1999) Testing the '+2 rule' for lipoprotein sorting in the Escherichia coli cell envelope with a new genetic selection. Mol Microbiol 34(4):810–821
Shanbhag VP, Stigbrand T, Jensen PE (1996) The conformational state of human alpha-2-macroglobulin influences its dissociation into half-molecules by sodium thiocyanate. Arch Biochem Biophys 333(1):35–41
Soker S, Svahn CM, Neufeld G (1993) Vascular endothelial growth factor is inactivated by binding to alpha-2-macroglobulin and the binding is inhibited by heparin. J Biol Chem 268(11):7685–7691
Sottrup-Jensen L (1989) alpha-Macroglobulin: structure, shape, and mechanism of proteinase complex formation. J Biol Chem 264(20):11539–11542
Sottrup-Jensen L (1994) Role of internal thiol esters in the alpha-macroglobulin-proteinase binding mechanism. Ann N Y Acad Sci 737:172–187
Sottrup-Jensen L, Folkersen J, Kristensen L, Tack BF (1984a) Partial primary structure of human pregnancy zone protein: extensive sequence homology with human alpha-2-macroglobulin. Proc Natl Acad Sci U S A 81:7353–7357
Sottrup-Jensen L, Stepanik TM, Kristensen T, Wierzbicki DM, Jones CM, Lonbland PB, Magnusson S, Petersen TE (1984b) Primary structure of human alpha-2-macroglobulin. V. The complete structure. J Biol Chem 259(13):8318–8327
Sottrup-Jensen L, Stepanik TM, Kristensen T, Lonbland PB, Jones CM, Wierzbicki DM, Magnusson S, Domdey H, Wetsel RA, Lundwall A, Tack BF, Fey GH (1985) Common evolu-

tionary origin of alpha-2-macroglobulin and complement components C3 and C4. Proc Natl Acad Sci U S A 82:9–13

Sottrup-Jensen L, Sand O, Kristensen T, Fey GH (1989) The alpha-macroglobulin bait region. Sequence diversity and localisation of cleavage sites for proteinases in five mammalian alpha-macroglobulins. J Biol Chem 264(27):15781–15789

Spycher SE, Arya S, Isenman DE, Painter RH (1987) A functional, thioester-containing alpha-2-macroglobulin homologue isolated from the hemolymph of the american lobster (Homarus americanus). J Biol Chem 262:14606–14611

Starkey PM, Barrett AJ (1982) Evolution of alpha-2-macroglobulin. The demonstration in a variety of vertebrate species of a protein resembling human alpha-2-macroglobulin. Biochem J 205:91–95

Stöcker W, Breit S, Sottrup-Jensen L, Zwilling R (1991) alpha-2-Macroglobulin from the haemolymph of the freshwater crayfish Astacus astacus. Comp Biochem Physiol 98B(4):501–509

Suda SA, Dolmer K, Gettins PG (1997) Critical role of asparagine 1065 of human alpha-2-macroglobulin in formation and reactivity of the thiol ester. J Biol Chem 272(49):31107–31112

Sunderic M, Malenkovic V, Nedic O (2015) Complexes between insulin-like growth factor binding proteins and alpha-2-macroglobulin in patients with tumor. Exp Mol Pathol 98(2):173–177

Swarnakar S, Asokan R, Quigley JP, Armstrong PB (2000) Binding of alpha-2-macrolgobulin and limulin: regulation of the plasma haemolytic system of the American horseshoe crab, Limulus. Biochem J 347:679–685

Thøgersen HC, Salvesen G, Brucato FH, Pizzo SV, Enghild JJ (1992) Purification and characterisation of an alpha-macroglobulin proteinase inhibitor from the mollusc Octopus vulgaris. Biochem J 285:521–527

Tortorella MD, Arner EC, Hills R, Easton A, Korte-Sarfaty J, Fok K, Wittwer AJ, Liu RQ, Malfait AM (2004) alpha-2-Macroglobulin is a novel substrate for ADAMTS-4 and ADAMTS-5 and represents an endogenous inhibitor of these enzymes. J Biol Chem 279(17):17554–17561

Tunstall AM, Merriman JML, Milne I, James K (1975) Normal and pathological serum levels of alpha-2-macroglobulins in men and mice. J Clin Pathol 28:133–139

Valnickova Z, Thøgersen IB, Christensen S, Chu CT, Pizzo SV, Enghild JJ (1996) Activate human plasma carboxypeptidase B is retained in the blood by binding to alpha-2-macroglobulin and pregnancy zone protein. J Biol Chem 271(22):12937–12943

Van Jaarsveld F, Naudé RJ, Oelofsen RJ, Travis J (1994) The isolation and partial characterisation of alpha-2-macroglobulin from the serum of the ostrich (Struthio camelus). Int J Biochem 26(1):97–110

Van Rompaey L, Proost P, Van den Berghe H, Marynen P (1995) Design of a new protease inhibitor by the manipulation of the bait region of alpha-2-macroglobulin: inhibition of the tobacco etch virus protease by mutant alpha-2-macroglobulin. Biochem J 312:191–195

Vanhoorelbeke K, Goossens A, Gielens C, Preaux G (1993) An alpha-2-macroglobulin-like proteinase inhibitor inthe haemolymph of the Decabrachia cephalopod Sepia officinalis. Arch Int Physiol Biochim Biophys 102:B25

Vilella E, Bengtsson-Olivecrona G, Stigbrand T, Jensen PE (1994) Binding of lipoprotein lipase to alpha-2-macroglobulin. Biochem J 300:443–448

Wacker M, Linton D, Hitchen PG, Nita-Lazar M, Haslam SM, North SJ, Panico M, Morris HR, Dell A, Wren BW, Aebi M (2002) N-linked glycosylation in Campylobacter jejuni and its functional transfer into E. coli. Science 298:1790–1793

Walter M, Sutton RM, Schechter NM (1999) Highly efficient inhibition of human chymase by alpha-2-macroglobulin. Arch Biochem Biophys 368(2):276–284

Wang S, Wei X, Zhou J, Zhang J, Li K, Chen Q, Terek R, Fleming BC, Goldring MB, Ehrlich MG, Zhang G, Wei L (2014) Identification of alpha-2-macroglobulin as a master inhibitor of cartilage-degrading factors that attenuates the progression of posttraumatic osteoarthritis. Arthritis Rheum 66(7):1843–1853

Webb DJ, Wen J, Karns LR, Kurilla MG, Gonias SL (1998) Localisation of the binding site for the transforming growth factor-beta in human alpha-2-macroglobulin to a 20 kDa peptide that also contains the bait region. J Biol Chem 273(21):13339–13346

Werb Z, Burleigh MC, Barrett AJ, Starkey PM (1973) The interaction of alpha-2-macroglobulin with proteinases. Binding and inhibitions of mammalian collagenases and other metal proteinases. Biochem J 139:359–368

Wolf BB, Gonias SL (1994) Neurotrophin binding to human alpha-2-macroglobulin under apparent equilibrium conditions. Biochemistry 33(37):11270–11277

Wollenberg G, LaMarre J, Rosendal S, Gonias SL, Hayes AM (1991) Binding of tumor necrosis factor alpha to activated forms of human plasma alpha-2-macroglobulin. Am J Pathol 138(2):265–272

Wong SG, Dessen A (2014) Structure of a bacterial alpha-2-macroglobulin reveals mimicry of eukaryotic innate immunity. Nat Commun 5:4917

Wyatt AR, Constantinescu P, Ecroyd H, Dobson CM, Wilson MR, Kumita JR, Yerbury JJ (2013) Protease-activated alpha-2-macroglobulin can inhibit amyloid formation via two distinct mechanisms. FEBS Lett 587(5):398–403

Wyatt AR, Kumita JR, Mifsud RW, Gooden CA, Wilson MR, Dobson CM (2014) Hypochlorite-induced structural modifications enhance the chaperone activity of human alpha-2-macroglobulin. Proc Natl Acad Sci U S A 111(20):E2081–E2090

Xiao T, Decamp DL, Sprang SR (2000) Structure of a rat alpha-1-macroglobulin receptor-binding domain dimer. Protein Sci 9:1889–1897

Yigzaw Y, Gielens C, Préaux G (2001) Isolation and characterisation of an alpha-macroglobulin from the gastropod mollusc Helix pomatia with tetrameric structure and preserved activity after methylamine treatment. Biochim Biophys Acta 1545:104–113

Chapter 7
The Structure and Function of the PRMT5:MEP50 Complex

Stephen Antonysamy

Abstract Protein arginine methyltransferase 5 (PRMT5) plays multiple roles in cellular processes at different stages of the cell cycle in a tissue specific manner. PRMT5 in complex with MEP50/p44/WDR77 associates with a plethora of partner proteins to symmetrically dimethylate arginine residues on target proteins in both the nucleus and the cytoplasm. Overexpression of PRMT5 has been observed in several cancers, making it an attractive drug target. The structure of the 453 kDa heterooctameric PRMT5:MEP50 complex bound to an S-adenosylmethionine analog and a substrate peptide provides valuable insights into this intriguing target.

Keywords Epigenetics • Development • Methyltransferases • PRMT5 • MEP50 • WD40 domain • PPI

7.1 Introduction

Epigenetic control of gene expression and protein function by posttranslational modifications of DNA, histones and non-histone proteins play critical roles in normal cell function, and is invariably perturbed in disease states (Yang and Bedford 2013; Arrowsmith et al. 2012; Sarkies and Sale 2012; Baylin and Jones 2011). Mono- and di-methylation of arginine sidechains is a common posttranslational modification, where a methyl group is transferred from S-adenosylmethionine (AdoMet or SAM) to the guanidyl sidechain of arginine (Bedford and Clarke 2009; Herrmann et al. 2009; Schapira and Ferreira de Freitas 2014; Fuhrmann et al. 2015; Morales et al. 2016). Protein arginine methyltransferases (PRMTs) are classified as Type I, Type II, Type III or Type IV, based on their pattern of arginine methylation.

- Type I PRMTs transfer a methyl group to a terminal nitrogen atom of the guanidyl sidechain to form monomethylarginine, and then a second methyl group to the same atom to form asymmetric dimethylarginine

S. Antonysamy (✉)
Structural Biology, Discovery Chemistry Research and Technologies, Eli Lilly and Company, Lilly Biotechnology Center, 10290 Campus Point Drive, San Diego, CA 92121, USA
e-mail: antonysamyst@lilly.com

J.R. Harris, J. Marles-Wright (eds.), *Macromolecular Protein Complexes*, Subcellular Biochemistry 83, DOI 10.1007/978-3-319-46503-6_7

- Type II PRMTs sequentially transfer methyl groups to each of the terminal atoms of the guanidyl side chain to form monomethylarginine and symmetric dimethylarginine
- Type III PRMTs are only able to monomethylate substrates
- Type IV PRMT identified in yeast, methylates the δ-nitrogen atom of the guanidyl sidechain

The nine human PRMTs (PRMT1-9) share the conserved methyltransferase catalytic domain, while each has additional domains that modulate their functions and substrates. PRMT5 is the primary Type II arginine methyltransferase that symmetrical dimethylates the terminal nitrogen atoms of the guanidyl sidechain (Branscombe et al. 2001; Karkhanis et al. 2011). PRMT5 forms a tight heterooctameric complex with the WD40 protein MEP50/p44/WDR77, to constitute a central core which associates with other binding partners to form larger multi-molecular complexes that perform a plethora of functions within the cell (Antonysamy et al. 2012).

7.2 Diverse Functions of PRMT5

PRMT5 plays a vital role in a wide spectrum of cellular processes, and its methyltransferase activity is carefully controlled by its binding partners, as well as its expression levels and subcellular localization at different stages of the cell cycle (Stopa et al. 2015; Karkhanis et al. 2011). As the predominant Type II arginine methyltransferase, there is increasing evidence that it methylates an ever expanding array of targets that encompass both histones and not-histone proteins. Its substrates include Sm proteins, nucleolin, p53, NF-κB p65, HOXA9, E2F-1, SPT5, MBD2, EGFR, CRAF, SREBP1a and histones H2A, H3 and H4 (Friesen et al. 2001; Hou et al. 2008; Kwak et al. 2003; Cho et al. 2012; Jansson et al. 2008; Hsu et al. 2011; Andreu-Perez et al. 2011; Liu et al. 2016; Bandyopadhyay et al. 2012; Harris et al. 2014).

PRMT5 is essential for early development and loss of PRMT function leads to embryonic lethality (Tee et al. 2010). Translocating between the cytoplasm and the nucleus, it plays multiple roles in germ cell development and embryonic stem cell pluripotency (Kim et al. 2014). Recent data suggests that while PRMT5 may not play a role in primordial germ cell (PGC) specification, it functions after the first global epigenetic reprogramming of PGCs to promote proliferation and survival (Li et al. 2015). PRMT5 represses transcription by the symmetric dimethylation of histone H3-Arg8 and histone H4-Arg3 (Pal et al. 2004). Overexpression of PRMT5 in cancer cells could lead to tumor growth through the repression of tumor suppressor genes (Wang et al. 2008; Pal et al. 2007). PRMT5 is a regulator of splicing in mammals, and the absence of PRMT5 results in reduced methylation of Sm proteins and aberrant constitutive splicing (Bezzi et al. 2013). In response to IL-1β, the p65 subunit of NF-κB is dimethylated by PRMT5 on R30, enhancing the activity of NF-κB (Wei et al. 2013; Lu and Stark 2015). Recent data indicates that concurrent stimulation of endothelial cells by both TNF and IFN-γ activates PRMT5 to methylate Arg174, Arg30 and Arg35 on p65, recruiting it to the CXCL11 promoter leading to the induction of CXCL11 (Harris et al. 2016). PRMT5 has been shown to interact with tumor suppressor RASSF1A (ras-association domain family 1, isoform A), leading to RASSF1A anchoring cytosolic PRMT5 to microtubules. (Sakai et al. 2015). PRMT5 symmetrically dimethylates sterol regulatory element-binding protein (SREBP) in cancer cells leading to accelerated growth both *in vivo* and *in vitro* (Liu et al. 2016). Deletion of 5-Methythioadenosine phosphorylase (MTAP), a frequent occurrence in several cancers due to its chromosomal proximity to tumor suppressor gene CDKN2A, makes the MTAP-deleted cancer cells sensitive to PRMT5 depletion (Mavrakis et al. 2016). Aberrant PRMT5 methylation and regulation of KLF4 plays a role in breast carcinogenesis, through the dysregulation of the levels of p21 and Bax, genes involved in determining cell fate (Hu et al. 2015).

The level of PRMT5 in cells is thought to be regulated by the chaperone-dependent E3 ubiquitin ligase CHIP (carboxy terminus of heat shock cognate 70-interacting protein). Interaction with CHIP leads to the polyubiquitination and proteasomal degradation of PRMT5 (Zhang et al. 2016). Studies with *C. elegans*, coupled with *in vitro* experiments with the human enzyme suggest that PRMT5 plays a role in GPCR signaling (Likhite et al. 2015). The crosstalk between the multiple pathways that control proliferation and differentiation is exemplified by the interplay between PKCδ/p38 signaling and PRMT5. Stimulation of the PKCδ/p38 signaling cascade that increases differentiation, leads to a reduction of PRMT5 and MEP50 levels and consequently histone methylation and gene silencing (Saha et al. 2015). The complex functions of PRMT5 is illustrated by its role in acute myeloid leukemia (AML) where it mediates both silencing and transcription of genes to support AML (Tarighat et al. 2015). Of the two Type II arginine methyltransferases that perform symmetric dimethylation of arginine residues, PRMT5 is the clearly the more prominent enzyme with a remarkably large set of known substrates, while the recently characterized PRMT9 which symmetrically dimethylates the splicing factor SAP145 appears to play a more limited role (Yang et al. 2015).

7.3 Role of MEP50

MEP50 is a WD40 protein that was initially identified as an essential component of the 20 S protein arginine methyltransferase complex, termed the methylosome (Friesen et al. 2001), and independently as p44, an androgen receptor coactivator (Hosohata et al. 2003). The large WD40 family of proteins play central roles in cellular interactions through protein binding and recognition (Smith et al. 1999; Stirnimann et al. 2010; Xu and Min 2011; Migliori et al. 2012). MEP50 binds PRMT5 to stabilize and increase its activity, and also functions to recruit substrate proteins for methylation (Antonysamy et al. 2012; Burgos et al. 2015). Independent of its role in PRMT5 activity, MEP50 is also thought to play a role in other networks in both the cytoplasm and nucleus through interaction with partner proteins. MEP50 has been implicated in prostrate, lung and ovarian cancers (Ligr et al. 2011; Gao and Wang 2012; Gu et al. 2013). MEP50 contributes to cellular proliferation during lung tumorigenesis, by down regulating the expression of TGFβ ligand and its receptor, leading to decreased TGFβ signaling (Yi et al. 2014). The subcellular localization of MEP50 and its role in androgen receptor driven gene expression is regulated by cGMP-dependent protein kinase 1β to regulate androgen receptor (Zhou et al. 2014). The function of MEP50 is dependent on its subcellular localization; while cytoplasmic MEP50 is needed for proliferation of prostrate epithelial cells in early prostrate development, nuclear MEP50 is essential for the differentiation and functionality.

7.4 Structure of the Heterooctameric PRMT5:MEP50 Complex

The structure of the 453 kDa heterooctameric complex of PRMT5 with MEP50 bound to A9145C (a S-adenosylmethionine analog (Pugh et al. 1978) and a peptide substrate derived from histone H4 was determined by X-ray crystallography to a resolution of 2.2 Å (Antonysamy et al. 2012). This first structure of a mammalian Type II arginine methyltransferase provided structural insights and detailed information about the cofactor and substrate binding site and has enabled the structure based design of potent and selective inhibitors against PRMT5 (Duncan et al. 2016).

The N-terminal domain of PRMT5 was revealed to be a TIM barrel, which plays an essential role in the assembly of the heterooctameric PRMT5:MEP50 complex by packing together with the C-terminal methyltransferase catalytic domain to form the central tetrameric PRMT5 inner core of the oligomer (Fig. 7.1). The MEP50 molecules bind on the outside of this core, interacting exclusively with the TIM barrel domains, stabilizing the complex and increasing the affinity for both cofactor and substrate. This is in contrast to the Type I arginine methyl-transferases, which form homo-dimers in head-to-tail arrangements, mediated by a dimerization arm that extends from the β-barrel domain of each monomer to interact with the AdoMet binding domain of the second monomer (Troffer-Charlier et al. 2007; Schapira and Ferreira de Freitas 2014). The catalytic domain adopts the canonical arginine methyltransferase structure, which is conserved between Type I, Type II and Type III PRMTs, and consists of an AdoMet binding domain containing the nucleotide binding Rossmann fold and a β-sandwich domain that contributes to substrate binding. The linker between the C-terminal catalytic domain and the N-terminal TIM-barrel domain (that is specific to PRMT5), also forms a large part of the peptide substrate and co-factor binding pocket. The PRMT5:MEP50 structure in complex with peptide revealed that the highly conserved catalytic site glutamate residues Glu435 and Glu444 form a pair of hydrogen bonds each, with the arginine residue of the substrate peptide (Fig. 7.2). These bidentate interactions that position the guanidyl

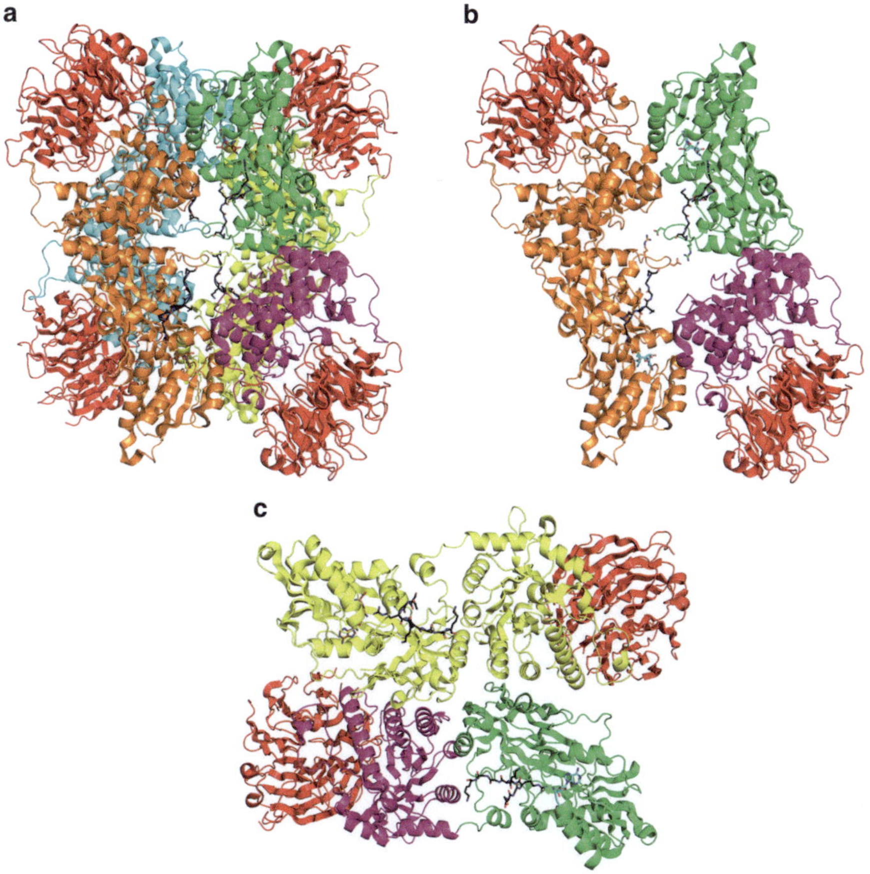

Fig. 7.1 (**a**) PRMT5:MEP50 heterooctamer in complex with S-adenosylmethionine analog A9145C, and a peptide substrate derived from histone H4. The PRMT5 catalytic domain is depicted as *green cartoons*, while the N-terminal TIM barrel domain is colored *magenta* in monomer A. PRMT5 monomers B, C and D are colored *orange cyan* and *yellow* respectively. All four MEP molecules are depicted as *red cartoons*, while the peptide substrate and AdoMet analog are show as *black* and *orange sticks* respectively. (**b**) One of the two PRMT5:MEP50 dimers that make up the $PRMT5_4$:$MEP50_4$ heterooctamer. Color scheme same as **a**. (**c**) The second dimer interaction within the $PRMT5_4$:$MEP50_4$ heterooctamer

sidechain for catalysis have now also been observed in Type 1 PRMTs, with the recent structure determination of CARM1 (PRMT4) with different substrate peptides (Boriack-Sjodin et al. 2016). Other than these two catalytic glutamates, the residues that line the substrate binding pocket exhibit significant variability among the different PRMTs. Analysis of the substrate pocket reveals that the majority of the interactions between the substrate peptide and PRMT5 are mediated by protein backbone interactions. Mutational studies on *C. elegans* PRMT5 has shown that, changing a phenylalanine in the substrate pocket to methionine, generates an

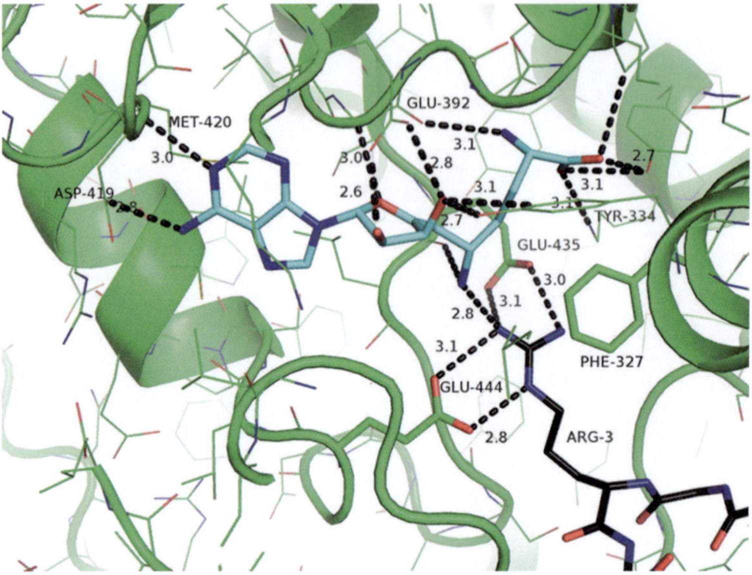

Fig. 7.2 Interactions of PRMT5 represented in *green cartoons* and *lines* with the peptide substrate and AdoMet analog (A9145C) in *black* and *cyan sticks* respectively. The structure clearly reveals the role of the conserved glutamate residues of the double-E loop (Glu435 and Glu444) in orienting the guanidyl sidechain for catalysis

enzyme that is now able to perform both symmetric and asymmetric dimethylation (Sun et al. 2011). Therefore, while the catalytic mechanism is conserved across the PRMT family, the ability to symmetrically or asymmetrically dimethylate, or to only monomethylated substrates seems to be determined by spatial and steric factors in the substrate binding site.

7.5 Summary

The complex role played by PRMT5 in cell function is accomplished by the diverse spectrum of substrates that it methylates to regulate varied biological processes. The lack of sequence specific side chain interactions in the substrate peptide binding pocket suggests that the necessary selectivity is likely achieved by protein-protein interactions between the complexes formed by PRMT5:MEP50 and its binding partners with the target of methylation. This view is supported by the fact that MEP50 is known to bind many of the substrates of PRMT5 (Antonysamy et al. 2012).

The critical role played by arginine methyltransferases in normal cell function, and their dysregulation in several disease states has led to their investigation as viable drug targets (Fuhrmann et al. 2015; Yang and Bedford 2013; Luo 2012; Yost et al. 2011; Arrowsmith et al. 2012). While much of the drug discovery efforts on arginine methyltransferases have focused on Type I PRMTs, the recent identification of a potent and selective inhibitor of PRMT5 with *in vivo* activity provides us with a valuable probe molecule, to help us unravel the biological roles of PRMT5 (Chan-Penebre et al. 2015).

Acknowledgements The author thanks Marijane Russell, Robert M Campbell and John E Munroe for valuable feedback on the manuscript.

References

Andreu-Perez P, Esteve-Puig R, de Torre-Minguela C, Lopez-Fauqued M, Bech-Serra JJ, Tenbaum S, Garcia-Trevijano ER, Canals F, Merlino G, Avila MA, Recio JA (2011) Protein arginine methyltransferase 5 regulates ERK1/2 signal transduction amplitude and cell fate through CRAF. Sci Signal 4(190):ra58. doi:10.1126/scisignal.2001936

Antonysamy S, Bonday Z, Campbell RM, Doyle B, Druzina Z, Gheyi T, Han B, Jungheim LN, Qian Y, Rauch C, Russell M, Sauder JM, Wasserman SR, Weichert K, Willard FS, Zhang A, Emtage S (2012) Crystal structure of the human PRMT5:MEP50 complex. Proc Natl Acad Sci U S A 109(44):17960–17965. doi:10.1073/pnas.1209814109

Arrowsmith CH, Bountra C, Fish PV, Lee K, Schapira M (2012) Epigenetic protein families: a new frontier for drug discovery. Nat Rev Drug Discov 11(5):384–400. doi:10.1038/nrd3674

Bandyopadhyay S, Harris DP, Adams GN, Lause GE, McHugh A, Tillmaand EG, Money A, Willard B, Fox PL, Dicorleto PE (2012) HOXA9 methylation by PRMT5 is essential for endothelial cell expression of leukocyte adhesion molecules. Mol Cell Biol 32 (7):1202–1213. doi:MCB.05977-11 [pii] 10.1128/MCB.05977-11

Baylin SB, Jones PA (2011) A decade of exploring the cancer epigenome – biological and translational implications. Nat Rev Cancer 11(10):726–734. doi:10.1038/nrc3130

Bedford MT, Clarke SG (2009) Protein arginine methylation in mammals: who, what, and why. Mol Cell 33(1):1–13. doi:10.1016/j.molcel.2008.12.013

Bezzi M, Teo SX, Muller J, Mok WC, Sahu SK, Vardy LA, Bonday ZQ, Guccione E (2013) Regulation of constitutive and alternative splicing by PRMT5 reveals a role for Mdm4 pre-mRNA in sensing defects in the spliceosomal machinery. Genes Dev 27(17):1903–1916. doi:10.1101/gad.219899.113

Boriack-Sjodin PA, Jin L, Jacques SL, Drew A, Sneeringer C, Scott MP, Moyer MP, Ribich S, Moradei O, Copeland RA (2016) Structural insights into ternary complex formation of human CARM1 with various substrates. ACS Chem Biol 11(3):763–771. doi:10.1021/acschembio.5b00773

Branscombe TL, Frankel A, Lee JH, Cook JR, Yang Z, Pestka S, Clarke S (2001) PRMT5 (Janus kinase-binding protein 1) catalyzes the formation of symmetric dimethylarginine residues in proteins. J Biol Chem 276 (35):32971–32976. doi:10.1074/jbc.M105412200 M105412200 [pii]

Burgos ES, Wilczek C, Onikubo T, Bonanno JB, Jansong J, Reimer U, Shechter D (2015) Histone H2A and H4 N-terminal tails are positioned by the MEP50 WD repeat protein for efficient methylation by the PRMT5 arginine methyltransferase. J Biol Chem 290(15):9674–9689. doi:10.1074/jbc.M115.636894

Chan-Penebre E, Kuplast KG, Majer CR, Boriack-Sjodin PA, Wigle TJ, Johnston LD, Rioux N, Munchhof MJ, Jin L, Jacques SL, West KA, Lingaraj T, Stickland K, Ribich SA, Raimondi A, Scott MP, Waters NJ, Pollock RM, Smith JJ, Barbash O, Pappalardi M, Ho TF, Nurse K, Oza KP, Gallagher KT, Kruger R, Moyer MP, Copeland RA, Chesworth R, Duncan KW (2015) A selective inhibitor of PRMT5 with in vivo and in vitro potency in MCL models. Nat Chem Biol 11(6):432–437. doi:10.1038/nchembio.1810

Cho EC, Zheng S, Munro S, Liu G, Carr SM, Moehlenbrink J, Lu YC, Stimson L, Khan O, Konietzny R, McGouran J, Coutts AS, Kessler B, Kerr DJ, Thangue NB (2012) Arginine methylation controls growth regulation by E2F-1. EMBO J 31(7):1785–1797. doi:10.1038/emboj.2012.17

Duncan KW, Rioux N, Boriack-Sjodin PA, Munchhof MJ, Reiter LA, Majer CR, Jin L, Johnston LD, Chan-Penebre E, Kuplast KG, Porter Scott M, Pollock RM, Waters NJ, Smith JJ, Moyer MP, Copeland RA, Chesworth R (2016) Structure and property guided design in the identification of PRMT5 tool compound EPZ015666. ACS Med Chem Lett 7(2):162–166. doi:10.1021/acsmedchemlett.5b00380

Friesen WJ, Paushkin S, Wyce A, Massenet S, Pesiridis GS, Van Duyne G, Rappsilber J, Mann M, Dreyfuss G (2001) The methylosome, a 20S complex containing JBP1 and pICln, produces dimethylarginine-modified Sm proteins. Mol Cell Biol 21(24):8289–8300. doi:10.1128/MCB.21.24.8289-8300.2001

Fuhrmann J, Clancy KW, Thompson PR (2015) Chemical biology of protein arginine modifications in epigenetic regulation. Chem Rev 115(11):5413–5461. doi:10.1021/acs.chemrev.5b00003

Gao S, Wang Z (2012) Subcellular localization of p44/WDR77 determines proliferation and differentiation of prostate epithelial cells. PLoS One 7(11):e49173. doi:10.1371/journal.pone.0049173

Gu Z, Zhang F, Wang ZQ, Ma W, Davis RE, Wang Z (2013) The p44/wdr77-dependent cellular proliferation process during lung development is reactivated in lung cancer. Oncogene 32(15):1888–1900. doi:10.1038/onc.2012.207

Harris DP, Bandyopadhyay S, Maxwell TJ, Willard B, DiCorleto PE (2014) Tumor necrosis factor (TNF)-alpha induction of CXCL10 in endothelial cells requires protein arginine methyltransferase 5 (PRMT5)-mediated nuclear factor (NF)-kappaB p65 methylation. J Biol Chem 289(22):15328–15339. doi:10.1074/jbc.M114.547349

Harris DP, Chandrasekharan UM, Bandyopadhyay S, Willard B, DiCorleto PE (2016) PRMT5-mediated methylation of NF-kappaB p65 at Arg174 is required for endothelial CXCL11 gene induction in response to TNF-alpha and IFN-gamma costimulation. PLoS One 11(2):e0148905. doi:10.1371/journal.pone.0148905

Herrmann F, Pably P, Eckerich C, Bedford MT, Fackelmayer FO (2009) Human protein arginine methyltransferases in vivo--distinct properties of eight canonical members of the PRMT family. J Cell Sci 122(Pt 5):667–677. doi:10.1242/jcs.039933

Hosohata K, Li P, Hosohata Y, Qin J, Roeder RG, Wang Z (2003) Purification and identification of a novel complex which is involved in androgen receptor-dependent transcription. Mol Cell Biol 23(19):7019–7029

Hou Z, Peng H, Ayyanathan K, Yan KP, Langer EM, Longmore GD, Rauscher FJ 3rd (2008) The LIM protein AJUBA recruits protein arginine methyltransferase 5 to mediate SNAIL-dependent transcriptional repression. Mol Cell Biol 28(10):3198–3207. doi:10.1128/MCB.01435-07

Hsu JM, Chen CT, Chou CK, Kuo HP, Li LY, Lin CY, Lee HJ, Wang YN, Liu M, Liao HW, Shi B, Lai CC, Bedford MT, Tsai CH, Hung MC (2011) Crosstalk between Arg 1175 methylation and Tyr 1173 phosphorylation negatively modulates EGFR-mediated ERK activation. Nat Cell Biol 13 (2):174–181. doi:ncb2158 [pii] 10.1038/ncb2158

Hu D, Gur M, Zhou Z, Gamper A, Hung MC, Fujita N, Lan L, Bahar I, Wan Y (2015) Interplay between arginine methylation and ubiquitylation regulates KLF4-mediated genome stability and carcinogenesis. Nat Commun 6:8419. doi:10.1038/ncomms9419

Jansson M, Durant ST, Cho EC, Sheahan S, Edelmann M, Kessler B, La Thangue NB (2008) Arginine methylation regulates the p53 response. Nat Cell Biol 10 (12):1431–1439. doi:ncb1802 [pii] 10.1038/ncb1802

Karkhanis V, Hu YJ, Baiocchi RA, Imbalzano AN, Sif S (2011) Versatility of PRMT5-induced methylation in growth control and development. Trends Biochem Sci 36 (12):633–641. doi:S0968-0004(11)00143-5 [pii] 10.1016/j.tibs.2011.09.001

Kim S, Gunesdogan U, Zylicz JJ, Hackett JA, Cougot D, Bao S, Lee C, Dietmann S, Allen GE, Sengupta R, Surani MA (2014) PRMT5 protects genomic integrity during global DNA demethylation in primordial germ cells and preimplantation embryos. Mol Cell 56(4):564–579. doi:10.1016/j.molcel.2014.10.003

Kwak YT, Guo J, Prajapati S, Park KJ, Surabhi RM, Miller B, Gehrig P, Gaynor RB (2003) Methylation of SPT5 regulates its interaction with RNA polymerase II and transcriptional elongation properties. Mol Cell 11 (4):1055–1066. doi:S1097276503001011 [pii]

Li Z, Yu J, Hosohama L, Nee K, Gkountela S, Chaudhari S, Cass AA, Xiao X, Clark AT (2015) The Sm protein methyltransferase PRMT5 is not required for primordial germ cell specification in mice. EMBO J 34(6):748–758. doi:10.15252/embj.201489319

Ligr M, Patwa RR, Daniels G, Pan L, Wu X, Li Y, Tian L, Wang Z, Xu R, Wu J, Chen F, Liu J, Wei JJ, Lee P (2011) Expression and function of androgen receptor coactivator p44/Mep50/WDR77 in ovarian cancer. PLoS One 6(10):e26250. doi:10.1371/journal.pone.0026250

Likhite N, Jackson CA, Liang MS, Krzyzanowski MC, Lei P, Wood JF, Birkaya B, Michaels KL, Andreadis ST, Clark SD, Yu MC, Ferkey DM (2015) The protein arginine methyltransferase PRMT5 promotes D2-like dopamine receptor signaling. Sci Signal 8(402):ra115. doi:10.1126/scisignal.aad0872

Liu L, Zhao X, Zhao L, Li J, Yang H, Zhu Z, Liu J, Huang G (2016) Arginine methylation of SREBP1a via PRMT5 promotes De Novo lipogenesis and tumor growth. Cancer Res 76(5):1260–1272. doi:10.1158/0008-5472.CAN-15-1766

Lu T, Stark GR (2015) NF-kappaB: regulation by methylation. Cancer Res 75(18):3692–3695. doi:10.1158/0008-5472.CAN-15-1022

Luo M (2012) Current chemical biology approaches to interrogate protein methyltransferases. ACS Chem Biol 7(3):443–463. doi:10.1021/cb200519y

Mavrakis KJ, ER MD 3rd, Schlabach MR, Billy E, Hoffman GR, deWeck A, Ruddy DA, Venkatesan K, Yu J, McAllister G, Stump M, deBeaumont R, Ho S, Yue Y, Liu Y, Yan-Neale Y, Yang G, Lin F, Yin H, Gao H, Kipp DR, Zhao S, JT MN, Sprague ER, Zheng B, Lin Y, Cho YS, Gu J, Crawford K, Ciccone D, Vitari AC, Lai A, Capka V, Hurov K, Porter JA, Tallarico J, Mickanin C, Lees E, Pagliarini R, Keen N, Schmelzle T, Hofmann F, Stegmeier F, Sellers WR (2016) Disordered methionine metabolism in MTAP/CDKN2A-deleted cancers leads to dependence on PRMT5. Science 351(6278):1208–1213. doi:10.1126/science.aad5944

Migliori V, Mapelli M, Guccione E (2012) On WD40 proteins: propelling our knowledge of transcriptional control? Epigenetics: Off J DNA Methylation Soc 7(8):815–822. doi:10.4161/epi.21140

Morales Y, Caceres T, May K, Hevel JM (2016) Biochemistry and regulation of the protein arginine methyltransferases (PRMTs). Arch Biochem Biophys 590:138–152. doi:10.1016/j.abb.2015.11.030

Pal S, Vishwanath SN, Erdjument-Bromage H, Tempst P, Sif S (2004) Human SWI/SNF-associated PRMT5 methylates histone H3 arginine 8 and negatively regulates expression of ST7 and NM23 tumor suppressor genes. Mol Cell Biol 24 (21):9630–9645. doi:24/21/9630 [pii] 10.1128/MCB.24.21.9630-9645.2004

Pal S, Baiocchi RA, Byrd JC, Grever MR, Jacob ST, Sif S (2007) Low levels of miR-92b/96 induce PRMT5 translation and H3R8/H4R3 methylation in mantle cell lymphoma. EMBO J 26 (15):3558–3569. doi:7601794 [pii] 10.1038/sj.emboj.7601794

Pugh CS, Borchardt RT, Stone HO (1978) Sinefungin, a potent inhibitor of virion mRNA(guanine-7-)-methyltransferase, mRNA(nucleoside-2′-)-methyltransferase, and viral multiplication. J Biol Chem 253(12):4075–4077

Saha K, Adhikary G, Eckert RL (2015) MEP50/PRMT5 reduces gene expression by histone arginine methylation and this is reversed by PKCdelta/p38delta signaling. J Invest dermatol. doi:10.1038/jid.2015.400

Sakai N, Saito Y, Fujiwara Y, Shiraki T, Imanishi Y, Koshimizu TA, Shibata K (2015) Identification of protein arginine N-methyltransferase 5 (PRMT5) as a novel interacting protein with the tumor suppressor protein RASSF1A. Biochem Biophys Res Commun 467(4):778–784. doi:10.1016/j.bbrc.2015.10.065

Sarkies P, Sale JE (2012) Cellular epigenetic stability and cancer. Trends Genet: TIG 28(3):118–127. doi:10.1016/j.tig.2011.11.005
Schapira M, Ferreira de Freitas R (2014) Structural biology and chemistry of protein arginine methyltransferases. Med Chem Commun 5(12):1779–1788. doi:10.1039/c4md00269e
Smith TF, Gaitatzes C, Saxena K, Neer EJ (1999) The WD repeat: a common architecture for diverse functions. Trends Biochem Sci 24(5):181–185
Stirnimann CU, Petsalaki E, Russell RB, Muller CW (2010) WD40 proteins propel cellular networks. Trends Biochem Sci 35(10):565–574. doi:10.1016/j.tibs.2010.04.003
Stopa N, Krebs JE, Shechter D (2015) The PRMT5 arginine methyltransferase: many roles in development, cancer and beyond. Cell Mol Life Sci 72(11):2041–2059. doi:10.1007/s00018-015-1847-9
Sun L, Wang M, Lv Z, Yang N, Liu Y, Bao S, Gong W, Xu RM (2011) Structural insights into protein arginine symmetric dimethylation by PRMT5. Proc Natl Acad Sci U S A 108 (51):20538–20543. doi:1106946108 [pii] 10.1073/pnas.1106946108
Tarighat SS, Santhanam R, Frankhouser D, Radomska HS, Lai H, Anghelina M, Wang H, Huang X, Alinari L, Walker A, Caligiuri MA, Croce CM, Li L, Garzon R, Li C, Baiocchi RA, Marcucci G (2015) The dual epigenetic role of PRMT5 in acute myeloid leukemia: gene activation and repression via histone arginine methylation. Leukemia. doi:10.1038/leu.2015.308
Tee WW, Pardo M, Theunissen TW, Yu L, Choudhary JS, Hajkova P, Surani MA (2010) Prmt5 is essential for early mouse development and acts in the cytoplasm to maintain ES cell pluripotency. Genes Dev 24 (24):2772–2777. doi:24/24/2772 [pii] 10.1101/gad.606110
Troffer-Charlier N, Cura V, Hassenboehler P, Moras D, Cavarelli J (2007) Functional insights from structures of coactivator-associated arginine methyltransferase 1 domains. EMBO J 26(20):4391–4401. doi:10.1038/sj.emboj.7601855
Wang L, Pal S, Sif S (2008) Protein arginine methyltransferase 5 suppresses the transcription of the RB family of tumor suppressors in leukemia and lymphoma cells. Mol Cell Biol 28(20):6262–6277. doi:10.1128/MCB.00923-08
Wei H, Wang B, Miyagi M, She Y, Gopalan B, Huang DB, Ghosh G, Stark GR, Lu T (2013) PRMT5 dimethylates R30 of the p65 subunit to activate NF-kappaB. Proc Natl Acad Sci U S A 110(33):13516–13521. doi:10.1073/pnas.1311784110
Xu C, Min J (2011) Structure and function of WD40 domain proteins. Protein Cell 2(3):202–214. doi:10.1007/s13238-011-1018-1
Yang Y, Bedford MT (2013) Protein arginine methyltransferases and cancer. Nat Rev Cancer 13(1):37–50. doi:10.1038/nrc3409
Yang Y, Hadjikyriacou A, Xia Z, Gayatri S, Kim D, Zurita-Lopez C, Kelly R, Guo A, Li W, Clarke SG, Bedford MT (2015) PRMT9 is a type II methyltransferase that methylates the splicing factor SAP145. Nat Commun 6:6428. doi:10.1038/ncomms7428
Yi P, Gao S, Gu Z, Huang T, Wang Z (2014) P44/WDR77 restricts the sensitivity of proliferating cells to TGFbeta signaling. Biochem Biophys Res Commun 450(1):409–415. doi:10.1016/j.bbrc.2014.05.125
Yost JM, Korboukh I, Liu F, Gao C, Jin J (2011) Targets in epigenetics: inhibiting the methyl writers of the histone code. Curr Chem Genomics 5(Suppl 1):72–84. doi:10.2174/1875397301005010072
Zhang HT, Zeng LF, He QY, Tao WA, Zha ZG, Hu CD (2016) The E3 ubiquitin ligase CHIP mediates ubiquitination and proteasomal degradation of PRMT5. Biochim Biophys Acta 1863(2):335–346. doi:10.1016/j.bbamcr.2015.12.001
Zhou L, Hosohata K, Gao S, Gu Z, Wang Z (2014) cGMP-dependent protein kinase Ibeta interacts with p44/WDR77 to regulate androgen receptor-driven gene expression. PLoS One 8(6):e63119. doi:10.1371/journal.pone.0063119

Chapter 8
Symmetry-Directed Design of Protein Cages and Protein Lattices and Their Applications

Aaron Sciore and E. Neil G. Marsh

Abstract The assembly of individual protein subunits into large-scale structures is important in many biological contexts. Proteins may assemble into geometrical cages or extended lattices that are characterized by a high degree of symmetry; examples include viral capsids and bacterial S-layers. The precisely defined higher order structure exhibited by these assemblies has inspired efforts to design such structures *de novo* by applying the principles of symmetry evident in natural protein assemblies. Here we discuss progress towards this goal and also examples of natural protein cages and lattices that have been engineered to repurpose them towards a diverse range of applications in materials science and nano-medicine.

Keywords Protein assembly • Protein cages • Protein lattices • Computational protein design • Symmetry • Synthetic biology • Biomaterials • Nanomedicine

8.1 Introduction

In recent years our understanding of protein-protein interactions has blossomed, driven by advances in techniques such as native mass spectrometry, x-ray crystallography, electron microscopy, and computational modeling. The assembly of proteins into highly ordered, large-scale oligomeric structures has been found to play an important role in an increasing number of biological phenomena. At the same time our understanding of the principles by which proteins oligomerize has advanced to the point that it is now possible to rationally design such structures (Jha et al. 2010; Huang et al. 2007). This is an exciting development, as it paves the way for the design of protein-based nanomaterials that assemble in a highly ordered manner.

A. Sciore
Department of Chemistry, University of Michigan, Ann Arbor, MI 48109, USA
e-mail: sciore@umich.edu

E.N.G. Marsh (✉)
Department of Chemistry, University of Michigan, Ann Arbor, MI 48109, USA

Department of Biological Chemistry, University of Michigan, Ann Arbor, MI 48109, USA
e-mail: nmarsh@umich.edu

J.R. Harris, J. Marles-Wright (eds.), *Macromolecular Protein Complexes*,
Subcellular Biochemistry 83, DOI 10.1007/978-3-319-46503-6_8

Such materials are very attractive for industrial and medicinal applications as they are biocompatible, retain their biological function and may exhibit new properties that emerge as a consequence of their higher order structure.

This chapter is focused on applying the principles of symmetry, exhibited in naturally occurring protein assemblies, to the design of new protein-based nanomaterials. Initially, we briefly review some examples of naturally formed symmetric protein assemblies, primarily to illustrate how their oligomeric structures relate to their biological functions. Next we discuss various approaches that have been taken to repurpose protein assemblies for therapeutic and materials science applications. In the last part of the chapter we will describe the basic theory underlying approaches to designing symmetrical protein assemblies, followed by a fairly comprehensive survey of the methods that have been used to create *de novo* designed protein assemblies.

8.2 Natural Protein Assemblies

The self-association of protein subunits is a ubiquitous feature of biology. It is essential to functions as diverse as catalysis and regulation of enzymes, cell signaling, cytoskeleton maintenance and cell motility, sequestering of metabolic pathways, iron storage and packaging of nucleic acids by viruses. Proteins generally self-associate through complementary and highly specific hydrophobic interfaces on their surfaces. Except for the case of a symmetrical dimer, each protein subunit will have two (or more) self-association interfaces. The precise orientation of these interfaces relative to each other defines whether the protein will self-assemble into either an extended structure, represented by one-dimensional fibers and two- and three-dimensional lattices, or a discrete structure, in which the subunits form either rings or a highly symmetrical 3-dimensional assembly, generally referred to as a 'protein cage'. In all of these cases the structures can be mathematically described by a combination of rotational and translational symmetry operations that map one subunit on to any of the others.

Naturally occurring one-dimensional protein fibers include the well-studied proteins actin and tubulin, which play vital structural roles in the cell. Actin (Fig. 8.1a) is one of the most common fiber-forming proteins, and is found in high abundance in nearly all eukaryotic cells, where it is the principle component of the cytoskeleton (Dominguez and Holmes 2011). The dynamic polymerization and de-polymerization of actin underpins important mechanical properties in the cell such as motility, and elongation during cell division. The cylindrical fibers formed by tubulin are the principle component of the microtubules that form the mitotic spindles responsible for the separation of chromosomes during mitosis. This protein also undergoes dynamic polymerization and de-polymerization (Abal et al. 2003). Inhibition of this process by taxanes forms the basis for these drugs' potent anti-cancer activity. Taxane-capped microtubules are unable to depolymerize, effectively inhibiting mitosis without causing immediate cell death.

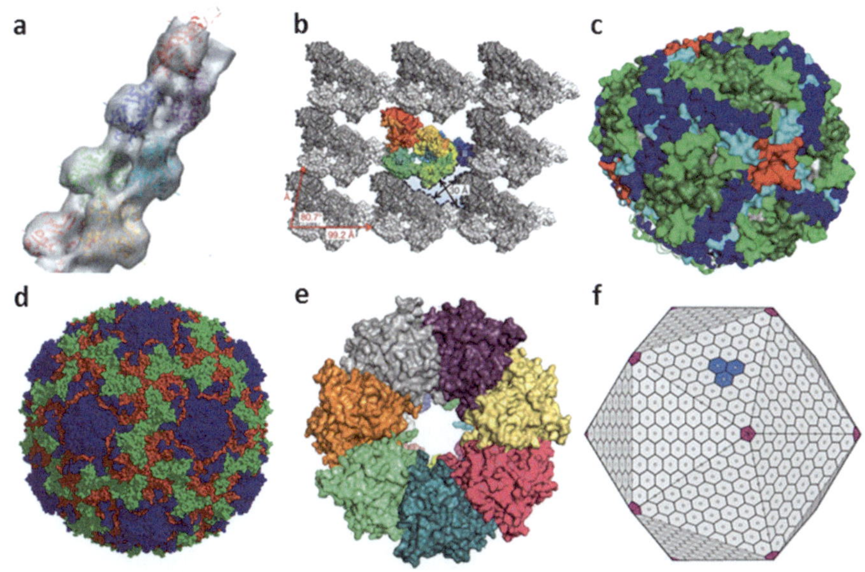

Fig. 8.1 Natural protein cages. (**a**) Cryo-EM reconstruction of fibrous actin (Adapted with permission from Oda et al. 2009, © Macmillan Publishers Ltd) (**b**) Crystal structure of an S-layer protein lattice with *P4* symmetry (Adapted with permission from Baranova et al. © Macmillan Publishers Ltd, 2012) (**c**) Crystal structure of octahedral bacterioferritin (PDB ID 3GVY) (**d**) Crystal structure of icosahedral rhinovirus capsid (PDB ID 4RHV) (**e**) View along the sevenfold axis of the crystal structure of GroEL (PDB ID 1GRL) (**f**) Cartoon of an assembled bacterial microcompartment consisting of hexameric (*blue*) and pentameric (*purple*) subunits (Adapted with permission from Yeates et al. 2011 © Elsevier)

Two-dimensional protein lattices are probably the least well-known and studied class of protein assemblies. They are found primarily in bacteria and archaea, where they form porous surface layers, or S-layers (Fig. 8.1b) that adhere to the outer layer of the cell wall and can comprise up to 20% of an organism's total protein content (Sára and Sleytr 2000). S-layers comprise multiple copies of one or a few glycoproteins, which can be made to self-assemble *in vitro* into lattices, rods, or spheres (Sleytr and Beveridge 1999). Each species has one or more unique S-layer proteins that form pores with sizes ranging from 2–8 nm. Different S-layers are characterized by different symmetries – so far S-layers have been identified with oblique (*P1* and *P2*), square (*P4*), and hexagonal (*P3* and *P6*) symmetries. These proteins are thought to play a role in such critical cellular functions such as cell adhesion, molecular sieving, and binding of toxic cations. Several species of bacteria have been found to synthesize different S-layer proteins in response to environmental stress, suggesting that S-layers may play a role in cellular adaptation as well (Sara et al. 1994).

Three-dimensional protein cages are characterized by a high degree of symmetry and are widely distributed in Nature. These play important roles in the storage of nucleic acids and iron, in protein folding and in confining reactive intermediates in metabolism. A key feature of protein cages is their ability to create a unique microenvironment within the cage interior that determines the cage's function. A good example is the iron storage protein ferritin (Theil 1987). This octahedral, 24-subunit protein cage is highly conserved and found in almost all organisms (Fig. 8.1c). The interior surface of the ferritin cage contains a large number of negatively charged residues. These bind ferrous iron atoms and catalyze their oxidation to the ferric form and thence to ferric oxide, which in turn serves as a nucleation site for other iron atoms. Additional nucleation of iron continues until the interior of the ferritin protein cage is completely filled with as many as 4500 iron atoms. This allows the cell to store iron, a biologically scarce element, in a highly efficient manner compared to a protein with discrete iron binding sites.

Many viruses encapsulate their DNA or RNA in an icosahedral protein capsid that is built from multiples of 60 protein subunits (Fig. 8.1d). In contrast to ferritin, the interior-facing residues of capsid subunits are positively-charged, thereby neutralizing the negatively charged nucleic acids and allowing them to be efficiently packaged within the capsid. Viral capsids also nicely illustrate another feature of protein cages – highly cooperative assembly of the subunits. Whereas binding interactions between individual subunits are usually much weaker than most protein-protein interactions, dissociating a single subunit from an assembled capsid requires significantly more energy (Singh and Zlotnick 2003). This cooperative binding effect imbues the capsid with a high degree of stability and protects the capsid proteins and enclosed nucleotides against environmental degradation (Ross et al. 2006).

Chaperonins, also known as heat shock proteins, illustrate the use of protein cages to create unique reaction environments, in this case to facilitate refolding of misfolded proteins (Vabulas et al. 2010). Most chaperonins form protein complexes with various symmetries including one of the only known tetrahedral protein complexes, but all with hydrophobic patches in the interior microenvironment that facilitates the refolding or degradation of misfolded proteins. The GroEL/GroES system is the best studied chaperonin, GroEL forms a barrel-shaped 14-subunit protein cage with a 4.5 nm wide hydrophobic interior channel (Fig. 8.1e), with heptameric GroES serving as a 'lid' for the barrel. Misfolded proteins bind to the hydrophobic interior of the barrel, followed by an ATP-driven conformational change in the GroEL subunits, which results in the interior surface becoming significantly more polar and pulling the misfolded protein apart. The unfolded protein then has the opportunity to refold correctly in the sequestered environment provided by the protein cage (Landry and Gierasch 1991).

Bacterial microcompartments (BMCs) provide another good example of a protein cage designed to facilitate biological reactions. BMCs comprise hexameric shell proteins that form a honeycomb-shaped lattice, interspersed with a small number of pentameric shell proteins that introduce kinks into the lattice (Yeates et al. 2008). The combination of hexamers and pentamers yields large (40–200 nm in diameter) pseudo-icosahedral protein complexes (Fig. 8.1f) capable of co-encapsulating multiple enzymes in a metabolic pathway to increase the efficiency of catalysis. The best-studied BMC is the carboxysome, found in photosynthetic cyanobacteria, which co-encapsulates carbonic anhydrase and RuBisCo. Carbonic anhydrase produces CO_2 from bicarbonate, which is quickly reacted with ribulose bisphosphate by RuBisCo as part of the Calvin cycle. Delocalizing carbonic anhydrase to the cytosol results in significant loss of RuBisCo efficiency (Price and Badger 1989), indicating that the carboxysome plays an important role in the efficient fixation of CO_2 by cyanobacteria.

8.3 Applications of Natural Protein Assemblies

8.3.1 Functionalization of Protein Cages

Naturally-occurring protein assemblies have been adapted for use in a diverse range of materials science and nano-medicine applications. The simplest functionalization of a protein assembly involves modifying the assembly to perform tasks similar to its cellular function. The natural ability of ferritin to biomineralize iron salts to form iron oxide nanoparticles of a defined size has been exploited to synthesize a wide variety of nanoparticles including silver, platinum, palladium, cobalt oxide, and cadmium sulfide (Flenniken et al. 2009). In the presence of a small peptide fragment a ferromagnetic nanoparticle of CoPt could also be synthesized (Klem et al. 2005). Biomineralization functionality could also be engineered into heat shock proteins (McMillan et al. 2005; Ishii et al. 2003) and viral capsids (Brumfield et al. 2004) by mutating the hydrophobic or cationic interior of these proteins to metal-binding or anionic residues. This allowed highly mono-disperse iron oxide nanoparticles to be synthesized with sizes determined by the interior diameter of these protein cages. These protein cage-derived iron-oxide nanoparticles could then be used as a nucleation site for the synthesis of single-walled carbon nanotubes, with the diameter of these nanotubes being proportional to the size of the seed iron oxide particle (Kramer et al. 2005).

Precisely sized nanoparticles, in particular iron and cobalt oxide, are of considerable interest in nanoelectronics research (Jutz et al. 2015). A single layer of protein cages containing nanoparticles can be deposited onto a pre-coated surface in a tight, hexagonal packing pattern, with a packing density close to theoretical values (Atsushi et al. 2006). In what is known as the bio-nano process, silicon wafers or other substrates are precisely patterned with hydrophobic or hydrophilic coatings, with nanoparticle-containing ferritin localizing only on the hydrophilic film. This is then exposed to heat, burning away the ferritin to leave only the iron oxide nanoparticle, which is then reduced to yield a precisely-patterned array of metallic iron spheres (Takuro et al. 2007). This has been used to generate semiconducting logic devices on the nano scale such as thin film transistor flash memory and floating nanodot gate memory devices (Kazunori et al. 2007). Parameters of the bio-nano process can be controlled with superior precision than existing nanoelectronics technology (Kiyohito et al. 2007) although this technology has not yet found a commercial application.

The microenvironment of the protein cage interior presents an attractive target for entrapping catalytic species, thereby creating a nano-scale bioreactor (Bode et al. 2011). Palladium nanoparticles formed on the interior of ferritin have been shown to catalyze the hydrogenation of olefins (Ueno et al. 2004). The kinetics of this reaction could be controlled by varying the size of the olefin substrate, indicating that this reaction is limited by diffusion of the olefin through the pores of ferritin. Similarly a rhodium (II) complex bound to the interior of ferritin was used to catalyze the polymerization of phenylacetylene (Abe et al. 2009). The resulting polymer, encapsulated within the ferritin cage, had a remarkably narrow size distribution. This level of precision in polymerization control may open up new avenues of research into designed smart materials.

The hollow interior of a protein cage has also sparked interest for entrapment and immobilization of enzymes, fashioning a bioreactor similar to the carboxysome. Protein cages can be assembled around enzymes in solution, trapping them, and these enzymes retain their catalytic activity (Comellas-Aragones et al. 2007). Using one of the larger viral capsid shells, multiple enzymes in a metabolic pathway were co-localized, but this had negligible effect on turnover rates (Patterson et al. 2014a), and there are indications that the catalytic activity is actually reduced due to crowding effects when a large number of enzymes are encapsulated in each protein cage (Minten et al. 2011).

Protein cages also hold promise for biomedical applications. Viral capsids, i.e. protein cages lacking the viral genome, are a particularly attractive target for therapeutic delivery systems, as the viral capsid has evolved to cross the cell membrane and deliver its payload. The sizeable interior cavity of a viral capsid has been adapted to encapsulate drug molecules and imaging agents, and can also accommodate plasmids as large as 17.6 kbp, thereby protecting them from nucleases (Kimchi-Sarfaty et al. 2003; Štokrová et al. 1999). An exotoxin-encoding plasmid encapsulated in a viral capsid was delivered in this way to tumor cells, resulting in a reduction of the size of the tumor *in vivo* (Kimchi-Sarfaty et al. 2006). Likewise, small molecules that are encapsulated in the capsid interior or covalently bound to the protein cage can be delivered to the cells with greater efficiency than unmodified small molecules (Aljabali et al. 2013), and can be released over an extended period of time (Flenniken et al. 2005).

Using protein cages as a delivery vehicle for therapeutics still has the drawback that their toxic payload is delivered indiscriminately. Therefore significant research has gone into decorating the capsid exterior with targeting ligands that localize the therapeutic protein cage to the cell type of interest. Drugs can be selectively targeted to specific cell types by decorating viral capsids with either small molecules (Zhao et al. 2011; Banerjee et al. 2010) or large biomolecules (Huang et al. 2011; Lockney et al. 2011) that bind receptors that are overexpressed by many cancer cell types. For example, by engineering appropriate receptor-targeting sequences onto the exterior of the capsid, therapeutic molecules could selectively target Jurkat leukemia T cells (Stephanopoulos et al. 2010) or human hepatocellular carcinoma cells (Ashley et al. 2011). In both cases, the protein cages delivered their cytotoxic payload exclusively to the targeted cells, inducing cell death in the majority of those cancer cells without affecting any of the control cells.

Protein cages also hold promise for therapeutic imaging *in vivo* by encapsulating positron emission tomography (PET) imaging agents or magnetic resonance imaging (MRI) contrast agents (Cormode et al. 2010). The common PET imaging agent ^{18}F was bioconjugated to viral capsids (Hooker et al. 2008) and the location of these capsids could be dynamically imaged *in vivo* (Flexman et al. 2008). Iron oxide nanoparticles, a simple and effective MRI contrast agent, could be inserted into the hollow cavity of ferritin (Uchida et al. 2008) or viral capsids (Ghosh et al. 2012) with similar results. Notably, incorporation of Gd^{3+} into the calcium-binding sites of the cowpea chlorotic mosaic virus led to a species with the highest relaxivity values measured to date, potentially leading to new applications with low-dose MRI contrast agents (Allen et al. 2005).

Viral capsids have a tendency to elicit strong immunogenic responses due to the display of multiple epitopes on the capsid surface (Kaiser et al. 2007). Whereas this is a disadvantage for many applications, it has attracted significant interest in the field of vaccine development. Recombinantly-produced viral capsids induce a strong immune response while offering a safe alternative to the use of attenuated or killed virus particles as they contain no genetic information from the virus; the major concern being that this immunogenicity must be modulated to avoid an adverse immune response (Rebeaud and Bachmann 2012; Jennings and Bachmann 2009). They also obviate the use of toxic adjuvants such as aluminium that are needed to boost immune response when low-copy epitopes are used as vaccines (Kawano et al. 2013). Currently over a dozen capsid-based vaccines are approved for clinical trials or clinical use, targeting Influenza, Hepatitis A & B, HPV, and others (Teunissen et al. 2013; Donaldson et al. 2015). Most interestingly, viral capsids have shown promise as vectors for therapeutic vaccinations against non-viral diseases, such as Alzheimer's disease (Wiessner et al. 2011), cancer (Garcia 2011; Speiser et al. 2010), arthritis (Spohn et al. 2008), and nicotine addiction (Maurer et al. 2005).

8.3.2 Functionalization of S-Layer proteins

Bacterial S-layer proteins have also been explored for functionalization, although the extended geometry results in different capabilities. Like protein cages, assembled S-layers have interfacial pores, which can be modified to biomineralize a broad range of metal ions (Mertig et al. 1999; Shenton et al. 1997). However, unlike protein cages, S-layers lack a defined interior or exterior, instead surface residues can be categorized based on which face of the assembled S-layer it resides. One face of the S-layer is strongly cytophilic due to the presence of several homologous 4-residue long sequences which bind to cell wall-associated peptidoglycans, while the other face of the S-layer is strongly cytophobic (Engelhardt and Peters 1998).

For one of the best-studied S-layer proteins, SbpA from *Lysinibacillus sphaericus*, it was discovered that the orientation of the faces was dependent on the pH of the solution during deposition: basic or neutral solutions induce SbpA to crystallize with the cytophobic face exposed, whereas acidic solutions induce crystallization with the cytophilic face exposed (Rothbauer et al. 2013). This property was used to design antifouling (cytophobic) ultrafiltration membranes (Weigert and Sára 1996). Co-immobilizing proteins such as glucose oxidase (Picher et al. 2013) or a prostate-specific antigen (PSA) antibody (Pleschberger et al. 2004) to the cytophobic face of the S-layer allowed metabolites such as glucose and prostate-specific antigen to be accurately detected. This method gave significantly lower rates of false positives due to nonspecific cell adhesion than standard methods such as adhesion to gold nanoparticles. Ligands that bind specific cell types have also been attached to the cytophobic face of an S-layer, and by controlling the deposition of several orthogonally mutated S-layers, co-cultures of different cell types could be obtained in a spatially defined manner (Rothbauer et al. 2015).

S-layers can also be functionalized for therapeutic aims, although they have been utilized much less extensively than protein cages. Due to their repetitive structure, S-layers induce a heightened immune response, but are generally regarded as less effective than other methods of epitope presentation (Ilk et al. 2011). However, purified S-layers can be used to inoculate against the species of bacteria that produces the S-layer. This approach was used to generate an effective vaccine against six virulent isolates of the fish pathogen *Aeromonas hydrophila*, which was otherwise very difficult to vaccinate against due to the wide variation between isolated strains (Poobalane et al. 2010).

Bacterial S-layers have also been explored for drug delivery applications. The cytophilic face of SbpA was found to bind to the outer layer of liposomes, encasing them as it would a cell wall (Ucisik et al. 2015). This dramatically increased the shelf life of liposome-based therapeutics by reducing the rate that the drug molecules diffused through the liposome, a common limitation of liposome-based therapies (Raza et al. 2013). The S-layers on encased liposomes are functionalizable – by genetically fusing an eGFP sequence to the terminus of an S-layer, individual liposomes could be observed as they diffused through the cell membrane (Ilk et al. 2004).

8.4 Design of *de novo* Protein Assemblies

Functionalizing natural protein assemblies represents one approach to developing new bio-inspired materials; unfortunately there are a relatively small number of well-characterized natural assemblies amenable to modification. An alternative approach that has gained popularity in recent years is to design protein assemblies *de novo* from multiple copies of smaller symmetric building blocks. The primary advantage of this approach is customizability: depending upon the choice of protein building block and the manner in which the blocks are assembled, one can introduce a wide range of functionality into the assembly, control the interior cavity size and potentially make its assembly responsive to environmental conditions or specific ligands.

8.4.1 Conceptual Approaches to Protein Assembly

The basic requirements for assembling a protein into a geometrically well-defined structure are deceptively simple: one must have two rotationally symmetric protein domains connected at the proper dihedral angle (Fig. 8.2a) (Yeates and Padilla 2002). For simplicity, we will only discuss the assembly of protein cages, but the principles described below apply to the design of extended protein assemblies as well.

Protein cages formed by connecting two symmetric protein domains can be found in four distinct point symmetries: tetrahedral (T or 332), octahedral (O or 432), icosahedral (I or 532), and prism (D_n). For the first three of these symmetries, any two of the three symmetry operators may be combined at the proper angle to yield a protein assembly with that symmetry. Thus, an icosahedral cage can, in principle, be formed from a subunit possessing domains that oligomerize with the following rotational symmetries: a threefold and a twofold oligomerization domain, a fivefold and a threefold domain, or a fivefold and a twofold domain, although the requisite dihedral angle between the two symmetry axes differs for each of these three "symmetry pairs". The binary permutations of symmetry pairs that can be used to assemble protein cages with tetrahedral, octahedral and icosahedral geometries are illustrated in Fig. 8.2b. In addition, any symmetry pair with a twofold symmetry element also has the potential to form a prism. GroEL, a 14-subunit protein cage formed from a sevenfold and a twofold symmetry pair, serves as an excellent example of this symmetry.

Although it is relatively easy to genetically fuse two protein domains that oligomerize with the desired symmetry to each other, it is the alignment of their respective symmetry axes that has posed the greatest challenge to the successful design of protein cages. This is because protein structures are both inherently asymmetric and flexible. Designing in the correct dihedral angle between two symmetry sites is difficult and requires a high level of precision, as a rigid, but improperly oriented dihedral angle will lead to aggregation. The alternative, allowing the dihedral angle to remain flexible, can lead to a wide variety of possible assemblies with irregular geometries, although the range of structures can be restricted, as discussed below.

For flexible self-assembling systems, in which the dihedral angle between symmetry elements is completely unconstrained, the only constraint is that there must be no unpaired symmetry elements. This means that the set of stable, discrete protein assemblies, in principle, contains almost all of the assemblies with a subunit number that is a multiple of the least common denominator of the two symmetry elements. Thus, a trimer-dimer symmetry pair forms assemblies with multiples of 6

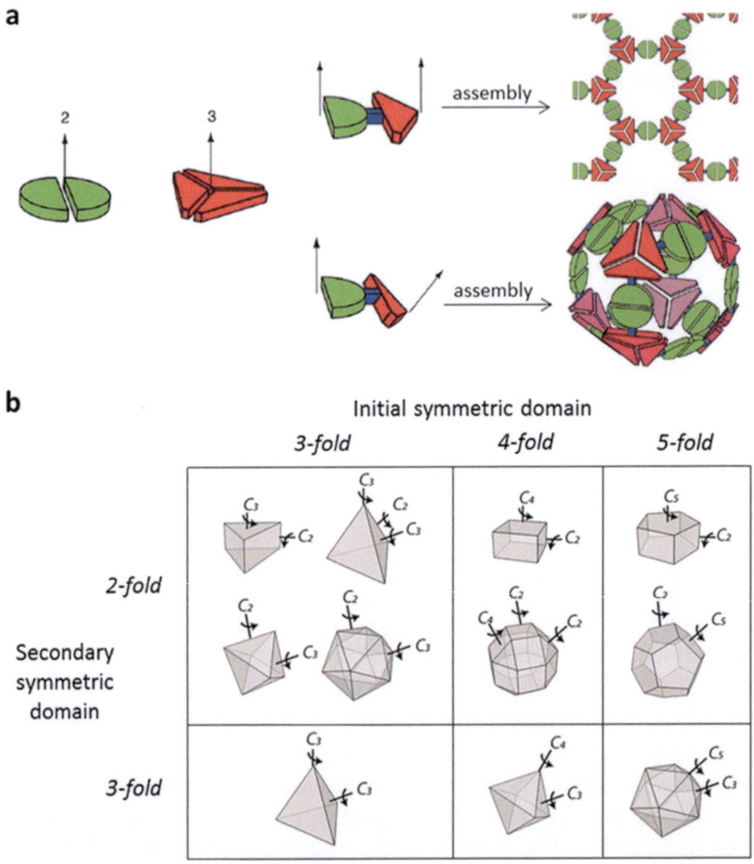

Fig. 8.2 Assembly of designed fusion protein assemblies (**a**) Rigidly linking a dimeric protein (*green*) with a trimeric protein (*red*) will result in different assemblies depending on the dihedral angle imparted by the rigid linker (*blue*) (Adapted from Padilla et al. 2001) (**b**) Six different symmetry pairs of fusion proteins connected at a proper dihedral angle are expected to result in the assembly of closed Euclidean solids

subunits (or 2 trimers), a trimer-trimer symmetry pair forms assemblies with multiples of 3 subunits (one trimer), and a trimer-tetramer symmetry pair forms assemblies with multiples of 12 subunits (4 trimers) (Fig. 8.3).

The specific complexes that can be formed for each symmetry pair is dependent on the range of allowed dihedral angles imparted by the flexibility in the system. For example, one of the possible 16-trimer assemblies formed from a subunit with a trimer-tetramer symmetry pair is arranged like an icosahedron but missing four nonadjacent trimers. If the range of allowed dihedral angles for the subunit of this trimer-tetramer system was restricted to one close to an icosahedron, this 16-mer pseudo-icosahedral species would be favored; whereas the ring-shaped 16-mer of trimers requires a much wider range of dihedral angles to form. While ostensibly,

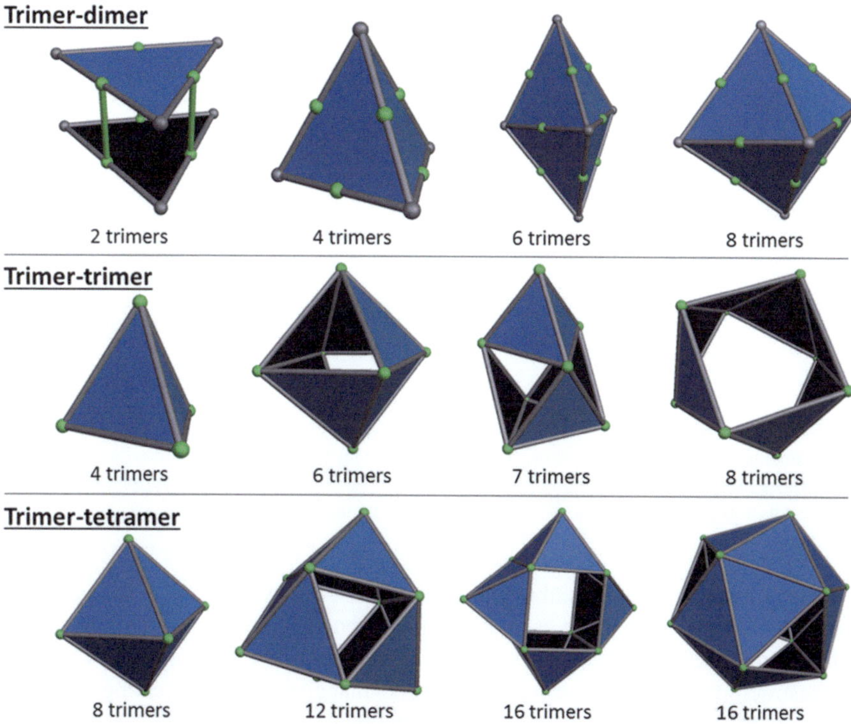

Fig. 8.3 Examples of possible assemblies that can be formed from fusion proteins with different symmetry pairs. *Blue triangles* represent a trimeric building block protein, fused with a second symmetric protein, represented as *green dots* at either the edges (dimeric symmetry domain) or vertices (trimeric or tetrameric symmetry domains) of the *blue triangles*. The assembled structures may be porous and/or require differing dihedral angles to form, but have the correct oligomerization state at every point of attachment and contain no unpaired symmetry elements

the analysis above implies that a flexibly-linked symmetry pair naturally makes an infinite assortment of complexes, in reality smaller assemblies are entropically favored (Boyle et al. 2012). Thus the assembly of the octahedron from a flexibly-linked trimer-tetramer symmetry pair is favored over either of the 16-trimer structures shown in Fig. 8.3.

8.4.2 *Extended Assemblies – Designed Fibers and Lattices*

The simplest *de novo* designed protein assemblies are one-dimensional protein fiber strands. These only require an asymmetric unit with two different self-associating dimeric interfaces. Since many natural proteins exist as dimers, it is relatively simple to design a fiber-forming asymmetric unit, with the earliest designs dating from the mid-1990s (Zhang et al. 1993; Woolfson 2010). This is because dimer-dimer

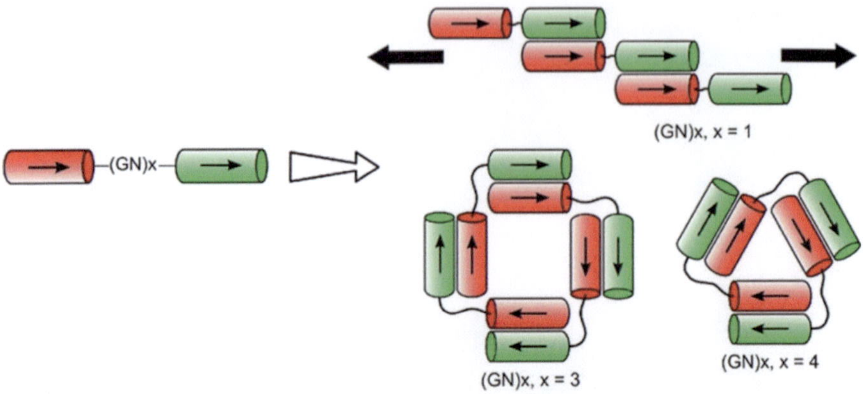

Fig. 8.4 A designed fusion protein with a flexibly-linked dimer-dimer symmetry pair assembles into filaments when this linker is short (*top*) but with longer linkers assembles into discrete oligomers whose size is inversely proportional to the length of the flexible linker (*bottom*) (Reprinted with permission from Boyle et al. 2012 © American Chemical Society)

symmetry pairs can still assemble into fibers even if the two dimeric domains aren't precisely oriented opposite to each other. In the general case, these assemblies form helical structures at the micro level, but still results in the formation of a fibrous macrostructure. These fibers can form networks that give rise to interesting solution-phase properties, such as hydrogelation (Rajagopal et al. 2009; Dong et al. 2008) and antibacterial activity (Salick et al. 2007; Ghadiri et al. 1993), and have been investigated as substrates for cell adhesion and growth (Haines-Butterick et al. 2007; Villard et al. 2006).

The formation of fibers requires a degree of rigidity between the oligomerization domains, otherwise the ends of the fiber will close on each other to form rings. This is illustrated clearly by a designed, rigid filament-forming fusion protein that was modified by adding a flexible linker between its two dimerization domains (Boyle et al. 2012). The protein assembly retained its filamentous structure after 2 or 4 flexible residues were added, but with the addition of a 6-residue linker the protein assembled into a tetramer-of-dimers, while an 8-residue linker yielded a trimer-of-dimers, and 10 residues led to a mixture of trimers-of-dimers and dimers-of-dimers (Fig. 8.4).

The fusion protein approach has also been used to design two-dimensional lattices; this is a harder problem due to the extra spatial dimension that must be aligned. In an early study a cysteine residue was introduced at each edge of a fourfold symmetric aldolase protein, allowing it to be tagged with biotin (Ringler and Schulz 2003). The dimeric, biotin-binding protein streptavidin was then mixed with the modified aldolase, resulting in a square lattice-like assembly. However, the lattice could only be propagated for a few repeating units before becoming disordered. A later study that examined the assembly of five different tetrameric proteins when fused to a streptavidin tag found that two of these five could be assembled into lattices that maintained order over a large number of subunit repeats (Fig. 8.5) (Sinclair

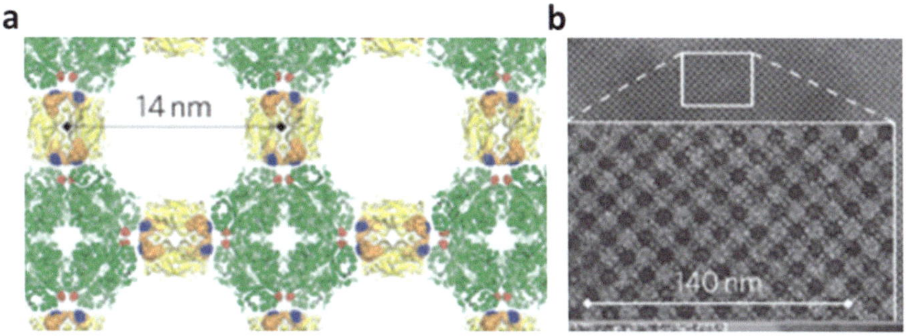

Fig. 8.5 A *de novo* designed square lattice (**a**) Designed lattice assembled by connecting a fourfold and a twofold-symmetric protein (**b**) TEM image of the assembled lattice (Adapted with permission from Sinclair et al. 2011, © Macmillan Publishers Ltd)

et al. 2011). For one protein lattice, assembly was found to be dependent on the linker length;: a linker of less than two residues resulted in nonspecific aggregation, but a two or three residue linker formed long-range lattices.

In a different approach, a two-component system based on *de novo* designed coiled-coils was constructed by linking a homo-trimeric coiled-coil to one of two heterodimer-forming α-helices through a disulfide bond (Fletcher et al. 2013). The locations of the two cysteine residues were chosen to orient the two oligomerization domains back-to-back, such that the resulting lattice would form a two dimensional honeycomb pattern as illustrated in Fig. 8.6. Instead, due to the slight inherent flexibility of proteins, these fusion proteins formed hollow spherical assemblies approximately 100 nm in diameter, similar in kind to bacterial microcompartments (Fletcher et al. 2013). Interestingly, a mutation of Asn to Ile introduced at the heterodimeric interface that increased the binding affinity of the heterodimeric coiled-coil resulted in particles forming with a 33% smaller diameter. This illustrates the interplay between thermodynamic forces in protein assembly – strengthening the coiled-coil interaction allows the proto-particle to overcome larger steric stresses associated with increasing the curvature of the honeycomb lattice and thus forms smaller particles.

Advances in computational methods have allowed more precisely designed protein lattices to be constructed. Lanci et al. generated a protein crystal with the rarely-observed space group *P6* by computationally designing a self-dimerizing interface into the outwardly-facing side-chains in a homo-trimeric coiled-coil (Fig. 8.7) (Lanci et al. 2012). A different hexagonal lattice was designed from a hexameric building block protein (Fig. 8.8) (Matthaei et al. 2015). After computationally aligning hexamers into the desired lattice, the inter-terminus distance was measured between subunits of adjacent hexagons. Using this as a guide, a linker was designed that was long enough to bridge between adjacent hexagons, but too short for both subunits of the fusion protein to fit into the same hexagon.

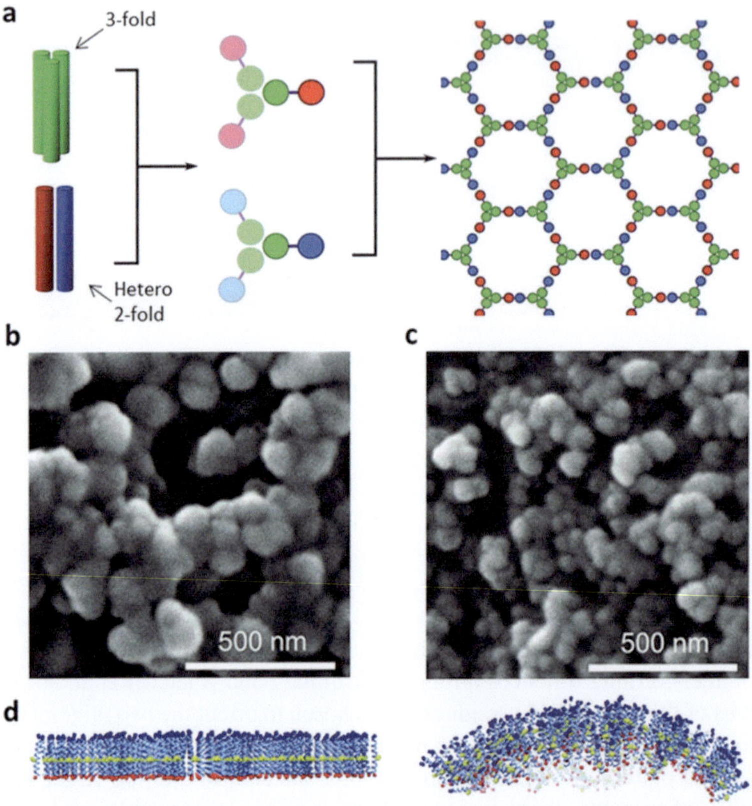

Fig. 8.6 Assembly of bacterial microcompartment-sized spheres from a flexible lattice-forming protein. (**a**) Design scheme: A homo-trimeric coiled-coil (*green*) was attached to one of two halves of a heterodimeric coiled-coil (*red & blue*), which combine to make a hexagonal lattice. (**b**, **c**) SEM images of the assembled superstructure with a weakly-associating (**b**) or a strongly-associating (**c**) heterodimer. (**d**) *Side view* of the molecular dynamics simulation of the potential curvature across 19 tessellated hexagons (Figure reproduced with permission from Fletcher et al. 2013, © American Academy for the Advancement of Science)

The Tezcan group has focused on the design of split metal-binding sites to mediate protein assembly. Each protein subunit supplies half of the ligands to the metal, typically Zn^{2+}, which bridges between the two subunits (Brodin et al. 2012). Computational design was used to position two Zn-mediated dimerization domains orthogonally to each other, with one domain having a significantly higher affinity than the second. Addition of one equivalent of Zn^{2+} resulted in dimerization in solution, as predicted, but the crystal structure of the dimeric complex revealed that there was a third, weak oligomerization site that connected two neighboring dimers together, such that they formed a zigzag-shaped chain (Fig. 8.9a). After a second equivalent of Zn^{2+} was added, dimerization occurred at the low-affinity site and the protein assembled into a 2-dimensional lattice. Interestingly, adding a large excess

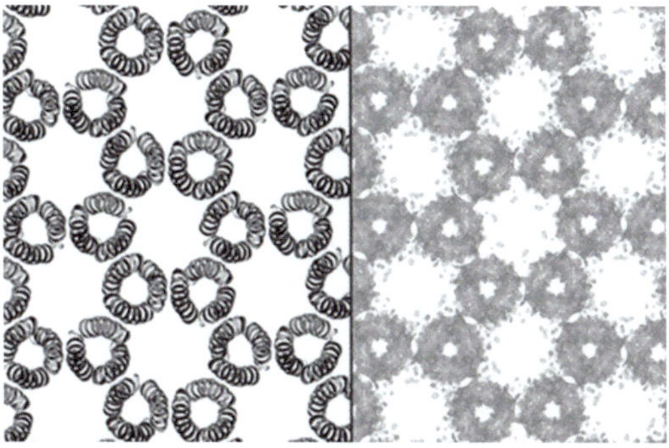

Fig. 8.7 Trimer-dimer system designed to form a protein crystal with the *P6* space group. Computational model (*left*) matches well with the cryo-EM reconstruction (*right*) (Figure adapted from Lanci et al. 2012)

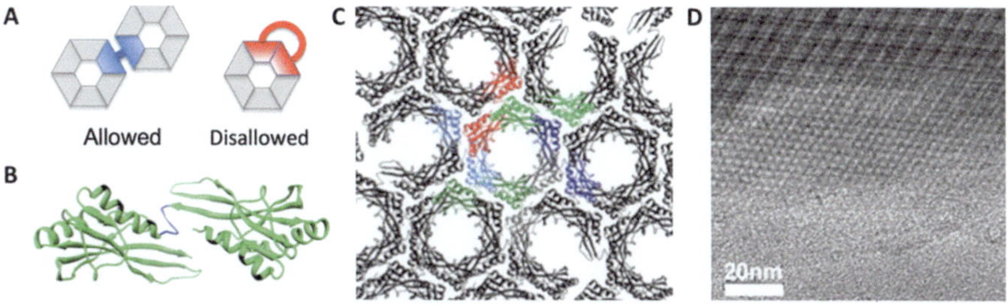

Fig. 8.8 Computational design of a flexibly-linked lattice. (**a**) Design scheme: The flexible linker must be long enough to connect two protein subunits on adjacent hexagons but not long enough to connect two protein subunits on the same hexagon. (**b**) The designed fusion protein comprising two subunits connected by a six glycine linker. (**c**) The predicted hexagonal lattice structure. Six fusion proteins comprising a single hexagon are colored separately. (**d**) TEM image of the assembled lattice (Figure adapted from Matthaei et al. 2015 © American Chemical Society)

of Zn^{2+} (>100 equivalents) or increasing the pH from 5.5 to 8.5 caused the formation of nanotubes (Fig. 8.10b). Most likely this was due to the increased binding affinity driving the assembly of smaller structures over the effects of steric hindrance, as discussed previously. Introduction of a cysteine residue at a position orthogonal to the two-dimensional lattice allowed these lattices to be further assembled into three-dimensional microcrystalline arrays using a dimaleimide cross-linker.

Interestingly, lattices have even been designed by using a natural protein cage as the building block. Yang et al. used ferritin as a building block to generate a square lattice, linking two adjacent ferritin nanoparticles together with a strand of poly(L) lysine, which was long enough to bind to the anionic interiors of two adjacent ferritin cages and bridge the gap between them (Fig. 8.10) (Yang et al. 2014). To accomplish this, the ferritin building blocks had the C-terminal α-helix removed, which widened the pore at the fourfold axis and allowed the poly(L)lysine connecting chain to pass into the cage interior. This ensured that the ferritin nanoparticles were aligned solely along their fourfold symmetry axes.

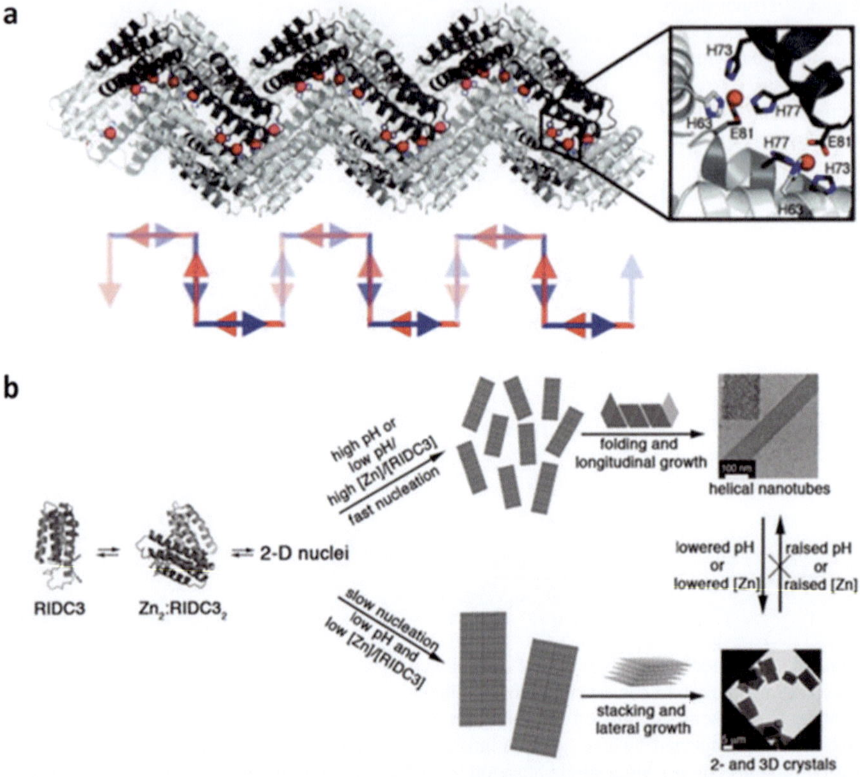

Fig. 8.9 Design of a lattice with multiple orthogonal Zn-mediated binding sites. (**a**) Crystal structure of the assembly after one equivalent of Zn was added. Instead of forming a homodimer as expected, 1-D arrays were generated due to weak dimer-dimer interactions. (**b**) The formation of different macrostructures is dependent on the ratio of excess Zn to protein (Adapted with permission from Brodin et al. 2012, © Macmillan Publishers Ltd)

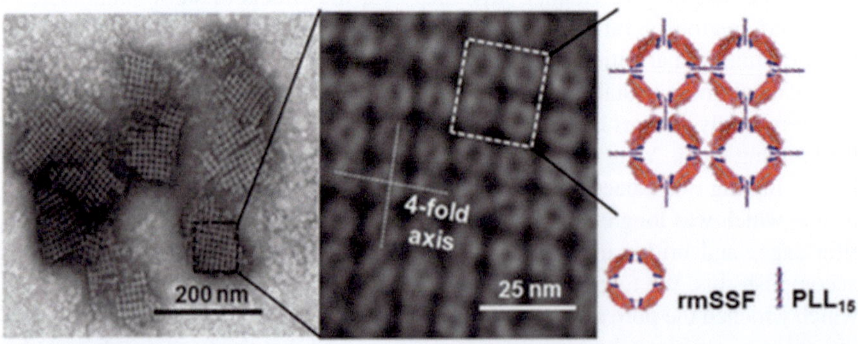

Fig. 8.10 A ferritin-based square lattice. TEM image depicts ferritin protein cages assembled into a *square lattice* by way of a poly(L)lysine linker (Reproduced from Yang et al. 2014 with permission of The Royal Society of Chemistry)

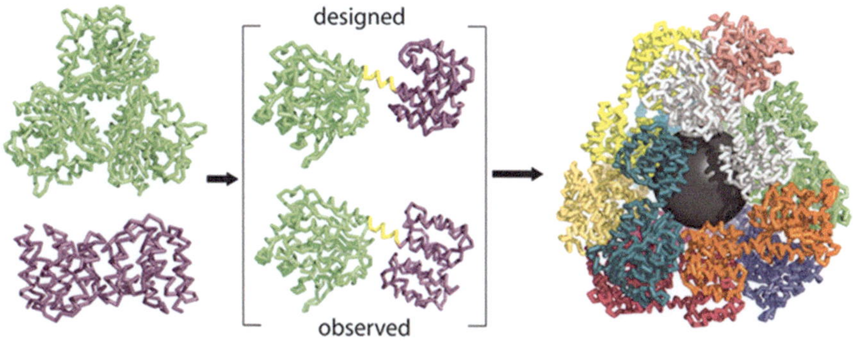

Fig. 8.11 Crystal structure of a *de novo* designed tetrahedral protein cage. While the designed subunits had a rigid α-helix connecting the trimeric and the dimeric domains, the crystal structure (*right*) shows considerable torsion between domains (Adapted with permission from Lai et al. © Lai et al. 2012 MacMillan Publishing Ltd)

8.4.3 Discrete Assemblies – Designed Protein Cages

In a pioneering study in 2001, Padilla et al. reported the design of a tetrahedral protein cage by linking a trimeric protein domain to a dimeric protein domain, through a rigid alpha helical linker sequence (Padilla et al. 2001). The linking α-helix extended between the C-terminal α-helix of the trimeric domain and the N-terminal α-helix of the dimeric domain, thereby maintaining a rigid connection. This allowed the dihedral angle between the two symmetric domains to be modified in a stepwise manner by adding or removing residues in the linking segment, with each additional residue twisting the dihedral angle by 100°. For the two symmetric protein domains used, a nine-residue α-helical linker was predicted to be close enough to orient the dihedral angle for tetrahedron formation. The resulting protein cages appeared tetrahedral when imaged by electron microscopy and had a molecular weight corresponding approximately 12 subunits, as judged by analytical ultracentrifugation. A later simulation of this assembly revealed that two residues from one domain interfered with the intended helix alignment (Lai et al. 2012). When these were mutated to helix-promoting residues the resulting assembly was significantly more homogeneous, and a crystal structure could be obtained. The crystal structure revealed significant torsion between the two oligomerization domains and a deviation of 8 Å from perfect tetrahedral geometry. Nevertheless the design assembled primarily as intended, unintentionally illustrating the benefits of retaining some flexibility in designing protein complexes (Fig. 8.11).

A variation of this design involved shortening the rigid α-helical linker from 9 to 4 residues, which aligned the dihedral angle between the trimeric and the dimeric protein domain to approximately 35°, the angle required for formation of a cubic structure (Lai et al. 2014). After optimization of this 4-residue linker region, the fusion protein assembled into 24-subunit cubes (40% of all assemblies), but also into 18-subunit trigonal bipyramids (50%) and 12-subunit tetrahedrons (10%) (Fig. 8.12).

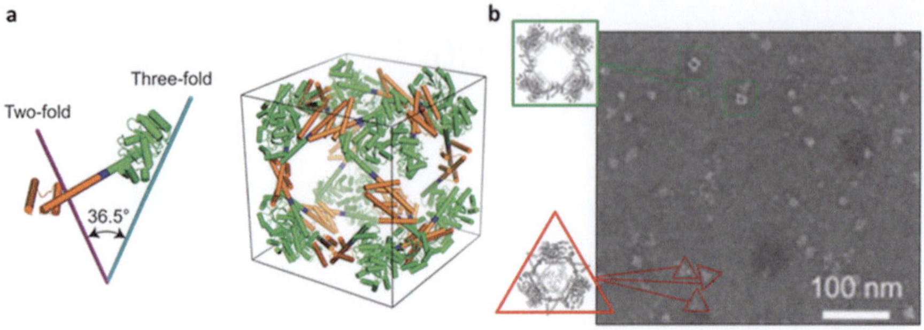

Fig. 8.12 A rigidly-linked trimer-dimer symmetry pair designed to assemble into a 24-subunit cube. (**a**) Model of the intended 24-subunit protein cube. (**b**) TEM of the heterogeneous mixture of assemblies formed. Complexes with apparent octahedral geometry are labeled with *green squares*, while smaller complexes are denoted with *red triangles* (Adapted with permission from Lai et al. 2014, © Macmillan Publishing Ltd)

This polymorphism appears to be a result of the innate flexibility of proteins as well as the tendency for tightly-associated super-symmetric protein subunits to favor smaller assemblies over larger ones, as discussed above. Despite this polymorphism, it was possible to obtain a single crystal of the assembled complex from a carefully purified sample that grew from the crystallization buffer after six months. The x-ray structure revealed that it formed a highly porous cubic structure that closely matched the intended design.

In a different approach, Patterson and coworkers designed two complementary building blocks based upon a trimeric protein, in which two complementary alpha-helical sequences, designed to form an anti-parallel, hetero-dimeric coiled-coil when mixed together, were appended to the N-termini of the protein through a long, flexible linker (Patterson et al. 2011, 2014b). The purpose of these studies was to explore the range of structures that could form in a system where the dihedral angle between the oligomerizing domains was completely unconstrained. Interestingly, upon mixing, the complementary trimeric building blocks assembled into a relatively small range of complexes, despite the unconstrained nature of the design. Analytical ultracentrifugation indicated that the system formed only six different complexes in significant concentrations, with the three major complexes having hydrodynamic properties consistent with the formation of protein cages with diameters appropriate for trigonal prism, tetrahedral or octahedral protein cages. Cryo EM analysis of the particles indicated that they varied in diameter in an almost continuous manner suggesting that their structures were extremely flexible, consistent with the design (Fig. 8.13).

A slightly different approach was taken by Kobayashi and coworkers, who attached a protein sequence comprising two helices of an antiparallel homo-dimeric 4-helix coiled-coil domain to a trimeric building block protein through a flexible linker (Kobayashi et al. 2015). In this case, because the coiled-coil was homo-dimeric, the proteins assembled into cages *in vivo*. This system also formed a mixture of assemblies expected for a trimer-dimer system, with the 6-, 12-, and

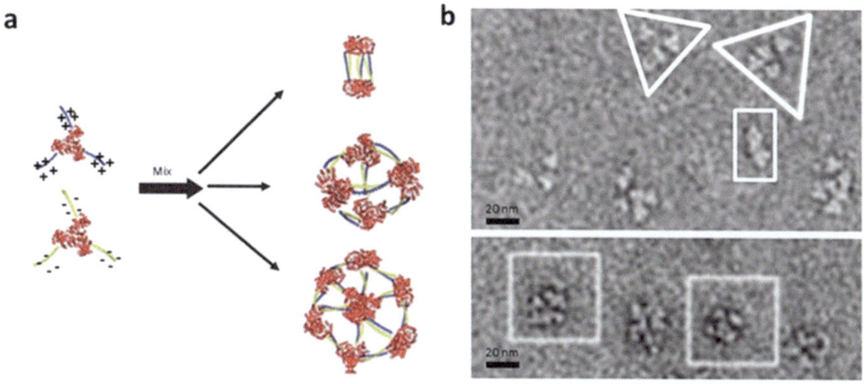

Fig. 8.13 Design of an extremely flexible self-assembling protein cage. (**a**) Design principle: the homo-trimeric building block was linked to an **α**-helix bearing either a strong positive or negative charge. The combination of the two should yield a variety of symmetries. (**b**) TEM of smaller assemblies (*top*) and of larger assemblies (*bottom*) after stoichiometric mixing and size exclusion purification (Adapted from Patterson et al. 2011, with permission from The Royal Society of Chemistry)

18-subunit assemblies able to be purified from the mixture by size exclusion chromatography.

Burkhard and co-workers have explored a cage-forming system that uses a pentameric symmetry element. This system comprises a pentamer-forming coiled-coil linked to a trimer-forming coiled-coil through a two glycine spacer, with the aim of assembling an icosahedral complex (Fig. 8.14) (Raman et al. 2006). To align the two oligomerizing coils with the correct dihedral angle for icosahedral formation, a cysteine residue was inserted on either side of the glycine linker that when oxidized to a disulfide bond would act as a staple. As expressed, this symmetric protein assembled into nonspecific complexes with a broad range of molecular weights. A more homogeneous preparation could be obtained by reducing and denaturing the protein with urea, followed by oxidation and a slow refolding. The refolded protein assembled into fairly homogeneous, spherical complexes of approximately 45 subunits at low concentrations (< 0.3 mg/ml), and approximately 60 subunits at higher concentrations. Interestingly, a variant of this design that substituted the two cysteine residues with alanine formed no 60-subunit icosahedral complexes. Instead, it formed a mixture of spherical assemblies with approximately 180, 240, 300, or 360 subunits, depending on the preparative conditions and the peptide sequence of the asymmetric unit (Indelicato et al. 2016; Yang et al. 2012). These studies clearly depict the challenge of designing well-defined assemblies with higher order symmetry elements.

Although structurally poorly defined, these protein assemblies were able to generate powerful immune responses similar to that of attenuated live viruses, and the intensity of this response could be modulated by changing the diameter of the assembled particles (Yang et al. 2012). This property was exploited to create

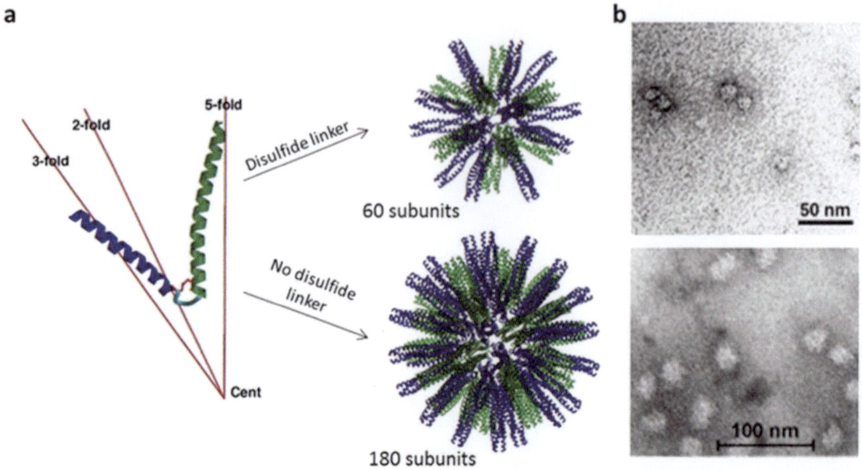

Fig. 8.14 Design of an icosahedral protein cage. (**a**) A protein subunit with a pentamer-trimer symmetry pair assembles into a 60-subunit icosahedron when a disulfide bond is present to lock the dihedral angle in place, but assembles into larger structures in its absence. (**b**) STEM images of the assemblies with (*top*) and without (*bottom*) the disulfide bond present (Adapted with permission from Yang et al. 2012, and Raman et al. 2006, © Elsevier)

functional mouse vaccines for malaria (Kaba et al. 2009), HIV (Wahome et al. 2012), influenza (Babapoor et al. 2011), and toxoplasma (El Bissati et al. 2014) by displaying the respective viral epitope at the terminus of each subunit, which induced long-lasting immune responses to each of these viruses without requiring the addition of an adjuvant.

Advances in computational methods, in particular the program Rosetta developed by Baker and coworkers, have resulted in some impressive advances in the design of *de novo* assembled protein cages. The Rosetta program rapidly and robustly models different docking conformations of protein-protein interfaces, using a built-in scoring function that assesses the energetic stability gained from burying hydrophobic surface residues and creating hydrogen bonds as well as the destabilizing effects of steric clashes and unfavorable Coulombic interactions. Analysis of docked protein interfaces with Rosetta has been used to predict the oligomerization state and binding surface of a self-associating protein from its crystal structure (André et al. 2007; Das et al. 2009). Additionally, new protein-protein interactions can be designed by remodeling a docked structure to add in inter-protein hydrogen bonds and hydrophobic interfaces (Huang et al. 2007; Jha et al. 2010; Stranges et al. 2011).

King and coworkers used Rosetta to design protein cages by symmetrically replicating the structures of 271 trimeric proteins with their symmetry axes centered perpendicular to the faces of either a tetrahedron or an octahedron (King et al. 2012). These assemblies were then analyzed for steric clashes (backbone atoms within 3 Å of each other) and close contacts (backbone atoms within 10 Å of each other) at the inter-trimer surface (Fig. 8.15a) (King et al. 2012). Each trimer was then simultane-

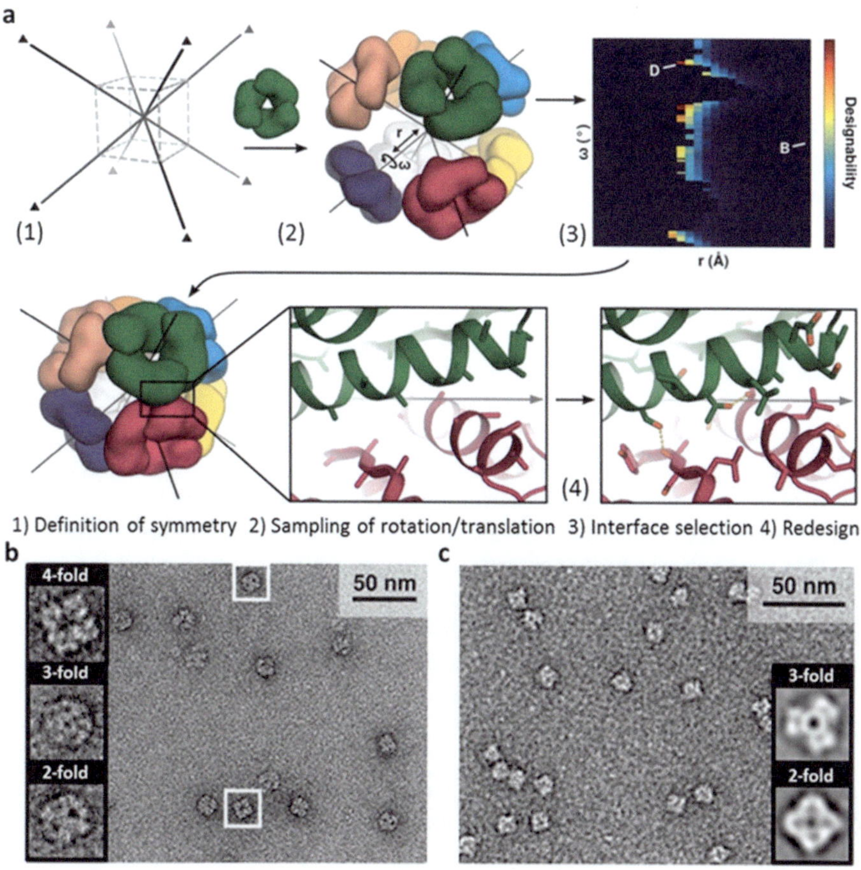

Fig. 8.15 Assembly of protein cages through *de novo* designed protein-protein interfaces (**a**) Outline of design strategy; for details see the text. (**b**, **c**) TEM of assembled octahedra (**b**) and tetrahedra (**c**) with averages of particles oriented at their respective symmetry axes (*inset*) (Figure adapted from King et al. 2012 with permission, © American Academy for the Advancement of Science)

ously translated 1 Å radially from the center of symmetry, until no two adjacent trimers had any close contacts. Next, each trimeric protein in the symmetrical assembly was simultaneously rotated 0.5° about its symmetry axis and the analysis repeated. This was performed 240 times to sample the entire set of rotational conformations. The 20 trimers that could be symmetrically docked into conformations with the most compact protein cages, i.e. the largest number of close contacts but without any steric clashes, were selected for interface redesign, and Rosetta was then used to design favorable interactions at the trimer-trimer interface.

35 *de novo* protein-protein interfaces were designed from these 20 proteins with an average of 9 mutations per design, of which 24 expressed as soluble proteins and 3 oligomerized into robust, symmetrical assemblies – one octahedral and two

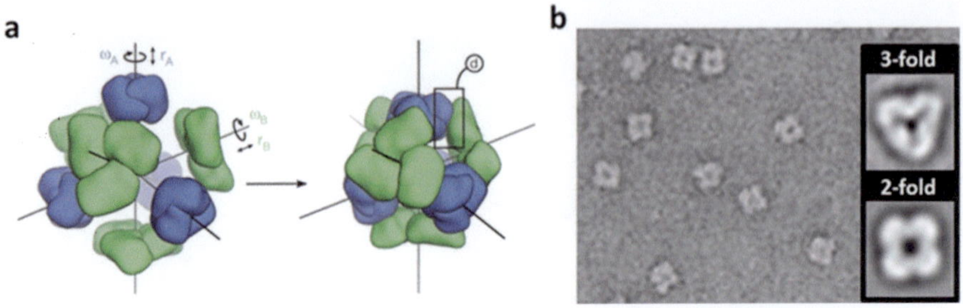

Fig. 8.16 Design of a tetrahedral heteroprotein cage. (**a**) Design strategy. Two oligomeric proteins were arranged at their respective symmetry axes for interface design, for details see the text. (**b**) TEM of assembled particles. *Inset* is averages of particles oriented at their respective axes (Adapted with permission from King et al. 2014, © Macmillan Publishing Ltd)

tetrahedral. Although the success rate of this approach was low, the crystal structures of those proteins that successfully assembled found that they were in close agreement with computational models, representing an impressive achievement.

This approach was further extended to design protein cages in which a trimeric protein was docked at the faces of a tetrahedron and either a dimeric or a trimeric protein was docked at the edges or vertices respectively (Fig. 8.16) (King et al. 2014). The rotational and translational space of both proteins was sampled simultaneously, resulting in a hetero-protein interface that was designed and optimized. In this case, 57 different designed hetero-protein pairs were expressed and characterized experimentally. Of these, four designs were characterized crystallographically. These four tetrahedral protein cages, three with a trimer-trimer symmetry pair and one with a trimer-dimer symmetry pair, were also found to overlay very closely with their computationally designed structures.

The examples above illustrate both the advantages and disadvantages associated with flexible and rigid approaches to assembling proteins. Our laboratory has recently focused on using higher-order symmetry elements to design protein assemblies represented by tetramer-trimer (Sciore et al. 2016), and trimer-trimer (Sciore et al., manuscript in preparation) symmetry pairs. In particular, avoiding the dimeric symmetry elements used in most of the studies discussed above greatly restricts the potential number of geometries that a protein cage can form without explicitly needing to design in a particular dihedral angle connecting the two symmetry elements. In these studies, we first selected a trimeric protein building block with its C-terminus located near the 'vertices' of the triangle formed from the three subunits. To assemble this building block into either octahedral or tetrahedral protein cages we attached at the C-terminus either tetrameric or trimeric *de novo* designed coiled-coils. We then used a modified version of the sampling algorithm developed by King et al. (2012, 2014) discussed above to determine the approximate minimum length of a flexible linker needed to connect the C-terminus of the trimeric building block pro-

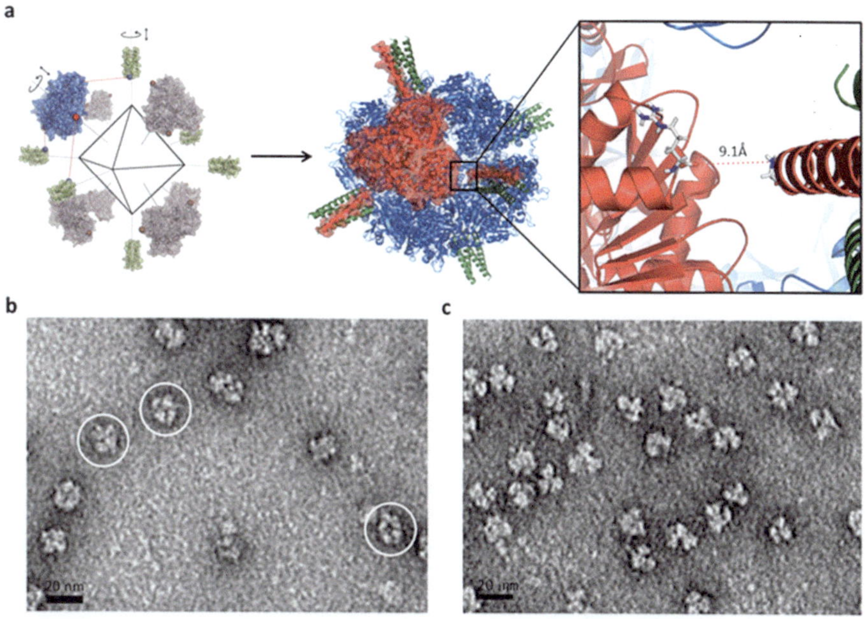

Fig. 8.17 Design of a flexibly-linked protein cage. (**a**) Design strategy: the two oligomeric proteins were aligned on their symmetry axes and the structure with the shortest inter-terminus distance was measured, for details see the text. This value was used to design the flexible linker length. (**b**, **c**) TEM images of the assembled octahedra (**b**) and tetrahedra (**c**)

tein with the N-terminus of the coiled-coils. The modeling indicated that the closest approach between the two design elements without unfavorable steric clashes was about 9 Å for the octahedron and 6 Å for the tetrahedron, which could in principle be bridged by a minimum of two to three residues.

From these computational models, we designed fusion proteins with trimer-trimer and trimer-tetramer symmetry pairs and a glycine-rich linker that varied between 2 and 8 residues in length (Fig. 8.17). In practice, a 4-residue linker proved optimal for the construction of an octahedral protein cage from a trimer-tetramer symmetry pair, whereas an 8-residue linker was required to form a tetrahedron from a trimer-trimer symmetry pair. Fusion proteins with fewer residues in the linker all assembled into a broad range of larger, spherical species. These studies demonstrate that cages of well-defined subunit composition and geometry can be constructed in a relatively simple manner that, in principle, should be highly generalizable. Although this "semi-flexible" approach does not produce as precisely defined or as rigid cages as by interface design, it requires neither sophisticated computational resources, nor extensive screening of many constructs to identify designs that assemble correctly.

8.5 Conclusions

The design of new symmetry-assembled protein nanomaterials will continue to present both interesting challenges and opportunities in the coming years. Research in this area continues to accelerate and there are still many potential avenues to be explored. Currently, research efforts have been primarily focused on exploring the principles of symmetry-directed protein assembly and in particular approaches to correctly orient symmetry elements to achieve the desired protein geometry. These fundamental studies are important, as the utility of assembling proteins into larger structures will only become apparent after methods are developed that allow this to be accomplished in a reliable and routine manner.

In the meantime, there is valuable work to be done investigating whether the current approaches to *de novo* designed protein assemblies can be adapted for, and more significantly, improve upon the range of applications that natural protein assemblies have been used for. For example, the design of protein cages in which enzymatic activity is integrated into the structure promises to be an especially interesting avenue of investigation that could lead to nano-devices that incorporate multiple catalytic activities in a spatially defined manner. Such applications will benefit from an expanded a library of *de novo* designed protein cages. One could imagine that the synthesis of a range of precisely-sized nanoparticles from functionalized protein cages with different interior cavities would be highly attractive for investigations into quantum phenomena.

In summary, the future is bright for symmetrical *de novo* designed protein assemblies. As the design principles described in this chapter, first articulated just fifteen years ago, are refined and perfected, there will be numerous potential applications for these new biological devices in nano-technology and medicine.

References

Abal M, Andreu JM, Barasoain I (2003) Taxanes: microtubule and centrosome targets, and cell cycle dependent mechanisms of action. Curr Cancer Drug Targets 3(3):193–203

Abe S, Hirata K, Ueno T, Morino K, Shimizu N, Yamamoto M, Takata M, Yashima E, Watanabe Y (2009) Polymerization of phenylacetylene by rhodium complexes within a discrete space of apo-ferritin. J Am Chem Soc 131(20):6958–6960. doi:10.1021/ja901234j

Aljabali AAA, Shukla S, Lomonossoff GP, Steinmetz NF, Evans DJ (2013) CPMV-DOX delivers. Mol Pharm 10(1):3–10. doi:10.1021/mp3002057

Allen M, Bulte JWM, Liepold L, Basu G, Zywicke HA, Frank JA, Young M, Douglas T (2005) Paramagnetic viral nanoparticles as potential high-relaxivity magnetic resonance contrast agents. Magn Reson Med 54(4):807–812. doi:10.1002/mrm.20614

André I, Bradley P, Wang C, Baker D (2007) Prediction of the structure of symmetrical protein assemblies. Proc Natl Acad Sci 104(45):17656–17661. doi:10.1073/pnas.0702626104

Ashley CE, Carnes EC, Phillips GK, Durfee PN, Buley MD, Lino CA, Padilla DP, Phillips B, Carter MB, Willman CL, Brinker CJ, Caldeira JC, Chackerian B, Wharton W, Peabody DS (2011) Cell-specific delivery of diverse cargos by bacteriophage MS2 virus-like particles. ACS Nano 5(7):5729–5745. doi:10.1021/nn201397z

Atsushi M, Takio H, Takashi M, Hiroshi Y, Tomoaki H, Yukiharu U, Takashi F, Shigeo Y, Ichiro Y (2006) Floating nanodot gate memory devices based on biomineralized inorganic nanodot array as a storage node. Jpn J Appl Phys 45(1L):L1

Babapoor S, Neef T, Mittelholzer C, Girshick T, Garmendia A, Shang H, Khan MI, Burkhard P (2011) A novel vaccine using nanoparticle platform to present immunogenic m2e against avian influenza infection. Influenza Res Treat 2011:12. doi:10.1155/2011/126794

Banerjee PS, Ostapchuk P, Hearing P, Carrico I (2010) Chemoselective attachment of small molecule effector functionality to human adenoviruses facilitates gene delivery to cancer cells. J Am Chem Soc 132(39):13615–13617. doi:10.1021/ja104547x

Baranova E, Fronzes R, Garcia-Pino A, Van Gerven N, Papapostolou D, Pehau-Arnaudet G, Pardon E, Steyaert J, Howorka S, Remaut H (2012) SbsB structure and lattice reconstruction unveil Ca2+ triggered S-layer assembly. Nature 487(7405):119–122. doi:10.1038/nature11155

Bode SA, Minten IJ, Nolte RJM, Cornelissen JJLM (2011) Reactions inside nanoscale protein cages. Nanoscale 3(6):2376–2389. doi:10.1039/C0NR01013H

Boyle AL, Bromley EHC, Bartlett GJ, Sessions RB, Sharp TH, Williams CL, Curmi PMG, Forde NR, Linke H, Woolfson DN (2012) Squaring the circle in peptide assembly: from fibers to discrete nanostructures by de Novo design. J Am Chem Soc 134(37):15457–15467. doi:10.1021/ja3053943

Brodin JD, Ambroggio XI, Tang C, Parent KN, Baker TS, Tezcan FA (2012) Metal-directed, chemically tunable assembly of one-, two- and three-dimensional crystalline protein arrays. Nat Chem 4(5):375–382

Brumfield S, Willits D, Tang L, Johnson JE, Douglas T, Young M (2004) Heterologous expression of the modified coat protein of Cowpea chlorotic mottle bromovirus results in the assembly of protein cages with altered architectures and function. J Gen Virol 85(Pt 4):1049–1053. doi:10.1099/vir.0.19688-0

Comellas-Aragones M, Engelkamp H, Claessen VI, Sommerdijk NAJM, Rowan AE, Christianen PCM, Maan JC, Verduin BJM, Cornelissen JJLM, Nolte RJM (2007) A virus-based single-enzyme nanoreactor. Nat Nano 2(10):635–639. doi:10.1038/nnano.2007.299

Cormode DP, Jarzyna PA, Mulder WJM, Fayad ZA (2010) Modified natural nanoparticles as contrast agents for medical imaging. Adv Drug Deliv Rev 62(3):329–338. doi:10.1016/j.addr.2009.11.005

Das R, André I, Shen Y, Wu Y, Lemak A, Bansal S, Arrowsmith CH, Szyperski T, Baker D (2009) Simultaneous prediction of protein folding and docking at high resolution. Proc Natl Acad Sci U S A 106(45):18978–18983. doi:10.1073/pnas.0904407106

Dominguez R, Holmes KC (2011) Actin structure and function. Annu Rev Biophys 40:169–186. doi:10.1146/annurev-biophys-042910-155359

Donaldson B, Al-Barwani F, Young V, Scullion S, Ward V, Young S (2015) Virus-like particles, a versatile subunit vaccine platform. In: Foged C, Rades T, Perrie Y, Hook S (eds) Subunit vaccine delivery. Advances in delivery science and technology. Springer, New York, pp 159–180. doi:10.1007/978-1-4939-1417-3_9

Dong H, Paramonov SE, Hartgerink JD (2008) Self-assembly of alpha-helical coiled coil nanofibers. J Am Chem Soc 130(41):13691–13695. doi:10.1021/ja8037323

El Bissati K, Zhou Y, Dasgupta D, Cobb D, Dubey JP, Burkhard P, Lanar DE, McLeod R (2014) Effectiveness of a novel immunogenic nanoparticle platform for Toxoplasma peptide vaccine in HLA transgenic mice. Vaccine 32(26):3243–3248. doi:10.1016/j.vaccine.2014.03.092

Engelhardt H, Peters J (1998) Structural research on surface layers: a focus on stability, surface layer homology domains, and surface layer–cell wall interactions. J Struct Biol 124(2–3):276–302. doi:10.1006/jsbi.1998.4070

Flenniken ML, Liepold LO, Crowley BE, Willits DA, Young MJ, Douglas T (2005) Selective attachment and release of a chemotherapeutic agent from the interior of a protein cage architecture. Chem Commun (Camb) 28(4):447–449. doi:10.1039/b413435d

Flenniken ML, Uchida M, Liepold LO, Kang S, Young MJ, Douglas T (2009) A library of protein cage architectures as nanomaterials. Curr Top Microbiol Immunol 327:71–93

Fletcher JM, Harniman RL, Barnes FRH, Boyle AL, Collins A, Mantell J, Sharp TH, Antognozzi M, Booth PJ, Linden N, Miles MJ, Sessions RB, Verkade P, Woolfson DN (2013) Self-assembling cages from coiled-coil peptide modules. Science 340(6132):595–599. doi:10.1126/science.1233936

Flexman JA, Cross DJ, Lewellen BL, Miyoshi S, Kim Y, Minoshima S (2008) Magnetically targeted viral envelopes: a PET investigation of initial biodistribution. IEEE Trans NanoBiosci 7(3):223–232. doi:10.1109/tnb.2008.2002288

Garcia JA (2011) Sipuleucel-T in patients with metastatic castration-resistant prostate cancer: an insight for oncologists. Therapeut Adv Med Oncol 3(2):101–108. doi:10.1177/1758834010397692

Ghadiri MR, Granja JR, Milligan RA, McRee DE, Khazanovich N (1993) Self-assembling organic nanotubes based on a cyclic peptide architecture. Nature 366(6453):324–327. doi:10.1038/366324a0

Ghosh D, Lee Y, Thomas S, Kohli AG, Yun DS, Belcher AM, Kelly KA (2012) M13-templated magnetic nanoparticles for targeted in vivo imaging of prostate cancer. Nat Nanotechnol 7(10):677–682. doi:10.1038/nnano.2012.146

Haines-Butterick L, Rajagopal K, Branco M, Salick D, Rughani R, Pilarz M, Lamm MS, Pochan DJ, Schneider JP (2007) Controlling hydrogelation kinetics by peptide design for three-dimensional encapsulation and injectable delivery of cells. Proc Natl Acad Sci 104(19):7791–7796. doi:10.1073/pnas.0701980104

Hooker JM, O'Neil JP, Romanini DW, Taylor SE, Francis MB (2008) Genome-free viral capsids as carriers for positron emission tomography radiolabels. Mol Imaging Biol: MIB: Off Publ Acad of Mol Imaging 10(4):182–191. doi:10.1007/s11307-008-0136-5

Huang P-S, Love JJ, Mayo SL (2007) A de novo designed protein–protein interface. Protein Sci 16(12):2770–2774. doi:10.1110/ps.073125207

Huang RK, Steinmetz NF, Fu C-Y, Manchester M, Johnson JE (2011) Transferrin-mediated targeting of bacteriophage HK97 nanoparticles into tumor cells. Nanomedicine (London, England) 6(1):55–68. doi:10.2217/nnm.10.99

Ilk N, Kupcu S, Moncayo G, Klimt S, Ecker RC, Hofer-Warbinek R, Egelseer EM, Sleytr UB, Sara M (2004) A functional chimaeric S-layer-enhanced green fluorescent protein to follow the uptake of S-layer-coated liposomes into eukaryotic cells. Biochem J 379(Pt 2):441–448. doi:10.1042/BJ20031900

Ilk N, Egelseer EM, Sleytr UB (2011) S-layer fusion proteins – construction principles and applications. Curr Opin Biotechnol 22(6):824–831. doi:10.1016/j.copbio.2011.05.510

Indelicato G, Wahome N, Ringler P, Müller Shirley A, Nieh M-P, Burkhard P, Twarock R (2016) Principles governing the self-assembly of coiled-coil protein nanoparticles. Biophys J 110(3):646–660. doi:dx.doi.org/10.1016/j.bpj.2015.10.057

Ishii D, Kinbara K, Ishida Y, Ishii N, Okochi M, Yohda M, Aida T (2003) Chaperonin-mediated stabilization and ATP-triggered release of semiconductor nanoparticles. Nature 423(6940):628–632

Jennings GT, Bachmann MF (2009) Immunodrugs: therapeutic VLP-based vaccines for chronic diseases. Annu Rev Pharmacol Toxicol 49(1):303–326. doi:10.1146/annurev-pharmtox-061008-103129

Jha RK, Leaver-Fay A, Yin S, Wu Y, Butterfoss GL, Szyperski T, Dokholyan NV, Kuhlman B (2010) Computational design of a PAK1 binding protein. J Mol Biol 400(2):257–270. doi:10.1016/j.jmb.2010.05.006

Jutz G, van Rijn P, Miranda BS, Boeker A (2015) Ferritin: a versatile building block for bionanotechnology. Chem Rev 115(4):1653–1701. doi:10.1021/cr400011b

Kaba SA, Brando C, Guo Q, Mittelholzer C, Raman S, Tropel D, Aebi U, Burkhard P, Lanar DE (2009) A nonadjuvanted polypeptide nanoparticle vaccine confers long-lasting protection against rodent malaria. J Immunol 183(11):7268–7277. doi:10.4049/jimmunol.0901957

Kaiser CR, Flenniken ML, Gillitzer E, Harmsen AL, Harmsen AG, Jutila MA, Douglas T, Young MJ (2007) Biodistribution studies of protein cage nanoparticles demonstrate broad tissue distribution and rapid clearance *in vivo*. Int J Nanomedicine 2(4):715–733

Kawano M, Matsui M, Handa H (2013) SV40 virus-like particles as an effective delivery system and its application to a vaccine carrier. Expert Rev Vaccines 12(2):199–210. doi:10.1586/erv.12.149

Kazunori I, Yukiharu U, Prakaipetch P, Hiroshi Y, Tomoaki H, Takashi F, Ichiro Y (2007) Low-temperature polycrystalline silicon thin film transistor flash memory with ferritin. Jpn J Appl Phys 46(9L):L804

Kimchi-Sarfaty C, Arora M, Sandalon Z, Oppenheim A, Gottesman MM (2003) High cloning capacity of in vitro packaged SV40 vectors with No SV40 virus sequences. Hum Gene Ther 14(2):167–177. doi:10.1089/104303403321070865

Kimchi-Sarfaty C, Vieira WD, Dodds D, Sherman A, Kreitman RJ, Shinar S, Gottesman MM (2006) SV40 Pseudovirion gene delivery of a toxin to treat human adenocarcinomas in mice. Cancer Gene Ther 13(7):648–657

King N, Sheffler W, Sawaya M, Vollmar B, Sumida J, Andre I, Gonen T, Yeates T, Baker D (2012) Computational design of self-assembling protein nanomaterials with atomic level accuracy. Science 6085(336):1171–1174

King NP, Bale JB, Sheffler W, McNamara DE, Gonen S, Gonen T, Yeates TO, Baker D (2014) Accurate design of co-assembling multi-component protein nanomaterials. Nature 510(7503):103–107. doi:10.1038/nature13404

Kiyohito Y, Shigeo Y, Shinya K, Atsushi M, Yukiharu U, Takashi F, Ichiro Y (2007) Effects of dot density and dot size on charge injection characteristics in nanodot array produced by protein supramolecules. Jpn J Appl Phys 46(11R):7549

Klem MT, Willits D, Solis DJ, Belcher AM, Young M, Douglas T (2005) Bio-inspired synthesis of protein-encapsulated CoPt nanoparticles. Adv Funct Mater 15(9):1489–1494. doi:10.1002/adfm.200400453

Kobayashi N, Yanase K, Sato T, Unzai S, Hecht MH, Arai R (2015) Self-assembling nano-architectures created from a protein nano-building block using an intermolecularly folded dimeric de Novo protein. J Am Chem Soc 137(35):11285–11293. doi:10.1021/jacs.5b03593

Kramer RM, Sowards LA, Pender MJ, Stone MO, Naik RR (2005) Constrained iron catalysts for single-walled carbon nanotube growth. Langmuir 21(18):8466–8470. doi:10.1021/la0506729

Lai Y-T, Cascio D, Yeates TO (2012) Structure of a 16-nm cage designed by using protein oligomers. Science 336(6085):1129–1129

Lai Y-T, Reading E, Hura GL, Tsai K-L, Laganowsky A, Asturias FJ, Tainer JA, Robinson CV, Yeates TO (2014) Structure of a designed protein cage that self-assembles into a highly porous cube. Nat Chem 6(12):1065–1071. doi:10.1038/nchem.2107

Lanci CJ, MacDermaid CM, S-g K, Acharya R, North B, Yang X, Qiu XJ, DeGrado WF, Saven JG (2012) Computational design of a protein crystal. Proc Natl Acad Sci U S A 109(19):7304–7309. doi:10.1073/pnas.1112595109

Landry SJ, Gierasch LM (1991) The chaperonin GroEL binds a polypeptide in an alpha-helical conformation. Biochemistry 30(30):7359–7362

Lockney DM, Guenther RN, Loo L, Overton W, Antonelli R, Clark J, Hu M, Luft C, Lommel SA, Franzen S (2011) The red clover necrotic mosaic virus capsid as a multifunctional cell targeting plant viral nanoparticle. Bioconjug Chem 22(1):67–73. doi:10.1021/bc100361z

Matthaei JF, DiMaio F, Richards JJ, Pozzo LD, Baker D, Baneyx F (2015) Designing two-dimensional protein arrays through fusion of multimers and interface mutations. Nano Lett 15(8):5235–5239. doi:10.1021/acs.nanolett.5b01499

Maurer P, Jennings GT, Willers J, Rohner F, Lindman Y, Roubicek K, Renner WA, Muller P, Bachmann MF (2005) A therapeutic vaccine for nicotine dependence: preclinical efficacy, and Phase I safety and immunogenicity. Eur J Immunol 35(7):2031–2040. doi:10.1002/eji.200526285

McMillan RA, Howard J, Zaluzec NJ, Kagawa HK, Mogul R, Li Y-F, Paavola CD, Trent JD (2005) A self-assembling protein template for constrained synthesis and patterning of nanoparticle arrays. J Am Chem Soc 127(9):2800–2801. doi:10.1021/ja043827s

Mertig M, Kirsch R, Pompe W, Engelhardt H (1999) Fabrication of highly oriented nanocluster arrays by biomolecular templating. Eur Phys J D-Atomic, Mol Opt Plasma Phys 9(1):45–48

Minten IJ, Claessen VI, Blank K, Rowan AE, Nolte RJM, Cornelissen JJLM (2011) Catalytic capsids: the art of confinement. Chem Sci 2(2):358–362. doi:10.1039/C0SC00407C

Oda T, Iwasa M, Aihara T, Maeda Y, Narita A (2009) The nature of the globular- to fibrous-actin transition. Nature 457(7228):441–445

Padilla JE, Colovos C, Yeates TO (2001) Nanohedra: using symmetry to design self assembling protein cages, layers, crystals, and filaments. Proc Natl Acad Sci U S A 98(5):2217–2221

Patterson DP, Desai AM, Holl MMB, Marsh ENG (2011) Evaluation of a symmetry-based strategy for assembling protein complexes. RSC Adv 1(6):1004–1012. doi:10.1039/c1ra00282a

Patterson DP, Schwarz B, Waters RS, Gedeon T, Douglas T (2014a) Encapsulation of an enzyme cascade within the bacteriophage P22 virus-like particle. ACS Chem Biol 9(2):359–365. doi:10.1021/cb4006529

Patterson DP, Su M, Franzmann TM, Sciore A, Skiniotis G, Marsh ENG (2014b) Characterization of a highly flexible self-assembling protein system designed to form nanocages. Protein Sci 23(2):190–199. doi:10.1002/pro.2405

Picher MM, Kupcu S, Huang C-J, Dostalek J, Pum D, Sleytr UB, Ertl P (2013) Nanobiotechnology advanced antifouling surfaces for the continuous electrochemical monitoring of glucose in whole blood using a lab-on-a-chip. Lab Chip 13(9):1780–1789. doi:10.1039/C3LC41308J

Pleschberger M, Saerens D, Weigert S, Sleytr UB, Muyldermans S, Sára M, Egelseer EM (2004) An S-Layer heavy chain camel antibody fusion protein for generation of a nanopatterned sensing layer to detect the prostate-specific antigen by surface plasmon resonance technology. Bioconjug Chem 15(3):664–671. doi:10.1021/bc049964w

Poobalane S, Thompson KD, Ardó L, Verjan N, Han H-J, Jeney G, Hirono I, Aoki T, Adams A (2010) Production and efficacy of an Aeromonas hydrophila recombinant S-layer protein vaccine for fish. Vaccine 28(20):3540–3547. doi:10.1016/j.vaccine.2010.03.011

Price GD, Badger MR (1989) Expression of human carbonic anhydrase in the cyanobacterium synechococcus PCC7942 creates a high CO(2)-requiring phenotype: evidence for a central role for carboxysomes in the CO(2) concentrating mechanism. Plant Physiol 91(2):505–513

Rajagopal K, Lamm MS, Haines-Butterick LA, Pochan DJ, Schneider JP (2009) Tuning the pH responsiveness of β-Hairpin peptide folding, self-assembly, and hydrogel material formation. Biomacromolecules 10(9):2619–2625. doi:10.1021/bm900544e

Raman S, Machaidze G, Lustig A, Aebi U, Burkhard P (2006) Structure-based design of peptides that self-assemble into regular polyhedral nanoparticles. Nanomedicine 2(2):95–102. doi:10.1016/j.nano.2006.04.007

Raza K, Katare OP, Setia A, Bhatia A, Singh B (2013) Improved therapeutic performance of dithranol against psoriasis employing systematically optimized nanoemulsomes. J Microencapsul 30(3):225–236. doi:10.3109/02652048.2012.717115

Rebeaud F, Bachmann M (2012) Virus-like particles as efficient delivery platform to induce a potent immune response. In: Baschieri S (ed) Innovation in vaccinology. Springer, Dordrecht, pp 87–122. doi:10.1007/978-94-007-4543-8_5

Ringler P, Schulz GE (2003) Self-assembly of proteins into designed networks. Science 302(5642):106–109

Ross PD, Conway JF, Cheng N, Dierkes L, Firek BA, Hendrix RW, Steven AC, Duda RL (2006) A free energy cascade with locks drives assembly and maturation of bacteriophage HK97 capsid. J Mol Biol 364(3):512–525. doi:10.1016/j.jmb.2006.08.048

Rothbauer M, Küpcü S, Sticker D, Sleytr UB, Ertl P (2013) Exploitation of S-layer anisotropy: pH-dependent nanolayer orientation for cellular micropatterning. ACS Nano 7(9):8020–8030

Rothbauer M, Ertl P, Theiler BA, Schlager M, Sleytr UB, Küpcü S (2015) Anisotropic crystalline protein nanolayers as multi-functional biointerface for patterned co-cultures of adherent and non-adherent cells in microfluidic devices. Adv Mater Interfaces 2(1):n/a–n/a. doi:10.1002/admi.201400309

Salick DA, Kretsinger JK, Pochan DJ, Schneider JP (2007) Inherent antibacterial activity of a peptide-based beta-hairpin hydrogel. J Am Chem Soc 129(47):14793–14799. doi:10.1021/ja076300z

Sára M, Sleytr UB (2000) S-Layer proteins. J Bacteriol 182(4):859–868. doi:10.1128/jb.182.4.859-868.2000

Sara M, Pum D, Kupcu S, Messner P, Sleytr UB (1994) Isolation of two physiologically induced variant strains of Bacillus stearothermophilus NRS 2004/3a and characterization of their S-layer lattices. J Bacteriol 176(3):848–860

Sciore A, Su M, Koldewey P, Eschweiler J, Diffley K, Linhares B, Ruotolo B, Bardwell J, Skiniotis G, Marsh ENG (2016) Flexible, symmetry-directed approach to assembling protein cages. Proc Natl Acad Sci U S A 113(31):8681–8686

Shenton W, Pum D, Sleytr UB, Mann S (1997) Synthesis of cadmium sulphide superlattices using self-assembled bacterial S-layers. Nature 389(6651):585–587

Sinclair JC, Davies KM, Venien-Bryan C, Noble MEM (2011) Generation of protein lattices by fusing proteins with matching rotational symmetry. Nat Nano 6(9):558–562. doi:10.1038/nnano.2011.122

Singh S, Zlotnick A (2003) Observed hysteresis of virus capsid disassembly is implicit in kinetic models of assembly. J Biol Chem 278(20):18249–18255. doi:10.1074/jbc.M211408200

Sleytr UB, Beveridge TJ (1999) Bacterial S-layers. Trends Microbiol 7(6):253–260. doi:10.1016/S0966-842X(99)01513-9

Speiser DE, Schwarz K, Baumgaertner P, Manolova V, Devevre E, Sterry W, Walden P, Zippelius A, Conzett KB, Senti G, Voelter V, Cerottini JP, Guggisberg D, Willers J, Geldhof C, Romero P, Kundig T, Knuth A, Dummer R, Trefzer U, Bachmann MF (2010) Memory and effector CD8 T-cell responses after nanoparticle vaccination of melanoma patients. J Immunother (Hagerstown, Md: 1997) 33(8):848–858. doi:10.1097/CJI.0b013e3181f1d614

Spohn G, Keller I, Beck M, Grest P, Jennings GT, Bachmann MF (2008) Active immunization with IL-1 displayed on virus-like particles protects from autoimmune arthritis. Eur J Immunol 38(3):877–887. doi:10.1002/eji.200737989

Stephanopoulos N, Tong GJ, Hsiao SC, Francis MB (2010) Dual-surface modified virus capsids for targeted delivery of photodynamic agents to cancer cells. ACS Nano 4(10):6014–6020. doi:10.1021/nn1014769

Štokrová J, Palková Z, Fischer L, Richterová Z, Korb J, Griffin BE, Forstová J (1999) Interactions of heterologous DNA with polyomavirus major structural protein, VP1. FEBS Lett 445(1):119–125. doi:10.1016/S0014-5793(99)00003-4

Stranges PB, Machius M, Miley MJ, Tripathy A, Kuhlman B (2011) Computational design of a symmetric homodimer using β-strand assembly. Proc Natl Acad Sci 108(51):20562–20567. doi:10.1073/pnas.1115124108

Takuro M, Nozomu M, Kenji I, Ken-Ichi S, Kiyotaka S, Ichiro Y (2007) Direct production of a two-dimensional ordered array of ferritin-nanoparticles on a silicon substrate. Jpn J Appl Phys 46(7L):L713

Teunissen EA, de Raad M, Mastrobattista E (2013) Production and biomedical applications of virus-like particles derived from polyomaviruses. J Control Release 172(1):305–321. doi:10.1016/j.jconrel.2013.08.026

Theil EC (1987) Ferritin: structure, gene regulation, and cellular function in animals, plants, and microorganisms. Annu Rev Biochem 56(1):289–315. doi:10.1146/annurev.bi.56.070187.001445

Uchida M, Terashima M, Cunningham CH, Suzuki Y, Willits DA, Willis AF, Yang PC, Tsao PS, McConnell MV, Young MJ, Douglas T (2008) A human ferritin iron oxide nano-composite magnetic resonance contrast agent. Magn Reson Med 60(5):1073–1081. doi:10.1002/mrm.21761

Ucisik MH, Sleytr UB, Schuster B (2015) Emulsomes meet S-layer proteins: an emerging targeted drug delivery system. Curr Pharm Biotechnol 16(4):392–405. doi:10.2174/138920101604150218112656

Ueno T, Suzuki M, Goto T, Matsumoto T, Nagayama K, Watanabe Y (2004) Size-selective olefin hydrogenation by a Pd nanocluster provided in an apo-ferritin cage. Angew Chem Int Ed 43(19):2527–2530. doi:10.1002/anie.200353436

Vabulas RM, Raychaudhuri S, Hayer-Hartl M, Hartl FU (2010) Protein folding in the cytoplasm and the heat shock response. Cold Spring Harb Perspect Biol 2(12). doi:10.1101/cshperspect.a004390

Villard V, Kalyuzhniy O, Riccio O, Potekhin S, Melnik TN, Kajava AV, Ruegg C, Corradin G (2006) Synthetic RGD-containing alpha-helical coiled coil peptides promote integrin-dependent cell adhesion. J Pept Sci 12(3):206–212. doi:10.1002/psc.707

Wahome N, Pfeiffer T, Ambiel I, Yang Y, Keppler OT, Bosch V, Burkhard P (2012) Conformation-specific display of 4E10 and 2F5 epitopes on self-assembling protein nanoparticles as a potential HIV vaccine. Chem Biol Drug Des 80(3):349–357. doi:10.1111/j.1747-0285.2012.01423.x

Weigert S, Sára M (1996) Ultrafiltration membranes prepared from crystalline bacterial cell surface layers as model systems for studying the influence of surface properties on protein adsorption. J Membr Sci 121(2):185–196. doi:10.1016/S0376-7388(96)00176-7

Wiessner C, Wiederhold KH, Tissot AC, Frey P, Danner S, Jacobson LH, Jennings GT, Luond R, Ortmann R, Reichwald J, Zurini M, Mir A, Bachmann MF, Staufenbiel M (2011) The second-generation active Abeta immunotherapy CAD106 reduces amyloid accumulation in APP transgenic mice while minimizing potential side effects. J Neurosci 31(25):9323–9331. doi:10.1523/jneurosci.0293-11.2011

Woolfson DN (2010) Building fibrous biomaterials from α-helical and collagen-like coiled-coil peptides. Pept Sci 94(1):118–127. doi:10.1002/bip.21345

Yang Y, Ringler P, Müller SA, Burkhard P (2012) Optimizing the refolding conditions of self-assembling polypeptide nanoparticles that serve as repetitive antigen display systems. J Struct Biol 177(1):168–176. doi:10.1016/j.jsb.2011.11.011

Yang R, Chen L, Yang S, Lv C, Leng X, Zhao G (2014) 2D square arrays of protein nanocages through channel-directed electrostatic interactions with poly([small alpha], l-lysine). Chem Commun 50(22):2879–2882. doi:10.1039/C3CC49306G

Yeates TO, Padilla JE (2002) Designing supramolecular protein assemblies. Curr Opin Struct Biol 12(4):464–470

Yeates TO, Kerfeld CA, Heinhorst S, Cannon GC, Shively JM (2008) Protein-based organelles in bacteria: carboxysomes and related microcompartments. Nat Rev Microbiol 6(9):681–691

Yeates TO, Thompson MC, Bobik TA (2011) The protein shells of bacterial microcompartment organelles. Curr Opin Struct Biol 21(2):223–231. doi:10.1016/j.sbi.2011.01.006

Zhang S, Holmes T, Lockshin C, Rich A (1993) Spontaneous assembly of a self-complementary oligopeptide to form a stable macroscopic membrane. Proc Natl Acad Sci 90(8):3334–3338. doi:10.1073/pnas.90.8.3334

Zhao Q, Chen W, Chen Y, Zhang L, Zhang J, Zhang Z (2011) Self-assembled virus-like particles from rotavirus structural protein VP6 for targeted drug delivery. Bioconjug Chem 22(3):346–352. doi:10.1021/bc1002532

Chapter 9
Structure and Function of RNA Polymerases and the Transcription Machineries

Joachim Griesenbeck, Herbert Tschochner, and Dina Grohmann

Abstract In all living organisms, the flow of genetic information is a two-step process: first DNA is transcribed into RNA, which is subsequently used as template for protein synthesis during translation. In bacteria, archaea and eukaryotes, transcription is carried out by multi-subunit RNA polymerases (RNAPs) sharing a conserved architecture of the RNAP core. RNAPs catalyse the highly accurate polymerisation of RNA from NTP building blocks, utilising DNA as template, being assisted by transcription factors during the initiation, elongation and termination phase of transcription. The complexity of this highly dynamic process is reflected in the intricate network of protein-protein and protein-nucleic acid interactions in transcription complexes and the substantial conformational changes of the RNAP as it progresses through the transcription cycle.

In this chapter, we will first briefly describe the early work that led to the discovery of multisubunit RNAPs. We will then discuss the three-dimensional organisation of RNAPs from the bacterial, archaeal and eukaryotic domains of life, highlighting the conserved nature, but also the domain-specific features of the transcriptional apparatus. Another section will focus on transcription factors and their role in regulating the RNA polymerase throughout the different phases of the transcription cycle. This includes a discussion of the molecular mechanisms and dynamic events that govern transcription initiation, elongation and termination.

Keywords RNA polymerase • Transcription • Transcription factors • Transcription cycle

---

J. Griesenbeck • H. Tschochner
Biochemie Zentrum Regensburg, Lehrstuhl Biochemie III, Universität Regensburg, Regensburg, Germany
e-mail: joachim.griesenbeck@ur.de; herbert.tschochner@ur.de

D. Grohmann (✉)
Institut für Biochemie, Genetik und Mikrobiologie, Lehrstuhl Mikrobiologie, Universität Regensburg, Regensburg, Germany
e-mail: dina.grohmann@ur.de

J.R. Harris, J. Marles-Wright (eds.), *Macromolecular Protein Complexes*, Subcellular Biochemistry 83, DOI 10.1007/978-3-319-46503-6_9

9.1 Multisubunit RNA Polymerases – Discovery of the Enzymes and Their Role Within the Central Dogma of Life

The description of the DNA structure in 1953 (Watson and Crick 1953b), together with early studies in genetics and bacterial transformation, stimulated the question how the repeating mononucleotide units that constitute DNA are assembled in the cell (Olby 2003). This led to the hypothesis that DNA replication would require DNA to function as a template for its own synthesis (Watson and Crick 1953a; Meselson and Stahl 1958). At this point, RNA was only a poorly understood molecule and, consequently, the search for a DNA replicating enzyme rather than an RNA synthesizing enzyme started soon after the double-helical structure of DNA was presented. Arthur Kornberg and colleagues were the first to describe DNA polymerase as the DNA synthesizing enzyme (Lehman et al. 1958). RNA only came into focus when the mechanism of protein synthesis was debated (Campbell and Work 1953; Dounce 1952). In 1955, Grunberg-Manago and Ochoa proposed that polynucleotide phosphorylase (PNP) is the enzyme responsible for RNA synthesis (Grunberg-Manago et al. 1955), an idea that proved wrong. The discovery of tRNAs as acceptor for amino acids (Hoagland et al. 1957) gave the first indication that RNA and protein synthesis are linked and that protein biosynthesis requires a high specificity of the tRNA. However, PNP could not fulfil this requirement for specificity as it only adds nucleotides at the RNA termini in a non-specific fashion. This led to the hypothesis that DNA could serve as a template for RNA synthesis and initiated the search for an enzyme capable of DNA-dependent RNA-polymerisation (Hurwitz 2005). RNA polymerase (RNAP) activity was first described by Weiss and Gladstone in 1959 who demonstrated that all four NTPs were incorporated into RNA when added to rat liver nuclear extracts (Weiss and Gladstone 1959). Shortly after, four laboratories independently reported on DNA-dependent RNA synthesising activity in cellular extracts of bacterial or eukaryotic origin (i.e. from *Escherichia coli* (Hurwitz and Bresler 1961; Stevens 1960), pea embryos (Huang et al. 1960) and *Micrococcus luteus* (Weiss and Nakamoto 1961)). Geiduschek et al. (1961) and Chamberlin et al. (1963) used purified RNAPs to demonstrate that *in vitro* transcribed RNA is complementary to the DNA substrate and that the DNA template remained intact after supporting RNA synthesis. Roeder and Rutter were the first to describe that not one but three different enzymes – RNAP I, II and III – transcribe the eukaryotic genome (Roeder and Rutter 1969, 1970). These early discoveries started a new research field and it became apparent that bacterial and eukaryotic RNAPs differ not only in their subunit composition, but also in their sensitivity towards small molecules, rendering RNAP a prime target for antibiotics. Moreover, the understanding of RNAP subunit composition was of critical importance to establish the archaea as an independent domain of life. Shortly after Woese suggested the existence of the third domain of life in 1977 (Woese and Fox 1977), Stetter and Zillig attempted the isolation and biochemical characterisation of RNAPs from different archaeal organisms (Schnabel et al. 1983; Zillig et al. 1978,

1979). Their work clearly demonstrated that archaeal RNAPs are composed of more than the four subunits prototypical for bacterial RNAPs and cannot be inhibited by the antibiotic Rifampicin, providing evidence that the archaeal RNAP shares more characteristics with eukaryotic than bacterial RNAPs. At this point, the subunit composition and the catalytic activity of RNAPs from all three domains could be explored in detail. But it took another 20 years until the first structure of a multisubunit (bacterial) RNAP was solved (Zhang et al. 1999), closely followed by a crystal structure of the eukaryotic RNAP enzyme from yeast (Gnatt et al. 2001; Cramer et al. 2000) and finally, in 2008, the first crystal structure of an archaeal RNAP was reported (Hirata et al. 2008).

9.2 Multisubunit RNA Polymerases and Transcription Factors in the Three Domains of Life

9.2.1 Overall Subunit Composition and Architecture of RNA Polymerase in Bacteria, Eukaryotes and Archaea

DNA-dependent RNA polymerases either belong to the family of single subunit or to the family of multisubunit RNAPs. Single subunit RNAPs can be found in chloroplasts, mitochondria and in bacteriophages, such as T7 or SP6, and are shaped like a right hand (Cheetham and Steitz 2000). In all living cells, however, DNA-dependent RNA synthesis is carried out by complex multisubunit RNAPs (Werner and Grohmann 2011). While bacteria and archaea employ a single type of RNAP, the eukaryotic genome is transcribed by at least three specialized RNAPs, which are dedicated to different subsets of genes. RNAP I transcribes the ribosomal RNA precursor for the mature 25/28S, 18S and 5.8S rRNAs, RNAP II is responsible for the transcription of messenger RNAs (mRNAs) and many non-coding RNAs, whereas RNAP III synthesises small structured RNAs like transfer RNAs (tRNA), spliceosomal U6 small nuclear RNA (snRNA), ribosomal 5S rRNA and 7 SL RNA (Sentenac 1985; Dieci et al. 2007). In plants, two additional multisubunit RNAPs, RNAP IV and V, are encoded. Both these RNAPs are involved in non-coding RNA-mediated gene silencing processes (Haag and Pikaard 2011).

The minimalistic bacterial RNAP is comprised of four different subunits (β, β', two copies of α and ω). Homologs of the bacterial subunits can be found in all multisubunit RNAPs (Fig. 9.1, panel a: homologous core subunits are coloured in blue). These subunits form the structurally conserved RNAP core that harbours the catalytic centre with coordinated magnesium ions at the interface of the two large bacterial subunits β/β' and archaeal-eukaryotic subunits Rpo1/2, A190/135, RBP1/2 and C160/128, respectively. Even though there is only a low sequence identity among the core subunits across the domains of life, there is a high degree of structural conservation (Fig. 9.1c). The core subunits are arranged like a “crab claw” forming the DNA cleft. The “jaws” are part of the large subunits and interact with the incoming

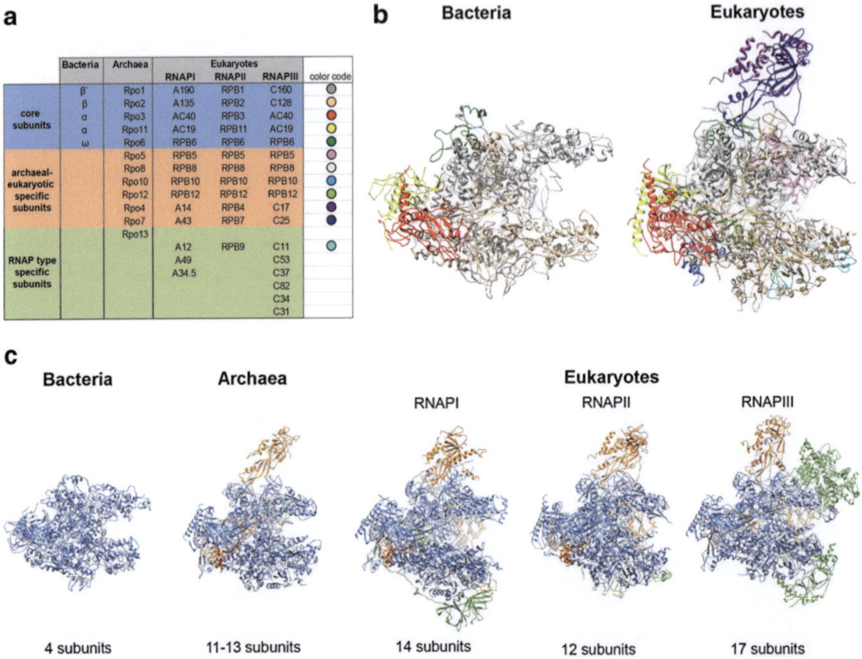

	Bacteria	Archaea	Eukaryotes RNAPI	RNAPII	RNAPIII	color code
core subunits	β'	Rpo1	A190	RPB1	C160	
	β	Rpo2	A135	RPB2	C128	
	α	Rpo3	AC40	RPB3	AC40	
	α	Rpo11	AC19	RPB11	AC19	
	ω	Rpo6	RPB6	RPB6	RPB6	
archaeal-eukaryotic specific subunits		Rpo5	RPB5	RPB5	RPB5	
		Rpo8	RPB8	RPB8	RPB8	
		Rpo10	RPB10	RPB10	RPB10	
		Rpo12	RPB12	RPB12	RPB12	
		Rpo4	A14	RPB4	C17	
		Rpo7	A43	RPB7	C25	
RNAP type specific subunits		Rpo13				
			A12	RPB9	C11	
			A49		C53	
			A34.5		C37	
					C82	
					C34	
					C31	

Fig. 9.1 Subunit composition and overall architecture of multisubunit RNA polymerases. (**a**) Subunit composition of multisubunit RNA polymerases in the three domains of life. Homologous core subunits are coloured in *light blue*, archaeal-eukaryotic specific subunits in *light orange* and subunits specific for the archaeal or specialized eukaryotic RNAPs I, II or II are highlighted in *light green*. (**b**) Structure of the bacterial (*left*, PDB:1I6V) and eukaryotic RNAP II (*right*, PDB: 2WAQ) with single subunits coloured according to the colour scheme given in the table. (**c**) The conserved crab-claw like architecture for multisubunit RNA polymerases from bacteria (*left*, PDB: 1I6V), archaea (PDB: 2WMZ) and eukaroytes (RNAP I: *middle*, PDB: 4C3I; RNAP II: *second right*, PDB: 2WAQ; RNAP III: *right*, PDB: 5FJ9). Universally conserved subunits are shown in *blue*, archaeal-eukaryotic specific subunits in *orange* and subunits unique of a specific RNAP type are shown in *green*

downstream DNA substrate (Fig. 9.1b). The large subunits also comprise a number of flexible elements like the bridge helix, the switch region and trigger loop, which are critically involved in the correct positioning of the next nucleotide and the translocation of RNA and DNA by RNAP (Zhang et al. 2010; Vassylyev et al. 2007; Brueckner et al. 2009a, b; Weinzierl 2010; Tan et al. 2008). Another mobile element of multisubunit RNAPs is the clamp domain, which can close over the DNA binding channel and adopts different conformational states in the different stages of the transcription cycle (Schulz et al. 2016; Chakraborty et al. 2012; Engel et al. 2013). The clamp represents an “interaction hotspot” as regulatory transcription factors like transcription factor E, and the elongation factor Spt4/5 bind the tip of the RNAP clamp region and adjust the position of the clamp (Grohmann et al. 2011; Grunberg et al. 2012; Klein et al. 2011; Schulz et al. 2016).

In addition to the core subunits, archaeal-eukaryotic RNAPs have an expanded set of subunits forming macromolecular assemblies up to 0.7 megadalton (MDa) in size (Fig. 9.1). Archaeal RNAPs typically contain 11–13 subunits (Grohmann et al. 2009) and the eukaryotic RNAPs 12–17 subunits (Vannini and Cramer 2012). Some of these subunits can be found in all archaeal-eukaryotic RNAPs (e.g. homologs of RNAP II subunits RPB4/5/7/8/10/12 highlighted in orange in Fig. 9.1a/c). Notably, subunits RPB5/6/8/10/12 are shared between the eukaryotic RNAPs I, II and III. Subunits 3/10/11/12 form the platform crucially important for the correct assembly of the catalytic subunits. A signature module of archaeal-eukaryotic RNAPs is the stalk domain composed of subunits Rpo4/7 (archaea), A14/A43 (RNAP I), RPB4/7 (RNAP II) and C17/C25 (RNAP III), respectively. The stalk protrudes from the core of the enzyme and is involved in a multitude of functions including the stabilisation of the initiation complex and the binding of the nascent RNA thereby increasing the elongation and termination efficiency of RNAPs. It furthermore serves as an interaction site for transcription factor E. In some archaea, subunit Rpo13 is part of the RNAP. No homologue of Rpo13 is encoded in eukaryotes or bacteria. Another example of a domain-specific subunit is RPB9, which is exclusively found in the eukaryotic domain (domain-specific subunits are color-coded in orange in Fig. 9.1c). RBP9 provides an interaction surface for transcription factor TFIIF (Ziegler et al. 2003) and is closely related to RNAP I and III subunits A12 and C11. However, A12 and C11 represent a fusion protein of the N-terminal RPB9 and the C-terminal part of transcript cleavage factor TFIIS (Ruan et al. 2011) and are considered to be "in-built" transcription factors. RNAP I and III contain additional, auxiliary subunits that are located at the surface of the RNAP thereby providing interaction platforms for factors that regulate RNAP activity. Subunits A49/A34.5 and C53/C37 are distantly related heterodimers that exhibit similarities to transcription factor F (TFIIF) (Fig. 9.2a/d) (Vannini and Cramer 2012). These TFIIF-like complexes are implicated to function in initiation complex stabilisation and occupy a location opposite of the stalk domain (Sainsbury et al. 2015). A49/34.5 furthermore enhances transcription elongation and transcript cleavage and C53/37 is important for termination. Just like A49/A34.5 and C53/C37, the protein complex C82/C34/C31 also represents an "in-built" transcription factor. C82/C34 is related to subunits alpha and beta of transcription factor E, which is associated with, but not integrated into, the archaeal RNAP and RNAP II (Fig. 9.2a/c) (Vannini and Cramer 2012; Blombach et al. 2015; Carter and Drouin 2010; Blombach et al. 2016). The archaeal TFE, TFIIE and C82/C34 all map to the RNAP clamp domain and fulfil functions during transcription initiation by stabilisation of the open DNA bubble (Blombach et al. 2016; Grohmann et al. 2011; Grunberg et al. 2012; Engel et al. 2013; Hoffmann et al. 2015). A unique feature of the largest core subunit of RNAP II is the C-terminal domain (CTD) composed of tandem heptad repeats (26 in yeast, 52 in vertebrates). The CTD is subject to extensive posttranslational modifications, most notably phosphorylation, throughout the transcription cycle, which allows a fine-tuned regulation of RNAP II activity and, among others, supports the coupling of transcription and post-transcriptional processing (for an overview see for example

a

Function	Bacteria	Archaea	Eukaryotes		
			RNAPI	RNAPII	RNAPIII
promoter DNA binding	σ	**TBP**	**TBP** UAF [UBF]	**TBP** (TFIID) TFIIA	**TBP** (TFIIIB) TFIIIA; TFIIIC; [SNAPc]
& polymerase binding	σ	**TFB**	**Rrn7** (CF), **[TAF1A** (SL1)**]**	**TFIIB**	**Brf1 [Brf2]** (TFIIIB)
polymerase binding & recruitment	σ		Rrn3 [TIF-IA]	Mediator	
initiation complex stabilisation & start site selection	σ		**A49 [PAF53]** (RNAP I) **A34.5 [PAF49, CAST]** (RNAP I)	**Tfg2 [TFIIFα]** (TFIIF) **Tfg2 [TFIIFβ]** (TFIIF)	**C37** (RNAP III); **Sfc1** (TFIIIC) **C53** (RNAP III); **Sfc7** (TFIIIC)
open complex formation	σ	**TFEα** (TFE) **TFEβ** (TFE)	**A49 [PAF53]** (RNAP I)	**Tfa1 (TFIIEα)** (TFIIE) **Tfa2 (TFIIEβ)** (TFIIE) TFIIH	**C82** (RNAP III) **C34** (RNAP III)
RNA cleavage	GreA; GreB	**TFS**	**A12.2** (RNAP I)	**TFIIS**	**C11** (RNAP III)
increasing processivity	**NusG**	**Spt5** **Spt4**		**Spt5** **Spt4**	

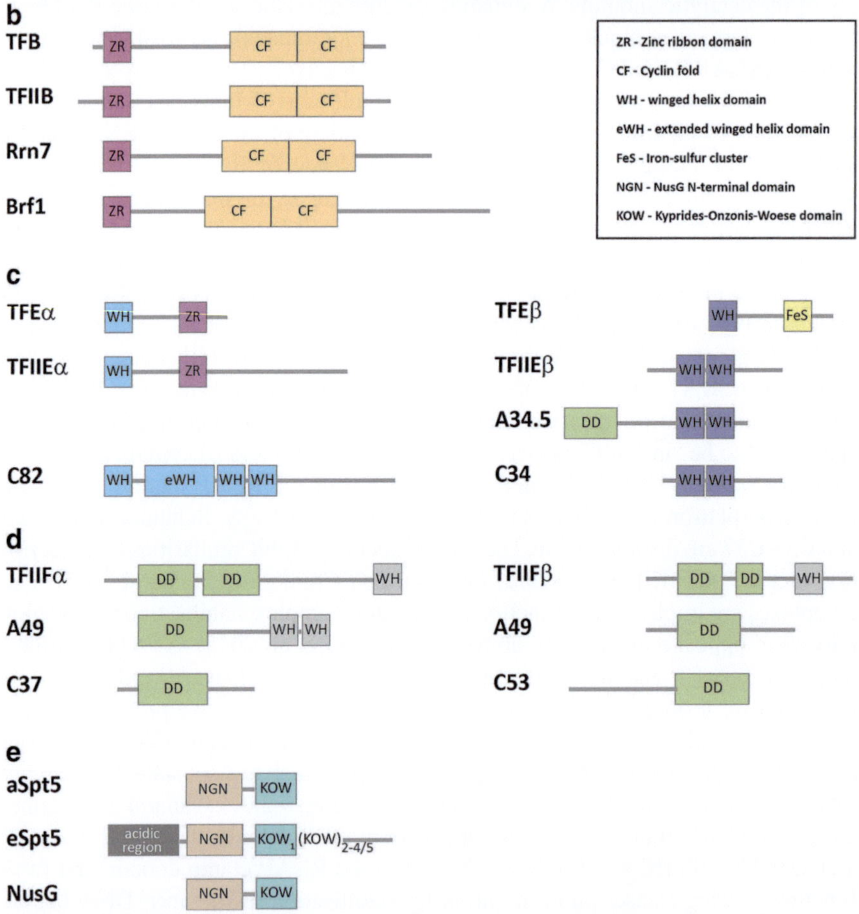

Fig. 9.2 Structure function relationship between transcription factors in bacteria, archaea and eukaryotes. (**a**) Overview of transcription factors implicated in transcription initiation and elongation in bacteria, archaea and eukaryotes. Factors with structural homologies are marked in *bold letters*. The yeast name and the mammalian name (in *squared brackets*) of the homologous protein are given. If these proteins are complex components, the name of the respective complex is depicted in *round brackets*. (**b–e**) Schematic representation of transcription factors and RNAP subunits sharing structural homologies. Structural domains are depicted as *boxes*, with similar domains having the same colour. Abbreviations are explained in the *legend on the right* and aSpt5 and eSpt5 refers to archaeal and eukaryotic Spt5, respectively

(Hsin and Manley 2012; Conaway and Conaway 2015). Taken together, multisubunit RNAPs are multiprotein complexes that are organised in conserved functional domains (e.g. core, assembly platform, stalk, clamp, jaws, DNA cleft) characterised by an increase in complexity and diversity in archaeal-eukaryotic RNAP variants.

9.2.2 Transcription Factors of RNA Polymerases in Bacteria, Eukaryotes and Archaea

Although RNAPs can synthesise RNAs on their own, they require additional factors guiding them to the promoter regions, as well as assisting transcription initiation, elongation and eventually termination. Initially, these factors were identified in different organisms owed to their ability to support different steps of transcription in highly purified *in vitro* systems. Results from these biochemical studies led to detailed models for the function of the individual factors within the transcription cycle. Many of the hypotheses derived from these biochemical studies were validated and extended by structural characterisation. Genetic analyses and the advent of technologies that allow the monitoring of the association status of the transcription machinery at virtually every gene locus *in vivo*, further helped to define a set of general factors supporting transcription in living cells.

As described before, RNAPs in the different domains of life share remarkable similarities with regard to their structure and function, without strong conservation at amino acid sequence level. This structure-function relationship is even more pronounced for transcription factors. Thus, proteins unrelated in their amino acid composition may exert similar functions, correlating with similar topology of structural elements. Consistently, in some cases, predicted structural similarities rather than sequence homologies have provided insights in a previously unknown role of a factor. This paragraph aims to put emphasis on the structure-function conservation within transcription factors in the three domains of life and the different transcription systems (Fig. 9.2). The focus will be on the so-called general transcription factors of transcription initiation, as well as on two factors implicated in transcription elongation. Details about the functional interaction of these factors with RNAPs in course of the transcription cycle are provided in paragraph 3.

9.2.2.1 Initiation Factors

RNAP needs transcription factors to: (i) Identify promoter regions, (ii) Mediate its interaction with the DNA, (iii) Determine the direction of transcription, (iv) Melt the double stranded template leading to the open complex, (v) Enter transcription elongation (Sainsbury et al. 2015; Werner and Grohmann 2011)

In bacteria, all of these tasks are accomplished by the protein family of the promoter specific sigma factors (σ factors) (Feklistov et al. 2014). σ factors were discovered as a biochemical fraction required by purified *E. coli* RNAP to transcribe selective regions of lambda DNA (Burgess et al. 1969). σ factor bound to RNAP recognises up to four out of five DNA motifs marking a prokaryotic promoter (see Sect. 9.3.2) (Bae et al. 2015). This is explained by the structure of σ proteins, organized in a variable number of folded protein domains (e.g. $\sigma_{1.1}$, σ_2, σ_3, σ_4 in the housekeeping σ^{70} in *E. coli*), which are connected by flexible linkers (Fig. 9.3a). Except $\sigma_{1.1}$, all other domains share the ability to bind to specific DNA elements. In solution, σ factors adopt a compact conformation, interfering with DNA-binding of the individual domains (Sorenson and Darst 2006; Sorenson et al. 2004). Additionally, σ^{70} has been suggested to bear an autoinhibitory N-terminal domain ($\sigma_{1.1}$)(Dombroski et al. 1992), which might stabilize the compact conformation. Upon interaction with core RNAP, σ factors undergo structural rearrangements yielding the initiation competent holoenzyme (Murakami et al. 2002a, b). After promoter binding of the holoenzyme, σ factors assist in formation and stabilisation of the nascent transcription bubble, owed to the ssDNA binding ability of the σ_2 domain (Bae et al. 2015) (see Sect. 9.3.2, Fig. 9.3a). The nascent RNA produced at the onset of transcription interferes with some of the σ-RNAP contacts resulting in stochastic dissociation of σ factors from the elongating RNAP after promoter escape (Nickels et al. 2005). The arrangement of structured domains and flexible linkers found in σ factors, which undergo functionally relevant structural changes in different phases of the transcription process, are a common feature of many transcription factors (see below).

The expression of σ variants with different promoter specificities provides prokaryotes with the possibility to express genes under specific environmental conditions (e.g. activation of genes induced under nitrogen stress conditions by σ^{54}) (Gruber and Gross 2003). Archaea and eukaryotes do not have homologs of σ factors for promoter recognition. Instead, they own the TATA-binding protein (TBP), and transcription factor B (TFB/TFIIB), which – similar to σ_4 in bacteria – bind to upstream DNA sequences in archaeal and eukaryotic promoters (Littlefield et al. 1999; Nikolov et al. 1995) (Fig. 9.3b/c). In contrast to bacterial σ factors, TBP and TFB/TFIIB may bind to promoter DNA in the absence of RNAPs, which are then recruited to the promoter-bound scaffold (Sainsbury et al. 2015). In some archaea and eukaryotes, the TBP gene underwent gene-duplication resulting in paralogs of TBP with distinct functions in gene regulation (Duttke 2015). Additionally, some archaea express multiple paralogs of TFB (Facciotti et al. 2007). Thus, in all three domains of life, gene duplication events resulting in variants of core promoter binding factors lead to an increase in the regulatory potential of gene expression.

Promoter-specific transcription by eukaryotic RNAPs I and III also strictly depends on TBP-interacting complexes containing TFB-like factors (Fig. 9.2a/b): RNAP I requires yeast core factor (CF), or its mammalian counterpart, selective factor 1 (SL1), whereas RNAP III depends on TFIIIB (Vannini and Cramer 2012). Homology between TFB/TFIIB, and the TFIIIB subunit Brf1 (and its paralog Brf2) have been described earlier (Colbert and Hahn 1992; López-De-León et al. 1992).

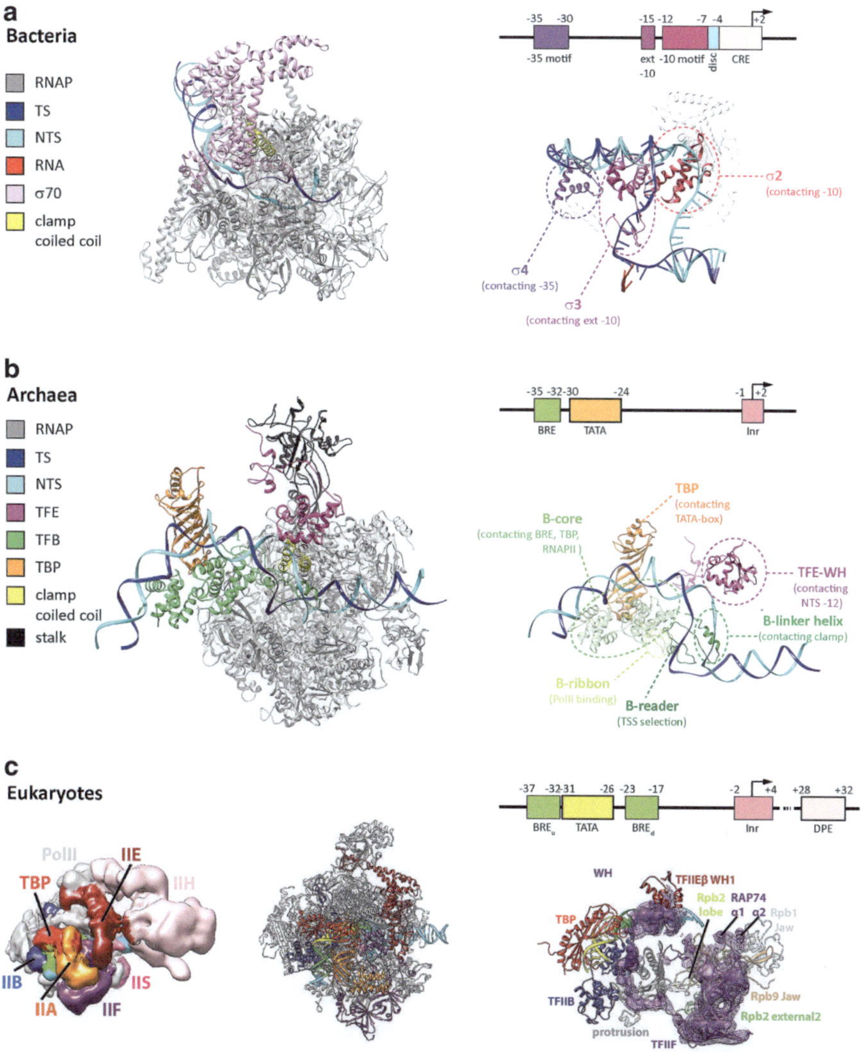

Fig. 9.3 Architecture of the open initiation complex in the three domains of life. In all cases, the template strand (TS) is inserted into the active site of the RNAP and the non-template strand (NTS) is accommodated outside the DNA binding cleft. (**a**) The bacterial initiation complex is composed of a sigma factor (here σ^{70}), the RNAP and DNA (PDB: 4YLO). σ^{70} is a multidomain protein that recognises the −35, −10 and discriminator (*disc*) motifs of the bacterial promoter and supports DNA melting. The RNAP contacts bases of the core recognition element (CRE) that surrounds the transcription start site (+1). (**b**) Model of the archaeal open pre-initiation complex (from (Nagy et al. 2015). The transcription initiation factors TBP and TFB recognise the TATA-box and B-recognition element (BRE) upstream of the transcription start site, respectively. The TFB-reader/helix domains are crucially important for promoter melting and transcription bubble stabilisation – a process furthermore supported by TFE. (**c**) Architecture of the human open pre-initiation complex (images kindly provided by Eva Nogales). EM-reconstruction of the holo-PIC (including TFIIH, *left*) and model of the core PIC (without TFIIH, *middle*). The arrangement of the transcription factors and RNAP motifs critically involved in promoter opening is shown (panel on the *right*). RAP74 is a subunit of TFIIF. Archaeal and eukaryotic promoters (shown here: human promoter) contain the initiator element (Inr) surrounding the transcription start site, and eukaryotic promoters additionally the downstream promoter element (DPE)

The homology between TFB/TFIIB and the yeast CF subunit Rrn7, or its human orthologue the SL1 subunit TAF1B, were only recently uncovered (Blattner et al. 2011; Naidu et al. 2011; Knutson and Hahn 2011). The latter was possible using a computational homology and structure prediction tool (Söding et al. 2005), in combination with functional assays. In agreement with the conserved function of σ-factors and TFB-like factors, the topology of secondary structure elements of these proteins within the different pre-initiation complexes (PICs), is remarkably similar (Burton and Burton 2014; Sekine et al. 2012) (Fig. 9.3a), despite the fact that they are unrelated in their amino acid sequence.

In addition to TBP and TFB, TFE is the third general transcription initiation factor common in the archaeal and eukaryotic transcription systems (Fig. 9.2a/c). Archaeal TFE occurs as either a monomeric TFEα (e.g. TFE from *Methanocaldococcus jannaschii*) or a dimeric TFEα/β variant (e.g. TFE from *Sulfolobus solfataricus*) (Blombach et al. 2009). TFEα is a homolog to the N-terminal part of the TFEIIEα subunit in the eukaryotic TFIIE heterodimer while the TFEβ subunit found in *Sulfolobus* is composed of one winged helix domain and one [4Fe-S] cluster containing domain that are homologous to the human RNAP III subunit RPC39 (Blombach et al. 2015, 2016) (Fig. 9.2c). TFE and TFIIE both contain winged helix (WH) domains (Meinhart et al. 2003). Interestingly, a tandem WH domain similar to a pair of WH domains in TFIIEβ has been found in the C-terminal region of the RNAP I subunit A49, and two adjacent WH domains exist in the RNAP III subunit C34 (Geiger et al. 2010). The human homolog of C82 has four WH domains, resembling the extended WH domain in TFE/TFIIEα (Lefèvre et al. 2011). As TFE/TFIIE, the WH domains of A49 and C34 may adopt a topologically similar positon over the cleft (Bischler et al. 2002; Vannini et al. 2010; Jennebach et al. 2012). As mentioned before (see Sect. 9.2.1), RNAP I and RNAP III subunits can structurally and functionally resemble RNAP II transcription factors, and may be considered as "in-built" transcription factors (Vannini and Cramer 2012).

Several transcription initiation factors are unique to the eukaryotic transcription machinery, some of which have paralogs in the different RNAP machineries. The transcription factor TFIIA is not required for RNAP II transcription *in vitro*, but it can stimulate RNA synthesis (Kang et al. 1995). This essential factor, composed of two subunits, has functions in stabilising the TBP-DNA interaction (Imbalzano et al. 1994). No TFIIA-like factor has to date been identified in the RNAP I and RNAP III transcription machineries.

Instead, RNAPs I and III contain subunits, structurally resembling TFIIF in the RNAP II system (Vannini and Cramer 2012) (see Sect. 9.2.1, Fig. 9.2a/d). Human TFIIF is a dimer of TFIIFα and TFIIFβ (Burton et al. 1988), with homologs in yeast (Chafin et al. 1991), acting together with an additional, non-essential TFIIF subunit (Henry et al. 1992). TFIIFα and TFIIFβ both contain a C-terminal WH domain (Groft et al. 1998; Kamada et al. 2001) (Chen et al. 2007; Eichner et al. 2010) (Fig. 9.2d). TFIIF binds to the lobe of RNAP II via a dimerisation module formed by the two subunits (Gaiser et al. 2000). A similar dimerization module is found in the A49/34.5 subunits of RNAP I and the C37/53 subunits of RNAP III (Geiger et al. 2010; Kuhn et al. 2007). In both polymerases, the respective dimerization modules bind close to the lobe (Engel et al. 2013; Fernandez-Tornero et al. 2013) (Fig. 9.3c) and – as TFIIF – might be implicated in diverse steps of transcription initiation. Interestingly, two subunits of the RNAP III-specific transcription factor TFIIIC establish a structure with a TFIIF-like dimerisation module, and a WH domain (Taylor et al. 2013). It has been hypothesised that these subunits assist PIC formation, while stabilising the single stranded non-template strand (NTS) DNA. There is evidence that the DNA binding mode of the TFIIIC subunit might resemble the one observed for the bacterial σ_2 domain to the single stranded −10 region (see also Sect. 9.3.2). A similar mechanism, supporting promoter opening, was suggested for TFIIF, which shows weak homology to σ_2 (Tan et al. 1994).

The core RNAP I and III machineries lack TFIIH-related factors. This 10-subunit transcription factor is recruited by TFIIE to the RNAP II PIC, where a translocase activity is required for OC formation in an ATP-consuming step (see Sect. 9.3.2) (Grunberg et al. 2012; Kim et al. 2000). Other enzymatic activities of holo-TFIIH include the phosphorylation of the CTD of the largest subunit of RNAP II (Feaver et al. 1991; Serizawa et al. 1995). A TFIIH subcomplex has an important role in DNA repair processes, mediated by two subunits with helicase activities (Coin et al. 2007). TFIIH has been visualized by cryo EM in both, the human and the yeast PIC, and the overall topology of the complex was consistent with its suggested functions during transcription initiation (He et al. 2016; Murakami et al. 2015; Murakami et al. 2013b).

Two other megadalton-sized regulatory complexes, TFIID and Mediator, are important for the regulation and modulation of RNAP II transcription initiation (Cler et al. 2009; Poss et al. 2013; Sikorski and Buratowski 2009; Thomas and Chiang 2006). TBP was originally identified as a component of TFIID, a large complex with 13 canonical TBP associated factors (TAFs), supporting activated transcription *in vitro. In vivo*, TBP in the context of TFIID can be recruited to genes lacking a TATA box, due to the recognition of additional promoter elements by TAFs (Baumann et al. 2010; Kadonaga 2012). Interestingly, interaction of the transcription factor TFIIA with TBP displaces the TAF1 N-terminal domain 1 from the TBP DNA-binding surface, enabling its interaction with promoter DNA (Kokubo et al. 1994). TFIID supports activated transcription from chromatin templates *in vitro*, suggesting that this might be important for its *in vivo* function in the chromosomal context (Wu and Hampsey 1999). Several high-resolution structures of individual TAF subcomplexes are available (Sainsbury et al. 2015) helping to interpret recent cryo EM structures of the full complex in solution or bound to a promoter DNA (Bieniossek et al. 2013; Cianfrocco et al. 2013; Elmlund et al. 2009; Louder et al. 2016; Papai et al. 2010).

Mediator was identified in search of a complex that transmits regulatory signals of DNA-binding factors to the RNAP II basal transcription machinery (Casamassimi and Napoli 2007; Thomas and Chiang 2006). Mediator also supports basal transcription by assisting PIC assembly *in vitro,* and regulates CTD phosphorylation by TFIIH (Sikorski and Buratowski 2009). As for TFIID, X-ray structures are available for individual Mediator subunits and sub-complexes and EM analyses have been performed with both the free complex and RNAP II-bound Mediator (Larivière et al. 2012). Recently, EM structures of the human Mediator-RNAP II-TFIIF complex (Bernecky et al. 2011), and a 15 subunit core Mediator interacting with an RNAP II core initiation complex, containing TFIIB, TBP, and the Pol II–TFIIF (Plaschka et al. 2015) have been obtained. The latter analysis suggested that Mediator might enhance CTD-phosphorylation by orienting the TFIIH kinase module and part of the CTD in the PIC. It also provided insights in how Mediator might interact with the upstream DNA and RNAP II to communicate signals from regulatory DNA binding proteins (Plaschka et al. 2016b).

Interestingly, the binding of the Mediator head module to RNAP II might be topologically related to the binding of the essential transcription factor Rrn3 (TIF-IA in mammals) to RNAP I (Blattner et al. 2011). Yeast Rrn3 and CF are sufficient to support RNAP I transcription *in vitro* even in absence of TBP (Bedwell et al. 2012; Keener et al. 1998; Merkl et al. 2014). The RNAP I-Rrn3 complex is the initiation-competent form of the polymerase, which might be recruited by CF bound to the RNAP I promoter (Peyroche et al. 2000). RNAP I-Rrn3 complex formation might be important to stabilise the interaction of the polymerase with Rrn7 driving transcription initiation (Blattner et al. 2011; Knutson et al. 2014).

TFIIIA represents a transcription factor, which is conserved in all eukaryotes investigated so far and is exclusively involved in RNAP III-dependent transcription initiation (Layat et al. 2013). It was the first eukaryotic transcription factor purified to homogeneity (Engelke et al. 1980; Segall et al. 1980). TFIIIA is needed for 5S rDNA transcription by RNAP III *in vitro* and *in vivo* (Andrews and Brown 1987; Rollins et al. 1993). The protein contains multiple C_2H_2 zinc finger repeats and binds to a specific DNA-sequence upstream of the 5S rDNA. Using its zinc fingers, TFIIIA might also bind to rRNA (Brow and Geiduschek 1987). Conservation between TFIIIAs from different species is largely restricted to the zinc finger domain (Huang and Maraia 2001), and the structural basis for its interaction with nucleic acids has been provided by X-ray crystallography (Lu et al. 2003).

9.2.2.2 Elongation Factors

After initiation, RNAPs start the elongation of the nascent transcript (see Sect. 9.3.3). Specific factors have evolved that directly interact with the RNAP and assist RNAPs to efficiently promote elongation. Interestingly, two general mechanisms of elongation factor operation appear to be conserved in the three domains of life (see Sect. 9.3.3):

I. Factors binding to the RNAP clamp domain, closing the RNAP active centre cleft and thereby stabilising the elongation complex (Fig. 9.5a)
II. Factors entering the RNAP active site via the secondary channel catalysing cleavage of backtracked transcripts and thus release arrested RNAPs (Fig. 9.5b)

Secondary channel binding proteins in bacteria belong to the GreA/B family and are unrelated in sequence and structure to the archaeal and eukaryotic secondary channel binding TFIIS-like proteins (Nickels and Hochschild 2004; Werner and Grohmann 2011). GreA/B and TFIIS stimulate elongation *in vitro* by releasing arrested, backtracked RNAPs (see Sect. 9.3.3) (Borukhov et al. 1992; Rutherford et al. 2007; Thomas et al. 1998). As observed for σ factors and TFB-like factors, those factors share striking similarity in their overall topology in complex with RNAP in good correlation with their related function (Cheung and Cramer 2011; Tagami et al. 2010; Wang et al. 2009). TFIIS-like proteins are conserved in archaea and in RNAP I and III subunits A12.2, and C11 (see Sect. 9.2.1) (Vannini and Cramer 2012; Werner and Grohmann 2011). Apart from their function in transcription elongation, secondary channel binding proteins in bacteria and eukaryotes may play a role in transcription initiation (Sikorski and Buratowski 2009).

The NusG/Spt5 RNAP clamp binding elongation factor is the only transcription factor conserved in sequence and structure in all three domains of life (Werner and Grohmann 2011; Yakhnin and Babitzke 2014). Bacterial NusG and archaeal Spt5 bear a NusG N-terminal domain (NGN) and a C-terminal Kyprides-Onzonis-Woese (KOW) domain (Knowlton et al. 2003; Mooney et al. 2009) (Figs. 9.2e and 9.5b). In eukaryotes, Spt5 may carry multiple KOW domains and extensions at the N- and C-termini. Archaeal and eukaryotic Spt5 further form a heterodimer with Spt4 (Hirtreiter et al. 2010a; Klein et al. 2011).

Taken together, transcription factors in the three domains of life seem to fall in at least three different categories:

I. Factors highly similar in sequence and structure (e.g. NusG/Spt5; TBP; TFB/TFIIB, TFE/TFIIE)
II. Factors with high structural homology (e.g. TFIIF/A34.5–49/C37–53/TFIIIC)
III. Factors, which do not share obvious sequence or structural similarities but are related in the topology of secondary structure elements in complex with RNAPs (e.g. GreA-B and TFS/TFIIS, σ-factor and TFB/TFIIB)

Thus, within the highly conserved RNAP active centre (see Sect. 9.2.1), transcription factors are required to adopt similar topologies to efficiently stimulate and regulate the catalytic process. The increasing complexity of the transcription machinery in eukaryotes, when compared to bacteria and archaea, culminates eventually in the evolution of complexes like TFIID and Mediator, comparable in size to RNAP II with associated general transcription factors. This complex network of protein-protein interactions may allow RNAP to integrate a large variety of different signals from DNA-bound regulators and from components of chromatin, the DNA-template of eukaryotic transcription.

9.3 Structural Dynamics of RNA Polymerase Throughout the Transcription Cycle

9.3.1 Overview

In the course of transcription, RNAPs repeatedly cycle through three distinct functional phases termed initiation, elongation and termination. Throughout the transcription cycle, RNAPs are associated with and regulated by transcription initiation, elongation and termination factors. During initiation, RNAP is specifically recruited to the promoter DNA located upstream of the transcription start site (TSS). Transcription initiation factors like σ factors (bacteria) or TBP and TFB-like factors (archaea/eukaryotes) specifically recognise and bind sequences in the promoter DNA, recruit the RNAP to the promoter and establish the directionality for transcription. Following RNAP recruitment, the DNA is locally melted at the TSS to allow loading of the template (TS) DNA strand into the active site of the RNAP. This process – called open complex formation – eventually yields the stable transcription bubble. RNAPs are able to start RNA synthesis *de novo* requiring no primer for RNA polymerisation. Once the RNAP escapes from the promoter, the phase of productive RNA synthesis (elongation) starts and the RNAP translocates along the TS with each new nucleotide added to the growing transcript chain. Elongation is a discontinuous process frequently interrupted by pausing events that can lead to backtracking of the RNAP. The elongation factors reduce pausing or recover backtracked arrested transcription complexes, thereby increasing the processivity of the RNAP. Eventually, the RNAP will encounter a termination signal leading to the dissociation of the elongation complex and release of the newly synthesized RNA.

9.3.2 Initiation

9.3.2.1 Bacteria

In bacteria, the RNAP core is transiently bound to a member of the σ factor family forming the bacterial RNAP holoenzyme. The number of encoded σ factors differs between bacterial species but one of the σ factor often serves a "housekeeping" factor that supports transcription initiation from the majority of promoters (Feklistov et al. 2014). *E.coli* encodes seven σ factors among which σ 70 (σ^{70}, also known as RpoD in *E.coli* and as σ^{A} in many other bacteria) is the "housekeeping" factor (Paget 2015). A well-studied alternative σ factors is σ^{54} (also known as RpoN or σ^{N}), which directs transcription under stress conditions and in response to environmental signals. σ^{70} and σ^{54} are structurally unrelated and belong to two distinct classes of σ factor that differ in their mode of action during transcription initiation (Paget 2015; Yang et al. 2015).

σ^{70} is organized in four functional domains termed $\sigma_{1.1}$, σ_2, σ_3 and σ_4 that are connected by flexible linkers (see Sect. 9.2.2). Sigma as part of the holoenzyme (but not by itself) undergoes specific interactions with conserved sequence elements of the bacterial core promoter on the one hand side and with the RNAP to perform sigma's three main functions:

I. Sequence-specific recognition of the promoter
II. Recruitment of the RNAP to the promoter
III. Promoting strand separation for initial transcription bubble formation (Paget 2015)

A helix-turn-helix motif in the highly conserved σ region 4 undergoes specific interactions with five nucleotides in the −35 region of the promoter and bends the DNA in a 30° angle (Basu et al. 2014; Murakami et al. 2002a; Zhang et al. 2012; Zuo and Steitz 2015). σ region 3 and 2 interact with the extended −10 motif and the −10 and discriminator region of the NTS, respectively (Fig. 9.3a). These interactions are critical for the site-specific recognition of the promoter DNA and only occur if the binding interfaces of sigma are exposed as a result of its interaction with the core RNAP. σ recognises the −35 and −10 motif in the double-stranded form and the nucleotides in the extended −10 and discriminator motif are "read" upon strand separation. In contrast, the bases of the NTS of the core recognition element (CRE) interact directly with the RNAP β subunit between positions −4 to +2. Here, a G at position +2 (G_{+2}) is unstacked and inserted into a deep pocket of the β subunit in a manner in which σ interacts with the bases A_{-11}, T_{-7} and G_{-6} (Zhang et al. 2012). It is thought that DNA opening starts at position A_{-11} followed by an extension of the initial melted region to yield an initial transcription bubble of 12–14 bases resulting in the open initiation complex (RP_o). The interaction of σ2 with the −10 motif results in a 90° bent in the DNA that directs the downstream DNA toward the RNAP active site cleft allowing the loading of the TS into the active site (Feklistov and Darst 2011). Upon sequence-specific recognition of the promoter DNA and loading of the DNA, $\sigma_{1.1}$ is displaced from the cleft. Relocation of $\sigma_{1.1}$ – a sigma domain that is only present in "housekeeping" sigma factors – does not occur if DNA is non-specifically bound to the RNAP rendering $\sigma_{1.1}$ a "gatekeeper" for the RNAP active site. Part of the linker region that connects σ_3 and σ_4 specifically interacts with the RNAP inserting a loop region (termed the sigma finger, $\sigma_{3.2}$) deep into the active site of the RNAP occupying the pathway of the growing RNA chain. Insertion of $\sigma_{3.2}$ stimulates binding of NTPs in the active site but when the RNA chain extends to a length of only four to five nucleotides, $\sigma_{3.2}$ contacts the nascent RNA. While this potentially stabilises short RNA-DNA hybrids (Zuo and Steitz 2015), it eventually results in a stressed intermediate state when further NTPs are added to the RNA. Single-molecule studies showed that during initial transcription, the RNAP does not translocate but remains connected to the promoter and downstream DNA is "scrunched" into the RNAP further adding to the accumulated stress (Kapanidis et al. 2006; Revyakin et al. 2006). The occurrence of the stressed intermediate state provides the mechanistic rational for an observation called "abortive initiation". Initially, RNAP repeatedly synthesises and releases short RNA transcript 2–15 nucleotides in length (Goldman et al. 2009) without entering the productive elongation phase, as σ blocks the RNA exit pathway. RNA release instead of productive initiation beyond nucleotide 15 is one possibility to escape from the stressed intermediate state while staying connected to the promoter DNA. Alternatively, the accumulated stress provides the driving force for promoter escape leading to productive initiation. The sigma finger and σ_4 are only displaced if the RNA reaches a length of 16 nucleotides and the RNA successfully competes for the space occupied by $\sigma_{3.2}$ (Nickels et al. 2005). Open complex formation is a spontaneous energy-independent process in σ^{70} containing holoenzymes, but σ^{54} holoenzymes require an additional transcriptional activator protein that belongs to the AAA+ ATPase family before DNA melting can commence (Wigneshweraraj et al. 2008; Saecker et al. 2011; Yang et al. 2015). The coupling of ATP hydrolysis to transcriptional activation allows a tighter control of the promoter and overall transcriptional output thereby allowing a swift and precise response to environmental change. Loading and unwinding of the DNA also leads to structural changes in the RNAP. The mobile RNAP clamp is predominantly in an open conformation in the DNA-free state of the bacterial RNAP. Upon open complex formation, the clamp closes securing the DNA in the nucleic acid cleft (Chakraborty et al. 2012).

9.3.2.2 Archaea

In archaea, transcription initiation is mediated by two transcription initiation factors, TBP and TFB. While the function of TBP and TFB – e.g. the sequence-specific positioning of the RNAP at the TSS – is comparable to that of the bacterial σ factors, TBP and TFB are not homologous to sigma. TBP recognises an AT-rich region termed the TATA-box centred at position −26/−27 with respect to the TSS (Soppa 1999). Upon binding, TBP induces a severe bend of approximately 90°in the DNA (Littlefield et al. 1999; Gietl et al. 2014). The B recognition element (BRE) is another archaeal promoter element located at the upstream end of the TATA-box (Fig. 9.3b). BRE is specifically recognised by the C-terminal cyclin domain of TFB (Werner and Weinzierl 2005). The TBP-DNA interaction is of transient nature and in some archaeal species (e.g. the crenarchaeon *Sulfolobus acidocaldarius*) TFB is required to stabilise the TBP-DNA interaction (Gietl et al. 2014) and recruits the RNAP via its N-terminal domains to the promoter DNA to form the functional pre-initiation complex (PIC) (Bell and Jackson 2000; Bell et al. 1999). As TBP is composed of two symmetric repeats (Brindefalk et al. 2013), TFBs orientation at the promoter determines the directionality of the transcription initiation complex on the DNA. TFB furthermore fulfils functions at the post-recruitment stage aiding in start site selection and promoter opening (Kostrewa et al. 2009; Wiesler et al. 2013; Wiesler and Weinzierl 2011). TBP and TFB are necessary and sufficient to drive promoter-directed transcription in the archaeal transcription system and transcription bubble formation does not necessitate ATP hydrolysis (Hausner et al. 1996;

Werner et al. 2006; Werner and Weinzierl 2002; Bell et al. 1998; Qureshi et al. 1997). However, TFE stabilises the PIC and aids open complex formation (Bell et al. 2001; Grohmann et al. 2011; Grünberg et al. 2007; Werner and Weinzierl 2005). Both, TFB and TFE are comprised of multiple domains connected by flexible linkers (Fig. 9.2b/c). A high-resolution structure of the archaeal PIC is not available but structures of archaeal and homologous eukaryotic subcomplexes exist, providing insights into the structural organisation of TFE and TFB and their respective RNAP binding sites. The N-terminal part of TFB is composed of the B-ribbon, −reader and –linker domains. The B-ribbon domain interacts with the RNAP dock domain while the reader and linker protrude deep into the cleft contacting the RNAP clamp and partly occupying the RNA exit channel (Fig. 9.3b) (Kostrewa et al. 2009; Liu et al. 2010; Sainsbury et al. 2013). The B-reader follows a similar path as σ3.2 that likewise results in a clash with the emerging RNA chain. TFB contacts the TS in a way that allows the organisation of the transcription bubble and prevents a tilting of short DNA-RNA hybrids thereby stimulating transcription (Werner and Weinzierl 2005; Sainsbury et al. 2013).

TFEα/β enhance open complex formation and stimulate productive initiation (Blombach et al. 2015; Grohmann et al. 2011). Archaeal TFEα is composed of an N-terminal WH domain that interacts with the RNAP clamp and a C-terminal zinc ribbon domain that contacts the RNAP at the base of the stalk – an interaction network conserved in the eukaryotic domain (Grohmann et al. 2011; Grunberg et al. 2012; Nagy et al. 2015; Engel et al. 2013). The stalk domain is fundamentally important for the recruitment and function of TFEα. The contribution of TFEβ to TFE function is less well understood. The C-terminal domain of TFEβ from *S. solfataricus* contains a [4Fe-S] cluster that is important for dimerization with TFEα and the interaction with the RNAP clamp. TFEα executes its function by contacting the NTS at position −12 thereby stabilising the upstream edge of the transcription bubble and it is likely that TFEβ (like its eukaryotic counterpart) reaches towards the cleft to assist with the handling of DNA strands.

Due to its high flexibility, the structure of the archaeal initiation complex could not be captured at high resolution. Nevertheless, using the distance constraints derived from single-molecule FRET measurements in combination with the known partial structures allowed the modelling of the archaeal open PIC (Nagy et al. 2015). The DNA is melted between position −12 and +2 and the TS is fully loaded into the active site while the NTS is deposited outside the cleft (Fig. 9.3b). The location of the downstream DNA is comparable to the bacterial and eukaryotic initiation complex. The upstream edge of the transcription bubble is located near the RNAP clamp and WH domain of TFEα suggesting that these structural elements collaborate to achieve the stabilisation of the transcription bubble. As observed for the bacterial RNAP clamp, a conformational change of the RNAP clamp accompanies the transition from the closed (dsDNA association without loading into the RNAP) to open complex. This transition is stimulated by TFE resulting in an equilibrium shift towards the open clamp if the PIC contains TFE (Schulz et al. 2016).

9.3.2.3 Eukaryotes

The basal transcription machineries of the archaeal and eukaryotic RNAP II transcription system are highly conserved. In both systems, TBP and TFIIB (homologous to the archaeal TBP and TFB) are sufficient to drive transcription from strong promoters and negatively supercoiled templates (Parvin and Sharp 1993). Archaeal TFE has the eukaryotic TFIIE as a functional counterpart. However, RNAP II initiation complexes contain additional general transcription factors TFIIA, TFIIF and TFIIH. Moreover, TBP is mainly part of the multiprotein complex TFIID (see Sect. 9.2.2). Like the archaeal PIC, the eukaryotic initiation complex most likely forms from individual factors in a stepwise manner *in vivo* (Fig. 9.3c). Initially, TBP (as part of TFIID) recognises the promoter DNA approximately 30 base pairs upstream of the TSS (in humans). While TBP is also required for transcription from all eukaryotic promoters, the presence of a TATA box is not mandatory for TBP binding. In fact, only 10–15% of the mammalian promoters contain a TATA-box (Huisinga and Pugh 2004; Lee et al. 2000). Nevertheless, TBP is found to be associated with the majority of TATA-containing and TATA-less promoters in yeast (Rhee and Pugh 2012; Juo et al. 2003) demonstrating the general role of TBP in transcription initiation. The mechanism of DNA-bending differs from the archaeal system, as eukaryotic TBP bends the DNA in a two-step process (Blair et al. 2012; Gietl et al. 2014; Tolic-Norrelykke et al. 2006; Wu and Hampsey 1999). Eukaryotic TBP forms long-lived complexes with TATA-containing promoter DNAs that are stable for minutes to hours. TFIIB stabilises the TBP-DNA complex in its fully bend form (Gietl et al. 2014). The auxiliary transcription factor TFIIA binds to the upstream side of the TATA-box and stabilises the TBP-DNA complex without changing the overall architecture of the complex (Blair et al. 2012). Like archaeal TFB, TFIIB associates with the TBP-DNA complex via sequence-specific interactions with the BRE element. In eukaryotes, BRE sequences line both sides of the TATA-box, termed upstream and downstream BRE (Deng and Roberts 2006). TFIIB is also responsible for the recruitment of RNAP in the eukaryotic system executing its role in a highly analogous manner to the archaeal system. When RNAP II engages with the TFIID-TFIIB-TFIIA-DNA complex it is already associated with TFIIF, a heterodimeric complex formed by an alpha and beta subunit (also known as Tfg1 and Tfg2 in yeast where a third protein, Tfg3, is part of the complex) (Chen et al. 2007). Around 50% of RNAP II are in complex with TFIIF (Rani et al. 2004), which may help to recruit the polymerase to the promoter by interfering with its non-specific DNA binding (Conaway et al. 1991). TFIIF fulfils various functions including the reinforcement of the PIC via a stabilisation of TFIIB within the PIC and stabilisation of the transcription bubble. The WH domain of TFIIFβ appears to be mobile and may adopt different positions to interact with the DNA during PIC formation (Chen et al. 2007; Eichner et al. 2010). TFIIF is also involved in TSS selection and stimulates early RNA synthesis (Khaperskyy et al. 2008; Ren et al. 1999; Tan et al. 1995; Yan et al. 1999; Rani et al. 2004; Fishburn and Hahn 2012). Association of TFIIE and TFIIH completes the PIC to form the closed initiation complex (Murakami et al. 2013a; Sainsbury et al. 2015). TFIIH and TFIIE seem to bind the initiation complex in a cooperative mode (He et al. 2016). No X-ray structure of the complete eukaryotic PIC is available but electron microscopy and crosslinking studies revealed the overall organisation of the human and yeast PIC (Fishburn and Hahn 2012; He et al. 2016; Murakami et al. 2013b, 2015; Plaschka et al. 2015) (Fig. 9.3c). TFIIE occupies a comparable position as archaeal TFE at the clamp coiled coil and another binding site close to the stalk domain. TFIIEβ stretches over the cleft. In contrast to σ^{70}-dependent bacterial initiation and initiation in archaea, open complex formation is an energy-dependent process in the RNAP II system. The ATPase activity resides in TFIIH, a 10-subunit factor that harbours an ATPase (XPB in humans, Ssl2 in yeast). TFIIH contacts the PIC in proximity to TFIIE and interacts with the downstream DNA but not with the transcription bubble. Mechanistically, TFIIH appears to act as an ATP-dependent dsDNA translocase that rotates the downstream DNA leading to torsional stress and the unwinding of dsDNA as the upstream DNA is fixed due to the tight interaction of the TATA/BRE-TFIIB-TBP complex with the RNAP (Fishburn et al. 2015). The RNAP II initiation complex unwinds approximately 11–15 bp of DNA. Recent cryo-EM reconstruction showed that the TFIIB linker is disordered in the closed human PIC but becomes ordered in the open PIC and directly contacts the NTS (He et al. 2016). A comparable scenario is found in the structures of PIC from *S. cerevisiae*. Here, the B-linker shows weak density and the B-reader is mobile (Plaschka et al. 2016a). An intricate network of TFIIB, the clamp coiled coil, the TFIIE "E-ribbon" domain and the RNAP rudder is likely to stabilise the transcription bubble. TFIIF and TFIIE bind the promoter from the opposite sites of the cleft working in concert to encircle, retain and open the DNA (He et al. 2016; Plaschka et al. 2016a). The RAP30 subunit of TFIIF, which contacts Rpb2 (external 2 and protrusion domain), TFIIB, TBP and the downstream BRE, is another factor critically involved in bubble stabilization. Just as observed for the bacterial and eukaryotic system, the clamp domain of RNAP II adopts different conformations at the different stages of PIC assembly (He et al. 2013, 2016; Plaschka et al. 2016a). In the closed complex, the clamp is found in an open state while the clamp is closed over the DNA cleft in the open PIC. Notably, the tip of the clamp coiled coil changes its interaction partner when progressing from the open to closed transition. In the closed PIC the tip contacts WH2 of TFIIEβ but relocates to interact with the WH domain of TFIIEα during open complex formation (personal communication Eva Nogales).

The process of initiation complex assembly and open complex formation is well described for the eukaryotic RNAP II system but less well understood for RNAP I and RNAP III transcription. TBP seems to be associated with all eukaryotic initiation complexes and TFIIB-like factors are integral parts of the RNAP I and III initiation complexes (Vannini and Cramer 2012; Knutson and Hahn 2013) (see Sect. 9.2.2). In the RNAP I system, Rrn7 (TAF1 B in humans) represents the TFIIB-like factor. Rrn7, together with Rrn11 and Rrn6, is part of the CF which recruits RNAP I to the core element (CE) of the promoter. In yeast, recruitment of RNAP I additionally requires the recognition of a sequence stretch upstream of the CE, the upstream element (UE), by the upstream activation factor (UAF). UAF is comprised of histones H3 and H4, UAF30 and the factors Rrn5, Rrn9 and Rrn10. In higher eukaryotes, the unrelated HMG-box protein upstream binding factor (UBF) might at least in part have a similar function (Sanij and Hannan 2009). RNAP I itself is associated with Rrn3 (TIF-IA in humans), which contacts subunit A43 and the CF fulfilling a bridging function.

In the RNAP III system, Brf1 represents the TFIIB-like factor when the RNAP III machinery is assembled at type I (e.g. 5S rRNA) and type II promoters (tRNAs). Here, Brf1 is found in complex with TBP and Bdp1, which together form the TFIIIB complex. Initiation at these promoter types also requires TFIIIC, a 6-subunit complex (Male et al. 2015). TFIIIC contains subunits Sfc1/Sfc7, which contains similar domains and an overall domain architecture comparable to the RNAP II-specific general transcription factor TFIIF Rap30/Rap74 (Taylor et al. 2013). At the 5S rRNA promoter, TFIIIA recruits the TFIIIC complex, whereas TFIIIC directly recognises the A and B box in tRNA promoters. RNAP III utilises a third promoter type represented for example by the human promoter for the U6 snRNA gene. Instead of Brf1, Brf2 is integral part of TFIIIB and the multisubunit SNAPc complex is crucially involved in promoter recognition upstream of the TATA-box. A structural characterisation of the Brf2/TBP/promoter DNA complex revealed the high degree of structural conservation as compared to the equivalent archaeal and RNAP II complex (Gouge et al. 2015).

Homologs of TFIIF (A49/A34.5 in RNAP I, C37/C53 in RNAP III) and TFIIE (A49 is homologous to TFIIEβ in RNAP I and C82/C34 in RNAP III) are stably integrated into RNAP I and RNAP III (see Sect. 9.2.2 and Fig. 9.2) and play a comparable role in open complex stabilisation (Hoffmann et al. 2015; Engel et al. 2013; Fernandez-Tornero et al. 2013). In the RNAP I and RNAP III system, open complex formation occurs without TFIIH-like factors rendering the melting of DNA independent of ATP hydrolysis.

9.3.3 *Elongation*

9.3.3.1 Promoter Clearance

Before a fully committed transcription elongation complex is established, an intermediate phase of transcription occurs in which RNAPs loose contact with initiation factors. Additionally, as noted above, the ternary RNAP-DNA-RNA complex is gradually stabilised when the first phosphodiester-bonds are synthesized. On the other hand, a few nucleotides (nt) long transcripts of bacterial RNAP are blocked by the $\sigma_{3.2}$ loop positioned at the RNA exit channel, which might lead to abortive transcription. To enter elongation, the $\sigma_{3.2}$ loop is displaced by the nascent RNA leading to its release when the DNA-RNA hybrid reaches a length of about 12 nt. The displacement of the $\sigma_{3.2}$ loop marks the initial state of promoter escape, triggering the release of RNAP from the promoter, which is required for RNAP translocation during transcription elongation (Murakami et al. 2002a).

The 5′end of eukaryotic RNAP II generated transcripts starts to enter the RNA exit channel when the DNA-RNA hybrid reaches a length of 8 nt (Westover et al. 2004b). While RNAP II scans for a transcription start site, the upstream edge of the separated strands remains fixed and the resulting transcription bubble extends downstream until it spans a length of 17–18 nt (Giardina and Lis 1993; Kuehner and Brow 2006). Immediately afterwards, the upstream region closes (Holstege et al. 1997; Pal et al. 2005) resulting in an approximately 10 nt long transcription bubble, which migrates with the translocating RNAP. Displacement of transcription factor TFIIB and initial RNA synthesis coincides with bubble collapse. It is a matter of discussion, how TFIIB destabilisation or the action of TFIIF and/or the elongation factor Spt4/Spt5 (see below) are implicated in this process (Andrecka et al. 2008; Blombach et al. 2013; Bushnell et al. 2004; Cabart et al. 2011; Fishburn and Hahn 2012; Kostrewa et al. 2009; Tran and Gralla 2008; Zawel et al. 1995). Furthermore, closing of the RNAP clamp accompanies the transition from the initiation to elongation phase (Chakraborty et al. 2012; Engel et al. 2013; Schulz et al. 2016; Gnatt et al. 2001). The stalk domain, which is present in archaea and eukaryotes but not in bacteria, is involved in this movement. It is formed by subunits homologous to the RNAP II subunits Rpb4/Rpb7 and promotes open complex formation and the processivity of the polymerase (Grohmann and Werner 2011; Hirtreiter et al. 2010b).

Transcription bubble collapse correlates with the dissociation of TFIIH and can occur after different RNA lengths have been reached, depending on the sequence of the promoter (Pal et al. 2005). Recent single-molecule analyses suggest that TFIIH-driven scanning for a TSS could result in largely extended transcription bubbles, which might either collapse back to a 10 nt containing DNA-RNA hybrid and allow promoter escape, or lead to repeated scanning, or cause dissociation of the RNAP machinery (Fazal et al. 2015).

Apparently, promoter escape can be modulated by different factors in the different RNAP systems, for example initiation factor UBF for RNAP I (Panov et al. 2006), or transcription factors TFIIF, FBP, FIR (Liu et al. 2001) and PC4 (Fukuda et al. 2004) for RNAP II. Promoter-proximal pausing represents an important regulatory step of transcription (Adelman and Lis 2012a; Jonkers and Lis 2015; Kugel and Goodrich 1998; Kwak and Lis 2013), which is relevant for a vast majority of genes in many organisms. Promoter-proximal nucleosomes or local DNA and/or RNA sequences might be implicated in this phenomenon, which might be overcome with the help of specific transcription (elongation) factors. Several detailed reviews give an overview about these important, but more specific gene-regulatory mechanisms (Liu et al. 2010, 2015, 2016; Adelman and Lis 2012b; Yamaguchi et al. 2013; Jonkers and Lis 2015).

9.3.3.2 Overview of the RNA Polymerisation Reaction

Once the promoter is cleared, RNA synthesis proceeds through repeating steps of nucleotide addition. Cooperation between RNAP subunits and precise structural refinements of the DNA-RNA hybrid in the active centre enables the RNAP to travel along the DNA, to incorporate the correct nucleotide, to remove wrongly added nucleotides and to re-enter productive transcription elongation if the enzyme is backtracked or stalled at a transcriptional barrier. Transcriptional barriers might be DNA-bound proteins like histones, or RNA-DNA hybrid structures, called R-loops. In this book chapter, we will mainly focus on the basic mechanism of RNAPs in RNA elongation and structural dynamics of RNAP domains during RNAP-movement along the naked DNA template. In addition, the function of three evolutionary conserved RNAP-associated elongation factors, (TFIIF, Spt4/5 like factors and TFIIS) will be described. Transcription elongation on a native chromatin template and the multiple interactions of RNAPs with many other factors involved in this process has been reviewed earlier in great detail (Dangkulwanich et al. 2014; Formosa 2013; Kwak and Lis 2013; Van Lijsebettens and Grasser 2014; Zhou et al. 2012).

9.3.3.3 The Nucleotide Addition Cycle

The active centre of all RNAPs is highly conserved and catalyses RNA chain elongation following a common basic mechanism, here described for RNAP II. The nucleotide addition cycle is defined as the condensation of a single nucleotide, which is incorporated as NMP in the growing RNA chain under release of pyrophosphate. Chain elongation is accompanied by the translocation of the enzyme along the DNA and by the movement of the transcription bubble. This involves both opening and closure at the 3′ and 5′ends of the 11–12 nucleotides long melted DNA strands, respectively, and annealing of 8–9 nucleotides of the nascent RNA.

In stable elongation complexes (EC), the incoming dsDNA unwinds in the downstream region, which allows the TS to contact the active centre (Gnatt et al. 2001). Positively charged residues in the RNAP II switch 2 region probably together with repellent negative charged amino acids of switch 1 and the fork loop 1 cooperate to separate the TS from the NTS (Kettenberger et al. 2004; Kireeva et al. 2011; Naji et al. 2008). Within the transcription bubble, eight to nine consecutive nucleotides hybridize with the nascent RNA, which is extended at its 3′end. Three RNAP loops-rudder, lid and fork loop – are considered to be involved in RNA-DNA strand separation (Gnatt et al. 2001; Kettenberger et al. 2004; Westover et al. 2004a). While the DNA-RNA hybrid is formed, the NTS makes a turn of about 90° near the catalytic centre and reanneals upstream with the TS to form the exiting DNA (Kornberg 2007).

Based on X-ray crystallography structures (Cheung and Cramer 2011, 2012; Kettenberger et al. 2004; Wang et al. 2006, 2009; Westover et al. 2004a; Zhang et al. 1999) six different stages can be distinguished in the nucleotide addition cycle

(NAC), which are accompanied or driven by structural rearrangements of the RNAP (Fig. 9.4) (see also Nudler 2009; Zhang et al. 2010; Cheung and Cramer 2012; Martinez-Rucobo and Cramer 2013; Dangkulwanich et al. 2014). The crucial conformational changes which are conserved through evolution (see also Bar-Nahum et al. 2005; Epshtein et al. 2002; Weinzierl 2011) concern especially two RNAP domains of the active centre, the trigger loop and the bridge helix (Fig. 9.4a). Before the nucleotide substrate (NTP) diffuses into the enzyme, RNAP II is in the post-translocation state (state I), which is characterised by an empty active site, an open trigger loop and a completely folded bridge helix (Fig. 9.4b). The NTP enters the active site between the RNA 3′end, the bridge helix and the mobile trigger loop and binds in a non-catalytic pre-insertion state (state II). A correct pairing NTP induces closure of the active centre because the trigger loop adopts an α-helical hairpin structure (state III). This conformational change brings the NTP in contact with all residues required for catalysis. Non-correct base pairing results in an equilibrium, which favours the open trigger loop formation and leads finally to dissociation of the non-pairing NTP. A two metal ion mechanism is used for catalysis (state IV). One metal ion A (Mg^{2+}) is permanently kept in the proximity of the RNA 3′end by three conserved aspartate side chains, whereas metal B (Mg^{2+}) contacts the NTP near conserved aspartate and glutamate residues of both the largest and second largest RNAP subunit. The RNA 3′hydroxyl group is deprotonated and attacks as a nucleophile the NTP-α-phosphate, which results in the formation of a new phosphodiester bond and the release of pyrophosphate (PPi) (Carvalho et al. 2011; Svetlov and Nudler 2013). Subsequently, the trigger loop opens and the elongation complex enters the pre-translocation state (state V). In this state, the nucleotide insertion site is occupied by the extended 3′end of the RNA. To incorporate the next nucleotide, the active site has to be emptied by translocation of the elongation complex by one nucleotide from the pre- to the post-translocation state (state I). Partial unfolding of the bridge helix in cooperation with trigger loop movement was suggested to push the nascent hybrid base pair out of the active centre first. In a second step, refolding of the bridge helix could guide the next DNA template base into the active centre by twisting it by 90°. However, the significance of the possible different bridge helix conformations for translocation is still discussed, especially since structural information confirming the predicted bridge helix movements is still missing (Svetlov and Nudler 2008).

To ensure transcriptional fidelity, RNAPs can move backwards (state VI). Backtracking of RNAP is a prerequisite to remove misincorporated nucleotides either by intrinsic RNAP-mediated cleavage or by cleavage supported by TFIIS-like factors.

9.3.3.4 Dynamics of Elongation

To unravel the exact molecular mechanism of transcription, crystallographic studies are not sufficient since X-ray structures are static and represent only single snapshots of the transcription cycle. Single-molecule analysis and molecular dynamics

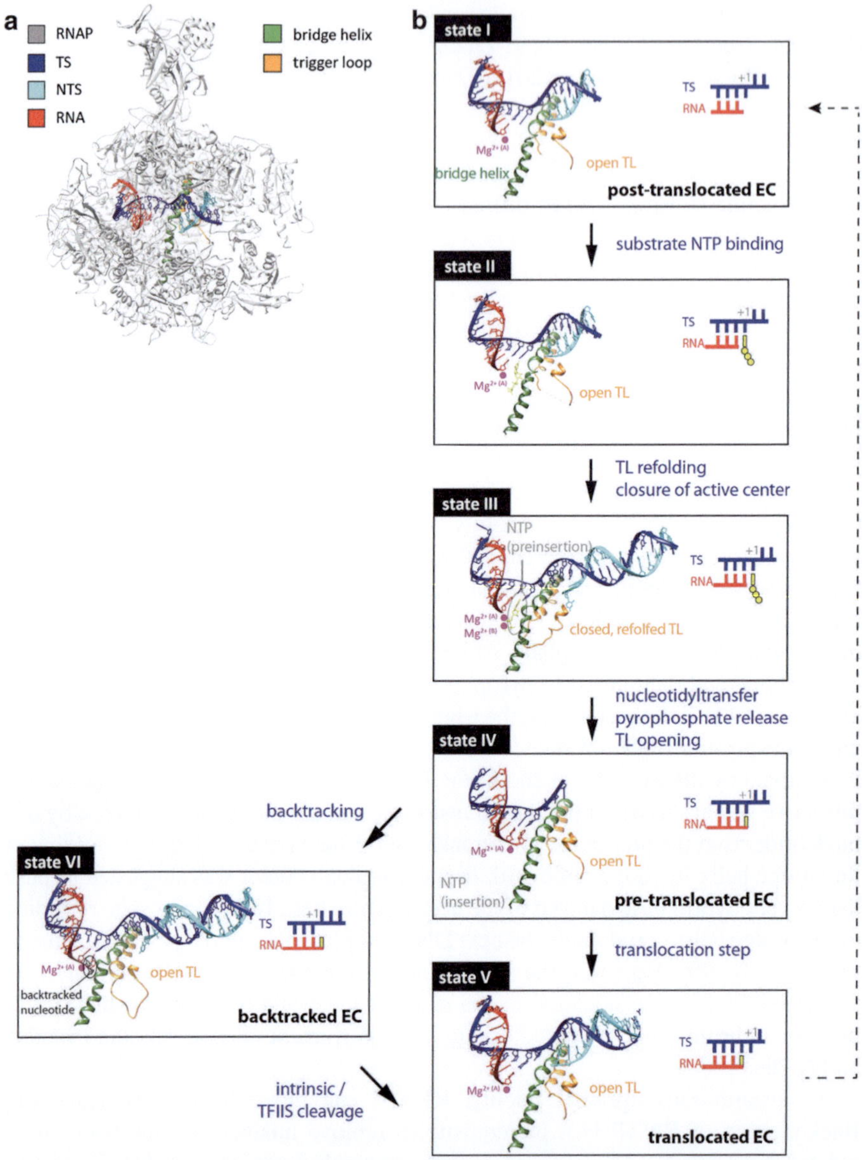

Fig. 9.4 Nucleotide addition cycle. (**a**) Overall organisation of the RNAP II transcription elongation complex (PDB: 1Y1W) including the TS (template strand), NTS (non-template strand), RNAP (RNA polymerase). Flexible elements important for the nucleotide addition cycle are highlighted. (**b**) Six distinguishable stages of the nucleotide addition cycle are shown (PDBs: 1Y1W (stage I), 1Y77 (II), 2E2H (III), 16IH (IV), 1VUM (V) and 3GTG (VI)). Dynamics of the trigger loop (TL) and the bridge helix (BH) according to published X-ray structures are indicated. Stage V represents a backtracked elongation complex. See text for details

simulation help to understand the temporal and spatial dynamics of biomolecules by considering both specific atomic interactions and the biological relevant kinetics (Zhang et al. 2016). Among other conclusions, these studies support a Brownian ratchet model to explain the dynamics of RNAP translocation. In principal, the mechano-chemical coupling that drives RNAP movement is still under debate and includes three models (Kireeva et al. 2010): a power-stroke mechanism, allosteric models and the Brownian ratchet model. Based on crystallographic analyses the power-stroke mechanism suggests that PPi release after nucleotide addition induces a structural change in the enzyme resulting in translocation and strand separation for 1 nt (Yin and Steitz 2004). Accordingly, the energy derived from the chemical reaction of NTP hydrolysis drives the motor forward. The allosteric model suggests that RNAP can exist in an activated and in a less activated state, which catalyse phosphodiester bond formation at different rates. Transition from the slow to the fast state is induced by binding of the templated NTP to the allosteric site (Foster et al. 2001; Holmes and Erie 2003; Nedialkov et al. 2003).

The Brownian ratchet model proposes that the enzyme oscillates between pre- and post-translocation states in a Brownian motion consuming thermal energy. NTP binding and formation of the phosphodiester bond favours the movement from the pre-to the post-translocation position (forward movement), but this step does not require a chemical reaction. In contrast, hydrolysis of both the pyrophosphate and the RNA transcript influences RNAP movement in the opposite direction (Abbondanzieri et al. 2005; Guajardo and Sousa 1997; Komissarova and Kashlev 1997a; Bar-Nahum et al. 2005). This model can explain the ability of RNAP to slide backwards on the DNA template without NTP utilisation when elongation is paused by specific DNA sequences or when NTPs become limiting. But it is still discussed whether such a back and forward movement occurs at all template positions during elongation. Investigating RNAP elongation kinetics via molecular dynamics simulation and single-molecule analyses using different experimental conditions clearly support the Brownian ratchet model (Kireeva et al. 2010; Zhang et al. 2016).

Another important aspect is that elongation speed is influenced by the underlying DNA sequence. RNAPs do not move at a constant speed over DNA templates with varying ATGC contents. Some sequences can cause a longer dwell time of the enzyme and two mechanisms – long and short pausing – can be distinguished. RNAP tends to backtrack during long pauses, which means that transcription is blocked until the 3′end of the nascent RNA is moved back in the active centre either by RNAP forward movement or by internal cleavage of the RNA. DNA sites provoking pauses have been identified for bacterial and human RNAPs (Imashimizu et al. 2013; Hawryluk et al. 2004; Hein et al. 2011; Herbert et al. 2006). Genome-wide analyses revealed almost 20,000 pausing sites on the *E. coli* chromosome, which are characterised by a distinct consensus sequence (Larson et al. 2014; Vvedenskaya et al. 2014).This sequence could cause pausing by interaction between the RNAP core enzyme and the 3′end of the RNA-DNA-hybrid. Accordingly, RNAPs might directly sense the shape or identity of basepairs, which could then delay enzyme translocation (Bochkareva et al. 2012).

9.3.3.5 Function of TFIIF-, Spt4/5- and TFIIS-Related Elongation Factors

One of the first factors found to be involved in RNAP II elongation was transcription initiation factor TFIIF, which associates with paused elongation complexes. During elongation, TFIIF stimulates the rate of elongation and suppresses pausing by stabilising the post-translocated elongation complex (Zhang and Burton 2004; Zhang et al. 2005). In RNAP I the heterodimer A49/34.5 and in RNAP III the subcomplex C37/53 share structural similarities, which correlates with their functional relatedness to TFIIF (Engel et al. 2013; Fernandez-Tornero et al. 2013; Geiger et al. 2010; Hoffmann et al. 2015).

Spt5/NusG-related factors are involved in elongation in all three domains of life indicating that they play an important role in supporting a basic mechanism of transcription. A common model emerged how the Spt4-Spt5-RNAP II architecture could influence transcription elongation (Fig. 9.5a). The NGN/Spt4 complex binds to the clamp, spans over the cleft and encircles the DNA in the enzyme. Binding of

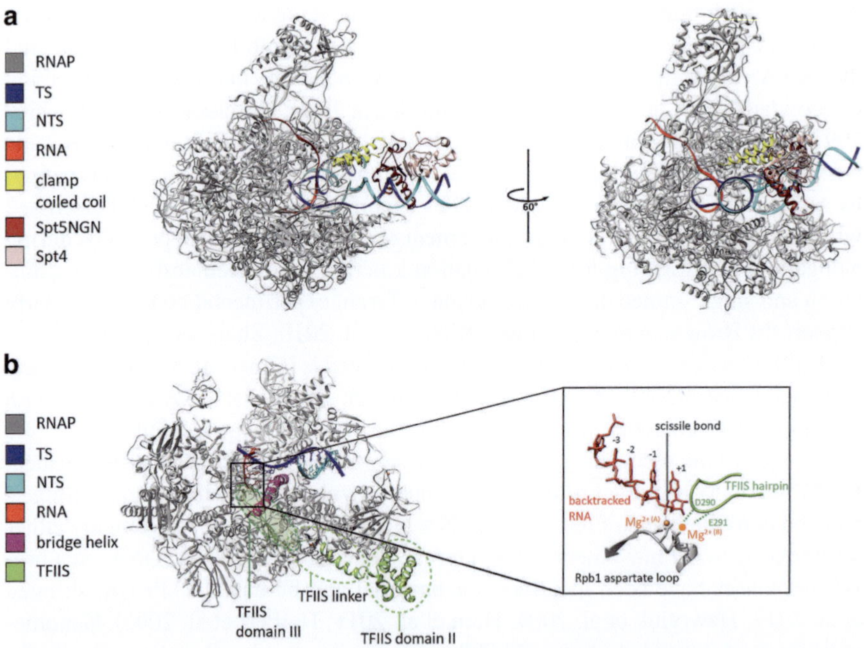

Fig. 9.5 RNAP II elongation complexes. (**a**) Structural model of a Spt5NGN/Spt4 containing RNAPII elongation complex (Model from Martinez-Rucobo et al. 2011). TS (template strand), NTS (non-template strand), RNAP (RNA polymerase), clamp coiled coil (Rpb1 domain interacting with Spt5NGN/Spt4). Spt5NGN/Spt4 bridges the cleft securing the DNA in the DNA binding channel, which contributes to RNAP processivity. (**b**) Structure of a TFIIS-containing elongation complex (PDB: 3PO3). The enlarged section shows components involved in the cleavage reaction of a backtracked RNA. See text for details

Spt4/5 induces closure of the clamp domain (Schulz et al. 2016). The downstream edge of the transcription bubble could be stabilised by interactions between the non-transcribed DNA strand and the NGN domain of Spt4/5. Biochemical and mutational analyses suggest that NusG speeds up the oscillation of RNAP between the pre- and post-translocation state (Fig. 9.4b) (Bar-Nahum et al. 2005; Borukhov and Nudler 2008). Taken together, these studies underline that the dissociation of the ternary transcription complex and release of the DNA template is prevented by Spt5/NusG elongation factors, which leads to an increase in polymerase processivity. Interestingly, cryo-EM studies indicated a shift of the Spt4/Spt5 location towards the opposite site of the cleft, interacting with the protrusion and lobe domains of RNAP II (Klein et al. 2011). This could mean that RNAP II domains (clamp and/or lobe) and the NGN domain of Spt5 interact in a dynamic fashion.

Recent structural investigations suggest that Spt4/Spt5/NusG proteins are associated with RNAP regions to which also transcription initiation factors bind. For instance, bacterial sigma factor or eukaryotic TFIIB binds to the coiled coil domain of the clamp (Arthur et al. 2000; Vassylyev et al. 2002; Kostrewa et al. 2009; Sevostyanova et al. 2008). Archaeal TFE and eukaryotic TFIIE interact with the clamp at a site that putatively overlaps with NGN domain-binding. Finally, the two large subunits of yeast TFIIF interact with RNAP domains that form one side of the cleft, but only at the periphery (Chen et al. 2007; Eichner et al. 2010). This raises the question whether competition of elongation and initiation factors for access at the clamp is a general mechanism to switch from initiation to elongation (Blombach et al. 2013). In fact, functional evidence was provided that archaeal TFE competes with Spt4/Spt5 for binding at the clamp coiled coil domain, and that this competition determines the transcription mode of the polymerase: TFE binding abolishes the inhibitory role of Spt4/Spt5 for initiation and Spt4/Spt5 removes TFE from the elongation complex (Grohmann et al. 2011). The latter could then facilitate the exchange between initiation factors and later acting RNAP-associated factors.

Pausing can lead to backtracking of the polymerase (Kireeva et al. 2005; Nudler et al. 1997; Komissarova and Kashlev 1997b). When the polymerase moves backwards along the DNA and RNA, the 3′RNA end dissociates from the DNA-RNA hybrid and is expelled from the active site through the pore underneath. Thus, RNA chain elongation is impaired. Transcription can only resume if the 3′end of the RNA is relocated into the active centre, which is mainly achieved through factor-stimulated RNA cleavage reactions. Arrested elongation complexes could be visualised by X-ray crystallography and revealed the exact position of the backtracked RNA (Cheung and Cramer 2011; Wang et al. 2009). In principal, a gating tyrosine residue Y769 can restrict the extent of backtracking, allowing the preferential cleavage at the level of dinucleotides. However, if RNAP is arrested, the DNA-RNA hybrid can become so instable that the RNA can easily backtrack beyond the gating tyrosine into the pore keeping the trigger loop in an inactive conformation (Fig. 9.4b, stage VI). Furthermore, the bending of the bridge helix, which apparently depends on mismatches in the RNA-DNA hybrid, was reported to promote RNAP II backtracking (Da et al. 2016). Backtracked RNAPs can be rescued by intrinsic RNA cleavage activity and with the help of factors that stimulate transcript cleavage at the active RNAP site. Upon RNA cleavage, 3–18 nucleotides long RNA stretches are released and the newly generated 3′OH of the RNA is aligned in the active site to become competent for RNA chain elongation. Bacteria, archaea and eukaryotic RNAP II have accessory factors Gre(B), TFS and TFIIS, respectively, to stimulate RNA cleavage of paused complexes (Fish and Kane 2002). Although GreB and TFIIS-related factors share no known structural homology (see Sect. 9.2.2), they function according to a similar principle (Martinez-Rucobo and Cramer 2013). In contrast to RNAP II, eukaryotic RNAP I and RNAP III employ the TFIIS-related polymerase subunits A12.2 and C11, respectively, to execute RNA cleavage (Chedin et al. 1998; Landrieux et al. 2006; Ruan et al. 2011; Jennebach et al. 2012; Kuhn et al. 2007; Lisica et al. 2016). Bacterial GreB binds to the jaw domain of the polymerase through its C-terminal globular domain while its N-terminal antiparallel α-helical coiled-coil domain approaches the active centre through the secondary channel and contacts the RNA (Opalka et al. 2003). TFIIS contains three domains. Domain II interacts with the polymerase at the RNAP jaw, and is connected via a linker domain with domain III, which is inserted into the active centre through the secondary channel (Fig. 9.5b) (Kettenberger et al. 2003, 2004). Apparently, the active site of RNAP II can be switched from polymerisation to cleavage mode and the transiently interacting TFIIS induces structural rearrangements of RNAP, which finally allow RNA processing and realign RNA in the active centre. The C-terminal parts of the RNAP I and III subunits A12.2 and C11 and the archaeal cleavage factor TFS, are homologous to the C-terminal domain of TFIIS, which is required for RNA processing. This molecular arrangement allows probably more efficient backtrack recovery and probably better proofreading (Lisica et al. 2016) for RNAP I and III. RNAP II and archaeal transcription might benefit from a dissociable cleavage factor since reversible association of TFIIS-related factors adds another level of transcriptional regulation (Ruan et al. 2011).

The mechanism of endonucleolytic cleavage is likely similar in all known TFIIS and GreB-related factors (Cheung and Cramer 2011; Kettenberger et al. 2003; Sosunov et al. 2003). Two conserved acidic residues (D290 and E291) are located at the foremost tip of the TFIIS-hairpin, which is inserted into the active centre of RNAP II. These residues position one of two Mg^{2+} ions involved in the cleavage reaction. The other Mg^{2+} ion is persistently bound to the RPB1 aspartate loop of the active site and binds the +1 RNA phosphate to align the hydrolysable phosphodiester bond (scissile bond, Fig. 9.5b). With the aid of the two acidic residues at the hairpin, the Mg^{2+} ion positions a water molecule, which then serves as a nucleophile. After RNA cleavage, the downstream RNA oligonucleotide is liberated, a new RNA 3′ end is associated to the active centre, the polymerase switches into the polymerisation mode and elongation can resume.

9.3.4 Termination

Correct transcription termination results in a length restricted transcript and in the release of RNAP and the transcript from the template. Termination of transcription requires common features for all transcription machineries, first, recognizing the end of the transcribed region, and second, overcoming related structural and energetic constraints to release the nascent transcript. Typically, terminator-complexes destabilise and release both, elongating RNA polymerases and associated transcripts. In addition, co-transcriptional RNA processing can also have a major impact on termination. No detailed structures of RNAP-containing termination complexes are yet available. Therefore, the possible mechanisms of termination are only briefly summarised.

In bacteria, two termination mechanisms are known: intrinsic termination induced by a palindromic sequence in the RNA that forms an RNA hairpin. A stem loop in the nascent RNA in which a series of U-residues follow a G-C rich element leads to pausing and destabilisation of transcription complexes (Larson et al. 2008; Peters et al. 2011; Washburn and Gottesman 2015). Mutations in the bridge helix can prevent termination, pointing at a crucial role of this RNAP domain also in the termination process (Nedialkov et al. 2013). The RNA hairpin has to be correctly folded when RNAP reaches the termination point. Furthermore, the sequence, size and length of the RNA loop influence termination efficiency.

Alternatively, termination in bacteria is mediated by the termination factor Rho (ρ). This termination factor is an RNA-dependent ATPase with helicase activity and forms a homohexameric ring. Two domains bind the nascent RNA and triggers dissociation of the RNAP from the DNA in an ATP-dependent manner. One RNA binding site binds RNA in the absence of ATP, the other site is stimulated to bind RNA transiently after the primary site is occupied. To reach the second binding site, the RNA is threaded to the central pore of Rho and the hexamer is closed when the RNA is attached. Subsequently, the transient interactions of RNA with the secondary binding site propel Rho along the RNA until it finally reaches the elongating RNAP at the release site. Dissociation of RNAP from the template can probably be achieved by unwinding of the RNA-DNA hybrid (Skordalakes and Berger 2003, 2006; Richardson 1982; Rabhi et al. 2011). The speed of elongation influences Rho dependent termination efficiency (Jin et al. 1992). In general, slowing down the elongation speed seems to be an important feature for most of the postulated termination mechanisms.

Termination in archaea and eukaryotes, however, is less well understood. Archaea and RNAP III terminate transcription after polymerizing a series of U residues (Campbell and Setzer 1992; Arimbasseri et al. 2013; Grohmann et al. 2010; Grohmann and Werner 2011; Hirtreiter et al. 2010b). This resembles the intrinsic bacteria-like termination mechanism. Whether the formation of an RNA hairpin upstream of the termination site of RNAP III is important for the mechanism of RNAP III termination is currently debated (Arimbasseri et al. 2014).

RNAP I termination involves a polymerase-specific termination factor that binds to a specific DNA sequence to pause elongation shortly upstream of its binding site, leading to transcript release (Németh et al. 2013; Jaiswal et al. 2016). In the yeast *S. cerevisiae,* this binding is sufficient to induce termination *in vitro* and *in vivo* (Merkl et al. 2014; Reiter et al. 2012).

Additionally, it was reported that factors involved in rRNA 3′end processing also participate in termination. Accordingly, co-transcriptional cleavage by rRNA processing factors generates a 5′RNA entry point for an exonuclease, which progressively degrades the nascent transcript associated with the elongating RNAP. After the exonuclease reaches the polymerase, which is slowed down at the termination site, it disrupts the ternary complex probably with the help of a helicase in a process termed the "torpedo model" (El Hage et al. 2008; Kawauchi et al. 2008; Rondon et al. 2009).

For RNAP II dependent termination either RNA 3′end processing factors or RNA-DNA helicases were suggested to disrupt the ternary transcription complexes (Kuehner et al. 2011; Mischo and Proudfoot 2013; Arndt and Reines 2015; Porrua and Libri 2015). RNAP II transcripts are often cleaved at a specific site (AAUAA) after the coding sequence and subsequently polyadenylated, which plays an important role in termination of protein-coding genes (Logan et al. 1987; Connelly and Manley 1988; Whitelaw and Proudfoot 1986). According to the "torpedo model" described above, the endoribonucleolytic cleavage allows entry of an exonuclease and rapid 5′-3′degradation, which leads to a clash between the exonuclease and the elongating RNAP II. This releases the polymerase from the template (Kuehner et al. 2011; Tollervey 2004). Alternatively, RNAP II transcription through the polyA-signal could induce structural changes of the elongation complex leading to its release from the template (Nag et al. 2006; Orozco et al. 2002). The involvement of additional factors is possible. Termination of RNAP II genes, which contain no polyA-tail, require factors containing RNA/DNA helicase activity. These factors destabilise the ternary elongation complex, thus, resembling bacterial Rho-dependent termination (Kuehner et al. 2011; Mischo and Proudfoot 2013; Porrua and Libri 2015).

Acknowledgements The authors apologise to those whose work could not be discussed owing to space limitation. The authors would like to thank Klaus Grasser for critical reading of the manuscript. We thank Eva Nogales for sharing and discussing unpublished information and for providing images used for the preparation of Fig. 9.3. We would like to thank the members of the Tschochner/Griesenbeck and Grohmann laboratory for their courage and hard work. D.G. would like to thank Finn Werner for passing on his enthusiasm for transcription and the many fruitful discussions.

References

Abbondanzieri EA, Greenleaf WJ, Shaevitz JW, Landick R, Block SM (2005) Direct observation of base-pair stepping by RNA polymerase. Nature 438(7067):460–465. doi:10.1038/nature04268

Adelman K, Lis JT (2012a) Promoter-proximal pausing of RNA polymerase II: emerging roles in metazoans. Nat Rev Genet 13(10):720–731. doi:10.1038/nrg3293

Adelman K, Lis JT (2012b) Promoter-proximal pausing of RNA polymerase II: emerging roles in metazoans. Nat Rev Genet 13(10):720–731. doi:10.1038/nrg3293

Andrecka J, Lewis R, Bruckner F, Lehmann E, Cramer P, Michaelis J (2008) Single-molecule tracking of mRNA exiting from RNA polymerase II. Proc Natl Acad Sci U S A 105(1):135–140. doi:10.1073/pnas.0703815105

Andrews MT, Brown DD (1987) Transient activation of oocyte 5S RNA genes in Xenopus embryos by raising the level of the trans-acting factor TFIIIA. Cell 51(3):445–453

Arimbasseri AG, Rijal K, Maraia RJ (2013) Transcription termination by the eukaryotic RNA polymerase III. Biochim Biophys Acta 1829(3–4):318–330. doi:10.1016/j.bbagrm.2012.10.006

Arimbasseri AG, Rijal K, Maraia RJ (2014) Comparative overview of RNA polymerase II and III transcription cycles, with focus on RNA polymerase III termination and reinitiation. Transcription 5(1):e27639. doi:10.4161/trns.27369

Arndt KM, Reines D (2015) Termination of transcription of short noncoding RNAs by RNA Polymerase II. Annu Rev Biochem 84:381–404. doi:10.1146/annurev-biochem-060614-034457

Arthur TM, Anthony LC, Burgess RR (2000) Mutational analysis of beta '260-309, a sigma 70 binding site located on Escherichia coli core RNA polymerase. J Biol Chem 275(30):23113–23119. doi:10.1074/jbc.M002040200

Bae B, Feklistov A, Lass-Napiorkowska A, Landick R, Darst SA (2015) Structure of a bacterial RNA polymerase holoenzyme open promoter complex. Elife 4. doi:10.7554/eLife.08504

Bar-Nahum G, Epshtein V, Ruckenstein AE, Rafikov R, Mustaev A, Nudler E (2005) A ratchet mechanism of transcription elongation and its control. Cell 120(2):183–193. doi:10.1016/j.cell.2004.11.045

Basu RS, Warner BA, Molodtsov V, Pupov D, Esyunina D, Fernandez-Tornero C, Kulbachinskiy A, Murakami KS (2014) Structural basis of transcription initiation by bacterial RNA polymerase holoenzyme. J Biol Chem 289(35):24549–24559. doi:10.1074/jbc.M114.584037

Baumann M, Pontiller J, Ernst W (2010) Structure and basal transcription complex of RNA polymerase II core promoters in the mammalian genome: an overview. Mol Biotechnol 45(3):241–247. doi:10.1007/s12033-010-9265-6

Bedwell GJ, Appling FD, Anderson SJ, Schneider DA (2012) Efficient transcription by RNA polymerase I using recombinant core factor. Gene 492(1):94–99. doi:10.1016/j.gene.2011.10.049

Bell SD, Jackson SP (2000) The role of transcription factor B in transcription initiation and promoter clearance in the archaeon Sulfolobus acidocaldarius. J Biol Chem 275(17):12934–12940

Bell SD, Jaxel C, Nadal M, Kosa PF, Jackson SP (1998) Temperature, template topology, and factor requirements of archaeal transcription. Proc Natl Acad Sci U S A 95(26):15218–15222

Bell SD, Kosa PL, Sigler PB, Jackson SP (1999) Orientation of the transcription preinitiation complex in archaea. Proc Natl Acad Sci U S A 96(24):13662–13667

Bell SD, Brinkman AB, van der Oost J, Jackson SP (2001) The archaeal TFIIEalpha homologue facilitates transcription initiation by enhancing TATA-box recognition. EMBO Rep 2(2):133–138. doi:10.1093/embo-reports/kve021

Bernecky C, Grob P, Ebmeier CC, Nogales E, Taatjes DJ (2011) Molecular architecture of the human Mediator-RNA polymerase II-TFIIF assembly. PLoS Biol 9(3):e1000603. doi:10.1371/journal.pbio.1000603

Bieniossek C, Papai G, Schaffitzel C, Garzoni F, Chaillet M, Scheer E, Papadopoulos P, Tora L, Schultz P, Berger I (2013) The architecture of human general transcription factor TFIID core complex. Nature 493(7434):699–702. doi:10.1038/nature11791
Bischler N, Brino L, Carles C, Riva M, Tschochner H, Mallouh V, Schultz P (2002) Localization of the yeast RNA polymerase I-specific subunits. EMBO J 21(15):4136–4144
Blair RH, Goodrich JA, Kugel JF (2012) Single-molecule fluorescence resonance energy transfer shows uniformity in TATA binding protein-induced DNA bending and heterogeneity in bending kinetics. Biochemistry 51(38):7444–7455. doi:10.1021/bi300491j
Blattner C, Jennebach S, Herzog F, Mayer A, Cheung ACM, Witte G, Lorenzen K, Hopfner K-P, Heck AJR, Aebersold R, Cramer P (2011) Molecular basis of Rrn3-regulated RNA polymerase I initiation and cell growth. Genes Dev 25(19):2093–2105. doi:10.1101/gad.17363311
Blombach F, Makarova KS, Marrero J, Siebers B, Koonin EV, van der Oost J (2009) Identification of an ortholog of the eukaryotic RNA polymerase III subunit RPC34 in Crenarchaeota and Thaumarchaeota suggests specialization of RNA polymerases for coding and non-coding RNAs in Archaea. Biol Direct 4:39. doi:10.1186/1745-6150-4-39
Blombach F, Daviter T, Fielden D, Grohmann D, Smollett K, Werner F (2013) Archaeology of RNA polymerase: factor swapping during the transcription cycle. Biochem Soc Trans 41(1):362–367. doi:10.1042/BST20120274
Blombach F, Salvadori E, Fouqueau T, Yan J, Reimann J, Sheppard C, Smollett KL, Albers SV, Kay CW, Thalassinos K, Werner F (2015) Archaeal TFEalpha/beta is a hybrid of TFIIE and the RNA polymerase III subcomplex hRPC62/39. Elife 4:e08378. doi:10.7554/eLife.08378
Blombach F, Smollett KL, Grohmann D, Werner F (2016) Molecular mechanisms of transcription initiation – structure, function and evolution of TFE/TFIIE-like factors and open complex formation. J Mol Biol. doi:10.1016/j.jmb.2016.04.016
Bochkareva A, Yuzenkova Y, Tadigotla VR, Zenkin N (2012) Factor-independent transcription pausing caused by recognition of the RNA-DNA hybrid sequence. EMBO J 31(3):630–639. doi:10.1038/emboj.2011.432
Borukhov S, Nudler E (2008) RNA polymerase: the vehicle of transcription. Trends Microbiol 16(3):126–134. doi:10.1016/j.tim.2007.12.006
Borukhov S, Polyakov A, Nikiforov V, Goldfarb A (1992) GreA protein: a transcription elongation factor from Escherichia coli. Proc Natl Acad Sci U S A 89(19):8899–8902
Brindefalk B, Dessailly BH, Yeats C, Orengo C, Werner F, Poole AM (2013) Evolutionary history of the TBP-domain superfamily. Nucleic Acids Res 41(5):2832–2845. doi:10.1093/nar/gkt045
Brow DA, Geiduschek EP (1987) Modulation of yeast 5 S rRNA synthesis in vitro by ribosomal protein YL3. A possible regulatory loop. J Biol Chem 262(29):13953–13958
Brueckner F, Armache KJ, Cheung A, Damsma GE, Kettenberger H, Lehmann E, Sydow J, Cramer P (2009a) Structure-function studies of the RNA polymerase II elongation complex. Acta Crystallogr D Biol Crystallogr 65(Pt 2):112–120. doi:10.1107/S0907444908039875
Brueckner F, Ortiz J, Cramer P (2009b) A movie of the RNA polymerase nucleotide addition cycle. Curr Opin Struct Biol 19(3):294–299. doi:10.1016/j.sbi.2009.04.005
Burgess RR, Travers AA, Dunn JJ, Bautz EKF (1969) Factor stimulating transcription by RNA polymerase. Nature 221(5175):43–46. doi:10.1038/221043a0
Burton SP, Burton ZF (2014) The σ enigma: bacterial σ factors, archaeal TFB and eukaryotic TFIIB are homologs. Transcription 5(4):e967599. doi:10.4161/21541264.2014.967599
Burton ZF, Killeen M, Sopta M, Ortolan LG, Greenblatt J (1988) RAP30/74: a general initiation factor that binds to RNA polymerase II. Mol Cell Biol 8(4):1602–1613
Bushnell DA, Westover KD, Davis RE, Kornberg RD (2004) Structural basis of transcription: an RNA polymerase II-TFIIB cocrystal at 4.5 Angstroms. Science 303(5660):983–988. doi:10.1126/science.1090838
Cabart P, Ujvari A, Pal M, Luse DS (2011) Transcription factor TFIIF is not required for initiation by RNA polymerase II, but it is essential to stabilize transcription factor TFIIB in early elongation complexes. Proc Natl Acad Sci U S A 108(38):15786–15791. doi:10.1073/pnas.1104591108

Campbell FE Jr, Setzer DR (1992) Transcription termination by RNA polymerase III: uncoupling of polymerase release from termination signal recognition. Mol Cell Biol 12(5):2260–2272

Campbell PN, Work TS (1953) Biosynthesis of proteins. Nature 171(4362):997–1001

Carter R, Drouin G (2010) The increase in the number of subunits in eukaryotic RNA polymerase III relative to RNA polymerase II is due to the permanent recruitment of general transcription factors. Mol Biol Evol 27(5):1035–1043. doi:10.1093/molbev/msp316

Carvalho AT, Fernandes PA, Ramos MJ (2011) The catalytic mechanism of RNA polymerase II. J Chem Theory Comput 7(4):1177–1188. doi:10.1021/ct100579w

Casamassimi A, Napoli C (2007) Mediator complexes and eukaryotic transcription regulation: an overview. Biochimie 89(12):1439–1446. doi:10.1016/j.biochi.2007.08.002

Chafin DR, Claussen TJ, Price DH (1991) Identification and purification of a yeast protein that affects elongation by RNA polymerase II. J Biol Chem 266(14):9256–9262

Chakraborty A, Wang D, Ebright YW, Korlann Y, Kortkhonjia E, Kim T, Chowdhury S, Wigneshweraraj S, Irschik H, Jansen R, Nixon BT, Knight J, Weiss S, Ebright RH (2012) Opening and closing of the bacterial RNA polymerase clamp. Science 337(6094):591–595. doi:10.1126/science.1218716

Chamberlin M, Baldwin RL, Berg P (1963) An enzymically synthesized RNA of alternating base sequence: physical and chemical characterization. J Mol Biol 7:334–349

Chedin S, Riva M, Schultz P, Sentenac A, Carles C (1998) The RNA cleavage activity of RNA polymerase III is mediated by an essential TFIIS-like subunit and is important for transcription termination. Genes Dev 12(24):3857–3871

Cheetham GM, Steitz TA (2000) Insights into transcription: structure and function of single-subunit DNA-dependent RNA polymerases. Curr Opin Struct Biol 10(1):117–123

Chen HT, Warfield L, Hahn S (2007) The positions of TFIIF and TFIIE in the RNA polymerase II transcription preinitiation complex. Nat Struct Mol Biol 14(8):696–703. doi:10.1038/nsmb1272

Cheung AC, Cramer P (2011) Structural basis of RNA polymerase II backtracking, arrest and reactivation. Nature 471(7337):249–253. doi:10.1038/nature09785

Cheung AC, Cramer P (2012) A movie of RNA polymerase II transcription. Cell 149(7):1431–1437. doi:10.1016/j.cell.2012.06.006

Cianfrocco MA, Kassavetis GA, Grob P, Fang J, Juven-Gershon T, Kadonaga JT, Nogales E (2013) Human TFIID binds to core promoter DNA in a reorganized structural state. Cell 152(1–2):120–131. doi:10.1016/j.cell.2012.12.005

Cler E, Papai G, Schultz P, Davidson I (2009) Recent advances in understanding the structure and function of general transcription factor TFIID. Cell Mol Life Sci 66(13):2123–2134. doi:10.1007/s00018-009-0009-3

Coin F, Oksenych V, Egly J-M (2007) Distinct roles for the XPB/p52 and XPD/p44 subcomplexes of TFIIH in damaged DNA opening during nucleotide excision repair. Mol Cell 26(2):245–256. doi:10.1016/j.molcel.2007.03.009

Colbert T, Hahn S (1992) A yeast TFIIB-related factor involved in RNA polymerase III transcription. Genes Dev 6(10):1940–1949

Conaway RC, Conaway JW (2015) Orchestrating transcription with the Pol II CTD. Nat Rev Mol Cell Biol 16(3):128. doi:10.1038/nrm3956

Conaway RC, Garrett KP, Hanley JP, Conaway JW (1991) Mechanism of promoter selection by RNA polymerase II: mammalian transcription factors alpha and beta gamma promote entry of polymerase into the preinitiation complex. Proc Natl Acad Sci U S A 88(14):6205–6209

Connelly S, Manley JL (1988) A functional mRNA polyadenylation signal is required for transcription termination by RNA polymerase II. Genes Dev 2(4):440–452

Cramer P, Bushnell DA, Fu J, Gnatt AL, Maier-Davis B, Thompson NE, Burgess RR, Edwards AM, David PR, Kornberg RD (2000) Architecture of RNA polymerase II and implications for the transcription mechanism. Science 288(5466):640–649

Da LT, Pardo-Avila F, Xu L, Silva DA, Zhang L, Gao X, Wang D, Huang X (2016) Bridge helix bending promotes RNA polymerase II backtracking through a critical and conserved threonine residue. Nat Commun 7:11244. doi:10.1038/ncomms11244

Dangkulwanich M, Ishibashi T, Bintu L, Bustamante C (2014) Molecular mechanisms of transcription through single-molecule experiments. Chem Rev 114(6):3203–3223. doi:10.1021/cr400730x

Deng W, Roberts SG (2006) Core promoter elements recognized by transcription factor IIB. Biochem Soc Trans 34(Pt 6):1051–1053. doi:10.1042/BST0341051

Dieci G, Fiorino G, Castelnuovo M, Teichmann M, Pagano A (2007) The expanding RNA polymerase III transcriptome. Trends Genet 23(12):614–622. doi:10.1016/j.tig.2007.09.001

Dombroski AJ, Walter WA, Record MT, Siegele DA, Gross CA (1992) Polypeptides containing highly conserved regions of transcription initiation factor sigma 70 exhibit specificity of binding to promoter DNA. Cell 70(3):501–512

Dounce AL (1952) Duplicating mechanism for peptide chain and nucleic acid synthesis. Enzymologia 15(5):251–258

Duttke SHC (2015) Evolution and diversification of the basal transcription machinery. Trends Biochem Sci 40(3):127–129. doi:10.1016/j.tibs.2015.01.005

Eichner J, Chen H-T, Warfield L, Hahn S (2010) Position of the general transcription factor TFIIF within the RNA polymerase II transcription preinitiation complex. EMBO J 29(4):706–716. doi:10.1038/emboj.2009.386

El Hage A, Koper M, Kufel J, Tollervey D (2008) Efficient termination of transcription by RNA polymerase I requires the 5′ exonuclease Rat1 in yeast. Genes Dev 22(8):1069–1081. doi:10.1101/gad.463708

Elmlund H, Baraznenok V, Linder T, Szilagyi Z, Rofougaran R, Hofer A, Hebert H, Lindahl M, Gustafsson CM (2009) Cryo-EM reveals promoter DNA binding and conformational flexibility of the general transcription factor TFIID. Structure (London, England: 1993) 17(11):1442–1452. doi:10.1016/j.str.2009.09.007

Engel C, Sainsbury S, Cheung AC, Kostrewa D, Cramer P (2013) RNA polymerase I structure and transcription regulation. Nature 502(7473):650–655. doi:10.1038/nature12712

Engelke DR, Ng SY, Shastry BS, Roeder RG (1980) Specific interaction of a purified transcription factor with an internal control region of 5S RNA genes. Cell 19(3):717–728

Epshtein V, Mustaev A, Markovtsov V, Bereshchenko O, Nikiforov V, Goldfarb A (2002) Swing-gate model of nucleotide entry into the RNA polymerase active center. Mol Cell 10(3):623–634

Facciotti MT, Reiss DJ, Pan M, Kaur A, Vuthoori M, Bonneau R, Shannon P, Srivastava A, Donohoe SM, Hood LE, Baliga NS (2007) General transcription factor specified global gene regulation in archaea. Proc Natl Acad Sci U S A 104(11):4630–4635. doi:10.1073/pnas.0611663104

Fazal FM, Meng CA, Murakami K, Kornberg RD, Block SM (2015) Real-time observation of the initiation of RNA polymerase II transcription. Nature 525(7568):274–277. doi:10.1038/nature14882

Feaver WJ, Gileadi O, Li Y, Kornberg RD (1991) CTD kinase associated with yeast RNA polymerase II initiation factor b. Cell 67(6):1223–1230

Feklistov A, Darst SA (2011) Structural basis for promoter-10 element recognition by the bacterial RNA polymerase sigma subunit. Cell 147(6):1257–1269. doi:10.1016/j.cell.2011.10.041

Feklistov A, Sharon BD, Darst SA, Gross CA (2014) Bacterial sigma factors: a historical, structural, and genomic perspective. Annu Rev Microbiol 68:357–376. doi:10.1146/annurev-micro-092412-155737

Fernandez-Tornero C, Moreno-Morcillo M, Rashid UJ, Taylor NM, Ruiz FM, Gruene T, Legrand P, Steuerwald U, Muller CW (2013) Crystal structure of the 14-subunit RNA polymerase I. Nature 502(7473):644–649. doi:10.1038/nature12636

Fish RN, Kane CM (2002) Promoting elongation with transcript cleavage stimulatory factors. Biochim Biophys Acta 1577(2):287–307

Fishburn J, Hahn S (2012) Architecture of the yeast RNA polymerase II open complex and regulation of activity by TFIIF. Mol Cell Biol 32(1):12–25. doi:10.1128/Mcb.06242-11

Fishburn J, Tomko E, Galburt E, Hahn S (2015) Double-stranded DNA translocase activity of transcription factor TFIIH and the mechanism of RNA polymerase II open complex formation. Proc Natl Acad Sci U S A 112(13):3961–3966. doi:10.1073/pnas.1417709112

Formosa T (2013) The role of FACT in making and breaking nucleosomes. Biochim Biophys Acta 1819(3–4):247–255

Foster JE, Holmes SF, Erie DA (2001) Allosteric binding of nucleoside triphosphates to RNA polymerase regulates transcription elongation. Cell 106(2):243–252

Fukuda A, Nakadai T, Shimada M, Tsukui T, Matsumoto M, Nogi Y, Meisterernst M, Hisatake K (2004) Transcriptional coactivator PC4 stimulates promoter escape and facilitates transcriptional synergy by GAL4-VP16. Mol Cell Biol 24(14):6525–6535. doi:10.1128/MCB.24.14.6525-6535.2004

Gaiser F, Tan S, Richmond TJ (2000) Novel dimerization fold of RAP30/RAP74 in human TFIIF at 1.7 A resolution. J Mol Biol 302(5):1119–1127. doi:10.1006/jmbi.2000.4110

Geiduschek EP, Nakamoto T, Weiss SB (1961) The enzymatic synthesis of RNA: complementary interaction with DNA. Proc Natl Acad Sci U S A 47(9):1405–1415

Geiger SR, Lorenzen K, Schreieck A, Hanecker P, Kostrewa D, Heck AJ, Cramer P (2010) RNA polymerase I contains a TFIIF-related DNA-binding subcomplex. Mol Cell 39(4):583–594. doi:10.1016/j.molcel.2010.07.028

Giardina C, Lis JT (1993) DNA melting on yeast RNA polymerase II promoters. Science 261(5122):759–762

Gietl A, Holzmeister P, Blombach F, Schulz S, von Voithenberg LV, Lamb DC, Werner F, Tinnefeld P, Grohmann D (2014) Eukaryotic and archaeal TBP and TFB/TF(II)B follow different promoter DNA bending pathways. Nucleic Acids Res 42(10):6219–6231. doi:10.1093/nar/gku273

Gnatt AL, Cramer P, Fu J, Bushnell DA, Kornberg RD (2001) Structural basis of transcription: an RNA polymerase II elongation complex at 3.3 A resolution. Science 292(5523):1876–1882. doi:10.1126/science.1059495

Goldman SR, Ebright RH, Nickels BE (2009) Direct detection of abortive RNA transcripts in vivo. Science 324(5929):927–928. doi:10.1126/science.1169237

Gouge J, Satia K, Guthertz N, Widya M, Thompson AJ, Cousin P, Dergai O, Hernandez N, Vannini A (2015) Redox signaling by the RNA polymerase III TFIIB-related factor Brf2. Cell 163(6):1375–1387. doi:10.1016/j.cell.2015.11.005

Groft CM, Uljon SN, Wang R, Werner MH (1998) Structural homology between the Rap30 DNA-binding domain and linker histone H5: implications for preinitiation complex assembly. Proc Natl Acad Sci U S A 95(16):9117–9122

Grohmann D, Werner F (2011) Cycling through transcription with the RNA polymerase F/E (RPB4/7) complex: structure, function and evolution of archaeal RNA polymerase. Res Microbiol 162(1):10–18. doi:10.1016/j.resmic.2010.09.002

Grohmann D, Hirtreiter A, Werner F (2009) Molecular mechanisms of archaeal RNA polymerase. Biochem Soc Trans 37(Pt 1):12–17. doi:10.1042/BST0370012

Grohmann D, Klose D, Klare JP, Kay CW, Steinhoff HJ, Werner F (2010) RNA-binding to archaeal RNA polymerase subunits F/E: a DEER and FRET study. J Am Chem Soc 132(17):5954–5955. doi:10.1021/ja101663d

Grohmann D, Nagy J, Chakraborty A, Klose D, Fielden D, Ebright RH, Michaelis J, Werner F (2011) The initiation factor TFE and the elongation factor Spt4/5 compete for the RNAP clamp during transcription initiation and elongation. Mol Cell 43(2):263–274. doi:10.1016/j.molcel.2011.05.030

Gruber TM, Gross CA (2003) Multiple sigma subunits and the partitioning of bacterial transcription space. Annu Rev Microbiol 57:441–466. doi:10.1146/annurev.micro.57.030502.090913

Grünberg S, Bartlett MS, Naji S, Thomm M (2007) Transcription factor E is a part of transcription elongation complexes. J Biol Chem 282(49):35482–35490. doi:10.1074/jbc.M707371200

Grunberg S, Warfield L, Hahn S (2012) Architecture of the RNA polymerase II preinitiation complex and mechanism of ATP-dependent promoter opening. Nat Struct Mol Biol 19(8):788–796. doi:10.1038/nsmb.2334

Grunberg-Manago M, Oritz PJ, Ochoa S (1955) Enzymatic synthesis of nucleic acidlike polynucleotides. Science 122(3176):907–910

Guajardo R, Sousa R (1997) A model for the mechanism of polymerase translocation. J Mol Biol 265(1):8–19. doi:10.1006/jmbi.1996.0707

Haag JR, Pikaard CS (2011) Multisubunit RNA polymerases IV and V: purveyors of non-coding RNA for plant gene silencing. Nat Rev Mol Cell Biol 12(8):483–492. doi:10.1038/nrm3152

Hausner W, Wettach J, Hethke C, Thomm M (1996) Two transcription factors related with the eucaryal transcription factors TATA-binding protein and transcription factor IIB direct promoter recognition by an archaeal RNA polymerase. J Biol Chem 271(47):30144–30148

Hawryluk PJ, Ujvari A, Luse DS (2004) Characterization of a novel RNA polymerase II arrest site which lacks a weak 3′ RNA-DNA hybrid. Nucleic Acids Res 32(6):1904–1916. doi:10.1093/nar/gkh505

He Y, Fang J, Taatjes DJ, Nogales E (2013) Structural visualization of key steps in human transcription initiation. Nature 495(7442):481–486. doi:10.1038/nature11991

He Y, Yan C, Fang J, Inouye C, Tjian R, Ivanov I, Nogales E (2016) Near-atomic resolution visualization of human transcription promoter opening. Nature 533(7603):359–365. doi:10.1038/nature17970

Hein PP, Palangat M, Landick R (2011) RNA transcript 3′-proximal sequence affects translocation bias of RNA polymerase. Biochemistry 50(32):7002–7014. doi:10.1021/bi200437q

Henry NL, Sayre MH, Kornberg RD (1992) Purification and characterization of yeast RNA polymerase II general initiation factor g. J Biol Chem 267(32):23388–23392

Herbert KM, La Porta A, Wong BJ, Mooney RA, Neuman KC, Landick R, Block SM (2006) Sequence-resolved detection of pausing by single RNA polymerase molecules. Cell 125(6):1083–1094. doi:10.1016/j.cell.2006.04.032

Hirata A, Klein BJ, Murakami KS (2008) The X-ray crystal structure of RNA polymerase from Archaea. Nature 451(7180):851–854. doi:10.1038/nature06530

Hirtreiter A, Damsma GE, Cheung AC, Klose D, Grohmann D, Vojnic E, Martin AC, Cramer P, Werner F (2010a) Spt4/5 stimulates transcription elongation through the RNA polymerase clamp coiled-coil motif. Nucleic Acids Res 38(12):4040–4051. doi:10.1093/nar/gkq135

Hirtreiter A, Grohmann D, Werner F (2010b) Molecular mechanisms of RNA polymerase--the F/E (RPB4/7) complex is required for high processivity in vitro. Nucleic Acids Res 38(2):585–596. doi:10.1093/nar/gkp928

Hoagland MB, Zamecnik PC, Stephenson ML (1957) Intermediate reactions in protein biosynthesis. Biochim Biophys Acta 24(1):215–216

Hoffmann NA, Jakobi AJ, Moreno-Morcillo M, Glatt S, Kosinski J, Hagen WJ, Sachse C, Muller CW (2015) Molecular structures of unbound and transcribing RNA polymerase III. Nature 528(7581):231–236. doi:10.1038/nature16143

Holmes SF, Erie DA (2003) Downstream DNA sequence effects on transcription elongation. Allosteric binding of nucleoside triphosphates facilitates translocation via a ratchet motion. J Biol Chem 278(37):35597–35608. doi:10.1074/jbc.M304496200

Holstege FC, Fiedler U, Timmers HT (1997) Three transitions in the RNA polymerase II transcription complex during initiation. EMBO J 16(24):7468–7480. doi:10.1093/emboj/16.24.7468

Hsin JP, Manley JL (2012) The RNA polymerase II CTD coordinates transcription and RNA processing. Genes Dev 26(19):2119–2137. doi:10.1101/gad.200303.112

Huang Y, Maraia RJ (2001) Comparison of the RNA polymerase III transcription machinery in Schizosaccharomyces pombe, Saccharomyces cerevisiae and human. Nucleic Acids Res 29(13):2675–2690

Huang RC, Maheshwari N, Bonner J (1960) Enzymatic synthesis of RNA. Biochem Biophys Res Commun 3:689–694

Huisinga KL, Pugh BF (2004) A genome-wide housekeeping role for TFIID and a highly regulated stress-related role for SAGA in Saccharomyces cerevisiae. Mol Cell 13(4):573–585

Hurwitz J (2005) The discovery of RNA polymerase. J Biol Chem 280(52):42477–42485. doi:10.1074/jbc.X500006200

Hurwitz J, Bresler AE (1961) The incorporation of ribonucleotides into ribonucleic acid. J Biol Chem 236:542–548

Imashimizu M, Kireeva ML, Lubkowska L, Gotte D, Parks AR, Strathern JN, Kashlev M (2013) Intrinsic translocation barrier as an initial step in pausing by RNA polymerase II. J Mol Biol 425(4):697–712. doi:10.1016/j.jmb.2012.12.002

Imbalzano AN, Zaret KS, Kingston RE (1994) Transcription factor (TF) IIB and TFIIA can independently increase the affinity of the TATA-binding protein for DNA. J Biol Chem 269(11):8280–8286

Jaiswal R, Choudhury M, Zaman S, Singh S, Santosh V, Bastia D, Escalante CR (2016) Functional architecture of the Reb1-Ter complex of Schizosaccharomyces pombe. Proc Natl Acad Sci U S A 113(16):E2267–E2276. doi:10.1073/pnas.1525465113

Jennebach S, Herzog F, Aebersold R, Cramer P (2012) Crosslinking-MS analysis reveals RNA polymerase I domain architecture and basis of rRNA cleavage. Nucleic Acids Res 40(12):5591–5601. doi:10.1093/nar/gks220

Jin DJ, Burgess RR, Richardson JP, Gross CA (1992) Termination efficiency at rho-dependent terminators depends on kinetic coupling between RNA polymerase and rho. Proc Natl Acad Sci U S A 89(4):1453–1457

Jonkers I, Lis JT (2015) Getting up to speed with transcription elongation by RNA polymerase II. Nat Rev Mol Cell Biol 16(3):167–177. doi:10.1038/nrm3953

Juo ZS, Kassavetis GA, Wang J, Geiduschek EP, Sigler PB (2003) Crystal structure of a transcription factor IIIB core interface ternary complex. Nature 422(6931):534–539. doi:10.1038/nature01534

Kadonaga JT (2012) Perspectives on the RNA polymerase II core promoter. Wiley Interdiscip Rev Dev Biol 1(1):40–51. doi:10.1002/wdev.21

Kamada K, De Angelis J, Roeder RG, Burley SK (2001) Crystal structure of the C-terminal domain of the RAP74 subunit of human transcription factor IIF. Proc Natl Acad Sci U S A 98(6):3115–3120. doi:10.1073/pnas.051631098

Kang JJ, Auble DT, Ranish JA, Hahn S (1995) Analysis of the yeast transcription factor TFIIA: distinct functional regions and a polymerase II-specific role in basal and activated transcription. Mol Cell Biol 15(3):1234–1243

Kapanidis AN, Margeat E, Ho SO, Kortkhonjia E, Weiss S, Ebright RH (2006) Initial transcription by RNA polymerase proceeds through a DNA-scrunching mechanism. Science 314(5802):1144–1147. doi:10.1126/science.1131399

Kawauchi J, Mischo H, Braglia P, Rondon A, Proudfoot NJ (2008) Budding yeast RNA polymerases I and II employ parallel mechanisms of transcriptional termination. Genes Dev 22(8):1082–1092. doi:10.1101/gad.463408

Keener J, Josaitis CA, Dodd JA, Nomura M (1998) Reconstitution of yeast RNA polymerase I transcription in vitro from purified components. TATA-binding protein is not required for basal transcription. J Biol Chem 273(50):33795–33802

Kettenberger H, Armache KJ, Cramer P (2003) Architecture of the RNA polymerase II-TFIIS complex and implications for mRNA cleavage. Cell 114(3):347–357

Kettenberger H, Armache KJ, Cramer P (2004) Complete RNA polymerase II elongation complex structure and its interactions with NTP and TFIIS. Mol Cell 16(6):955–965. doi:10.1016/j.molcel.2004.11.040

Khaperskyy DA, Ammerman ML, Majovski RC, Ponticelli AS (2008) Functions of Saccharomyces cerevisiae TFIIF during transcription start site utilization. Mol Cell Biol 28(11):3757–3766. doi:10.1128/mcb.02272-07

Kim TK, Ebright RH, Reinberg D (2000) Mechanism of ATP-dependent promoter melting by transcription factor IIH. Science (New York, NY) 288(5470):1418–1422

Kireeva ML, Hancock B, Cremona GH, Walter W, Studitsky VM, Kashlev M (2005) Nature of the nucleosomal barrier to RNA polymerase II. Mol Cell 18(1):97–108. doi:10.1016/j.molcel.2005.02.027

Kireeva M, Kashlev M, Burton ZF (2010) Translocation by multi-subunit RNA polymerases. Biochim Biophys Acta 1799(5–6):389–401. doi:10.1016/j.bbagrm.2010.01.007

Kireeva ML, Domecq C, Coulombe B, Burton ZF, Kashlev M (2011) Interaction of RNA polymerase II fork loop 2 with downstream non-template DNA regulates transcription elongation. J Biol Chem 286(35):30898–30910. doi:10.1074/jbc.M111.260844

Klein BJ, Bose D, Baker KJ, Yusoff ZM, Zhang X, Murakami KS (2011) RNA polymerase and transcription elongation factor Spt4/5 complex structure. Proc Natl Acad Sci U S A 108(2):546–550. doi:10.1073/pnas.1013828108

Knowlton JR, Bubunenko M, Andrykovitch M, Guo W, Routzahn KM, Waugh DS, Court DL, Ji X (2003) A spring-loaded state of NusG in its functional cycle is suggested by X-ray crystallography and supported by site-directed mutants. Biochemistry 42(8):2275–2281. doi:10.1021/bi0272508

Knutson BA, Hahn S (2011) Yeast Rrn7 and human TAF1B are TFIIB-related RNA polymerase I general transcription factors. Science (New York, NY) 333(6049):1637–1640. doi:10.1126/science.1207699

Knutson BA, Hahn S (2013) TFIIB-related factors in RNA polymerase I transcription. Biochim Biophys Acta 1829(3–4):265–273. doi:10.1016/j.bbagrm.2012.08.003

Knutson BA, Luo J, Ranish J, Hahn S (2014) Architecture of the Saccharomyces cerevisiae RNA polymerase I Core Factor complex. Nat Struct Mol Biol 21(9):810–816. doi:10.1038/nsmb.2873

Kokubo T, Yamashita S, Horikoshi M, Roeder RG, Nakatani Y (1994) Interaction between the N-terminal domain of the 230-kDa subunit and the TATA box-binding subunit of TFIID negatively regulates TATA-box binding. Proc Natl Acad Sci U S A 91(9):3520–3524

Komissarova N, Kashlev M (1997a) RNA polymerase switches between inactivated and activated states by translocating back and forth along the DNA and the RNA. J Biol Chem 272(24):15329–15338

Komissarova N, Kashlev M (1997b) Transcriptional arrest: Escherichia coli RNA polymerase translocates backward, leaving the 3′ end of the RNA intact and extruded. Proc Natl Acad Sci U S A 94(5):1755–1760

Kornberg RD (2007) The molecular basis of eukaryotic transcription. Proc Natl Acad Sci U S A 104(32):12955–12961. doi:10.1073/pnas.0704138104

Kostrewa D, Zeller ME, Armache KJ, Seizl M, Leike K, Thomm M, Cramer P (2009) RNA polymerase II-TFIIB structure and mechanism of transcription initiation. Nature 462(7271):323–330. doi:10.1038/nature08548

Kuehner JN, Brow DA (2006) Quantitative analysis of in vivo initiator selection by yeast RNA polymerase II supports a scanning model. J Biol Chem 281(20):14119–14128. doi:10.1074/jbc.M601937200

Kuehner JN, Pearson EL, Moore C (2011) Unravelling the means to an end: RNA polymerase II transcription termination. Nat Rev Mol Cell Biol 12(5):283–294. doi:10.1038/nrm3098

Kugel JF, Goodrich JA (1998) Promoter escape limits the rate of RNA polymerase II transcription and is enhanced by TFIIE, TFIIH, and ATP on negatively supercoiled DNA. Proc Natl Acad Sci U S A 95(16):9232–9237

Kuhn CD, Geiger SR, Baumli S, Gartmann M, Gerber J, Jennebach S, Mielke T, Tschochner H, Beckmann R, Cramer P (2007) Functional architecture of RNA polymerase I. Cell 131(7):1260–1272. doi:10.1016/j.cell.2007.10.051

Kwak H, Lis JT (2013) Control of transcriptional elongation. Annu Rev Genet 47:483–508. doi:10.1146/annurev-genet-110711-155440

Landrieux E, Alic N, Ducrot C, Acker J, Riva M, Carles C (2006) A subcomplex of RNA polymerase III subunits involved in transcription termination and reinitiation. EMBO J 25(1):118–128. doi:10.1038/sj.emboj.7600915

Larivière L, Seizl M, Cramer P (2012) A structural perspective on Mediator function. Curr Opin Cell Biol 24(3):305–313. doi:10.1016/j.ceb.2012.01.007

Larson MH, Greenleaf WJ, Landick R, Block SM (2008) Applied force reveals mechanistic and energetic details of transcription termination. Cell 132(6):971–982. doi:10.1016/j.cell.2008.01.027

Larson MH, Mooney RA, Peters JM, Windgassen T, Nayak D, Gross CA, Block SM, Greenleaf WJ, Landick R, Weissman JS (2014) A pause sequence enriched at translation start sites drives transcription dynamics in vivo. Science 344(6187):1042–1047. doi:10.1126/science.1251871

Layat E, Probst AV, Tourmente S (2013) Structure, function and regulation of Transcription Factor IIIA: From Xenopus to Arabidopsis. Biochim Biophys Acta 1829(3–4):274–282. doi:10.1016/j.bbagrm.2012.10.013

Lee TI, Causton HC, Holstege FC, Shen WC, Hannett N, Jennings EG, Winston F, Green MR, Young RA (2000) Redundant roles for the TFIID and SAGA complexes in global transcription. Nature 405(6787):701–704. doi:10.1038/35015104

Lefèvre S, Dumay-Odelot H, El-Ayoubi L, Budd A, Legrand P, Pinaud N, Teichmann M, Fribourg S (2011) Structure-function analysis of hRPC62 provides insights into RNA polymerase III transcription initiation. Nat Struct Mol Biol 18(3):352–358. doi:10.1038/nsmb.1996

Lehman IR, Bessman MJ, Simms ES, Kornberg A (1958) Enzymatic synthesis of deoxyribonucleic acid. I. Preparation of substrates and partial purification of an enzyme from Escherichia coli. J Biol Chem 233(1):163–170

Lisica A, Engel C, Jahnel M, Roldan E, Galburt EA, Cramer P, Grill SW (2016) Mechanisms of backtrack recovery by RNA polymerases I and II. Proc Natl Acad Sci U S A. doi:10.1073/pnas.1517011113

Littlefield O, Korkhin Y, Sigler PB (1999) The structural basis for the oriented assembly of a TBP/TFB/promoter complex. Proc Natl Acad Sci U S A 96(24):13668–13673

Liu J, Akoulitchev S, Weber A, Ge H, Chuikov S, Libutti D, Wang XW, Conaway JW, Harris CC, Conaway RC, Reinberg D, Levens D (2001) Defective interplay of activators and repressors with TFIH in xeroderma pigmentosum. Cell 104(3):353–363

Liu X, Bushnell DA, Wang D, Calero G, Kornberg RD (2010) Structure of an RNA polymerase II-TFIIB complex and the transcription initiation mechanism. Science 327(5962):206–209. doi:10.1126/science.1182015

Liu X, Kraus WL, Bai X (2015) Ready, pause, go: regulation of RNA polymerase II pausing and release by cellular signaling pathways. Trends Biochem Sci 40(9):516–525. doi:10.1016/j.tibs.2015.07.003

Liu B, Zuo Y, Steitz TA (2016) Structures of E. coli sigmaS-transcription initiation complexes provide new insights into polymerase mechanism. Proc Natl Acad Sci U S A. doi:10.1073/pnas.1520555113

Logan J, Falck-Pedersen E, Darnell JE Jr, Shenk T (1987) A poly(A) addition site and a downstream termination region are required for efficient cessation of transcription by RNA polymerase II in the mouse beta maj-globin gene. Proc Natl Acad Sci U S A 84(23):8306–8310

López-De-León A, Librizzi M, Puglia K, Willis IM (1992) PCF4 encodes an RNA polymerase III transcription factor with homology to TFIIB. Cell 71(2):211–220

Louder RK, He Y, Lopez-Blanco JR, Fang J, Chacon P, Nogales E (2016) Structure of promoter-bound TFIID and model of human pre-initiation complex assembly. Nature. doi:10.1038/nature17394

Lu D, Searles MA, Klug A (2003) Crystal structure of a zinc-finger-RNA complex reveals two modes of molecular recognition. Nature 426(6962):96–100. doi:10.1038/nature02088

Male G, von Appen A, Glatt S, Taylor NM, Cristovao M, Groetsch H, Beck M, Muller CW (2015) Architecture of TFIIIC and its role in RNA polymerase III pre-initiation complex assembly. Nat Commun 6:7387. doi:10.1038/ncomms8387

Martinez-Rucobo FW, Cramer P (2013) Structural basis of transcription elongation. Biochim Biophys Acta 1829(1):9–19. doi:10.1016/j.bbagrm.2012.09.002

Martinez-Rucobo FW, Sainsbury S, Cheung AC, Cramer P (2011) Architecture of the RNA polymerase-Spt4/5 complex and basis of universal transcription processivity. EMBO J 30(7):1302–1310. doi:10.1038/emboj.2011.64

Meinhart A, Blobel J, Cramer P (2003) An extended winged helix domain in general transcription factor E/IIE alpha. J Biol Chem 278(48):48267–48274. doi:10.1074/jbc.M307874200

Merkl P, Perez-Fernandez J, Pilsl M, Reiter A, Williams L, Gerber J, Bohm M, Deutzmann R, Griesenbeck J, Milkereit P, Tschochner H (2014) Binding of the termination factor Nsi1 to its cognate DNA site is sufficient to terminate RNA polymerase I transcription in vitro and to induce termination in vivo. Mol Cell Biol 34(20):3817–3827. doi:10.1128/MCB.00395-14

Meselson M, Stahl FW (1958) The replication of DNA in Escherichia Coli. Proc Natl Acad Sci U S A 44(7):671–682

Mischo HE, Proudfoot NJ (2013) Disengaging polymerase: terminating RNA polymerase II transcription in budding yeast. Biochim Biophys Acta 1829(1):174–185. doi:10.1016/j.bbagrm.2012.10.003

Mooney RA, Schweimer K, Rösch P, Gottesman M, Landick R (2009) Two structurally independent domains of E. coli NusG create regulatory plasticity via distinct interactions with RNA polymerase and regulators. J Mol Biol 391(2):341–358. doi:10.1016/j.jmb.2009.05.078

Murakami KS, Masuda S, Campbell EA, Muzzin O, Darst SA (2002a) Structural basis of transcription initiation: an RNA polymerase holoenzyme-DNA complex. Science (New York, NY) 296(5571):1285–1290. doi:10.1126/science.1069595

Murakami KS, Masuda S, Darst SA (2002b) Structural basis of transcription initiation: RNA polymerase holoenzyme at 4 A resolution. Science (New York, NY) 296(5571):1280–1284. doi:10.1126/science.1069594

Murakami K, Calero G, Brown CR, Liu X, Davis RE, Boeger H, Kornberg RD (2013a) Formation and fate of a complete 31-protein RNA polymerase II transcription preinitiation complex. J Biol Chem 288(9):6325–6332. doi:10.1074/jbc.M112.433623

Murakami K, Elmlund H, Kalisman N, Bushnell DA, Adams CM, Azubel M, Elmlund D, Levi-Kalisman Y, Liu X, Gibbons BJ, Levitt M, Kornberg RD (2013b) Architecture of an RNA polymerase II transcription pre-initiation complex. Science 342(6159):1238724. doi:10.1126/science.1238724

Murakami K, Tsai KL, Kalisman N, Bushnell DA, Asturias FJ, Kornberg RD (2015) Structure of an RNA polymerase II preinitiation complex. Proc Natl Acad Sci U S A 112(44):13543–13548. doi:10.1073/pnas.1518255112

Nag A, Narsinh K, Kazerouninia A, Martinson HG (2006) The conserved AAUAAA hexamer of the poly(A) signal can act alone to trigger a stable decrease in RNA polymerase II transcription velocity. RNA 12(8):1534–1544. doi:10.1261/rna.103206

Nagy J, Grohmann D, Cheung AC, Schulz S, Smollett K, Werner F, Michaelis J (2015) Complete architecture of the archaeal RNA polymerase open complex from single-molecule FRET and NPS. Nat Commun 6:6161. doi:10.1038/ncomms7161

Naidu S, Friedrich JK, Russell J, Zomerdijk JCBM (2011) TAF1B is a TFIIB-like component of the basal transcription machinery for RNA polymerase I. Science (New York, NY) 333(6049):1640–1642. doi:10.1126/science.1207656

Naji S, Bertero MG, Spitalny P, Cramer P, Thomm M (2008) Structure-function analysis of the RNA polymerase cleft loops elucidates initial transcription, DNA unwinding and RNA displacement. Nucleic Acids Res 36(2):676–687. doi:10.1093/nar/gkm1086

Nedialkov YA, Gong XQ, Hovde SL, Yamaguchi Y, Handa H, Geiger JH, Yan H, Burton ZF (2003) NTP-driven translocation by human RNA polymerase II. J Biol Chem 278(20):18303–18312. doi:10.1074/jbc.M301103200

Nedialkov YA, Opron K, Assaf F, Artsimovitch I, Kireeva ML, Kashlev M, Cukier RI, Nudler E, Burton ZF (2013) The RNA polymerase bridge helix YFI motif in catalysis, fidelity and translocation. Biochim Biophys Acta 1829(2):187–198. doi:10.1016/j.bbagrm.2012.11.005

Németh A, Perez-Fernandez J, Merkl P, Hamperl S, Gerber J, Griesenbeck J, Tschochner H (2013) RNA polymerase I termination: where is the end? Biochim Biophys Acta 1829(3–4):306–317. doi:10.1016/j.bbagrm.2012.10.007

Nickels BE, Hochschild A (2004) Regulation of RNA polymerase through the secondary channel. Cell 118(3):281–284. doi:10.1016/j.cell.2004.07.021

Nickels BE, Garrity SJ, Mekler V, Minakhin L, Severinov K, Ebright RH, Hochschild A (2005) The interaction between sigma70 and the beta-flap of Escherichia coli RNA polymerase inhibits extension of nascent RNA during early elongation. Proc Natl Acad Sci U S A 102(12):4488–4493. doi:10.1073/pnas.0409850102

Nikolov DB, Chen H, Halay ED, Usheva AA, Hisatake K, Lee DK, Roeder RG, Burley SK (1995) Crystal structure of a TFIIB-TBP-TATA-element ternary complex. Nature 377(6545):119–128. doi:10.1038/377119a0

Nudler E (2009) RNA polymerase active center: the molecular engine of transcription. Annu Rev Biochem 78:335–361. doi:10.1146/annurev.biochem.76.052705.164655

Nudler E, Mustaev A, Lukhtanov E, Goldfarb A (1997) The RNA-DNA hybrid maintains the register of transcription by preventing backtracking of RNA polymerase. Cell 89(1):33–41

Olby R (2003) Quiet debut for the double helix. Nature 421(6921):402–405. doi:10.1038/nature01397

Opalka N, Chlenov M, Chacon P, Rice WJ, Wriggers W, Darst SA (2003) Structure and function of the transcription elongation factor GreB bound to bacterial RNA polymerase. Cell 114(3):335–345

Orozco IJ, Kim SJ, Martinson HG (2002) The poly(A) signal, without the assistance of any downstream element, directs RNA polymerase II to pause in vivo and then to release stochastically from the template. J Biol Chem 277(45):42899–42911. doi:10.1074/jbc.M207415200

Paget MS (2015) Bacterial sigma factors and anti-sigma factors: structure, function and distribution. Biomolecules 5(3):1245–1265. doi:10.3390/biom5031245

Pal M, Ponticelli AS, Luse DS (2005) The role of the transcription bubble and TFIIB in promoter clearance by RNA polymerase II. Mol Cell 19(1):101–110. doi:10.1016/j.molcel.2005.05.024

Panov KI, Friedrich JK, Russell J, Zomerdijk JC (2006) UBF activates RNA polymerase I transcription by stimulating promoter escape. EMBO J 25(14):3310–3322. doi:10.1038/sj.emboj.7601221

Papai G, Tripathi MK, Ruhlmann C, Layer JH, Weil PA, Schultz P (2010) TFIIA and the transactivator Rap1 cooperate to commit TFIID for transcription initiation. Nature 465(7300):956–960. doi:10.1038/nature09080

Parvin JD, Sharp PA (1993) DNA topology and a minimal set of basal factors for transcription by RNA polymerase II. Cell 73(3):533–540

Peters JM, Vangeloff AD, Landick R (2011) Bacterial transcription terminators: the RNA 3′-end chronicles. J Mol Biol 412(5):793–813. doi:10.1016/j.jmb.2011.03.036

Peyroche G, Milkereit P, Bischler N, Tschochner H, Schultz P, Sentenac A, Carles C, Riva M (2000) The recruitment of RNA polymerase I on rDNA is mediated by the interaction of the A43 subunit with Rrn3. EMBO J 19(20):5473–5482. doi:10.1093/emboj/19.20.5473

Plaschka C, Lariviere L, Wenzeck L, Seizl M, Hemann M, Tegunov D, Petrotchenko EV, Borchers CH, Baumeister W, Herzog F, Villa E, Cramer P (2015) Architecture of the RNA polymerase II-Mediator core initiation complex. Nature 518(7539):376–380. doi:10.1038/nature14229

Plaschka C, Hantsche M, Dienemann C, Burzinski C, Plitzko J, Cramer P (2016a) Transcription initiation complex structures elucidate DNA opening. Nature 533(7603):353–358. doi:10.1038/nature17990

Plaschka C, Nozawa K, Cramer P (2016b) Mediator architecture and RNA polymerase II interaction. J Mol Biol. doi:10.1016/j.jmb.2016.01.028

Porrua O, Libri D (2015) Transcription termination and the control of the transcriptome: why, where and how to stop. Nat Rev Mol Cell Biol 16(3):190–202. doi:10.1038/nrm3943

Poss ZC, Ebmeier CC, Taatjes DJ (2013) The Mediator complex and transcription regulation. Crit Rev Biochem Mol Biol 48(6):575–608. doi:10.3109/10409238.2013.840259

Qureshi SA, Bell SD, Jackson SP (1997) Factor requirements for transcription in the Archaeon Sulfolobus shibatae. EMBO J 16(10):2927–2936. doi:10.1093/emboj/16.10.2927
Rabhi M, Gocheva V, Jacquinot F, Lee A, Margeat E, Boudvillain M (2011) Mutagenesis-based evidence for an asymmetric configuration of the ring-shaped transcription termination factor Rho. J Mol Biol 405(2):497–518. doi:10.1016/j.jmb.2010.11.006
Rani PG, Ranish JA, Hahn S (2004) RNA polymerase II (Pol II)-TFIIF and Pol II-mediator complexes: the major stable Pol II complexes and their activity in transcription initiation and reinitiation. Mol Cell Biol 24(4):1709–1720
Reiter A, Hamperl S, Seitz H, Merkl P, Perez-Fernandez J, Williams L, Gerber J, Nemeth A, Leger I, Gadal O, Milkereit P, Griesenbeck J, Tschochner H (2012) The Reb1-homologue Ydr026c/Nsi1 is required for efficient RNA polymerase I termination in yeast. EMBO J 31(16):3480–3493. doi:10.1038/emboj.2012.185
Ren D, Lei L, Burton ZF (1999) A region within the RAP74 subunit of human transcription factor IIF is critical for initiation but dispensable for complex assembly. Mol Cell Biol 19(11):7377–7387
Revyakin A, Liu C, Ebright RH, Strick TR (2006) Abortive initiation and productive initiation by RNA polymerase involve DNA scrunching. Science 314(5802):1139–1143. doi:10.1126/science.1131398
Rhee HS, Pugh BF (2012) Genome-wide structure and organization of eukaryotic pre-initiation complexes. Nature 483(7389):295–301. doi:10.1038/nature10799
Richardson JP (1982) Activation of rho protein ATPase requires simultaneous interaction at two kinds of nucleic acid-binding sites. J Biol Chem 257(10):5760–5766
Roeder RG, Rutter WJ (1969) Multiple forms of DNA-dependent RNA polymerase in eukaryotic organisms. Nature 224(5216):234–237
Roeder RG, Rutter WJ (1970) Specific nucleolar and nucleoplasmic RNA polymerases. Proc Natl Acad Sci U S A 65(3):675–682
Rollins MB, Del Rio S, Galey AL, Setzer DR, Andrews MT (1993) Role of TFIIIA zinc fingers in vivo: analysis of single-finger function in developing Xenopus embryos. Mol Cell Biol 13(8):4776–4783
Rondon AG, Mischo HE, Kawauchi J, Proudfoot NJ (2009) Fail-safe transcriptional termination for protein-coding genes in S. cerevisiae. Mol Cell 36(1):88–98. doi:10.1016/j.molcel.2009.07.028
Ruan W, Lehmann E, Thomm M, Kostrewa D, Cramer P (2011) Evolution of two modes of intrinsic RNA polymerase transcript cleavage. J Biol Chem 286(21):18701–18707. doi:10.1074/jbc.M111.222273
Rutherford ST, Lemke JJ, Vrentas CE, Gaal T, Ross W, Gourse RL (2007) Effects of DksA, GreA, and GreB on transcription initiation: insights into the mechanisms of factors that bind in the secondary channel of RNA polymerase. J Mol Biol 366(4):1243–1257. doi:10.1016/j.jmb.2006.12.013
Saecker RM, Record MT Jr, Dehaseth PL (2011) Mechanism of bacterial transcription initiation: RNA polymerase – promoter binding, isomerization to initiation-competent open complexes, and initiation of RNA synthesis. J Mol Biol 412(5):754–771. doi:10.1016/j.jmb.2011.01.018
Sainsbury S, Niesser J, Cramer P (2013) Structure and function of the initially transcribing RNA polymerase II-TFIIB complex. Nature 493(7432):437–440. doi:10.1038/nature11715
Sainsbury S, Bernecky C, Cramer P (2015) Structural basis of transcription initiation by RNA polymerase II. Nat Rev Mol Cell Biol 16(3):129–143. doi:10.1038/nrm3952
Sanij E, Hannan RD (2009) The role of UBF in regulating the structure and dynamics of transcriptionally active rDNA chromatin. Epigenetics 4(6):374–382
Schnabel R, Thomm M, Gerardy-Schahn R, Zillig W, Stetter KO, Huet J (1983) Structural homology between different archaebacterial DNA-dependent RNA polymerases analyzed by immunological comparison of their components. EMBO J 2(5):751–755

Schulz S, Kramm K, Werner F, Grohmann D (2015) Fluorescently labeled recombinant RNAP system to probe archaeal transcription initiation. Methods 86:10–18. doi:10.1016/j.ymeth.2015.04.017

Schulz S, Gietl A, Smollett K, Tinnefeld P, Werner F, Grohmann D (2016) TFE and Spt4/5 open and close the RNA polymerase clamp during the transcription cycle. Proc Natl Acad Sci U S A. doi:10.1073/pnas.1515817113

Segall J, Matsui T, Roeder RG (1980) Multiple factors are required for the accurate transcription of purified genes by RNA polymerase III. J Biol Chem 255(24):11986–11991

Sekine S-i, Tagami S, Yokoyama S (2012) Structural basis of transcription by bacterial and eukaryotic RNA polymerases. Curr Opin Struct Biol 22(1):110–118. doi:10.1016/j.sbi.2011.11.006

Sentenac A (1985) Eukaryotic RNA polymerases. CRC Crit Rev Biochem 18(1):31–90

Serizawa H, Mäkelä TP, Conaway JW, Conaway RC, Weinberg RA, Young RA (1995) Association of Cdk-activating kinase subunits with transcription factor TFIIH. Nature 374(6519):280–282. doi:10.1038/374280a0

Sevostyanova A, Svetlov V, Vassylyev DG, Artsimovitch I (2008) The elongation factor RfaH and the initiation factor sigma bind to the same site on the transcription elongation complex. Proc Natl Acad Sci U S A 105(3):865–870. doi:10.1073/pnas.0708432105

Sikorski TW, Buratowski S (2009) The basal initiation machinery: beyond the general transcription factors. Curr Opin Cell Biol 21(3):344–351. doi:10.1016/j.ceb.2009.03.006

Skordalakes E, Berger JM (2003) Structure of the Rho transcription terminator: mechanism of mRNA recognition and helicase loading. Cell 114(1):135–146

Skordalakes E, Berger JM (2006) Structural insights into RNA-dependent ring closure and ATPase activation by the Rho termination factor. Cell 127(3):553–564. doi:10.1016/j.cell.2006.08.051

Söding J, Biegert A, Lupas AN (2005) The HHpred interactive server for protein homology detection and structure prediction. Nucleic Acids Res 33(Web Server issue):W244–W248. doi:10.1093/nar/gki408

Soppa J (1999) Normalized nucleotide frequencies allow the definition of archaeal promoter elements for different archaeal groups and reveal base-specific TFB contacts upstream of the TATA box. Mol Microbiol 31(5):1589–1592

Sorenson MK, Darst SA (2006) Disulfide cross-linking indicates that FlgM-bound and free sigma28 adopt similar conformations. Proc Natl Acad Sci U S A 103(45):16722–16727. doi:10.1073/pnas.0606482103

Sorenson MK, Ray SS, Darst SA (2004) Crystal structure of the flagellar sigma/anti-sigma complex sigma(28)/FlgM reveals an intact sigma factor in an inactive conformation. Mol Cell 14(1):127–138

Sosunov V, Sosunova E, Mustaev A, Bass I, Nikiforov V, Goldfarb A (2003) Unified two-metal mechanism of RNA synthesis and degradation by RNA polymerase. EMBO J 22(9):2234–2244. doi:10.1093/emboj/cdg193

Stevens A (1960) Incorporation of the adenine ribonucleotide into RNA by cell fractions from E. coli. Biochem Biophys Res Commun 3:92–96

Svetlov V, Nudler E (2008) Jamming the ratchet of transcription. Nat Struct Mol Biol 15(8):777–779. doi:10.1038/nsmb0808-777

Svetlov V, Nudler E (2013) Basic mechanism of transcription by RNA polymerase II. Biochim Biophys Acta 1829(1):20–28. doi:10.1016/j.bbagrm.2012.08.009

Tagami S, Sekine S-I, Kumarevel T, Hino N, Murayama Y, Kamegamori S, Yamamoto M, Sakamoto K, Yokoyama S (2010) Crystal structure of bacterial RNA polymerase bound with a transcription inhibitor protein. Nature 468(7326):978–982. doi:10.1038/nature09573

Tan S, Garrett KP, Conaway RC, Conaway JW (1994) Cryptic DNA-binding domain in the C terminus of RNA polymerase II general transcription factor RAP30. Proc Natl Acad Sci U S A 91(21):9808–9812

Tan S, Conaway RC, Conaway JW (1995) Dissection of transcription factor TFIIF functional domains required for initiation and elongation. Proc Natl Acad Sci U S A 92(13):6042–6046

Tan L, Wiesler S, Trzaska D, Carney HC, Weinzierl RO (2008) Bridge helix and trigger loop perturbations generate superactive RNA polymerases. J Biol 7(10):40. doi:10.1186/jbiol98

Taylor NMI, Baudin F, von Scheven G, Müller CW (2013) RNA polymerase III-specific general transcription factor IIIC contains a heterodimer resembling TFIIF Rap30/Rap74. Nucleic Acids Res 41(19):9183–9196. doi:10.1093/nar/gkt664

Thomas MC, Chiang C-M (2006) The general transcription machinery and general cofactors. Crit Rev Biochem Mol Biol 41(3):105–178. doi:10.1080/10409230600648736

Thomas MJ, Platas AA, Hawley DK (1998) Transcriptional fidelity and proofreading by RNA polymerase II. Cell 93(4):627–637

Tolic-Norrelykke SF, Rasmussen MB, Pavone FS, Berg-Sorensen K, Oddershede LB (2006) Stepwise bending of DNA by a single TATA-box binding protein. Biophys J 90(10):3694–3703. doi:10.1529/biophysj.105.074856

Tollervey D (2004) Molecular biology: termination by torpedo. Nature 432(7016):456–457. doi:10.1038/432456a

Tran K, Gralla JD (2008) Control of the timing of promoter escape and RNA catalysis by the transcription factor IIb fingertip. J Biol Chem 283(23):15665–15671. doi:10.1074/jbc.M801439200

Van Lijsebettens M, Grasser KD (2014) Transcript elongation factors: shaping transcriptomes after transcript initiation. Trends Plant Sci 19(11):717–726. doi:10.1016/j.tplants.2014.07.002

Vannini A, Cramer P (2012) Conservation between the RNA polymerase I, II, and III transcription initiation machineries. Mol Cell 45(4):439–446. doi:10.1016/j.molcel.2012.01.023

Vannini A, Ringel R, Kusser AG, Berninghausen O, Kassavetis GA, Cramer P (2010) Molecular basis of RNA polymerase III transcription repression by Maf1. Cell 143(1):59–70. doi:10.1016/j.cell.2010.09.002

Vassylyev DG, Sekine S, Laptenko O, Lee J, Vassylyeva MN, Borukhov S, Yokoyama S (2002) Crystal structure of a bacterial RNA polymerase holoenzyme at 2.6 A resolution. Nature 417(6890):712–719. doi:10.1038/nature752

Vassylyev DG, Vassylyeva MN, Perederina A, Tahirov TH, Artsimovitch I (2007) Structural basis for transcription elongation by bacterial RNA polymerase. Nature 448(7150):157–162. doi:10.1038/nature05932

Vvedenskaya IO, Vahedian-Movahed H, Bird JG, Knoblauch JG, Goldman SR, Zhang Y, Ebright RH, Nickels BE (2014) Interactions between RNA polymerase and the "core recognition element" counteract pausing. Science 344(6189):1285–1289. doi:10.1126/science.1253458

Wang D, Bushnell DA, Westover KD, Kaplan CD, Kornberg RD (2006) Structural basis of transcription: role of the trigger loop in substrate specificity and catalysis. Cell 127(5):941–954. doi:10.1016/j.cell.2006.11.023

Wang D, Bushnell DA, Huang X, Westover KD, Levitt M, Kornberg RD (2009) Structural basis of transcription: backtracked RNA polymerase II at 3.4 angstrom resolution. Science 324(5931):1203–1206. doi:10.1126/science.1168729

Washburn RS, Gottesman ME (2015) Regulation of transcription elongation and termination. Biomolecules 5(2):1063–1078. doi:10.3390/biom5021063

Watson JD, Crick FH (1953a) Genetical implications of the structure of deoxyribonucleic acid. Nature 171(4361):964–967

Watson JD, Crick FH (1953b) Molecular structure of nucleic acids; a structure for deoxyribose nucleic acid. Nature 171(4356):737–738

Weinzierl RO (2010) The nucleotide addition cycle of RNA polymerase is controlled by two molecular hinges in the Bridge Helix domain. BMC Biol 8:134. doi:10.1186/1741-7007-8-134

Weinzierl RO (2011) The Bridge Helix of RNA polymerase acts as a central nanomechanical switchboard for coordinating catalysis and substrate movement. Archaea 2011:608385. doi:10.1155/2011/608385

Weiss SB, Gladstone L (1959) A mammalian system for the incorporation of cytidine triphosphate into ribonucleic ACID1. J Am Chem Soc 81(15):4118–4119. doi:10.1021/ja01524a087

Weiss SB, Nakamoto T (1961) Net synthesis of ribonucleic acid with a microbial enzyme requiring deoxyribonucleic acid and four ribonucleoside triphosphates. J Biol Chem 236:PC18–PC20

Werner F, Grohmann D (2011) Evolution of multisubunit RNA polymerases in the three domains of life. Nat Rev Microbiol 9(2):85–98. doi:10.1038/nrmicro2507

Werner F, Weinzierl RO (2002) A recombinant RNA polymerase II-like enzyme capable of promoter-specific transcription. Mol Cell 10(3):635–646

Werner F, Weinzierl RO (2005) Direct modulation of RNA polymerase core functions by basal transcription factors. Mol Cell Biol 25(18):8344–8355. doi:10.1128/MCB.25.18.8344-8355.2005

Werner F, Wiesler S, Nottebaum S, Weinzierl RO (2006) Modulation of RNA polymerase core functions by basal transcription factor TFB/TFIIB. Biochem Soc Symp 73:49–58

Westover KD, Bushnell DA, Kornberg RD (2004a) Structural basis of transcription: nucleotide selection by rotation in the RNA polymerase II active center. Cell 119(4):481–489. doi:10.1016/j.cell.2004.10.016

Westover KD, Bushnell DA, Kornberg RD (2004b) Structural basis of transcription: separation of RNA from DNA by RNA polymerase II. Science 303(5660):1014–1016. doi:10.1126/science.1090839

Whitelaw E, Proudfoot N (1986) Alpha-thalassaemia caused by a poly(A) site mutation reveals that transcriptional termination is linked to 3′ end processing in the human alpha 2 globin gene. EMBO J 5(11):2915–2922

Wiesler SC, Weinzierl RO (2011) The linker domain of basal transcription factor TFIIB controls distinct recruitment and transcription stimulation functions. Nucleic Acids Res 39(2):464–474. doi:10.1093/nar/gkq809

Wiesler SC, Werner F, Weinzierl RO (2013) Promoter independent abortive transcription assays unravel functional interactions between TFIIB and RNA polymerase. Methods Mol Biol 977:217–227. doi:10.1007/978-1-62703-284-1_17

Wigneshweraraj S, Bose D, Burrows PC, Joly N, Schumacher J, Rappas M, Pape T, Zhang X, Stockley P, Severinov K, Buck M (2008) Modus operandi of the bacterial RNA polymerase containing the sigma54 promoter-specificity factor. Mol Microbiol 68(3):538–546. doi:10.1111/j.1365-2958.2008.06181.x

Woese CR, Fox GE (1977) Phylogenetic structure of the prokaryotic domain: the primary kingdoms. Proc Natl Acad Sci U S A 74(11):5088–5090

Wu W-H, Hampsey M (1999) An activation-specific role for transcription factor TFIIB in vivo. PNAS 96(6):2764–2769. doi:10.1073/pnas.96.6.2764

Yakhnin AV, Babitzke P (2014) NusG/Spt5: are there common functions of this ubiquitous transcription elongation factor? Curr Opin Microbiol 18:68–71. doi:10.1016/j.mib.2014.02.005

Yamaguchi Y, Shibata H, Handa H (2013) Transcription elongation factors DSIF and NELF: promoter-proximal pausing and beyond. Biochim Biophys Acta 1829(1):98–104. doi:10.1016/j.bbagrm.2012.11.007

Yan Q, Moreland RJ, Conaway JW, Conaway RC (1999) Dual roles for transcription factor IIF in promoter escape by RNA polymerase II. J Biol Chem 274(50):35668–35675

Yang Y, Darbari VC, Zhang N, Lu D, Glyde R, Wang YP, Winkelman JT, Gourse RL, Murakami KS, Buck M, Zhang X (2015) TRANSCRIPTION. Structures of the RNA polymerase-sigma54 reveal new and conserved regulatory strategies. Science 349(6250):882–885. doi:10.1126/science.aab1478

Yin YW, Steitz TA (2004) The structural mechanism of translocation and helicase activity in T7 RNA polymerase. Cell 116(3):393–404

Zawel L, Kumar KP, Reinberg D (1995) Recycling of the general transcription factors during RNA polymerase II transcription. Genes Dev 9(12):1479–1490

Zhang C, Burton ZF (2004) Transcription factors IIF and IIS and nucleoside triphosphate substrates as dynamic probes of the human RNA polymerase II mechanism. J Mol Biol 342(4):1085–1099. doi:10.1016/j.jmb.2004.07.070

Zhang G, Campbell EA, Minakhin L, Richter C, Severinov K, Darst SA (1999) Crystal structure of Thermus aquaticus core RNA polymerase at 3.3 A resolution. Cell 98(6):811–824

Zhang C, Zobeck KL, Burton ZF (2005) Human RNA polymerase II elongation in slow motion: role of the TFIIF RAP74 alpha1 helix in nucleoside triphosphate-driven translocation. Mol Cell Biol 25(9):3583–3595. doi:10.1128/MCB.25.9.3583-3595.2005

Zhang J, Palangat M, Landick R (2010) Role of the RNA polymerase trigger loop in catalysis and pausing. Nat Struct Mol Biol 17(1):99–104. doi:10.1038/nsmb.1732

Zhang Y, Feng Y, Chatterjee S, Tuske S, Ho MX, Arnold E, Ebright RH (2012) Structural basis of transcription initiation. Science 338(6110):1076–1080. doi:10.1126/science.1227786

Zhang L, Pardo-Avila F, Unarta IC, Cheung PP, Wang G, Wang D, Huang X (2016) Elucidation of the dynamics of transcription elongation by RNA polymerase II using kinetic network models. Acc Chem Res. doi:10.1021/acs.accounts.5b00536

Zhou Q, Li T, Price DH (2012) RNA polymerase II elongation control. Annu Rev Biochem 81:119–143. doi:10.1146/annurev-biochem-052610-095910

Ziegler LM, Khaperskyy DA, Ammerman ML, Ponticelli AS (2003) Yeast RNA polymerase II lacking the Rpb9 subunit is impaired for interaction with transcription factor IIF. J Biol Chem 278(49):48950–48956. doi:10.1074/jbc.M309656200

Zillig W, Stetter KO, Tobien M (1978) DNA-dependent RNA polymerase from Halobacterium halobium. Eur J Biochem 91(1):193–199

Zillig W, Stetter KO, Janekovic D (1979) DNA-dependent RNA polymerase from the archaebacterium Sulfolobus acidocaldarius. Eur J Biochem 96(3):597–604

Zuo Y, Steitz TA (2015) Crystal structures of the E. coli transcription initiation complexes with a complete bubble. Mol Cell 58(3):534–540. doi:10.1016/j.molcel.2015.03.010

Chapter 10
Dihydrodipicolinate Synthase: Structure, Dynamics, Function, and Evolution

F. Grant Pearce, André O. Hudson, Kerry Loomes, and Renwick C.J. Dobson

Abstract Enzymes are usually comprised of multiple subunits and more often than not they are made up of identical subunits. In this review we examine lysine biosynthesis and focus on the enzyme dihydrodipicolinate synthase in terms of its structure, function and the evolution of its varied number of subunits (quaternary structure). Dihydrodipicolinate synthase is the first committed step in the biosynthesis of lysine, which occurs naturally in plants, bacteria, archaea and fungi, but is not synthesized in mammals. In bacteria, there have been four separate pathways identified from tetrahydrodipicolinate to *meso*-diaminopimelate, which is the immediate precursor to lysine. Dihydrodipicolinate synthases from many bacterial and plant species have been structurally characterised and the results show considerable variability with respect to their quaternary structure, hinting at their evolution. The oligomeric state of the enzyme plays a key role, both in catalysis and in the allosteric regulation of the enzyme by lysine. While most bacteria and plants have tetrameric enzymes, where the structure of the dimeric building blocks is conserved, the

F. Grant Pearce
Biomolecular Interaction Centre and School of Biological Sciences, University of Canterbury, Private Bag 4800, Christchurch 8041, New Zealand
e-mail: grant.pearce@canterbury.ac.nz

A.O. Hudson
Thomas H. Gosnell School of Life Sciences, Rochester Institute of Technology, Rochester, NY, USA
e-mail: aohsbi@rit.edu

K. Loomes
School of Biological Sciences & Maurice Wilkins Centre for Molecular Biodiscovery, University of Auckland, Auckland, New Zealand
e-mail: k.loomes@auckland.ac.nz

R.C.J. Dobson (✉)
Biomolecular Interaction Centre and School of Biological Sciences, University of Canterbury, Private Bag 4800, Christchurch 8041, New Zealand

Department of Biochemistry and Molecular Biology, Bio21 Molecular Science and Biotechnology Institute, University of Melbourne, 30 Flemington Road, Parkville, VIC 3010, Australia
e-mail: renwick.dobson@canterbury.ac.nz

J.R. Harris, J. Marles-Wright (eds.), *Macromolecular Protein Complexes*, Subcellular Biochemistry 83, DOI 10.1007/978-3-319-46503-6_10

arrangement of the dimers differs. We also review a key development in the field, namely the discovery of a human dihydrodipicolinate synthase-like enzyme, now known as 4-hydroxy-2-oxoglutarate aldolase. This discovery complicates the rationale underpinning drug development against bacterial dihydrodipicolinate synthases, since genetic errors in 4-hydroxy-2-oxoglutarate aldolase cause the disease Primary Hyperoxaluria Type 3 and therefore compounds that are geared towards the inhibition of bacterial dihydrodipicolinate synthase may be toxic to mammalian cells.

Keywords Lysine biosynthesis • Dihydrodipicolinate synthase • 4-Hydroxy-2-oxoglutarate aldolase

10.1 Background

10.1.1 Enzymes Are Often Oligomeric

Proteins are usually comprised of multiple subunits and more often than not they are comprised of identical subunits (Goodsell and Olson 2000). However, the reasons why evolution has selected to assemble proteins (quaternary complexes) is less clear, especially compared to our understanding of protein tertiary structure. There are many reviews on the topic, for example see (Goodsell and Olson 2000; Devenish and Gerrard 2009; Griffin and Gerrard 2012) and equally many hypotheses that attempt to provide a general explanation for this observation.

Our initial understanding came from the first crystal structures of heamoglobin, which demonstrated that its tetrameric architecture is necessary for regulation through changes in the conformation of the tetramer (Perutz 1989). Others also noted that enzyme regulation can be accomplished through changes in the oligomeric state, e.g. the association of less active monomers to active dimers (Burgess et al. 2008). Another level of regulation involves the rate of encounter of an enzyme with a substrate, which is proportional to the effective radius of the enzyme. As such, the elimination of extraneous protein surface through oligomerisation will increase the frequency of productive encounters (Phillips et al. 2012; Lynch 2013). Other proposals include the efficient folding of multiple small proteins compared to a single large protein, the stability of larger (multimeric) proteins compared to smaller (monomeric) proteins (Goodsell and Olson 2000), and the fact that osmotic pressure is reduced when monomeric units aggregate (Perham 1975). Given the appreciable rates of error in transcription and translation, it is more efficient to generate a large protein from smaller components (Goodsell and Olson 2000). Moreover, the formation of quaternary structure could be an 'editing' step—erroneous polypeptides are less likely to form an active quaternary structure (Perham 1975).

In this review, we note recent developments in the enzymology of plant and bacterial lysine biosynthesis and focus on the enzyme dihydrodipicolinate synthase, relating its function and evolution to its varied quaternary structure. We also note the recent discovery of a hitherto unknown dihydrodipicolinate synthase like enzyme in humans, which has now been annotated as a 4-hydroxy-2-oxoglutarate aldolase. As we shall see, the quaternary complex of these protein complexes is critical for their catalytic function and regulation.

10.1.2 Lysine Biosynthesis – A Model System for Quaternary Structure

The biosynthesis of lysine occurs naturally in plants, bacteria, archaea and some fungi, it is not synthesized in mammals. As such, lysine is one of the nine proteogenic amino acids that is deemed essential for humans (Rose 1949; Hutton et al. 2007; Dogovski et al. 2009). For this reason, this pathway has received much attention from two major perspectives:

1. The design of antibiotics to inhibit this pathway for the control of bacterial infections (Scapin et al. 1998; Dobson et al. 2004a; Turner et al. 2005; Hutton et al. 2007; Boughton et al. 2008a; Mitsakos et al. 2008; Kefala et al. 2008; Burgess et al. 2008; Boughton et al. 2008b; Domigan et al. 2009; Evans et al. 2011; Skovpen et al. 2016)
2. The up-regulation of lysine biosynthesis in plants to increase the levels of free lysine in crop plants given that it is limited in crop plants especially cereal crops (Jander and Joshi 2009, 2010; Galili 2011)

Two independent evolutionary lysine biosynthesis pathways have been identified to date:

- The α-aminoadipate pathway, found in euglenoids, fungi and a few archeal species
- The diaminopimelate pathway which is employed predominantly by bacteria, photosynthetic organisms and fungi (Zabriskie and Jackson 2000).

In bacteria, there have been four separate pathways identified from tetrahydrodipicolinate to *meso*-diaminopimelate, which is the immediate precursor to lysine (Fig. 10.1). These pathways can be most readily differentiated as follows:

1. The dehydrogenase pathway found in *Bacillus sphaericus*, *Corynecbacterium glutamicum* and *Brevibacterium sp.* (Cox 1996) is the most chemically direct and involves the direct reductive amination of tetrahydrodipicolinate to *meso*-diaminopimelate by *meso*-diaminopimelate dehydrogenase (Scapin and Blanchard 1998)

2. The succinylase pathway, employed by *Escherichia coli* (Kindler and Gilvarg 1960) and *Staphylococcus aureus* (Barnes et al. 1969), involves the enzymatic succinylation of tetrahydrodipicolinate to succinyl-ε-keto-α-aminopimelate by tetrahydrodipicolinate succinylase (Simms et al. 1984). Tetrahydrodipicolinate is then aminated (Peterkofsky and Gilvarg 1961) and subsequently desuccinylated (Lin et al. 1988) to *S,S*-diaminopimelate before being epimerised to *meso*-diaminopimelate by diaminopimelate epimerase (Wiseman and Nichols 1984)
3. The acetylase pathway, limited to some bacterial species of the genus *Bacillus* (Weinberger and Gilvarg 1970), differs from the most common succinylase pathway in only the protecting group utilised
4. A recently discovered pathway, discussed in more detail below, is the aminotransferase pathway, which is commonly found in photosynthetic organisms (Hudson et al. 2006), *Chlamydia trachomatis* (McCoy et al. 2006; Watanabe et al. 2011), *Bacteriodes fragilis,* and *Clostridium thermocellum* (Hudson et al. 2011)

At the convergence point of the four known bacterial and plant lysine biosynthesis pathways, *meso*-diaminopimelate is then decarboxylated by *meso*-diaminopimelate decarboxylase to yield *S*-lysine in the final step of the lysine biosynthesis pathway in bacteria (White and Kelly 1965) (Fig. 10.1).

An alternative diaminopimelate pathway has recently been proposed with the identification of an *S,S*-diaminopimelate aminotransferase from *Arabidopsis thaliana* and the photosynthetic bacterium *Synechocystis sp.,* which catalyses the direct transamination of tetrahydrodipicolinate to *S,S*-diaminopimelate (Hudson et al. 2006) (Fig. 10.1). Although this enzyme operates almost 60 times more efficiently in the reverse direction, the high glutamate to α-ketoglutarate ratio in the chloroplast stroma is believed to favour the forward reaction (Hudson et al. 2006). In addition to bacterial species, this enzyme has recently been identified and characterised from the alga *Chlamydomonas reinhardtii* (Dobson et al. 2011), suggesting a wider prevalence of this pathway than previously understood. The identification of *S,S*-diaminopimelate aminotransferase orthologues across plant, chlamydial and cyanobacterial species is one line of evidence indicative of the endosymbiotic origin of the chloroplast and suggests that the aminotransferase pathway is common amongst higher plants (McCoy et al. 2006). The final steps of the aminotransferase pathway are shared with the other bacterial pathways and consist of the epimerisation and decarboxylation of *S,S*-diaminopimelate, catalysed sequentially by diaminopimelate epimerase (Tyagi et al. 1982) and *meso*-diaminopimelate decarboxylase (Kelland et al. 1985) (Fig. 10.1).

Dihydrodipicolinate synthase serves as a regulatory point of lysine biosynthesis in plants, bacteria and some fungi, and has been shown to have differing levels of lysine feedback-inhibition across plant and bacterial species. Furthermore, the quaternary architectures between plant and bacterial dihydrodipicolinate synthase are different, but essential for regulation and catalysis.

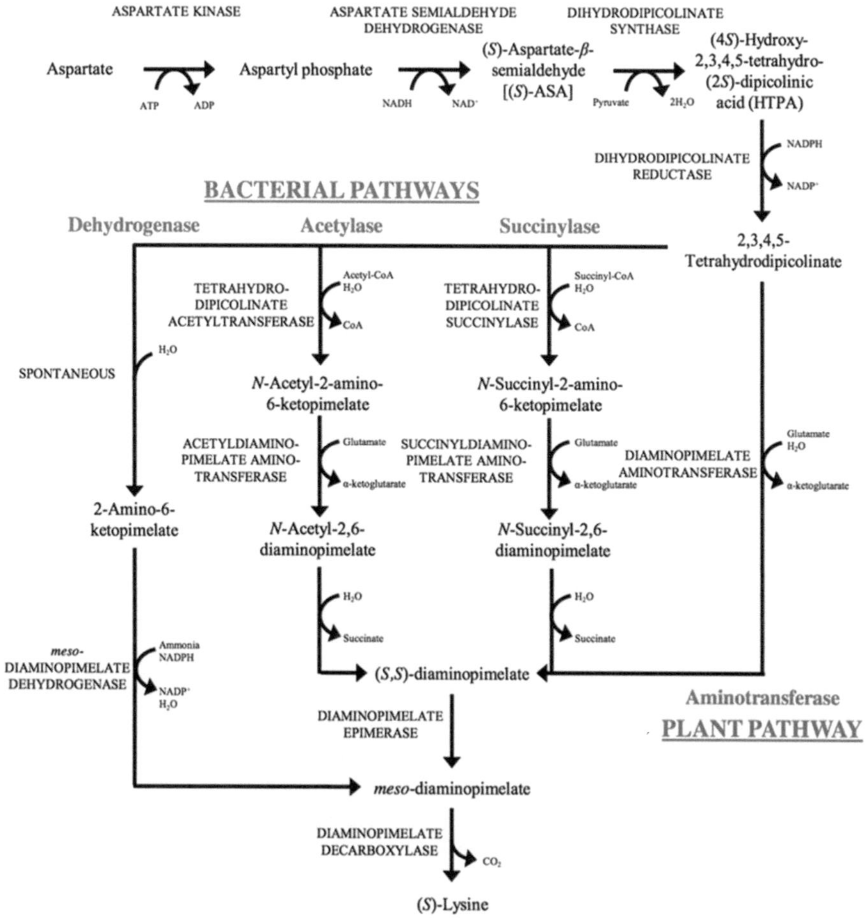

Fig. 10.1 A schematic of the alternate pathways to lysine biosynthesis in plants and bacteria

10.2 Dihydrodipicolinate Synthase: Structure and Mechanism

10.2.1 Structure

Dihydrodipicolinate synthases from many bacterial and plant species have now been structurally characterised and these structures, as will be describe later, show some variability with respect to their quaternary assemblies that hints at their evolution.

The archetypal structure is from *Escherichia coli*, which was the first to be characterized (Mirwaldt et al. 1995; Dobson et al. 2005; Devenish et al. 2008). It is a homotetramer of four identical $(\beta/\alpha)_8$-barrel monomers with a C-terminal α-helical

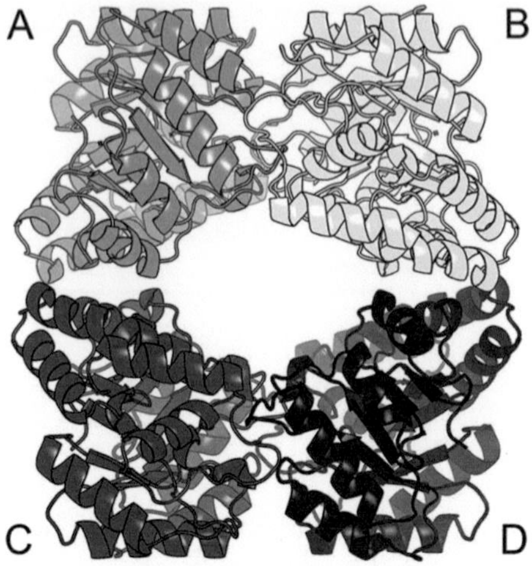

Fig. 10.2 Tetrameric structure of *E. coli* dihydrodipicolinate synthase (PDB id: 1yxc)

domain that is important for the maintenance of the tetramer (Guo et al. 2009). The quaternary structure can be described as a dimer of dimers (monomers A and B and monomers C and D, Fig. 10.2) (Dobson et al. 2005). The tetramer contains strong interactions between monomers A and B (C and D) at the so-called 'tight dimer interface' and weaker interactions between the dimers at the 'weak dimer interface' (A and C) (Fig. 10.2). Active site residues reach across the tight-dimer interface and contribute to the active site of the neighbouring monomer. The association of the four monomers creates a large water-filled cavity at the centre of the tetramer, such that each monomer has contacts with two neighbouring monomers only. The active site is located at the C-terminal end of the $(\beta/\alpha)_8$-barrel and each active site faces into the water filled cavity. The two allosteric binding sites are located at each 'tight dimer interface' (Dobson et al. 2005) (Fig. 10.2), away from the four active sites. Thus, the oligomeric state of the enzyme plays a key role in allosteric regulation, in addition to catalysis.

10.2.2 Catalytic Mechanism

Kinetic analyses demonstrate that dihydrodipicolinate synthase enzymes proceed via a 'ping-pong' kinetic scheme, which necessitates an ordered mechanism of catalysis (Karsten 1997; Dobson et al. 2004b). The first catalytic step is a Schiff base formation with pyruvate (Fig. 10.3), which proceeds *via* the nucleophilic attack of the ε-amino group of Lys161 to the keto carbon of pyruvate, resulting in the formation of an imine moiety (Laber et al. 1992; Soares da Costa et al. 2010). Mutagenic analysis suggests that a catalytic triad of residues, Tyr133, Thr44 and Tyr107 (*E. coli* dihydrodipicolinate synthase numbering), acts as a proton relay to transfer protons to and from the active site *via* a water-filled channel that leads to the bulk

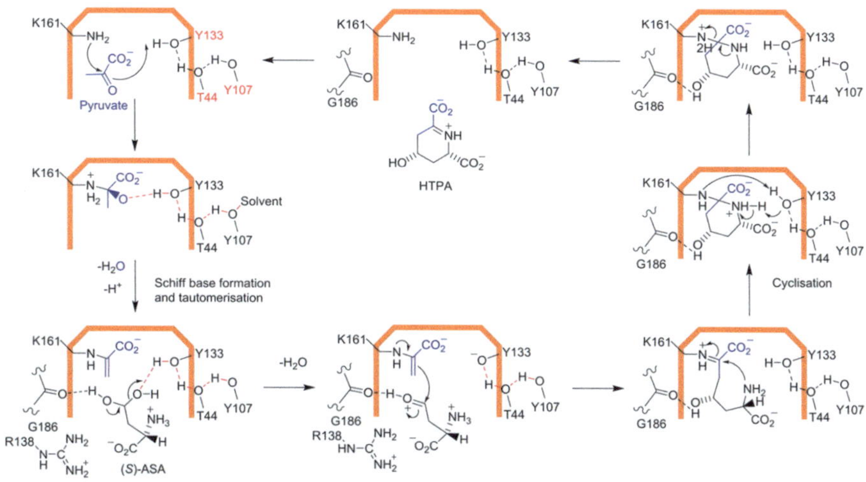

Fig. 10.3 Proposed catalytic mechanism for dihydrodipicolinate synthase

solvent (Dobson et al. 2004c). Tyr107 contributes the other monomer to the active site form across the tight dimer interface. The imine moiety is converted to its enamine form through a nucleophilic attack of *S*-aspartate semialdehyde. *S*-Aspartate semialdehyde adds its aldehyde group to the enamine group of the Schiff base (Boughton et al. 2012), with the displacement of a water molecule. The backbone oxygen of Ile203, which is in an unusual configuration, having a non-planar peptide bond, may be important for this step (Blickling et al. 1997b; Dobson et al. 2008). In aqueous solution, *S*-aspartate semialdehyde is known to exist in the hydrated form rather than in the aldehyde form (Coulter et al. 1996). The product, 4-hydroxy-tetrahydrodipicolinate, is formed by the nucleophilic attack of the amino group of the hydrated *S*-aspartate semialdehyde to the Schiff base, leading to cyclisation and the detachment of the product from the enzyme. It is worth noting here that the product is not dihydrodipicolinate, which has resulted the re-naming of this enzyme to 4-hydroxy-tetrahydrodipicolinate synthase (EC 4.3.3.7).

The catalytic triad, consisting of Tyr133, Thr44 and Tyr107, acts as a proton relay to transfer protons to and from the active site *via* a water-filled channel leading to the lysine binding site and the bulk solvent (Dobson et al. 2004c, 2005). Interestingly, the high resolution structures of pyruvate and lysine bound, and pyruvate bound and unbound *AgT*-dihydrodipicolinate synthase each show two distinct conformations of Tyr133 (Atkinson et al. 2014). The two conformations are well supported by electron density and Tyr133 from the lower resolution *E. coli* dihydrodipicolinate synthase structure is situated between these two conformations, perhaps an average of the two. Given that Tyr133 has been shown to play an important role in substrate binding, donating a proton to the Schiff base hydroxyl (Dobson et al. 2005), having multiple subtle conformations in the catalytic triad may have mechanistic importance, or be important for feedback regulation by lysine.

10.2.3 Mechanism of Allosteric Regulation

The activity of dihydrodipicolinate synthase is inhibited by lysine through a negative feedback loop. This phenomenon has been investigated in several plants, in addition to several Gram-negative and Gram-positive bacterial species.

The kinetics and inhibition of dihydrodipicolinate synthase from the plant species *Triticum aestivium* (Kumpaisal et al. 1987), *Daucus carota sativa* (Matthews and Widholm 1979), *Spinacia oleracea* (Wallsgrove and Mazelis 1980), *Zea mays* (Frisch et al. 1991), and *Pisum sativum* (Dereppe et al. 1992) have been studied. In general, it is found that plant enzymes are strongly inhibited by lysine (IC_{50} = 0.01–0.05 mM).

In contrast, bacterial dihydrodipicolinate synthase enzymes are significantly less sensitive to lysine than those from plants. Dihydrodipicolinate synthase from Gram-negative bacteria such as *E. coli* (Yugari and Gilvarg 1965), *Neisseria meningitidis* (Devenish et al. 2009), and *Methanobacterium thermoautotrophicum* (Bakhiet et al. 1984) are weakly inhibited, with an IC_{50} of between 0.25 and 1.0 mM. Dihydrodipicolinate synthase from Gram-positive bacteria such as *Bacillus lichenformis* (Stahly 1969), *Bacillus megaterium* (Webster and Lechowich 1970), *Bacillus subtillis* (Yamakura et al. 1974), *Corynebacterium glutamicum* (Cremer et al. 1988), *Bacillus cereus* (Hoganson and Stahly 1975), *Lactobacillus plantarum* (Cahyanto et al. 2006), and methicillin-resistant *Staphylococcus aureus* (Burgess et al. 2008) show no inhibition by lysine, with an IC_{50} that is often as high as 50 mM and thus not biologically relevant.

The crystal structure of dihydrodipicolinate synthase from *E. coli* in the presence of lysine shows that two lysine molecules bind the lysine binding site. Isothermal titration calorimetry experiments (Phenix and Palmer 2008; Muscroft-Taylor et al. 2010) have shown that inhibition is cooperative: the second molecule of lysine binds 105 times more tightly than the first. The mechanism by which lysine exerts its regulatory control over bacterial dihydrodipicolinate synthase is not well understood, although kinetic and structural studies suggest that it is an allosteric inhibitor, causing partial inhibition (90%) at saturating concentrations. Isothermal titration calorimetry experiments have also illustrated that the substrate has a substantial effect on the nature of enzyme-inhibitor association (Phenix and Palmer 2008; Muscroft-Taylor et al. 2010).

The mechanisms of lysine inhibition of dihydrodipicolinate synthase from plants has been investigated to some extent in wheat (Kumpaisal et al. 1987). It was found that lysine is a non-competitive inhibitor relative to pyruvate and a competitive inhibitor relative to *S*-aspartate semialdehyde. A structural study of dihydrodipicolinate synthase from the plant *N. sylvestris* suggests a significant change in conformation upon lysine binding (Blickling et al. 1997a). Several of the residues involved in contacts at the tight dimer interface are displaced when lysine is bound, dislocating the dimers in relation to each other. This translocation displaces the active site residues Arg138 and Tyr133, hindering coordination of the carboxyl group of *S*-aspartate semialdehyde during the cyclisation of the substrate. Such an altered conformation was not observed in the *E. coli* dihydrodipicolinate synthase structure (Dobson et al. 2005).

There have been various reports that suggest that high levels of *S*-aspartate semialdehyde may also inhibit *E. coli* dihydrodipicolinate synthase (Karsten 1997), but this is not consistently observed, perhaps indicative of differences in *S*-aspartate semialdehyde preparation (Dobson et al. 2004a). Nevertheless, the inhibition of dihydrodipicolinate synthase activity from *Thermoanaerobacter tengcongensis* (Wolterink-van Loo et al. 2008) and *S. aureus* (Burgess et al. 2008) has been observed at *S*-aspartate semialdehyde concentrations higher than 2 mM and 1.2 mM, respectively, despite using *S*-aspartate semialdehyde synthesised by cleaner methods (Roberts et al. 2003).

10.3 4-Hydroxy-2-oxoglutarate Aldolase – The New Kid on the Block

4-Hydroxy-2-oxoglutarate aldolase is a mammalian class 1 aldolase, which has been placed within the dihydrodipicolinate synthase/*N*-acetyl neuraminate lyase superfamily (Dekker and Kitson 1992; Riedel et al. 2011) and was previously annotated as dihydrodipicolinate synthase-like, until its biological function was realised (Belostotsky et al. 2010). 4-Hydroxy-2-oxoglutarate aldolase is of special biomedical interest, as inborn genetic errors in the *HOGA1* gene are responsible for primary hyperoxaluria type-3 in humans.

Primary hyperoxaluria is a rare disease characterised by increased endogenous production of oxalate, primarily in the liver, leading to urinary calculi formation and end stage renal failure in severe cases (Cochat and Rumsby 2013). Primary hyperoxaluria is classified into three types, each characterised by genetic mutations that affect the function of enzymes involved in glyoxylate metabolism (Fig. 10.4). Overall, defects in glyoxylate processing lead to increased oxalate levels through lactate dehydrogenase. Of the three types, primary hyperoxaluria type-3 is the least severe form, where clinical remission can occur despite the persistence of primary hyperoxaluria (Beck et al. 2013).

4-Hydroxy-2-oxoglutarate aldolase is distantly related to bacterial dihydrodipicolinate synthase (22% sequence identity with *E. coli* dihydrodipicolinate synthase) (Riedel et al. 2011), but both share similar tertiary and quaternary architectures. 4-Hydroxy-2-oxoglutarate aldolase has a similar folding topology and tetrameric structure that can be viewed as a dimer of dimers, characterised by weak and major subunit interfaces (Bunker 2010; Riedel et al. 2011) (Fig. 10.5). The respective active site geometries are also similar with conservation of the active site lysine

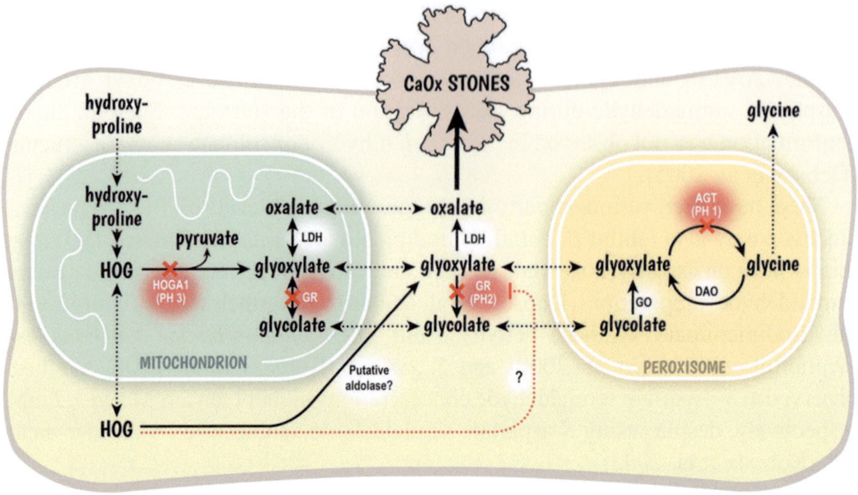

Fig. 10.4 Primary hyperoxaluria type 3 and glyoxylate metabolism. Abbreviations: 4-hydroxy-2-oxoglutarate aldolase (HOGA1); 4-hydroxy-2-oxoglutarate (HOG); lactate dehydrogenase (LDH); calcium oxalate (CaOx); glyoxylate reductase (GR); GO; alanine-glyoxylate aminotransferase (AGT); D-amino acid oxidase (DAO); Glycolate oxidase (GO); primary hyperoxaluria type 1 (PH1); primary hyperoxaluria type 12 (PH2); primary hyperoxaluria type 3 (PH3)

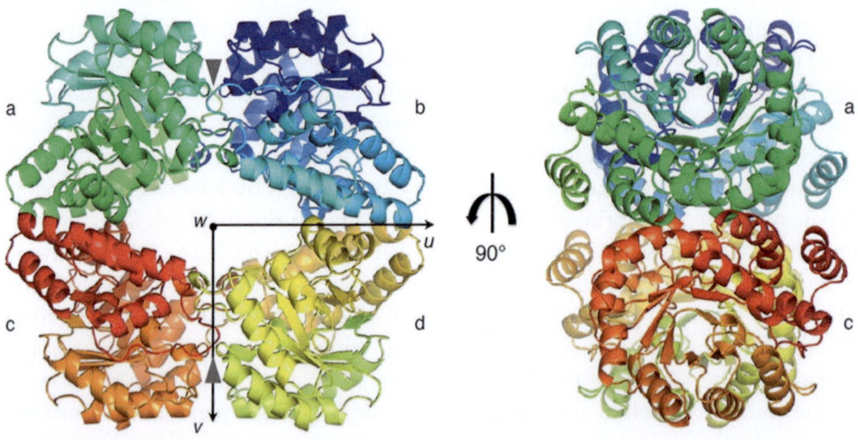

Fig. 10.5 The HOGA1 tetramer. The major dimer interface is formed between monomers a/b and c/d. The minor dimer interface is formed between monomers a/c and b/d. The principle axes, *w*, *v* and *u* are indicated. The positions corresponding to the lysine allosteric sites for some bacterial dihydrodipicolinate synthase enzymes are indicated by *grey* arrowheads (Figure taken from Bunker 2010)

residue involved in Schiff base formation together with the proton-relay network comprising Ser77, Tyr168 and Tyr140 (hydroxy-2-oxoglutarate aldolase numbering) (Riedel et al. 2011). For both dihydrodipicolinate synthase and 4-hydroxy-2-oxoglutarate aldolase, one of the residues in this network (Tyr140) originates from an adjacent monomer across the major dimer interface. Nevertheless, there are functional differences, as Tyr140 is not essential for 4-hydroxy-2-oxoglutarate aldolase activity, in contrast to dihydrodipicolinate synthase (Riedel et al. 2011). Residues involved in lysine binding and inhibition of dihydrodipicolinate synthase (His53, His56 and Glu84, *E. coli* dihydrodipicolinate synthase numbering) are also not conserved in hydroxy-2-oxoglutarate aldolase. Of potential interest is the space within the hydroxy-2-oxoglutarate aldolase structure that corresponds to the dihydrodipicolinate synthase lysine allosteric site, where a molecule could conceivably bind (Fig. 10.5) (Bunker 2010).

To date at least 25 clinical mutations in the HOGA1 gene have been identified in association with type-3 hyperoxaluria (Belostotsky et al. 2010; Monico et al. 2011; Williams et al. 2012; Beck et al. 2013; Wang et al. 2015; Hopp et al. 2015). These mutants result in a variety of missense substitutions, in-frame amino acid deletions, premature truncation products, and amino acid insertions. There are two prevalent mutations that in one study accounted for 74.2% of the total allelic mutations identified (Hopp et al. 2015). The first is an in-frame deletion of Glu-315 (Δ315), which located within the final C-terminal helix and two amino acids downstream of a salt bridge (Arg-144/Glu-313) that stabilises the major dimer interface. The second (c.700 + 5G>T) is a splice-site mutation, which leads to the addition of 17 amino acids within the minor interface region. Both of these sites are not close to the active site and are at positions that would presumably disrupt the major and minor interfaces.

Several lines of evidence show that these *HOGA1* gene mutations lead to a loss of function. First, various recombinant hydroxy-2-oxoglutarate aldolase mutants expressed in *E. coli* have a propensity to produce aggregate forms that are devoid of activity (Riedel et al. 2012). In the case of the Δ315 and c.700 + 5G>T mutant proteins, their purified recombinant forms are thermally unstable and are rapidly denatured when compared to wild type hydroxy-2-oxoglutarate aldolase (MacDonald et al. 2016). Nevertheless, the Δ315 mutant is partially stabilised against thermal denaturation by the presence of pyruvate, indicating the preservation of some level of functional capacity (MacDonald et al. 2016). Second, transient expression of some recombinant mutant hydroxy-2-oxoglutarate aldolase forms in mammalian cells result in the inability to produce detectable protein (Riedel et al. 2012; Beck et al. 2013). Using a cell line that was stably-transformed with the Δ315 mutant, it was shown that proteasome mediated degradation was responsible for a complete lack of functional protein (MacDonald et al. 2016). Third, nonsense mutations have been identified that result in premature termination products that presumably could not be active (Williams et al. 2012; Beck et al. 2013). Finally, knockout mice lacking 4-hydroxy-2-oxoglutarate aldolase expression replicate the build-up of the 4-hydroxy-2-oxoglutarate aldolase substrate, hydroxy-2-oxoglutarate, and other metabolites also observed in human hyperoxaluria type-3 patients (Li et al. 2015).

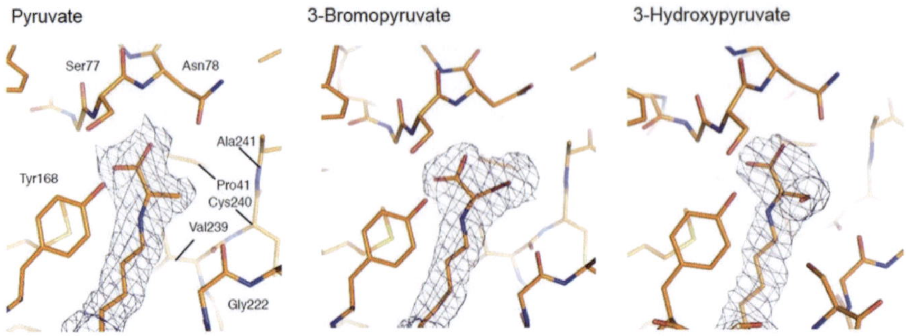

Fig. 10.6 4-Hydroxy-2-oxoglutarate aldolase active site Lys196 adducts. The initial, unbiased 2mFo–DFc electron density map calculated with Lys196 omitted and contoured at 1 σ (blue mesh) is displayed with the final structural model (Figure taken from Bunker 2010)

It is still not clear exactly how a loss of 4-hydroxy-2-oxoglutarate aldolase activity leads to increased oxalate production *in vivo* (Fig. 10.4). Unlike the loss-of-function mutations that occur in primary hyperoxaluria-type 1 and primary hyperoxaluria-type 2, a lack of hydroxy-2-oxoglutarate aldolase activity would intuitively lead to reduced glyoxylate and oxalate production. Two possible mechanisms have been proposed to account for the endogenous increase in oxalate in the face of hydroxy-2-oxoglutarate aldolase inhibition. The first is that the hydroxy-2-oxoglutarate aldolase substrate, 4-hydroxy-2-oxoglutarate, accumulates and migrates from the mitochondria to the cytosol where it is converted to glyoxylate by a putative aldolase (Monico et al. 2011). Alternatively, increased 4-hydroxy-2-oxoglutarate levels inhibit glyoxylate reductase, in effect mimicking the primary hyperoxaluria type-2 phenotype by impairing glyoxylate catabolism (Riedel et al. 2012).

The role of 4-hydroxy-2-oxoglutarate aldolase in hyperoxaluria type-3 raises potential off-target considerations for the development of new antibiotics, based on the inhibition of bacterial dihydrodipicolinate synthase. The similarities of the two active site geometries are illustrated with the pyruvate analogues, 3-bromopyruvate and 3-hydroxypyruvate, which are both inhibitors of dihydrodipicolinate synthase (Borthwick et al. 1995; Devenish et al. 2008; Boughton et al. 2012). Similar to pyruvate, 3-bromopyruvate interacts with the active site in hydroxy-2-oxoglutarate aldolase, forming a Schiff base adduct with Lys196 as shown in Fig. 10.6 (Bunker 2010). The presence of the bromine atom induces the displacement of the Asn78 side chain away from the pyruvate moiety. Similarly, 3-hydroxypyruvate forms a Schiff base adduct with Lys196, but in this case the 3-hydroxypyruvate moiety is in an opposite orientation to that seen in the dihydrodipicolinate synthase structure (Dobson et al. 2008). Consequently, inhibitors targeting the active site of dihydrodipicolinate synthase would have to be developed with special design constraints to avoid interactions with hydroxy-2-oxoglutarate aldolase.

10.4 Evolution of Dihydrodipicolinate Synthase Quaternary Structure

It would appear that the structure and function of dihydrodipicolinate synthase has all the elements of an exciting molecular evolutionary story. While most bacteria and plants have tetrameric enzymes, and the structure of dimeric building block is conserved, the arrangements of the dimers differ. Bacterial dihydrodipicolinate synthase enzymes adopt a 'back-to-back' arrangement that is the same as for other members of the *N*-acetyl neuraminate lyase family, including 4-hydroxy-2-oxoglutarate aldolase, while plant enzymes adopt a 'head-to-head' arrangement (Atkinson et al. 2012; Griffin et al. 2012) (Fig. 10.7). Additionally, some bacteria have a dimeric rather than tetrameric structure (Girish et al. 2008; Burgess et al. 2008), suggesting that a tetrameric structure is not necessarily essential for full catalytic activity.

In order to probe the way in which the oligomeric state plays a role in protein function, complexes of different oligomeric orders can be engineered through mutations at the interface regions. The structure and function of these engineered proteins can then be compared to the wild-type protein, possibly revealing insight into the relationship between the oligomeric state and protein function. To this end, several studies have used mutagenesis to change the oligomeric state of different dihydrodipicolinate synthase enzymes.

Initial studies generated a dimeric variant of the *E. coli* dihydrodipicolinate synthase enzyme in which the active site structure was not substantially altered, but the activity of the enzyme was reduced due to the irreversible inhibition by α-ketoglutarate (Griffin et al. 2008, 2010). A similar trend was observed for a

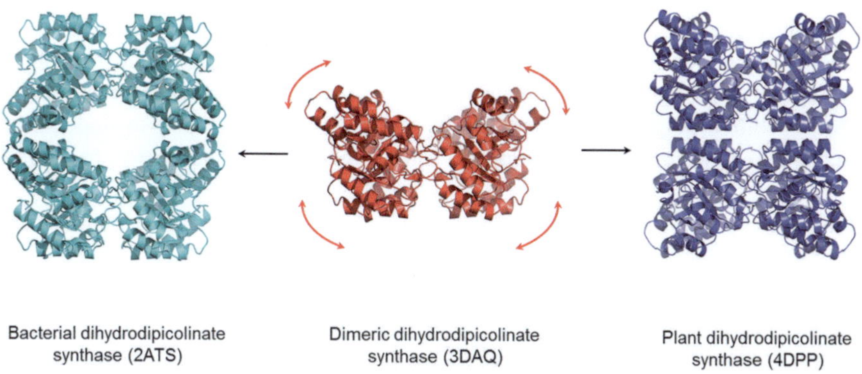

Fig. 10.7 Dihydrodipicolinate synthase can exist primarily in three different forms. In some bacteria (including *S. aureus*), the enzyme has a dimeric arrangement. It has been suggested that this dimeric enzyme form may have been flexible, reducing the efficiency of catalysis. In order to reduce flexibility, many dihydrodipicolinate synthase enzymes have a higher order oligomeric structure. Most bacteria have a "back-to-back" arrangement, while plants adopt a 'head-to-head" arrangement

dimeric variant of the *Bacillus anthracis* enzyme, although in this case the enzyme was not inhibited, simply less active (Voss et al. 2010). The decrease in activity has been attributed to increased flexibility of the enzyme, which in the *E. coli* enzyme resulted in inhibition by a natural metabolite and in the *B. anthracis* enzyme resulted in a defective active site. It has been proposed that the tetrameric structure may have evolved to optimise dynamics within the tight-dimer (Griffin et al. 2008). Molecular dynamics simulation studies have shown striking contrast between the dynamics of the dimeric and tetrameric forms, with the dimeric enzyme showing high flexibility, resulting in deformation of key residues in the active site (Reboul et al. 2012). In contrast, dimeric mutants of *Mycobacterium tuberculosis* and *Thermotoga maritima* dihydrodipicolinate synthase maintained high levels of catalytic activity (Evans et al. 2011; Pearce et al. 2011), suggesting there may be alternative mechanisms for stabilising the active site and that additional selection pressures may be driving conservation of the dihydrodipicolinate synthase tetrameric architecture.

In all cases, the dimeric mutants had reduced thermal stability compared to the original tetrameric enzyme, suggesting a role for quaternary structure in providing protein stability. The intricate link between quaternary structure and activity make dihydrodipicolinate synthase an excellent model to study the catalysis and regulation of oligomeric enzymes.

10.5 Conclusion

The past two decades has seen significant advances in our understanding of the enzymes of the lysine biosynthetic pathway. In particular, a new biosynthetic pathway has been discovered, a human dihydrodipicolinate synthase like enzyme has been discovered (which has implications with regards to the design of inhibitors against its bacterial counterparts), and our understanding of role of quaternary structure in enzyme function has expanded. However, there still lacks a general understanding of how dihydrodipicolinate synthase enzymes are allosterically regulated. Overall, dihydrodipicolinate synthase has provided fertile ground to understand enzyme mechanism, allosteric function, and the evolution of quaternary structure.

Acknowledgments AOH acknowledges the College of Science and the Thomas H. Gosnell School of Life Sciences at the Rochester Institute of Technology for ongoing support. RCJD acknowledges the following for funding support, in part: (1) the Ministry of Business, Innovation and Employment (contract UOCX1208); (2) the New Zealand Royal Society Marsden Fund (contract UOC1013) and (3) the US Army Research Laboratory and US Army Research Office under contract/grant number W911NF-11-1-0481.

References

Atkinson SC, Dogovski C, Downton MT et al (2012) Crystal, solution and *in silico* structural studies of dihydrodipicolinate synthase from the common grapevine. PLoS One 7:e38318–e38319
Atkinson SC, Hor L, Dogovski C et al (2014) Identification of the bona fide DHDPS from a common plant pathogen. Proteins 82:1869–1883
Bakhiet N, Forney FW, Stahly DP, Daniels L (1984) Lysine biosynthesis in *Methanobacterium thermoautotrophicum* is by the diaminopimelic acid pathway. Curr Microbiol 10:195–198
Barnes IJ, Bondi A, Moat AG (1969) Biochemical characterization of lysine auxotrophs of *Staphylococcus aureus*. J Bacteriol 99:169–174
Beck BB, Baasner A, Buescher A et al (2013) Novel findings in patients with primary hyperoxaluria type III and implications for advanced molecular testing strategies. Eur J Hum Genet 21:162–172
Belostotsky R, Seboun E, Idelson GH et al (2010) Mutations in DHDPSL are responsible for primary hyperoxaluria type III. Am J Hum Genet 87:392–399
Blickling S, Beisel HG, Bozic D et al (1997a) Structure of dihydrodipicolinate synthase of *Nicotiana sylvestris* reveals novel quaternary structure. J Mol Biol 274:608–621
Blickling S, Renner C, Laber B et al (1997b) Reaction mechanism of *Escherichia coli* dihydrodipicolinate synthase investigated by X-ray crystallography and NMR spectroscopy. Biochemistry 36:24–33
Borthwick EB, Connell SJ, Tudor DW et al (1995) *Escherichia coli* dihydrodipicolinate synthase: characterization of the imine intermediate and the product of bromopyruvate treatment by electrospray mass spectrometry. Biochem J 305(Pt 2):521–524
Boughton BA, Dobson RCJ, Gerrard JA, Hutton CA (2008a) Conformationally constrained diketopimelic acid analogues as inhibitors of dihydrodipicolinate synthase. Bioorg Med Chem Lett 18:460–463
Boughton BA, Griffin MDW, O'Donnell PA et al (2008b) Irreversible inhibition of dihydrodipicolinate synthase by 4-oxo-heptenedioic acid analogues. Bioorg Med Chem 16:9975–9983
Boughton BA, Dobson RCJ, Hutton CA (2012) The crystal structure of dihydrodipicolinate synthase from *Escherichia coli* with bound pyruvate and succinic acid semialdehyde: unambiguous resolution of the stereochemistry of the condensation product. Proteins 80:2117–2122
Bunker RD (2010) Enzymes associated with the complications of diabetes mellitus. Ph.D. thesis, School of Biological Sciences, University of Auckland. https://researchspace.auckland.ac.nz/handle/2292/7142
Burgess BR, Dobson RCJ, Bailey MF et al (2008) Structure and evolution of a novel dimeric enzyme from a clinically important bacterial pathogen. J Biol Chem 283:27598–27603
Cahyanto MN, Kawasaki H, Nagashio M et al (2006) Regulation of aspartokinase, aspartate semialdehyde dehydrogenase, dihydrodipicolinate synthase and dihydrodipicolinate reductase in *Lactobacillus plantarum*. Microbiology (Reading, England) 152:105–112
Cochat P, Rumsby G (2013) Primary hyperoxaluria. N Engl J Med 369:649–658
Coulter CV, Gerrard JA, Kraunsoe JAE et al (1996) (S)-Aspartate semi-aldehyde: Synthetic and structural studies. Tetrahedron 52:7127–7136
Cox RJ (1996) The DAP pathway to lysine as a target for antimicrobial agents. Nat Prod Rep 13:29–43
Cremer J, Treptow C, Eggeling L, Sahm H (1988) Regulation of enzymes of lysine biosynthesis in *Corynebacterium glutamicum*. Microbiology (Reading, England) 134:3221–3229
Dekker EE, Kitson RP (1992) 2-Keto-4-hydroxyglutarate aldolase: purification and characterization of the homogeneous enzyme from bovine kidney. J Biol Chem 267:10507–10514
Dereppe C, Bold G, Ghisalba O et al (1992) Purification and characterization of dihydrodipicolinate synthase from pea. Plant Physiol 98:813–821
Devenish SRA, Gerrard JA (2009) The role of quaternary structure in $(\beta/\alpha)8$-barrel proteins: evolutionary happenstance or a higher level of structure-function relationships? Org Biomol Chem 7:833–839

Devenish SRA, Gerrard JA, Jameson GB, Dobson RCJ (2008) The high-resolution structure of dihydrodipicolinate synthase from *Escherichia coli* bound to its first substrate, pyruvate. Acta Crystallogr Sect F: Struct Biol Cryst Commun 64:1092–1095

Devenish SRA, Huisman FHA, Parker EJ et al (2009) Cloning and characterisation of dihydrodipicolinate synthase from the pathogen *Neisseria meningitidis*. Biochim Biophys Acta 1794:1168–1174

Dobson RCJ, Gerrard JA, Pearce FG (2004a) Dihydrodipicolinate synthase is not inhibited by its substrate, (*S*)-aspartate β-semialdehyde. Biochem J 377:757–762

Dobson RCJ, Griffin MDW, Roberts SJ, Gerrard JA (2004b) Dihydrodipicolinate synthase (DHDPS) from *Escherichia coli* displays partial mixed inhibition with respect to its first substrate, pyruvate. Biochimie 86:311–315. doi:10.1016/j.biochi.2004.03.008

Dobson RCJ, Valegård K, Gerrard JA (2004c) The crystal structure of three site-directed mutants of *Escherichia coli* dihydrodipicolinate synthase: further evidence for a catalytic triad. J Mol Biol 338:329–339

Dobson RCJ, Griffin MDW, Jameson GB, Gerrard JA (2005) The crystal structures of native and (*S*)-lysine-bound dihydrodipicolinate synthase from *Escherichia coli* with improved resolution show new features of biological significance. Acta Crystallogr D Biol Crystallogr 61:1116–1124

Dobson RCJ, Griffin MDW, Devenish SRA et al (2008) Conserved main-chain peptide distortions: a proposed role for Ile203 in catalysis by dihydrodipicolinate synthase. Protein Sci 17:2080–2090

Dobson RCJ, Girón I, Hudson AO (2011) L,L-diaminopimelate aminotransferase from *Chlamydomonas reinhardtii*: a target for algaecide development. PLoS One 6:e20439–e20413

Dogovski C, Atkinson SC, Dommaraju SR et al (2009) Lysine biosynthesis in bacteria: an unchartered pathway for novel antibiotic design. Encycl Life Support Syst 11:116–136

Domigan LJ, Scally SW, Fogg MJ et al (2009) Characterisation of dihydrodipicolinate synthase (DHDPS) from *Bacillus anthracis*. Biochim Biophys Acta 1794:1510–1516

Evans G, Schuldt L, Griffin MDW et al (2011) A tetrameric structure is not essential for activity in dihydrodipicolinate synthase (DHDPS) from *Mycobacterium tuberculosis*. Arch Biochem Biophys 512:154–159

Frisch DA, Gengenbach BG, Tommey AM et al (1991) Isolation and characterization of dihydrodipicolinate synthase from maize. Plant Physiol 96:444–452

Galili G (2011) The aspartate-family pathway of plants: linking production of essential amino acids with energy and stress regulation. Plant Signal Behav 6:192–195

Girish TS, Sharma E, Gopal B (2008) Structural and functional characterization of *Staphylococcus aureus* dihydrodipicolinate synthase. FEBS Lett 582:2923–2930

Goodsell DS, Olson AJ (2000) Structural symmetry and protein function. Annu Rev Biophys Biomol Struct 29:105–153

Griffin MDW, Gerrard JA (2012) The relationship between oligomeric state and protein function. Adv Exp Med Biol 747:74–90

Griffin MDW, Dobson RCJ, Pearce FG et al (2008) Evolution of quaternary structure in a homotetrameric enzyme. J Mol Biol 380:691–703

Griffin MDW, Dobson RCJ, Gerrard JA, Perugini MA (2010) Exploring the dihydrodipicolinate synthase tetramer: how resilient is the dimer–dimer interface? Arch Biochem Biophys 494:58–63

Griffin MDW, Billakanti JM, Wason A et al (2012) Characterisation of the first enzymes committed to lysine biosynthesis in *Arabidopsis thaliana*. PLoS One 7:e40318–e40312

Guo BBB, Devenish SRA, Dobson RCJ et al (2009) The C-terminal domain of *Escherichia coli* dihydrodipicolinate synthase (DHDPS) is essential for maintenance of quaternary structure and efficient catalysis. Biochem Biophys Res Commun 380:802–806

Hoganson DA, Stahly DP (1975) Regulation of dihydrodipicolinate synthase during growth and sporulation of *Bacillus cereus*. J Bacteriol 124:1344–1350

Hopp K, Cogal AG, Bergstralh EJ et al (2015) Phenotype-genotype correlations and estimated carrier frequencies of primary hyperoxaluria. J Am Soc Nephrol 26:2559–2570

Hudson AO, Singh BK, Leustek T, Gilvarg C (2006) An *LL*-diaminopimelate aminotransferase defines a novel variant of the lysine biosynthesis pathway in plants. Plant Physiol 140:292–301

Hudson AO, Klartag A, Gilvarg C et al (2011) Dual diaminopimelate biosynthesis pathways in *Bacteroides fragilis* and *Clostridium thermocellum*. Biochim Biophys Acta 1814:1162–1168

Hutton CA, Perugini MA, Gerrard JA (2007) Inhibition of lysine biosynthesis: an evolving antibiotic strategy. Mol BioSyst 3:458–465

Jander G, Joshi V (2009) Aspartate-derived amino acid biosynthesis in *Arabidopsis thaliana*. Arabidopsis Book 7:e0121

Jander G, Joshi V (2010) Recent progress in deciphering the biosynthesis of aspartate-derived amino acids in plants. Mol Plant 3:54–65

Karsten WE (1997) Dihydrodipicolinate synthase from *Escherichia coli*: pH dependent changes in the kinetic mechanism and kinetic mechanism of allosteric inhibition by L-lysine. Biochemistry 36:1730–1739

Kefala G, Evans GL, Griffin MDW et al (2008) Crystal structure and kinetic study of dihydrodipicolinate synthase from *Mycobacterium tuberculosis*. Biochem J 411:351–310

Kelland JG, Palcic MM, Pickard MA, Vederas JC (1985) Stereochemistry of lysine formation by *meso*-diaminopimelate decarboxylase from wheat germ: use of proton-carbon-13 NMR shift correlation to detect stereospecific deuterium labeling. Biochemistry 24:3263–3267

Kindler SH, Gilvarg C (1960) *N*-Succinyl-*L*-2,6-diaminopimelic acid deacylase. J Biol Chem 235:3532–3535

Kumpaisal R, Hashimoto T, Yamada Y (1987) Purification and characterization of dihydrodipicolinate synthase from wheat suspension cultures. Plant Physiol 85:145–151

Laber B, Gomis-Rüth FX, Romão MJ, Huber R (1992) *Escherichia coli* dihydrodipicolinate synthase. Identification of the active site and crystallization. Biochem J 288(Pt 2):691–695

Li X, Knight J, Todd Lowther W, Holmes RP (2015) Hydroxyproline metabolism in a mouse model of Primary Hyperoxaluria Type 3. Biochim Biophys Acta 1852:2700–2705

Lin YK, Myhrman R, Schrag ML, Gelb MH (1988) Bacterial *N*-succinyl-*L*-diaminopimelic acid desuccinylase. Purification, partial characterization, and substrate specificity. J Biol Chem 263:1622–1627

Lynch M (2013) Evolutionary diversification of the multimeric states of proteins. Proc Natl Acad Sci U S A 110:E2821–E2828

MacDonald JR, Huang AD, Loomes K (2016) Cellular degradation of 4-hydroxy-2-oxoglutarate aldolase leads to absolute deficiency in Primary Hyperoxaluria Type 3. FEBS Lett 590(10):1467–1476

Matthews BF, Widholm JM (1979) Expression of aspartokinase, dihydrodipicolinic acid synthase and homoserine dehydrogenase during growth of carrot cell suspension cultures on lysine- and threonine-supplemented media. Zeitschrift für Naturforschung Section C: Biosciences 34:1177–1185

McCoy AJ, Adams NE, Hudson AO et al (2006) L,L-diaminopimelate aminotransferase, a trans-kingdom enzyme shared by Chlamydia and plants for synthesis of diaminopimelate/lysine. Proc Natl Acad Sci 103:17909–17914. doi:10.1073/pnas.0608643103

Mirwaldt C, Korndörfer I, Huber R (1995) The crystal structure of dihydrodipicolinate synthase from *Escherichia coli* at 2.5 A resolution. J Mol Biol 246:227–239

Mitsakos V, Dobson RCJ, Pearce FG et al (2008) Inhibiting dihydrodipicolinate synthase across species: towards specificity for pathogens? Bioorg Med Chem Lett 18:842–844

Monico CG, Rossetti S, Belostotsky R et al (2011) Primary hyperoxaluria type III gene HOGA1 (formerly DHDPSL) as a possible risk factor for idiopathic calcium oxalate urolithiasis. Clin J Am Soc Nephrol 6:2289–2295

Muscroft-Taylor AC, Soares da Costa TP, Gerrard JA (2010) New insights into the mechanism of dihydrodipicolinate synthase using isothermal titration calorimetry. Biochimie 92:254–262

Pearce FG, Dobson RCJ, Jameson GB et al (2011) Characterization of monomeric dihydrodipicolinate synthase variant reveals the importance of substrate binding in optimizing oligomerization. Biochim Biophys Acta Protein Proteomics 1814:1900–1909
Perham RN (1975) Self-assembly of biological macromolecules. Philos Trans R Soc Lond Ser B Biol Sci 272:123–136
Perutz MF (1989) Mechanisms of cooperativity and allosteric regulation in proteins. Q Rev Biophys 22:139–237
Peterkofsky B, Gilvarg C (1961) *N*-Succinyl-*L*-diaminopimelic-glutamic transaminase. J Biol Chem 236:1432–1438
Phenix CP, Palmer DRJ (2008) Isothermal titration microcalorimetry reveals the cooperative and noncompetitive nature of inhibition of *Sinorhizobium meliloti* L5-30 dihydrodipicolinate synthase by (*S*)-lysine. Biochemistry 47:7779–7781
Phillips R, Kondev J, Theriot J, Garcia H (2012) Physical biology of the cell, 2nd edn. New York, Garland Science
Reboul CF, Porebski BT, Griffin MDW et al (2012) Structural and dynamic requirements for optimal activity of the essential bacterial enzyme dihydrodipicolinate synthase. PLoS Comput Biol 8:e1002537
Riedel TJ, Johnson LC, Knight J et al (2011) Structural and biochemical studies of human 4-hydroxy-2-oxoglutarate aldolase: implications for hydroxyproline metabolism in primary hyperoxaluria. PLoS One 6:e26021
Riedel TJ, Knight J, Murray MS et al (2012) 4-Hydroxy-2-oxoglutarate aldolase inactivity in primary hyperoxaluria type 3 and glyoxylate reductase inhibition. Biochim Biophys Acta 1822:1544–1552
Roberts SJ, Morris JC, Dobson RCJ, Gerrard JA (2003) The preparation of (*S*)-aspartate semialdehyde appropriate for use in biochemical studies. Bioorg Med Chem Lett 13:265–267
Rose WC (1949) Amino acid requirements of man. Fed Proc 8:546–552
Scapin G, Blanchard JS (1998) Enzymology of bacterial lysine biosynthesis. Adv Enzymol Relat Areas Mol Biol 72:279–324
Scapin G, Cirilli M, Reddy SG et al (1998) Substrate and inhibitor binding sites in *Corynebacterium glutamicum* diaminopimelate dehydrogenase. Biochemistry 37:3278–3285
Simms SA, Voige WH, Gilvarg C (1984) Purification and characterization of succinyl-CoA: tetrahydrodipicolinate *N*-succinyltransferase from *Escherichia coli*. J Biol Chem 259:2734–2741
Skovpen YV, Conly CJT, Sanders DAR, Palmer DRJ (2016) Biomimetic design results in a potent allosteric inhibitor of dihydrodipicolinate synthase from *Campylobacter jejuni*. J Am Chem Soc 138:2014–2020
Soares da Costa TP, Muscroft-Taylor AC, Dobson RCJ et al (2010) How essential is the "essential" active-site lysine in dihydrodipicolinate synthase? Biochimie 92:837–845
Stahly DP (1969) Dihydrodipicolinic acid synthase of *Bacillus licheniformis*. Biochim Biophys Acta Enzymol 191:439–451
Turner JJ, Healy JP, Dobson RCJ et al (2005) Two new irreversible inhibitors of dihydrodipicolinate synthase: diethyl (*E,E*)-4-oxo-2,5-heptadienedioate and diethyl (*E*)-4-oxo-2-heptenedioate. Bioorg Med Chem Lett 15:995–998
Tyagi VVS, Henke RR, Farkas WR (1982) Occurrence of diaminopimelic epimerase in maize. Biochim Biophys Acta Gen Subj 719:363–369
Voss JE, Scally SW, Taylor NL et al (2010) Substrate-mediated stabilization of a tetrameric drug target reveals Achilles heel in anthrax. J Biol Chem 285:5188–5195
Wallsgrove RM, Mazelis M (1980) The enzymology of lysine biosynthesis in higher plants: complete localization of the regulatory enzyme dihydrodipicolinate synthase in the chloroplasts of spinach leaves. FEBS Lett 116:189–192
Wang X, Zhao X, Wang X et al (2015) Two novel HOGA1 splicing mutations identified in a Chinese patient with primary hyperoxaluria type 3. Am J Nephrol 42:78–84

Watanabe N, Clay MD, van Belkum MJ et al (2011) The structure of *LL*-diaminopimelate aminotransferase from *Chlamydia trachomatis*: implications for its broad substrate specificity. J Mol Biol 411:649–660

Webster FH, Lechowich RV (1970) Partial purification and characterization of dihydrodipicolinic acid synthetase from sporulating *Bacillus megaterium*. J Bacteriol 101:118–126

Weinberger S, Gilvarg C (1970) Bacterial distribution of the use of succinyl and acetyl blocking groups in diaminopimelic acid biosynthesis. J Bacteriol 101:323–324

White PJ, Kelly B (1965) Purification and properties of diaminopimelate decarboxylase from *Escherichia coli*. Biochem J 96:75–84

Williams EL, Bockenhauer D, van't Hoff WG et al (2012) The enzyme 4-hydroxy-2-oxoglutarate aldolase is deficient in primary hyperoxaluria type 3. Nephrol Dial Transplant 27:3191–3195

Wiseman JS, Nichols JS (1984) Purification and properties of diaminopimelic acid epimerase from *Escherichia coli*. J Biol Chem 259:8907–8914

Wolterink-van Loo S, Levisson M, Cabrières MC et al (2008) Characterization of a thermostable dihydrodipicolinate synthase from *Thermoanaerobacter tengcongensis*. Extremophiles 12:461–469

Yamakura F, Ikeda Y, Kimura K, Sasakawa T (1974) Partial purification and some properties of pyruvate-aspartic semialdehyde condensing enzyme from sporulating *Bacillus subtilis*. J Biochem 76:611–621

Yugari Y, Gilvarg C (1965) The condensation step in diaminopimelate synthesis. J Biol Chem 240:4710–4716

Zabriskie TM, Jackson MD (2000) Lysine biosynthesis and metabolism in fungi. Nat Prod Rep 17:85–97

Chapter 11
"Pyruvate Carboxylase, Structure and Function"

Mikel Valle

Abstract Pyruvate carboxylase is a metabolic enzyme that fuels the tricarboxylic acid cycle with one of its intermediates and also participates in the first step of gluconeogenesis. This large enzyme is multifunctional, and each subunit contains two active sites that catalyze two consecutive reactions that lead to the carboxylation of pyruvate into oxaloacetate, and a binding site for acetyl-CoA, an allosteric regulator of the enzyme. Pyruvate carboxylase oligomers arrange in tetramers and covalently attached biotins mediate the transfer of carboxyl groups between distant active sites. In this chapter, some of the recent findings on pyruvate carboxylase functioning are presented, with special focus on the structural studies of the full length enzyme. The emerging picture reveals large movements of domains that even change the overall quaternary organization of pyruvate carboxylase tetramers during catalysis.

Keywords Multifunctional enzyme • Biotin-dependent carboxylase • Pyruvate carboxylase • Allosteric regulation • Acetyl-CoA

11.1 Introduction

Pyruvate carboxylase (PC) is a biotin-dependent enzyme that catalyses the carboxylation of pyruvate into oxaloacetate. PC fuels the tricarboxylic acid (TCA) cycle with one of its intermediates and also participates in the first step of gluconeogenesis. At these metabolic hubs, the activity of PC is regulated by acetyl coenzyme A (acetyl-CoA). PC is multifunctional, and each subunit contains two active sites that catalyze the two consecutive reactions that take place during carboxylation of pyruvate. A covalently attached biotin mediates the transfer of carboxyl groups between distant active sites. PC oligomers arrange in tetramers that are organized into two layers, with two opposing monomers in each layer. The biotin cofactor is tethered to a swinging domain that travels long distances between catalytic sites located at

M. Valle (✉)
Structural Biology Unit, Center for Cooperative Research in Biosciences, CIC bioGUNE, 48160 Derio, Spain
e-mail: mvalle@cicbiogune.es

J.R. Harris, J. Marles-Wright (eds.), *Macromolecular Protein Complexes*, Subcellular Biochemistry 83, DOI 10.1007/978-3-319-46503-6_11

opposite subunits. Thus, the mechanism of action of PC requires large movements of domains, conformational changes that are under allosteric regulation. Indeed, available crystal structures reveal two different configurations for PC tetramers, the so-called symmetric and asymmetric, and these conformations were understood as characteristic molecular architectures for PC from different organisms. Recent cryoEM studies suggest, however, that PC tetramers map both conformations during its catalytic cycle, and that each of the structural states is linked to one of the two sequential reactions.

11.2 Biotin-Dependent Carboxylases

Biotin-dependent carboxylases are metabolic enzymes that are widely distributed in all organisms, where they catalyze the transfer of carboxyl groups to several substrates that are crucial in fatty acid, amino acid and carbohydrate metabolism (Waldrop et al. 2013; Tong 2013). They also play different roles in a variety of metabolic routes specific to some microorganisms. This group of enzymes is multifunctional, carries a covalently linked biotin, and is usually active in homo- or hetero-oligomeric forms. Biotin-dependent carboxylases perform their activity in two sequential steps at two active sites (Tong 2013; Attwood and Wallace 2002). The first reaction involves the carboxylation of the biotin using bicarbonate as the carboxyl donor, this requires Mg-ATP and takes place at the biotin carboxylase (BC) active site (Fig. 11.1). The second step is carried out in the carboxyltransferase (CT) component, where the carboxyl group is transferred from the carboxybiotin to the substrate specific for each enzyme. The BC and CT activities work in a sequential fashion and the biotin moves between reaction centers to transfer the activated carboxyl group (Fig. 11.1). Biotin is linked to the side chain of a lysine residue within the biotin carboxyl carrier protein (BCCP) domain. The BC and BCCP domains are well conserved between biotin-dependent carboxylases, but the CT component is diverse and specialized in the carboxylation of the specific substrate (Waldrop et al. 2013; Tong 2013; Attwood and Wallace 2002).

The biotin residue must visit the BC and CT reaction centers and the structures of several biotin-dependent carboxylases (Wei and Tong 2015; Tran et al. 2015; Jurado et al. 2015; Huang et al. 2012; Fan et al. 2012; Huang et al. 2010; Xiang and Tong 2008; St Maurice et al. 2007) have shown that the distance between the two sites is in the range of 45–85 Å. The prosthetic biotin is linked to the BCCP domain through an arm contributed by the lysine and biotin itself. This arm is, however, too short to cover the distance between the BC and CT reaction sites, and thus, the enzymes need the BCCP domains to be translocated across long distances during catalysis. The available crystallographic structures have significantly contributed to unveiling the chemical catalytic mechanisms for biotin-dependent carboxylases, but the data have not yet explained the dynamic behavior of BCCP domains in any of the studied systems. It is clear that understanding of the communication between

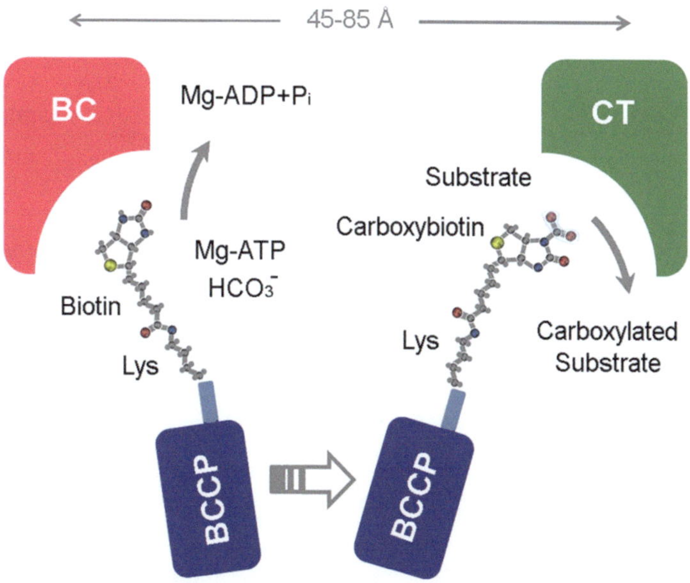

Fig. 11.1 Sequential enzymatic reactions in biotin-dependent carboxylases. Prosthetic biotin is attached to a lysine residue in the BCCP domain. In the first step biotin is carboxylated at the BC active site. Then, BCCP and attached carboxybiotin are translocated to the CT active center, where the carboxyl group is transferred to the specific substrate

reaction centers is crucial to fully comprehend the functioning and regulation of this group of enzymes.

11.3 Physiological Role of Pyruvate Carboxylase

11.3.1 Metabolic Activity of PC

Eukaryotic PC is a mitochondrial enzyme that carboxylates pyruvate into oxaloacetate, an essential metabolite in the TCA cycle (Utter and Keech 1960). Oxaloacetate is a key intermediate in metabolism, linking carbohydrate, lipid, amino acid, and nucleotide metabolism (Gray et al. 2014). Thus, the activity of the enzyme is involved in a large variety of metabolic routes (Fig. 11.2) whose significance depends on the organism, tissue and cell type. Pyruvate is the end-product of glycolysis and a major input of carbon flux into the TCA cycle. Cytosolic pyruvate crosses the inner mitochondrial membrane transported by the mitochondrial pyruvate carrier (MPC). Once in the mitochondrial matrix, pyruvate can be converted into acetyl-CoA by the pyruvate dehydrogenase (PDH) complex or into oxaloacetate by PC (Fig. 11.2). These two paths are interconnected by the positive regulation of PC activity by acetyl-CoA. So, high levels of acetyl-CoA stimulate PC to

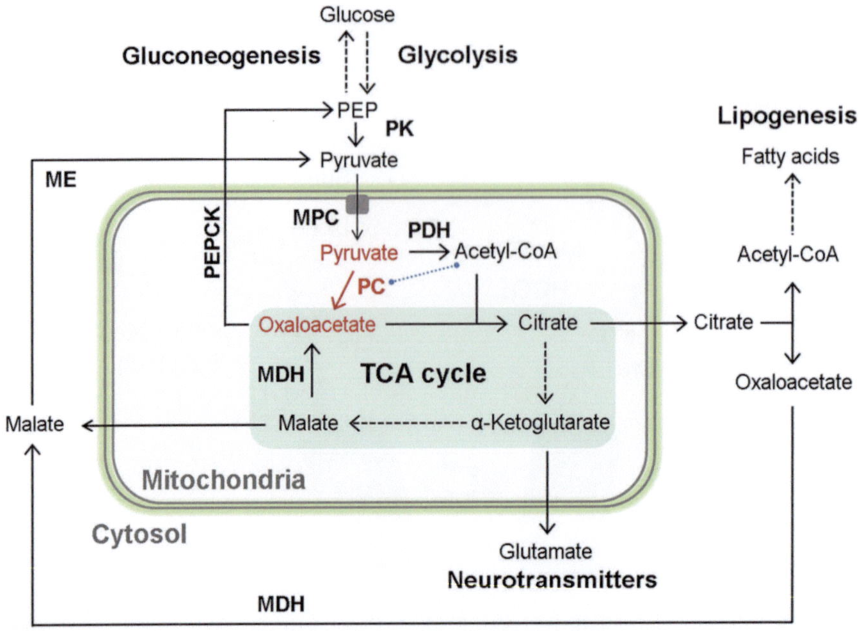

Fig. 11.2 Schematic view of eukaryotic PC metabolic network. The carboxylation of pyruvate into oxaloacetate catalyzed by PC (*colored red*) is an anaplerotic reaction that feeds the TCA cycle in mitochondria. TCA cycle intermediates participate in several metabolic routes. Alternatively, oxaloacetate can be diverted to gluconeogenesis, where PC reaction is the first step. PC enzymatic activity is stimulated by acetyl-CoA (*blue line*). *Black* and *unbroken arrows* portrait single step enzymatic reactions, and *dashed arrows* summarize multiple steps. The acronyms in the figure stand for: phosphoenolpyruvate (*PEP*); pyruvate kinase (*PK*); mitochondrial pyruvate carrier (*MPC*); phosphoenolpyruvate carboxykinase (*PEPCK*); pyruvate dehydrogenase (*PDH*); pyruvate carboxylase (*PC*); tricarboxylic acid (*TCA*) cycle; malic enzyme (*ME*); and malate dehydrogenase (*MDH*)

synthesize oxaloacetate and feed the TCA cycle or divert the carbon flux to gluconeogenesis. PDH and PC share the same substrate, and the product of the PDH reaction stimulates PC, this way the interplay between these two enzymes controls the carbon flux (details regarding PDH can be found in the Chap. 19 “The Pyruvate dehydrogenase complex and related assemblies in health and disease” by Olwyn Byron and Gordon Lindsay). PC is stimulated by acetyl-CoA or derivatives to varying degrees by activating the ATP-cleavage and biotin carboxylation at BC site (Adina-Zada et al. 2012c, b).

The PC enzyme has a critical role in gluconeogenesis, since it catalyzes the first step by channeling pyruvate (Fig. 11.2), and accordingly, PC expression levels are high in liver and kidney (Gray et al. 2014). Fasting increases pyruvate flux through gluconeogenesis by increased activities of both PC and phosphoenolpyruvate carboxykinase (PEPCK) (Bizeau et al. 2001). On the other hand, starvation induces an increase of PC expression in liver and kidney cortex, and non-carbohydrate precursors linked to the TCA cycle are used in glucose production (Salto et al. 1996).

Thus, PC expression levels are used to divert the excess of carbon flux to glucose production, or to generate fuel for organs that rely on glucose during scarcity of nutrients. PC levels are also high in adipose tissue. Here, using the TCA cycle, oxaloacetate production is converted to citrate. The citrate is transported to the cytoplasm and is cleaved to generate oxaloacetate and acetyl-CoA, which is now available for the synthesis of fatty acids in lipogenesis (Fig. 11.2).

The exchange of pyruvate with TCA cycle intermediates is also correlated with glucose-stimulated insulin secretion from pancreatic cells (Lu et al. 2002). The cycling between cytosolic and mitochondrial pyruvate involves the export of citrate and malate from the mitochondria to the cytoplasm, where they can be converted again to pyruvate (Fig. 11.2). This flux is controlled by the conversion of mitochondrial pyruvate into oxaloacetate by PC, and contributes to the production of NADPH within the TCA cycle, and to the conversion of malate to pyruvate by malic enzyme (ME). The reducing power of NADPH is then used for the production of ATP, sustaining the ATP/ADP ratio required for insulin secretion (Jitrapakdee et al. 2008). Another relevant role of the catalytic activity of PC takes place in astrocytes. Here, the TCA cycle provides α-ketoglutarate as a precursor for the neurotransmitter glutamate (Gamberino et al. 1997).

Recently, another physiological role for PC has been proposed. In human cell lines, PC is seen to enhance the virus-triggered activation of the immune response, including the expression of interferons and pro-inflammatory cytokines, to hinder viral replication and infection (Cao et al. 2016). This function seems unrelated to the catalytic activity of PC and appears to be caused by the direct interaction of the enzyme with several proteins of the innate immune response network. Further work is needed to substantiate this function of PC.

11.3.2 PC and Human Disease

Since PC function is embedded in several central metabolic pathways, its deficiency in humans triggers a variety of metabolic disorders. In most of the cases, these diseases predominantly manifest via developmental delay, neurological disorders, and metabolic acidosis (Marin-Valencia et al. 2010). PC deficiency causes the accumulation of pyruvate in the plasma, which is subsequently converted into lactate by the enzyme lactate dehydrogenase, causing an elevated plasma concentration of lactic acid. Low production of oxaloacetate burdens gluconeogenesis, promoting hypoglycemia, and prevents the liver from oxidizing acetyl-CoA derived from pyruvate and fatty acids. The increment of the acetyl-CoA pool results in hepatic ketone body synthesis or ketoacidosis (DeVivo et al. 1977). Impaired PC activity disrupts the TCA cycle and related biosynthetic pathways and might also cause abnormal levels of several amino acids, or affect other routes, such as the urea cycle causing hyperammonemia (Wang and De Vivo 1993). Delayed mental development, recurrent seizures, and psychomotor retardation are neurological manifestations also linked to PC dysfunction. The failure to produce the required levels of neurotransmitters and/or its combination with other metabolic imbalances might cause these observed clinical manifestations (Garcia-Cazorla et al. 2006).

PC deficiency is an autosomal recessive syndrome caused exclusively by mutations in the gene of the enzyme (Wang and De Vivo 1993). It is a rare disease with an incidence of 1:250,000 (Robinson 2006). There are three forms of clinical manifestations:

- Type A (infantile form) or North American type is characterized by an infantile onset with mild to moderate lactic acidemia, psychomotor retardation, and convulsions. Most of the affected children die in infancy or early childhood (Robinson 2006; Wang and De Vivo 1993). Type A patients show residual levels of PC and low activity of the enzyme
- Type B (severe neonatal form) or French type (Saudubray et al. 1976) patients usually show no presence of PC or PC mRNA, suffer severe biochemical abnormalities, intellectual disability, and abnormal brain development. The majority of affected infants die within the first three months of life (Garcia-Cazorla et al. 2006)
- Type C (benign form) is characterized by normal or mildly delayed neurologic development and episodic lactic acidosis (Wang and De Vivo 1993).

It is worth to mention that during the last years, several reports have explored the role of PC during tumor development and growth. The metabolic re-programming of cancer cells, known as Warburg effect (Warburg 1956), increases the involvement of TCA cycle intermediates into biosynthetic pathways. In this scenario, anaplerotic reactions, as the one catalyzed by PC, become essential to replenish the consumed metabolites. Indeed, it has been shown that PC-mediated anaplerosis is required for tumor survival and proliferation in non-small-cell lung cancer (Sellers et al. 2015). In breast cancer, PC protein and PC mRNA levels are increased 2–3 fold, and siRNA-mediated knockdown of PC expression resulted in a 50% reduction of cell proliferation, migration and *in vitro* invasion ability (Phannasil et al. 2015). Also, PC seems to sustain cell growth in succinate dehydrogenase (SDH) deficient cells, a frequent loss-of-function in cancer due mutations in SDH genes (Cardaci et al. 2015). These studies suggest that anaplerosis via PC plays a key role in supporting the metabolic requirements of cancer cells, and that targeting the enzyme could provide additional tools for tumor clearance.

11.4 Enzymatic Reactions Catalyzed by PC

In the first step of the PC reaction (Fig. 11.3), the BC domain catalyzes the carboxylation of the biotin cofactor using bicarbonate as the carboxyl donor in a reaction that requires Mg-ATP (Attwood and Wallace 2002; Knowles 1989). The second step is carried out by the CT component that promotes the transfer of CO_2^- from carboxybiotin to pyruvate. PC can catalyze full and partial reactions in forward and reverse modes (Menefee and Zeczycki 2014).

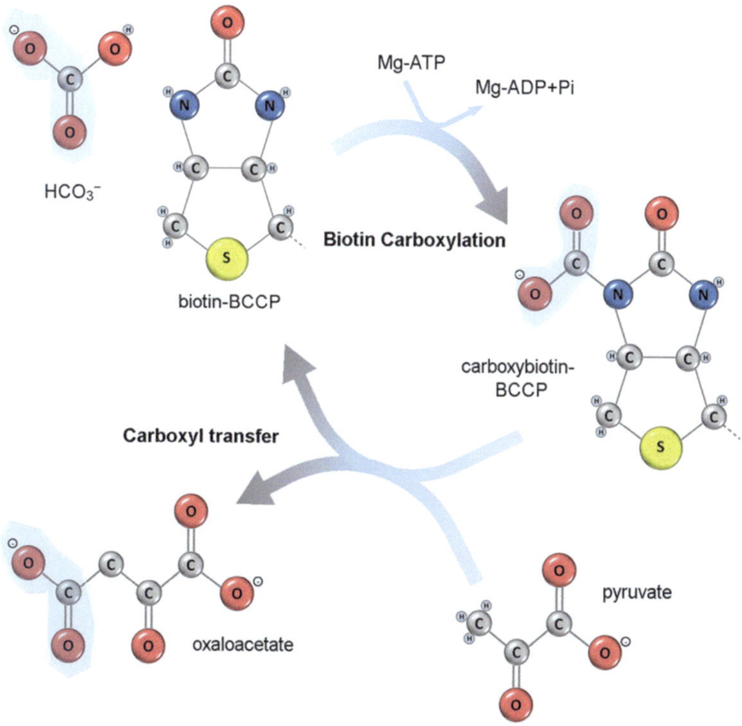

Fig. 11.3 Reaction catalyzed by PC. In the first reaction biotin is carboxylated at the N_1-position using bicarbonate (HCO_3^-) as carboxyl donor and Mg-ATP as cofactor. At the second step, the CO_2^- is transferred from carboxybiotin to pyruvate, oxaloacetate is produced and biotin is again available for a new round. The atoms of the incorporated CO_2^- group are traced (*highlighted*) along the process

Biotin-dependent carboxylases use HCO_3^- and not CO_2 as carboxyl donor (Kaziro et al. 1962). In PC, the reaction requires two Mg^{2+} equivalents, one free metal cation bound to the active site, and another in complex with ATP (Keech and Barritt 1967). Mg^{2+} and Mg-ATP bind to the BC site first, and promote a closure of the active site that enables the proper orientation of amino acid residues and retention of both ligands (Menefee and Zeczycki 2014). The biotin cofactor binds after Mg^{2+}, Mg-ATP, and HCO_3^-, and further closes the active site in PC (Lietzan et al. 2011). Despite BC being common to biotin-dependent carboxylases, the exact nature of the reaction intermediates is not well resolved. Activation of HCO_3^- seems to progress through a carboxyphospate intermediate by the nucleophilic attack on the ɣ-phosphate of ATP (Tipton and Cleland 1988; Ogita and Knowles 1988; Ashman and Keech 1975). This putative intermediate is labile and there is no direct evidence of its presence during the reaction. Once formed, carboxyphophate is thought to decompose into CO_2^- and PO_4^{3-}, PO_4^{3-} deprotonates biotin at N_1-position, and the biotin enolate reacts with CO_2^- resulting in carboxybiotin (Menefee and

Zeczycki 2014). At this BC step free biotin can be utilized, however, the CT reaction strictly requires a tethered biotin (Adina-Zada et al. 2008).

The CT reaction is specific for each type of carboxylase, and in PC this second step of the enzyme reactions is well understood. The active site needs a dicationic metal (Zn^{2+} or Mn^{2+}), and binding of pyruvate substrate induces a binding pocket for carboxybiotin (Lietzan and St Maurice 2013b). Pyruvate is then enolized, carboxybiotin decarboxylates, and the pyruvate-enolate intermediate reacts with the released CO_2^- to produce oxaloacetate (Menefee and Zeczycki 2014).

Numerous kinetic studies on PC activity have explored the coupling between BC and CT active sites. Carboxylation of pyruvate is thermodynamically non-favored, but the hydrolysis of ATP during the biotin carboxylation can compensate it. There is not, however, a full coordination between BC and CT reactions, since low pyruvate concentrations lead to non-productive ATP hydrolysis. The binding of pyruvate to CT induces the translocation of the carboxybiotin to the CT site (Goodall et al. 1981), and the subsequent release of Mg-ADP and Pi (Easterbrook-Smith et al. 1978) from a BC domain free of the interaction with the BCCP domain (Zeczycki et al. 2011a,b). Thus, the full forward reaction of PC proceeds in a sequential manner, where binding of pyruvate precedes the release of products from the first reaction (Menefee and Zeczycki 2014). At sub-saturating concentrations of pyruvate, this substrate might be liberated without the formation of oxaloacetate, promoting a non-productive decarboxylation of the biotin moiety and a new round of ATP hydrolysis. At concentrations of pyruvate near saturating levels, BC and CT activities are fully coupled and the ATP hydrolytic leak is avoided (Goodall et al. 1981). How pyruvate binding to CT site triggers the release of carboxybiotin from BC is not known.

11.5 Atomic Structure

11.5.1 Structural Domains in PC Subunit

Typically, PC is found as tetramers of four identical subunits with monomers of around 120–130 kDa in size, the α_4 type of PC (Jitrapakdee et al. 2008). In some archea and bacteria PC is composed by two polypeptide chains, an $\alpha_4\beta_4$ type, where the largest subunit (~65–75 kDa) carries the biotin and the CT activity, while the smallest (~50–55 kDa) contains the BC domain (Goss et al. 1981; Kondo et al. 2004; Lai et al. 2006; Mukhopadhyay et al. 1998, 2000, 2001). The nomenclature of the α/β subunits is unclear, and in some cases the α subunit is the largest one (Goss et al. 1981; Kondo et al. 2004), while other authors name this subunit as the β one (Lai et al. 2006; Mukhopadhyay et al. 1998, 2000, 2001). There are several available atomic structures for α_4 PC constructs from different organisms and for the BC-containing subunit of a couple of the $\alpha_4\beta_4$ PC type (Table 11.1). The first crystallographic studies for full length PC in tetrameric organization were reported for

Table 11.1 Atomic structures reported for PC

PDB code	PC construct	S/A	BCCP on	Ligand	References
4QSH	LmPC	S	exo3 (4)	c-di-AMP (CT) citrate (AL)	Sureka et al. (2014)
4QSK	LmPC	S	exo3 (4)	c-di-AMP (CT) citrate (AL)	
4QSL	LmPC	A	flex (4)	–	
4LOC	RePC (CT + AL)	S	–	oxamate (CT) free biotin (exo2)	Lietzan et al. (2014)
4M6V	RePC (CT + AL)	S	–	pyruvate (CT) free byocitin (exo2)	
4MIM	RePC (CT + AL)	S	–	3-bromopyruvate (CT)	Lietzan and St Maurice (2013a)
4MFD	RePC (CT + AL)	S	–	oxalate (CT)	
4MFE	RePC (CT + AL)	S	–	3-hydroxypyruvate (CT)	
4JX4	RePC (CT + AL)	S	–	–	Lietzan and St Maurice (2013b)
4JX5	RePC (CT + AL)	S	–	pyruvate (CT)	
4JX6	RePC Y628A (CT + AL)	S	–	pyruvate (CT)	
4HNT	SaPC F403A	S	flex (4)	–	Yu et al. (2013)
4HNU	SaPC K442E	S	flex (4)	–	
4HNV	SaPC R54E	S	CT(1) exo2 (2) flex (1)	–	
3TW6	RePC T882A	A	BC (2) flex (2)	acetyl-CoA (2) (AL)	Lietzan et al. (2011)
3TW7	RePC T882A	A	flex (4)	L-Aspartate	
3HB9	SaPC A610T	S	exo2 (2) flex (2)	–	Yu et al. (2009)
3HBL	SaPC T908A	S	CT (1 flex (1) exo2 (2)	–	
3HO8	SaPC	S	flex (4)	acetyl-CoA (4) (AL)	

(continued)

Table 11.1 (continued)

PDB code	PC construct	S/A	BCCP on	Ligand	References
3BG3	HsPC (ΔBC)	S	CT (2)	pyruvate (CT)	Xiang and Tong (2008)
			flex (2)		
3BG9	HsPC (ΔBC)	Dimeric	flex (4)	–	
3BG5	SaPC	S	CT (1)	pyruvate (CT)	
			exo2 (2)		
			flex(1)		
2QF7	RePC	A	exo1 (2)	ethyl-CoA (2) (AL)	St Maurice et al. (2007)
			flex (2)		
2DZD	BtPC (BC)	–	–	–	Kondo et al. (2007)
1ULZ	AePC (BC)	–	–	–	Kondo et al. (2004)

The table lists the crystallographic structures available in the Protein Data Bank (PDB) with the corresponding pdb code. There are PC from several sources: *Listeria monocytogenes* (LmPC); *Rhizobium etli* (RePC); *Staphylococcus aureus* (SaPC); *Homo sapiens* (HsPC); *Bacillus thermodenitrificans* (BtPC); and *Aquifex aeolicus* (AePC). The "A/S" column indicates the symmetric or asymmetric organization of PC tetramers. The "BCCP on" accounts for the position of BCCP domains in the atomic structures. Individual BCCP domains have been observed bound to BC and CT domains, but also to different "exo" binding sites outside the catalytic centers. Whenever BCCP domains are not visible in the crystal structures, they are considered flexible (flex). In the column of ligands, only some cofactors or substrates are highlighted (excluding ATP/ADP or derivatives and metal ions)

Rhizobium etli (RePC) (St Maurice et al. 2007) and *Staphylococcus aureus* (SaPC) (Xiang and Tong 2008). The PC oligomer is a tetrameric rhombohedron organized in two layers, with two opposing monomers in each layer (Fig. 11.4). Each subunit contains all the three aforementioned domains, the catalytic BC and CT, and the BCCP domain where biotin is covalently attached. In the tetramers, the layers are connected through BC-BC and CT-CT contact surfaces, and there is little direct interaction between subunits in the same layer of the tetramer. The crystal structures revealed an additional structural domain that was not inferred from the protein sequence where its two constitutive regions are far apart from each other. This region was termed allosteric domain (referred as AL along this chapter) and serves as mediator between the other protein domains. The AL region connects BC and CT domains and is the anchor site for a flexible loop (about 20 amino acids) where BCCP domain is attached. The BC and CT sites are in far distance (around 75 Å), and it is clear that the linker between BCCP and AL facilitates the translocation of the prosthetic biotin for catalysis.

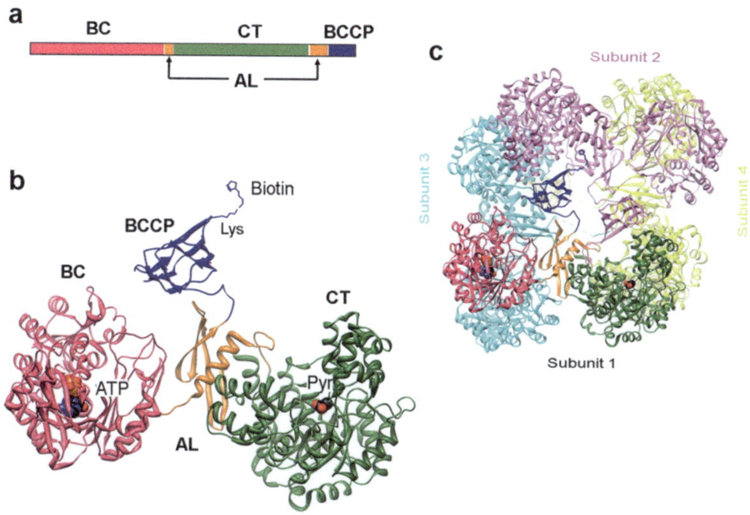

Fig. 11.4 Atomic structure and domains of PC. (**a**) Schematic representation of structural domains in the primary structure of PC. Domains are colored: BC, *magenta*; CT, *green*; BCCP, *blue*; and AL domain, *golden*. The color code for domains is kept along the figures unless stated otherwise. (**b**) One subunit of SaPC is depicted in ribbons taken from the crystallographic structure of SaPC (3BG5) (Xiang and Tong 2008). One ATP molecule residing in the BC active site, and one molecule of pyruvate in the CT site are shown. Biotin is represented attached to the conserved Lys residue of the BCCP domain. (**c**) Crystal structure of tetrameric full-length SaPC (3BG5) (Xiang and Tong 2008). One of the subunits is colored domain wise (subunit 1), while subunits 2–4 are seen *pink*, *blue* and *yellow* respectively

11.5.2 Quaternary Structures and the Allosteric (AL) Domain

There is a large difference in the quaternary structure of the reported RePC and SaPC tetramers (Fig. 11.5). In SaPC the tetramer shows a mostly symmetrical configuration (Fig. 11.5a) where all four subunits have similar structures (Fig. 11.5b). In RePC, however, the two layers show a considerable asymmetry (Fig. 11.5c), and when monomers are aligned by the CT domains, the BC regions are seen to display a large relative rotation of about 60–70^0 with the hinge located between BC and AL domains (Fig. 11.5d). In RePC the contacts between subunits involve exclusively CT-CT and BC-BC surfaces, but in the SaPC tetramer all four AL domains have a similar arrangement and are found to mediate contacts between subunits in different layers. In fact, the mutations in the AL region were seen to impair oligomerization of SaPC (Xiang and Tong 2008), and the domain was termed as tetramerization domain (PT). The tetrameric organization and activity of SaPC was restored in AL (or PT) mutants in the presence of acetyl-CoA. Interestingly, the RePC crystal structure revealed the binding site for ethyl-CoA (a nonhydrolyzable analog of the activator acetyl-CoA) located at the interface between AL and BC regions, exactly at the hinge that separates the two types of subunits in RePC (Fig. 11.5d). Initially, the occurrence of symmetric and asymmetric architectures was linked to the presence

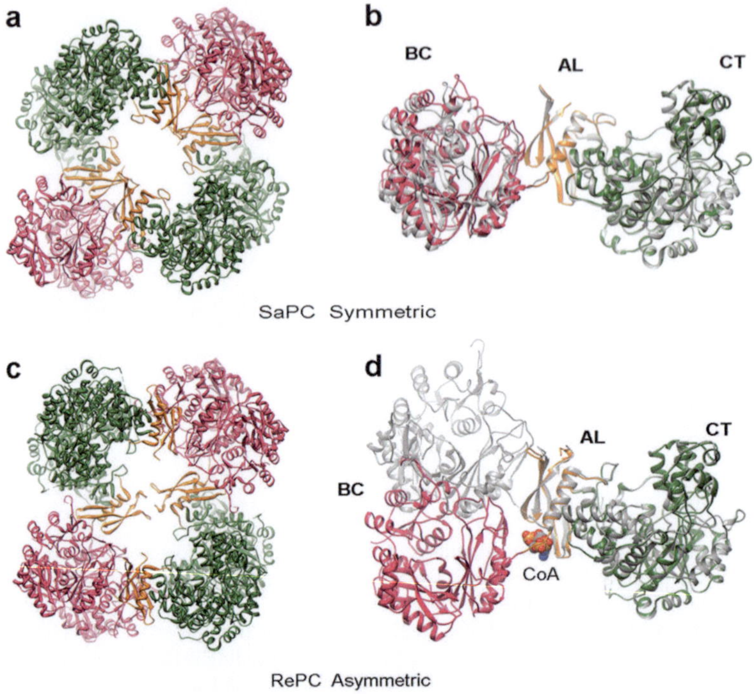

Fig. 11.5 Symmetric and asymmetric PC tetramers. (**a**) Ribbons representation for the crystal structure of SaPC tetramer (3BG5) (Xiang and Tong 2008) showing the symmetric arrangement and (**b**) comparison between subunits from opposite layers aligned by the CT domain. (**c**) Depiction of the crystal structure of RePC tetramer (2QF7) (St Maurice et al. 2007) displaying the asymmetric architecture and (**d**) comparison between subunits at *top* and *bottom layers* from the same structure. In the panels comparing subunits, one is *colored* domain-wise and other is depicted *grey*. In panel (**d**) the CoA molecule is bound to the monomer *colored* domain-wise. For clarity, the ribbons renderings do not include the BCCP domains

of ethyl-CoA in RePC structure, where only two of the subunits show bound cofactor. Puzzling, the RePC monomers carrying the CoA molecule are more similar in conformation to SaPC symmetric subunits in the absence of CoA. Later on, SaPC tetramers with four acetyl-CoA molecules bound showed again a symmetric structure (Yu et al. 2009), and RePC without CoA compound remained in the asymmetric mode (Lietzan et al. 2011). Thus, there is no evident correlation between the presence of CoA and the two structural designs observed for PC, and there was an assumption that the overall symmetry of PC tetramers was species-specific (Lietzan et al. 2011; St Maurice et al. 2007). However, a recent crystal structure of RePC lacking BC and BCCP domains showed a tetramer with a symmetrical architecture (Lietzan and St Maurice 2013b), where as in SaPC, the AL domains were also found to mediate inter-layer contacts. PC tetramers from *Listeria monocytogenes* (LmPC) are symmetric while bound to cyclic di-adenosine monophosphate (c-di-AMP), a conserved second messenger in bacteria and an allosteric regulator of PC (Sureka et al. 2014). But LmPC tetramers organize in asymmetric fashion in the apo form. This leaves open the possibility that symmetrical and asymmetrical structures represent a range of conformations for general PC. Since the enzyme is subjected to allosteric regulation and follows a sequential mode of action, it is clear that the possibility to arrange in asymmetric mode might be crucial for catalysis.

11.5.3 Structural Insights of BCCP Domain Interaction with Active Sites

The variety of atomic structures for PC constructs have delivered some snapshots of the interaction of the BCCP domain and the attached biotin with enzymatic reaction centers and other regions (Table 11.1). Although most of the published crystal structures show BCCP domains located in exo binding sites far from the catalytic centers, or missing due to their flexible nature, there are some exceptions. The BCCP domain has been observed next to the BC active site in just one RePC structure, and after mutation of the CT site (Lietzan et al. 2011). The interaction between BCCP and CT domains has been, however, described in several structures for SaPC (Yu et al. 2013, 2009; Xiang and Tong 2008) and HsPC (Xiang and Tong 2008).

11.5.3.1 Interaction of Biotin-BCCP with the BC Active Site

The BC domain contains and ATP-grasp module that directly binds the Mg-ATP molecule. Proteins of the ATP-grasp family have an overall structural design containing 3 conserved domains commonly termed the A (the N-terminal region), B (or central domain), and C (C-terminal domain). Regions A and C assemble together and form a large central core that houses the essential substrate binding sites. The B or central domain contributes to the ATP binding site via the conserved phosphate-binding P-loop. The B-domain constructs a flexible lid over the active site, and undergoes a transition to an ordered configuration upon ATP binding (Fawaz et al. 2011).

In RePC, the T882A mutant shows no catalytic activity for CT reactions but an increase in ADP phosphorylation and bicarbonate-dependent ATPase reactions at the BC site (Zeczycki et al. 2009). The crystal structure of RePC T882A revealed the interaction between BCCP and BC domains (Lietzan et al. 2011), being the only high resolution description of this interaction for any biotin-dependent carboxylase. In the tetramer, two BCCP domains are visible and bound to the BC site of their own subunits. The BCCP domain docks between the B-subdomain and a loop in the C-subdomain, essentially through non-specific and not conserved contacts (Lietzan et al. 2011), which could facilitate the dissociation and translocation of the BCCP domain during catalysis. In the structure, the BCCP domain inserts the region with the lysine residue carrying the prosthetic biotin within the BC domain (Fig. 11.6a).

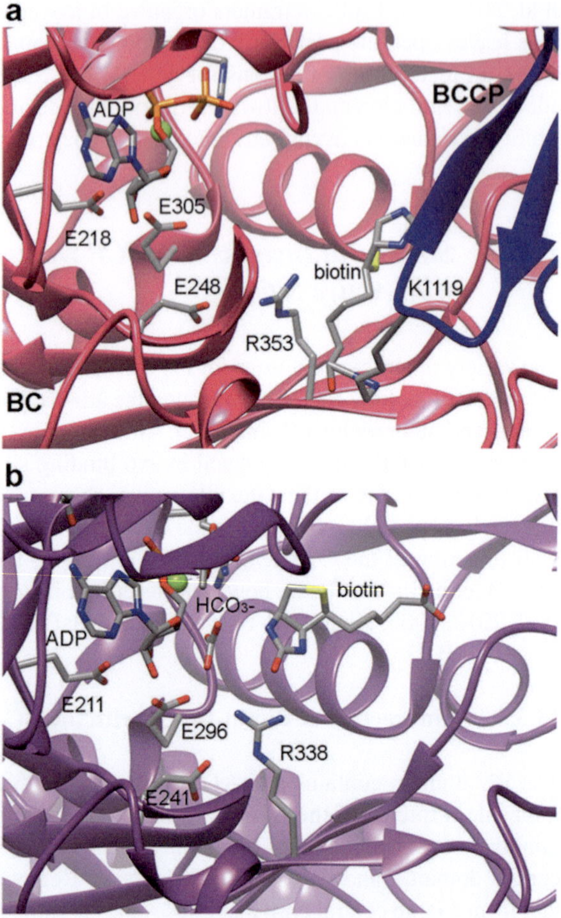

Fig. 11.6 Structural insights in biotin carboxylation (BC). (**a**) Close view of the BCCP-BC interaction observed in the crystallographic structure for RePC T882A mutant (3TW6) (Lietzan et al. 2011). (**b**) Details from the structure of the *E. coli* BC subunit from ACC in the presence of free biotin, MgADP and HCO_3^- (3G8C) (Chou et al. 2009). Both panels display similar views of the BC active site. Some of the relevant catalytic residues are labeled. The *green spheres* represent Mg^{2+} ions

The electron density and the B-factors for the tethered biotin suggest high mobility, and the biotin does not reach the BC active site. This suggests that this BCCP-BC interaction does not represent a catalytically active state. The entrance for the biotin moiety is precluded by a salt bridge between Arg353 and Glu248 (RePC numbering). The proposed catalytic residues in the BC domain of RePC are involved in MgATP-binding and cleavage (Glu218 and Lys245), HCO_3^- deprotonation (Glu305 and Arg301), and biotin enolization (Arg353) (Zeczycki et al. 2011a). The most complete atomic structure of a BC active site was reported for biotin carboxylase (BC) subunit from *Escherichia coli* acetyl-CoA carboxylase (ACC) in complex with biotin, bicarbonate, and Mg-ADP (Chou et al. 2009). This BC subunit can catalyze the carboxylation of free biotin in the absence of the CT and BCCP subunits of the ACC holoenzyme. The structure revealed clear signals for substrate and cofactors showing that the bicarbonate resides in a binding pocket between Mg-ADP and the free biotin (Fig. 11.6b). The structure, together with mutagenic and kinetic studies, points out that in BC enzyme Glu296 is the general base that deprotonates the bicarbonate, and that Arg338 stabilizes enolized biotin prior to its carboxylation (Chou et al. 2009). By direct comparison between atomic structures (Fig. 11.6) it is concluded that Glu305 (HCO_3^- deprotonation) and Arg353 (biotin enolization) are the key catalytic residues in RePC involved in those crucial steps. The position and orientation for the free biotin in BC (Fig. 11.6b) are significantly different from the one shown for the tethered biotin in RePC (Fig. 11.6a), reinforcing a catalytically non-competent configuration in RePC. Interestingly, Arg353 participates in the salt bridge that closes the active site for the BCCP-attached biotin (Fig. 11.6a). It has been proposed that after the biotin carboxylation, a reorientation of Arg353 toward Glu248 precludes carboxybiotin from reentering the BC active site, thus preventing its premature decarboxylation prior to the binding of a carboxyl acceptor in the CT domain (Chou et al. 2009). Thus, the RePC crystallographic structure might represent the state where the BCCP-biotin is ready for translocation to the BC site.

The catalytic residues at the BC site are well conserved in biotin-dependent carboxylases including PC. Regarding human diseases linked to PC deficiency, none of the reported missense HsPC mutations are directly involved in catalysis, although the available data is very limited. Point mutations in conserved Arg205 (human PC numbering) in type B patient (Ostergaard et al. 2013), and Arg270 and Tyr304 in type A patients (Monnot et al. 2009) are thought to promote local conformational changes near the BC active site. The Arg270W HsPC mutation would correspond to Arg242 in RePC, an amino acid that resides in a motif near the biotin and bicarbonate binding sites, and close to the interface between BCCP and BC domains.

11.5.3.2 Interaction of Biotin-BCCP with the CT Active Site

The carboxylation of pyruvate by the transfer of CO_2^- from carboxybiotin is specific for PC. The CT domain folds in an $\alpha_8\beta_8$ TIM barrel opened in a funnel that leads to the active site that contains a dicationic metal (Zn^{2+} or Mn^{2+}) (St Maurice et al. 2007).

The first atomic structures for HsPC and SaPC revealed biotin-BCCP bound to CT active sites from opposite subunits (Xiang and Tong 2008). The crystals for both PC constructs were grown in the presence of oxaloacetate, but its rapid spontaneous decarboxylation left pyruvate bound within the active site. In both structures some BCCP domains (two in HsPC and one in SaPC tetramers, see Table 11.1) are seen to interact with the C-terminal extension of the core of the CT domain at the entrance of the funnel, and the biotin moiety protrudes away from the BCCP domain reaching the vicinity of pyruvate bound in the CT active site (Fig. 11.7). Again, weaker electron density for the biotin (which is decarboxylated in the structure) suggests partial flexibility. The N_1 atom of biotin is within 4.7 Å of the methyl group of pyruvate suggesting a catalytically competent conformation (Xiang and Tong 2008). The binding of pyruvate promotes local rearrangements and a binding pocket for biotin (Lietzan and St Maurice 2013b). In the active site, strictly conserved Thr908 (HsPC numbering) resides in a location where it can transfer a proton from pyruvate to the

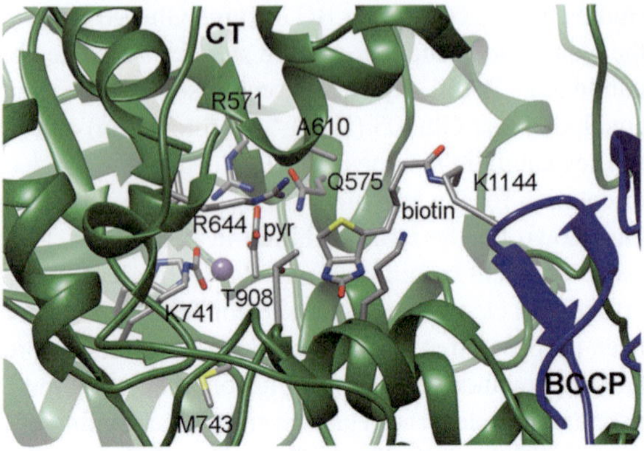

Fig. 11.7 Structural insights in carboxyl transferase (CT) reaction. Close-up view of the interaction between the BCCP-biotin and the BC active site observed in HsPC (3BG3) (Xiang and Tong 2008). Some amino acids are labeled. The *purple sphere* represents a divalent metal cation

biotin enolate (Zeczycki et al. 2009). The enolpyruvate intermediate thus generated attacks the liberated CO_2^- to form oxaloacetate. The enolpyruvate is stabilized by coordination with the Lewis metal cation present at the active site, Mn^{2+} or Zn^2, and with residues Arg571, Gln575 and Arg644, as confirmed by site-directed mutagenesis (Duangpan et al. 2010). The central role of Thr908 in the CT catalytic activity has been explored with a SaPC T908A mutant that shows more than 30-fold loss in k_{cat}, although its atomic structure does not reveal significant conformational changes (Yu et al. 2009). The catalytic reaction at the CT active site has been further explored in several atomic structures containing bound inhibitors and analogs of the enolpyruvate intermediate (Lietzan et al. 2014; Lietzan and St Maurice 2013a).

Single point mutations in the CT domain have been reported for residues in its active site. Missense mutations Ala610Thr and Met743Ile have been found in several type A patients (Carbone et al. 1998). Ala610 is at the biotin binding site, and the change to a larger side chain could impair proper binding. The crystal structure for SaPC A610T mutant (following HsPC numbering) shows that the Thr side chain would cause steric clashes with a bound biotin, and the tetramer does not contain biotin bound at the CT active site (Yu et al. 2009). The substitution of Met743 at the active site by Ile could also interfere with proper positioning of substrates. More recently, the Gly869Asp mutation has been observed in a patient affected by type B deficiency (Ostergaard et al. 2013). Gly869 resides next to the ureido group of biotin, and the change from the conserved nonpolar Gly to a polar Asp could also disrupt biotin binding.

11.5.3.3 BCCP Domain Traveling Between Reaction Centers

The first published structure for full-length PC tetramers (St Maurice et al. 2007) clearly suggested that the movement of the BCCP domain between reaction centers on opposing subunits was more plausible than the performance of the sequential reactions within the same polypeptide. The use of hybrid mutants for PC tetramers confirmed this mode of action (St Maurice et al. 2007). The BCCP domain has also

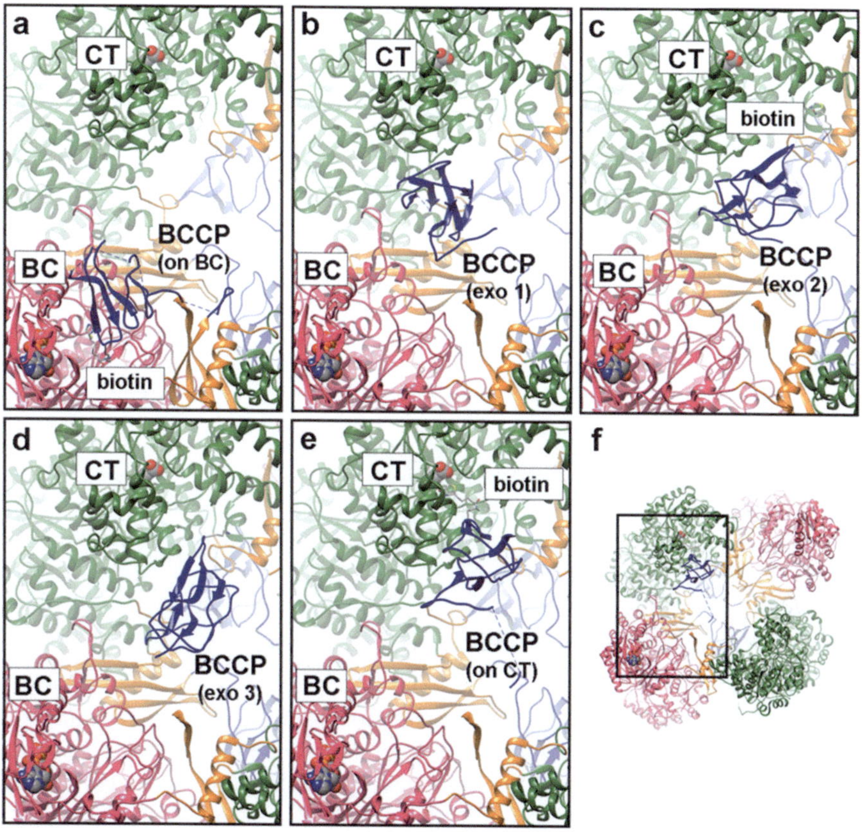

Fig. 11.8 Positions observed for BCCP domain. Ribbon representations for PC tetramers focused on the reported positions for BCCP domains at BC and CT active sites and between them, in exo sites. The panels show a combination of the atomic coordinates for RePC mutant T882A (3TW6) (Lietzan et al. 2011) and SaPC (3BG5) (Xiang and Tong 2008) to illustrate the traveling of BCCP domain between reaction centers from opposite subunits within PC tetramers. Also, the panels include BCCP domains from several structures after the alignment of tetramers: (**a**) BCCP domain from RePC T882A mutant (3TW6) engaged in interaction with the BC site (Lietzan et al. 2011); (**b**) BCCP at exo1 site seen in RePC structure (2QF7) (St Maurice et al. 2007); (**c**) the BBCP domain at exo2 site observed in SaPC (3BG5) (Xiang and Tong 2008); (**d**) BCCP bound in exo3 site in LmPC (4QSH) (Sureka et al. 2014); (**e**) the BBCP domain engaged with the CT active site in SaPC (3BG5) (Xiang and Tong 2008). In panel (**f**) the overall structure of a PC tetramer is displayed in the same orientation as a guide. In all the panels atomic models for ADP (at BC site) and pyruvate (at CT site) locate the active centers of the enzyme. Whenever the biotin moiety was visible in the atomic structures, this is indicated in the panels

been observed bound outside BC and CT reaction centers, in the so-called "exo" binding sites. Although the relevance of these sites is not clear, the set of exo sites depict a suggestive path for the movement of the BCCP domain (Fig. 11.8).

The most frequent position for BCCP in crystal structures is the exo2 site (Table 11.1), and it is the only one (out of the three exo sites) where the prosthetic biotin is well ordered in the crystallographic structures (Yu et al. 2009, 2013; Xiang and Tong 2008). The biotin is located in a small pocket between CT and AL domains (Fig. 11.8c) a region that includes conserved residues in bacterial PC. The BCCP-biotin at the exo2 site has been reported only for SaPC, but RePC crystallographic structures showed free biotin and biocytin (the amide made by biotin and Lys) in the same site, although with the biotin at different orientation (Lietzan et al. 2014). It is possible that this exo2 site represents an intermediate position for the BCCP domain during catalysis. However, the atomic interactions at this binding pocket for the biotin suggest that a carboxylated biotin could not be accommodated (Xiang and Tong 2008). Alternatively, the binding of biotin to the exo2 site might be a mechanism to increase the local concentration of free biotin and promote the biotinylation of the BCCP domain by biotin protein ligase in bacterial PC (Lietzan et al. 2014). From the overall view of the exo sites (Fig. 11.8) it seems that BCCP domain at exo1 corresponds to a true intermediate during the journey between BC and CT reaction centers.

11.6 Allosteric Control of PC

11.6.1 Communication Between BC Domains

In PC, the coordination between catalytic centers, the movements of domains, and the effects of activators and inhibitors, play in allosteric mode. However, there are few insights into how this allosteric control takes place. Studies with the BC subunit of the *E. coli* ACC have provided a model for the BC reaction (as shown in Fig. 11.6b) and for the communication between subunits. These BC subunits associate in dimers (Fig. 11.9a) where their active sites are about 50 Å away (Waldrop et al. 1994), but mutations in the active site of one of the monomers result in full inactivation of the dimer (Janiyani et al. 2001). Surprisingly, BC mutants at the dimer interface that are essentially monomeric in solution retain significant catalytic activity (Shen et al. 2006), thus, the dimeric arrangement is not required for BC function. The configuration of residues at the interface seems to be responsible of the control of BC activity. This specific conformation might also occur in the monomeric construct, as shown for BC mutants (Shen et al. 2006), and, in the context of the dimer, it seems to regulate the allosteric control of distant active sites. Although not all the BC containing subunits from different organisms associate in dimers by themselves, it is assumed that the interplay between BC domains is common between biotin-dependent carboxylases. Importantly, the organization of BC dimer from *E. coli* ACC (Fig. 11.9a) is similar to the dimer of BC domains within tetrameric PC (Fig. 11.9b).

In the atomic structure of BC in the presence of an ATP analog (ADPCF$_2$P) the ligand is bound to one subunit of BC while the other remains unligated (Fig. 11.9a) (Mochalkin et al. 2008). In the subunit bearing the nucleotide, the B-subdomain lid

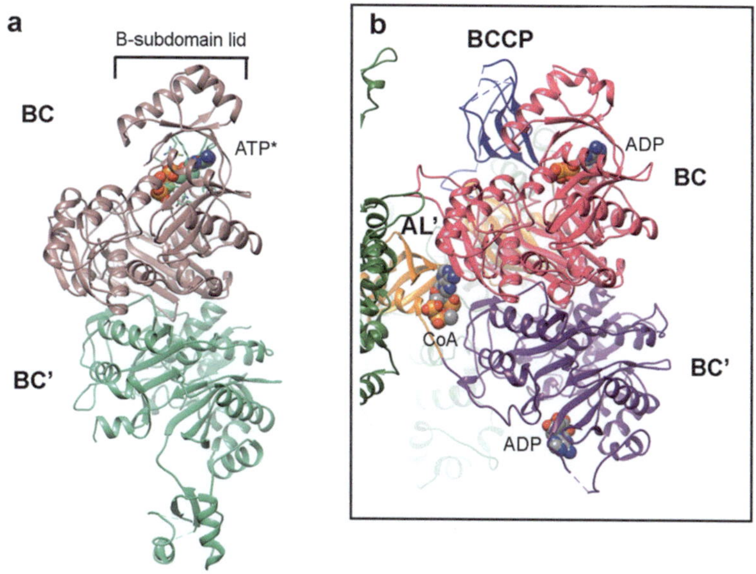

Fig. 11.9 BC dimer interplay and CoA binding site. (**a**) rendering of the atomic structure of the BC dimer from *E. coli* ACC in the presence of ATP analog AMPPCF$_2$P (2VR1) (Mochalkin et al. 2008). One BC subunit is *colored brown* (*top*) and the other is seen *green* (*bottom*). The ATP analog ADPCF$_2$P is rendered with *spheres* and labeled ATP*. (**b**) Focus on the BC-BC interaction in the context of full length RePC T882A mutant (3TW6) (Lietzan et al. 2011). Bound ADP and acetyl-CoA are indicated. The ribbons are *colored* following the domain-wise color code, but the BC domain at the *bottom layer* is depicted *purple*

rearranges and closes upon the active site. This conformational change in BC (Thoden et al. 2000) is a feature of the ATP-grasp superfamily of enzymes (Galperin and Koonin 1997), where this subdomain rotates around a hinge region by 40–50° to trap the nucleotide. Thus, the presence of one BC subunit with bound nucleotide and a closed lid, and one subunit unligated with the lid in an open conformation, goes along with an alternate mode of action, where while one monomer undergoes catalysis, the counterpart is releasing the product (de Queiroz and Waldrop 2007; Janiyani et al. 2001). This negative cooperativity, a half-of-sites play mode, requires communication through the interface between subunits to avoid simultaneous catalytic reactions.

In PC, isolated BC domains are monomeric in solution and inactive (Yu et al. 2013). Specific mutants in the BC-BC interface still arrange in tetramers in SaPC, but are catalytically inactive (Yu et al. 2013). This is in clear agreement with a communication via an unknown mechanism between residues at the interacting surface and the BC active sites. In the large majority of PC crystallographic structures, the B-subdomain lid is disordered in some of the subunits, and the asymmetry between neighbor BC domains regarding ligated nucleotides has never been observed. However the behavior of BC domains follows some asymmetry in PC. For instance, in the RePC T882A mutant structure showing the BCCP-BC interaction (Lietzan et al. 2011) there is a molecule of ADP in each BC domain, however the structure shows well defined and closed B-subdomains only in the BC regions of one of the layers, while in the opposite face, the lids are disordered (Fig. 11.9b). This type of top-bottom asymmetry might be related to the ability of interacting BC subunits to modulate their function, and in the current crystallographic structure the BCCP-BC interaction takes place only in one of the layers, the one with closed B-subdomains (Fig. 11.9b). An important question regards the role of the putative asymmetric behavior of BC dimers in the observed symmetric/asymmetric quaternary structures of PC (Fig. 11.5). All the crystal structures for full length RePC show asymmetric tetramers (Lietzan et al. 2011; St Maurice et al. 2007), however when the constructs include only CT and AL domains (ΔBCΔBCCP construct for RePC) the four subunits are almost identical and the tetramer is highly symmetric (Lietzan et al. 2014; Lietzan and St Maurice 2013a, b). The atomic structures for SaPC tetramers display the symmetric architecture (Yu et al. 2009, 2013; Xiang and Tong 2008), but a cryo-electron microscopy (cryoEM) work with full length SaPC suggests that the tetramers also map the asymmetric organization during catalysis (Lasso et al. 2014). Although there is no definitive insight into the causative role of BC domains in defining the quaternary structure of PC, it is clear that the asymmetric configuration is seen only when the BC domain is included in the construct (Table 11.1). Also, the polarity of the asymmetry between subunits at top and bottom layers in PC tetramers is coincident with the observed for BC dimers. Therefore, the communication between BC subunits within PC tetramers might be related to the distinct quaternary structures.

11.6.2 Allosteric Activator Acetyl-CoA

Among the allosteric modulators of PC activity there are several TCA cycle intermediates and derivatives that play as feedback inhibitors. This way L-aspartate is an inhibitor of bacterial and fungal α_4 PC, and L-glutamate and 2-oxoglutarate inhibit PC in vertebrates (Zeczycki et al. 2010). There are no structural insights into their binding sites in PC, and the atomic structure for RePC in the presence of L-aspartate showed high B-factors and no electron density attributable to this ligand (Lietzan et al. 2011). In the other hand, the majority of the α_4-type of PC enzymes are activated by acetyl-CoA to varying degrees, while $\alpha_4\beta_4$ PCs seem unaffected (Adina-Zada et al. 2012b). This way, tested avian PCs strictly require acetyl-CoA (Dugal 1973; Scrutton and Utter 1967; Scrutton et al. 1965) but mammalian PC retains between 2% and 30% of the maximal activity in the absence of acetyl-Coa (Adina-Zada et al. 2012b) being the PC from pig liver the one that keeps the higher catalytic action without the cofactor (Warren and Tipton 1974). The activation of the mammalian PCs follows a sigmoidal dependence on acetyl-CoA concentrations, with Hill coefficients ranging 1.9–2.5 values, implying the binding of more than one molecule of acetyl-CoA to the holenzyme in cooperative mode. The effect of acetyl CoA in PC catalysis is related to the BC reaction, i.e., the ATP-cleavage and the biotin carboxylation (Legge et al. 1996; Attwood and Graneri 1992; Scrutton et al.

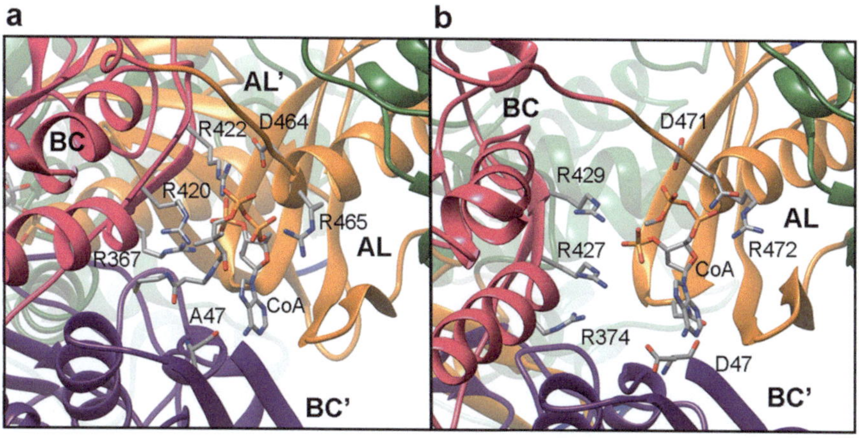

Fig. 11.10 Binding site for acetyl-CoA. (**a**) Close up view of the binding pocket for acetyl-CoA in the SaPC tetramer structure (3HO8) (Yu et al. 2009). (**b**) the binding site for acetyl-CoA activator in RePC T882A mutant (3TW6) (Lietzan et al. 2011). In both panels the ribbons are *colored* domain-wise. The BC subunit at the *bottom* is seen *purple* to differentiate BC from BC' domains. Acetyl-CoA molecule and some of PC amino acids are labeled

1965; Freedman and Kohn 1964). Comparison with pre-steady state studies of the formation of the enzyme-carboxybiotin complex indicate that acetyl-CoA reduces the uncoupling between ATP cleavage and biotin carboxylation (Easterbrook-Smith et al. 1979). The first structure for PC holoenzyme in the presence of ethyl-CoA revealed two molecules bound to tetrameric RePC (St Maurice et al. 2007). Interestingly, the binding site of the activator acetyl-CoA is located between the AL domain and the boundary between BC domains (Fig. 11.9b), opening the possibility for modulated communication between BC domains by the action of this allosteric activator. As mentioned earlier, in the asymmetric architecture there are two available binding sites for acetyl-CoA as observed for RePC (Lietzan et al. 2011; St Maurice et al. 2007), while the symmetric tetramer display four CoA binding pockets in SaPC (Yu et al. 2009). Whatever the role of acetyl-CoA, it does not determine the symmetry of PC tetramers, although some still unclear relationship is suspected. Intriguingly, in the crystal structure of RePC mutant T882A (Lietzan et al. 2011) the subunits with bound acetyl-CoA do not have the BCCP-BC interaction, which happens only in the opposite face (Fig. 11.9b). If the allosteric ligand plays in the activation of the BC reaction, the atomic structure suggests that acetyl-CoA stimulates the catalytic activity in the opposite and unligated layer.

The binding site for acetyl-CoA at the interface between BC and AL domains has been described at atomic level for RePC (Lietzan et al. 2011; St Maurice et al. 2007) and SaPC (Yu et al. 2009), and the binding pocket is rich in evolutionary conserved Arg residues (Fig. 11.10). In SaPC the 3′-phosphate of CoA is surrounded by interactions with arginines at positions 367, 420 and 422 (SaPC numbering) from the BC domain (Fig. 11.10a). In RePC, however, despite the conservation of these amino acids (arginines 374, 427 and 429 in RePC numbering), only Arg427 is seen to bind

CoA, where the other two amino acids are far from interaction (Fig. 11.10b). Since SaPC is a symmetric tetramer, AL domains from different layers are interacting with each other, and the second AL domain (AL' in Fig. 11.10a) seems to pack the entire region resulting in a higher number of interactions between BC domain and acetyl-CoA in SaPC compared with RePC. This difference suggests that the structure and the nature of atomic interactions between CoA and PC might vary depending on the symmetric/asymmetric arrangement of PC tetramers. The rest of the interactions are very similar between SaPC and RePC, and another arginine, now from the AL domain, is seen to contact the 5′-α-phosphate of the CoA (Arg465 in SaPC and Arg472 in RePC). The importance of residues Arg427 and Arg472 in RePC were tested with mutants showing greatly increased activation constants (Adina-Zada et al. 2012a). Likewise, SaPC R420C mutant is much less sensitive to acetyl-CoA activation (Xiang and Tong 2008). Also, in both structures, acetyl-CoA is in interaction distance with the main chain carbonyl of one amino acid from the BC' domain, i.e., with the BC domain at the other layer of the tetramer (Ala47 in SaPC and Asp47 in RePC). Thus, acetyl-CoA binding site resides in the crossroad between the dimer of BC domains and the AL domain, pointing to some role modulating the communication between them.

There is a reported disease related mutant in PC at this allosteric binding site for acetyl-CoA. The R451C mutation found in a family with siblings affected with type A deficiency (Wexler et al. 1998) would correspond to arginines 420 (SaPC) and 427 (RePC) that are seen in direct contact with the ligand and which mutants show impaired activation (Adina-Zada et al. 2012a; Xiang and Tong 2008).

In one of the crystallographic structures reported for LmPC, four citrate molecules were seen bound in the acetyl-CoA binding sites (Table 11.1) (Sureka et al. 2014), probably due to the high concentration of the metabolite in the solution where crystals were grown. Although this TCA cycle intermediate could have some feed-back regulatory role for PC (Fig. 11.2), kinetic studies show only small effects on PC catalytic activity. A possible competition between citrate and acetyl-CoA in PC regulation has not been explored.

11.7 Quaternary Structure of PC in Solution

11.7.1 CryoEM of SaPC

The structure of PC tetramers has been also explored by cryoEM. All the publicly available 3D maps have been obtained for SaPC and at moderate resolution in the 8–13 Å range (Lasso et al. 2010, 2014; Yu et al. 2009). At this level of structural detail the overall quaternary structure and the relative position of domains is unveiled, but no direct insights regarding catalytic reactions at atomic level can be drawn. The cryoEM data, however, depict PC tetramers outside the constraints of 3D crystals, and might provide clues about the symmetric *vs* asymmetric organizations. The molecular weight of PC tetramers (around 500 kDa) is very well suited for cryoEM studies, and their single particles display good contrast and typical

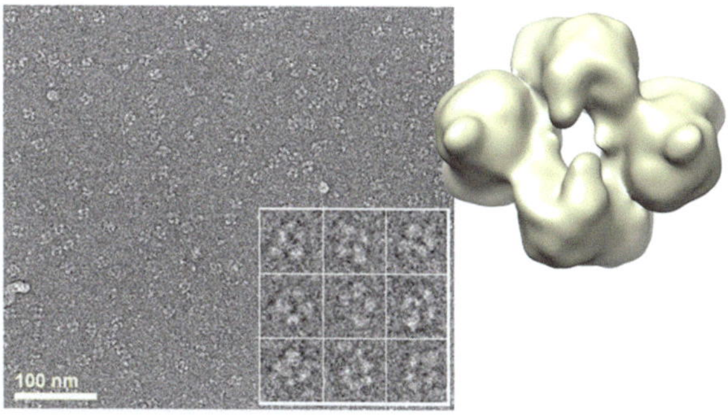

Fig. 11.11 SaPC tetramers seen by cryoEM. The figure shows the field of an electron micrograph with single particles of SaPC tetramers in a cryoEM experiment. The inset highlights some individual tetramers. At the *right side* of the figure, a cryoEM map (emd-1736) for SaPC (Lasso et al. 2010) is rendered *yellow*

organization in four lobes when observed in frontal views (Fig. 11.11). The close packed dimers of BC-BC and CT-CT domains in the context of the tetramer are apparent and singled out.

The first cryoEM analysis of SaPC tetramers revealed a symmetric organization in the presence of activator acetyl-CoA (Fig. 11.12) (Lasso et al. 2010; Yu et al. 2009). This is a clear correlation between the oligomeric architecture of tetramers shown in crystallographic experiments and the one uncovered for free particles. Therefore, the packing in 3D crystals does not induce the observed quaternary structure for SaPC. Although subunits at opposite layers are very similar and symmetric in the relative position for BC, AL and CT domains, in the cryoEM map for SaPC there is a local difference in the B-subdomain lid of BC domains between layers (Fig. 11.12b). Also, the BBCP domains are visible only in one of the layers and involved in BCCP-CT interactions (Fig. 11.12a), in similar position and orientation as described in the crystal structures.

Further cryoEM works on SaPC in the presence of different combination of substrates and cofactors brought similar results (Lasso et al. 2010). The assays included acetyl-CoA, the ATP analog AMP-PNP, $KHCO_3$, pyruvate, and even oxaloacetate. In all the cryoEM maps, the BBCP domains were visible in just one of the layers and engaged in interaction with the CT active sites. Essentially, the tested conditions did not revealed any BCCP-BC interaction. In the presence of oxaloacetate, PC performs the partial reverse reaction of oxaloacetate decarboxylation. In the cryoEM sample of SaPC in the presence of oxaloacetate the BBCP domains were seen liberated from the binding with CT active sites, but far from a close interaction with the BC domains (Lasso et al. 2010). Thus, the attempts to characterize by cryoEM the conformational landscape in PC functioning did not reveal any large structural change of the oligomer (Lasso et al. 2010; Yu et al. 2009). These works tried to halt SaPC tetramers in distinct steps of its enzymatic reaction in a way that all the molecules were synchronized. However, none of the 3D maps displayed the BCCP domain close to the BC active site, the first reaction of the sequence for pyruvate carboxylation.

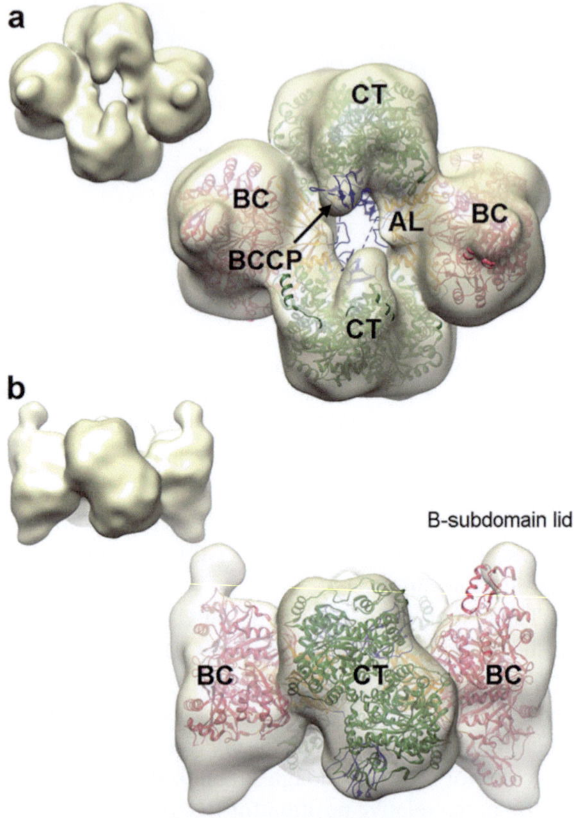

Fig. 11.12 Symmetric SaPC tetramers in the presence of acetyl-CoA. (**a**) (*top*) and (**b**) (*side*) views of the 3D density map for SaPC tetramer in the presence of acetyl-CoA (emd-1736) (Lasso et al. 2010). The map is depicted semi-transparent and the fitted coordinates for SaPC crystal structure (3BG5) (Xiang and Tong 2008) are seen in ribbons

11.7.2 Observing SaPC by cryoEM During Catalysis

SaPC tetramers were observed by cryoEM in the presence of all the substrates and cofactors required for its catalytic activity (Lasso et al. 2014). The sample contained $KHCO_3$, ATP, pyruvate, $MgCl_2$, and the activator acetyl-CoA, and the production of oxaloacetate was monitored. In the tested conditions, PC tetramers were active for few minutes, and their structure was explored before consumption of the substrates. The experiment was not in time-resolved mode, and individual tetramers were expected to perform several catalytic cycles. The analysis by cryoEM was conducted using unsupervised classification techniques (Scheres 2012, 2010). Essentially, the technique allows to separate homogenous subsets from an heterogeneous population of single particles without imposing any constrain in the nature of the structural variation. Thus, the study of freely working SaPC was intended to fish several conformational states within a single sample of active enzyme. The results revealed two dramatically different structures for SaPC tetramers (Fig. 11.13), and the flexible fitting of atomic coordinates located PC domains and described their relative arrangement. Essentially, the work showed that SaPC tetramers display

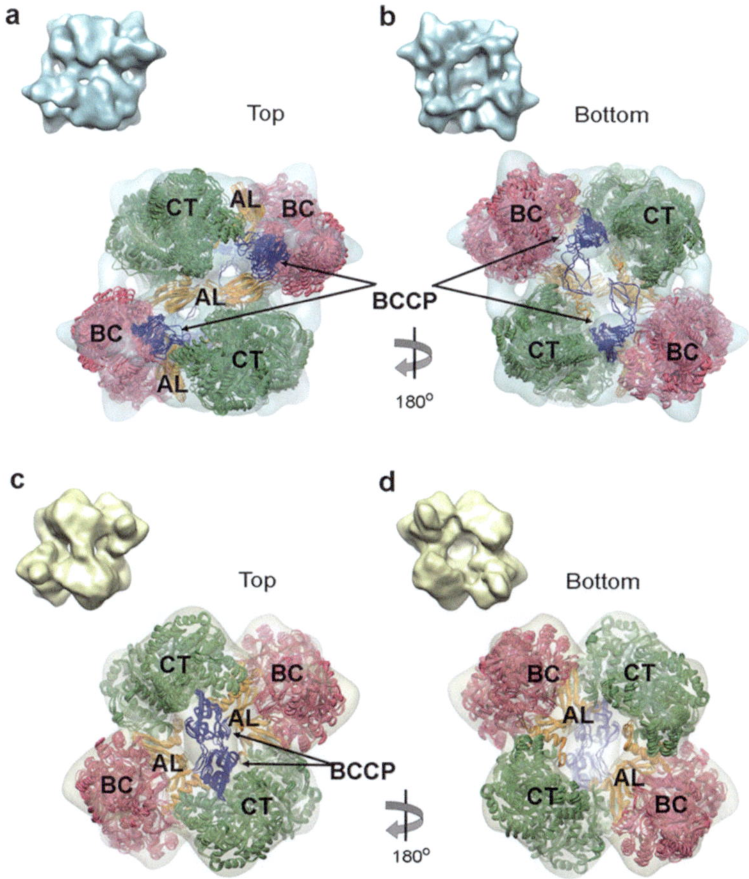

Fig. 11.13 Symmetric and asymmetric SaPC tetramers during catalysis. CryoEM maps for SaPC calculated for tetramers engaged in catalytic activity and after classification. The 3D maps are seen semi-transparent together with ensembles of calculated atomic models. The *blue* map in panels (**a**) and (**b**) shows an asymmetric architecture, and the atomic models reveal BC-BCCP interactions at the *top layer* (**a**), and BCCP domains residing in exo site at the *bottom layer* (**b**). The map depicted in panels (**c**) and (**d**), follows a symmetric organization, with BCCP domains engaged in interaction with CT sites at the *top layer* (**c**) but without density for BCCP regions in the *bottom layer* (**d**). Thumbnails of the cryoEM maps depicted in solid mode are also included in each panel (Reproduced with permission from Lasso et al. (2014) with modifications)

both, symmetric and asymmetric quaternary structures, similar to the ones seen in crystal structures for PC form different organisms, but in this case for the same PC enzyme, SaPC. Furthermore, the position of BCCP domains was clearly linked to the architecture of the tetramer. This way, BCCP domains in one layer of the asymmetric SaPC were seen bound to the BC domains (Fig. 11.13a), while the symmetric SaPC had BCCP domains engaged in interactions with the CT domain in one of the layers (Fig. 11.13c).

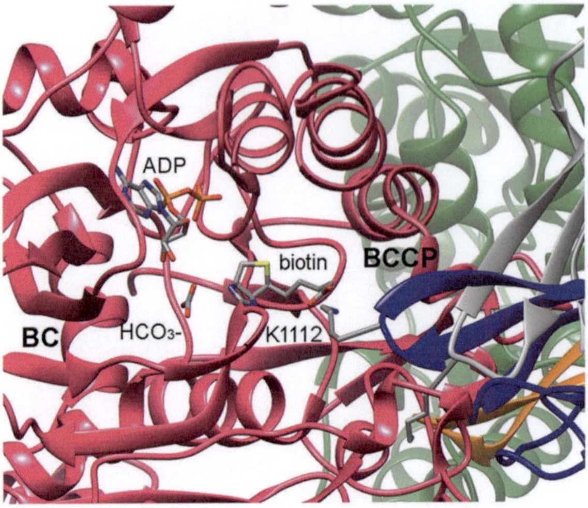

Fig. 11.14 Interaction of BCCP domain with BC active site observed by cryoEM. The figure portrays one atomic model calculated for the asymmetric SaPC tetramer (as shown in Fig. 11.13a) derived from the cryoEM analysis of SaPC engaged in catalysis (Lasso et al. 2014). The position of the biotin, ADP and bicarbonate are taken from the *E. coli* BC subunit from ACC (3G8C) (Chou et al. 2009) shown in Fig. 11.6b, and after alignment of the BC active site between the atomic structures. The BCCP domain depicted *grey* belongs to the RePC crystal structure that shows BCCP-BC interaction (3TW6) (Lietzan et al. 2011)

The cryoEM results clearly have two main implications: first, that asymmetric and symmetric configurations seem linked to the location of BCCP domains during PC function; and that BCCP domains are seen bound to catalytic domains only in one of the faces, what supports a half-site reactivity mode for PC tetramers, where only one of the layers at a time is engaged in enzymatic reactions. Regarding the relationship between the symmetry of PC tetramers and the sequential reactions in PC, it seems that the BC reaction needs an asymmetric tetramer, and that the CT step occurs in the symmetric one. This observation in SaPC by cryoEM, could have been also anticipated by the analysis of the several PC crystal structures (Table 11.1), where BCCP domain has only been observed bound to the BC domain in an asymmetric tetramer (Lietzan et al. 2011), and to the CT site within a symmetric one (Yu et al. 2009, 2013; Xiang and Tong 2008).

Interestingly, the position and orientation of the BCCP domain bound to the BC site revealed in the cryoEM assay of SaPC (Lasso et al. 2014) differs from the one described in the crystal structure for RePC (Lietzan et al. 2011), and it is compatible with the location of the free biotin within the *E. coli* BC subunit from ACC (Fig. 11.14) (Chou et al. 2009). Thus, despite the moderate resolution of the data, the configuration of the BCCP-BC interaction in the cryoEM map for SaPC is fully compatible with a catalytically competent interaction.

11.8 A Model for PC Enzyme Functioning

In the current understanding of PC structure, there are two architectures, the symmetric and asymmetric, which have been observed in crystal structures for different organisms. A putative link between both configurations and the sequential catalytic reactions performed by PC is very suggestive (Fig. 11.15). Thus, in a full reaction, PC tetramers start in the asymmetric state with BCCP domains bound to the BC site (Fig. 11.15a). Next, a transition to a symmetric configuration allows for the binding of BCCP domains to the CT active site (Fig. 11.15b) to fulfill the enzymatic reaction in one of the layers. The symmetric-asymmetric transition requires relative movements and rotation between domains. Noteworthy, the LmPC structure bound to inhibitor c-di-AMP shows a highly symmetric tetramer where the ligand binds to the CT-CT interface playing as a molecular staple that locks the enzyme in a single conformation (Sureka et al. 2014), which is not compatible with catalysis. In the putative structural transformations the AL domain might play a central role since it

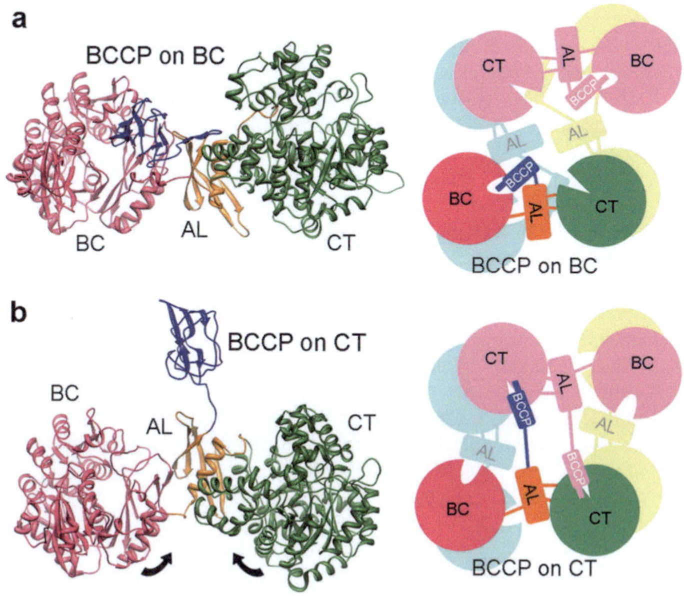

Fig. 11.15 Model for PC functioning during catalytic activity. Panels show in the *left side ribbons* representations for distinct atomic models for SaPC subunits based in cryoEM data (Lasso et al. 2014). The models for the subunits are taken from: (**a**) *top layer* of emd-5945; (**b**) *top layer* of emd-5944. The BCCP domains establish their interactions within their layer, and their location is indicated. Structural transitions related to the relative rotation between CT and BC domains within the subunits are indicated by *arrows*. In the *right side* of the panels, the arrangement of SaPC tetramers is depicted in *cartoons*. The architecture of the tetramers has been observed in: emd-5945 (**a**); emd-5944 (**b**) (Reproduced with permission from Lasso et al. (2014) with modifications)

might integrate the relative changes between BC and CT regions and is the anchor point for the linker of the BCCP domain. Also, the AL domain participates in the binding site for the allosteric activator acetyl-CoA. Unraveling the exact role of the activator, the long range communication between active sites, the local changes that trigger the translocation of BCCP domains, and the overall transition between symmetric and asymmetric architecture, will require higher resolution structures for PC tetramers in solution and engaged in catalysis.

Acknowledgements This work was supported by grant BFU2012-34873 from the Spanish Ministry of Economy and Competitiveness, and RGP0062/2016 from the HFSP.

References

Adina-Zada A, Jitrapakdee S, Surinya KH, McIldowie MJ, Piggott MJ, Cleland WW, Wallace JC, Attwood PV (2008) Insights into the mechanism and regulation of pyruvate carboxylase by characterisation of a biotin-deficient mutant of the Bacillus thermodenitrificans enzyme. Int J Biochem Cell Biol 40(9):1743–1752. doi:10.1016/j.biocel.2008.01.001

Adina-Zada A, Sereeruk C, Jitrapakdee S, Zeczycki TN, St Maurice M, Cleland WW, Wallace JC, Attwood PV (2012a) Roles of Arg427 and Arg472 in the binding and allosteric effects of acetyl CoA in pyruvate carboxylase. Biochemistry 51(41):8208–8217

Adina-Zada A, Zeczycki TN, Attwood PV (2012b) Regulation of the structure and activity of pyruvate carboxylase by acetyl CoA. Arch Biochem Biophys 519(2):118–130. doi:10.1016/j.abb.2011.11.015

Adina-Zada A, Zeczycki TN, St Maurice M, Jitrapakdee S, Cleland WW, Attwood PV (2012c) Allosteric regulation of the biotin-dependent enzyme pyruvate carboxylase by acetyl-CoA. Biochem Soc Trans 40(3):567–572. doi:10.1042/BST20120041

Ashman LK, Keech DB (1975) Sheep kidney pyruvate carboxylase. Studies on the coupling of adenosine triphosphate hydrolysis and CO2 fixation. J Biol Chem 250(1):14–21

Attwood PV, Graneri BD (1992) Bicarbonate-dependent ATP cleavage catalysed by pyruvate carboxylase in the absence of pyruvate. Biochem J 287(Pt 3):1011–1017

Attwood PV, Wallace JC (2002) Chemical and catalytic mechanisms of carboxyl transfer reactions in biotin-dependent enzymes. Acc Chem Res 35(2):113–120

Bizeau ME, Short C, Thresher JS, Commerford SR, Willis WT, Pagliassotti MJ (2001) Increased pyruvate flux capacities account for diet-induced increases in gluconeogenesis in vitro. Am J Phys Regul Integr Comp Phys 281(2):R427–R433

Cao Z, Zhou Y, Zhu S, Feng J, Chen X, Liu S, Peng N, Yang X, Xu G, Zhu Y (2016) Pyruvate Carboxylase Activates the RIG-I-like Receptor-Mediated Antiviral Immune Response by Targeting the MAVS signalosome. Sci Report 6:22002. doi:10.1038/srep22002

Carbone MA, MacKay N, Ling M, Cole DE, Douglas C, Rigat B, Feigenbaum A, Clarke JT, Haworth JC, Greenberg CR, Seargeant L, Robinson BH (1998) Amerindian pyruvate carboxylase deficiency is associated with two distinct missense mutations. Am J Hum Genet 62(6):1312–1319

Cardaci S, Zheng L, MacKay G, van den Broek NJ, MacKenzie ED, Nixon C, Stevenson D, Tumanov S, Bulusu V, Kamphorst JJ, Vazquez A, Fleming S, Schiavi F, Kalna G, Blyth K, Strathdee D, Gottlieb E (2015) Pyruvate carboxylation enables growth of SDH-deficient cells by supporting aspartate biosynthesis. Nat Cell Biol 17(10):1317–1326. doi:10.1038/ncb3233

Chou CY, Yu LP, Tong L (2009) Crystal structure of biotin carboxylase in complex with substrates and implications for its catalytic mechanism. J Biol Chem 284(17):11690–11697. doi:10.1074/jbc.M805783200 de Queiroz MS, Waldrop GL (2007) Modeling and numerical simulation of biotin carboxylase kinetics: implications for half-sites reactivity. J Theor Biol 246(1):167–175

DeVivo DC, Haymond MW, Leckie MP, Bussman YL, McDougal DB Jr, Pagliara AS (1977) The clinical and biochemical implications of pyruvate carboxylase deficiency. J Clin Endocrinol Metab 45(6):1281–1296. doi:10.1210/jcem-45-6-1281

Duangpan S, Jitrapakdee S, Adina-Zada A, Byrne L, Zeczycki TN, St Maurice M, Cleland WW, Wallace JC, Attwood PV (2010) Probing the catalytic roles of Arg548 and Gln552 in the carboxyl transferase domain of the Rhizobium etli pyruvate carboxylase by site-directed mutagenesis. Biochemistry 49(15):3296–3304

Dugal BS (1973) Apparent co-operative effect of acetyl-CoA on pigeon liver pyruvate carboxylase. FEBS Lett 30(2):181–184

Easterbrook-Smith SB, Wallace JC, Keech DB (1978) A reappraisal of the reaction pathway of pyruvate carboxylase. Biochem J 169(1):225–228

Easterbrook-Smith SB, Campbell AJ, Keech DB, Wallace JC (1979) The atypical velocity response by pyruvate carboxylase to increasing concentrations of acetyl-coenzyme A. Biochem J 179(3):497–502

Fan C, Chou CY, Tong L, Xiang S (2012) Crystal structure of urea carboxylase provides insights into the carboxyltransfer reaction. J Biol Chem 287(12):9389–9398. doi:10.1074/jbc.M111.319475

Fawaz MV, Topper ME, Firestine SM (2011) The ATP-grasp enzymes. Bioorg Chem 39(5–6):185–191. doi:10.1016/j.bioorg.2011.08.004

Freedman AD, Kohn L (1964) Pyruvate metabolism and control: factors affecting pyruvic carboxylase activity. Science (New York, NY) 145(3627):58–60

Galperin MY, Koonin EV (1997) A diverse superfamily of enzymes with ATP-dependent carboxylate-amine/thiol ligase activity. Protein Sci 6(12):2639–2643. doi:10.1002/pro.5560061218

Gamberino WC, Berkich DA, Lynch CJ, Xu B, LaNoue KF (1997) Role of pyruvate carboxylase in facilitation of synthesis of glutamate and glutamine in cultured astrocytes. J Neurochem 69(6):2312–2325

Garcia-Cazorla A, Rabier D, Touati G, Chadefaux-Vekemans B, Marsac C, de Lonlay P, Saudubray JM (2006) Pyruvate carboxylase deficiency: metabolic characteristics and new neurological aspects. Ann Neurol 59(1):121–127. doi:10.1002/ana.20709

Goodall GJ, Baldwin GS, Wallace JC, Keech DB (1981) Factors that influence the translocation of the N-carboxybiotin moiety between the two sub-sites of pyruvate carboxylase. Biochem J 199(3):603–609

Goss JA, Cohen ND, Utter MF (1981) Characterization of the subunit structure of pyruvate carboxylase from Pseudomonas citronellolis. J Biol Chem 256(22):11819–11825

Gray LR, Tompkins SC, Taylor EB (2014) Regulation of pyruvate metabolism and human disease. Cell Mol Life Sci 71(14):2577–2604. doi:10.1007/s00018-013-1539-2

Huang CS, Sadre-Bazzaz K, Shen Y, Deng B, Zhou ZH, Tong L (2010) Crystal structure of the alpha(6)beta(6) holoenzyme of propionyl-coenzyme A carboxylase. Nature 466(7309):1001–1005. doi:10.1038/nature09302

Huang CS, Ge P, Zhou ZH, Tong L (2012) An unanticipated architecture of the 750-kDa alpha6beta6 holoenzyme of 3-methylcrotonyl-CoA carboxylase. Nature 481(7380):219–223. doi:10.1038/nature10691

Janiyani K, Bordelon T, Waldrop GL, Cronan JE Jr (2001) Function of *Escherichia coli* biotin carboxylase requires catalytic activity of both subunits of the homodimer. J Biol Chem 276(32):29864–29870. doi:10.1074/jbc.M104102200

Jitrapakdee S, St Maurice M, Rayment I, Cleland WW, Wallace JC, Attwood PV (2008) Structure, mechanism and regulation of pyruvate carboxylase. Biochem J 413(3):369–387

Jurado AR, Huang CS, Zhang X, Zhou ZH, Tong L (2015) Structure and substrate selectivity of the 750-kDa alpha6beta6 holoenzyme of geranyl-CoA carboxylase. Nat Commun 6:8986. doi:10.1038/ncomms9986

Kaziro Y, Hass LF, Boyer PD, Ochoa S (1962) Mechanism of the propionyl carboxylase reaction. II. Isotopic exchange and tracer experiments. J Biol Chem 237:1460–1468

Keech B, Barritt GJ (1967) Allosteric activation of sheep kidney pyruvate carboxylase by the magnesium ion (Mg2+) and the magnesium adenosine triphosphate ion (MgATP2-). J Biol Chem 242(9):1983–1987

Knowles JR (1989) The mechanism of biotin-dependent enzymes. Annu Rev Biochem 58:195–221

Kondo S, Nakajima Y, Sugio S, Yong-Biao J, Sueda S, Kondo H (2004) Structure of the biotin carboxylase subunit of pyruvate carboxylase from Aquifex aeolicus at 2.2 A resolution. Acta Crystallogr D Biol Crystallogr 60(Pt 3):486–492. doi:10.1107/S0907444904000423

Kondo S, Nakajima Y, Sugio S, Sueda S, Islam MN, Kondo H (2007) Structure of the biotin carboxylase domain of pyruvate carboxylase from Bacillus thermodenitrificans. Acta Crystallogr D Biol Crystallogr 63(Pt 8):885–890. doi:10.1107/S0907444907029423

Lai H, Kraszewski JL, Purwantini E, Mukhopadhyay B (2006) Identification of pyruvate carboxylase genes in Pseudomonas aeruginosa PAO1 and development of a P. aeruginosa-based overexpression system for alpha4- and alpha4beta4-type pyruvate carboxylases. Appl Environ Microbiol 72(12):7785–7792. doi:10.1128/AEM.01564-06

Lasso G, Yu LP, Gil D, Xiang S, Tong L, Valle M (2010) Cryo-EM analysis reveals new insights into the mechanism of action of pyruvate carboxylase. Structure 18(10):1300–1310

Lasso G, Yu LP, Gil D, Lazaro M, Tong L, Valle M (2014) Functional conformations for pyruvate carboxylase during catalysis explored by cryoelectron microscopy. Structure 22(6):911–922. doi:10.1016/j.str.2014.04.011

Legge GB, Branson JP, Attwood PV (1996) Effects of acetyl CoA on the pre-steady-state kinetics of the biotin carboxylation reaction of pyruvate carboxylase. Biochemistry 35(12):3849–3856. doi:10.1021/bi952797q

Lietzan AD, Menefee AL, Zeczycki TN, Kumar S, Attwood PV, Wallace JC, Cleland WW, St Maurice M (2011) Interaction between the biotin carboxyl carrier domain and the biotin carboxylase domain in pyruvate carboxylase from Rhizobium etli. Biochemistry 50(45):9708–9723

Lietzan AD, St Maurice M (2013a) Insights into the carboxyltransferase reaction of pyruvate carboxylase from the structures of bound product and intermediate analogs. Biochem Biophys Res Commun 441(2):377–382. doi:10.1016/j.bbrc.2013.10.066

Lietzan AD, St Maurice M (2013b) A substrate-induced biotin binding pocket in the carboxyl transferase domain of pyruvate carboxylase. J Biol Chem 288(27):19915–19925

Lietzan AD, Lin Y, St Maurice M (2014) The role of biotin and oxamate in the carboxyltransferase reaction of pyruvate carboxylase. Arch Biochem Biophys 562:70–79. doi:10.1016/j.abb.2014.08.008

Lu D, Mulder H, Zhao P, Burgess SC, Jensen MV, Kamzolova S, Newgard CB, Sherry AD (2002) 13C NMR isotopomer analysis reveals a connection between pyruvate cycling and glucose-stimulated insulin secretion (GSIS). Proc Natl Acad Sci U S A 99(5):2708–2713. doi:10.1073/pnas.052005699

Marin-Valencia I, Roe CR, Pascual JM (2010) Pyruvate carboxylase deficiency: mechanisms, mimics and anaplerosis. Mol Genet Metab 101(1):9–17

Menefee AL, Zeczycki TN (2014) Nearly 50 years in the making: defining the catalytic mechanism of the multifunctional enzyme, pyruvate carboxylase. FEBS J 281(5):1333–1354. doi:10.1111/febs.12713

Mochalkin I, Miller JR, Evdokimov A, Lightle S, Yan C, Stover CK, Waldrop GL (2008) Structural evidence for substrate-induced synergism and half-sites reactivity in biotin carboxylase. Protein Sci 17(10):1706–1718

Monnot S, Serre V, Chadefaux-Vekemans B, Aupetit J, Romano S, De Lonlay P, Rival JM, Munnich A, Steffann J, Bonnefont JP (2009) Structural insights on pathogenic effects of novel mutations causing pyruvate carboxylase deficiency. Hum Mutat 30(5):734–740. doi:10.1002/humu.20908

Mukhopadhyay B, Stoddard SF, Wolfe RS (1998) Purification, regulation, and molecular and biochemical characterization of pyruvate carboxylase from Methanobacterium thermoautotrophicum strain deltaH. J Biol Chem 273(9):5155–5166

Mukhopadhyay B, Patel VJ, Wolfe RS (2000) A stable archaeal pyruvate carboxylase from the hyperthermophile Methanococcus jannaschii. Arch Microbiol 174(6):406–414

Mukhopadhyay B, Purwantini E, Kreder CL, Wolfe RS (2001) Oxaloacetate synthesis in the methanarchaeon Methanosarcina barkeri: pyruvate carboxylase genes and a putative *Escherichia coli*-type bifunctional biotin protein ligase gene (bpl/birA) exhibit a unique organization. J Bacteriol 183(12):3804–3810. doi:10.1128/JB.183.12.3804-3810.2001

Ogita T, Knowles JR (1988) On the intermediacy of carboxyphosphate in biotin-dependent carboxylations. Biochemistry 27(21):8028–8033

Ostergaard E, Duno M, Moller LB, Kalkanoglu-Sivri HS, Dursun A, Aliefendioglu D, Leth H, Dahl M, Christensen E, Wibrand F (2013) Novel Mutations in the PC Gene in Patients with Type B Pyruvate Carboxylase Deficiency. JIMD Rep 9:1–5. doi:10.1007/8904_2012_173

Phannasil P, Thuwajit C, Warnnissorn M, Wallace JC, MacDonald MJ, Jitrapakdee S (2015) Pyruvate carboxylase is up-regulated in breast cancer and essential to support growth and invasion of MDA-MB-231 cells. PLoS One 10(6):e0129848. doi:10.1371/journal.pone.0129848

Robinson BH (2006) Lactic acidemia and mitochondrial disease. Mol Genet Metab 89(1–2):3–13. doi:10.1016/j.ymgme.2006.05.015

Salto R, Sola M, Oliver FJ, Vargas AM (1996) Effects of starvation, diabetes and carbon tetrachloride intoxication on rat kidney cortex and liver pyruvate carboxylase levels. Arch Physiol Biochem 104(7):845–850. doi:10.1076/apab.104.7.845.13111

Saudubray JM, Marsac C, Cathelineau CL, Besson Leaud M, Leroux JP (1976) Neonatal congenital lactic acidosis with pyruvate carboxylase deficiency in two siblings. Acta Paediatr Scand 65(6):717–724

Scheres SH (2010) Classification of structural heterogeneity by maximum-likelihood methods. Methods Enzymol 482:295–320

Scheres SH (2012) RELION: implementation of a Bayesian approach to cryo-EM structure determination. J Struct Biol 180(3):519–530

Scrutton MC, Utter MF (1967) Pyruvate carboxylase IX. Some properties of the activation by certain acyl derivatives of coenzyme A. J Biol Chem 242(8):1723–1735

Scrutton MC, Keech DB, Utter MF (1965) Pyruvate carboxylase. IV. Partial reactions and the locus of activation by acetyl coenzyme A. J Biol Chem 240:574–581

Sellers K, Fox MP, Bousamra M 2nd, Slone SP, Higashi RM, Miller DM, Wang Y, Yan J, Yuneva MO, Deshpande R, Lane AN, Fan TW (2015) Pyruvate carboxylase is critical for non-small-cell lung cancer proliferation. J Clin Invest 125(2):687–698. doi:10.1172/JCI72873

Shen Y, Chou CY, Chang GG, Tong L (2006) Is dimerization required for the catalytic activity of bacterial biotin carboxylase? Mol Cell 22(6):807–818

St Maurice M, Reinhardt L, Surinya KH, Attwood PV, Wallace JC, Cleland WW, Rayment I (2007) Domain architecture of pyruvate carboxylase, a biotin-dependent multifunctional enzyme. Science (New York, NY) 317(5841):1076–1079

Sureka K, Choi PH, Precit M, Delince M, Pensinger DA, Huynh TN, Jurado AR, Goo YA, Sadilek M, Iavarone AT, Sauer JD, Tong L, Woodward JJ (2014) The cyclic dinucleotide c-di-AMP is an allosteric regulator of metabolic enzyme function. Cell 158(6):1389–1401. doi:10.1016/j.cell.2014.07.046

Thoden JB, Blanchard CZ, Holden HM, Waldrop GL (2000) Movement of the biotin carboxylase B-domain as a result of ATP binding. J Biol Chem 275(21):16183–16190

Tipton PA, Cleland WW (1988) Catalytic mechanism of biotin carboxylase: steady-state kinetic investigations. Biochemistry 27(12):4317–4325

Tong L (2013) Structure and function of biotin-dependent carboxylases. Cell Mol Life Sci 70(5):863–891

Tran TH, Hsiao YS, Jo J, Chou CY, Dietrich LE, Walz T, Tong L (2015) Structure and function of a single-chain, multi-domain long-chain acyl-CoA carboxylase. Nature 518(7537):120–124. doi:10.1038/nature13912

Utter MF, Keech DB (1960) Formation of oxaloacetate from pyruvate and carbon dioxide. J Biol Chem 235:PC17–PC18

Waldrop GL, Rayment I, Holden HM (1994) Three-dimensional structure of the biotin carboxylase subunit of acetyl-CoA carboxylase. Biochemistry 33(34):10249–10256

Waldrop GL, Holden HM, St Maurice M (2013) The enzymes of biotin dependent CO(2) metabolism: what structures reveal about their reaction mechanisms. Protein Sci 21(11):1597–1619

Wang D, De Vivo D (1993) Pyruvate Carboxylase Deficiency. In: Pagon RA, Adam MP, Ardinger HH et al. (eds) GeneReviews(R). Seattle (WA)

Warburg O (1956) On the origin of cancer cells. Science (New York, NY) 123(3191):309–314

Warren GB, Tipton KF (1974) The role of acetyl-CoA in the reaction pathway of pig-liver pyruvate carboxylase. Eur J Biochem 47(3):549–554

Wei J, Tong L (2015) Crystal structure of the 500-kDa yeast acetyl-CoA carboxylase holoenzyme dimer. Nature 526(7575):723–727. doi:10.1038/nature15375

Wexler ID, Kerr DS, Du Y, Kaung MM, Stephenson W, Lusk MM, Wappner RS, Higgins JJ (1998) Molecular characterization of pyruvate carboxylase deficiency in two consanguineous families. Pediatr Res 43(5):579–584. doi:10.1203/00006450-199805000-00004

Xiang S, Tong L (2008) Crystal structures of human and *Staphylococcus aureus* pyruvate carboxylase and molecular insights into the carboxyltransfer reaction. Nat Struct Mol Biol 15(3):295–302

Yu LP, Xiang S, Lasso G, Gil D, Valle M, Tong L (2009) A symmetrical tetramer for *S. Aureus* pyruvate carboxylase in complex with coenzyme A. Structure 17(6):823–832

Yu LP, Chou CY, Choi PH, Tong L (2013) Characterizing the importance of the biotin carboxylase domain dimer for *Staphylococcus aureus* pyruvate carboxylase catalysis. Biochemistry 52(3):488–496

Zeczycki TN, St Maurice M, Jitrapakdee S, Wallace JC, Attwood PV, Cleland WW (2009) Insight into the carboxyl transferase domain mechanism of pyruvate carboxylase from Rhizobium etli. Biochemistry 48(20):4305–4313

Zeczycki TN, Maurice MS, Attwood PV (2010) Inhibitors of Pyruvate Carboxylase. Open Enzym Inhib J 3:8–26. doi:10.2174/1874940201003010008

Zeczycki TN, Menefee AL, Adina-Zada A, Jitrapakdee S, Surinya KH, Wallace JC, Attwood PV, St Maurice M, Cleland WW (2011a) Novel insights into the biotin carboxylase domain reactions of pyruvate carboxylase from Rhizobium etli. Biochemistry 50(45):9724–9737

Zeczycki TN, Menefee AL, Jitrapakdee S, Wallace JC, Attwood PV, St Maurice M, Cleland WW (2011b) Activation and inhibition of pyruvate carboxylase from Rhizobium etli. Biochemistry 50(45):9694–9707

Chapter 12
Cullin-RING E3 Ubiquitin Ligases: Bridges to Destruction

Henry C. Nguyen, Wei Wang, and Yong Xiong

Abstract Ubiquitination is a highly conserved post-translational modification in eukaryotes, well known for targeting proteins for degradation by the 26S proteasome. Proteins destined for proteasomal degradation are selected by E3 ubiquitin ligases. Cullin-RING E3 ubiquitin ligases (CRLs) are the largest superfamily of E3 ubiquitin ligases, with over 400 members known in mammals. These modular complexes are tightly regulated in the cell. In this chapter, we highlight recent structural and biochemical advances shedding light on the assembly and architecture of cullin-RING ligases, their dynamic regulation by a variety of host factors, and their manipulation by viral pathogens and small molecules.

Keywords Ubiquitin • Ubiquitination • Ubiquitylation • E3 ligase • Cullin • Cullin-RING ligase • NEDD8 • Neddylation • Protein-protein interaction • Viral hijacking

12.1 Introduction: Overview and Function of CRLs

Ubiquitin (Ub) is a small, 8.5 kDa protein that is conserved across all eukaryotes and that regulates important cellular processes (Hershko and Ciechanover 1998). Indeed ubiquitous, it is expressed at very high concentrations *in vivo*, at approximately 85 μM in human embryonic kidney (HEK) cells for example (Kaiser et al. 2011). The covalent attachment of Ub to a target protein, termed ubiquitylation or ubiquitination, has a variety of effects on the regulation and fate of the target

H.C. Nguyen
Department of Molecular Biophysics and Biochemistry, Yale University, New Haven, CT 06511, USA

Department of Biochemistry and Biophysics, University of California, San Francisco, CA 94158, USA
e-mail: Henry.Nguyen@ucsf.edu

W. Wang • Y. Xiong (✉)
Department of Molecular Biophysics and Biochemistry, Yale University, New Haven, CT 06511, USA
e-mail: w.wang@yale.edu; yong.xiong@yale.edu

J.R. Harris, J. Marles-Wright (eds.), *Macromolecular Protein Complexes*, Subcellular Biochemistry 83, DOI 10.1007/978-3-319-46503-6_12

protein. Ub-mediated protein degradation by the 26S proteasome was discovered in the late 1970s and early 1980s. This work led to the receiving of the 2004 Nobel Prize in Chemistry by Aaron Ciechanover, Avram Hershko, and Irwin Rose (Ciechanover et al. 2004). Besides its role in proteasomal degradation, ubiquitination has been shown to mediate many other processes, such as protein-protein interactions, subcellular localization, and DNA repair. This process requires a cascade of three enzymes: E1 ubiquitin activating enzymes, E2 ubiquitin conjugating enzymes, and E3 ubiquitin ligases (reviewed in Pickart 2001). E1 activating enzymes transfer Ub in an ATP-dependent manner to the E2 conjugating enzyme. The E3 Ub ligase bridges the E2 and the target protein (the substrate) to facilitate the transfer of Ub to a lysine or the N-terminus of the target.

E3 ligases are responsible for selecting substrates to be ubiquitinated. While humans only have two E1 activating enzymes, UBA1 and UBA6, and ~30 different E2 conjugating enzymes, there are over 600 different E3 ligases (van Wijk and Timmers 2010). E3 ligases are divided into three main families: the homologous to E6AP carboxyl terminus (HECT), the really interesting new gene and U box (RING and U box), and the RING-between-RING (RBR) families (Berndsen and Wolberger 2014). The HECT and RBR families catalyze ubiquitination in two steps. First, Ub is catalytically transferred from the E2 to the E3, and then from the E3 to the substrate. In contrast, the RING family mediates direct transfer of ubiquitin from the E2 to the substrate, bypassing an E3-Ub intermediate. While not catalyzing a reaction, the RING family is still considered an enzyme (Bulatov and Ciulli 2015). The majority of E3 ligases belong to the RING family (Deshaies and Joazeiro 2009). Within that family, cullin-RING E3 ligases (CRLs) are the largest superfamily, with over 200 members and are responsible for ~20% of all ubiquitination in cells (Soucy et al. 2009; Petroski and Deshaies 2005). The core CRL consists of four components: a cullin protein that serves to scaffold the CRL, a RING finger protein that binds to an E2 ubiquitin conjugating enzyme, a substrate receptor that recognize the target protein, and adaptor proteins that bridge the substrate receptor to the cullin (Fig. 12.1a).

Mammals express seven canonical cullin proteins (Cul1, Cul2, Cul3, Cul4A, Cul4B, Cul5, and Cul7) that form modular, multisubunit CRLs, designated CRL1–7. Cullins are named because they "cull" (originally meaning to selectively slaughter a herd of wild animals) proteins for proteasomal degradation (Sarikas et al. 2011). Bioinformatic analysis suggests that the canonical eukaryotic cullins arose from three ancestral genes (Culα, Culβ, Culγ) where Cul1, Cul2, Cul5, and Cul7 descend from Culα, Cul3 from Culβ, and Cul4A/Cul4B from Culγ (Marin 2009). Cul4A and Cul4B mainly differ by an additional 149 amino acids at the Cul4B N-terminus, a portion of which harbors a nuclear localization sequence (Zou et al. 2009). Cul7 is an atypical cullin as it is much larger than the average cullin (~1700 amino acids compared to ~760 amino acids). Mammals also encode two other proteins that contain cullin homology domains, Cul9/PARkin-like cytoplasmic protein (PARC), and subunit 2 of the anaphase promoting complex/cyclosome (APC2), but these two proteins are overall highly divergent from canonical cullins as they contain additional domains (Skaar et al. 2007; Zachariae et al. 1998).

Within a CRL, the cullin protein serves as the central scaffold of the E3 Ub ligase (Fig. 12.1a). The C-terminal domain of a cullin anchors the RING protein, either

A

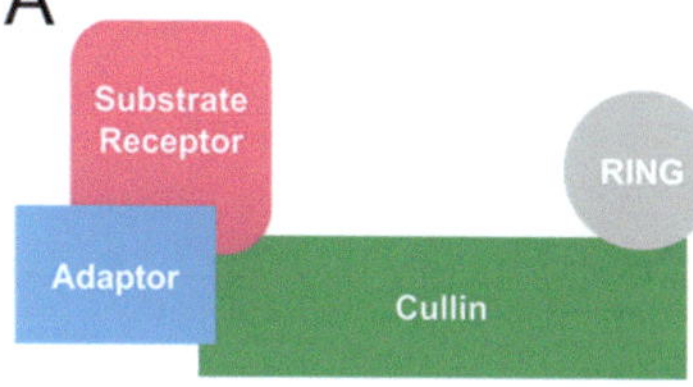

B

Cullin	Adaptor	Substrate Receptor	RING protein
Cul1	Skp1	F-box proteins	Rbx1
Cul2	EloBC	VHL box protein	Rbx1
Cul3	BTB proteins	BTB proteins	Rbx1
Cul4A/B	DDB1	DCAF proteins	Rbx1
Cul5	EloBC	SOCS box proteins	Rbx2
Cul7	Skp1	Fbw8	Rbx1

C D

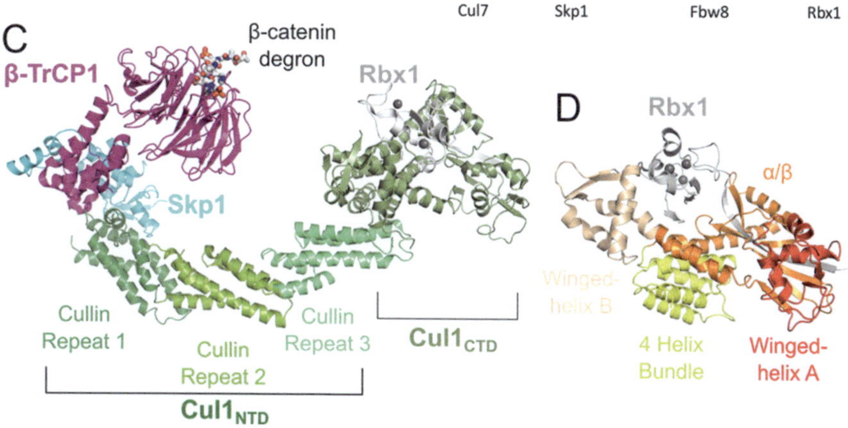

Fig. 12.1 Architecture of cullin-RING ligases. (**a**) Schematic representation of the components of a CRL. (**b**) Specific protein factors for each CRL family. (**c**) Structural architecture of a CRL1 in ribbon representation generated from PDB IDs 1LDK and 1P22. (**d**) Organization of Cul1 CTD–Rbx1. The domains of Cul1 are labeled in (**c**) and (**d**). The Zn^{2+} ions in (**c**) and (**d**) are shown as *gray spheres*

Rbx1 or Rbx2, which recruit the Ub-charged E2 enzymes. At the opposite end of the cullin, the N-terminal domain interacts with the adaptor component of the CRL. Adaptors share one of two common folds that are used to interact with cullins. Skp1, Elongin C, and Bric-a-brac, Tramtrack and Broad Complex (BTB) proteins share a common α-helical BTB fold to interact with their CRL, while the DDB1–Cul4-associated factor (DCAF) family uses a WD40 β–propeller fold.

The adaptor component of a CRL helps link the cullin to a family of substrate receptors (Fig. 12.1b). Typically, each substrate receptor family contains anywhere from approximately 30 to 70 proteins (Sarikas et al. 2011). For example, the Skp1 adaptor protein connects the F-box substrate receptor protein family, with over 60 representatives, to a CRL1 (Skaar et al. 2013). However, the atypical Cul7, to date, has only one identified substrate receptor protein, Fbw8 (Dias et al. 2002). A specific CRL complex can be denoted with its substrate receptor, such as $CRL7^{Fbw8}$. The substrate receptor has an additional domain that is responsible for recognizing a substrate of the CRL through a 'degron' (Fig. 12.1c), a specific amino acid sequence that is often post-translationally modified (Bulatov and Ciulli 2015; Lydeard et al. 2013). Thus, the substrate receptor component is responsible for dictating the specificity of a particular CRL towards its cellular target proteins. Since various substrate receptors can target a repertoire of substrates, CRLs are involved

Table 12.1 List of some diseases associated with cullin-RING ligases

CRL	Cellular function	Related diseases
CRL1$^{\beta\text{-TrCP1}}$	Cell adhesion and signaling	Gastric cancer; split hand-split foot malformation
CRL1$^{\text{Skp2}}$	Cell division	Lung cancer; squamous cell carcinoma
CRL1$^{\text{Atrogin1}}$	Muscle differentiation	Muscle wasting; dilated cardiomyopathy
CRL2$^{\text{LRR-1}}$	Cell migration	
CRL2^{VHL}	Hypoxic response	von Hippel-Lindau disease; renal cell carcinomas
CRL3$^{\text{KLHL9}}$	Cytokinesis	Early onset distal myopathy
CRL3^{SPOP}	Cell growth and development	
CRL3$^{\text{Keap1}}$	Oxidative stress response	Pulmonary papillary; adenocarcinoma
CRL4$^{\text{DDB2}}$	DNA damage response	Xeroderma pigmentosum; cockayne syndrome
CRL4$^{\text{Cdt2}}$	Cell division	
CRL5$^{\text{SOCS3}}$	Cytokine signaling	Diabetes
CRL5$^{\text{ASB3}}$	T-cell signaling regulation	

in a diverse array of biological processes and are associated with a multitude of diseases (Table 12.1).

12.1.1 CRL1 Defines the Prototypical CRL

The prototypical and best-characterized CRL is CRL1 or Skp1–Cul1–F-box (SCF) E3 ligase. Much of our understanding of CRLs comes from CRL1. The structure of CRL1$^{\text{Skp2}}$, containing Skp1, the F-box domain of Skp2 (F-box$^{\text{Skp2}}$), Cul1, and Rbx1 (Zheng et al. 2002) was the first to provide insight toward the overall CRL arrangement. The complex has an arch or banana-shaped architecture, with the curved Cul1 at the center, scaffolding Skp1 and Skp2 at its N-terminus and Rbx1 at its C-terminus (Fig. 12.1c). The spatial organization and overall architecture of CRL1 is representative for all CRLs.

Cul1 is composed of a helical N-terminal domain (NTD) and a globular C-terminal domain (CTD) (Fig. 12.1c). The N-terminus can be divided into three cullin repeat domains, each composed of 5-helix bundles, with the first cullin repeat containing an additional helix inserted after helix α4 (Fig. 12.1c). The first cullin repeat interacts with Skp1 and F-box$^{\text{Skp2}}$ via hydrophobic and electrostatic interactions with helices α2 and α5, which is a conserved mechanism across all CRLs. The F-box motif is a three-helix bundle that packs along the helices of Skp1 (Fig. 12.2b)

Fig. 12.2 (continued) Charge complementary at the binary cullin–substrate receptor interface between VHL and Cul2 helix α5 (*top*) or between SOCS2 and Cul5 helix α5 (*bottom*). Cullins are shown in electrostatic surface representation (*Blue*: positive; *Red*: negative). Each region is marked in (**b**) with a *blue asterisk* and with the point of view depicted by the *arrow*. (**d**) Structure of the CRL4$^{\text{DDB2}}$–UV-damaged DNA complex (PDB ID: 4A0K) in ribbon representation. The domains of DDB1 are marked

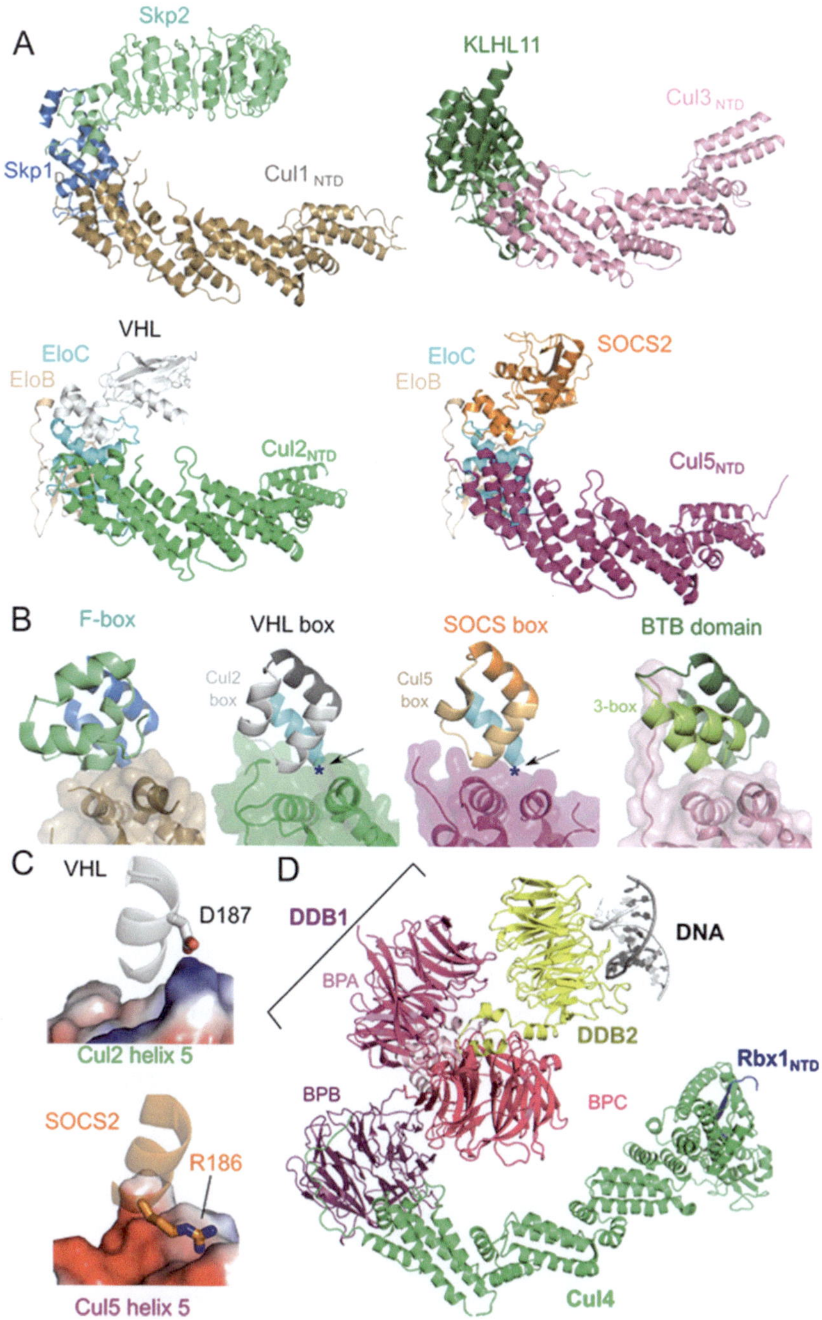

Fig. 12.2 Similar architecture of cullin-RING ligases. (**a**) Structures of VHL–EloBC–$Cul2_{NTD}$ (PDB ID: 4WQO), SOCS2–EloBC–$Cul5_{NTD}$ (PDB ID: 4JGH), Skp2–Skp1–$Cul1_{NTD}$ (PDB ID: 1LDK), and KLHL11-$Cul3_{NTD}$ (PDB ID: 4AP2) show a common spatial organization of components. Cullin repeats 2 and 3 of Cul2 were modeled from Cul1. (**b**) The cullin–adaptor–substrate receptor binding regions for each CRL shown in (**a**) are in ribbon representations with semitransparent surfaces shown for the cullins. The Cul2 box, the Cul5 box, and the 3-box are marked. (**c**)

(Schulman et al. 2000). The Cul1 C-terminus is globular, containing a 4-helix bundle, an α/β domain, and two winged-helix domains, A and B (Fig. 12.1d). The CTD creates a V-shaped binding-pocket for Rbx1. Rbx1 contains three segments: an N-terminal β-strand that inserts into the Cul1 α/β domain, a flexible linker region, and a C-terminal RING finger domain. The Rbx1 RING domain contains a RING motif, which binds two zinc ions, but with an additional 20-residue insertion, forming an additional zinc ion coordinating site (Zheng et al. 2002). In the crystal structure, Rbx1 packs against the winged-helix B of Cul1, in an inactive conformation (discussed in Sect. 12.2.1). The Cul1 NTD and CTD are rigid with respect to each other, acting as a stiff platform for the proteins at each end.

The two domains of Cul1 can be expressed *in vitro* independently of one another. With solubility-enhancing mutations at the interfaces, either the Cul1 NTD or the Cul1 CTD with Rbx1 can be expressed and purified from *E. coli*. This technique has been utilized for the expression and crystallization of fragments of Cul3 and Cul5 (Errington et al. 2012; Canning et al. 2013; Muniz et al. 2013). The two wild-type halves can be co-expressed together to reconstitute a 'full-length' Cul1 in a 'split-n-coexpress' method (Zheng et al. 2002).

12.1.2 CRL2 and CRL5: The Same Adaptor, Different Substrate Receptors

Among the CRL families, CRL2 and CRL5 are the most similar to CRL1 (Fig. 12.2a). While Skp1 is the adaptor protein for CRL1, the obligate heterodimer of Elongin B and Elongin C (EloBC) is the adaptor component for CRL2 and CRL5. Even with the same adaptor, Cul2 and Cul5 recruit different substrate receptors. VHL box substrate receptors interact with Cul2/EloBC while SOCS box proteins bind Cul5/EloBC. The VHL box motif is similar in sequence and structure to the SOCS box motif, making it difficult to predict their assembly with either Cul2 or Cul5 (Fig. 12.2b) (Mahrour et al. 2008). Both the VHL box and the SOCS box motifs are composed of a BC box and a cullin box. The BC box mediates the association with EloBC while the cullin box is important for Cul2/Cul5 specificity. Swapping of cullin boxes can switch the substrate receptor preference for Cul2 or Cul5 (Kamura et al. 2004; Mahrour et al. 2008).

Charge complementary helps dictate substrate receptor preference for Cul2 or Cul5. The structures of $CRL2^{VHL}$ and $CRL5^{SOCS2}$ reveal that the interactions between substrate receptor and cullin are centered around cullin helices α2 and α5 (Guo et al. 2014; Kim et al. 2013; Nguyen et al. 2015). The surface charge polarity between VHL and Cul2 helix α5 and between SOCS2 and Cul5 helix α5 are opposite, with the Cul2 helix α5 interface being basic and the Cul5 counterpart being acidic (Fig. 12.2c). In addition, the $CRL2^{VHL}$ structure revealed that the determinants for Cul2 selection can lie outside of the cullin box (Nguyen et al. 2015). The residues used by the substrate receptors to recruit the respective cullin are not conserved across the other VHL or SOCS box proteins. Therefore, even with the knowledge of charge complementarity at the substrate receptor-cullin interface, a more explicit rule to predict Cul2/5 preference is lacking and will require additional CRL2/5 structures.

12.1.3 CRL3: Integrating Adaptors and Substrate Receptors to Target One or Two Copies of a Substrate

CRL3 fuses the functions of the substrate receptor and adaptor into a single polypeptide, a BTB protein. BTB proteins contain the BTB domain, which is responsible for binding to Cul3, and a protein-protein interaction domain to target a substrate. The BTB domain is structurally homologous to the adaptors Skp1 and EloC in CRL1 and CRL2/5, respectively (Fig. 12.2b). However, not all proteins containing BTB domains bind to Cul3 (Genschik et al. 2013). Only those that contain an additional motif, termed the 3-box, are recruited to CRL3. The 3-box is analogous to the other cullin-binding motifs such as the F-box, VHL box, and SOCS box, as it aids in the binding of Cul3 (Zhuang et al. 2009). The 3-box creates a hydrophobic groove into which the 22 amino acid N-terminal extension of Cul3 is inserted. Deletion of the Cul3 N-terminal extension results in a 30-fold loss of binding (Canning et al. 2013). Overall, assembly of CRL3s is similar in architecture to CRL1/2/5 (Fig. 12.2a) (Canning et al. 2013; Ji and Prive 2013; Errington et al. 2012).

BTB proteins can dimerize to control substrate ubiquitination. This leads to an assembled CRL3 that has two substrate receptors and two catalytic RING domains, which can either act independently and target two substrates total (one for each Rbx1), or the dimer can act to target a single protein. The BTB protein Keap1 falls into the latter category and is thought to recognize two distinct degrons present in the same target substrate, Nrf2 (Tong et al. 2006, 2007). In addition, multimeric CRL3 assemblies can be formed through BTB proteins containing the BACK domains, such as SPOP. $CRL3^{SPOP}$ has a much higher E3 ligase activity than that of a dimerization-deficient mutant, which could be due to either enhanced avidity or increased local concentration of CRLs (Errington et al. 2012). While some CRL1 F-box proteins, such as Fbw7, have been reported to dimerize (Welcker and Clurman 2007), the ability to form dimeric CRL is a hallmark characteristic of the CRL3 family.

12.1.4 CRL4: Propellers Drive Assembly

CRL4s are highly distinct from the other CRLs as they use the adaptor protein damaged DNA binding protein 1 (DDB1), which does not have the BTB fold found in the adaptors of CRL1/2/3/5. Instead, DDB1 consists of three WD40 β-propeller domains (BPA, BPB, BPC) and a helical C-terminal domain (Fig. 12.2d). Only the BPB domain interacts with Cul4, and while it is structurally distinct from the other adaptors, it binds to Cul4 in an equivalent manner as other adaptor-cullins interactions. BPB recognizes Cul4 helices α2 and α5 of the first cullin repeat and is wrapped around by a long N-terminal extension of Cul4. BPA and BPC form a cleft that is the binding site for the DCAF substrate receptors. The DDB1 BPA-BPC double propeller and the BPB propeller arrangement can be quite flexible, with a 150° range of rotational freedom (Angers et al. 2006; Zimmerman et al. 2010). This flexibility may allow binding of substrates of various sizes and shapes or to scan a large area centered around the DCAF–substrate complex (Scrima et al. 2008). DCAF proteins utilize a short helix, termed the H-box motif, or two conserved Asp-x-Arg motifs located at the bottom surface of the WD40 propeller (WDXR motif) to interact with DDB1 (Angers et al. 2006; Jin et al. 2006; Li et al. 2010). While CRL4 exhibits vast structural differences in its subunit composition, the overall architecture is surprisingly similar to the other CRLs.

CRL4s have been shown to recognize non-canonical substrates. They are the only CRLs known to date where the substrate may not be a protein, but DNA. $CRL4^{DDB2}$ detects UV-induced lesions in DNA and ubiquitinates nearby proteins to help direct nucleotide excision repair (Chen et al. 2001; Fischer et al. 2011; Scrima et al. 2008). Small molecules can also modulate the substrate specificity of a CRL. The immune modulatory drugs thalidomide and its derivatives lenalidomide and pomalidomide bind to the CRL4 substrate receptor Cereblon, prevent targeting of its native substrates, and subvert it to target casein kinase 1α for polyubiquitination (Petzold et al. 2016). These examples demonstrate that CRL can recognize non-protein substrates and that they can be altered to target non-canonical proteins (discussed in Sects. 12.3 and 12.4).

12.2 Regulation of CRLs

The active state of CRLs must be tightly regulated to ensure proper cellular fitness. This is achieved by post-translational modifications of cullins and controlled alterations to the subunit composition of the complexes. These are highly dynamic processes that require the interplay of many other proteins. Here we describe how neddylation affects CRLs, how CAND1 allows for remodeling of substrate receptors on CRLs, and how E3 activity can be downregulated by the protein Glomulin.

12.2.1 Neddylation Activates CRLs Through Conformational Change at the Cullin-RING Interface

All cullins are post-translationally modified by covalent attachment of the protein Neural precursor cell Expressed Developmentally Down-regulated protein 8 (NEDD8) to activate ligase activity (Hori et al. 1999; Pan et al. 2004). The attachment of NEDD8, termed neddylation, requires a cascade of enzymes similar to those required for ubiquitination. NAE, a heterodimer of NAE1 and UBA3, acts as the E1 for neddylation while UBC12 and UBE2F act as the NEDD8 E2s. Interestingly and distinct from ubiquitination, a NEDD8 E3 and a 'co-E3' are required for neddylation. The NEDD8 E3 is the RING protein of the CRLs, either Rbx1 or Rbx2, while the 'co-E3' is the protein Defective in Cullin Neddylation 1 (DCN1) (Kurz et al. 2005). Both Rbx1/Rbx2 and DCN1 are required for the NEDD8-charged E2 to neddylate a cullin. DCN1 helps recruit the E2 by using its Potentiating Neddylation (PONY) domain to clasp onto both the cullin and the acetylated N-terminal helix of the E2 UBC12, while the RING correctly positions the E2 via contacts with both UBC12 and NEDD8 (Scott et al. 2014).

Neddylation occurs on the cullin CTD (Lys720 in the winged-helix B motif of Cul1) and leads to a conformation change of the cullin winged helix B motif and the cullin–RING interface (Fig. 12.3a). NEDD8 interacts with the winged-helix B motif in the cullin CTD, leading to the displacement of Rbx1 from the cullin. In this conformation, Rbx1 can sample multiple conformations to bring the E2 closer to the target substrate tethered at the other end of the cullin by the substrate receptor (Fig. 12.3a). Small angle X-ray scattering and biochemical studies validated that NEDD8-induced conformational change results in the catalytically active CRL state (Duda et al. 2008).

Deneddylation is regulated by the COP9 Signalosome (CSN), which is a ~ 350 kDa, eight-subunit complex that harbors similarity to the 19S lid of the proteasome (Pick et al. 2009; Lyapina et al. 2001). It is composed of CSN1-8 where CSN5 is the catalytic subunit, a zinc metalloproteinase that cleaves NEDD8. The CRL-free CSN is catalytically inactive and has three organizational centers: a bottom base, a central box, and a cap. The base is a hand-like structure that is composed of CSN1-4 and CSN7-8. The central box sits at the palm of the hand and is a large helical bundle containing the C-terminal helices of every CSN subunit. The cap sits at the top of the box and consists of a CSN5–CSN6 heterodimer (Fig. 12.3b) (Deshaies 2014). CSN4 senses binding of a neddylated CRL and leads to a conformational rearrangement of the CSN that activates CSN5 to cleave NEDD8 (Lingaraju et al. 2014).

Besides deneddylation, CSN has additional regulatory roles that are not well understood. *in vivo*, 10–20% of CRLs are associated with CSN at steady-state level, with only a two-fold decrease when neddylation is inhibited, revealing that the association is NEDD8-independent (Bennett et al. 2010). This is supported by experiments *in vitro*, as electron microscopy structures of CSN bound to either CRL1^{Skp2} or CRL4^{DDB2} revealed that CSN interacts with unneddylated CRLs and blocks the binding sites for both the substrate receptor and the E2 (Fig. 12.3c) (Enchev et al.

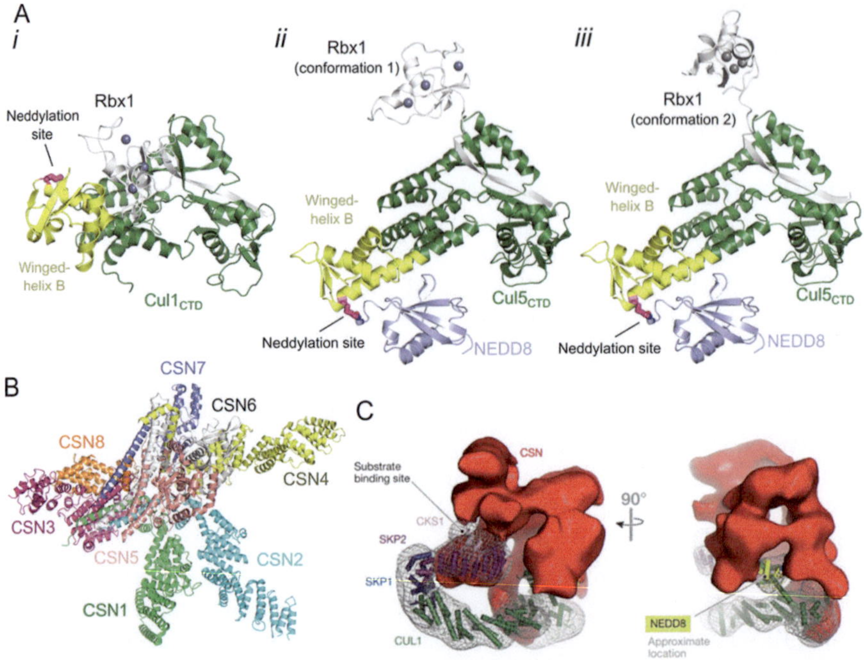

Fig. 12.3 Neddylation activates cullin-RING ligases. (**a**) *i*, Structure of unneddylated, inactive $Cul1_{CTD}$–Rbx1 (PDB ID: 1LDK). Lys720, the neddylation site, is shown in *magenta* as sticks. *ii and iii*, Neddylation of Cul_{CTD} causes a change in orientation of the winged-helix B motif and rearrangement of Rbx1 (PDB ID: 4P5O). Shown are two conformations of Rbx1 observed in the crystal structure. The isopeptide bond between Cul5 and NEDD8 is shown as sticks. (**b**) Crystal structure of the CSN (PDB ID: 4D10). (**c**) Negative staining electron microscopy reconstruction of CSN bound to $CRL1^{Skp2}$ (EMD-2173) (Reprinted from Lydeard et al. 2013)

2012; Cavadini et al. 2016). In addition, deneddylation does not always occur even when CSN binds neddylated CRLs (Bennett et al. 2010; Emberley et al. 2012). This suggests that either binding of CSN to CRL itself is not sufficient for deneddylation or another regulatory step is required. However, these studies could be biased by the epitope-tagged CSN that may lack the catalytic subunit (Lydeard et al. 2013). Moreover, when preformed CSN–CRL complexes are incubated with a substrate *in vitro*, CSN dissociates from CRL while the substrate associates (Bennett et al. 2010; Fischer et al. 2011; Emberley et al. 2012). This suggests a mode of CRL regulation in which high substrate levels out-compete CSN for CRL. Binding between CSN and CRL can also be regulated by inositol hexakisphosphate (Scherer et al. 2016). Further experiments are needed to completely delineate the deneddylase-independent roles of CSN in regulating CRLs.

12.2.2 *CAND1 Acts as an Adaptor Protein Exchange Factor*

Cullin-associated NEDD8-dissociated factor 1 (CAND1) was first identified as an inhibitor of CRL activity, but further studies revealed that it promotes the dissociation of the bound substrate receptor from a cullin to allow the binding of a new substrate receptor. CAND1 is a ~ 136 kDa HEAT (huntingtin, elongation factor 3, protein phosphatase 2A, TOR) – repeat protein that reversibly binds to deneddylated CRL by wrapping around both the NTD and CTD of the cullin (Goldenberg et al. 2004). CAND1 binding occludes both the substrate receptor binding site and the NEDD8 acceptor lysine site to inhibit neddylation, supporting its role as a negative regulator of CRLs (Fig. 12.4a) (Goldenberg et al. 2004). This is in contrast to genetic studies that showed it as a positive regulator, promoting CRL activity (Zhang et al. 2008). These conflicting views were resolved by a Förster resonance energy transfer (FRET) study, which revealed that CAND1 significantly enhances (by ~one million fold) the dissociation of incorporated substrate receptors from CRLs. The addition of CAND1 to a mixture of pre-assembled $CRL1^{\beta\text{-TRCP}}$ and a free F-box protein-adaptor, Skp1–FBXW7, led to the removal of β-TRCP and the exchange of FBXW7 onto CRL1 (Pierce et al. 2013). CAND1 seems to behave like a substrate exchange factor, analogous to guanine nucleotide exchange factors that promote exchange of GTP for GDP in GTPases.

Figure 12.4b summarizes a model for CAND1-mediated substrate receptor exchange (Lydeard et al. 2013). A cullin-RING binary complex (step 1) engages the first substrate receptor–adaptor, SR1 (step 2). This assembled complex could either encounter CAND1 to replace SR1 with a different substrate receptor–adaptor, SR2, (step 3) or could become neddylated, recruit substrates, and start ubiquitin-mediated degradation (steps 4 and 5). As the substrate concentration decreases, CSN deneddylates the CRL and then dissociates (step 6), freeing the CRL complex to repeat the cycle in step 2. CAND1 exchanges SR1 with SR2 through a proposed intermediate where CAND1 and SR2 bind simultaneously, but transiently, with the cullin–RING binary complex (step 3). Exchange of SR1 for SR2 would change which substrates are targeted in the cell.

12.2.3 *Glomulin Sterically Blocks Rbx1-E2 Interactions*

The Rbx1 utilizing activity of CRLs can be negatively regulated by another HEAT-repeat protein called Glomulin, which specifically interacts with Rbx1, but not Rbx2 (Tron et al. 2012). This interaction inhibits CRLs by obscuring the E2-binding site, preventing both ubiquitination and neddylation (Fig. 12.4c). This interaction is independent of the neddylation state of the cullin or the presence of CAND1 *in vitro* (Duda et al. 2012). However, the *in vivo* role of Glomulin is less understood as it

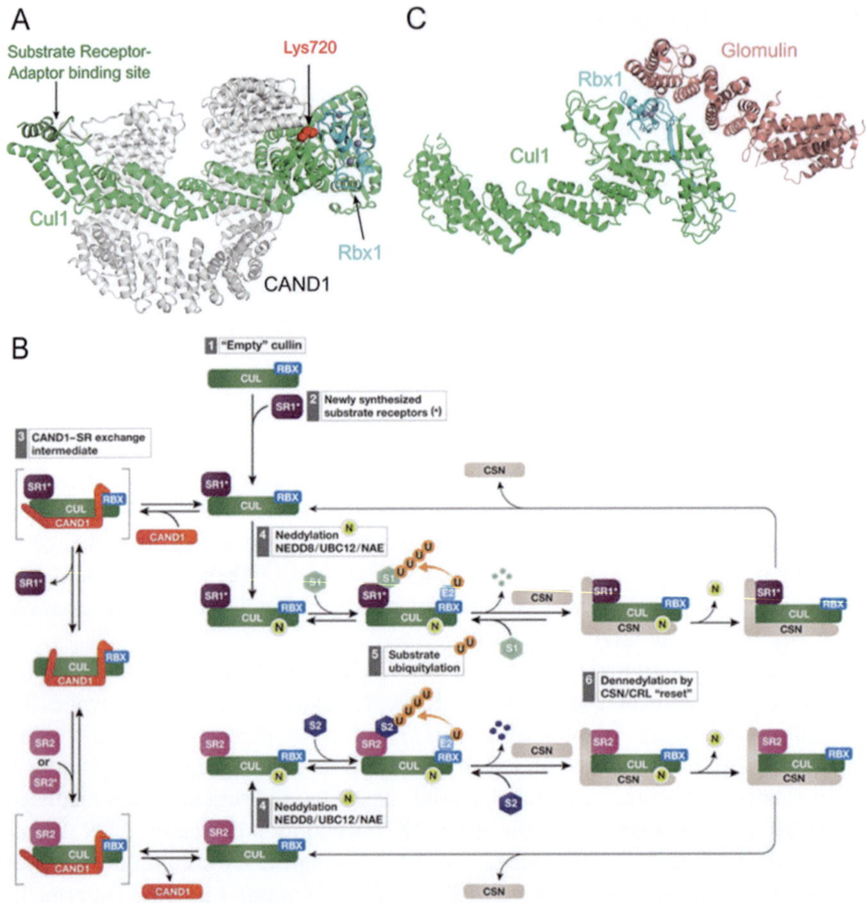

Fig. 12.4 Regulation of cullin-RING ligases by CAND1 and Glomulin. (**a**) Structure of Cul1–Rbx1 inhibited by CAND1 (PDB ID: 1U6G) in ribbon representation. CAND1 sterically blocks the neddylation site on the cullin (shown in *red spheres*) and the substrate receptor binding site at the cullin NTD (highlighted in *dark green*). (**b**) Model of remodeling of CRL composition by CAND1 and CSN (Reprinted from Lydeard et al. 2013. See main text for description). (**c**) Structural inhibition of Rbx1 by glomulin (PDB ID: 4F52) show in ribbon representation. Glomulin inhibits CRL activity by binding to Rbx1 and blocking the E2 interacting site. The Zn^{2+} ions in (**a**) and (**c**) are shown as *gray spheres*

only affects CRL1^{Fbxw7} activity, yet its deletion leads to decreased cellular levels of Cul1, Rbx1, and Fbxw7 (Tron et al. 2012). Genetic analysis reveals that Glomulin mutations cause glomuvenous malformation disease, characterized by cutaneous venous lesions (Brouillard et al. 2002). The mechanism of this Glomulin-associated disease is currently unknown.

12.3 Viral Hijacking of CRLs: Turning the Cellular Machinery Against Itself

Viruses often hijack host E3 ubiquitin ligases by mimicking the substrate receptors to evade the immune system and promote replication (Barry and Früh 2006) (Fig. 12.5a). For example, Epstein-Barr virus expresses BZLF1 (also known as EB1, Zta, or ZEBRA) to target p53 for degradation via both CRL2 and CRL5, promoting viral proliferation during the lytic phase (Sato et al. 2009). Similarly, paramyxoviruses, Hepatitis B virus (HBV), and human immunodeficiency virus (HIV) have evolved various accessory proteins to target host factors for polyubiquitination and degradation (Barry and Früh 2006; Jia et al. 2015). For the purpose of this review, we will focus on some recent examples of structural studies on viral hijacking of CRLs.

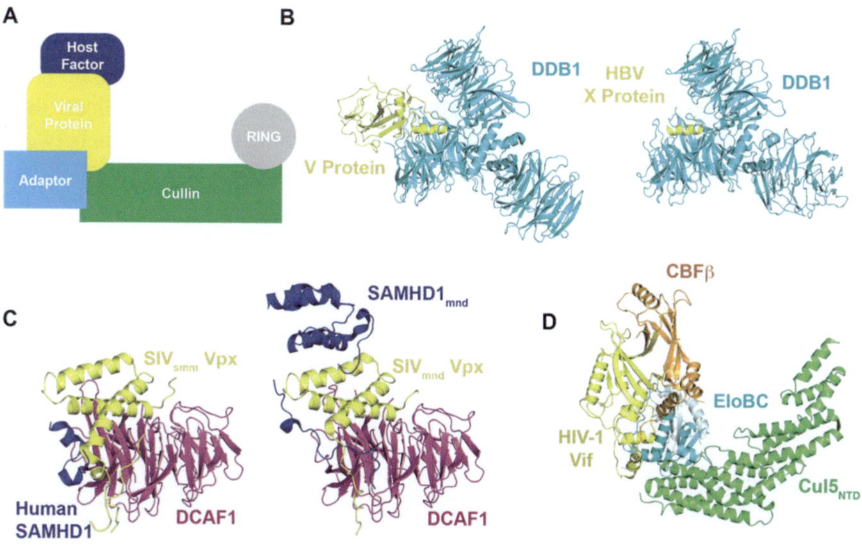

Fig. 12.5 Viral hijacking of cullin-RING ligases. (**a**) Schematic of how viral proteins recruit host factors to CRL for ubiquitination and degradation. (**b**) Paramyxoviruses V protein (PDB ID: 2B5L) and HBV X protein (PDB ID: 3I7H) bind to DDB1 using a short helix, the H-Box. (**c**) SIV Vpx-DCAF1 complexes interact with the human SAMHD1 C-terminus (PDB ID: 4CC9) and the mandrill SAMHD1 N-terminus (PDB ID: 5AJA) using similar binding modes. (**d**) HIV-1 Vif interacts with Cul5 with the help of CBFβ (PDB ID 4N9F)

12.3.1 Paramyxoviruses and Hepatitus B Virus Utilizes CRL4 for Successfully Infection

Paramyxoviruses V protein and HBV X protein function as substrate receptors to target host factors to CRL4 for polyubiquitination and degradation. Paramyxoviruses V protein targets the host Signal Transducer and Activator of Transcription (STAT) protein, which is involved in the interferon signaling pathway, thus overcoming the host interferon antiviral responses (Ulane and Horvath 2002; Precious et al. 2005). In HBV, the X protein (HBx), which is essential for virus replication *in vivo* (Bouchard and Schneider 2004), also interacts with DDB1 (Sitterlin et al. 1997). It has recently been reported that HBx functions by targeting the structural maintenance of chromosomes (Smc) complex Smc5/6 to CRL4 for degradation (Decorsière et al. 2016).

The paramyxoviruses V protein and HBx directly interact with the CRL4 adaptor protein DDB1 and they do so in a very similar manner (Fig. 12.5b) (Angers et al. 2006; Li et al. 2006, 2010). Paramyxoviruses V protein and HBx both present a 3-turn helix, which docks into the binding pocket enclosed by BPA and BPC. The helix mainly contacts the top of BPC, using hydrophobic residues on one helical face and forming hydrogen bonds with BPC residues. Although there are no common residues shared between paramyxoviruses V protein and HBx helices, their DBB1 interaction modes are remarkably similar and mimic the interaction between DBB1 and its cellular substrate receptors (e.g. DCAF9, DDB2, etc) (Li et al. 2010). This short helical domain was therefore named the H-box.

Paramyxoviruses V protein makes additional interactions with DDB1 outside the H-box (Li et al. 2006), like other DBB1 substrate receptors. It has been suggested that the DDB1 and H-box interaction alone is not sufficient for productive CRL formation. The CRLs are activated when additional interactions between the substrate receptors (such as V protein, DCAF9 and DDB2) and DDB1 are formed. This bipartite interaction may allow CRL4 to switch between productive and non-productive forms without disassembling the complex (Li et al. 2010).

12.3.2 SIVs Employ Vpx to Suppress SAMHD1 Via CRL4

Simian immunodeficiency viruses (SIVs) express the accessory protein Vpx to hijack CRL4 to evade the host antiviral protein SAMHD1 (Hrecka et al. 2011; Laguette et al. 2011). SAMHD1 represses lentiviral infection in non-dividing cells by depleting cellular dNTP levels, inhibiting reverse transcription (Kim et al. 2012; St Gelais et al. 2012). Interestingly, Vpx from different lineages of SIVs targets different regions of SAMHD1 to a CRL4. Vpx from HIV-2 and SIV_{smm} (SIV infecting sooty mangabey monkey) lineages interact with the SAMHD1 C-terminus, whereas Vpx from $SIV_{mnd\text{-}2}$ (mandrill-infecting) lineage targets the N-terminus of SAMHD1. When expressed in human cell lines, Vpx of SIV_{smm} can recognize the C-terminal tail of human SAMHD1 for degradation (Schwefel et al. 2014).

Although Vpx proteins from different species recognize distinct regions of SAMHD1, they interact with the substrate receptor DCAF1 in a remarkably similar manner (Fig. 12.5c) (Schwefel et al. 2014, 2015). Vpx consists of a flexible N-terminal region and a 3-helix globular domain, which interact with the side and the top regions of DCAF1 respectively. The C-terminal tail of human SAMHD1 is located at the SIV_{smm} Vpx–DCAF1 interaction interface, making contacts with both components. In contrast, the non-structured N-terminal region of mandrill SAMHD1 is sandwiched between SIV_{mnd-2} Vpx N-terminus and DCAF1, while the SAM domain of mandrill SAMHD1 makes additional contacts with Vpx helices to stabilize the complex.

The two variable regions (VR1 and VR2) of Vpx are responsible for the ability of the architecturally similar Vpx–DCAF1 complexes to bind different regions of SAMHD1. The Vpx of HIV-2 and SIV_{smm} feature a GEET-motif in VR1 located just N-terminal of helix 1. In contrast, the SIV_{mnd-2} Vpx VR1 comprises a proline followed by several other conserved residues, P(x)GAG[D/E]V/A, which is conserved among SIV Vpx that recognize SAMHD1 N-terminus (Schwefel et al. 2015). Moreover, the VR2 of Vpx has two conserved tyrosine residues, facilitating the complex formation with SAMHD1 and DCAF1. On the SAMHD1 side, two critical residues Phe52 and Phe15 also control the interaction between mandrill SAMHD1 N-terminus and Vpx-DCAF1 complex.

HIV-1 encodes a Vpx homolog protein in its genome, Vpr, which induces host G2/M cell cycle arrest, but does not target human SAMHD1 for degradation. Presumably, this is due to Vpr's role in targeting an unknown factor for degradation by CRL4. Recently, the crystal structure of HIV-1 Vpr bound to DDB1–DCAF1 and a substrate, uracil DNA glycosylase 2 was solved. This revealed that Vpr engages DCAF1 in a similar mode as Vpx does, but Vpr uses different structural regions to recruit its substrate (Wu et al. 2016).

12.3.3 HIV-1 Vif Downregulates the Immune Response by Targeting APOBEC3 Proteins to a CRL5

HIV express virion infectivity factor (Vif) to target the host antiviral proteins APOBEC3G (A3G) and APOBEC3F (A3F) to a CRL5 for polyubiquitination and subsequent proteasomal degradation (Sheehy et al. 2002; Yu et al. 2003). A3G and A3F are cytidine deaminases that lethally hypermutate the viral genome to suppress HIV infection (Harris et al. 2003; Lecossier et al. 2003; Mangeat et al. 2003). Unlike any other known CRL, cellular or viral, Vif recruits an additional cofactor, core binding factor β (CBFβ), to form the CRL5 (Jager et al. 2012; Zhang et al. 2012). CBFβ is a transcription factor that normally interacts with the RUNX family of proteins to promote cell differentiation (de Bruijn and Speck 2004; Ito 2008). In CRL5, CBFβ acts as a molecular chaperone to stabilize the Vif structure and allow it to interact with Cul5 (Fribourgh et al. 2014).

The crystal structure of the Vif–CBFβ–EloBC–$Cul5_{NTD}$ complex reveals how Vif recruits components of the E3 ligase (Fig. 12.5d) (Guo et al. 2014). Vif is at the center of the complex, buttressing CBFβ, EloC, and $Cul5_{NTD}$. Vif is composed of two domains, a larger α/β and a smaller α-helical domain with a Zn^{2+} binding motif serving to organize the two domains. The α/β domain has no known structural homology to other proteins. CBFβ makes extensive interactions with the α/β domain as well as the Zn^{2+} binding motif, serving as a molecular chaperone to help Vif fold into a proper conformation. The Vif α-helical domain interacts with EloBC through the conserved BC box motif as shown previously (Stanley et al. 2008), but uses a non-canonical motif to engage Cul5. Interestingly, Vif still interacts with the same general interface of Cul5 as other substrate receptors, making the overall assembly of $CRL5^{Vif}$ similar to the other CRLs.

While HIV-1 Vif was originally found to recruit CRL5, lentiviral Vifs have shown some promiscuity or ambiguity on the exact CRL they recruit. HIV-1 Vif has also been shown to recruit CRL2 depending on the cell line (Jager et al. 2012; Kane et al. 2015). Likewise, Vif from bovine immunodeficiency virus (BIV) and maedi-visna virus (MVV), a sheep-specific lentivirus, have been shown to associate with both CRL2 and CRL5 (Zhang et al. 2014; Kane et al. 2015). The composition of these CRLs also differs from that of HIV-1 Vif CRL5, as BIV Vif does not require an additional cellular cofactor while MVV Vif recruits cyclophilin A instead of CBFβ (Kane et al. 2015). Given the similarities between CRL2 and CRL5, it could be to the advantage of the virus to recruit both CRLs to achieve its goal.

12.4 Therapeutic Targeting of CRL

The misregulation of protein levels by the ubiquitin-proteasome system (UPS) causes a large number of disorders, including cancer, diabetes, and neurodegenerative diseases (Table 12.1). While some proteasome inhibitors are effective at treating certain diseases, such as Bortezomib treatment for multiple myeloma and mantle cell lymphoma, these drugs are broadly cytotoxic due to nonspecific blocking of all UPS-dependent protein degradation (Orlowski and Kuhn 2008). Interfering with CRL functions represents an alternative strategy to modulate the UPS in a more specific way. While small-molecule inhibitors of specific substrate receptors have been reviewed (Bulatov and Ciulli 2015), we will focus on two more general strategies targeting CRLs.

12.4.1 Inactivation of Neddylation by MLN4924

Currently MLN4924 (also known as pevonedistat or TAK-924) is the only general, small-molecule inhibitor of CRLs, targeting the neddylation reaction (Soucy et al. 2009). It is a sulfamate analog of AMP and is a mechanism-based inhibitor of NAE,

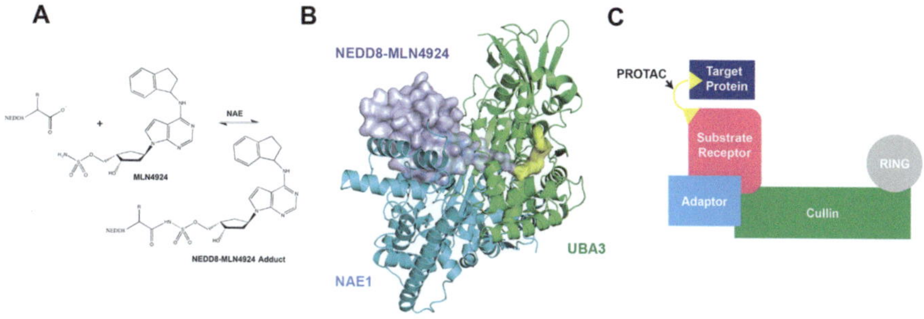

Fig. 12.6 Therapeutic manipulation of the cullin-RING ligase activity. (**a**) MLN4924 is covalently linked to the C-terminus of NEDD8 through catalysis by NAE. (**b**) The NAE heterodimer, composed of NAE1 and UBA3, forms a tight complex with the NEDD8–MLN4924 adduct. NEDD8–MLN4924 is shown in surface representation, with MLN4924 in *yellow* (PDB ID: 3GZN). (**c**) A PROTAC bridges target protein to CRL for ubiquitination and degradation

the E1 for neddylation (Fig. 12.6a). Specifically, MLN4924 binds to the E1 nucleotide binding pocket on the NAE1 subunit of NAE, which then catalyzes the addition of MLN4924 to the C-terminus of NEDD8, forming a stable NEDD8–MLN4924 adduct (Fig. 12.6b). The modified NEDD8 adduct remains tightly bound to NAE. The crystal structure of the NEDD8–MLN4924 adduct with NAE reveals that NEDD8–MLN4924 mimics the charged NEDD8–AMP intermediate, but cannot be processed further by enzymes in the NEDD8 pathway (Brownell et al. 2010; Petroski 2010; Huang et al. 2007; Walden et al. 2003). Thus, MLN4924, through the actions of NAE, generates a null NEDD8 that cannot be conjugated onto cullins. This leads to an overall decrease in active CRLs and an accumulation of substrates that would normally be targeted for degradation.

This general CRL inhibition has been shown to be an effective treatment for some cancers. In HCT-116 cells, a human colon cancer cell line, treatment with MLN4924 leads to S-phase arrest and eventually apoptosis (Soucy et al. 2009). While the MLN4923-triggered apoptosis is not completely understood, these cells accumulate multiple copies of DNA without undergoing mitosis, termed re-replication. This leads to an accumulation of DNA damage (Archambault et al. 2005), but it is unknown whether the DNA damage or accumulation of CRL substrates is responsible for induction of apoptosis. This phenotype is also observed *in vivo*, as MLN4924 suppressed the growth of xenografted tumors in various mouse models (Kuo et al. 2015; Milhollen et al. 2010; Soucy et al. 2009). It is unclear why MLN4924 affects tumors more strongly than normal cells. Nonetheless, it is undergoing Phase I clinical trials in patients suffering from advanced nonhematologic malignancies, relapsed/refractory lymphoma or multiple myeloma (Shah et al. 2016; Sarantopoulos et al. 2015).

12.4.2 Small-Molecule Subversion of CRL Targeting by PROTACs

By taking advantage of knowledge of substrate receptor binding sites, PROTACs (proteolysis targeting chimeras) have been developed to modulate specific protein levels by CRLs. PROTACs are bifunctional small molecules that simultaneously bind a target protein and the substrate receptor of a ubiquitin ligase, thus leading to the ubiquitination and degradation of the target (Fig. 12.6c). In essence, they bridge novel proteins to CRLs for polyubiquitination and degradation. The first PROTAC, Protac-1, was composed of a phosphopeptide fragment of IκBα, a ligand for the F-box protein β-TRCP, that was fused via a linker to the small molecule ovalicin, which binds to methionine aminopeptidase-2 (MetAP-2), which is involved in G_1 phase cell cycle arrest and is not normally targeted to any CRL. *in vitro*, Protac-1 has been shown to target MetAP-2 to the $CRL1^{\beta\text{-TRCP}}$ for polyubiquitination (Sakamoto et al. 2001). Furthermore, in Xenopus egg extracts, Protac-1 led to the degradation of MetAP-2 within 30 min (Sakamoto et al. 2001). However, Protac-1 is not cell permeable, so it cannot act *in vivo*, limiting its therapeutic potential. Recently, many PROTACs have been developed with improved pharmacological properties (Toure and Crews 2016). Many of these take advantage of small molecules in place of phosphopeptides to recruit either the substrate receptors VHL or Cereblon to target oncogenic proteins for degradation and are functional *in vivo* (Bondeson et al. 2015; Winter et al. 2015). Unlike traditional small-molecule inhibitors, PROTACs can act catalytically to induce multiple rounds of targeting polyubiquitination and degradation (Bondeson et al. 2015), allowing for lower doses to be effective. While no PROTACs have made it to clinical trials yet, there is great potential for the PROTAC technology as a viable therapeutic pathway to reprogram a CRL to target any protein of interest.

12.5 Future Perspectives

CRLs are the largest superfamily of E3 ubiquitin ligase and play critical roles in a diverse range of cellular processes. The extensive studies discussed above have shed enormous insight into the architecture and regulatory mechanisms of these macromolecular complexes. Yet while the general architecture of CRLs are known, much remain to be answered regarding their assembly. While we know surface charge complementarity contribute to the selection of VHL box or SOCS box proteins in CRL2/5s, there is no explicit rule to define the preference of a VHL box or SOCS box protein for Cul2 or Cul5. In addition, even with the knowledge of assembled and NEDD8-activated CRLs, it is unknown how an E2 enzyme targets specific lysines on the substrate or how an E2 accommodates the initial target site and then elongates the ubiquitin chain during polyubiquitination. In the complex case of the dimeric $CRL3^{Keap1}$, with two binding sites for a single substrate, it will be interesting to see how the two E2s are positioned relative to the single Nrf2 substrate. Structural biology, in particular the rapidly developing field of cryo-electron microscopy (cryo-EM), will likely play a major role in answering these questions.

The regulation of CRL assembly and activity is very dynamic and more complex than previously thought. A recent study revealed that the RBR E3 ARIH1 and CRL1/2/3s work in concert to polyubiquitinate substrates. ARIH1 binds to neddylated CRLs to add the first Ub to the substrate while the CRL and its E2 elongate the Ub chain (Scott et al. 2016). However, it will be important to determine the dynamics of the ARIH1-CRL interactions with respect to ARIH1 monoubiquitin kinetics (Kleiger and Deshaies 2016). While the kinetics of CAND1 association with CRL and its effect on substrate receptor exchange have been revealed, the kinetics of CSN association with CRL and the subsequent deneddylation are unknown. Of particular interest is the interplay of these regulatory factors *in vivo*. How the combined action of CSN, CAND1, substrates, and substrate receptors, at relevant *in vivo* protein concentrations, affects CRL assembly needs to be addressed. To address these questions, fluorescence microscopy of endogenous proteins and single molecule techniques may be necessary.

Finally, with the structural knowledge of how viruses hijack CRLs for their own survival, the road to rational drug design at the viral protein-CRL interface is now open. An alternative avenue focusing on CRL regulation, PROTACs or other approaches to manipulate CRLs can be used to combat viral hijacking of CRLs or other CRL related diseases. The development of these new therapeutics will likely depend on the collaboration between continued research in basic biology and efforts in pharmaceutical sciences.

Acknowledgements We thank Olga Buzovetsky, Xiaofei Jia, Kirsten Knecht, April Rose, Sarah Smaga, and Samantha Ziegler for critical comments and discussion. This work was supported in part by NIH grants AI078831 and AI16313 to Y.X.

References

Angers S, Li T, Yi XH, MacCoss MJ, Moon RT, Zheng N (2006) Molecular architecture and assembly of the DDB1-CUL4A ubiquitin ligase machinery. Nature 443(7111):590–593. doi:10.1038/Nature05175

Archambault V, Ikui AE, Drapkin BJ, Cross FR (2005) Disruption of mechanisms that prevent rereplication triggers a DNA damage response. Mol Cell Biol 25(15):6707–6721. doi:10.1128/MCB.25.15.6707-6721.2005

Barry M, Früh K (2006) Viral modulators of cullin RING ubiquitin ligases: culling the host defense. Sci Signal 2006(335):pe21–pe21. doi:10.1126/stke.3352006pe21

Bennett EJ, Rush J, Gygi SP, Harper JW (2010) Dynamics of cullin-RING ubiquitin ligase network revealed by systematic quantitative proteomics. Cell 143(6):951–965. doi:10.1016/j.cell.2010.11.017

Berndsen CE, Wolberger C (2014) New insights into ubiquitin E3 ligase mechanism. Nat Struct Mol Biol 21(4):301–307. doi:10.1038/nsmb.2780

Bondeson DP, Mares A, Smith IED, Ko E, Campos S, Miah AH, Mulholland KE, Routly N, Buckley DL, Gustafson JL, Zinn N, Grandi P, Shimamura S, Bergamini G, Faelth-Savitski M,

Bantscheff M, Cox C, Gordon DA, Willard RR, Flanagan JJ, Casillas LN, Votta BJ, den Besten W, Famm K, Kruidenier L, Carter PS, Harling JD, Churcher I, Crews CM (2015) Catalytic *in vivo* protein knockdown by small-molecule PROTACs. Nat Chem Biol 11(8):611–U120. doi:10.1038/Nchembio.1858

Bouchard MJ, Schneider RJ (2004) The enigmatic X Gene of hepatitis B virus. J Virol 78(23):12725–12734. doi:10.1128/JVI.78.23.12725-12734.2004

Brouillard P, Boon LM, Mulliken JB, Enjolras O, Ghassibe M, Warman ML, Tan OT, Olsen BR, Vikkula M (2002) Mutations in a novel factor, glomulin, are responsible for glomuvenous malformations ("glomangiomas"). Am J Hum Genet 70(4):866–874. doi:10.1086/339492

Brownell JE, Sintchak MD, Gavin JM, Liao H, Bruzzese FJ, Bump NJ, Soucy TA, Milhollen MA, Yang X, Burkhardt AL, Ma J, Loke HK, Lingaraj T, Wu D, Hamman KB, Spelman JJ, Cullis CA, Langston SP, Vyskocil S, Sells TB, Mallender WD, Visiers I, Li P, Claiborne CF, Rolfe M, Bolen JB, Dick LR (2010) Substrate-assisted inhibition of ubiquitin-like protein-activating enzymes: the NEDD8 E1 inhibitor MLN4924 forms a NEDD8-AMP mimetic in situ. Mol Cell 37(1):102–111. doi:10.1016/j.molcel.2009.12.024

Bulatov E, Ciulli A (2015) Targeting Cullin-RING E3 ubiquitin ligases for drug discovery: structure, assembly and small-molecule modulation. Biochem J 467:365–386. doi:10.1042/Bj20141450

Canning P, Cooper CDO, Krojer T, Murray JW, Pike ACW, Chaikuad A, Keates T, Thangaratnarajah C, Hojzan V, Ayinampudi V, Marsden BD, Gileadi O, Knapp S, von Delft F, Bullock AN (2013) Structural basis for Cul3 protein assembly with the BTB-Kelch family of E3 ubiquitin ligases (vol 288, pg 7803, 2013). J Biol Chem 288(39):28304–28304. doi:10.1074/Jbc.A112.437996

Cavadini S, Fischer ES, Bunker RD, Potenza A, Lingaraju GM, Goldie KN, Mohamed WI, Faty M, Petzold G, Beckwith RE, Tichkule RB, Hassiepen U, Abdulrahman W, Pantelic RS, Matsumoto S, Sugasawa K, Stahlberg H, Thoma NH (2016) Cullin-RING ubiquitin E3 ligase regulation by the COP9 signalosome. Nature 531(7596):598–603. doi:10.1038/nature17416

Chen X, Zhang Y, Douglas L, Zhou P (2001) UV-damaged DNA-binding proteins are targets of CUL-4A-mediated ubiquitination and degradation. J Biol Chem 276(51):48175–48182. doi:10.1074/jbc.M106808200

Ciechanover A, Hershko A, Rose I (2004) The Nobel Prize in Chemistry 2004. Nobel Media AB. http://www.nobelprize.org/nobel_prizes/chemistry/laureates/2004/. Accessed 15 Mar 2016

de Bruijn MF, Speck NA (2004) Core-binding factors in hematopoiesis and immune function. Oncogene 23(24):4238–4248. doi:10.1038/sj.onc.1207763

Decorsière A, Mueller H, van Breugel PC, Abdul F, Gerossier L, Beran RK, Livingston CM, Niu C, Fletcher SP, Hantz O, Strubin M (2016) Hepatitis B virus X protein identifies the Smc5/6 complex as a host restriction factor. Nature 531(7594):386–380. doi:10.1038/nature17170

Deshaies RJ (2014) Structural biology: corralling a protein-degradation regulator. Nature 512(7513):145–146. doi:10.1038/nature13644

Deshaies RJ, Joazeiro CAP (2009) RING domain E3 ubiquitin ligases. Annu Rev Biochem 78:399–434. doi:10.1146/Annurev.Biochem.78.101807.093809

Dias DC, Dolios G, Wang R, Pan ZQ (2002) CUL7: a DOC domain-containing cullin selectively binds Skp1.Fbx29 to form an SCF-like complex. Proc Natl Acad Sci U S A 99(26):16601–16606. doi:10.1073/pnas.252646399

Duda DM, Borg LA, Scott DC, Hunt HW, Hammel M, Schulman BA (2008) Structural insights into NEDD8 activation of cullin-RING ligases: conformational control of conjugation. Cell 134(6):995–1006. doi:10.1016/j.cell.2008.07.022

Duda DM, Olszewski JL, Tron AE, Hammel M, Lambert LJ, Waddell MB, Mittag T, DeCaprio JA, Schulman BA (2012) Structure of a glomulin-RBX1-CUL1 complex: inhibition of a RING E3 ligase through masking of its E2-binding surface. Mol Cell 47(3):371–382. doi:10.1016/j.molcel.2012.05.044

Emberley ED, Mosadeghi R, Deshaies RJ (2012) Deconjugation of Nedd8 from Cul1 Is directly regulated by Skp1-F-box and substrate, and the COP9 signalosome inhibits deneddylated SCF

by a noncatalytic mechanism. J Biol Chem 287(35):29679–29689. doi:10.1074/jbc.M112.352484
Enchev RI, Scott DC, da Fonseca PCA, Schreiber A, Monda JK, Schulman BA, Peter M, Morris EP (2012) Structural basis for a reciprocal regulation between SCF and CSN. Cell Rep 2(3):616–627. doi:10.1016/J.Celrep.2012.08.019
Errington WJ, Khan MQ, Bueler SA, Rubinstein JL, Chakrabartty A, Prive GG (2012) Adaptor protein self-assembly drives the control of a cullin-RING ubiquitin ligase. Structure 20(7):1141–1153. doi:10.1016/J.Str.2012.04.009
Fischer ES, Scrima A, Bohm K, Matsumoto S, Lingaraju GM, Faty M, Yasuda T, Cavadini S, Wakasugi M, Hanaoka F, Iwai S, Gut H, Sugasawa K, Thoma NH (2011) The molecular basis of CRL4(DDB2/CSA) ubiquitin ligase architecture, targeting, and activation. Cell 147 (5):1024–1039. doi:10.1016/J.Cell.2011.10.035
Fribourgh JL, Nguyen HC, Wolfe LS, Dewitt DC, Zhang W, Yu XF, Rhoades E, Xiong Y (2014) Core binding factor beta plays a critical role by facilitating the assembly of the Vif-cullin 5 E3 ubiquitin ligase. J Virol 88(6):3309–3319. doi:10.1128/JVI.03824-13
Genschik P, Sumara I, Lechner E (2013) The emerging family of CULLIN3-RING ubiquitin ligases (CRL3s): cellular functions and disease implications. EMBO J 32(17):2307–2320. doi:10.1038/emboj.2013.173
Goldenberg SJ, Cascio TC, Shumway SD, Garbutt KC, Liu J, Xiong Y, Zheng N (2004) Structure of the Cand1-Cul1-Roc1 complex reveals regulatory mechanisms for the assembly of the multisubunit cullin-dependent ubiquitin ligases. Cell 119(4):517–528. doi:10.1016/j.cell.2004.10.019
Guo Y, Dong L, Qiu X, Wang Y, Zhang B, Liu H, Yu Y, Zang Y, Yang M, Huang Z (2014) Structural basis for hijacking CBF-beta and CUL5 E3 ligase complex by HIV-1 Vif. Nature 505(7482):229–233. doi:10.1038/nature12884
Harris RS, Bishop KN, Sheehy AM, Craig HM, Petersen-Mahrt SK, Watt IN, Neuberger MS, Malim MH (2003) DNA deamination mediates innate immunity to retroviral infection. Cell 113(6):803–809. doi:10.1016/s0092-8674(03)00423-9
Hershko A, Ciechanover A (1998) The ubiquitin system. Annu Rev Biochem 67:425–479. doi:10.1146/annurev.biochem.67.1.425
Hori T, Osaka F, Chiba T, Miyamoto C, Okabayashi K, Shimbara N, Kato S, Tanaka K (1999) Covalent modification of all members of human cullin family proteins by NEDD8. Oncogene 18(48):6829–6834. doi:10.1038/sj.onc.1203093
Hrecka K, Hao C, Gierszewska M, Swanson SK, Kesik-Brodacka M, Srivastava S, Florens L, Washburn MP, Skowronski J (2011) Vpx relieves inhibition of HIV-1 infection of macrophages mediated by the SAMHD1 protein. Nature 474(7353):658–661. doi:10.1038/nature10195
Huang DT, Hunt HW, Zhuang M, Ohi MD, Holton JM, Schulman BA (2007) Basis for a ubiquitin-like protein thioester switch toggling E1-E2 affinity. Nature 445(7126):394–398. doi:10.1038/nature05490
Ito Y (2008) RUNX genes in development and cancer: regulation of viral gene expression and the discovery of RUNX family genes. Adv Cancer Res 99:33–76. doi:10.1016/S0065-230X(07)99002-8
Jager S, Cimermancic P, Gulbahce N, Johnson JR, McGovern KE, Clarke SC, Shales M, Mercenne G, Pache L, Li K, Hernandez H, Jang GM, Roth SL, Akiva E, Marlett J, Stephens M, D'Orso I, Fernandes J, Fahey M, Mahon C, O'Donoghue AJ, Todorovic A, Morris JH, Maltby DA, Alber T, Cagney G, Bushman FD, Young JA, Chanda SK, Sundquist WI, Kortemme T, Hernandez RD, Craik CS, Burlingame A, Sali A, Frankel AD, Krogan NJ (2012) Global landscape of HIV-human protein complexes. Nature 481(7381):365–370. doi:10.1038/nature10719
Ji AX, Prive GG (2013) Crystal structure of KLHL3 in complex with Cullin3. PLoS One 8(4):e60445. doi:10.1371/journal.pone.0060445
Jia X, Zhao Q, Xiong Y (2015) HIV suppression by host restriction factors and viral immune evasion. Curr Opin Struct Biol 31:106–114. doi:10.1016/j.sbi.2015.04.004

Jin JP, Arias EE, Chen J, Harper JW, Walter JC (2006) A family of diverse Cul4-Ddb1-interacting proteins includes Cdt2, which is required for S phase destruction of the replication factor Cdt1. Mol Cell 23(5):709–721. doi:10.1016/j.molcel.2006.08.010

Kaiser SE, Riley BE, Shaler TA, Trevino RS, Becker CH, Schulman H, Kopito RR (2011) Protein standard absolute quantification (PSAQ) method for the measurement of cellular ubiquitin pools. Nat Methods 8(8):691–696. doi:10.1038/nmeth.1649

Kamura T, Maenaka K, Kotoshiba S, Matsumoto M, Kohda D, Conaway RC, Conaway JW, Nakayama KI (2004) VHL-box and SOCS-box domains determine binding specificity for Cul2-Rbx1 and Cul5-Rbx2 modules of ubiquitin ligases. Genes Dev 18(24):3055–3065. doi:10.1101/gad.1252404

Kane JR, Stanley DJ, Hultquist JF, Johnson JR, Mietrach N, Binning JM, Jonsson SR, Barelier S, Newton BW, Johnson TL, Franks-Skiba KE, Li M, Brown WL, Gunnarsson HI, Adalbjornsdottir A, Fraser JS, Harris RS, Andresdottir V, Gross JD, Krogan NJ (2015) Lineage-specific viral hijacking of non-canonical E3 ubiquitin ligase cofactors in the evolution of Vif Anti-APOBEC3 activity. Cell Rep 11(8):1236–1250. doi:10.1016/j.celrep.2015.04.038

Kim B, Nguyen LA, Daddacha W, Hollenbaugh JA (2012) Tight interplay among SAMHD1 protein level, cellular dNTP levels, and HIV-1 proviral DNA synthesis kinetics in human primary monocyte-derived macrophages. J Biol Chem 287(26):21570–21574. doi:10.1074/jbc.C112.374843

Kim YK, Kwak MJ, Ku B, Suh HY, Joo K, Lee J, Jung JU, Oh BH (2013) Structural basis of intersubunit recognition in elongin BC-cullin 5-SOCS box ubiquitin-protein ligase complexes. Acta Crystallogr D Biol Crystallogr 69(Pt 8):1587–1597. doi:10.1107/S0907444913011220

Kleiger G, Deshaies R (2016) Tag team ubiquitin ligases. Cell 166(5):1080–1081. doi:10.1016/j.cell.2016.08.014

Kuo KL, Ho IL, Shi CS, Wu JT, Lin WC, Tsai YC, Chang HC, Chou CT, Hsu CH, Hsieh JT, Chang SC, Pu YS, Huang KH (2015) MLN4924, a novel protein neddylation inhibitor, suppresses proliferation and migration of human urothelial carcinoma: *in vitro* and *in vivo* studies. Cancer Lett 363(2):127–136. doi:10.1016/j.canlet.2015.01.015

Kurz T, Ozlu N, Rudolf F, O'Rourke SM, Luke B, Hofmann K, Hyman AA, Bowerman B, Peter M (2005) The conserved protein DCN-1/Dcn1p is required for cullin neddylation in C. elegans and S. cerevisiae. Nature 435(7046):1257–1261. doi:10.1038/nature03662

Laguette N, Sobhian B, Casartelli N, Ringeard M, Chable-Bessia C, Segeral E, Yatim A, Emiliani S, Schwartz O, Benkirane M (2011) SAMHD1 is the dendritic- and myeloid-cell-specific HIV-1 restriction factor counteracted by Vpx. Nature 474(7353):654–657. doi:10.1038/nature10117

Lecossier D, Bouchonnet F, Clavel F, Hance AJ (2003) Hypermutation of HIV-1 DNA in the absence of the Vif protein. Science 300 (5622):1112–1112. doi:10.1126/Science.1083338

Li T, Chen X, Garbutt KC, Zhou P, Zheng N (2006) Structure of DDB1 in complex with a paramyxovirus V protein: viral hijack of a propeller cluster in ubiquitin ligase. Cell 124(1):105–117. doi:10.1016/j.cell.2005.10.033

Li T, Robert EI, van Breugel PC, Strubin M, Zheng N (2010) A promiscuous alpha-helical motif anchors viral hijackers and substrate receptors to the CUL4-DDB1 ubiquitin ligase machinery. Nat Struct Mol Biol 17(1):105–111. doi:10.1038/nsmb.1719

Lingaraju GM, Bunker RD, Cavadini S, Hess D, Hassiepen U, Renatus M, Fischer ES, Thoma NH (2014) Crystal structure of the human COP9 signalosome. Nature 512(7513):161–165. doi:10.1038/nature13566

Lyapina S, Cope G, Shevchenko A, Serino G, Tsuge T, Zhou C, Wolf DA, Wei N, Shevchenko A, Deshaies RJ (2001) Promotion of NEDD-CUL1 conjugate cleavage by COP9 signalosome. Science 292(5520):1382–1385. doi:10.1126/science.1059780

Lydeard JR, Schulman BA, Harper JW (2013) Building and remodelling Cullin-RING E3 ubiquitin ligases. EMBO Rep 14(12):1050–1061. doi:10.1038/embor.2013.173

Mahrour N, Redwine WB, Florens L, Swanson SK, Martin-Brown S, Bradford WD, Staehling-Hampton K, Washburn MP, Conaway RC, Conaway JW (2008) Characterization of Cullin-box sequences that direct recruitment of Cul2-Rbx1 and Cul5-Rbx2 modules to elongin BC-based ubiquitin ligases. J Biol Chem 283(12):8005–8013. doi:10.1074/Jbc.M706987200

Mangeat B, Turelli P, Caron G, Friedli M, Perrin L, Trono D (2003) Broad antiretroviral defence by human APOBEC3G through lethal editing of nascent reverse transcripts. Nature 424(6944):99–103. doi:10.1038/Nature01709

Marin I (2009) Diversification of the cullin family. BMC Evol Biol 9:267. doi:10.1186/1471-2148-9-267

Milhollen MA, Traore T, Adams-Duffy J, Thomas MP, Berger AJ, Dang L, Dick LR, Garnsey JJ, Koenig E, Langston SP, Manfredi M, Narayanan U, Rolfe M, Staudt LM, Soucy TA, Yu J, Zhang J, Bolen JB, Smith PG (2010) MLN4924, a NEDD8-activating enzyme inhibitor, is active in diffuse large B-cell lymphoma models: rationale for treatment of NF-{kappa} B-dependent lymphoma. Blood 116(9):1515–1523. doi:10.1182/blood-2010-03-272567

Muniz JR, Guo K, Kershaw NJ, Ayinampudi V, von Delft F, Babon JJ, Bullock AN (2013) Molecular architecture of the ankyrin SOCS box family of Cul5-dependent E3 ubiquitin ligases. J Mol Biol 425(17):3166–3177. doi:10.1016/j.jmb.2013.06.015

Nguyen HC, Yang H, Fribourgh JL, Wolfe LS, Xiong Y (2015) Insights into cullin-RING E3 ubiquitin ligase recruitment: structure of the VHL-EloBC-Cul2 complex. Structure. doi:10.1016/j.str.2014.12.014

Orlowski RZ, Kuhn DJ (2008) Proteasome inhibitors in cancer therapy: lessons from the first decade. Clin Cancer Res 14(6):1649–1657. doi:10.1158/1078-0432.Ccr-07-2218

Pan ZQ, Kentsis A, Dias DC, Yamoah K, Wu K (2004) Nedd8 on cullin: building an expressway to protein destruction. Oncogene 23(11):1985–1997. doi:10.1038/sj.onc.1207414

Petroski MD (2010) Mechanism-based neddylation inhibitor. Chem Biol 17(1):6–8. doi:10.1016/j.chembiol.2010.01.002

Petroski MD, Deshaies RJ (2005) Function and regulation of Cullin-RING ubiquitin ligases. Nat Rev Mol Cell Biol 6(1):9–20. doi:10.1038/Nrm1547

Petzold G, Fischer ES, Thoma NH (2016) Structural basis of lenalidomide-induced CK1alpha degradation by the CRL4 ubiquitin ligase. Nature. doi:10.1038/nature16979

Pick E, Hofmann K, Glickman MH (2009) PCI complexes: beyond the proteasome, CSN, and eIF3 Troika. Mol Cell 35(3):260–264. doi:10.1016/j.molcel.2009.07.009

Pickart CM (2001) Mechanisms underlying ubiquitination. Annu Rev Biochem 70:503–533. doi:10.1146/Annurev.Biochem.70.1.503

Pierce NW, Lee JE, Liu X, Sweredoski MJ, Graham RLJ, Larimore EA, Rome M, Zheng N, Clurman BE, Hess S, Shan SO, Deshaies RJ (2013) Cand1 promotes assembly of new SCF complexes through dynamic exchange of F Box proteins. Cell 153 (1):206–215. doi:10.1016/J.Cell.2013.02.024

Precious B, Childs K, Fitzpatrick-Swallow V, Goodbourn S, Randall RE (2005) Simian virus 5 V protein acts as an adaptor, linking DDB1 to STAT2, to facilitate the ubiquitination of STAT1. J Virol 79(21):13434–13441. doi:10.1128/jvi.79.21.13434-13441.2005

Sakamoto KM, Kim KB, Kumagai A, Mercurio F, Crews CM, Deshaies RJ (2001) Protacs: chimeric molecules that target proteins to the Skp1-Cullin-F box complex for ubiquitination and degradation. Proc Natl Acad Sci U S A 98(15):8554–8559. doi:10.1073/pnas.141230798

Sarantopoulos J, Shapiro GI, Cohen RB, Clark JW, Kauh JS, Weiss GJ, Cleary JM, Mahalingam D, Pickard MD, Faessel HM, Berger AJ, Burke K, Mulligan G, Dezube BJ, Harvey RD (2015) Phase I study of the investigational NEDD8-activating enzyme inhibitor Pevonedistat (TAK-924/MLN4924) in patients with advanced solid tumors. Clin Cancer Res. doi:10.1158/1078-0432.CCR-15-1338

Sarikas A, Hartmann T, Pan ZQ (2011) The cullin protein family. Genome Biol 12(4). doi:10.1186/Gb-2011-12-4-220

Sato Y, Kamura T, Shirata N, Murata T, Kudoh A, Iwahori S, Nakayama S, Isomura H, Nishiyama Y, Tsurumi T (2009) Degradation of phosphorylated p53 by viral protein-ECS E3 ligase complex. PLoS Path 5(7). doi:10.1371/Journal.Ppat.1000530

Scherer PC, Ding Y, Liu Z, Xu J, Mao H, Barrow JC, Wei N, Zheng N, Snyder SH, Rao F (2016) Inositol hexakisphosphate (IP6) generated by IP5K mediates cullin-COP9 signalosome interactions and CRL function. Proc Natl Acad Sci U S A 113(13):3503–3508. doi:10.1073/pnas.1525580113

Schulman BA, Carrano AC, Jeffrey PD, Bowen Z, Kinnucan ERE, Finnin MS, Elledge SJ, Harper JW, Pagano M, Pavietich NP (2000) Insights into SCF ubiquitin ligases from the structure of the Skp1-Skp2 complex. Nature 408(6810):381–386. doi:10.1038/35042620

Schwefel D, Groom HCT, Boucherit VC, Christodoulou E, Walker PA, Stoye JP, Bishop KN, Taylor IA (2014) Structural basis of lentiviral subversion of a cellular protein degradation pathway. Nature 505(7482):234–238. doi:10.1038/nature12815

Schwefel D, Boucherit Virginie C, Christodoulou E, Walker Philip A, Stoye Jonathan P, Bishop Kate N, Taylor Ian A (2015) Molecular determinants for recognition of divergent SAMHD1 proteins by the lentiviral accessory protein Vpx. Cell Host Microbe 17(4):489–499. doi:10.1016/j.chom.2015.03.004

Scott DC, Rhee DY, Duda DM, Kelsall IR, Olszewski JL, Paulo JA, de Jong A, Ovaa H, Alpi AF, Harper JW, Schulman BA (2016) Two distinct types of E3 ligases work in unison to regulate substrate ubiquitylation. Cell 166(5):1198–1214 e1124. doi:10.1016/j.cell.2016.07.027

Scott DC, Sviderskiy VO, Monda JK, Lydeard JR, Cho SE, Harper JW, Schulman BA (2014) Structure of a RING E3 trapped in action reveals ligation mechanism for the ubiquitin-like protein NEDD8. Cell 157(7):1671–1684. doi:10.1016/j.cell.2014.04.037

Scrima A, Konickova R, Czyzewski BK, Kawasaki Y, Jeffrey PD, Groisman R, Nakatani Y, Iwai S, Pavletich NP, Thoma NH (2008) Structural basis of UV DNA-damage recognition by the DDB1-DDB2 complex. Cell 135(7):1213–1223. doi:10.1016/j.cell.2008.10.045

Shah JJ, Jakubowiak AJ, O'Connor OA, Orlowski RZ, Harvey RD, Smith MR, Lebovic D, Diefenbach C, Kelly K, Hua Z, Berger AJ, Mulligan G, Faessel HM, Tirrell S, Dezube BJ, Lonial S (2016) Phase I study of the novel investigational NEDD8-activating enzyme inhibitor pevonedistat (MLN4924) in patients with relapsed/refractory multiple myeloma or lymphoma. Clin Cancer Res 22(1):34–43. doi:10.1158/1078-0432.CCR-15-1237

Sheehy AM, Gaddis NC, Choi JD, Malim MH (2002) Isolation of a human gene that inhibits HIV-1 infection and is suppressed by the viral Vif protein. Nature 418(6898):646–650. doi:10.1038/nature00939

Sitterlin D, Lee TH, Prigent S, Tiollais P, Butel JS, Transy C (1997) Interaction of the UV-damaged DNA-binding protein with hepatitis B virus X protein is conserved among mammalian hepadnaviruses and restricted to transactivation-proficient X-insertion mutants. J Virol 71(8):6194–6199

Skaar JR, Florens L, Tsutsumi T, Arai T, Tron A, Swanson SK, Washburn MP, DeCaprio JA (2007) PARC and CUL7 form atypical cullin RING ligase complexes. Cancer Res 67(5):2006–2014. doi:10.1158/0008-5472.CAN-06-3241

Skaar JR, Pagan JK, Pagano M (2013) Mechanisms and function of substrate recruitment by F-box proteins. Nat Rev Mol Cell Biol 14(6):369–381. doi:10.1038/nrm3582

Soucy TA, Smith PG, Milhollen MA, Berger AJ, Gavin JM, Adhikari S, Brownell JE, Burke KE, Cardin DP, Critchley S, Cullis CA, Doucette A, Garnsey JJ, Gaulin JL, Gershman RE, Lublinsky AR, McDonald A, Mizutani H, Narayanan U, Olhava EJ, Peluso S, Rezaei M, Sintchak MD, Talreja T, Thomas MP, Traore T, Vyskocil S, Weatherhead GS, Yu J, Zhang J, Dick LR, Claiborne CF, Rolfe M, Bolen JB, Langston SP (2009) An inhibitor of NEDD8-activating enzyme as a new approach to treat cancer. Nature 458(7239):732–736. doi:10.1038/nature07884

Stanley BJ, Ehrlich ES, Short L, Yu Y, Xiao Z, Yu XF, Xiong Y (2008) Structural insight into the human immunodeficiency virus Vif SOCS box and its role in human E3 ubiquitin ligase assembly. J Virol 82(17):8656–8663. doi:10.1128/JVI.00767-08

St Gelais C, de Silva S, Amie SM, Coleman CM, Hoy H, Hollenbaugh JA, Kim B, Wu L (2012) SAMHD1 restricts HIV-1 infection in dendritic cells (DCs) by dNTP depletion, but its expression in DCs and primary CD4 + T-lymphocytes cannot be upregulated by interferons. Retrovirology 9(1):1–15. doi:10.1186/1742-4690-9-105

Tong KI, Katoh Y, Kusunoki H, Itoh K, Tanaka T, Yamamoto M (2006) Keap1 recruits Neh2 through binding to ETGE and DLG motifs: characterization of the two-site molecular recognition model. Mol Cell Biol 26(8):2887–2900. doi:10.1128/MCB.26.8.2887-2900.2006

Tong KI, Padmanabhan B, Kobayashi A, Shang C, Hirotsu Y, Yokoyama S, Yamamoto M (2007) Different electrostatic potentials define ETGE and DLG motifs as hinge and latch in oxidative stress response. Mol Cell Biol 27(21):7511–7521. doi:10.1128/MCB.00753-07

Toure M, Crews CM (2016) Small-molecule PROTACS: new approaches to protein degradation. Angew Chem Int Ed Eng 55(6):1966–1973. doi:10.1002/anie.201507978

Tron AE, Arai T, Duda DM, Kuwabara H, Olszewski JL, Fujiwara Y, Bahamon BN, Signoretti S, Schulman BA, DeCaprio JA (2012) The glomuvenous malformation protein Glomulin binds Rbx1 and regulates cullin RING ligase-mediated turnover of Fbw7. Mol Cell 46(1):67–78. doi:10.1016/j.molcel.2012.02.005

Ulane CM, Horvath CM (2002) Paramyxoviruses SV5 and HPIV2 assemble STAT protein ubiquitin ligase complexes from cellular components. Virology 304(2):160–166. doi:10.1006/viro.2002.1773

van Wijk SJL, Timmers HTM (2010) The family of ubiquitin-conjugating enzymes (E2s): deciding between life and death of proteins. FASEB J 24(4):981–993. doi:10.1096/fj.09-136259

Walden H, Podgorski MS, Huang DT, Miller DW, Howard RJ, Minor DL, Holton JM, Schulman BA (2003) The structure of the APPBP1-UBA3-NEDD8-ATP complex reveals the basis for selective ubiquitin-like protein activation by an E1. Mol Cell 12(6):1427–1437. doi:10.1016/S1097-2765(03)00452-0

Welcker M, Clurman BE (2007) Fbw7/hCDC4 dimerization regulates its substrate interactions. Cell Div 2:7. doi:10.1186/1747-1028-2-7

Winter GE, Buckley DL, Paulk J, Roberts JM, Souza A, Dhe-Paganon S, Bradner JE (2015) Phthalimide conjugation as a strategy for *in vivo* target protein degradation. Science 348(6241):1376–1381. doi:10.1126/science.aab1433

Wu Y, Zhou X, Barnes CO, DeLucia M, Cohen AE, Gronenborn AM, Ahn J, Calero G (2016) The DDB1-DCAF1-Vpr-UNG2 crystal structure reveals how HIV-1 Vpr steers human UNG2 toward destruction. Nat Struct Mol Biol 23(10):933–940 doi:10.1038/nsmb.3284

Yu X, Yu Y, Liu B, Luo K, Kong W, Mao P, Yu XF (2003) Induction of APOBEC3G ubiquitination and degradation by an HIV-1 Vif-Cul5-SCF complex. Science 302(5647):1056–1060. doi:10.1126/science.1089591

Zachariae W, Shevchenko A, Andrews PD, Ciosk R, Galova M, Stark MJ, Mann M, Nasmyth K (1998) Mass spectrometric analysis of the anaphase-promoting complex from yeast: identification of a subunit related to cullins. Science 279(5354):1216–1219. doi:10.1126/science.279.5354.1216

Zhang W, Ito H, Quint M, Huang H, Noel LD, Gray WM (2008) Genetic analysis of CAND1-CUL1 interactions in Arabidopsis supports a role for CAND1-mediated cycling of the SCFTIR1 complex. Proc Natl Acad Sci U S A 105(24):8470–8475. doi:10.1073/pnas.0804144105

Zhang W, Du J, Evans SL, Yu Y, Yu XF (2012) T-cell differentiation factor CBF-beta regulates HIV-1 Vif-mediated evasion of host restriction. Nature 481(7381):376–379. doi:10.1038/nature10718

Zhang WY, Wang H, Li ZL, Liu X, Liu GC, Harris RS, Yu XF (2014) Cellular requirements for bovine immunodeficiency virus Vif-mediated inactivation of bovine APOBEC3 proteins. J Virol 88(21):12528–12540. doi:10.1128/Jvi.02072-14

Zheng N, Schulman BA, Song LZ, Miller JJ, Jeffrey PD, Wang P, Chu C, Koepp DM, Elledge SJ, Pagano M, Conaway RC, Conaway JW, Harper JW, Pavletich NP (2002) Structure of the Cul1-Rbx1-Skp1-F box(Skp2) SCF ubiquitin ligase complex. Nature 416(6882):703–709. doi:10.1038/416703a

Zhuang M, Calabrese MF, Liu J, Waddell MB, Nourse A, Hammel M, Miller DJ, Walden H, Duda DM, Seyedin SN, Hoggard T, Harper JW, White KP, Schulman BA (2009) Structures of SPOP-substrate complexes: insights into molecular architectures of BTB-Cul3 ubiquitin ligases. Mol Cell 36(1):39–50. doi:10.1016/j.molcel.2009.09.022

Zimmerman ES, Schulman BA, Zheng N (2010) Structural assembly of cullin-RING ubiquitin ligase complexes. Curr Opin Struct Biol 20(6):714–721. doi:10.1016/J.Sbi.2010.08.010

Zou Y, Mi J, Cui J, Lu D, Zhang X, Guo C, Gao G, Liu Q, Chen B, Shao C, Gong Y (2009) Characterization of nuclear localization signal in the N terminus of CUL4B and its essential role in cyclin E degradation and cell cycle progression. J Biol Chem 284(48):33320–33332. doi:10.1074/jbc.M109.050427

Chapter 13
The Ccr4-Not Complex: Architecture and Structural Insights

Martine A. Collart and Olesya O. Panasenko

Abstract The Ccr4-Not complex is an essential multi-subunit protein complex that plays a fundamental role in eukaryotic mRNA metabolism and has a multitude of different roles that impact eukaryotic gene expression. It has a conserved core of three Not proteins, the Ccr4 protein, and two Ccr4 associated factors, Caf1 and Caf40. A fourth Not protein, Not4, is conserved, but is only a stable subunit of the complex in yeast. Certain subunits have been duplicated during evolution, with functional divergence, such as Not3 in yeast, and Ccr4 or Caf1 in human. However the complex includes only one homolog for each protein. In addition, species-specific subunits are part of the complex, such as Caf130 in yeast or Not10 and Not11 in human. Two conserved catalytic functions are associated with the complex, deadenylation and ubiquitination. The complex adopts an L-shaped structure, in which different modules are bound to a large Not1 scaffold protein. In this chapter we will summarize our current knowledge of the architecture of the complex and of the structure of its constituents.

Keywords Ccr4-Not complex • Eukaryotic gene expression • Deadenylation • Ubiquitination • RING E3 ligase • mRNA decay • Transcription

13.1 Introduction

The Ccr4-Not complex is a functionally diverse machine essential and conserved in all eukaryotes (for review *see* Collart 2016). It performs two different and conserved catalytic functions, as well as essential scaffolding functions that are key for the regulation of eukaryotic gene expression. The former are well understood, enzymatically and structurally, while we are only gradually grasping at the complexity and importance of the Ccr4-Not scaffolding platform. Ccr4 and Caf1 are the subunits responsible for the first catalytic activity, deadenylation, which involves the shortening of the cytoplasmic mRNA poly(A) tail. This represses mRNA translation

M.A. Collart (✉) • O.O. Panasenko
Department of Microbiology and Molecular Medicine, Faculty of Medicine, University of Geneva, 1 rue Michel Servet, Geneva, Switzerland
e-mail: martine.collart@unige.ch; olesya.panasenko@unige.ch

J.R. Harris, J. Marles-Wright (eds.), *Macromolecular Protein Complexes*, Subcellular Biochemistry 83, DOI 10.1007/978-3-319-46503-6_13

and is the rate-limiting step that initiates mRNA degradation. Appropriate control of this step in the gene expression pathway is essential in many biological processes. Not4 is the subunit responsible for the second catalytic activity, ubiquitination, which involves covalent attachment of one or multiple 8 kDa ubiquitin moieties at lysine residues of the target protein. Poly-ubiquitination generally targets proteins for degradation by the proteasome. In contrast, mono-ubiquitination serves multiple functions, ranging from modification of protein function to interaction with partners for targeting of proteins to degradation by the lysosome. Appropriate control of ubiquitination is essential for many biological processes and for cellular proteostasis. Finally, Not1 is the major protein relevant for the scaffolding function of the Ccr4-Not complex. The other Ccr4-Not complex subunits dock onto this very large scaffold protein, which is tethered to different mRNA substrates. The tethering to mRNAs is mediated by Not1 itself, by other Ccr4-Not subunits, or by other factors such as RNA binding proteins or the microRNA machinery that interact with different complex subunits (for review see Collart 2016). Hence, Not1 brings together components interacting with different Ccr4-Not complex interfaces and mRNA targets. It seems likely that the Ccr4-Not subunits that have no identified catalytic function play important roles as interaction surfaces on the Not1 scaffold.

Genes encoding subunits of the Ccr4-Not complex were first identified by genetic selections in yeast designed to identify transcription factors (for review see Collart 2003). Since these early studies implicating the complex in transcriptional regulation, many studies have demonstrated that the complex is important for transcription, mRNA quality control and export, mRNA translatability, co-translational quality control, translation regulation and mRNA decay. It also regulates post-translational protein modifications and protein decay (reviewed in Collart 2016). The implication of the complex at all stages of gene expression makes it a key player in gene expression homeostasis and in response to extracellular signaling. Not surprisingly, the complex has a role in many diverse physiological processes including cell proliferation, apoptosis, gametogenesis and embryogenesis, heart function, bone formation, energy metabolism and many more (for review *see* Shirai et al. 2014).

Our current knowledge of the Ccr4-Not complex has been summarized recently in a large number of different reviews (Collart et al. 2012; Collart and Panasenko 2012; Doidge et al. 2012; Miller and Reese 2012; Panepinto et al. 2013; Reese 2013; Wahle and Winkler 2013; Chapat and Corbo 2014; Inada and Makino 2014; Panasenko 2014; Shirai et al. 2014; Temme et al. 2014; Xu et al. 2014; Villanyi and Collart 2015). A major recent achievement has been the resolution of many structures of Ccr4-Not constituents and the definition of the global architecture of the complex by cryo-EM as will be outlined below.

13.2 Composition of the Ccr4-Not Complex

The composition of the complex in different organisms where it has mostly been studied is summarized in Table 13.1. There are subunits that are common to the different eukaryotes and can be thought of as the core subunits. These are Not1, Not2, Not4, Not5, Caf1, Ccr4 and Caf40. Some confusion exists concerning Not5 in yeast, named CNOT3 in human and/or Not3 in other eukaryotes. In budding yeast there are 2 genes, *NOT3* and *NOT5*, which encode similar proteins. When the human gene was first isolated it was considered to be more similar to yeast Not3 than yeast Not5. However Not5 is the most relevant functional subunit in *S. cerevisiae*. Not4 is a core subunit since the gene is present in all eukaryotes and the human protein can complement the absence of the protein in *S. cerevisiae*. But it is a stable subunit of the complex only in budding and fission yeast. There are additional species-specific subunits of the complex. The main ones are Caf130 in budding yeast, the RNA binding protein Mmi1 in fission yeast, and Not10 and Not11 in human and flies.

The Ccr4-Not complex is built around Not1, a large scaffold protein of more than 200 kDa. The other subunits are organized in at least 4 different functional and physical separable modules that dock onto the Not1 scaffold (Fig. 13.1). Ccr4 and Caf1 constitute a deadenylase module, Not4 a ubiquitnation module, a third NOT module is defined by the Not2-Not3-Not5 proteins, and finally Caf40 is module 4. It is likely that Caf130 in yeast, and Not10-Not11 in human or flies, define a fifth functional module. Consistent with different functions mediated the different modules, these dock onto different areas of the Not1 scaffold (Fig. 13.1). This docking is essential for the recruitment of the different modules to the target mRNAs.

Table 13.1 Name of *CCR4-NOT* genes in yeast, flies and human

Core name	Drosophila gene name	*S. cerevisiae* gene name	*S. pombe* gene name	*H. sapiens* gene name
Not1	Not1	NOT1/CDC39	NOT1	CNOT1
Not2	Regena (Rga)	NOT2/CDC36	NOT2	CNOT2
Not5	Not3	NOT5	NOT3	CNOT3
Caf1	Pop2	CAF1/POP2	CAF1	Caf1a/CNOT7/CAF1
				Caf1b/CNOT8/POP2/CALIF
Ccr4	twin	CCR4	CCR4	CCR4a/CNOT6C
				CCR4b/CNOT6L
Caf40	Rcd1	CAF40	CAF40	CNOT9/Rcd1/CAF40/RQCD1
Not4		NOT4/MOT2/SIG1	NOT4	CNOT4
Not10	Not10			CNOT10
Not11	Not11			CNOT11/C2orf29
Caf130		CAF130		
Not3		NOT3		
Mmi1			MMI1	

Collart (2016)

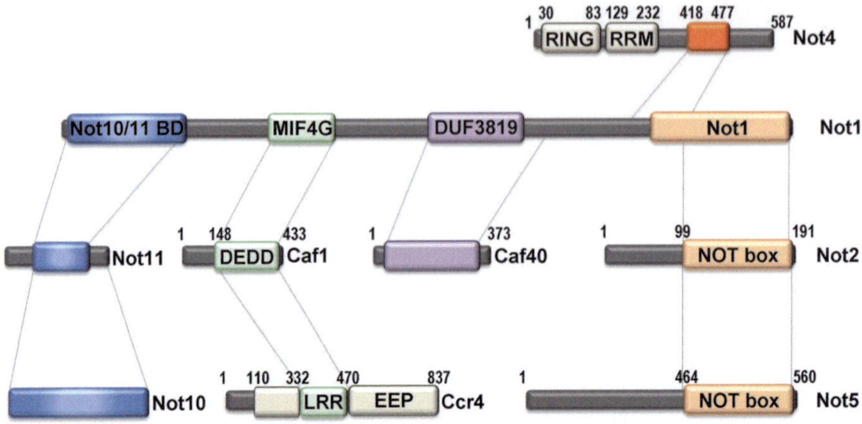

Fig. 13.1 Schematic representation of the Not1 scaffold of the Ccr4-Not complex with docking sites for the other Ccr4-Not subunits (Adapted from Bawankar et al. (2013), Collart (2016)). The numbering refers to amino acids of the yeast subunits that are discussed in the chapter. The domains of interaction between the subunits are indicated, as are additional domains of the proteins that have been identified. The human-specific Not10, Not11 subunits and their site of docking on the human Not1 scaffold are included. For each subunit, the known signature motifs are indicated by their given names (RING, RRM, MIF4G, DUF3819, DEDD, EEP, LRR and NOT box)

13.3 Architecture of the Ccr4-Not Complex

The global architecture of the complex was determined first in budding yeast in 2011 (Nasertorabi et al. 2011) and more recently in fission yeast (Ukleja et al. 2016). In the first study (Nasertorabi et al. 2011), the scaffold of the complex, Not1, was tagged at its C-terminus with a Tap-tag (consisting of a calmodulin binding peptide, a cleavage site for the Tobacco Etch Virus nuclear-inclusion-a endopeptidase (TEV) and finally a protein A moiety). Not1 was purified from total extracts on IgG sepharose beads and eluted with TEV cleavage. The Not1 eluate was then cross-linked by glutaraldehyde and fractionated on a glycerol gradient. This yielded a complex of all nine Ccr4-Not subunits with a size somewhat greater than 1 MDa. The mild cross-linking was necessary because the complex shows an inherent instability. Electron microscopy of negatively stained particles were obtained at a resolution of 33 Å. Projections of the three-dimensional map obtained revealed an L-shaped particle with 2 arms of similar lengths (180 and 190 Å) (Fig. 13.2a). The shorter arm was evaluated to account for approximately 600 kDa, the longer thinner arm for 300 kDa and the hinge for 100 kDa. The relative arrangement of the arms was observed to be variable, indicating flexibility of the whole assembly. Moreover the arms defined also a cavity in the center that could provide a platform for controlled interaction with RNA and accessory factors.

In the more recent study, the Not2 subunit was tagged at its C-terminus with the same tag used previously for Not1. The tagged Not2 was purified first by IgG sepharose and eluted by TEV cleavage. Second it was separated on a glycerol gradient.

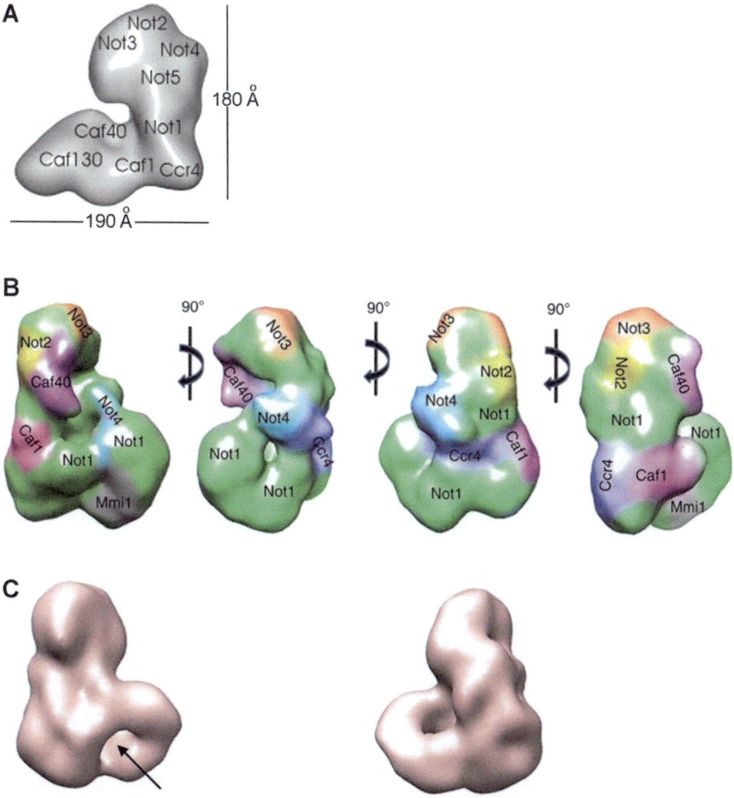

Fig. 13.2 **EM structures of the Ccr4-Not complex**. (**a**) EM structure of the *S. cerevisiae* Ccr4-Not complex with a proposed location of the nine subunits (Nasertorabi et al. 2011). The structure in the EMBD data base is ID 1901. (**b**) EM structure of the *S. pombe* Ccr4-Not complex with the mapped positions of the Ccr4-Not subunits and the RNA binding protein Mmi1 (Ukleja et al. 2016). The 3D reconstruction code in EMBD is EMD3232. (**c**) Structure of the *S. pombe* Ccr4-Not complex without Mmi1. The "hole" from lack of Mmi1 is indicated by an *arrow* in the left structure. 2 orientations are shown

High salt conditions were used to avoid co-purification of non-specific components. This procedure yielded the core Ccr4-Not components and an additional fission yeast-specific subunit Mmi1. Negative staining and 3D–reconstruction defined the same L-shaped structure described previously, with slightly shorter arms (140 and 150 Å). For cryo-EM the authors repeated the procedure with low glutaraldehyde concentrations in the glycerol gradient as used in the first study. They obtained a volume at 20 Å resolution very similar to the negative staining EM but with additional details (Fig. 13.2b).

Mmi1 is an RNA binding protein that is important for removal of meiosis-specific transcripts during vegetative growth. Its presence in the purified fission yeast Ccr4-Not complex seems to indicate a role of Ccr4-Not in regulation of meiosis in *S. pombe*. A complex without the essential Mmi1 protein was obtained using a biochemical trick. The 3D reconstruction rendered a volume that was very similar, but with a small channel in the shorter mass (Fig. 13.2c).

The EM structures of the Ccr4-Not complex from fission and budding yeast are very similar, indicating that the subunit organization within the complex is conserved. For the *S. pombe* complex, immuno-EM was used to localize each subunit tagged with the green fluorescent protein (GFP) within the 3D reconstruction. The protrusions in the different images enabled the localization of the different subunits on the L-shaped front view (Fig. 13.2b).

13.4 The Ccr4-Not Complex Consists of Different Functional Modules

13.4.1 The Not2-Not5 Heterodimer

13.4.1.1 Physiological Relevance of the NOT Module

The Not2 and Not5 proteins form a heterodimer that associates with a C-terminal region of Not1 (Fig. 13.1). This forms the NOT module very important for the integrity of the entire Ccr4-Not complex. Budding yeast lacking either one of these 2 subunits have a doubling time more than 3 times longer than wild-type cells (Maillet et al. 2000) and deletions of either one of the genes encoding these 2 subunits is embryonic lethal in all animals (Maeda et al. 2001; Kamath and Ahringer 2003). In addition specific functions have been attributed to Not2 and Not5. In mouse and in human this module is important for self-renewal of ES cells, and not surprisingly mutations in CNOT3, the human ortholog of yeast Not5 (Table 13.1) have been identified in several cancers. For instance *CNOT3* was identified as a tumor suppressor in T-cell acute lymphoblastic leukemia and its knockdown in a sensitized *Drosophila melanogaster* model caused cancer (De Keersmaecker et al. 2013). Mutations in *CNOT3* have also been defined as a modifier of mutations in *PRPF31* responsible for retinitis pigmentosa, preventing retinal degeneration (Venturini et al. 2012). Along the same lines, a recent study modeled changes in expression of *CNOT2* and showed that this was predictive for tumor cell metastatic potential *in vivo* (Faraji et al. 2014).

An important role of the NOT module for the ability of animals to stock and mobilize nutrients has been suggested by the observation that heterozygous CNOT3+/− mice are born smaller than wild type mice with a particular reduction of liver and adipose tissues and a higher metabolic rate (Morita et al. 2011). Such heterozygous mice show osteopenia and exacerbated ageing-induced osteoporosis (Watanabe et al. 2014). They have also revealed that *CNOT3* is a critical regulator of heart function (Neely et al. 2010). Conditional knock out of *CNOT3* has provided knowledge about additional specific functions. In B lineage cells, it revealed that CNOT3 is important for B cell differentiation because it controls *Igh* rearrangements and it is important for p53 mRNA destabilization (Inoue et al. 2015). In mouse embryonic fibroblasts instead, it promotes necroptosis, and this correlates with induction of mRNAs encoding cell-death inducing proteins (Suzuki et al. 2015). Clearly the Not2-Not5 module is essential for a wide-range of physiological phenomena.

13.4.1.2 Molecular Mechanism of NOT Module Action

At the cellular level, the function of the Not2-Not5 module is still unclear. Neither protein has a catalytic domain. It seems likely that this module serves as a protein interaction surface that can regulate the enzymatic activities of the Ccr4-Not complex. Indeed in several studies down-regulation of either protein was shown to have consequences on specific mRNA levels, or less frequently on length of poly(A) tails, probably because it is important for the recruitment of the Ccr4-Not complex to specific mRNAs (Morita et al. 2011). It has also been reported to contribute to decapping (Muhlrad and Parker 2005; Alhusaini and Coller 2016). One has to keep in mind that any function identified by phenotypes of cells lacking Not2 or Not5 could be indirectly due to decreased Ccr4-Not complex integrity. There is nevertheless at least one example in *Drosophila* where Not3 binds an RNA-binding protein Bicaudal-C and thereby contributes to the recruitment of the entire Ccr4-Not complex to mRNAs for their deadenylation (Chicoine et al. 2007). Moreover the structural characterization of the NOT module has shown that Not2 and Not5 NOT boxes create a V-shaped surface that displays patches of positively charged residues and have the ability within the NOT module to bind poly(U) RNA (Bhaskar et al. 2013).

Mostly this module has been connected to transcription. It interacts with a diversity of general and specific transcription factors, is present at sites of transcription and has an impact on the presence of transcription factors at promoters and on global histone acetylation levels (for review see Collart 2016). The Not2-Not5 heterodimer is also important during translation, since it is needed during synthesis of the largest subunit of RNA polymerase II to prevent its aggregation (Villanyi et al. 2014).

13.4.1.3 Structure of the Not Module

A 2.8 Å resolution crystal structure of a core of Not proteins from budding yeast (Bhaskar et al. 2013) and a corresponding structure from a core of human Not proteins at 3.2 Å resolution (Boland et al. 2013) are available (Fig. 13.3a, b). In both cases the stoichiometry is 1:1:1. The C-terminal region of Not1 that interacts with the Not2-Not5 heterodimer adopts an L-shape with two perpendicularly arranged stacks of alpha helices (shown for human CNOT1 on Fig. 13.3c with a cartoon representation on Fig. 13.3d). The topology is typical of that observed in HEAT repeat proteins where the repeats are characterized by a helix A-turn-helix B motif and are arranged in tandem in an almost parallel way. The Not1 C-terminal domain organization does not change in the trimeric NOT module (cartoon on Fig. 13.3e).

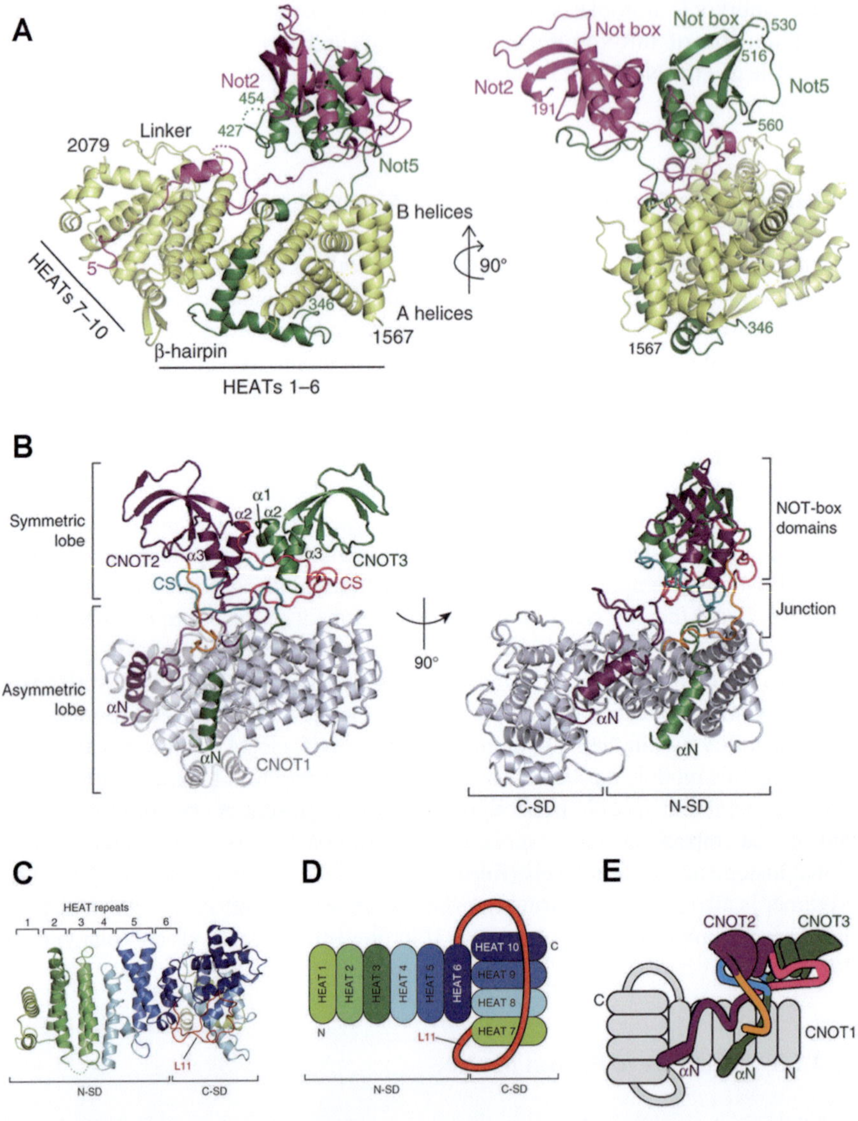

Fig. 13.3 **Structure of the NOT module** of the Ccr4-Not complex. (**a**) Structure of the *S. cerevisiae* NOT module (accession code PDB 4B46) (Boland et al. 2013). Not1 is depicted in *yellow* (residues 1567–2079), Not2 in *pink* (residues 5–191) and Not5 in *green* (residues 346 to 560). Disordered regions of Not2 (14–29 and 44–48) and two loops of Not5 (428–453) are excluded. Different structural elements are indicated. (**b**) Structure of the human NOT module (PDB 4C0D) (Bhaskar et al. 2013). The human hetero-trimeric complex contains CNOT1 indicated in *silver* (1842–2353), CNOT2 in *pink* (350–540) and CNOT3 in *green* (607–748). The symmetric lobe defined by the NOT box domains and the asymmetric lobe comprising the N-terminal extensions of CNOT2 and CNOT3 interacting with CNOT1 are indicated. (**c**) Structure of the C-terminal HEAT domain of CNOT1 that associates with the Not heterodimer (PDB 4C0E). (**d**) Cartoon scheme of the structure of the C-terminal Not1 HEAT domain presented in (**c**) (Bhaskar et al. 2013). (**e**) Cartoon scheme of the human NOT module showing that the structure of the CNOT1 domain does not change upon CNOT2-CNOT3 binding (Bhaskar et al. 2013)

Not5 and Not2 share a conserved C-terminal domain called the Not box (Fig. 13.1). This domain is a homo- or hetero-dimerization domain, and it is very similar for both proteins, human or yeast. The Not5 and Not2 homo-dimers differ in the conformation adopted by their N-terminal helices (compare Fig. 13.4a for CNOT3 homo-dimers and Fig. 13.4b for CNOT2 homo-dimers). The hetero-dimer adopts the same arrangement as the Not5 homo-dimer (Fig. 13.3a, b). The Not box is located between residues 99 and 191 for yeast Not2 and residues 464–560 for yeast Not5. Extensive interactions between the 2 proteins are centered on Phe114 and Leu115 of Not2, including Glu177, Trp176, Thr111, Phe107 and Leu106 and centered on the corresponding Phe479 and Ile480 of Not5, including Lys471, Phe472, Thr476 and Trp541 (Fig. 13.4c).

A 3D model obtained by negative EM staining of the Not2-Not5 dimer from *S. cerevisiae* shows 2 connected masses, one of about 110 Å long and one of about 70 Å long (Ukleja et al. 2016). The central part of the volume encompasses a spacious cavity most likely occupied by the C-terminal part of Not1 (Fig. 13.4d, upper row). This volume was located within the cryo-EM reconstruction of the entire Ccr4-Not complex with some manual fitting (Fig. 13.4d, lower row).

Besides their C-terminal Not box, Not2 and Not5 have extended N-terminal domains (residues 5–75 for yeast Not2 and 346–404 for yeast Not5) that span more than 100 Å and wrap around an important surface on the Not1 interaction domain (cartoon on Fig. 13.3e). The N-terminal extension of Not2 is organized in 3 segments. The first segment of Not2 (residues 5–13) binds the alpha helices of HEAT repeats 9 and 10 of the Not1 C-terminal domain (Fig. 13.5a). Leu9 of Not2 inserts into a hydrophobic pocket of Not1 defined by a number of residues including F2064, I2071, L2027, L2013 and Y1982. Residues Leu6 and Pro11 of Not2 also make important contacts. The second segment of Not2 (residues 31–64) formed by a short helix and a hairpin docks back and forth between the HEAT repeats 4–6 of the Not1 C-terminal domain (Fig. 13.5b). The helix positions itself on a hydrophobic surface centered around Arg1811 and Leu1814 of HEAT 5 with residues Leu34, Met37 and Leu41 contributing to dock the helix. The hairpin fits into a groove of HEAT repeats 4 and 5 defined by residues Phe1751 to Ile1812, such that residues Trp60, Pro59 and Phe56 can make relevant contacts (Fig. 13.5b). Finally, the third segment of Not2 between residues 65–75 interacts over the beta helices of HEAT 3 of the Not1 C-terminal domain (Fig. 13.5c). The extended N-terminal domain of Not5 is also divided in 3 relevant segments that cover HEAT repeats 1–5 of the Not1 C-terminal domain. First apolar interactions occur between residues 346–373 within an alpha helix of Not5, in particular residues Trp362, Phe350, Pro353 and Phe354, and an alpha helix of Not1 above Val 1673 and Val 1671 (Fig. 13.5d). The second segment (residues 374–391), in particular residues Met380, Leu384 and Leu388 of an alpha helix of Not5, makes hydrophobic contacts with an intra-repeat region of HEATs 3–5 of the Not1 C-terminal domain, around residues Phe1788, Pro1743 and Phe1751 (Fig. 13.5e). Finally the third segment of Not5 (residues 392–404) makes both polar and apolar contacts, involving in particular Asp393, Asp396, Pro400 and Leu402, in the vicinity of Arg65 of Not2, and on top of beta helices between HEAT repeats 1–3 of the Not1 C-terminal domain (Fig. 13.5c).

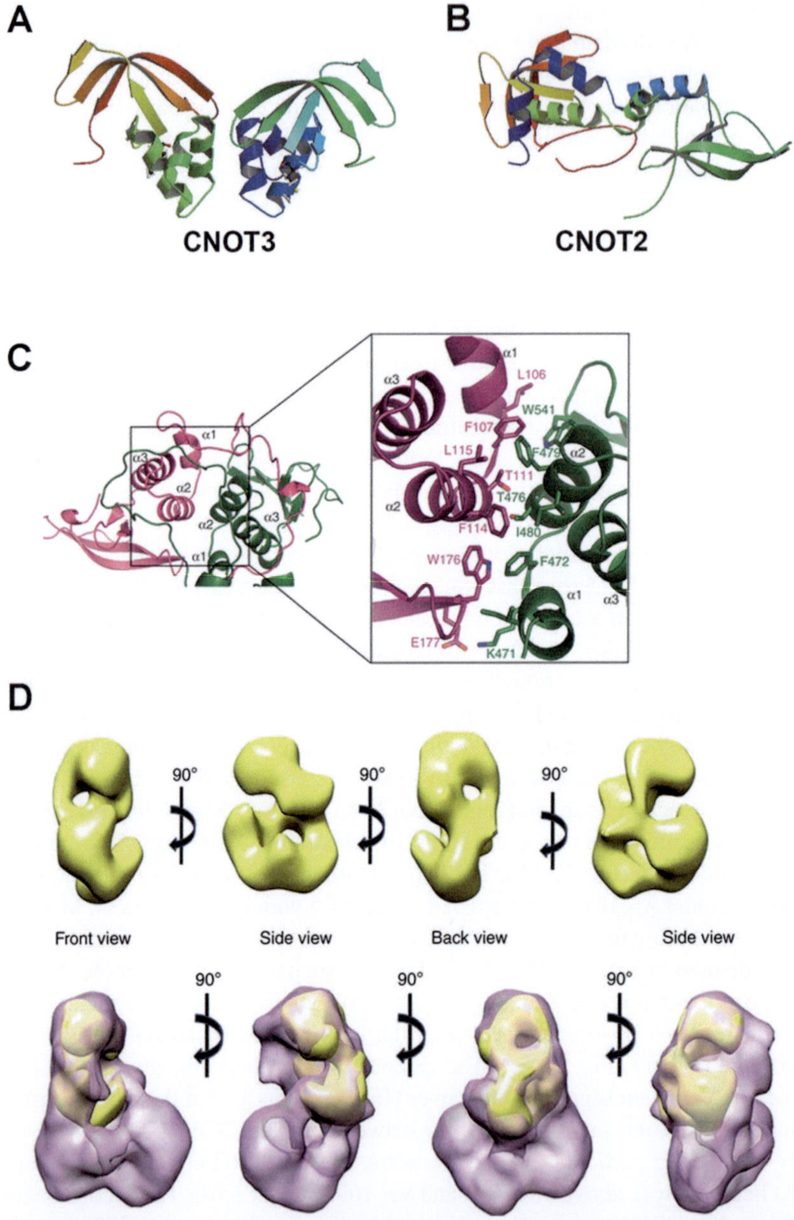

Fig. 13.4 Structures of the NOT box dimers. (**a**) The human CNOT3 homo-dimer is depicted (PDB 4C0G) (Boland et al. 2013). (**b**) The human CNOT2 homo-dimer is depicted (PDB 4C0F). (**c**) The Not box interaction surface between yeast Not2 and yeast Not5 is depicted (Bhaskar et al. 2013) (PDB 4BY6). Not2 is indicated in *pink* and Not5 in *green*. (**e**) 3D reconstruction of the *S. cerevisiae* Not2-Not5 heterodimer at 16 Å resolution (*upper row*) and manual docking of this reconstruction onto the *S. pombe* cryo-EM 3D reconstruction shown above in Fig. 13.3 (Ukleja et al. 2016)

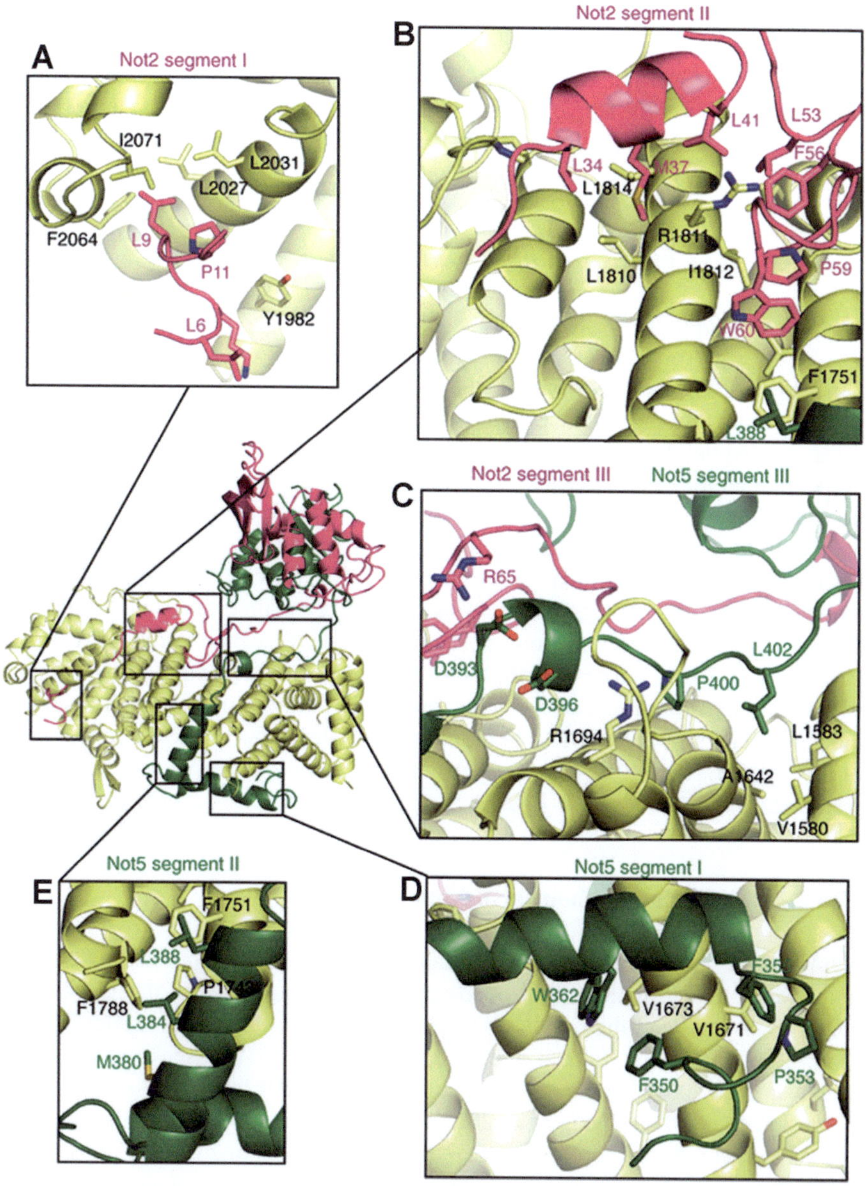

Fig. 13.5 Expanded view of the interaction surfaces between yeast Not1, Not2 and Not5 (Bhaskar et al. 2013). Conserved residues that participate in interactions are indicated. Not1 is in *yellow*, Not2 in *pink* and Not5 in *green*. (**a**) The Not2 segment 1 that interacts with Not1 is expanded. (**b**) The Not2 segment 2 that interacts with Not1 is expanded. (**c**) The Not2 and Not5 segments 3 that interact with each other and with Not1 are expanded. (**d**) The Not5 segment 1 that interacts with Not1 is expanded. (**e**) The Not5 segment 2 that interacts with Not1 is expanded

These structures reveal that the formation of the NOT module requires the cooperative binding of the N-terminal domains of Not2 and Not5 that extend from the Not box to the C-terminal domain of Not1. The Not2-Not5 Not box heterodimer forms a symmetric lobe above the conserved helical surface of the more N-terminal portion of the Not1 interacting domain (Fig. 13.3a, b).

13.4.2 *The Ubiquitylation Module*

13.4.2.1 Pysiological Role of Not4

Our knowledge of Not4's physiological role is less extensive than for the other core Ccr4-Not subunits, and it concerns mostly yeasts. In *Candida albicans* Not4 is necessary for hyphal development and pathogenicity (Krueger et al. 2004). In *K. lactis* it is important for carbon metabolism (Mazzoni et al. 2005). In *S. cerevisiae* Not4 is necessary for resistance to DNA replication stress induced by hydroxyurea (Mulder et al. 2005) probably because it is needed for destabilization of Cdc17 itself needed for resistance to hydroxyurea (Haworth et al. 2010). Not4 is also needed during oxidative stress in budding yeast for degradation of cyclin C, that with the Cdk8 kinase represses a subset of stress genes needed for resistance to oxidative stress (Cooper et al. 2012). It also contributes to degrade the oxidative stress-induced transcriptional activator Yap1, and thereby limits the duration of the stress response induction (Gulshan et al. 2012). In human and in flies, Not4 is a positive regulator of the JAK/STAT pathway essential for innate immunity, organogenesis and stress responses (Gronholm et al. 2012).

13.4.2.2 Molecular Mechanism of Not4 Action

Not4 contains an N-terminal RING domain (Fig. 13.1) and harbors E3 ligase activity in yeast and in human. It depends upon its interaction with the Ubc4/5 E2 enzymes in yeast and the ortholog UbcH5 in humans (Albert et al. 2002; Mulder et al. 2007). It ubiquitylates a wide range of substrates (Panasenko et al. 2006; Mersman et al. 2009; Cooper et al. 2012; Gulshan et al. 2012; Panasenko and Collart 2012; Sun et al. 2015; Zhang et al. 2015), in the nucleus and in the cytoplasm, and as indicated above, this activity is certainly responsible for many of the physiological roles assigned to Not4. Curiously however, mutations that abolish Not4's interaction with its partner E2 enzymes in budding yeast has only very mild phenotypes compared to deletion of the entire RING domain, or C-terminal truncation of the protein that abolishes Not4's binding to Not1 (reviewed in Collart 2013). Thus the importance of Not4's docking onto Not1 and function as E3 ligase still needs to be clarified in the context of global Not4 function.

Recent data has linked Not4 to translation and in particular to co-translational quality control (reviewed in Collart 2013; Panasenko 2014). However the function of Not4 during translation is rather controversial. It is reported to poly-ubiquitinate proteins at stalled ribosomes (Dimitrova et al. 2009), repress translation initiation from mRNAs carrying stalled ribosomes (Preissler et al. 2015) or finally preserve translation from mRNAs carrying transiently stalled ribosome complexes (Halter et al. 2014). A model to reconcile these different observations suggests that Not4's action may occur in 2 steps, first prevent repression of translation initiation in case the ribosome can resume elongation, and second let repression occur and allow mRNA decay to proceed if the ribosome is definitively stalled (Collart 2016).

A ribosome-associated chaperone and a ribosomal protein are Not4 substrates, and interestingly these substrates are not de-stabilized by ubiquitination but accumulate instead as stable mono-ubiquitinated proteins. However whether these ubiquitination events mediate the roles of Not4 in translation is not known. Not4 also has a putative RNA-Recognition-Motif (RRM) and it docks onto the Not1 scaffold via a C-terminal domain (Fig. 13.1).

13.4.2.3 Structure of the Ubiquitination Module

The structure of the RING domain of CNOT4 was solved first (Hanzawa et al. 2001) and is unusual in the sense that it folds via a C4C4 motif in which zinc is coordinated by 8 cysteine residues. A structure of the RING domain of yeast Not4 with the yeast Ubc4 E2 enzyme (Fig. 13.6a) became available recently (Bhaskar et al. 2015). It is very similar to the human Not4 RING domain in isolation and also to the previously determined structure of Ubc4 in isolation (Cook et al. 1993). This structure allowed for the suggestion that the specificity of Not4 for the Ubc4 E2 enzyme is due to subtle differences in the more extensive interaction network possible between Ubc4 and Not4 compared to what could occur between Not4 and other E2 enzymes.

A complex between the C-terminal region of yeast Not4 (residues 418–477) and residues 1541–2093 of yeast Not1 led to crystals that diffracted to a resolution of 3.6 Å (Bhaskar et al. 2015). The HEAT repeat structure of Not1 in this structure is similar to that in the NOT module (see above). The C-terminal domain of Not4 that binds Not1 folds into an alpha helix that is flanked by regions lacking defined secondary structural elements (Fig. 13.6b). Several amino acids are critical for three different points of contact between Not1 and Not4 (Fig. 13.6c). The mutations of 2 residues in the first interacting segment of Not4, L463E, F464E or F464E, W466E, is sufficient to disrupt the association of the 2 proteins. The resolution of this structure has provided some explanation for the curious observation that Not4 is a stable subunit of the complex in budding and fission yeast but not in metazoa (Lau et al. 2009; Temme et al. 2010). Indeed, the third segment of Not4, essential for stable binding to Not1 in yeast, is not present in CNOT4.

The NOT module and the ubiquitnation module bind to a similar C-terminal domain of Not1. However, a superposition of the yeast NOT structure and the structure of the ubiquitnation module suggests that the binding of the Not2-Not5 heterodimer and Not4 enzyme onto Not1 are independent (Fig. 13.6d).

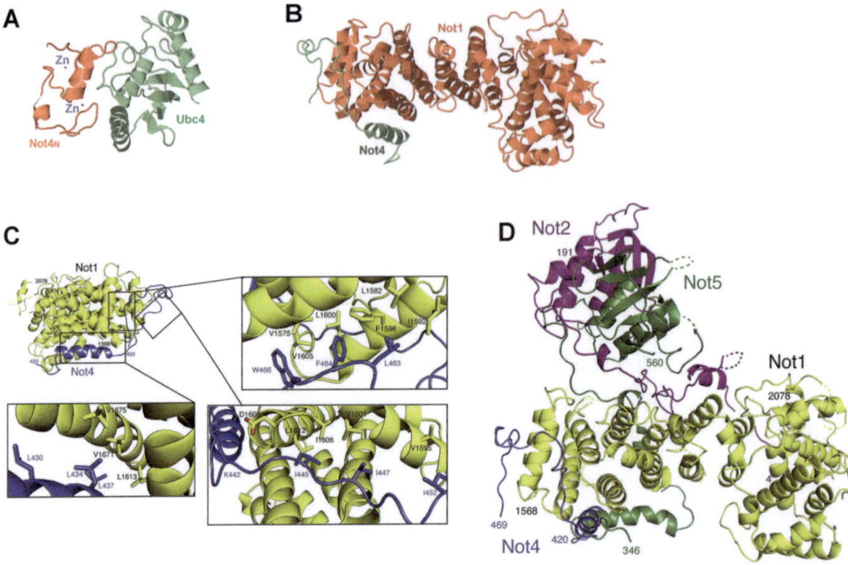

Fig. 13.6 Structure of the ubiquitination module of the Ccr4-Not complex (Bhaskar et al. 2015). (**a**) Structure of the complex between the *S. cerevisiae* N-terminal Not4 RING E3 domain (residues 30–83) depicted in *orange* and the *S. cerevisiae* Ubc4 E2 (residues 1–148) depicted in *green* (PDB 5AIE). 2 ions of Zn^{2+} are indicated. (**b**) Structure of the C-terminal domain of *S. cerevisiae* Not1 (residues 1541–2093) depicted in *orange* and the *S. cerevisiae* Not4 C-terminal domain (residues 418–477) depicted in *green* (PDB 5AJD). (**c**) Expanded view of the interaction interface between Not1 and Not4. Not1 is in *yellow* and Not4 is in *purple*. The residues involved in interactions are shown as *sticks* and their positions are labeled. (**d**) Superposition of the NOT and ubiquitination modules determines that Not4 C-terminal domain and the Not2-Not5 heterodimer bind independently. Not1 is indicated in *yellow*, Not2 in *pink*, Not5 in *green* and Not4 in *green* (Not1 interaction domain) and *blue* (Bhaskar et al. 2015)

13.4.3 Caf40

13.4.3.1 Physiological Relevance of Caf40

Caf40 was first known under the name Rcd1 (Required for Cell Differentiation) when it was identified as a factor essential for nitrogen-starvation-induced sexual differentiation in fission yeast (Okazaki et al. 1998). In mammalian cells it was described as a critical transcriptional co-factor for retinoic acid-induced differentiation of teratocarcinoma cells and for lung development (Hiroi et al. 2002). It is frequently up-regulated in breast cancer bioposies and cell lines, and its knock down suppresses their proliferation (Ajiro et al. 2009). Very recently a recurrent mutation in Caf40, P131L, was identified in cutaneous melanoma (Wong et al. 2015). In budding yeast, Caf40 has not been specifically characterized by mutant phenotypes, and it was identified biochemically as a component of the Ccr4-Not complex (Chen et al. 2001).

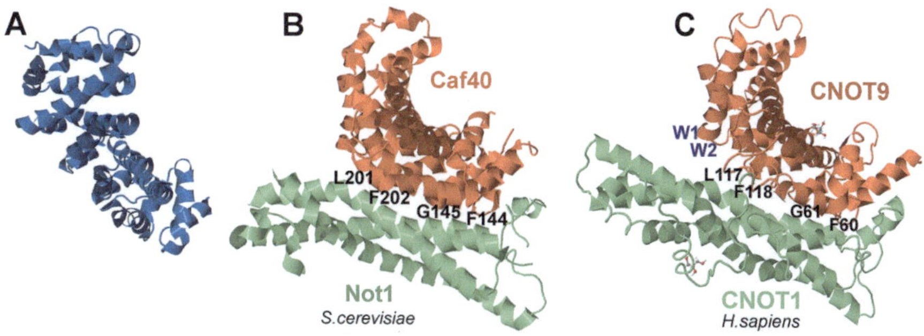

Fig. 13.7 Structure of the Caf40 module of the Ccr4-Not complex. (**a**) Cartoon representation of human CNOT9 (PDB 2FV2). (**b**) Structure of the *S. cerevisiae* Caf40 in complex with the DUF Not1 binding domain (PDB 4CV5). Not1 is in *green* and Caf40 is in *orange* (Mathys et al. 2014). Caf40 resides that contribute to interaction are indicated on the interface in a cartoon way. (**c**) Structure of human CNOT9 in complex with the CNOT1 DUF domain (PDB 4CT7) in the same color code as for panel B (Mathys et al. 2014). CNOT9 resides that contribute to the interaction are indicated above the interface. W1 corresponds to CNOT1 Trp1603 that is included in the structure despite being within an unstructured region and W2 is an additional Trp that gets included in the structure upon addition of an excess of free L-Trp (Chen et al. 2014). These Trp residues reveal that Caf40 has Trp-binding pockets

13.4.3.2 Molecular Mechanism of Caf40 Action

Until recently very little was understood about the mechanism of Caf40 action. It interacts with transcription factors and alters their activation potential (Hiroi et al. 2002; Haas et al. 2004; Garapaty et al. 2008). It can bind single stranded DNA polymers, but surprisingly not poly(A) sequences (Garces et al. 2007). Finally, it is important for the interaction of an RNA helicase Mtr4 and an exoncuclease Rrp6, which are co-factors of the nuclear exosome (Azzouz et al. 2009). A breakthrough has been the recent finding that the interaction of Caf40 with Not1 provides binding sites for GW182/TNRC6 proteins of the microRNA machinery and thereby contributes to gene silencing by microRNAs (Chen et al. 2014; Mathys et al. 2014). A common theme between these different observations is that Caf40, through its association with Not1 and ability to bind single stranded nucleic acids, might provide surfaces that tether together proteins and nucleic acids. The recent structural resolution of Caf40-Not1 complexes makes it now possible to test Caf40 derivatives that do not associate with Not1 and progress in our understanding of this functional module of Ccr4-Not.

13.4.3.3 Structure of the Caf40 Module

A first crystal structure of human Caf40 revealed that the protein has an Armadillo repeat formed by a bundle of alpha helices arranged in a clockwork spiral (Garces et al. 2007) (Fig. 13.7a). A positively charged cleft was suggested to be important for nucleotide binding while the super-helix produced by the folded Armadillo repeats thought to provide a versatile platform for protein-protein interactions.

Caf40 and its orthologs dock onto a DUF3819 domain in the middle of the Not1 protein (Fig. 13.1). A complex of yeast Caf40 (residues 102–370) and yeast Not1 (residues 1071–1282) was solved at 3.8 Å resolution (Fig. 13.7b) (Mathys et al. 2014). A similar complex of the human orthologs, CNOT9 (residues 16–284) and CNOT1 (residues 1351–1588), was obtained and refined at 2 Å resolution (Fig. 13.7c) (Mathys et al. 2014). Another structure at 1.65 Å resolution of a complex between CNOT9 and a fragment of CNOT1 composed of residues 1356–1607 was reported independently (Chen et al. 2014). The new structures of Caf40 and CNOT9 in complex with Not1 and CNOT1 are virtually identical to the previous one of CNOT9 in isolation, suggesting that CNOT9 does not undergo major rearrangements upon CNOT1 binding. It was observed that CNOT9 in isolation forms homo-dimers, but the interaction surface between CNOT9 and CNOT1 partially overlaps with the homo-dimerization domain of CNOT9 and it is more extensive, indicating that CNOT9 and CNOT1 complexes are probably favored over the homodimer in vivo. Multiple mutations are needed to disrupt this interface.

The CNOT9-binding domain of CNOT1 is composed of 7 alpha helices. It differs from the other docking domains of Not1 organized into HEAT repeats and it is quite a unique folding domain. The specific region of CNOT1 that contacts CNOT9 is very much disordered. It participates in an important way in the complex however, because it inserts Trp1603 into a Trp-binding pocket of CNOT9 and it is therefore required for crystallization. This Trp residue is clearly defined in the structure (indicated W1 on Fig. 13.7c). A second Trp-binding pocket in CNOT9 is visible when excess of L-Trp is added in the crystallization solution (W2 on Fig. 13.7c). The distance between the two Trp pockets is 20–25 Å. They are not relevant for CNOT9 homo-dimerization. However, if the 2 pockets are mutated (A248F, Y203A, R244A for pocket 1 and P165G, R205A, H208A for pocket 2), CNOT9 in complex with its CNOT1 docking domain fails to bind TNRC6 in vitro. This would suggest that CNOT9 has the primary Trp-binding sites yet in vivo CNOT9 does not bind TNRC6 if it is defective for interaction with CNOT1. Moreover full-length CNOT1 mutants that no longer interact with CNOT9 can bind TNRC6. These observations indicate that CNOT1 must carry additional sites for TNRC6 besides those available on CNOT9. Similar observations were made with the *Drosophila* proteins indicating that these interactions are conserved. They are consistent with many studies indicating that GW182/TNRC6 proteins bind CNOT1 through multiple W-containing motifs. Moreover binding redundancy correlates also with functional redundancy for mRNA repression.

13.4.4 *The Ccr4-Caf1 Nuclease Module*

13.4.4.1 Physiological Roles of Ccr4 and Caf1

Most eukaryotic mRNAs are heterogeneously polyadenylated. This ensures their stability in the cytoplasm and their translatability. Ccr4 and Caf1 are 2 different deadenylating enzymes. They come into action after an initial shortening of poly(A) tails to a rather homogenous length of approximately 110 nucleotides by the Pan2-Pan3 deadenylases. They function at the rate-limiting step, to deadenylate mRNAs further to a rather heterogenous length between 20 and 110 nucleotides. This initiates mRNA degradation and translational repression. This process concerns all polyadenylated mRNAs and is the major route for mRNA degradation in eukaryotes.

Since Ccr4 and Caf1 contribute to decay of all mRNAs, one might expect that these proteins are essential. However, this is not the case either in higher eukaryotes or in yeast, where they are encoded by single copy genes. Hence one has to conclude that deadenylation by Ccr4 and Caf1, even though it is the major route for mRNA decay of all polyadenylated mRNAs in eukaryotes, is only essential under certain circumstances. In most cases, either cells use alternative routes to degrade mRNAs in the absence of Ccr4 and Caf1, such as the 3′ to 5′ exonuclease called the exosome, or cells adapt to increased mRNA stability by reducing transcription, or more likely both mechanisms co-exist.

Ccr4

Genetic and RNA interference experiments have defined the role of Ccr4 in many different species ((Goldstrohm and Wickens 2008) and for review see (Shirai et al. 2014)). In yeast only Ccr4 but not Caf1 is active as an enzyme, whereas in contrast in *Trypanosoma brucei* no ortholog of Ccr4 has been identified. In *Drosophila* cell lines, depletion of Ccr4 has no major effect on bulk poly(A). However adult flies bearing a null allelic combination of its encoding gene *twin* have an increased steady-state bulk poly(A) tail length, and decay of specific mRNAs is slower than in wild type. The females are affected in their reproductive capacity. Moreover, embryos derived from *twin* mutant mothers have a reduced viability (for review see Temme et al. 2014). Self-renewal of germ line stem cells needs recruitment of Ccr4 by the specific RNA binding proteins Nanos and Pumilio. In *C. elegans* germ cells, depletion of *CCR4* leads to a global increased poly(A) tail length (Nousch et al. 2013). In mammals, the 2 Ccr4 orthologs have partially overlapping functions, but distinctive tissue-specific distribution, with CNOT6L being ubiquitous and CNOT6 restricted to testis ovary, thymus and spleen (Chen et al. 2011). Consistently, CNOT6L seems to play a more major role. Nevertheless, mice lacking either ortholog are viable and fertile. Both human orthologs contribute to growth of the MCF-7 breast cancer cell line, but only CNOT6L contributes to growth of murine NIH3T3 fibroblasts (Morita et al. 2007). Moreover depletion of CNOT6L has more important effects on mRNA levels than CNOT6 (Morita et al. 2007; Mittal et al. 2011). Recent data has linked changes in expression of the Ccr4 orthologs to several cancers. Loss of CNOT6L copy number in human colon adenocarcinoma samples (Tay et al. 2011) or reduced expression of CNOT6L in ALL or AML leukemia cells with slight up-regulation of CNOT6 (Maragozidis et al. 2012) have been reported, as well as a significant association of SNPs in CNOT6 with B-cell ALL susceptibility (Gutierrez-Camino et al. 2014).

Caf1

The biological significance of Caf1 varies among eukaryotes. In *S. cerevisiae* Caf1 is essential for deadenylation because it tethers Ccr4 to Not1, but it does not play a catalytic role (Tucker et al. 2002). In *Drosophila* the catalytic activity of Caf1 is thought to be more potent than that of Ccr4 (Temme et al. 2004, 2010) and in *Trypanosoma brucei* it is the only deadenylase (Schwede et al. 2008). In mammals the two orthologs are ubiquitously expressed. They have overlapping roles in repressing expression of anti-proliferative genes and allowing cell proliferation (Aslam et al. 2009). CNOT7 is a downstream effector in interferon pathways and degrades STAT1-regulated mRNAs (Chapat et al. 2013). Its knockout in mice results in severe infertility associated with a defect in spermatogenesis (Berthet et al. 2004; Nakamura et al. 2004). The mice also show increased bone mass because CNOT7 functions as a suppressor of bone morphogenetic protein signaling (Washio-Oikawa et al. 2007). Like CNOT6L, CNOT7 is down-regulated in ALL and AML leukemia cells. CNOT8 on the other hand is elevated in colorectal carcinoma and metastatic legions compared to the normal mucosa (Seiden-Long et al. 2006).

13.4.4.2 Molecular Mechanism of Ccr4 and Caf1 Action

Ccr4 and Caf1 are both deadenylases, but they belong to distinct families based upon their nuclease domain signature. Both hydrolyze the RNA in the 3′ to 5′ direction with the concomitant release of 5′AMP. Caf1 in yeast, or the CNOT7 and CNOT8 human orthologs, belong to the DEDD-type nucleases that have conserved catalytic Asp and Glu residues. Ccr4 in yeast, or CNOT6L and CNOT6 in human, are members of the exonuclease-endonuclease-phosphatase (EEP) family. The catalytic roles of Ccr4 and Caf1 differ in different organisms. In addition, Caf1 serves in all eukaryotes an additional scaffolding or tethering function.

Deadenylation is catalyzed by two metal ions coordinated by acidic residues. Deadenylating enzymes require an active catalytic site and must be tethered to their substrates, because they do not degrade cytoplasmic mRNA poly(A) tails indiscriminately. They get tethered to their target mRNAs in several different ways. As mentioned above, Ccr4 is tethered to the Ccr4-Not complex via Caf1. In turn the complex will be recruited to mRNAs. Components of the antiproliferative BTG-

TOB family of proteins serve as one set of adapter proteins that interact with CNOT7 or CNOT8 to tether the Ccr4-Not complex to mRNAs. Tob1 for instance, interacts with the Poly(A) binding protein PABP for general recruitment of the deadenylase to mRNAs. The Caf1 proteins can also interact with CEBP3 bound to a cis element upstream of the poly(A) site for specific recruitment (for review see Doidge et al. 2012). RNA binding proteins, independently of BTG-TOB proteins, also recruit the deadenylase to mRNAs. One particularly well-studied example is tristetraprolin (TTP) that recognizes AU-rich elements and mediates rapid mRNA decay (Sandler et al. 2011). Another example is Nanos2, an RNA binding protein required for development of germ cells and male fertility (Suzuki et al. 2010) (see further) or the pumilio protein Mpt5 in yeast (Goldstrohm et al. 2006). In several cases, different subunits of the Ccr4-Not complex have been shown to associate with RNA binding proteins to tether Not1 to mRNAs. One example is CNOT3 that interacts with Biaudal-C (Chicoine et al. 2007). Finally the miRNA machinery recruits the Ccr4-Not deadenylase to mRNAs, by interacting with Not1 and Caf40 that both have Trp-binding pockets (see above, and reviewed in Collart 2016).

Deadenylation activity in vivo requires the Ccr4-Not complex even though in vitro the enzymes are active without being integrated into the complex (Basquin et al. 2012). This seems curious, since in many cases CNOT7 and CNOT8 are at the same time the catalytic subunit and the Ccr4-Not subunit that is tethered to the mRNA. Maybe the Ccr4-Not complex is important in vivo for stability of the deadenylases, but this is probably not the only explanation. Indeed, in *Drosophila*, a Caf1 mutant that does not interact with Not1 is dominant-negative (Petit et al. 2012).

13.4.4.3 Structures of the Ccr4 and Caf1 Deadenylases

The crystal structure of Caf1 was first defined for the *S. cerevisiae* protein in 2003 (Thore et al. 2003). The overall structure displays 13 alpha helices and 6 beta strands (Fig. 13.8a). It has the common fold of DEDD-type nucleases, but has instead a SEDQ motif and lacks essential residues to be catalytically active. Caf1 from the fission yeast *S. pombe* is very similar (Fig. 13.8b), but reveals an intact DEDD motif that can chelate the metal ions (Fig. 13.8c). The metal ions have a great importance on the rate of deadenylation. The presence of Zn^{2+} in the catalytic site for instance slows the reaction whereas high activity is observed with Mn^{2+} (Andersen et al. 2009). A structure of the mammalian ortholog, CNOT7, in complex with a BTG-TOB protein was solved (Horiuchi et al. 2009) (Fig. 13.8d), and the CNOT7 within this structure has high structural similarity to the yeast Caf1 proteins (reviewed in Bartlam and Yamamoto 2010). Like for the yeast protein the highest activity for CNOT7 was observed with Mn^{2+}.

Ccr4 is the major cytoplasmic deadenylase in yeast. It has a non-conserved N-terminal region that is rich in glutamines and asparagines and was originally described to interact with the transcription machinery. The central region of Ccr4 contains tandem copies of a leucine-rich-repeat domain (LRR) that interacts with Caf1 and docks Ccr4 onto the Not1 scaffold (Fig. 13.1) (Malvar et al. 1992; Draper

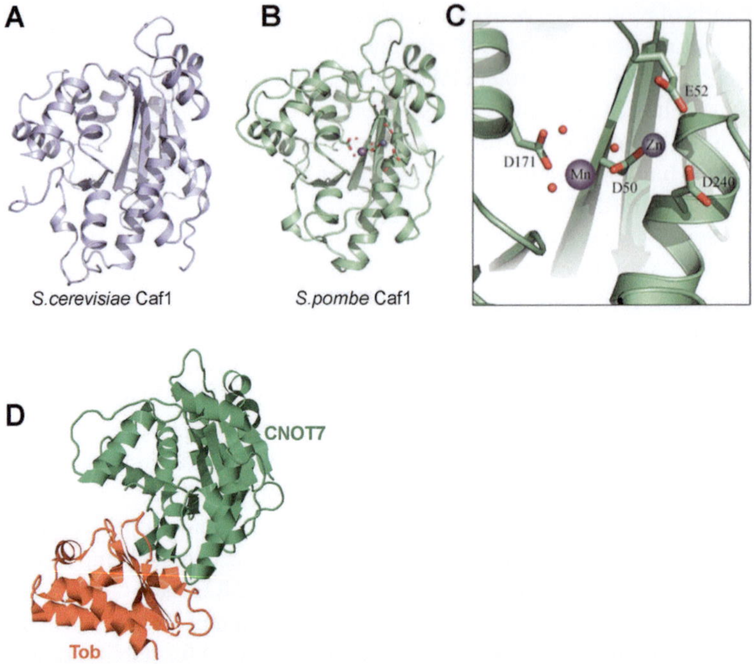

Fig. 13.8 Cartoon representations of Caf1 structures (Bartlam and Yamamoto 2010). (**a**) The structure of *S. cerevisiae* is represented (PDB 1U0C). (**b**) The structure of *S. pombe* is represented (PDB 2P51). (**c**) Expanded view of the catalytic site with Mn^{2+} is represented. (**d**) Structure of human CNOT7 (*green*) in complex with Tob1 (*orange*) is depicted (PDB 3DJU)

et al. 1995; Liu et al. 2001). The C-terminal region contains the deadenylase domain typical of the EEP superfamily, with conserved catalytic Asp and His residues in the active site. An intact Ccr4 nuclease domain structure for human CNOT6L is available (Wang et al. 2010). It has a heart-like shape in which outer layers of alpha helices sandwich 2 interior beta sheets (Fig. 13.9a). It reveals a prominent active-site pocket at the top of the molecule with 2 bound Mg^{2+} ions. Glu240 coordinates the first ion and is essential, whereas Asp410, Asn412, and His529 coordinate more weakly the second one (Fig. 13.9b). Activity is Mg^{2+}-dependent and targets strictly poly(A) RNA substrates. Structural studies have indicated that residues Pro365, Asn412, and Phe484 are important for substrate recognition and have explained substrate specificity. Indeed, even the purine base G, which shares the most similarity with A, has a carbonyl oxygen in position 6′ of the base that would clash with Asn412, while the pyrimidine bases could not be accommodated within the recognition pocket.

The deadenylase module docks onto the Not1 MIF4G domain via the Caf1 subunit. A structure of the human CNOT1 MIF4G domain (residues 1093–1317) alone (Fig. 13.10a) and in complex with CNOT7 (Fig. 13.10b) was solved. The very N- and C-terminal residues of CNOT7 are not visible in the structure and are probably flexible or disordered. The Not1 MIF4G domain comprises five helical hairpins

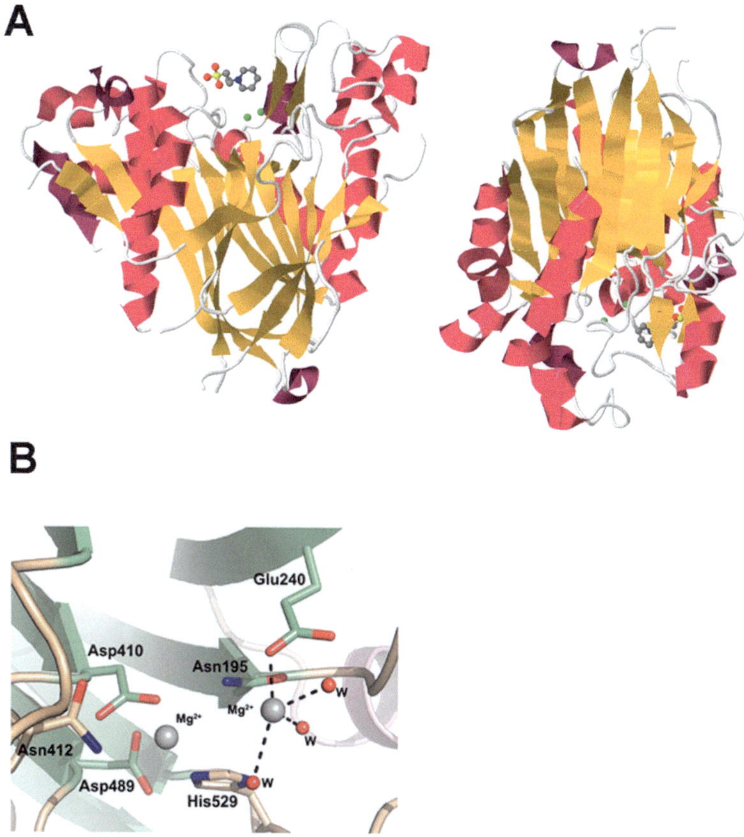

Fig. 13.9 Structure of the Ccr4 nuclease domain. (**a**) Cartoon representation of CNOT6L (PDB 3NGQ) in 2 different orientations. 3-pyridinium-1 YLPropane-1-sulfonate (PPS) used for the crystallization is visible in the structure close to 2 Mg^{2+} ions shown in *green*. (**b**) Expanded view of the catalytic site with 2 Mg^{2+} ions (Wang et al. 2010)

organized in HEAT-like repeats that are followed by a C-terminal extension that wraps around alpha helices 9 and 7 in a structurally conserved manner (Fig. 13.10a). An alignment of Not1 MIF4G domains shows a conservation of 41 residues throughout the domain. Many of these conserved residues lie within the hydrophobic core, or within internal interaction regions between the helices, indicating that these conserved residues are needed for the global integrity of the domain. Some conserved residues are also in loops between helices, solvent exposed, and thus are likely to engage in protein-protein interactions. The loops between helices 6 and 7 and 8 and 9 are the ones that are involved in Caf1 binding. Many important proline residues are also conserved (see further).

A structural view of a complete deadenylase module has been reported for the yeast complex (Fig. 13.10c). This structure contains only a partial Ccr4 nuclease domain, but the structure of the nuclease domain of CNOT6L can be superimposed

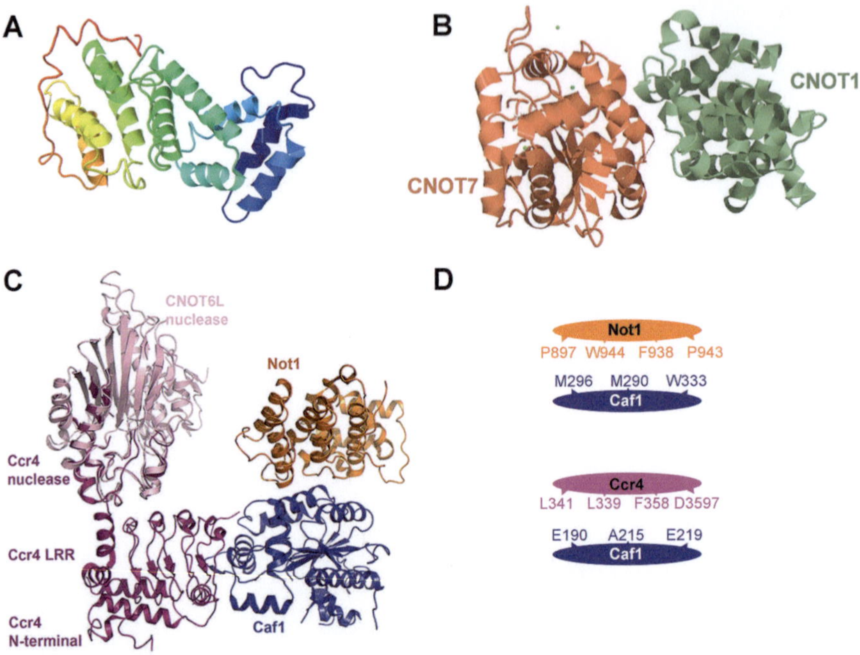

Fig. 13.10 Structure of the Ccr4-Not nuclease module. (**a**) Structure of the MIF4G domain of CNOT1 (PDB 4GML). (**b**) Structure of the complex of human CNOT7 (*orange*) with the CNOT1 MIG4G binding site (*green*) (PDB 4GMJ). (**c**) Structure of the entire nuclease module of *S. cerevisiae* (PDB 4B8C) (Basquin et al. 2012) with the additional superposition of the complete CNOT6L nuclease domain to complete the structure (Xu et al. 2014). The Not1 MIF4G domain is *orange*, Caf1 is in *blue*, yeast Ccr4 is in *dark pink* and the CNOT6L nuclease domain is in *light pink*. The different domains of Ccr4 are indicated. (**d**) Cartoon indication of the important residues within the interaction interfaces between Not1 and Caf1, as well as between Caf1 and Ccr4 (Basquin et al. 2012)

to give a complete view (Xu et al. 2014). The N-terminal arm of Not1 (residues 754–1000) shows the same HEAT motif as the human protein, with its two antiparallel helices and successive HEAT motifs packed together nearly parallel. The domain of Caf1 that associates with Not1 is opposite to its active site pocket, such that, like for CNOT7 in complex with the MIF4G domain of CNOT1, very little of the total accessible solvent area is masked within the Not1-Caf1 interface. Ccr4 interacts with Caf1 via its LRR domain. It makes no direct interaction with Not1 and its nuclease domain remains accessible. There is a disordered 70 amino acid linker between the yeast-specific N-terminal domain of Ccr4 and the LRR domain, and inter-molecular interactions build a single structural unit. The part of the Ccr4 nuclease domain that is resolved in this nuclease module is the helices that interact with the LRR domain. Hydrophobic and polar contacts are defined in the surface between the LRR and nuclease domains. A cartoon of the important residues defining interactions between the 3 subunits of the nuclease module is shown (Fig. 13.10d)

(Basquin et al. 2012). The L339E, L341E double mutation in Ccr4 disrupts the interaction between Ccr4 and Caf1 but does not touch the interaction between Caf1 and Not1, while a M290K, M296K double mutation in Caf1 disrupts the interaction with Not1 but does not affect the interaction with Ccr4 (Basquin et al. 2012).

13.4.5 Non-conserved Ccr4-Not Subunits

There is no structural information concerning the non-conserved subunits, except for the RNA binding protein Mmi1, which is present in the cryo EM structure of *S. pombe* Ccr4-Not. In the case of Not10 and Not11, only the subunits' docking site onto the very first amino acids of Not1 have been defined (Fig. 13.1) (Mauxion et al. 2013).

13.5 Additional Structural Information

Several additional structures between Not1 and other proteins are available. For instance there are structures for CNOT1 and the RNA binding protein TTP (Fabian et al. 2013) or Nanos (Bhandari et al. 2014), or for the DEAD-box protein DDX6 and CNOT1, with or without additionally the 4E-T protein (Chen et al. 2014; Mathys et al. 2014; Ozgur et al. 2015).

The Nanos proteins interact with the same C-terminal domain of CNOT1 that interacts with the CNOT2-CNOT3 heterodimer, within the so-called superfamily homology domain (SHD). They insert conserved aromatic residues into a hydrophobic pocket on the CNOT1 surface. A crystal structure between the CNOT1 SHD (residues 1833–2361) and the interaction peptide of Nanos 1 (amino acids 40–56) was resolved at 2.8 Å (Fig. 13.11a, b). The overall arrangement of the CNOT1 domain in the CNOT1-Nanos 1 structure does not change compared to its arrangement in the NOT module, except in an alpha helix that displays 2 additional turns in the CNOT1-Nanos 1 complex. This observation suggests that the Nanos binding pocket is not preformed, but instead is induced by binding.

The binding site for TTP on the Not1 scaffold is located between amino acids 800–1015 of CNOT1. A structure at 1.5 Å resolution of this domain with a peptide of human TTP covering residues Ala312 to Glu326 was obtained (Fig. 13.11c). The N-terminal 20 residues of CNOT1 of this domain are disordered, while the region between residues 820 and 999 consists of 8 alpha helices stacked as a series of helix-turn-helix motifs, similar to the typical HEAT repeats found throughout Not1. The TTP binding site is formed by a highly conserved hydrophobic groove between 2 alpha helices flanked by negatively charged patches (Fig. 13.11c).

Dhh1 in yeast, or DDX6 in human and flies, interacts with Not1 through the MIF4G domain. This protein is thought to contribute to gene repression induced by tethering of the Ccr4-Not complex to mRNAs. A co-crystal of the CNOT1 MIF4G

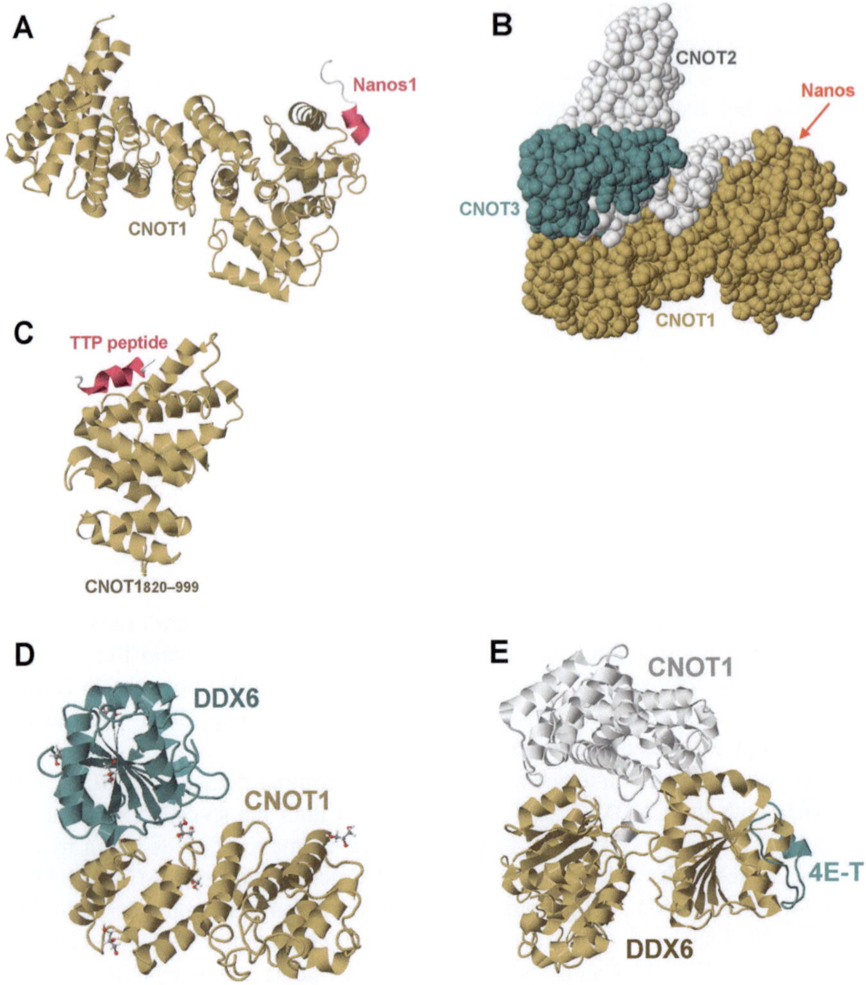

Fig. 13.11 (**a**) Structure of the Nanos peptide with the C-terminal domain of CNOT1 (PDB 4CQ0). (**b**) Representation of the NOT module with an *arrow* to indicate where the Nanos peptide binds (Bhandari et al. 2014). (**c**) Structure of the complex between TTP and CNOT1. CNOT1 (residues 820–999) indicated in *yellow* in complex with the TTP peptide (depicted in *pink*) is shown (PDB 4J8S) (Fabian et al. 2013). (**d**) Structure of the complex between the DDX6 RecA domain and the CNOT1 MIF4G domain. Glycerol used for the crystallization is visible in *red*. (**e**) Cartoon representation of the structure between the human CNOT1 MIF4G domain, DDX6 RecA1 and A2 domains, and the 4E-T CHD domain (PDB 5ANR)

domain with the C-terminal RecA domain of DDX6 was solved at 1.75 Å (Fig. 13.11d). The interaction of the RecA domain of DDX6 and the concave surface of the MIF4G domain of CNOT1 is similar to the eIF4G and eiF4A interaction, with helix 1 and loops L3, 5 and 7 of the RecA DDX6 domain interacting with the first two N-terminal HEAT repeats of the MIF4G domain. This structure reveals that the binding of DDX6 is compatible with the simultaneous binding of Caf1 occurring on an opposite surface of the MIF4G domain.

Several studies have pointed to 4E-T (eiF4E transporter) as a key translational repressor connected to DDX6. A structure between the MIF4G domain of CNOT1, DDX6 and 4E–T was solved at 2.1 Å. It includes residues 216–238 of 4E–T, residues 95–462 of DDX6 and residues 1065–1309 of CNOT1. The DDX6-CNOT1 portion of the ternary complex is similar to the binary complex described above. The concave surface of CNOT1 MIF4G faces DDX6 and interacts via its first and last HEAT repeats to contact the RecA2 and RecA1 domains of DDX6 (Fig. 13.11e).

13.6 Perspectives

The eukaryotic Ccr4-Not complex is an extremely sophisticated regulator of gene expression. The multitude of site and type of actions of the complex has been truly tantalizing. With the available structures we can start to visualize how the complex is tethered to its mRNA substrates and can act to modify these substrates at different functional sites in different ways. We are still far from a complete integrated view, but we can address specific questions by mutating selected interfaces. The need to use cross-linking for the cryo EM structures has confirmed that the complex is very dynamic, and hence we will need to consider that there may be alternative conformations from the ones for which we have structures. Moreover, the assembly of the complex in time and space within the cell is a question that has to be tackled to truly understand Ccr4-Not function in the nucleus and cytoplasm and how these are connected. This is certainly the biggest and exciting challenge ahead.

References

Ajiro M, Katagiri T, Ueda K, Nakagawa H, Fukukawa C, Lin ML, Park JH, Nishidate T, Daigo Y, Nakamura Y (2009) Involvement of RQCD1 overexpression, a novel cancer-testis antigen, in the Akt pathway in breast cancer cells. Int J Oncol 35(4):673–681

Albert TK, Hanzawa H, Legtenberg YI, de Ruwe MJ, van den Heuvel FA, Collart MA, Boelens R, Timmers HT (2002) Identification of a ubiquitin-protein ligase subunit within the CCR4-NOT transcription repressor complex. Embo J 21(3):355–364

Alhusaini N, Coller J (2016) The deadenylase components Not2p, Not3p, and Not5p promote mRNA decapping. RNA 22(5):709–721

Andersen KR, Jonstrup AT, Van LB, Brodersen DE (2009) The activity and selectivity of fission yeast Pop2p are affected by a high affinity for Zn2+ and Mn2+ in the active site. RNA 15(5):850–861

Aslam A, Mittal S, Koch F, Andrau JC, Winkler GS (2009) The Ccr4-NOT deadenylase subunits CNOT7 and CNOT8 have overlapping roles and modulate cell proliferation. Mol Biol Cell 20(17):3840–3850

Azzouz N, Panasenko OO, Colau G, Collart MA (2009) The CCR4-NOT complex physically and functionally interacts with TRAMP and the nuclear exosome. PLoS One 4(8):e6760

Bartlam M, Yamamoto T (2010) The structural basis for deadenylation by the CCR4-NOT complex. Protein Cell 1(5):443–452
Basquin J, Roudko VV, Rode M, Basquin C, Seraphin B, Conti E (2012) Architecture of the nuclease module of the yeast Ccr4-not complex: the Not1-Caf1-Ccr4 interaction. Mol Cell 48(2):207–218
Bawankar P, Loh B, Wohlbold L, Schmidt S, Izaurralde E (2013) NOT10 and C2orf29/NOT11 form a conserved module of the CCR4-NOT complex that docks onto the NOT1 N-terminal domain. RNA Biol 10(2):228–244
Berthet C, Morera AM, Asensio MJ, Chauvin MA, Morel AP, Dijoud F, Magaud JP, Durand P, Rouault JP (2004) CCR4-associated factor CAF1 is an essential factor for spermatogenesis. Mol Cell Biol 24(13):5808–5820
Bhandari D, Raisch T, Weichenrieder O, Jonas S, Izaurralde E (2014) Structural basis for the Nanos-mediated recruitment of the CCR4-NOT complex and translational repression. Genes Dev 28(8):888–901
Bhaskar V, Roudko V, Basquin J, Sharma K, Urlaub H, Seraphin B, Conti E (2013) Structure and RNA-binding properties of the Not1-Not2-Not5 module of the yeast Ccr4-Not complex. Nat Struct Mol Biol 20(11):1281–1288
Bhaskar V, Basquin J, Conti E (2015) Architecture of the ubiquitylation module of the yeast Ccr4-Not complex. Structure 23(5):921–928
Boland A, Chen Y, Raisch T, Jonas S, Kuzuoglu-Ozturk D, Wohlbold L, Weichenrieder O, Izaurralde E (2013) Structure and assembly of the NOT module of the human CCR4-NOT complex. Nat Struct Mol Biol 20(11):1289–1297
Chapat C, Corbo L (2014) Novel roles of the CCR4-NOT complex. Wiley Interdiscip Rev RNA 5(6):883–901
Chapat C, Kolytcheff C, Le Romancer M, Auboeuf D, De La Grange P, Chettab K, Sentis S, Corbo L (2013) hCAF1/CNOT7 regulates interferon signalling by targeting STAT1. EMBO J 32(5):688–700
Chen J, Rappsilber J, Chiang YC, Russell P, Mann M, Denis CL (2001) Purification and characterization of the 1.0 MDa CCR4-NOT complex identifies two novel components of the complex. J Mol Biol 314(4):683–694
Chen C, Ito K, Takahashi A, Wang G, Suzuki T, Nakazawa T, Yamamoto T, Yokoyama K (2011) Distinct expression patterns of the subunits of the CCR4-NOT deadenylase complex during neural development. Biochem Biophys Res Commun 411(2):360–364
Chen Y, Boland A, Kuzuoglu-Ozturk D, Bawankar P, Loh B, Chang CT, Weichenrieder O, Izaurralde E (2014) A DDX6-CNOT1 complex and W-binding pockets in CNOT9 reveal direct links between miRNA target recognition and silencing. Mol Cell 54(5):737–750
Chicoine J, Benoit P, Gamberi C, Paliouras M, Simonelig M, Lasko P (2007) Bicaudal-C recruits CCR4-NOT deadenylase to target mRNAs and regulates oogenesis, cytoskeletal organization, and its own expression. Developmental Cell 13(5):691–704
Collart MA (2003) Global control of gene expression in yeast by the Ccr4-Not complex. Gene 313:1–16
Collart MA (2013) The NOT4 RING E3 ligase: a relevant player in co-translational quality control. ISRN Mol Biol 2013:1–19
Collart MA (2016) The Ccr4-Not complex is a key regulator of eukaryotic gene expression. Wiley Interdiscip Rev RNA 7(4):438–454
Collart MA, Panasenko OO (2012) The Ccr4--not complex. Gene 492(1):42–53
Collart M, Panasenko O, Nikolaev S (2012) The Not3/5 subunit of the Ccr4-Not complex: a central regulator of gene expression that integrates signals between the cytoplasm and the nucleus in eukaryotic cells. Cell Signal 25(4):743–751
Cook WJ, Jeffrey LC, Xu Y, Chau V (1993) Tertiary structures of class I ubiquitin-conjugating enzymes are highly conserved: crystal structure of yeast Ubc4. Biochemistry 32(50):13809–13817
Cooper KF, Scarnati MS, Krasley E, Mallory MJ, Jin C, Law MJ, Strich R (2012) Oxidative-stress-induced nuclear to cytoplasmic relocalization is required for Not4-dependent cyclin C destruction. J Cell Sci 125(Pt 4):1015–1026

De Keersmaecker K, Atak ZK, Li N, Vicente C, Patchett S, Girardi T, Gianfelici V, Geerdens E, Clappier E, Porcu M, Lahortiga I, Luca R, Yan J, Hulselmans G, Vranckx H, Vandepoel R, Sweron B, Jacobs K, Mentens N, Wlodarska I, Cauwelier B, Cloos J, Soulier J, Uyttebroeck A, Bagni C, Hassan BA, Vandenberghe P, Johnson AW, Aerts S, Cools J (2013) Exome sequencing identifies mutation in CNOT3 and ribosomal genes RPL5 and RPL10 in T-cell acute lymphoblastic leukemia. Nat Genet 45(2):186–190

Dimitrova LN, Kuroha K, Tatematsu T, Inada T (2009) Nascent peptide-dependent translation arrest leads to Not4p-mediated protein degradation by the proteasome. J Biol Chem 284(16):10343–10352

Doidge R, Mittal S, Aslam A, Winkler GS (2012) Deadenylation of cytoplasmic mRNA by the mammalian Ccr4-Not complex. Biochem Soc Trans 40(4):896–901

Draper MP, Salvadore C, Denis CL (1995) Identification of a mouse protein whose homolog in *Saccharomyces cerevisiae* is a component of the CCR4 transcriptional regulatory complex. Mol. Cell. Biol. 15(7):3487–3495

Fabian MR, Frank F, Rouya C, Siddiqui N, Lai WS, Karetnikov A, Blackshear PJ, Nagar B, Sonenberg N (2013) Structural basis for the recruitment of the human CCR4-NOT deadenylase complex by tristetraprolin. Nat Struct Mol Biol 20(6):735–739

Faraji F, Hu Y, Wu G, Goldberger NE, Walker RC, Zhang J, Hunter KW (2014) An integrated systems genetics screen reveals the transcriptional structure of inherited predisposition to metastatic disease. Genome Res 24(2):227–240

Garapaty S, Mahajan MA, Samuels HH (2008) Components of the CCR4-NOT complex function as nuclear hormone receptor coactivators via association with the NRC-interacting Factor NIF-1. J Biol Chem 283(11):6806–6816

Garces RG, Gillon W, Pai EF (2007) Atomic model of human Rcd-1 reveals an armadillo-like-repeat protein with in vitro nucleic acid binding properties. Protein Sci 16(2):176–188

Goldstrohm AC, Wickens M (2008) Multifunctional deadenylase complexes diversify mRNA control. Nat Rev Mol Cell Biol 9(4):337–344

Goldstrohm AC, Hook BA, Seay DJ, Wickens M (2006) PUF proteins bind Pop2p to regulate messenger RNAs. Nat Struct Mol Biol 13(6):533–539

Gronholm J, Kaustio M, Myllymaki H, Kallio J, Saarikettu J, Kronhamn J, Valanne S, Silvennoinen O, Ramet M (2012) Not4 enhances JAK/STAT pathway-dependent gene expression in Drosophila and in human cells. FASEB J 26(3):1239–1250

Gulshan K, Thommandru B, Moye-Rowley WS (2012) Proteolytic degradation of the Yap1 transcription factor is regulated by subcellular localization and the E3 ubiquitin ligase Not4. J Biol Chem 287(32):26796–26805

Gutierrez-Camino A, Lopez-Lopez E, Martin-Guerrero I, Pinan MA, Garcia-Miguel P, Sanchez-Toledo J, Carbone Baneres A, Uriz J, Navajas A, Garcia-Orad A (2014) Noncoding RNA-related polymorphisms in pediatric acute lymphoblastic leukemia susceptibility. Pediatr Res 75(6):767–773

Haas M, Siegert M, Schürmann A, Sodeik B, Wolfes H (2004) c-Myb protein interacts with Rcd-1, a component of the CCR4 transcription mediator complex. Biochem. 43:8152–8159

Halter D, Collart MA, Panasenko OO (2014) The Not4 E3 ligase and CCR4 deadenylase play distinct roles in protein quality control. PLoS One 9(1):e86218

Hanzawa H, de Ruwe MJ, Albert TK, van Der Vliet PC, Timmers HT, Boelens R (2001) The structure of the C4C4 ring finger of human NOT4 reveals features distinct from those of C3HC4 RING fingers. J Biol Chem 276(13):10185–10190

Haworth J, Alver RC, Anderson M, Bielinsky AK (2010) Ubc4 and Not4 regulate steady-state levels of DNA polymerase-alpha to promote efficient and accurate DNA replication. Mol Biol Cell 21(18):3205–3219

Hiroi N, Ito T, Yamamoto H, Ochiya T, Jinno S, Okayama H (2002) Mammalian Rcd1 is a novel transcriptional cofactor that mediates retinoic acid-induced cell differentiation. EMBO J. 21:5235–5244

Horiuchi M, Takeuchi K, Noda N, Muroya N, Suzuki T, Nakamura T, Kawamura-Tsuzuku J, Takahasi K, Yamamoto T, Inagaki F (2009) Structural basis for the antiproliferative activity of the Tob-hCaf1 complex. J Biol Chem 284(19):13244–13255

Inada T, Makino S (2014) Novel roles of the multi-functional CCR4-NOT complex in post-transcriptional regulation. Front Genet 5:135

Inoue T, Morita M, Hijikata A, Fukuda-Yuzawa Y, Adachi S, Isono K, Ikawa T, Kawamoto H, Koseki H, Natsume T, Fukao T, Ohara O, Yamamoto T, Kurosaki T (2015) CNOT3 contributes to early B cell development by controlling Igh rearrangement and p53 mRNA stability. J Exp Med 212(9):1465–1479

Kamath RS, Ahringer J (2003) Genome-wide RNAi screening in *Caenorhabditis elegans*. Methods 30:313–321

Krueger KE, Ghosh AK, Krom BP, Cihlar RL (2004) Deletion of the NOT4 gene impairs hyphal development and pathogenicity in *Candida albicans*. Microbiology 150(Pt 1):229–240

Lau NC, Kolkman A, van Schaik FM, Mulder KW, Pijnappel WW, Heck AJ, Timmers HT (2009) Human Ccr4-Not complexes contain variable deadenylase subunits. Biochem J 422(3):443–453

Liu HY, Chiang YC, Pan J, Chen J, Salvadore C, Audino DC, Badarinarayana V, Palaniswamy V, Anderson B, Denis CL (2001) Characterization of CAF4 and CAF16 reveals a functional connection between the CCR4-NOT complex and a subset of SRB proteins of the RNA polymerase II holoenzyme. J Biol Chem 276(10):7541–7548

Maeda I, Kohara Y, Yamamoto M, Sugimoto A (2001) Large-scale analysis of gene function in *Caenorhabditis elegans* by high-throughput RNAi. Current Biol. 11:171–176

Maillet L, Tu C, Hong YK, Shuster EO, Collart MA (2000) The essential function of Not1 lies within the Ccr4-Not complex. J Mol Biol 303(2):131–143

Malvar T, Biron RW, Kaback DB, Denis CL (1992) The CCR4 protein from *Saccharomyces cerevisiae* contains a leucine-rich repeat region which is required for its control of ADH2 gene expression. Genetics 132(4):951–962

Maragozidis P, Karangeli M, Labrou M, Dimoulou G, Papaspyrou K, Salataj E, Pournaras S, Matsouka P, Gourgoulianis KI, Balatsos NA (2012) Alterations of deadenylase expression in acute leukemias: evidence for poly(a)-specific ribonuclease as a potential biomarker. Acta Haematol 128(1):39–46

Mathys H, Basquin J, Ozgur S, Czarnocki-Cieciura M, Bonneau F, Aartse A, Dziembowski A, Nowotny M, Conti E, Filipowicz W (2014) Structural and biochemical insights to the role of the CCR4-NOT complex and DDX6 ATPase in microRNA repression. Mol Cell 54(5):751–765

Mauxion F, Preve B, Seraphin B (2013) C2ORF29/CNOT11 and CNOT10 form a new module of the CCR4-NOT complex. RNA Biol 10(2):267–276

Mazzoni C, Serafini A, Falcone C (2005) The inactivation of KlNOT4, a Kluyveromyces lactis gene encoding a component of the CCR4-NOT complex, reveals new regulatory functions. Genetics 170(3):1023–1032

Mersman DP, Du HN, Fingerman IM, South PF, Briggs SD (2009) Polyubiquitination of the demethylase Jhd2 controls histone methylation and gene expression. Genes Dev 23(8):951–962

Miller JE, Reese JC (2012) Ccr4-Not complex: the control freak of eukaryotic cells. Critical reviews in biochemistry and molecular biology 47(4):315–333

Mittal S, Aslam A, Doidge R, Medica R, Winkler GS (2011) The Ccr4a (CNOT6) and Ccr4b (CNOT6L) deadenylase subunits of the human Ccr4-Not complex contribute to the prevention of cell death and senescence. Molecular biology of the cell 22(6):748–758

Morita M, Suzuki T, Nakamura T, Yokoyama K, Miyasaka T, Yamamoto T (2007) Depletion of mammalian CCR4b deadenylase triggers elevation of the p27Kip1 mRNA level and impairs cell growth. Mol Cell Biol 27(13):4980–4990

Morita M, Oike Y, Nagashima T, Kadomatsu T, Tabata M, Suzuki T, Nakamura T, Yoshida N, Okada M, Yamamoto T (2011) Obesity resistance and increased hepatic expression of catabolism-related mRNAs in Cnot3+/− mice. EMBO J 30(22):4678–4691

Muhlrad D, Parker R (2005) The yeast EDC1 mRNA undergoes deadenylation-independent decapping stimulated by Not2p, Not4p, and Not5p. Embo J 24(5):1033–1045

Mulder KW, Winkler GS, Timmers HT (2005) DNA damage and replication stress induced transcription of RNR genes is dependent on the Ccr4-Not complex. Nucleic Acids Res 33(19):6384–6392

Mulder KW, Brenkman AB, Inagaki A, van den Broek NJ, Timmers HT (2007) Regulation of histone H3K4 tri-methylation and PAF complex recruitment by the Ccr4-Not complex. Nucleic Acids Res 35(7):2428–2439

Nakamura T, Yao R, Ogawa T, Suzuki T, Ito C, Tsunekawa N, Inoue K, Ajima R, Miyasaka T, Yoshida Y, Ogura A, Toshimori K, Noce T, Yamamoto T, Noda T (2004) Oligo-astheno-teratozoospermia in mice lacking Cnot7, a regulator of retinoid X receptor beta. Nat Genet 36(5):528–533

Nasertorabi F, Batisse C, Diepholz M, Suck D, Bottcher B (2011) Insights into the structure of the CCR4-NOT complex by electron microscopy. FEBS Letters 585:2182–2186

Neely GG, Kuba K, Cammarato A, Isobe K, Amann S, Zhang L, Murata M, Elmen L, Gupta V, Arora S, Sarangi R, Dan D, Fujisawa S, Usami T, Xia CP, Keene AC, Alayari NN, Yamakawa H, Elling U, Berger C, Novatchkova M, Koglgruber R, Fukuda K, Nishina H, Isobe M, Pospisilik JA, Imai Y, Pfeufer A, Hicks AA, Pramstaller PP, Subramaniam S, Kimura A, Ocorr K, Bodmer R, Penninger JM (2010) A global in vivo Drosophila RNAi screen identifies NOT3 as a conserved regulator of heart function. Cell 141(1):142–153

Nousch M, Techritz N, Hampel D, Millonigg S, Eckmann CR (2013) The Ccr4-Not deadenylase complex constitutes the main poly(A) removal activity in *C. elegans*. J Cell Sci 126(Pt 18):4274–4285

Okazaki N, Okazaki K, Watanabe Y, Kato-Hayashhi M, Yamamoto M, Okayama H (1998) Novel factor highly conserved among eukaryotes controls sexual development in fission yeast. Mol. Cell. Biol. 18:887–895

Ozgur S, Basquin J, Kamenska A, Filipowicz W, Standart N, Conti E (2015) Structure of a Human 4E-T/DDX6/CNOT1 Complex Reveals the Different Interplay of DDX6-Binding Proteins with the CCR4-NOT Complex. Cell Rep 13(4):703–711

Panasenko OO (2014) The role of the E3 ligase Not4 in cotranslational quality control. Front Genet 5:141

Panasenko OO, Collart MA (2012) Presence of Not5 and ubiquitinated Rps7A in polysome fractions depends upon the Not4 E3 ligase. Molecular microbiology 83(3):640–653

Panasenko O, Landrieux E, Feuermann M, Finka A, Paquet N, Collart MA (2006) The yeast Ccr4-Not complex controls ubiquitination of the nascent-associated polypeptide (NAC-EGD) complex. J Biol Chem 281(42):31389–31398

Panepinto JC, Heinz E, Traven A (2013) The cellular roles of Ccr4-NOT in model and pathogenic fungi-implications for fungal virulence. Front Genet 4:302

Petit AP, Wohlbold L, Bawankar P, Huntzinger E, Schmidt S, Izaurralde E, Weichenrieder O (2012) The structural basis for the interaction between the CAF1 nuclease and the NOT1 scaffold of the human CCR4-NOT deadenylase complex. Nucleic Acids Res 40(21):11058–11072

Preissler S, Reuther J, Koch M, Scior A, Bruderek M, Frickey T, Deuerling E (2015) Not4-dependent translational repression is important for cellular protein homeostasis in yeast. EMBO J 34(14):1905–1924

Reese JC (2013) The control of elongation by the yeast Ccr4-not complex. Biochim Biophys Acta 1829(1):127–133

Sandler H, Kreth J, Timmers HT, Stoecklin G (2011) Not1 mediates recruitment of the deadenylase Caf1 to mRNAs targeted for degradation by tristetraprolin. Nucleic acids research 39(10):4373–4386

Schwede A, Ellis L, Luther J, Carrington M, Stoecklin G, Clayton C (2008) A role for Caf1 in mRNA deadenylation and decay in trypanosomes and human cells. Nucleic Acids Res 36(10):3374–3388

Seiden-Long IM, Brown KR, Shih W, Wigle DA, Radulovich N, Jurisica I, Tsao MS (2006) Transcriptional targets of hepatocyte growth factor signaling and Ki-ras oncogene activation in colorectal cancer. Oncogene 25(1):91–102

Shirai YT, Suzuki T, Morita M, Takahashi A, Yamamoto T (2014) Multifunctional roles of the mammalian CCR4-NOT complex in physiological phenomena. Front Genet 5:286

Sun HY, Kim N, Hwang CS, Yoo JY (2015) Protein Degradation of RNA Polymerase II-Association Factor 1(PAF1) Is Controlled by CNOT4 and 26S Proteasome. PLoS One 10(5):e0125599

Suzuki A, Igarashi K, Aisaki K, Kanno J, Saga Y (2010) NANOS2 interacts with the CCR4-NOT deadenylation complex and leads to suppression of specific RNAs. Proceedings of the National Academy of Sciences of the United States of America 107(8):3594–3599

Suzuki T, Kikuguchi C, Sharma S, Sasaki T, Tokumasu M, Adachi S, Natsume T, Kanegae Y, Yamamoto T (2015) CNOT3 suppression promotes necroptosis by stabilizing mRNAs for cell death-inducing proteins. Sci Rep 5:14779

Tay Y, Kats L, Salmena L, Weiss D, Tan SM, Ala U, Karreth F, Poliseno L, Provero P, Di Cunto F, Lieberman J, Rigoutsos I, Pandolfi PP (2011) Coding-independent regulation of the tumor suppressor PTEN by competing endogenous mRNAs. Cell 147(2):344–357

Temme C, Zaessinger S, Meyer S, Simonelig M, Wahle E (2004) A complex containing the CCR4 and CAF1 proteins is involved in mRNA deadenylation in Drosophila. Embo J 23(14):2862–2871

Temme C, Zhang L, Kremmer E, Ihling C, Chartier A, Sinz A, Simonelig M, Wahle E (2010) Subunits of the Drosophila CCR4-NOT complex and their roles in mRNA deadenylation. RNA 16(7):1356–1370

Temme C, Simonelig M, Wahle E (2014) Deadenylation of mRNA by the CCR4-NOT complex in Drosophila: molecular and developmental aspects. Front Genet 5:143

Thore S, Mauxion F, Seraphin B, Suck D (2003) X-ray structure and activity of the yeast Pop2 protein: a nuclease subunit of the mRNA deadenylase complex. EMBO Rep 4(12):1150–1155

Tucker M, Staples RR, Valencia-Sanchez MA, Muhlrad D, Parker R (2002) Ccr4p is the catalytic subunit of a Ccr4p/Pop2p/Notp mRNA deadenylase complex in *Saccharomyces cerevisiae*. EMBO J. 21:1427–1436

Ukleja M, Cuellar J, Siwaszek A, Kasprzak JM, Czarnocki-Cieciura M, Bujnicki JM, Dziembowski A, Valpuesta JM (2016) The architecture of the Schizosaccharomyces pombe CCR4-NOT complex. Nat Commun 7:10433

Venturini G, Rose AM, Shah AZ, Bhattacharya SS, Rivolta C (2012) CNOT3 is a modifier of PRPF31 mutations in retinitis pigmentosa with incomplete penetrance. PLoS Genet 8(11):e1003040

Villanyi Z, Collart MA (2015) Ccr4-Not is at the core of the eukaryotic gene expression circuitry. Biochem Soc Trans 43(6):1253–1258

Villanyi Z, Ribaud V, Kassem S, Panasenko OO, Pahi Z, Gupta I, Steinmetz L, Boros I, Collart MA (2014) The Not5 subunit of the ccr4-not complex connects transcription and translation. PLoS Genet 10(10):e1004569

Wahle E, Winkler GS (2013) RNA decay machines: deadenylation by the Ccr4-Not and Pan2-Pan3 complexes. Biochim Biophys Acta 1829(6–7):561–570

Wang H, Morita M, Yang X, Suzuki T, Yang W, Wang J, Ito K, Wang Q, Zhao C, Bartlam M, Yamamoto T, Rao Z (2010) Crystal structure of the human CNOT6L nuclease domain reveals strict poly(A) substrate specificity. The EMBO journal 29(15):2566–2576

Washio-Oikawa K, Nakamura T, Usui M, Yoneda M, Ezura Y, Ishikawa I, Nakashima K, Noda T, Yamamoto T, Noda M (2007) Cnot7-null mice exhibit high bone mass phenotype and modulation of BMP actions. J Bone Miner Res 22(8):1217–1223

Watanabe C, Morita M, Hayata T, Nakamoto T, Kikuguchi C, Li X, Kobayashi Y, Takahashi N, Notomi T, Moriyama K, Yamamoto T, Ezura Y, Noda M (2014) Stability of mRNA influences osteoporotic bone mass via CNOT3. Proc Natl Acad Sci U S A 111(7):2692–2697

Wong SQ, Behren A, Mar VJ, Woods K, Li J, Martin C, Sheppard KE, Wolfe R, Kelly J, Cebon J, Dobrovic A, McArthur GA (2015) Whole exome sequencing identifies a recurrent RQCD1 P131L mutation in cutaneous melanoma. Oncotarget 6(2):1115–1127

Xu K, Bai Y, Zhang A, Zhang Q, Bartlam MG (2014) Insights into the structure and architecture of the CCR4-NOT complex. Front Genet 5:137

Zhang L, Tran NT, Su H, Wang R, Lu Y, Tang H, Aoyagi S, Guo A, Khodadadi-Jamayran A, Zhou D, Qian K, Hricik T, Cote J, Han X, Zhou W, Laha S, Abdel-Wahab O, Levine RL, Raffel G, Liu Y, Chen D, Li H, Townes T, Wang H, Deng H, Zheng YG, Leslie C, Luo M, Zhao X (2015) Cross-talk between PRMT1-mediated methylation and ubiquitylation on RBM15 controls RNA splicing. Elife 4:e07938

Chapter 14
Higher-Order Structure in Bacterial VapBC Toxin-Antitoxin Complexes

Kirstine L. Bendtsen and Ditlev E. Brodersen

Abstract Toxin-antitoxin systems are widespread in the bacterial kingdom, including in pathogenic species, where they allow rapid adaptation to changing environmental conditions through selective inhibition of key cellular processes, such as DNA replication or protein translation. Under normal growth conditions, type II toxins are inhibited through tight protein-protein interaction with a cognate antitoxin protein. This toxin-antitoxin complex associates into a higher-order macromolecular structure, typically heterotetrameric or heterooctameric, exposing two DNA binding domains on the antitoxin that allow auto-regulation of transcription by direct binding to promoter DNA. In this chapter, we review our current understanding of the structural characteristics of type II toxin-antitoxin complexes in bacterial cells, with a special emphasis on the staggering variety of higher-order architecture observed among members of the VapBC family. This structural variety is a result of poor conservation at the primary sequence level and likely to have significant and functional implications on the way toxin-antitoxin expression is regulated.

Keywords DNA-binding protein • Ribonuclease • RNA-binding protein • Toxin-antitoxin • Transcriptional regulation

14.1 Introduction

Toxin-antitoxin (TA) systems are characterised by being non-essential for normal growth of bacterial cells, but essential for survival during stress, where they provide a means of rapid adaptation via an adjustment of the overall metabolic rate. Stress conditions include changes in humidity, temperature, nutrient supply, oxidative

K.L. Bendtsen
Faculty of Health and Medical Sciences, Department of Drug Design and Pharmacology ,
University of Copenhagen, Jagtvej 162, DK-2100 Copenhagen, Denmark
e-mail: kirstine.bendtsen@sund.ku.dk

D.E. Brodersen (✉)
Centre for Bacterial Stress Response and Persistence, Department of Molecular Biology and Genetics, Aarhus University, Gustav Wieds Vej 10c, 8000 Aarhus C, Denmark
e-mail: deb@mbg.au.dk

J.R. Harris, J. Marles-Wright (eds.), *Macromolecular Protein Complexes*,
Subcellular Biochemistry 83, DOI 10.1007/978-3-319-46503-6_14

state, the presence of antibiotics and bacteriophages or other competing organisms (Hayes 2003; Sat et al. 2003; Hazan et al. 2004; Sat et al. 2001; Hazan and Engelberg-Kulka 2004). TA systems are widespread in both bacteria and archaea and are often found in large numbers, a surprising fact given the small size of most prokaryotic genomes (Pandey and Gerdes 2005; Sevin and Barloy-Hubler 2007; Makarova et al. 2009). Free-living prokaryotes appear to contain an especially large number of TA systems, whereas obligate intracellular prokaryotes contain few or even no TA systems (Pandey and Gerdes 2005; Sevin and Barloy-Hubler 2007; Makarova et al. 2009). First discovered in the 1980s as loci that conferred plasmid maintenance (Gerdes et al. 1986b; Ogura and Hiraga 1983), TA systems were later identified also on chromosomes of free-living prokaryotes (Masuda et al. 1993; Gerdes 2000), and in the genomes of obligate intracellular organisms (Ogata et al. 2005). TA systems are characterised by two components, a stable toxin and an unstable antitoxin that inhibits the toxin under normal cellular conditions by a range of different mechanisms (Pandey and Gerdes 2005; Gerdes et al. 2005; Van Melderen and Saavedra De Bast 2009). During cellular stress, the antitoxin is inactivated, in some cases by proteases, releasing the active toxin. Once activated, toxins inhibit essential cellular processes, including DNA replication, protein synthesis, peptidoglycan formation and cell division, but the full range of targets has not yet been elucidated as many putative TA systems remain functionally uncharacterised (Gerdes et al. 2005; Unterholzner et al. 2013).

The overall physiological roles of TA systems range widely, from plasmid maintenance, programmed cell death, and stress-response, to the formation of persister cells (Magnuson 2007). The *ccdAB* locus was initially identified on the mini-F plasmid of *Escherichia coli* as a segment promoting plasmid maintenance during cell division (Ogura and Hiraga 1983). Shortly after, other TA systems with similar functions were identified and the observed function was coined *post-segregational killing* (Gerdes et al. 1986b; Jaffe et al. 1985). According to this principle, daughter cells stochastically receive either a plasmid containing the TA system during cell division, or not. In daughter cells without plasmid, toxin and antitoxin molecules will still be present, but will not be replenished. Therefore, due to the instability of the antitoxin, toxin molecules will gradually be released and activated, resulting in cell death (Van Melderen and Saavedra De Bast 2009). Consequently, plasmids containing the TA system (TA+) are able to outcompete plasmids devoid of the TA system (TA-), since a loss of the TA+ plasmid will lead to cell death (Van Melderen and Saavedra De Bast 2009). A similar function was described for some chromosome-encoded TA systems and termed *programmed cell death* (Arcus et al. 2011). In this situation, a population of cells is believed to benefit from suicide of a specific subpopulation by activation of the toxin (Yarmolinsky 1995; Jensen and Gerdes 1995). For example, the *E. coli* MazEF system (Aizenman et al. 1996) was shown to play a part in mediating programmed cell death under a variety of stress conditions, such as phage infection, high temperature, DNA damage, oxidative stress, and the presence of antibiotics (Hazan and Engelberg-Kulka 2004; Hazan et al. 2004; Sat et al. 2001). However, other studies have found that the MazF toxin improves survival of cells by inducing a static condition where cells are viable, but unable to proliferate during unfavourable conditions. This *bacteriostatic* effect (as opposed to *bactericidal*) is reversible and can be relieved upon expression of MazE antitoxin at any later time (Pedersen et al. 2002). TA systems have also been suggested to act as quality control elements at the post-translational level: Under conditions of limited growth, a cell can save energy by optimising its gene expression quality so only functional transcripts are expressed (Makarova et al. 2009). Both the *E. coli* RelBE and MazEF TA systems have been shown to inhibit translation by cleavage of mRNA during nutritional stress and could thus be thought of as quality control factors (Van Melderen and Saavedra De Bast 2009; Gerdes and Wagner 2007; Hayes 2003).

Finally, it has recently become clear that TA systems likely play a role in the formation of bacterial *persister cells*. Persister cells are a subset of a bacterial population that have entered a dormant state and remain viable after treatment with e.g. antibiotics (Dorr et al. 2010; Brantl and Jahn 2015; Wang and Wood 2011). Characteristic of persister cells is that they do not acquire inheritable resistance and revert to normal growing cells upon further culturing in absence of the selective pressure (Singh et al. 2009). The detailed mechanism of persister cell formation remains somewhat of an enigma, but it has been suggested that TA systems play an important role by down-regulating essential metabolic functions during stress, which could cause a dormant state (Wang and Wood 2011). The first TA system to be implicated in persister cell formation was the *E. coli* HipBA system (Keren et al. 2004), where overexpression of the HipA toxin can be shown to correlate with increased persister cell formation and, conversely, deletion of the entire HipBA locus leads to a decrease in persister cell formation. Since this discovery, numerous other TA systems have been associated with persister cell formation, including TisAB, MqsRA, RelBE, HigAB, MazEF, DinJ-YafQ, and YefM-YoeB (Kim and Wood 2010; Dorr et al. 2010; Wang and Wood 2011; Lewis 2010; Brantl and Jahn 2015).

14.2 Functional and Genetic Organisation of TA Systems

A TA system is defined by the presence of both a toxin and an antitoxin component, and the group has been subdivided into six distinct types (I-VI) based on the nature of the antitoxin, which can be either RNA (types I and III) or protein (types II, IV, V and VI), and the mode of action of the toxin, which is always protein (Fig. 14.1) (Unterholzner et al. 2013; Pandey and Gerdes 2005). By definition, the antitoxin blocks the cellular function of the toxin under normal growth conditions in what is called the *inhibited state*. The toxin and antitoxin are often encoded in a bicistronic locus, nearly always with the antitoxin preceding the toxin and under the control of a unique TA promoter. This genetic organisation presumably ensures a high antitoxin:toxin ratio and a solid inhibition of the toxin under normal growth conditions (Pandey and Gerdes 2005). Finally, the antitoxin often regulates the transcription from the TA operon by binding directly to the promoter region at a pseudo-palindromic site, which provides a means of auto-regulating TA levels in the cell for any given locus (see Fig. 14.1 for an overview). In the following, we will

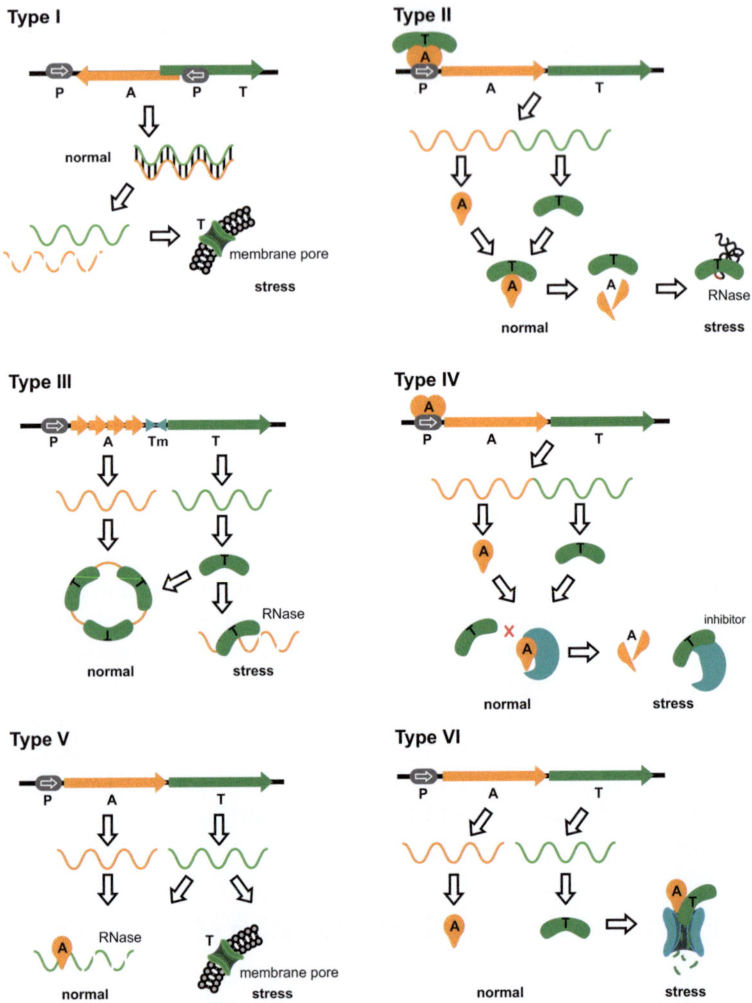

Fig. 14.1 Overview of the six types of TA systems found in bacteria. In all cases, the toxin (*T*) and antitoxin (*A*) are expressed from a common promoter (*P*), but in some cases (type III) interrupted by a terminator (*Tm*). **Type I.** The *T* and *A* transcripts are partly complementary and base-pair in the inhibited state. Upon degradation of the *A* transcript, *T* is expressed and forms a membrane pore. **Type II.** Both *T* and *A* are expressed as proteins which form a tight complex in the inhibited state that regulates transcription through operator binding. Upon degradation of *A*, *T* is released and is active as an RNA endonuclease with specific cellular targets. **Type III.** The *A* transcript consists of a number of repeats and is expressed as RNA only. *T* is expressed as a protein which binds to the repeats of the *A* RNA in the inhibited state. The activated *T* is an RNase capable of degrading *A*. **Type IV.** Both *T* and *A* are expressed as proteins as in Type II, however, *A* functions by binding to the target of *T* in the inhibited state, thus preventing toxin action (*red cross*). Upon degradation of *A*, *T* becomes active and can bind its target. **Type V.** During the inhibited state, *A* is expressed as a protein that is able to cleave the *T* transcript. During stress, *T* is expressed and forms a membrane pore like in Type I. **Type VI.** Both *T* and *A* are expressed as proteins as in Type II, but *A* has the ability to guide *T* to the cellular protein degradation machinery, thus inactivating the toxin by promoting its degradation. When *A* is not active, *T* is not degraded and can exert its function as a toxin

describe the characteristics of each type of TA system, highlighting in each case the regulation and function of the active toxin.

Type I TA Systems In type I TA systems, transcription of the antitoxin results in a small antisense RNA, which is complementary to a part of the toxin mRNA. Under normal growth conditions, the antitoxin transcript anneals to the toxin mRNA, resulting in a very stable dsRNA complex that is readily degraded, thus preventing expression of the toxin (Fig. 14.1, Type I) (Fozo et al. 2008a; Gerdes and Wagner 2007; Fozo et al. 2008b). The toxin protein product is often a small, hydrophobic polypeptide that folds into a single α-helix which locates to the inner cell membrane. The transmembrane helix then polymerises, forming a pore that destroys the membrane potential and inhibits ATP synthesis (Fozo et al. 2008a; Fozo et al. 2008b; Yamaguchi et al. 2011). The *Hok-Sok system* was the first type I TA system to be characterised, identified in 1986 in a screen for plasmid-stabilising genes (Gerdes et al. 1986a). The *hok-sok* locus contains three genes: *hok*, *mok*, and *sok*, with *sok* in the antisense direction preceding the *hok* open reading frame. Translation of the gene encoding the toxin, *hok*, results in production of a 52-residue transmembrane polypeptide that irreversibly damages the cell membrane through loss of the membrane potential. The *mok* open reading frame overlaps with both *sok* and *hok* and is required for translation of *hok*, while the *sok* gene specifies a small antisense RNA that blocks translation of *mok*, thereby indirectly inhibiting *hok* translation (Gerdes et al. 1986a; Thisted and Gerdes 1992). During normal growth, the *sok* antisense RNA, which is transcribed from a downstream promoter on the reverse strand, is expressed in excess over toxin mRNA and therefore inhibits its translation by formation of a dsRNA complex, which is rapidly degraded by RNase III (Gerdes et al. 1990; Gerdes et al. 1992). However, the *sok* transcript is unstable and when degraded, the *hok* mRNA is translated producing the toxin protein product. Several other type I loci have since been discovered, including *rdlD-ldrA* (Kawano et al. 2002), *symR-symE* (Kawano et al. 2007), and *istR-tisB* (Vogel et al. 2004). Due to their small size, type I TA systems are very difficult to identify and their distribution in the bacterial kingdom is therefore not known in detail.

Type II TA Systems The type II TA systems are by far the most well-characterised and most abundant in prokaryotes (Pandey and Gerdes 2005; Sevin and Barloy-Hubler 2007; Makarova et al. 2009). In type II systems, both toxin and antitoxin are proteins and form a tight complex under normal growth conditions thereby keeping the toxin inactive (Fig. 14.1, Type II). At the genomic level, the TA genes are arranged in a bicistronic operon, usually with the antitoxin gene preceding the toxin gene. An exception to this rule is the *higAB* and *hicAB* families, where the toxin precedes the antitoxin (Tian et al. 1996a; Jorgensen et al. 2009; Makarova et al. 2006). In operons where the antitoxin gene precedes the toxin gene, overlaps of 1–4 base pairs between the open reading frames of the antitoxin and toxin are common, reflecting a translational coupling that presumably helps secure a stable 1:1 expression ratio of the proteins (Pandey and Gerdes 2005). The operon is tightly controlled by negative auto-regulation of transcription via a DNA-binding domain on the antitoxin (Gerdes et al. 1986a), except for the ω-ε-ζ and *paaR-paaA-parE* families, which are three-component systems for which the additional ω (or PaaR, respectively) protein is the transcriptional repressor (de la Hoz et al. 2000; Hallez et al. 2010). The antitoxin binds to the promoter region of the operon either in complex with the toxin or alone, and both toxin and DNA binding stabilises the antitoxin, which is otherwise disordered and prone to proteolysis (Hayes 2003). The type II TA complex often binds to the promoter region more tightly than the isolated antitoxin, and in these cases the toxin is a co-repressor of its own transcription (Gerdes et al. 2005). During stress, the antitoxin is degraded by cellular proteases and the toxin is released and activated (Gerdes et al. 2005; Yamaguchi et al. 2011; Unterholzner et al. 2013; Tian et al. 1996b). The type II TA systems have been subdivided into families (VapBC, CcdAB, MazEF, Phd/Doc, ParDE, HigAB, RelBE, HipBA, HicAB and ω-ε-ζ) based on the mode of action of the toxin as well as conserved structural domains (Pandey and Gerdes 2005; Ogura and Hiraga 1983; Masuda et al. 1993; Lehnherr et al. 1993; Roberts et al. 1994; Tian et al. 1996b; Gotfredsen and Gerdes 1998; Black et al. 1991; de la Hoz et al. 2000; Jorgensen et al. 2009).

Type III TA Systems For type III TA systems, the antitoxin gene is expressed as an RNA that directly binds and inhibits the toxin protein (Fig. 14.1, Type III). At the genomic level, the operon is bicistronic with the antitoxin gene, which consists of a tandem array of direct repeats, preceding the toxin gene. Two short, inverted repeats are located between the two genes where they function as transcriptional terminators to regulate the relative amounts of antitoxin and toxin RNA (Fineran et al. 2009). Binding of antitoxin RNA to the toxin protein results in the formation of a heterohexameric complex consisting of three toxin proteins and three antitoxin RNAs (Unterholzner et al. 2013). The free toxin is an active RNase targeting the bicistronic TA mRNA at the sites of the direct repeats, thus abolishing the function of the antitoxin (Blower et al. 2012). The first characterised type III system was ToxIN from the *Pectobacterium carotovoum* plasmid pECA1039 (Fineran et al. 2009). Recent studies have shown that the 125 type III TA systems identified to date can be assigned to three families based on sequence similarity, ToxIN, CptIN, and TenpIN (Blower et al. 2012).

Type IV TA Systems For type IV TA systems, the antitoxin is a protein that suppresses the toxicity of the toxin by functioning as an antagonist that blocks the toxin's target (Fig. 14.1, Type IV) (Masuda et al. 2012a). At the genomic level, the two proteins are expressed from a single operon where the antitoxin precedes the toxin. At the protein level, the antitoxin does not interfere with the toxin, but binds directly to the target of it, thereby preventing binding of the toxin. Only two type IV TA systems are known, *E. coli yeeU-cbtA* and *Streptococcus agalactiae abiEi-abiEii* (Masuda et al. 2012b; Dy et al. 2014). In the *yeeU-cbtA* system, the genes are separated by 89 base pairs in the bicistronic operon, which might play a role in protein expression. The toxin, CbtA, targets cell division and morphology by inhibiting polymerisation of the proteins MreB and FtsZ (Masuda et al. 2012a), homologues of the eukaryotic cytoskeleton proteins, actin and tubulin (van den Ent et al.

2001; Erickson 1997). The antitoxin, YeeU, acts as an antagonist of CbtA by binding MreB and FtsZ, enhancing their polymerisation, thus counteracting the effect of CbtA (Masuda et al. 2012a). In the *abiEi-abiEii* system, the two genes overlap by four base pairs at the 5′-end of the *abiEii*, reflecting translational coupling. The toxin, AbiEii, is a GTP-specific NTase that transfers a nucleotide to a yet unknown target, resulting in growth inhibition (Dy et al. 2014). The antitoxin, AbiEi, negatively auto-regulates transcription via binding to operator DNA and counteracts the toxicity of the toxin, probably by interfering with the toxin target. However, the exact mechanism of how AbiEi inhibits AbiEii is still unclear and the designation of the TA system as type IV is thus somewhat tentative (Dy et al. 2014).

Type V TA Systems In type V TA systems, the antitoxin is a protein that masks the toxicity of the toxin by cleaving its mRNA (Fig. 14.1, Type V). To date only one type V TA system has been identified, namely *E. coli ghoS-ghoT* (Wang et al. 2012). At the genomic level, the antitoxin and the toxin are encoded by a single operon, again with the antitoxin preceding the toxin, and a distance of 27 base pairs between the open reading frames in the case of *ghoS-ghoT*. The protein products are the antitoxin GhoS and the toxin GhoT for which the antitoxin is a small protein with sequence-specific endoribonuclease activity related to the CAS2 CRISPR RNase family. GhoS directly targets the GhoT mRNA, thus preventing translation (Wang et al. 2012). Unusually, however, the GhoS antitoxin is stable and remains in the cell during stressful periods and moreover, the antitoxin does not regulate its own transcription through DNA binding (Wang et al. 2012). The toxin is a small (57 residues), hydrophobic protein with two transmembrane regions (Hofmann and Stoffel 1993) that causes persistence and ghost cell formation by disrupting the membrane, upon activation, hence the name (Wang et al. 2012).

Type VI TA Systems A putative type VI TA system has been described relatively recently in which the antitoxin is a protein that promotes the degradation of the toxin (Fig. 14.1, Type VI) (Aakre et al. 2013; Markovski and Wickner 2013). To date, only one type VI TA system has been identified, namely *Caulobacter crescentus socAB*. At the genomic level, the two proteins are expressed from a single operon where the antitoxin precedes the toxin. At the protein level, the unstable toxin, SocB, is constitutively degraded by the ClpXP protease by a mechanism where the antitoxin, SocA, acts as a proteolytical adaptor by bringing the toxin to the protease. During stress, when the toxin protein is not degraded, it functions to disrupt DNA replication elongation by binding to the β-sliding clamp of DNA polymerase and thereby outcompeting the other clamp-binding proteins.

TA TA loci, even of different types, tend to cluster in the bacterial genomes, probably reflecting a tendency towards horizontal gene transfer and intra-genomic recombination, and for this reason they are generally considered mobile genetic elements (Pandey and Gerdes 2005; Sevin and Barloy-Hubler 2007). Genome-wide prediction of TA loci has further suggested that some bacteria contain toxin genes that are located in proximity to unrelated antitoxins (Gerdes et al. 2005; Hayes 2003). Furthermore, some toxins (including RelE, VapC, and MazF) have also been found as solitary genes, not accompanied by antitoxins (Pandey and Gerdes 2005). Some organisms, including major pathogens, have been found to contain a surprisingly large number of TA systems in their genomes (Pandey and Gerdes 2005; Sevin and Barloy-Hubler 2007; Makarova et al. 2009). This is particularly true for *Mycobacterium tuberculosis*, which harbours at least 88 TA systems (Ramage et al. 2009), and *E. coli* K12, which has at least 33 known TA systems (Yamaguchi et al. 2011). On the other hand, there are also prokaryotes that have few or even no TA systems. For instance *M. smegmatis*, a non-pathogenic *Mycobacterium* species related to *M. tuberculosis*, only harbours three identified TA systems (Pandey and Gerdes 2005; Robson et al. 2009). It is still debated what the benefits of such a high number of TA systems are and several studies have investigated possible correlations between a large number of loci and various phenotypic features, such as life style, phylum, growth rate, cell shape, respiratory system, GC content, presence of one or several replicons, generation time, and genome size, but no correlation has been found yet (Pandey and Gerdes 2005; Sevin and Barloy-Hubler 2007). As mentioned above, a link between TA loci and the ability to form persister cells, which are critical to pathogenicity, is currently being investigated. Presumably, a large number of TA loci would be particularly important for pathogenic species that have to survive in a challenging host-pathogen environment. It is possible that the plethora of toxins allow pathogenic bacteria to specifically target several cellular pathways at the same time and thus allow for a greater variety in adaption to stressful situations. Finally, it is worth noting that the cellular levels of both toxins and antitoxins are kept at stable steady-state levels during normal growth, which allows cells to very quickly respond to environmental changes through toxin activation, also under conditions where gene expression is slow or inhibited.

14.3 Structural Hierarchy of the VapBC TA Systems

The VapBC (Virulence associated proteins B and C) system constitutes the largest subgroup of the type II TA systems (Pandey and Gerdes 2005), and *vapBC* loci are found in surprisingly large numbers in some prokaryotes and archaea. For example, *M. tuberculosis* contains at least 47 *vapBC* loci and these loci have also been found in large numbers in most hyper thermophilic archaea (Pandey and Gerdes 2005; Sevin and Barloy-Hubler 2007). The specific, biological advantage of having so many homologous loci is presently unknown, but it is possible that it allows organisms to deliver a very fine-grained response towards external stress conditions, like antibiotic pressure or the immune system of a host organism in the case of human pathogens. The *vapBC* operon is an archetypical type II TA system, and as such bicistronic at the genomic level, with the *vapB* antitoxin gene preceding the *vapC* toxin gene, allowing expression levels of the antitoxin to exceed those of the toxin (Gerdes et al. 2005) (Fig. 14.1, Type II). During normal growth conditions, the VapC and VapB proteins form a tight protein complex, in which the VapB antitoxin wraps its C-terminal tail around and inhibits VapC (Gerdes et al. 2005). This inhibited VapBC complex negatively auto-regulates transcription of the operon by binding directly to the promoter region using a DNA-binding domain located in the N-terminus of VapB (Gerdes et al. 2005; Wilbur et al. 2005; Robson et al. 2009). During cellular stress, the labile and highly flexible VapB antitoxin is degraded, releasing the active VapC. Due to the high prevalence of *vapBC* loci in pathogenic bacteria and the prospects of understanding human infectious diseases at the molecular level, a lot of research has recently gone into deciphering the molecular architecture and mechanisms of the VapBC TA systems. To date, crystal structures of seven distinct VapBC-type complexes have been determined, and together these highlight the great structural diversity of the family (Mattison et al. 2006; Miallau et al. 2009; Dienemann et al. 2011; Mate et al. 2012; Min et al. 2012; Das et al. 2014; Lee et al. 2015) (Table 14.1).

The VapC Toxin All VapC toxins belong to the conserved PIN (PilT N-terminal domain) domain family, which is a relatively small (app. 130 residues), compact domain with *bona fide* Mg^{2+}/Mn^{2+}-dependent endoribonuclease activity (Arcus et al. 2011; Levin et al. 2004). PIN domains are found in all domains of life and are involved in cleavage of a range of specific RNA targets in both sequence and structure specific fashions (Arcus et al. 2011). However, in prokaryotes, the vast majority of PIN domains are associated with TA systems (Arcus et al. 2011). While the over-

Table 14.1 Overview of the available VapBC complex structures

TA pair	PDB ID	TA architecture	Inhibition mode	Active site inhibition	DNA-binding domain
Ngo FitAB	2H1O	$VapB_4C_4$ octamer	1:1 proximal	Arg	RHH
Sfl VapBC	3TND	$VapB_4C_4$ octamer	1:1 proximal	Arg + Gln	AbrB
Rfe VapBC2	3ZVK	$VapB_4C_4$ octamer	1:2	Arg or Tyr	AbrB
Mtu VapBC-3	3H87	$VapB_4C_4$ octamer	1:1 distal	Arg	RHH
Mtu VapBC-30	4XGQ	$VapB_2C_2$ tetramer	1:1 distal	No specific interactions	N/A[a]
Mtu VapBC-15	4CHG	$VapB_2C_2$ tetramer	1:1 proximal	Glu	N/A[a]
	4CHG	$VapBC_2$ heterotrimer	1:2	Glu or Arg + Lys	N/A[a]
Mtu VapBC-15	3DBO	$VapB_2C_2$ tetramer	1:1 proximal	Arg	N/A[a]

TA architecture, mode of antitoxin inhibition and active site inhibition, and DNA binding domain for TA structures determined so far: *Neisseria gonorrhoeae* FitAB (Mattison et al. 2006), *Shigella flexneri* VapBC (Dienemann et al. 2011); *Rickettsia felis* VapBC2 (Mate et al. 2012); *Mycobacterium tuberculosis* VapBC-3 (Min et al. 2012), VapBC-30 (Lee et al. 2015), VapBC-15 (Das et al. 2014), and VapBC-15 (Das et al. 2014)

[a]DNA-binding domain not part of the structure

all PIN domain fold is highly conserved there is surprisingly low overall primary sequence conservation and alignments are most precise when carried out structurally with tools like PROMALS3D (Pei and Grishin 2014). There are currently 11 different structures of prokaryotic and archaeal PIN domains in the PDB that have been functionally classified as VapC toxins, many of which have been determined in complex with their cognate VapB antitoxin. These structures include *A. fulgidus* VapC-9 (DAF0591) (Levin et al. 2004), *M. tuberculosis* VapC-3, VapC-5, VapC-15, and VapC-30 (Min et al. 2012; Lee et al. 2015; Das et al. 2014; Miallau et al. 2009), *N. gonorrhoeae* FitB, a VapC homologue (Mattison et al. 2006), *P. aerophilum* VapC-3 and Vap-C9 (Bunker et al. 2008; Arcus et al. 2004), *P. horikoshii* VapC-4 (Jeyakanthan et al. 2005), *R. felis* VapC2 (Mate et al. 2012), and a VapC from *S. flexneri* virulence plasmid pMYSH6000 (Dienemann et al. 2011). In addition to these, the PDB currently contains two structures of homologous proteins from *A. fulgidus* (AF1683, PDB ID 1W8I) and *P. furiosus* (Pfu-367848-001, PDB ID 1Y82) with no associated publications, which are not considered here due to their putative assignment as VapC toxins. Structural alignment of the 11 VapC proteins using PROMALS3D reveals that the sequence identity varies between 10 and 36% among the structurally characterised VapC toxins, with the *R. felis* and *S. flexneri* orthologues being the most similar proteins (Fig. 14.2). When grouping amino acids based on similar functionality (F, Y, W = aromatic; V, I, L = aliphatic; R, K, H = charged positive; D, E = charged negative; S, T = alcohol, and N, Q = polar), the VapC orthologues are between 23 and 61% similar, again with the *R. felis* and *S. flexneri* toxins as the most closely related proteins. Most PIN domain proteins are active as Mg^{2+}/Mn^{2+}-dependent endoribonucleases and share between three and five conserved, acidic residues that are required for metal ion binding and consequently RNase activity. Of these, three (D-E-D) are universally conserved (Fig. 14.2, dashed boxes) and these residues can be used as a signature sequence for locating PIN domain proteins, which can otherwise be difficult to identify. For bacterial TA systems, the genetic organisation with the antitoxin preceding the PIN domain toxin is also a strong constraint for locating such loci (Pandey and Gerdes 2005).

The PIN Domain The PIN domain fold itself consists of a α/β/α sandwich core with alternating α-helices and β-strands that organise the conserved acidic residues in close proximity in the active site. In this chapter, we will use a standard nomenclature to denote the architecture of toxins, antitoxins, and their complexes: T_nA_m, where n is the number of toxin (T) molecules and *m* the number of antitoxin (A) molecules in any given complex. Using this nomenclature, the toxin monomer can be referred to as T_1 and its core PIN domain fold is shown in Fig. 14.3. Accumulating evidence indicates that VapC PIN domain RNases are able to target specific RNAs in bacterial cells, including unique tRNA and rRNA species (Winther et al. 2013; Winther and Gerdes 2011; Cruz et al. 2015; Sharp et al. 2012). However, there are yet no structures of PIN domains bound to their cognate RNA targets that would allow us to establish a detailed molecular basis of recognition. The mechanism of cleavage, on the other hand, is likely related to T4 RNase H, which is a nuclease specific for RNA-DNA duplexes (Xu et al. 2016). As such structures exhibit a

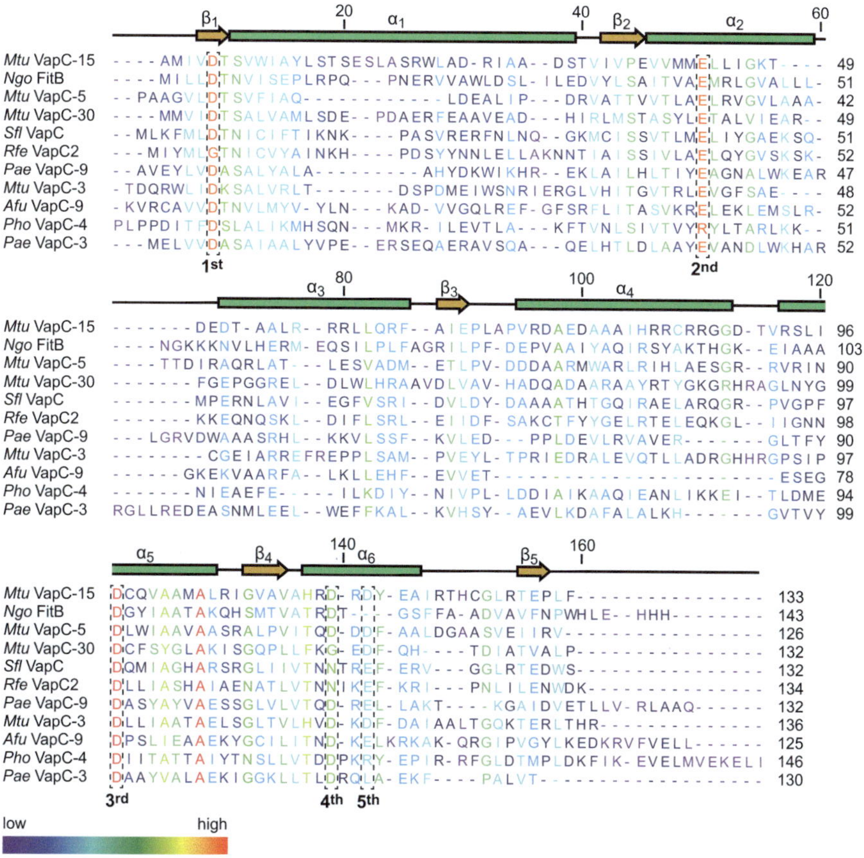

Fig. 14.2 Structural alignment of bacterial and archaeal VapC toxins. Structural alignment of *M. tuberculosis* VapC-15 (PDB ID 4CHG) (Das et al. 2014), VapC-5 (PDB ID 3DBO) (Miallau et al. 2009), VapC-30 (PDB ID 4XGQ) (Lee et al. 2015), and VapC-3 (PDB ID 3H87) (Min et al. 2012), *N. gonorrhoeae* FitB (PDB ID 1H1O) (Mattison et al. 2006), *S. flexneri* pMYSH6000 VapC (PDB ID 3TND) (Dienemann et al. 2011), *R. felis* VapC2 (PDB ID 3ZVK) (Mate et al. 2012), *P. aerophilum* VapC-9 (PDB ID 1V8P) (Arcus et al. 2004) and VapC-3 (PDB ID 2FE1) (Bunker et al. 2008), *A. fulgidus* VapC-9 (PDB ID 1O4W) (Levin et al. 2004), and *P. horikoshii* VapC-5 (PDB ID 1 V96) (Jeyakanthan et al. 2005) coloured by conservation from low (*purple*) to high (*red*) as shown by the inset *colour bar*. Secondary structure elements are shown and annotated and the five conserved active site residues are indicated (1st through 5th)

conformation similar to A-form RNA they may be quite similar to the structured RNA targets of the PIN domains. An intriguing observation is that even though each PIN domain appears to contain a complete active site, all structures of bacterial and archaeal VapCs determined so far have shown the protein as a homodimer. This is unlikely to be a result of crystallisation as several of the VapC toxins display the same behaviour in solution as judged by gel filtration chromatography (Dienemann et al. 2011; Xu et al. 2013; Xu et al. 2016). The VapC dimer is formed by tight pack-

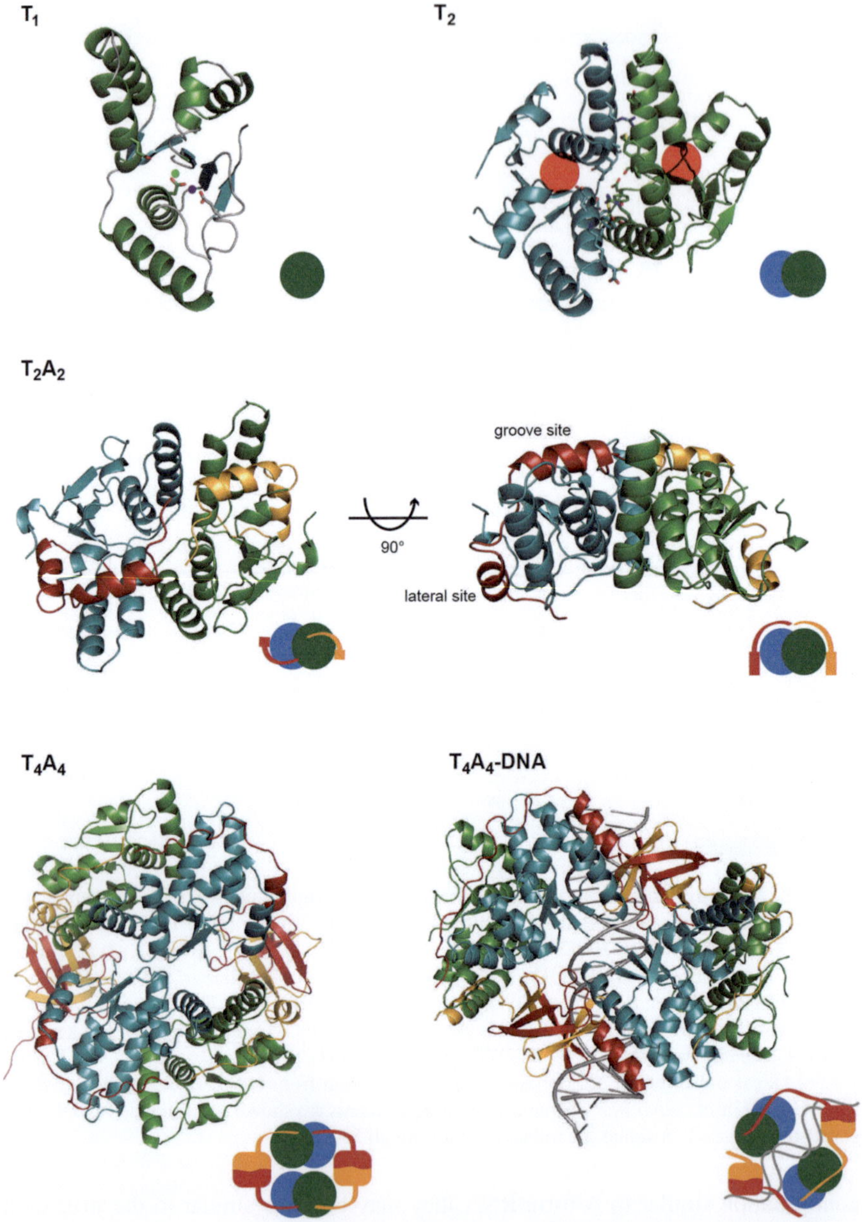

Fig. 14.3 Hierarchical architecture of TA systems. Overview of the hierarchical structure of the TA systems. The figure uses the T_nA_m nomenclature to indicate the complex architecture, where *n* is the number of toxin (T) molecules and *m* the number of antitoxin (A) molecules in the complex. The small schematics illustrate the overall architecture of each complex, with *green* and *blue spheres* representing toxin molecules, *red* and *orange* the antitoxin DNA-binding domain (*rounded squares*) and C-terminal extension (*lines*), and *grey lines* representing DNA. $\mathbf{T_1}$,

ing of the third α-helix (α_3, Fig. 14.2), the β-strand following (β_3) and the penultimate α-helix (α_5), and sometimes the fourth and ultimate α-helices (α_4 and α_6). Although there are typically both a number of strong hydrogen bonds as well as hydrophobic interactions between the chains, dimerisation appears to be driven largely by shape complementarity between the two PIN domains folds. This mode of dimerisation is observed among all structures of bacterial and archaeal VapC proteins determined to date, either as a direct dimer in the crystallographic asymmetric unit or in a few instances, as the results of a crystallographic two-fold symmetry (such as in the cases of *M. tuberculosis* VapC-5, PDB ID 3DBO (Miallau et al. 2009) and *P. aerophilum* VapC-3, PDB ID 2FE1 (Bunker et al. 2008)). The only departure from this rule is the putative VapC homologue from *A. fulgidus* (AF1683, PDB ID 1W8I), for which the dimer is similar, but skewed compared to the canonical structures. Since there are presently no structures available of bacterial PIN domains bound to their cognate RNA targets, it is not known if the dimer represents the active conformation of the toxin or somehow mimics the higher-order structure of the antitoxin-inhibited state (see below). Nevertheless, two active sites are placed in relative proximity in the T_2 dimer (Fig. 14.3, T_2, red spheres), so if the dimer does indeed represent the active conformation of the toxin it would lead to a whole new range of questions concerning substrate specificity and activity such as whether both active sites are engaged simultaneously in RNA cleavage or both bind RNA, but only one is catalytically active.

The VapB Antitoxin The VapB antitoxin serves two main purposes in VapBC TA complexes, (1) it inhibits the activity of the toxin by binding directly to its active site, and (2) it carries a DNA-binding domain that promotes binding of the TA complex to operator DNA and confers auto-inhibition of transcription from the TA locus. These functions are maintained by two distinct domains in the antitoxins, an N-terminal DNA-binding domain and a C-terminal intrinsically disordered region. VapB toxins are typically around 80 amino acids in length, of which the first approx. 50 dimerise to form a DNA-binding domain while the last approx. 30 amino acids contain a flexible "tail" that wraps around the toxin and physically inhibits it by

Fig. 14.3 (continued) the isolated toxin structure showing the PIN domain fold (PDB ID 5ECW) in *green cartoon* (Xu et al. 2016). Acidic residues in the active site are shown in *sticks* and the expected positions of the bound Mn^{2+} and Mg^{2+} ions are shown with *purple* and *green spheres*, respectively, according to their position in *M. tuberculosis* VapC-15 (4CHG) (Das et al. 2014). **T_2**, the VapC toxin dimer (*blue* and *green cartoons*) as observed in nearly all crystal structures with residues interacting at the interface (*sticks*) and the location of the two active sites (*red circles*) shown (PDB ID 5ECW) (Xu et al. 2016). **T_2A_2**, two orthogonal views of the heterotetramer formed in the absence of the antitoxin DNA-binding domains. Colouring as above with the antitoxin molecules in *red and orange* (PDB ID 3DBO) (Miallau et al. 2009). **T_4A_4**, the heterooctameric structure formed upon dimerisation of the antitoxin DNA-binding domains (*red and orange*) (PDB ID 3TND) (Dienemann et al. 2011). **T_4A_4-DNA**, interaction of the heterooctamer with duplex DNA at the operator binding site (PDB ID 3ZVK) (Mate et al. 2012). Organisms abbreviations: *Ngo, Neisseria gonorrhoeae; Sfl, Shigella flexneri; Rfe, Rickettsia felis; Mtu, Mycobacterium tuberculosis*. All structure figures in this chapter were created with PyMol (Schrodinger 2010)

directly binding the active site. Four types of DNA-binding domains have been identified so far in VapB toxins, namely the AbrB-type, the ribbon-helix-helix (RHH) domain, the helix-turn-helix (HTH), and the Phd/YefM-type domain (Gerdes et al. 2005). The VapB C-terminal tail is intrinsically disordered in the isolated antitoxin and prone to proteolysis, and presumably only forms correct secondary structure upon binding to its cognate toxin molecule. Consequently, all VapB structures have been determined in the context of the VapBC TA complex, except in a few cases, such as isolated DNA-binding domains (*R. equi* VapB, PDB ID 4CV7 (Geerds et al. 2014)) or putative VapB antitoxins (*M. tuberculosis* VapB-49, Rv2018, PDB ID 5AF3). As for the VapBC complexes, there are currently seven unique structures available in the PDB (Table 14.1). In all these cases the flexible C-terminal domain of the VapB antitoxin is visible. The N-terminal DNA binding domain is visible in some of these structures, but has also been removed in others (Mattison et al. 2006; Miallau et al. 2009; Dienemann et al. 2011; Mate et al. 2012; Min et al. 2012; Das et al. 2014; Lee et al. 2015).

The Mechanism of VapB Inhibition A conserved feature of VapBC TA systems (and type II TA systems in general) is that the antitoxin blocks the cellular function of the toxin by formation of a tight protein complex. This function is maintained by the flexible C-terminal domain of VapB which contains functional residues that appear to be widely conserved within the VapBC family (Dienemann et al. 2011). The seven available VapBC structures reveal that the C-terminal domain of the VapB proteins most often consists of an α-helix followed by an extended tail with no secondary structure that wraps around the toxin (Fig. 14.3, T_2A_2). Exceptions to this include *M. tuberculosis* VapB-3, which has of a loop followed by an α-helix (Min et al. 2012) and *M. tuberculosis* VapB-5 (Miallau et al. 2009) and VapB-15 (Das et al. 2014), which both contain two α-helices. The interaction of VapB with VapC can be separated into two sites, which we term the *lateral* and *groove* sites (Fig. 14.3, T_2A_2). The first site, the lateral site, occurs on the side of the TA complex and usually involves an α-helix of the antitoxin that makes hydrophobic interactions with residues of nearby α-helices (α_2 and α_4) of the toxin, as well as a few specific hydrogen bonds involving both side chains and protein backbone of both partners (Mattison et al. 2006; Dienemann et al. 2011; Mate et al. 2012; Lee et al. 2015) (Fig. 14.3, T_2A_2, and Fig. 14.5d). The remaining, extended tail of the antitoxin is located in a large continuous groove formed by the dimerisation of the toxins on the "top" of the complex, which also contains the two actives sites of the T_2 dimer. Outside the active sites, this groove is lined with hydrophobic residues from the core of the protein as well as some charged residues (Mate et al. 2012; Min et al. 2012). The tail of the antitoxin runs along the groove, where hydrophobic residues from both proteins interact with each other, tightly anchoring the antitoxin to the surface of the toxin (Fig. 14.3, T_2A_2, and Fig. 14.5d). Although the antitoxin tails show a large degree of variability, both with respect to length and amino acid content, some of these interactions appear to be conserved across the VapBC family (Dienemann et al. 2011).

14.4 Overall TA Complex Architecture

Despite the relatively small size and conserved folds of their components, intact TA complexes display a surprising structural and architectural variety. In all VapBC and isolated VapC structures determined to date, the VapC toxin exhibits the conserved PIN domain structure, which appears as a rigid domain with only minor conformational changes observed between the VapB-bound and free toxin states (Xu et al. 2016). The structural complexity of TA pairs therefore appears to be largely due to the presence of one or more VapB antitoxin molecules, which contain both flexible regions and a variable DNA-binding domain. Since both toxins and antitoxins form homodimers, most intact VapBC complexes have a heterooctameric T_4A_4 architecture that can be described as $(TA\text{-}AT)_2$ or in other words, homodimers of homodimers of heterodimers (Fig. 14.3, T_4A_4). This complex presents two antitoxin DNA-binding homodimers (A_2) on one side of the complex, which are able to interact with two adjacent major grooves on operator DNA (Fig. 14.3, T_4A_4-DNA). There are presently seven unique VapBC-type TA structures in the PDB, which include *N. gonorrhoeae* (*N. gonorrhoeae*) FitAB (a VapBC orthologue) (Mattison et al. 2006), *S. flexneri* (*S. flexneri*) pMYSH6000 VapBC (Dienemann et al. 2011), *R. felis* (*R. felis*) VapBC2 (Mate et al. 2012), and four complexes from *M. tuberculosis* (*M. tuberculosis*), VapBC-3 (Min et al. 2012), VapBC-5 (Miallau et al. 2009), VapBC-30 (Lee et al. 2015), and VapBC-15 (Das et al. 2014) (Fig. 14.4). Two of these structures, *N. gonorrhoeae* FitAB and *R. felis* VapBC2 where determined as bound to their cognate operator DNA sequence, but not in isolation. Since much structural variation is observed between various VapBC orthologues this means that we still do not know exactly whether DNA binding induces conformational changes in the TA complex. *R. felis* VapBC2 was reported to form a heterohexamer in solution, however, this result is based on size exclusion chromatography making it somewhat uncertain (Mate et al. 2012). In fact, the most commonly observed stoichiometry of TA complexes in absence of DNA is a heterooctamer with four antitoxin molecules and four toxin molecules in a T_4A_4 architecture (Table 14.1 and Fig. 14.4, *N. gonorrhoeae* FitAB, *S. flexneri* VapBC, *R. felis* VapBC2, and *M. tuberculosis* VapBC-3) (Mattison et al. 2006; Dienemann et al. 2011; Mate et al. 2012; Min et al. 2012; Lee et al. 2015). *M. tuberculosis* VapBC-30 was also reported as a heterooctamer, however, the interaction between T_2A_2 heterotetramers in the crystallographic asymmetric unit is weak and not likely to be physiological (Fig. 14.4e) (Lee et al. 2015). Likewise, *M. tuberculosis* VapBC-5 was reported as a heterodimer, but appears more likely to form a heterotetrameric assembly in the crystal (Fig. 14.4g) (Miallau et al. 2009). Adding to this complexity, the structure of *M. tuberculosis* VapBC-15 contains both a heterotrimeric T_2A_1 and a heterotetrameric T_2A_2 complex in the crystallographic asymmetric unit (Fig. 14.4f). Finally, even for identical stoichiometries, large differences are observed in complex organisation, such as *N. gonorrhoeae* FitAB, for example, which displays a very loose arrangement of domains, whereas most of the other structures adopt a much more compact arrangement (Fig. 14.4). Due to these discrepancies and since apparent oligomeric

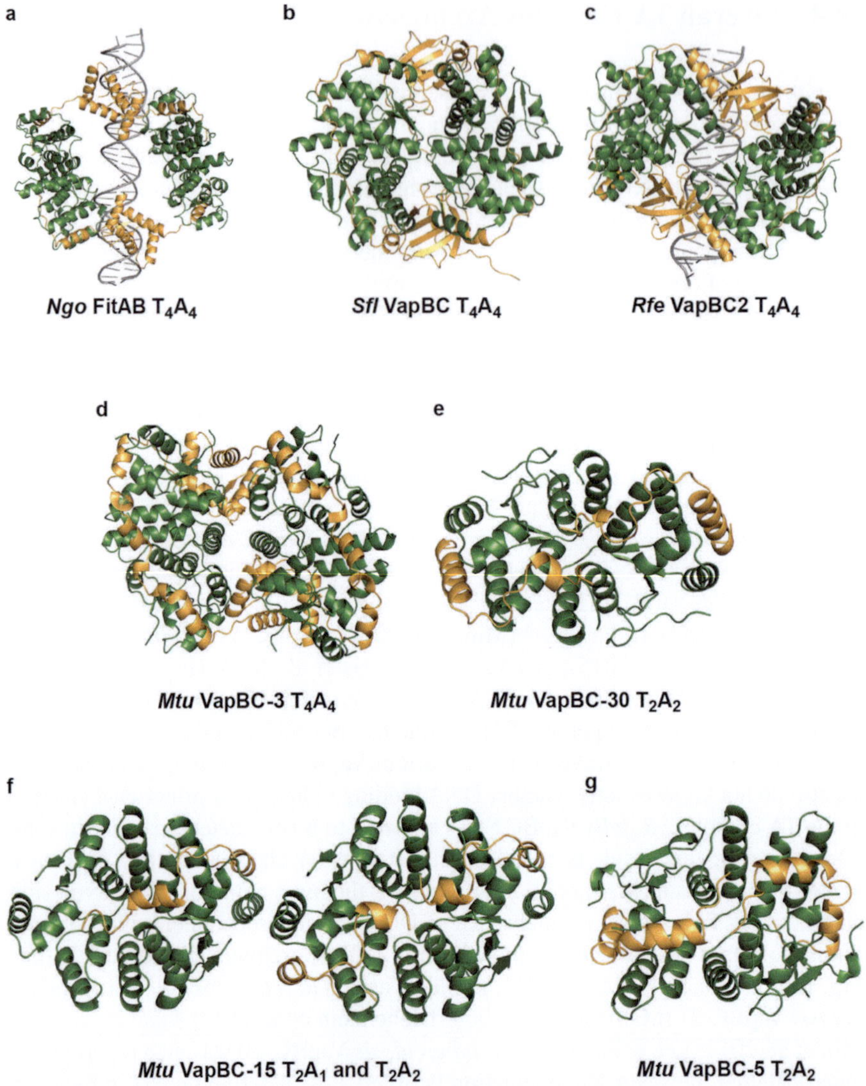

Fig. 14.4 VapBC higher-order complex structure. Overview of known VapBC complex structures shown as cartoon with toxins in *green* and antitoxins in *orange*. (**a**) *N. gonorrhoeae* FitAB heterooctamer bound to DNA (PDB ID 2H1O) (Mattison et al. 2006). (**b**) *S. flexneri* pMYSH6000 VapBC heterooctamer (PDB ID 3TND) (Dienemann et al. 2011). (**c**) *R. felis* VapBC2 heterooctamer bound to DNA (PDB ID 3ZVK) (Mate et al. 2012). (**d**) *M. tuberculosis* VapBC-3 heterooctamer (PDB ID 3H87) (Min et al. 2012). (**e**) *M. tuberculosis* VapBC-30 heterotetramer (PDB ID 4XGR) (Lee et al. 2015). (**f**) *M. tuberculosis* VapBC-15 T_2A_1 heterotrimer (*left*) and heterotetramer (*right*) (PDB ID 4CHG) (Das et al. 2014). (**g**) *M. tuberculosis* VapBC-5 heterotetramer (PDB ID 3DBO) (Miallau et al. 2009)

state can be affected by crystal packing interactions, we decided to reinvestigate all known VapBC (and homologous) complex structures using the protein interaction server PISA to assess the extent and strength of reported and observed homodimer and heterotetramer interaction interfaces (Krissinel and Henrick 2007). Tables 14.1 and 14.2 provide an overview of all crystal structures of VapBC-type T_nA_m complexes, along with their respective PDB codes and their overall architecture, as predicted by PISA analysis. Most correspond well with what was reported in the original papers, except *M. tuberculosis* VapBC-5 and *M. tuberculosis* VapBC-30, which differ from the reported as mentioned above. Such characteristics become evident when investigating the heterotetramer interaction areas and energies ($\Delta G_{interaction}$) listed in Table 14.2 (T_2A_2:T_2A_2). We find that stable heterooctamers typically display interface sizes in the 3–4000 Å^2 range, while it for *M. tuberculosis* VapBC-30, for example, is only 542 Å^2 suggesting this complex is more likely on heterotetrameric form. Likewise, $\Delta G_{interaction}$ values tend to lie in the −20 to −40 kcal/mol range for heterooctamers while for VapBC-30 we get a value of only −9.5 kcal/mol. The other columns of the table list the corresponding values for the TA-interactions ("T_2A (high)" and "T_2A (low)", see below), and the toxin (T:T) and antitoxin (A:A) homodimers. It is interesting to note that toxin and antitoxin dimerisation take place using comparable interaction areas and interaction energies, possibly reflecting that they are both equally important for formation of the heterooctamer during DNA-binding. Values for the antitoxin dimerisation are missing for those structures where the DNA-binding domain was left out of the structure or not visible (*M. tuberculosis* VapBC-5, VapBC-15, and VapBC-30).

14.5 The Stoichiometry of VapB Inhibition

In the canonical VapBC complexes, each VapC toxin is inhibited by a single VapB antitoxin, which wraps around it and inhibits its active site by directly interfering with the conserved, charged residues (see *Active site inhibition* below). Both VapC and VapB form homodimers giving rise to a circular $(AT\text{-}TA)_2$ structure, which is most clearly observed for *N. gonorrhoeae* FitAB due to the distance between the heterotetramers (Fig. 14.4a). In the most common case, the VapB antitoxin interferes with the active site of its nearby or cognate VapC toxin in a binding mode, which we term *1:1 proximal binding*. This mode of inhibition has been observed in the structures of *N. gonorrhoeae* FitAB, *S. flexneri* VapBC, and *M. tuberculosis* VapBC-5 (Table 14.1) (Mattison et al. 2006; Miallau et al. 2009; Dienemann et al. 2011). Only about ten residues of VapB (typically residues 60–70 out of about 80) are required for the interaction at the groove site, where the antitoxin polypeptide has a fully extended conformation (Fig. 14.5a). This leaves about ten residues in the C-terminus of VapB unaccounted for, which are presumed to be pointing into solution as the last few visible residues tend to point away from VapC (e.g. in *S. flexneri* VapBC) (Dienemann et al. 2011). However, as first observed in the structure of *R. felis* VapBC2, these residues can also in some instances continue across the combined

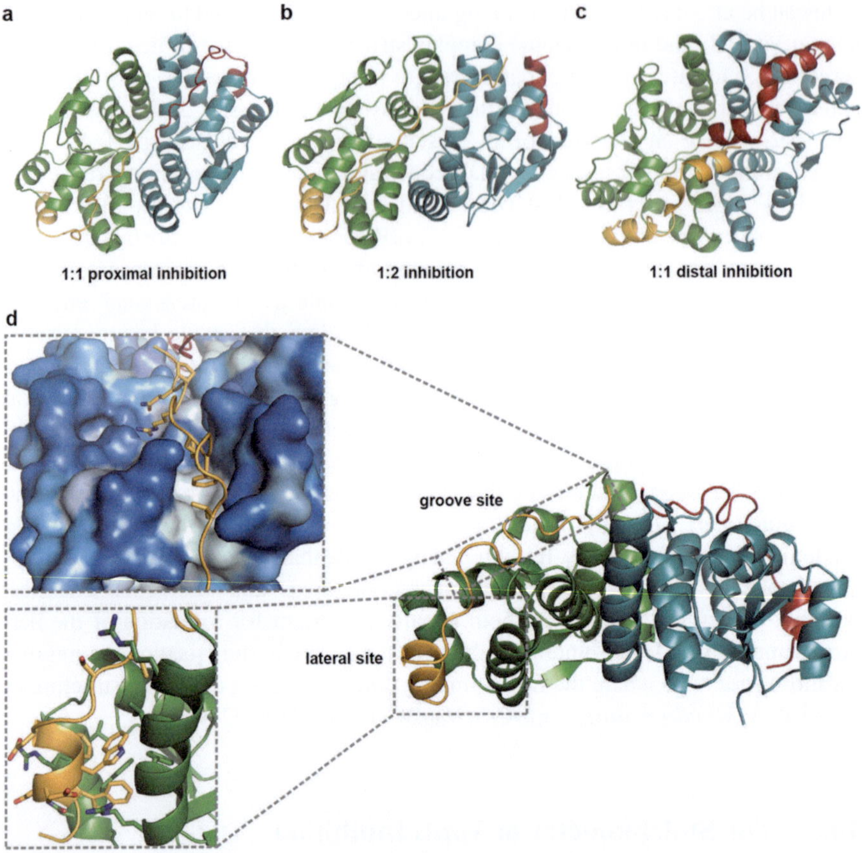

Fig. 14.5 Inhibition modes of the VapB antitoxin. (**a**) 1:1 inhibition of the *proximal* toxin as observed in the *S. flexneri* pMYSH6000 VapBC complex (PDB ID 3TND) (Dienemann et al. 2011). (**b**) 1:2 inhibition of the toxin dimer as observed in the *R. felis* VapBC2 complex (PDB ID 3ZVK) (Mate et al. 2012). (**c**) 1:1 inhibition of the *distal* toxin in the *M. tuberculosis* VapBC-3 complex (PDB ID 3H87) (Min et al. 2012). (**d**) Overview and details of the toxin-antitoxin interactions at the "lateral" and "groove" sites exemplified by the *S. flexneri* VapBC complex (PDB ID 3TND) (Dienemann et al. 2011). Interactions at the groove site (*upper inset*) are both hydrophilic at the active site (*top*) and hydrophobic (*bottom*). The surface of VapC is shown coloured by hydrophobicity, with *white* representing the most hydrophobic areas. Interactions at the lateral site (*lower inset*) are primarily hydrophobic

groove of the VapC homodimer and block the adjacent active site, thus one antitoxin interferes with two VapC active sites, a phenomenon that has been termed *1:2 binding* of antitoxin to toxin (Fig. 14.5b and Table 14.1) (Mate et al. 2012). Consequently, the VapB tail associated with the other VapC of the toxin dimer in the canonical interaction must detach and in such cases only interacts with VapC at the lateral site. This results in two distinct modes of toxin-antitoxin interaction in a single, two-fold symmetrical heterooctamer, which we term T_2A *(high)* and T_2A *(low)*, corresponding to high affinity (VapB bound at the lateral site and across the groove sites of both toxins) and low affinity (VapB bound exclusively to the lateral site), respectively. This distinction becomes especially clear when looking at the interaction areas and energies as listed in Table 14.2, where complexes displaying 1:2 binding display larger interaction areas for the high affinity interaction (e.g. *R. felis* VapBC2) as well as a much larger difference between the values for the high and low affinity sites. *M. tuberculosis* VapBC-15 is an interesting case as the crystal contains both a T_2A_2 heterotetramer and a T_2A_1 heterotrimer in the asymmetric unit for which the heterotetramer displays 1:1 binding and the heterotrimer 1:2 binding. Based solely on energetics, we cannot say whether the observed 1:2 interaction in the trimer is a result of the missing VapC molecule or vice versa, that the VapC molecule is missing due to the 1:2 interaction, as the T_2:A energies are comparable to the T:T energies for this complex. It is also not known at present whether there are any significant, biological consequences of the ability of VapB to bind two VapC molecules simultaneously, nor whether it interaction is affected by (or affects) DNA binding by the TA complex. Another variation in VapBC toxin-antitoxin interaction has been observed for *M. tuberculosis* VapBC-3 and VapBC-30, which both on the surface display 1:1 binding in that the VapB tails only cover the nearby VapC toxin groove. However, closer inspection of the structures revealed that the very last part of the antitoxin that is structurally organised actually reaches across in both cases and interferes with the active site of the adjacent VapC rather than the cognate one to which it is bound. We term the resulting crossed-over or swapped inactivation mode *1:1 distal binding* (Fig. 14.5c and Table 14.1) (Min et al. 2012; Lee et al. 2015).

14.6 Active Site Inhibition

The VapC active sites are located in two negatively charged cavities along a continuous groove on the toxin homodimer surface. In the inhibited state, electrostatic and hydrogen bond interactions between specific residues of the antitoxin and the toxin active site(s) effectively prevent binding of potential substrates (Table 14.1 and Fig. 14.6). In *N. gonorrhoeae* FitAB and *M. tuberculosis* VapBC-5, which both exhibit the canonical 1:1 proximal binding, and VapBC-3, in which the distal VapB is involved, a positively charged arginine residue from the antitoxin points directly into the active site, where its guanidinium group interacts electrostatically with the carboxyl groups of several of the conserved acidic residues of the toxin (Fig. 14.6a) (Mattison et al. 2006; Miallau et al. 2009; Min et al. 2012). In addition, in *M. tuberculosis* VapBC-3, a bulky tryptophan side chain from the antitoxin buries itself in the crevice between VapC monomers, which runs perpendicular to the groove, thus anchoring the antitoxin across to the next monomer (Fig. 14.6e) (Min et al. 2012). In the *S. flexneri* VapBC complex, which also displays 1:1 proximal binding, both an arginine (Arg64) and a glutamine residue (Gln66) from VapB interferes with the active site where they interact with the conserved acidic residues via electrostatic and hydrogen bond interactions (Fig. 14.6b) (Dienemann et al. 2011). Intriguingly,

Table 14.2 Interface areas and interaction energies for known TA complexes

			Interface area (Å²)					$\Delta G_{interaction}$ (kcal/mol)				
TA pair	PDB ID	T_nA_m	T_2:A (high)	T_2:A (low)	T:T	A:A	T_2A_2:T_2A_2	T_2:A (high)	T_2:A (low)	T:T	A:A	T_2A_2:T_2A_2
Ngo FitAB	2H1O	T_4A_4	1336.3	1027.2	953.0	1372.4	3088.2	−17.7	−16.3	−18.2	−16.2	−38.9
Sfl VapBC	3TND	T_4A_4	1513.2	1414.6	1050.4	1534.9	3892.4	−7.6	−5.7	−20.3	−10.4	−20.0
Rfe VapBC2	3ZVK[a]	T_4A_4	2398.2	764.4	1086.4	1625.4	3807.9	−24.7	−12.9	−11.7	−16.6	−35.4
			1877.8	1173.3	–	–	–	−23.7	−21.8	–	–	–
Mtu VapBC-3	3H87	T_4A_4	1758.6	1359.6	1419.5	1331.8	3407.1	−21.4	−21.1	−14.7	−13.6	−24.6
Mtu VapBC-30	4XGQ	T_2A_2	1193.9	1191.0	1128.3	N/A[b]	541.6	−17.1	−15.0	−13.1	N/A[b]	−9.5
Mtu VapBC-15	4CHG[c]	T_2A_2	1131.0	1137.6	1263.6	N/A[b]	N/A[c]	−15.4	−14.5	−18.1	N/A[b]	N/A[c]
	4CHG[d]	T_2A_1	1769.7	N/A[d]	1279.4	N/A[b]	N/A[d]	−16.2	N/A[d]	−19.1	N/A[b]	N/A[d]
Mtu VapBC-5	3DBO	T_2A_2	1819.1	1819.2	969.7	N/A[b]	N/A[c]	−10.3	−10.2	−18.2	N/A[b]	N/A[c]

For each available TA structure, the table lists the stoichiometry as T_nA_m, where *n* is the number of toxin molecules and *m* is the number of antitoxin molecules in each discrete complex observed structurally. The structures are: *Neisseria gonorrhoeae* FitAB (Mattison et al. 2006), *Shigella flexneri* VapBC (Dienemann et al. 2011); *Rickettsia felis* VapBC2 (Mate et al. 2012); *Mycobacterium tuberculosis* VapBC-3 (Min et al. 2012), VapBC-30 (Lee et al. 2015), VapBC-15 (Das et al. 2014), and VapBC-15 (Das et al. 2014). Average interface areas and interaction energies ($\Delta G_{interaction}$) as calculated using PISA (Krissinel and Henrick 2007) are shown for the indicated interfaces, where T_2:A (high) and T_2:A (low) represent the high and low affinity antitoxin interactions with a complete toxin dimer, respectively, T:T the toxin-toxin dimer interaction, A:A the antitoxin-antitoxin interaction (primarily through the DNA-binding domain), and T_2A_2:T_2A_2 the interaction between tetramers for heterooctameric complexes

[a]The structure is not two-fold symmetrical, so values for both tetramers are given rather than the average

[b]Structures determined without the DNA-binding domain

[c]Heterotetrameric T_2A_2 complex only

[d]Heterotrimeric T_2A complex only

the VapB interactions differ between the four VapC active sites in the *R. felis* VapBC2 structure, which displays 1:2 binding as described above, where one antitoxin straddles across both VapC molecules of the toxin homodimer. In one VapC dimer of the heterooctamer, two VapB arginine residues (Arg66/Arg74) block the two active sites through electrostatic interactions with the conserved acidic residues, and this is also the case for one VapC of the other homodimer (Arg66). However, in the last active site, the Arg has shifted away and a tyrosine residue (Tyr76) appears to have taken its place (Mate et al. 2012) (Fig. 14.6d). In the *M. tuberculosis* VapBC-15 structure, several configurations of the active site were also observed in that two active sites were found to bind divalent metal ions (one Mg^{2+} and one Mn^{2+}) while the two other sites were empty, despite all VapC toxins being in complex with antitoxin. In the active site containing bound ions, a glutamate residue (Glu67) from the antitoxin interacts with both ions (Fig. 14.6c), whereas in the active site without divalent cations, in analogy with the previously described pattern, an arginine residue (Arg74) and a lysine residue (Lys75) point into the active site where they interact with the conserved, acidic residues (Das et al. 2014). As if to highlight the great diversity in VapB inhibition modes, no direct interactions take place between the conserved, acidic residues in active site and the C-terminal tail of VapB in *M. tuberculosis* VapBC-30. Rather, several, likely specific hydrogen bonds are formed between residues close to the active site of VapC and VapB (Fig. 14.6f). In this case, the TA complex is further stabilised by a few hydrophobic interactions between the C-terminal tails of two VapB antitoxins (Lee et al. 2015).

14.7 DNA Binding by VapBC Complexes

All characterised type II TA systems, including VapBC, display auto-regulation of transcription through direct binding of the TA complex to operator DNA. As described above, this is achieved through presentation of the two DNA-binding antitoxin dimers on one side of the TA heterooctamer, where they are in a position to interact with two adjacent major grooves on DNA (Fig. 14.3, T_4A_4-DNA) (Gerdes et al. 2005). Interestingly, VapB antitoxins contain one of at least four different types of DNA binding domains, namely the *helix-turn-helix* (HTH) motif, the *ribbon-helix-helix* (RHH) motif, the *Phd/YefM* domain, or the *AbrB* domain (Table 14.1) (Gerdes et al. 2005). This supports the general notion that TA systems have evolved through active gene shuffling by swapping modules required for DNA binding and transcriptional regulation. Apart from the VapBC complex, the VapB antitoxin can also bind to the promoter region on its own, but biochemical analyses have shown that the TA complex binds with much higher affinity than the antitoxin alone due to the stabilisation conferred by effective cross-linking of VapB dimers by VapC in the heterooctamer. Consequently, the toxin can therefore be regarded as a co-repressor of its own transcription (Wilbur et al. 2005; Gerdes et al. 2005). Only two structures of VapBC complexes bound to their promoter region exist to date, namely the *N. gonorrhoeae* FitAB (PDB IDs 2BSQ, 2H1O, and 2H1C) (Mattison et al. 2006;

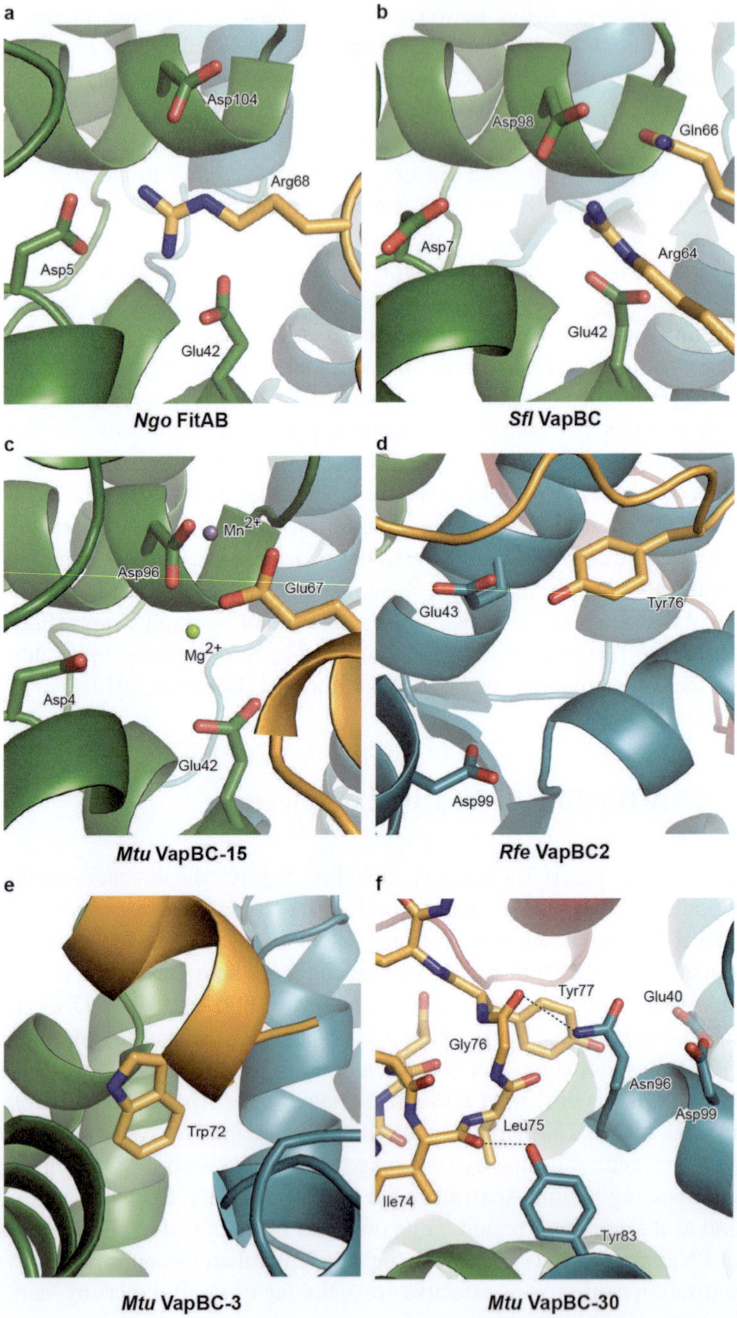

Fig. 14.6 Details of VapB-VapC active site interactions. (**a**) In *N. gonorrhoeae* FitA, Arg68 inhibits the FitB active site through electrostatic interactions and possibly displacement of divalent metal ions (PDB ID 2H1O) (Mattison et al. 2006). (**b**) In *S. flexneri*, VapB inhibits the

Wilbur et al. 2005) and *R. felis* VapBC2 (PDB ID 3ZVK) complexes (Mattison et al. 2006; Mate et al. 2012). Furthermore, despite DNA being absent, the VapB DNA-binding domains are present in what appears to be approximately the correct orientation in both the *S. flexneri* VapBC (PDB ID 3TND) and *M. tuberculosis* VapB-3 (PDB ID 3H87) heterooctamer structures (Dienemann et al. 2011; Min et al. 2012).

Among the known VapBC structures, two of the four types of DNA binding domains are represented; the RHH motif, which is found in *N. gonorrhoeae* FitA and *M. tuberculosis* VapB-3 (Mattison et al. 2006; Min et al. 2012), and the AbrB motif, found in *S. flexneri* VapB and *R. felis* VapB2 (Dienemann et al. 2011; Mate et al. 2012). The RHH domain was originally identified in the phage P22 Arc repressor and consists of a β/α motif with a β-strand followed by two α-helices (Raumann et al. 1994b, 1994a). The DNA-binding unit is a homodimer of the β/α motif and thus has an antiparallel β-sheet surrounded by four α-helices (Somers and Phillips 1992). Of these, the β-sheet inserts into the DNA major groove and is responsible for specific sequence recognition, whereas the α-helices primarily are responsible for dimerisation (Somers and Phillips 1992; Raumann et al. 1994a). The transcription of many RHH-containing proteins is found to be auto-regulated by the protein itself through binding to pseudo-palindromic, inverted repeats in the promoter region (Wilbur et al. 2005). The *N. gonorrhoeae* FitAB antitoxin, FitA, belongs to the RHH family (Wilbur et al. 2005) and consequently adopts the conserved β/α motif (Fig. 14.7a) (Mattison et al. 2006). As for other type II TA systems, the FitAB complex was found to bind to its promoter DNA with higher affinity than FitA alone due to the formation of the heterooctameric superstructure, which results in increased stability (Wilbur et al. 2005). Two RHH DNA-binding domains were observed in the crystal structure of FitAB bound to a 36 base pair cognate operator DNA fragment (Fig. 14.7b) (Mattison et al. 2006). The two RHH domains in the heterooctamer bind to two adjacent inverted repeat sequences on operator DNA. The domains bind on the same side of the DNA and make few specific contacts, and those there are, are all mediated through the β-sheet. The two binding sites on DNA are separated by a 14 base pair thymine/adenine-rich sequence, which adopts a straight and rigid structure with a compressed minor groove and shorter helical repeat compared to standard B-DNA (Mattison et al. 2006). In the crystal structure of *M. tuberculosis* VapBC-3, the VapB-3 antitoxin was also found to contain a RHH motif and although DNA was not present in this crystal form, two complete DNA-binding domains were found to protrude from the heterooctameric structure (Fig. 14.7c) (Min et al. 2012).

Fig. 14.6 (continued) VapC active site using both Arg64 and Gln66 (PDB ID 3TND) (Dienemann et al. 2011). (**c**) In *M. tuberculosis* VapBC-15, VapB-15 inhibits the active site of VapC-15 using a single glutamate Glu67 (PDB ID 4CHG) (Das et al. 2014). (**d**) In *R. felis* VapBC2, VapB2 inhibits the active site of VapC2 by placing Tyr76 in the active site (PDB ID 3ZVK) (Mate et al. 2012). (**e**) In *M. tuberculosis* VapBC-3, Trp72 of the antitoxin locates in the toxin dimer interface (PDB ID 3H87) (Min et al. 2012). (**f**) Inhibition of the active site of *M. tuberculosis* VapC-30 (PDB ID 4XGQ) (Das et al. 2014). Toxins are shown as *green* and *teal cartoons* and antitoxins are in *orange* and *red*. Relevant residues are shown as *sticks*

The AbrB-type DNA binding domain was first identified in *B. subtilis* as a transition-state regulator that binds promoter DNA via its N-terminal domain (Xu and Strauch 2001; Strauch et al. 1989). It consists of a β/α/β motif with two β-strands forming a β-hairpin followed by a short α-helix and two more β-strands, which form another hairpin. The two hairpins interleave each other in the dimer, and curve upwards to form a deep DNA binding cleft (Coles et al. 2005). Like seen for the RHH domain, dimerisation is important for stability and binding to DNA (Xu and Strauch 2001; Sullivan et al. 2008). Adjacent to the β-sheet, which locates in the DNA major groove, the two short α-helices interact with DNA on the rim next to the major groove (Sullivan et al. 2008). Four arginine residues located throughout the protein sequence appear to be critical for the ability to bind DNA, which involves both electrostatic and hydrogen bond interactions (Sullivan et al. 2008). Like the RHH domain, the AbrB domain binds to pseudo-palindromic thymine/adenine-rich sequences in promoter regions. For TA operator sequences, this feature represents a hallmark (Klein and Marahiel 2002; Bailey and Hayes 2009). The *R. felis* VapB2 antitoxin contains an AbrB-type motif in its N-terminal domain that dimerises with a neighbouring VapB monomer using the β-strands to form a complete DNA binding domain. However, the VapB DNA-binding domains do not contain the short α-helices seen in the original AbrB domain. Due to the heterooctameric VapBC architecture, there are two DNA binding sites in the intact structure (Fig. 14.7d, e) (Mate et al. 2012). Sequence analysis identified two potential pseudo-palindromic binding sites in the promoter region of the *R. felis vapBC2* operon that subsequently led to the design of a 27 base pair pseudo-palindromic dsDNA sequence that could be stably bound to the protein. The crystal structure of *R. felis* VapBC2 bound to its cognate operator DNA revealed that binding of the protein complex in two adjacent major grooves results in perturbation of the widths of both minor and major grooves, with major grooves spanning 15.5 Å, somewhat expanded compared to normal B-DNA (11.7 Å) (Neidle 2002). Inside the DNA, the *R. felis* VapB2 AbrB domain interacts with the concave surface of the major groove floor, but the two domains in the heterooctamer approach the DNA double helix from opposite sides. Contacts to the backbone are thought to open up the major groove where there are electrostatic interactions with the phosphates of the DNA backbone (Mate et al. 2012). DNA is specifically recognised in two ways, both through direct base readout, involving direct hydrogen bonds between protein and specific atoms of the nucleobases of DNA, and indirect readout, where the affinity is based on shape complementarity between the AbrB domain and the DNA major groove (Mate et al. 2012). *S. flexneri* VapB from the pMYSH6000 virulence plasmid VapBC complex also contains an AbrB-type motif in its N-terminus, but the only available structure was determined in absence of nucleic acids (Dienemann et al. 2011). The *S. flexneri* VapB displays the classical swapped β-hairpin fold but lacks the two short α-helices like seen for *R. felis* VapBC2 (Fig. 14.7f). The *S. flexneri* VapBC complex was shown biochemically to bind specifically to two binding sites within the promoter region of the pMYSH6000 plasmid from where it is expressed, of the operon but the DNA-bound structure has not yet been determined (Dienemann et al. 2011).

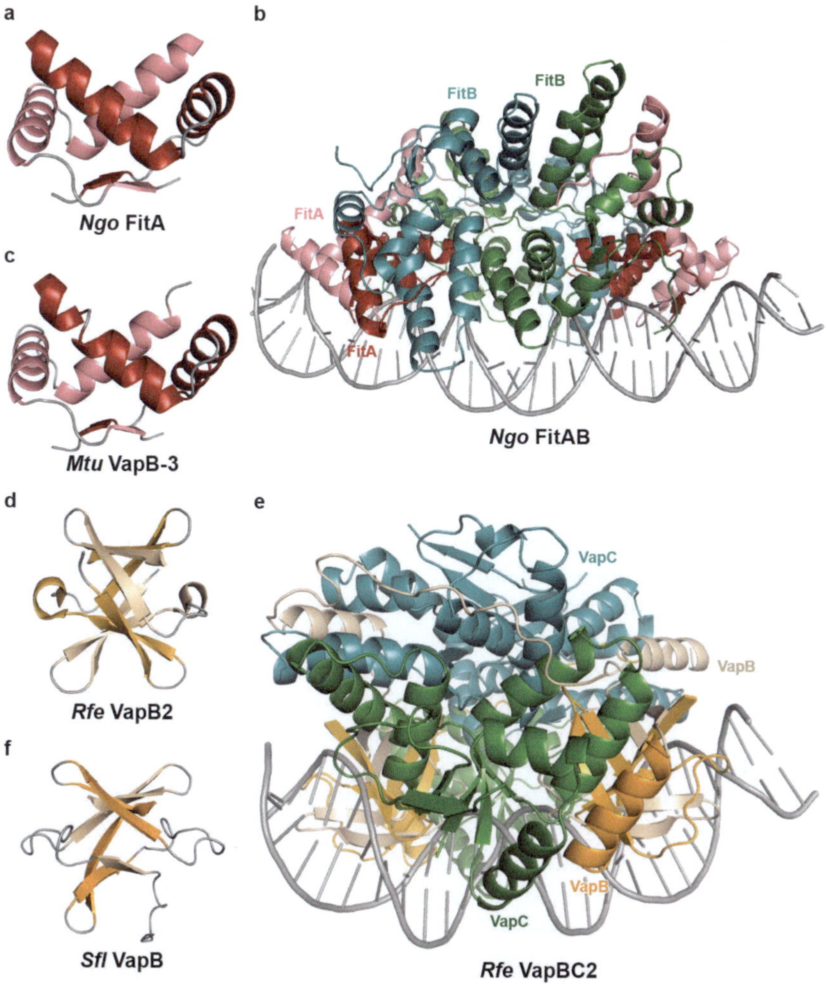

Fig. 14.7 DNA binding by VapBC complexes. (**a**) The ribbon-helix-helix DNA binding domain of *N. gonorrhoeae* FitA (PDB ID 2H1O). (**b**). The *N. gonorrhoeae* FitAB complex forms two DNA binding sites. (**c**) The ribbon-helix-helix DNA binding domain of *M. tuberculosis* VapB-3 (PDB ID 3H87). (**d**) The AbrB DNA binding domain of *R. felis* VapB2 (PDB ID 3ZVK). (**e**) The *R. felis* VapBC2 complex forms two DNA binding sites. (**f**) The AbrB DNA binding domain of *Shigella flexneri* VapB (PDB ID 3TND). For (**a**), (**b**) and (**c**) toxins are coloured *green* and *teal*, and antitoxins are coloured *red* and *pink*. For (**d**), (**e**) and (**f**), toxins are coloured *green* and *teal*, and antitoxins are coloured *orange* and *sand*

14.8 Discussion

Despite their relatively small size, bacterial toxin-antitoxin complexes display an impressive array of large, higher-order structures. Through dimerisation of both the toxin and antitoxin components; dimers, tetramers and even octamers are frequently observed and these structures are known to be central for maintaining steady cellular TA levels during normal growth. Although some TA complexes have been reported as heterotetramers, structural and energetic considerations reveal that this conclusion is usually either due to the DNA-binding domain being absent from the antitoxin or incomplete analysis of crystal packing. Therefore, for the type II TA complexes as exemplified here by the VapBC group, the canonical architecture is a heterooctamer consisting of four toxin and four antitoxin molecules. The heterooctamer is obviously required for placing two DNA-binding motifs into adjacent major grooves, so the question arises of whether the TA complexes are heterotetramers off DNA and only assemble when bound to the operator region? Several lines of evidence speak against the free TA complex as being present on tetramer form. Firstly, the interaction energies involved in toxin-toxin and antitoxin-antitoxin dimerisation are comparable (Table 14.2) and secondly, all intact VapBC structures determined so far, i.e. in presence of the DNA-binding domain, show the complex on heterooctameric form. Together these data suggest that the heterooctamer exists in solution and approaches the DNA operator as such.

Intriguingly, several types of DNA-binding domains are observed in type II TA systems, the HTH motif, the RHH motif, the *Phd/YefM* domain, and the *AbrB* domain (Table 14.1) (Gerdes et al. 2005). Two of these, the RHH and AbrB domains, are observed among the available VapBC structures. Likewise, several types of antitoxin inhibition, including 1:1 proximal, 1:1 distal, and 1:2 have been described for members of the VapBC family. This has led to speculations as to whether there is a correlation between the type of DNA-binding domain and the mode of inhibition, for example, it has been proposed that the AbrB-type DNA-binding domain would correlate with 1:2 binding as observed in the *R. felis* VapBC-2 structure (Mate et al. 2012). However, such a simple correlation does not seem to be the case as the structure of the *S. flexneri* VapB also has the AbrB domain but interacts with VapC via 1:1 proximal binding (Table 14.1). Furthermore, there is some evidence that the toxin-antitoxin interaction mode might be flexible and vary depending on the cellular context of the complex. For example, in *M. tuberculosis* VapBC-3, an example of 1:1 distal binding, both antitoxins lie in the continuous groove formed by the dimerization of the toxins and appear to physically clash in the crystal structure near their termini. The authors therefore modelled the structure with static main chain disorder at this location reflecting a situation whereby the VapB tail can be either in one or the other conformation at any given time (Min et al. 2012). Currently, no VapBC structures are available both on and off DNA, so it is possible that some of this inherent flexibility relates to DNA binding.

When wrapping around their cognate VapC toxins, the VapB antitoxins in most cases place one or more charged residues in the active site, presumably to inactive the toxin. As described in this chapter, both positively charged residues (arginine and lysine) and negatively charged residues (glutamate) have been observed. The VapC PIN domain is known to function as divalent ion-depending endonucleases when activated, but usually no metal ions are observed in the inhibited, VapB-bound state. This suggests that the charged residues, in particular arginine and lysine residues, may function to displace ions from the active site by compensating for the loss of positive charge. The involvement of a glutamate residue is much more debatable and somewhat enigmatic. In the structure of *M. tuberculosis* VapBC-15 structure, a glutamate residue from VapB-15 was found in one of four active sites, which also contained two metal ions. This suggests that VapB inactivation functions by a charge-reversal principle, in which the presence of ions dictates the charge of the residue required for inactivation, or vice versa. However, neither the extent of this phenomenon, not its functional and regulatory consequences are known at present. However, the observations suggest that active site inhibition of VapC toxins by VapB antitoxins follows a complex and multipronged approach, in which both acidic and basic residues can take part. It is likely that in those cases where acidic residues of VapB are used, that these compete with the RNA substrate for binding to the active site. It is not known whether the metal ions are present in the VapC active sites in the absence of RNA, but if that is the case, it is likely that interaction with the positive charged arginine or lysine would physically displace the ions (Dienemann et al. 2011).

Acknowledgements We are thankful to Kristoffer Winther and Kenn Gerdes for insightful comments and critical reading of the manuscript. This work was supported by Lundbeck Foundation (grant no. R173-2014-1182) and the Danish National Research Foundation Centre for Bacterial Stress Response and Persistence (DNRF120).

References

Aakre CD, Phung TN, Huang D, Laub MT (2013) A bacterial toxin inhibits DNA replication elongation through a direct interaction with the beta sliding clamp. Mol Cell 52(5):617–628. doi:10.1016/j.molcel.2013.10.014

Aizenman E, Engelberg-Kulka H, Glaser G (1996) An Escherichia coli chromosomal "addiction module" regulated by guanosine [corrected] 3′,5′-bispyrophosphate: a model for programmed bacterial cell death. Proc Natl Acad Sci U S A 93(12):6059–6063

Arcus VL, Backbro K, Roos A, Daniel EL, Baker EN (2004) Distant structural homology leads to the functional characterization of an archaeal PIN domain as an exonuclease. J Biol Chem 279(16):16471–16478. doi:10.1074/jbc.M313833200

Arcus VL, McKenzie JL, Robson J, Cook GM (2011) The PIN-domain ribonucleases and the prokaryotic VapBC toxin-antitoxin array. Protein Eng Des Sel 24(1–2):33–40. doi:10.1093/protein/gzq081

Bailey SE, Hayes F (2009) Influence of operator site geometry on transcriptional control by the YefM-YoeB toxin-antitoxin complex. J Bacteriol 191(3):762–772. doi:10.1128/JB.01331-08

Black DS, Kelly AJ, Mardis MJ, Moyed HS (1991) Structure and organization of hip, an operon that affects lethality due to inhibition of peptidoglycan or DNA synthesis. J Bacteriol 173(18):5732–5739

Blower TR, Short FL, Rao F, Mizuguchi K, Pei XY, Fineran PC, Luisi BF, Salmond GP (2012) Identification and classification of bacterial Type III toxin-antitoxin systems encoded in chromosomal and plasmid genomes. Nucleic Acids Res 40(13):6158–6173. doi:10.1093/nar/gks231

Brantl S, Jahn N (2015) sRNAs in bacterial type I and type III toxin-antitoxin systems. FEMS Microbiol Rev 39(3):413–427. doi:10.1093/femsre/fuv003

Bunker RD, McKenzie JL, Baker EN, Arcus VL (2008) Crystal structure of PAE0151 from Pyrobaculum aerophilum, a PIN-domain (VapC) protein from a toxin-antitoxin operon. Proteins 72(1):510–518. doi:10.1002/prot.22048

Coles M, Djuranovic S, Soding J, Frickey T, Koretke K, Truffault V, Martin J, Lupas AN (2005) AbrB-like transcription factors assume a swapped hairpin fold that is evolutionarily related to double-psi beta barrels. Structure 13(6):919–928. doi:10.1016/j.str.2005.03.017

Cruz JW, Sharp JD, Hoffer ED, Maehigashi T, Vvedenskaya IO, Konkimalla A, Husson RN, Nickels BE, Dunham CM, Woychik NA (2015) Growth-regulating Mycobacterium tuberculosis VapC-mt4 toxin is an isoacceptor-specific tRNase. Nat Commun 6:7480. doi:10.1038/ncomms8480

Das U, Pogenberg V, Subhramanyam UK, Wilmanns M, Gourinath S, Srinivasan A (2014) Crystal structure of the VapBC-15 complex from Mycobacterium tuberculosis reveals a two-metal ion dependent PIN-domain ribonuclease and a variable mode of toxin-antitoxin assembly. J Struct Biol 188(3):249–258. doi:10.1016/j.jsb.2014.10.002

de la Hoz AB, Ayora S, Sitkiewicz I, Fernandez S, Pankiewicz R, Alonso JC, Ceglowski P (2000) Plasmid copy-number control and better-than-random segregation genes of pSM19035 share a common regulator. Proc Natl Acad Sci U S A 97(2):728–733

Dienemann C, Boggild A, Winther KS, Gerdes K, Brodersen DE (2011) Crystal structure of the VapBC toxin-antitoxin complex from Shigella flexneri reveals a hetero-octameric DNA-binding assembly. J Mol Biol 414(5):713–722. doi:10.1016/j.jmb.2011.10.024

Dorr T, Vulic M, Lewis K (2010) Ciprofloxacin causes persister formation by inducing the TisB toxin in Escherichia coli. PLoS Biol 8(2):e1000317. doi:10.1371/journal.pbio.1000317

Dy RL, Przybilski R, Semeijn K, Salmond GP, Fineran PC (2014) A widespread bacteriophage abortive infection system functions through a Type IV toxin-antitoxin mechanism. Nucleic Acids Res 42(7):4590–4605. doi:10.1093/nar/gkt1419

Erickson HP (1997) FtsZ, a tubulin homologue in prokaryote cell division. Trends Cell Biol 7(9):362–367. doi:10.1016/S0962-8924(97)01108-2

Fineran PC, Blower TR, Foulds IJ, Humphreys DP, Lilley KS, Salmond GP (2009) The phage abortive infection system, ToxIN, functions as a protein-RNA toxin-antitoxin pair. Proc Natl Acad Sci U S A 106(3):894–899. doi:10.1073/pnas.0808832106

Fozo EM, Hemm MR, Storz G (2008a) Small toxic proteins and the antisense RNAs that repress them. Microbiol Mol Biol Rev 72(4):579–589. Table of Contents. doi:10.1128/MMBR.00025-08

Fozo EM, Kawano M, Fontaine F, Kaya Y, Mendieta KS, Jones KL, Ocampo A, Rudd KE, Storz G (2008b) Repression of small toxic protein synthesis by the Sib and OhsC small RNAs. Mol Microbiol 70(5):1076–1093. doi:10.1111/j.1365-2958.2008.06394.x

Geerds C, Wohlmann J, Haas A, Niemann HH (2014) Structure of Rhodococcus equi virulence-associated protein B (VapB) reveals an eight-stranded antiparallel beta-barrel consisting of two Greek-key motifs. Acta Crystallogr F Struct Biol Commun 70(Pt 7):866–871. doi:10.1107/S2053230X14009911

Gerdes K (2000) Toxin-antitoxin modules may regulate synthesis of macromolecules during nutritional stress. J Bacteriol 182(3):561–572

Gerdes K, Wagner EG (2007) RNA antitoxins. Curr Opin Microbiol 10(2):117–124. doi:10.1016/j.mib.2007.03.003

Gerdes K, Bech FW, Jorgensen ST, Lobner-Olesen A, Rasmussen PB, Atlung T, Boe L, Karlstrom O, Molin S, von Meyenburg K (1986a) Mechanism of postsegregational killing by the hok gene product of the parB system of plasmid R1 and its homology with the relF gene product of the E. coli relB operon. EMBO J 5(8):2023–2029

Gerdes K, Rasmussen PB, Molin S (1986b) Unique type of plasmid maintenance function: post-segregational killing of plasmid-free cells. Proc Natl Acad Sci U S A 83(10):3116–3120
Gerdes K, Thisted T, Martinussen J (1990) Mechanism of post-segregational killing by the hok/sok system of plasmid R1: sok antisense RNA regulates formation of a hok mRNA species correlated with killing of plasmid-free cells. Mol Microbiol 4(11):1807–1818
Gerdes K, Nielsen A, Thorsted P, Wagner EG (1992) Mechanism of killer gene activation. Antisense RNA-dependent RNase III cleavage ensures rapid turn-over of the stable hok, srnB and pndA effector messenger RNAs. J Mol Biol 226(3):637–649
Gerdes K, Christensen SK, Lobner-Olesen A (2005) Prokaryotic toxin-antitoxin stress response loci. Nat Rev Microbiol 3(5):371–382. doi:10.1038/nrmicro1147
Gotfredsen M, Gerdes K (1998) The Escherichia coli relBE genes belong to a new toxin-antitoxin gene family. Mol Microbiol 29(4):1065–1076
Hallez R, Geeraerts D, Sterckx Y, Mine N, Loris R, Van Melderen L (2010) New toxins homologous to ParE belonging to three-component toxin-antitoxin systems in Escherichia coli O157:H7. Mol Microbiol 76(3):719–732. doi:10.1111/j.1365-2958.2010.07129.x
Hayes F (2003) Toxins-antitoxins: plasmid maintenance, programmed cell death, and cell cycle arrest. Science 301(5639):1496–1499. doi:10.1126/science.1088157
Hazan R, Engelberg-Kulka H (2004) Escherichia coli mazEF-mediated cell death as a defense mechanism that inhibits the spread of phage P1. Mol Genet Genomics 272(2):227–234. doi:10.1007/s00438-004-1048-y
Hazan R, Sat B, Engelberg-Kulka H (2004) Escherichia coli mazEF-mediated cell death is triggered by various stressful conditions. J Bacteriol 186(11):3663–3669. doi:10.1128/JB.186.11.3663-3669.2004
Hofmann K, Stoffel W (1993) TMBASE – a database of membrane spanning protein segments. Biol Chem 374(166)
Jaffe A, Ogura T, Hiraga S (1985) Effects of the ccd function of the F plasmid on bacterial growth. J Bacteriol 163(3):841–849
Jensen RB, Gerdes K (1995) Programmed cell death in bacteria: proteic plasmid stabilization systems. Mol Microbiol 17(2):205–210
Jeyakanthan J, Inagaki E, Kuroishi C, Tahirov TH (2005) Structure of PIN-domain protein PH0500 from Pyrococcus horikoshii. Acta Crystallogr Sect F Struct Biol Cryst Commun 61(Pt 5):463–468. doi:10.1107/S1744309105012406
Jorgensen MG, Pandey DP, Jaskolska M, Gerdes K (2009) HicA of Escherichia coli defines a novel family of translation-independent mRNA interferases in bacteria and archaea. J Bacteriol 191(4):1191–1199. doi:10.1128/JB.01013-08
Kawano M, Oshima T, Kasai H, Mori H (2002) Molecular characterization of long direct repeat (LDR) sequences expressing a stable mRNA encoding for a 35-amino-acid cell-killing peptide and a cis-encoded small antisense RNA in Escherichia coli. Mol Microbiol 45(2):333–349
Kawano M, Aravind L, Storz G (2007) An antisense RNA controls synthesis of an SOS-induced toxin evolved from an antitoxin. Mol Microbiol 64(3):738–754. doi:10.1111/j.1365-2958.2007.05688.x
Keren I, Shah D, Spoering A, Kaldalu N, Lewis K (2004) Specialized persister cells and the mechanism of multidrug tolerance in Escherichia coli. J Bacteriol 186(24):8172–8180. doi:10.1128/JB.186.24.8172-8180.2004
Kim Y, Wood TK (2010) Toxins Hha and CspD and small RNA regulator Hfq are involved in persister cell formation through MqsR in Escherichia coli. Biochem Biophys Res Commun 391(1):209–213. doi:10.1016/j.bbrc.2009.11.033
Klein W, Marahiel MA (2002) Structure-function relationship and regulation of two Bacillus subtilis DNA-binding proteins, HBsu and AbrB. J Mol Microbiol Biotechnol 4(3):323–329
Krissinel E, Henrick K (2007) Inference of macromolecular assemblies from crystalline state. J Mol Biol 372(3):774–797. doi:10.1016/j.jmb.2007.05.022
Lee IG, Lee SJ, Chae S, Lee KY, Kim JH, Lee BJ (2015) Structural and functional studies of the Mycobacterium tuberculosis VapBC30 toxin-antitoxin system: implications for the design of novel antimicrobial peptides. Nucleic Acids Res 43(15):7624–7637. doi:10.1093/nar/gkv689

Lehnherr H, Maguin E, Jafri S, Yarmolinsky MB (1993) Plasmid addiction genes of bacteriophage P1: doc, which causes cell death on curing of prophage, and phd, which prevents host death when prophage is retained. J Mol Biol 233(3):414–428. doi:10.1006/jmbi.1993.1521

Levin I, Schwarzenbacher R, Page R, Abdubek P, Ambing E, Biorac T, Brinen LS, Campbell J, Canaves JM, Chiu HJ, Dai X, Deacon AM, Di Donato M, Elsliger MA, Floyd R, Godzik A, Grittini C, Grzechnik SK, Hampton E, Jaroszewski L, Karlak C, Klock HE, Koesema E, Kovarik JS, Kreusch A, Kuhn P, Lesley SA, McMullan D, McPhillips TM, Miller MD, Morse A, Moy K, Ouyang J, Quijano K, Reyes R, Rezezadeh F, Robb A, Sims E, Spraggon G, Stevens RC, van den Bedem H, Velasquez J, Vincent J, von Delft F, Wang X, West B, Wolf G, Xu Q, Hodgson KO, Wooley J, Wilson IA (2004) Crystal structure of a PIN (PilT N-terminus) domain (AF0591) from Archaeoglobus fulgidus at 1.90 A resolution. Proteins 56(2):404–408. doi:10.1002/prot.20090

Lewis K (2010) Persister cells. Annu Rev Microbiol 64:357–372. doi:10.1146/annurev.micro.112408.134306

Magnuson RD (2007) Hypothetical functions of toxin-antitoxin systems. J Bacteriol 189(17):6089–6092. doi:10.1128/JB.00958-07

Makarova KS, Grishin NV, Koonin EV (2006) The HicAB cassette, a putative novel, RNA-targeting toxin-antitoxin system in archaea and bacteria. Bioinformatics 22(21):2581–2584. doi:10.1093/bioinformatics/btl418

Makarova KS, Wolf YI, Koonin EV (2009) Comprehensive comparative-genomic analysis of type 2 toxin-antitoxin systems and related mobile stress response systems in prokaryotes. Biol Direct 4:19. doi:10.1186/1745-6150-4-19

Markovski M, Wickner S (2013) Preventing bacterial suicide: a novel toxin-antitoxin strategy. Mol Cell 52(5):611–612. doi:10.1016/j.molcel.2013.11.018

Masuda Y, Miyakawa K, Nishimura Y, Ohtsubo E (1993) chpA and chpB, Escherichia coli chromosomal homologs of the pem locus responsible for stable maintenance of plasmid R100. J Bacteriol 175(21):6850–6856

Masuda H, Tan Q, Awano N, Wu KP, Inouye M (2012a) YeeU enhances the bundling of cytoskeletal polymers of MreB and FtsZ, antagonizing the CbtA (YeeV) toxicity in Escherichia coli. Mol Microbiol 84(5):979–989. doi:10.1111/j.1365-2958.2012.08068.x

Masuda H, Tan Q, Awano N, Yamaguchi Y, Inouye M (2012b) A novel membrane-bound toxin for cell division, CptA (YgfX), inhibits polymerization of cytoskeleton proteins, FtsZ and MreB, in Escherichia coli. FEMS Microbiol Lett 328(2):174–181. doi:10.1111/j.1574-6968.2012.02496.x

Mate MJ, Vincentelli R, Foos N, Raoult D, Cambillau C, Ortiz-Lombardia M (2012) Crystal structure of the DNA-bound VapBC2 antitoxin/toxin pair from Rickettsia felis. Nucleic Acids Res 40(7):3245–3258. doi:10.1093/nar/gkr1167

Mattison K, Wilbur JS, So M, Brennan RG (2006) Structure of FitAB from Neisseria gonorrhoeae bound to DNA reveals a tetramer of toxin-antitoxin heterodimers containing pin domains and ribbon-helix-helix motifs. J Biol Chem 281(49):37942–37951. doi:10.1074/jbc.M605198200

Miallau L, Faller M, Chiang J, Arbing M, Guo F, Cascio D, Eisenberg D (2009) Structure and proposed activity of a member of the VapBC family of toxin-antitoxin systems. VapBC-5 from Mycobacterium tuberculosis. J Biol Chem 284(1):276–283. doi:10.1074/jbc.M805061200

Min AB, Miallau L, Sawaya MR, Habel J, Cascio D, Eisenberg D (2012) The crystal structure of the Rv0301-Rv0300 VapBC-3 toxin-antitoxin complex from M. tuberculosis reveals a Mg(2)(+) ion in the active site and a putative RNA-binding site. Protein Sci 21(11):1754–1767. doi:10.1002/pro.2161

Neidle S (2002) Nucleic acid structure and recognition, 1st edn. Oxford University Press, Oxford

Ogata H, Renesto P, Audic S, Robert C, Blanc G, Fournier PE, Parinello H, Claverie JM, Raoult D (2005) The genome sequence of Rickettsia felis identifies the first putative conjugative plasmid in an obligate intracellular parasite. PLoS Biol 3(8):e248. doi:10.1371/journal.pbio.0030248

Ogura T, Hiraga S (1983) Mini-F plasmid genes that couple host cell division to plasmid proliferation. Proc Natl Acad Sci U S A 80(15):4784–4788

Pandey DP, Gerdes K (2005) Toxin-antitoxin loci are highly abundant in free-living but lost from host-associated prokaryotes. Nucleic Acids Res 33(3):966–976. doi:10.1093/nar/gki201

Pedersen K, Christensen SK, Gerdes K (2002) Rapid induction and reversal of a bacteriostatic condition by controlled expression of toxins and antitoxins. Mol Microbiol 45(2):501–510

Pei J, Grishin NV (2014) PROMALS3D: multiple protein sequence alignment enhanced with evolutionary and three-dimensional structural information. Methods Mol Biol 1079:263–271. doi:10.1007/978-1-62703-646-7_17

Ramage HR, Connolly LE, Cox JS (2009) Comprehensive functional analysis of Mycobacterium tuberculosis toxin-antitoxin systems: implications for pathogenesis, stress responses, and evolution. PLoS Genet 5(12):e1000767. doi:10.1371/journal.pgen.1000767

Raumann BE, Brown BM, Sauer RT (1994a) Major groove DNA recognition by β-sheets: the ribbon-helix-helix family of gene regulatory proteins. Curr Opin Struct Biol 4(1):36–43. doi:10.1016/s0959-440x(94)90057-4

Raumann BE, Rould MA, Pabo CO, Sauer RT (1994b) DNA recognition by beta-sheets in the Arc repressor-operator crystal structure. Nature 367(6465):754–757. doi:10.1038/367754a0

Roberts RC, Strom AR, Helinski DR (1994) The parDE operon of the broad-host-range plasmid RK2 specifies growth inhibition associated with plasmid loss. J Mol Biol 237(1):35–51. doi:10.1006/jmbi.1994.1207

Robson J, McKenzie JL, Cursons R, Cook GM, Arcus VL (2009) The vapBC operon from Mycobacterium smegmatis is an autoregulated toxin-antitoxin module that controls growth via inhibition of translation. J Mol Biol 390(3):353–367. doi:10.1016/j.jmb.2009.05.006

Sat B, Hazan R, Fisher T, Khaner H, Glaser G, Engelberg-Kulka H (2001) Programmed cell death in Escherichia coli: some antibiotics can trigger mazEF lethality. J Bacteriol 183(6):2041–2045. doi:10.1128/JB.183.6.2041-2045.2001

Sat B, Reches M, Engelberg-Kulka H (2003) The Escherichia coli mazEF suicide module mediates thymineless death. J Bacteriol 185(6):1803–1807

Schrodinger, LLC (2010) The PyMOL molecular graphics system, version 1.3r1

Sevin EW, Barloy-Hubler F (2007) RASTA-Bacteria: a web-based tool for identifying toxin-antitoxin loci in prokaryotes. Genome Biol 8(8):R155. doi:10.1186/gb-2007-8-8-r155

Sharp JD, Cruz JW, Raman S, Inouye M, Husson RN, Woychik NA (2012) Growth and translation inhibition through sequence-specific RNA binding by Mycobacterium tuberculosis VapC toxin. J Biol Chem 287(16):12835–12847. doi:10.1074/jbc.M112.340109

Singh R, Ray P, Das A, Sharma M (2009) Role of persisters and small-colony variants in antibiotic resistance of planktonic and biofilm-associated Staphylococcus aureus: an in vitro study. J Med Microbiol 58(Pt 8):1067–1073. doi:10.1099/jmm.0.009720-0

Somers WS, Phillips SE (1992) Crystal structure of the met repressor-operator complex at 2.8 A resolution reveals DNA recognition by beta-strands. Nature 359(6394):387–393. doi:10.1038/359387a0

Strauch MA, Spiegelman GB, Perego M, Johnson WC, Burbulys D, Hoch JA (1989) The transition state transcription regulator abrB of Bacillus subtilis is a DNA binding protein. EMBO J 8(5):1615–1621

Sullivan DM, Bobay BG, Kojetin DJ, Thompson RJ, Rance M, Strauch MA, Cavanagh J (2008) Insights into the nature of DNA binding of AbrB-like transcription factors. Structure 16(11):1702–1713. doi:10.1016/j.str.2008.08.014

Thisted T, Gerdes K (1992) Mechanism of post-segregational killing by the hok/sok system of plasmid R1. Sok antisense RNA regulates hok gene expression indirectly through the overlapping mok gene. J Mol Biol 223(1):41–54

Tian QB, Hayashi T, Murata T, Terawaki Y (1996a) Gene product identification and promoter analysis of hig locus of plasmid Rts1. Biochem Biophys Res Commun 225(2):679–684. doi:10.1006/bbrc.1996.1229

Tian QB, Ohnishi M, Tabuchi A, Terawaki Y (1996b) A new plasmid-encoded proteic killer gene system: cloning, sequencing, and analyzing hig locus of plasmid Rts1. Biochem Biophys Res Commun 220(2):280–284. doi:10.1006/bbrc.1996.0396

Unterholzner SJ, Poppenberger B, Rozhon W (2013) Toxin-antitoxin systems: biology, identification, and application. Mob Genet Elements 3(5):e26219. doi:10.4161/mge.26219

van den Ent F, Amos LA, Lowe J (2001) Prokaryotic origin of the actin cytoskeleton. Nature 413(6851):39–44. doi:10.1038/35092500

Van Melderen L, Saavedra De Bast M (2009) Bacterial toxin-antitoxin systems: more than selfish entities? PLoS Genet 5(3):e1000437. doi:10.1371/journal.pgen.1000437

Vogel J, Argaman L, Wagner EG, Altuvia S (2004) The small RNA IstR inhibits synthesis of an SOS-induced toxic peptide. Curr Biol 14(24):2271–2276. doi:10.1016/j.cub.2004.12.003

Wang X, Wood TK (2011) Toxin-antitoxin systems influence biofilm and persister cell formation and the general stress response. Appl Environ Microbiol 77(16):5577–5583. doi:10.1128/AEM.05068-11

Wang X, Lord DM, Cheng HY, Osbourne DO, Hong SH, Sanchez-Torres V, Quiroga C, Zheng K, Herrmann T, Peti W, Benedik MJ, Page R, Wood TK (2012) A new type V toxin-antitoxin system where mRNA for toxin GhoT is cleaved by antitoxin GhoS. Nat Chem Biol 8(10):855–861. doi:10.1038/nchembio.1062

Wilbur JS, Chivers PT, Mattison K, Potter L, Brennan RG, So M (2005) Neisseria gonorrhoeae FitA interacts with FitB to bind DNA through its ribbon-helix-helix motif. Biochemistry 44(37):12515–12524. doi:10.1021/bi0511080

Winther KS, Gerdes K (2011) Enteric virulence associated protein VapC inhibits translation by cleavage of initiator tRNA. Proc Natl Acad Sci U S A 108(18):7403–7407. doi:10.1073/pnas.1019587108

Winther KS, Brodersen DE, Brown AK, Gerdes K (2013) VapC20 of Mycobacterium tuberculosis cleaves the sarcin-ricin loop of 23S rRNA. Nat Commun 4:2796. doi:10.1038/ncomms3796

Xu K, Strauch MA (2001) DNA-binding activity of amino-terminal domains of the Bacillus subtilis AbrB protein. J Bacteriol 183(13):4094–4098. doi:10.1128/JB.183.13.4094-4098.2001

Xu K, Dedic E, Cob-Cantal P, Dienemann C, Boggild A, Winther KS, Gerdes K, Brodersen DE (2013) Protein expression, crystallization and preliminary X-ray crystallographic analysis of the isolated Shigella flexneri VapC toxin. Acta Crystallogr Sect F Struct Biol Cryst Commun 69(Pt 7):762–765. doi:10.1107/S1744309113014012

Xu K, Dedic E, Brodersen DE (2016) Structural analysis on the active site architecture of the VapC toxin from Shigella flexneri. Proteins. doi:10.1002/prot.25002

Yamaguchi Y, Park JH, Inouye M (2011) Toxin-antitoxin systems in bacteria and archaea. Annu Rev Genet 45:61–79. doi:10.1146/annurev-genet-110410-132412

Yarmolinsky MB (1995) Programmed cell death in bacterial populations. Science 267(5199):836–837

Chapter 15
D-Glyceraldehyde-3-Phosphate Dehydrogenase Structure and Function

Michael R. White and Elsa D. Garcin

Abstract Aside from its well-established role in glycolysis, glyceraldehyde-3-phosphate dehydrogenase (GAPDH) has been shown to possess many key functions in cells. These functions are regulated by protein oligomerization, biomolecular interactions, post-translational modifications, and variations in subcellular localization. Several GAPDH functions and regulatory mechanisms overlap with one another and converge around its role in intermediary metabolism. Several structural determinants of the protein dictate its function and regulation. GAPDH is ubiquitously expressed and is found in all domains of life. GAPDH has been implicated in many diseases, including those of pathogenic, cardiovascular, degenerative, diabetic, and tumorigenic origins. Understanding the mechanisms by which GAPDH can switch between its functions and how these functions are regulated can provide insights into ways the protein can be modulated for therapeutic outcomes.

Keywords Glyceraldehyde-3-phosphate dehydrogenase • GAPDH • Multifunctional • Localization • Post-translational modifications • Structural • Oligomerization • Interfaces • NAD$^+$

15.1 Introduction

Glyceraldehyde-3-phosphate dehydrogenase (GAPDH, EC 1.2.1.12) is a dynamic, multifunctional protein that is ubiquitously expressed and highly conserved across all kingdoms of life (Stone et al. 1985). In humans, GAPDH is encoded by a single structural gene, located on chromosome twelve, of somatic cells (Bruns et al. 1979). An additional transcript, located on chromosome nineteen, encodes spermatogenic

M.R. White (✉)
Department of Structural Biology, St. Jude Children's Research Hospital,
262 Danny Thomas Place, Memphis, TN 28105, USA
e-mail: whitem2@umbc.edu

E.D. Garcin
Department of Chemistry and Biochemistry, University of Maryland Baltimore County,
1000 Hilltop Circle, Baltimore, MD 21250, USA
e-mail: egarcin@umbc.edu

J.R. Harris, J. Marles-Wright (eds.), *Macromolecular Protein Complexes*,
Subcellular Biochemistry 83, DOI 10.1007/978-3-319-46503-6_15

GAPDH, referred to as GAPDS (Miki et al. 2004). Though no known alternate transcripts have been identified in somatic cells (Mezquita et al. 1998), several non-functional pseudogenes (Piechaczyk et al. 1984) are known. Additionally, various isoforms of GAPDH with altered physicochemical properties are proposed to coordinate the multiplicity of this dynamic protein.

GAPDH is a highly abundant enzyme and, as such, is often regarded as a static marker protein and is frequently used as a reference gene or loading control. However, GAPDH expression has been observed to be highly dynamic (Jacob et al. 2013) and sensitive to a variety of cell conditions, including calcium influx (Chao et al. 1990), hypoxia (Graven et al. 1994), iron concentration (Polati et al. 2012; Quail and Yeoh 1995), cell cycle progression (Mansur et al. 1993), and growth-accelerated states (Persons et al. 1989; Tokunaga et al. 1987). GAPDH has also been implicated in a wide array of cellular processes beyond metabolism, reviewed in Sect. 15.2.

With such diversity in expression and function, GAPDH activity must be highly regulated (reviewed in (Sirover 2012; Sirover 2014; Tristan et al. 2011)). To date, the mechanisms that regulate its activity remain enigmatic. However, several cases of biomolecular interactions, post-translational modifications, and alterations in localization have been elucidated, as discussed in Sects. 15.2 and 15.3. GAPDH is an important cellular target for study as it is increasingly shown to be involved in a variety of human pathologies, some of which include amyotrophic lateral sclerosis (Pierce et al. 2008), diabetes (Yego and Mohr 2010; Yego et al. 2009), neurodegenerative diseases (Mazzola and Sirover 2002; Mazzola and Sirover 2003; Nakajima et al. 2009; Shiozawa et al. 2003; Tatton 2000; Tsuchiya et al. 2004), several types of cancer (Colell et al. 2009; Zhou et al. 2008), and the propagation of infective agents (Boradia et al. 2014a; Daubenberger et al. 2003; Modun and Williams 1999; Modun et al. 2000). Increased knowledge of the structure (reviewed in Sect. 15.4; note that reference to specific amino acids will be done with respect to human GAPDH amino acid numbering throughout the text), function, and relationship therein of this ubiquitous multifunctional protein has significant therapeutic potential. GAPDH has been shown many times to be involved in a variety of cellular processes mediating normal cell function as well as pathological progression.

15.2 GAPDH Functional Diversity

15.2.1 Intermediary Metabolism

The most prominent role of GAPDH is as an oxidoreductase in cellular metabolism. In the cytosol, GAPDH catalyzes the sixth step of glycolysis, the reversible conversion of glyceraldehyde-3-phosphate (GAP) to 1,3-bisphosphoglycerate in the presence of the cofactor nicotinamide adenine dinucleotide (NAD^+) and inorganic phosphate. The mechanism of this reaction involves the formation of a hemi-acetal intermediate via an active site cysteine residue (C152, see Sect 15.4.3). Being at the intersection of multiple metabolic pathways, glycolysis, gluconeogenesis, fructose catabolism, glycerol metabolism, and the pentose phosphate pathway, GAPDH serves, in part, as a master regulator of carbon flux (Hildebrandt et al. 2015). The majority of cells are heavily reliant on glycolysis, the principle role of which is making high-energy bonds. Glycolysis additionally contributes to cellular redox homeostasis via the generation of NADH, which also serves as an electron source for cellular respiration. As we will discuss, this ability to switch metabolic flux could be important for cellular stress responses as well as dictating GAPDH function.

15.2.2 Membrane Fusion

GAPDH has been shown to play key roles in membrane fusion, linking to it functions in cell growth, regulation, and communication. First, GAPDH binds *in vitro* to phosphatidyl serine (PS) in phospholipid vesicles with high affinity, via residues 70–94 (Fig. 15.1a) in its N-terminal domain (Kaneda et al. 1997). Second, GAPDH stimulates nuclear membrane fusion by binding to PS in the membrane. This function is inhibited by a GAPDH-derived peptide spanning residues 70–94 and anti-GAPDH antibodies (Nakagawa et al. 2003). Both antibody and peptide blocking of GAPDH fusogenic activity resulted in the formation of a reticular network on the majority of chromatin, suggesting GAPDH plays a role in the second step of nuclear membrane fusion (Nakagawa et al. 2003).

15.2.2.1 Vesicle Fusion in Exocytosis

In human neutrophils, GAPDH promotes the fusion of liposomes (Hessler et al. 1998). GAPDH fusogenic activity is not mitigated by differences in bi-layer curvature between vesicle populations (with sizes ranging from 30 nm to 100 nm (De Arcuri et al. 1999)), an important facet of exocytosis. Additionally, fusion activity was shown to occur on the millisecond time scale, and was independent of GAPDH oxidoreductase activity (Glaser and Gross 1995).

Rabbit brain GAPDH has been shown to mediate the fusion of plasmalogen-containing vesicles, dependent on the presence of plasmenylethanolamine (PEA), cholesterol, and PS (Glaser and Gross 1995). In addition, GAPDH was able to promote fusion between islet plasma membranes and secretory granules, indicating a role in insulin exocytosis (Han et al. 1998). The fusogenic activity was again seen to be dependent on the lipid composition of the membranes. These results suggest an unexpected link between glucose metabolism and insulin secretion, and bring to light the theme of the balance between GAPDH glycolytic and other functions in the cell, as will be discussed throughout.

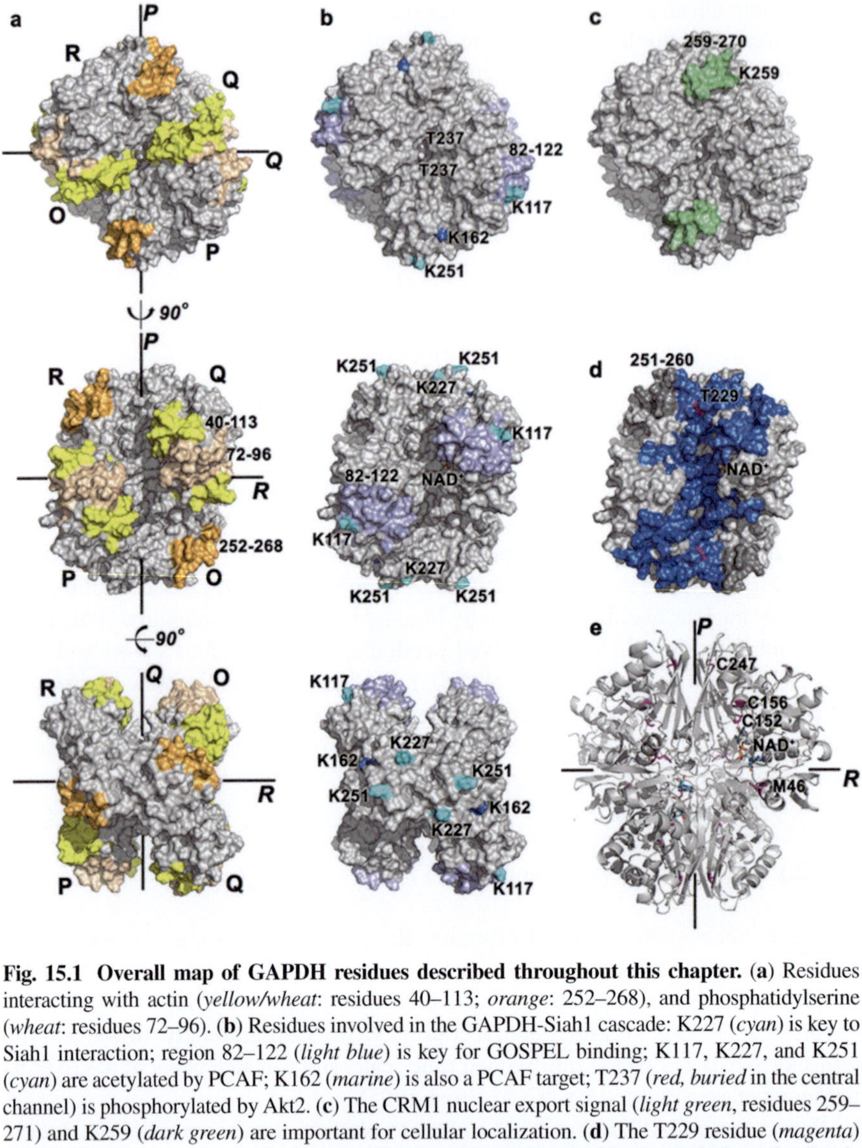

Fig. 15.1 Overall map of GAPDH residues described throughout this chapter. (**a**) Residues interacting with actin (*yellow/wheat*: residues 40–113; *orange*: 252–268), and phosphatidylserine (*wheat*: residues 72–96). (**b**) Residues involved in the GAPDH-Siah1 cascade: K227 (*cyan*) is key to Siah1 interaction; region 82–122 (*light blue*) is key for GOSPEL binding; K117, K227, and K251 (*cyan*) are acetylated by PCAF; K162 (*marine*) is also a PCAF target; T237 (*red, buried* in the central channel) is phosphorylated by Akt2. (**c**) The CRM1 nuclear export signal (*light green*, residues 259–271) and K259 (*dark green*) are important for cellular localization. (**d**) The T229 residue (*magenta*) proposed to be a site for O-Glc-Nac modification. The regions in *blue* located along the *P* axis showed structural and dynamic changes by HDXMS upon T229K mutation. The 251–260 region (*dark grey*) was shown to cross-link to RNA. (**e**) Residues that are sensitive to oxidative modifications include active site C152, nearby residue C156, dimer interface residue C247, and M46

Mutant CHO cells (FD1.3.25) exhibited increased accumulation and altered distribution of endocytic elements along with decreased exocytosis. This phenotype was shown to be due to a single point mutation P236S in a subpopulation of GAPDH in these cells (Robbins et al. 1995). Mutant GAPDH showed increased binding to microtubules and altered microtubule-dependent movement of endocytic vesicles leading to decreased exocytosis. This Proline residue is highly conserved across species and is found at the tetrameric interface. Therefore, it is possible that the mutation affects GAPDH oligomerization. Collectively, these results demonstrate the role of GAPDH in exocytosis mediated by its membrane fusion activity.

15.2.2.2 Vesicle Trafficking in Exocytosis

GAPDH is involved in the early secretory pathway as well and was shown to be involved in trafficking between the endoplasmic reticulum (ER) and the Golgi apparatus. In normal rat kidney cells, Rab2 directly interacts with GAPDH, stimulating its binding to vesicle membrane (Tisdale 2001). While GAPDH was not required for vesicle formation or budding, it was required for trafficking (Tisdale 2001), a function independent of its oxidoreductase activity (Tisdale et al. 2004). The process of retrograde trafficking involving GAPDH is now well established (reviewed in (Tisdale et al. 2009)). Rab2-GAPDH interaction and recruitment were also observed in *Plasmodium falciparum*. These results indicate that the role of GAPDH in Rab2-dependent vesicular trafficking is conserved throughout evolution (Daubenberger et al. 2003).

15.2.3 Interactions with Cytoskeletal Components

GAPDH was shown to associate with tubulin and actin to mediate cytoskeletal functional capacity; much of which is dictated by the higher-order architecture that is adopted by the cross-linking or bundling capacity of effector components.

15.2.3.1 Interactions with Tubulin and Microtubules

Several studies have shown that GAPDH interacts with microtubules as well as with tubulin, the main component of microtubules, in an ATP- and ionic strength-dependent manner (Durrieu et al. 1987a; 1987b; Huitorel and Pantaloni 1985; Kumagai and Sakai 1983; Launay et al. 1989; Muronetz et al. 1994). GAPDH, and other glycolytic enzymes, were shown to induce microtubule bundling via binding to a common site in the α-tubulin C-terminal region, residues 408–451 (Volker and Knull 1993, 1997; Walsh et al. 1989). GAPDH binding to tubulin has been shown to either decrease or increase GAPDH activity, depending on the cellular redox state (Durrieu et al. 1987a, 1987b; Landino et al. 2014; Muronetz et al. 1994).

The affinity and stoichiometry of the interaction between GAPDH and tubulin dimers or microtubules varies greatly. GAPDH:tubulin complexes with stoichiometries of 1:10 (Durrieu et al. 1987a) and 1:2 (Somers et al. 1990) have been described, with dissociation constants in the low μM range (Durrieu et al. 1987a; Somers et al.

1990; Walsh et al. 1989). Interestingly, these values varied depending whether GAPDH was free in solution or immobilized, leading to the concept of high- and low-affinity binding sites (Muronetz et al. 1994; Somers et al. 1990).

More recently, interactions between several proteins and microtubules have been shown to be facilitated by GAPDH, including the Ca^{2+}-binding protein p22 involved in microtubule-membrane interactions (Andrade et al. 2004); calsequestrin in skeletal muscle microsomes (Caswell and Corbett 1985); the heavy chain of microtubule associated protein 1B (Cueille et al. 2007); and the membrane trafficking protein Rab2 and the motor protein dynein, both involved in microtubule dynamics (Tisdale et al. 2009). Overall, it is still not known what the exact role of GAPDH in promoting the formation and/or stability of microtubules in cells may be.

15.2.3.2 Interactions with Actin and Microfilaments

It is well known that glycolytic enzymes bind actin (Arnold and Pette 1968) and that their oligomeric state is key to binding (Lowe et al. 2003). GAPDH binding to actin has been shown to occur between positively charged residues in the GAPDH N-terminal region (residues 40–113) and acidic residues in actin (Méjean et al. 1989; Waingeh et al. 2004), while other studies proposed GAPDH residues 252–268 (Fig. 15.1a) as the docking site for actin (Sheedy and Clarke 2001). Detailed modeling analysis suggested that GAPDH could use its quaternary structure to span the distance between neighboring actin subunits (Ouporov et al. 2000; Ouporov et al. 2001).

Due to the fact that multiple studies have found glycolytic enzymes, including GAPDH, associated with either microtubules, microfilaments, or both, it is tempting to speculate that a role of these cytoskeletal interactions is in the compartmentalization of glycolytic activities. However, the intracellular distribution of GAPDH under varying metabolic conditions has been ascertained, wherein the interaction of GAPDH with either cytoskeletal structure was not detected under normal culture conditions. These results suggest that GAPDH association with cytoskeletal components occurred in response to cellular conditions and as a means to conduct specialized functions, such as the trafficking of compartmentalized cargo or the induction or mediation of apoptotic pathways, as opposed to a more general function like the compartmentalization of glycolysis (Schmitz and Bereiter-Hahn 2002).

15.2.4 GAPDH as a Redox Sensor

Many cellular processes are influenced by redox reactions, key to which are reactive oxygen, sulfur, and nitrogen species (ROS, RSS, and RNS respectively). GAPDH is prone to several redox modifications (Hildebrandt et al. 2015) due to its highly reactive catalytic cysteine (C152, see Sect. 15.4 for structural overview), with a predicted pKa of about 6, owing to the microenvironment surrounding the active site.

Other GAPDH residues including C247, C156, and M46 are also targeted by redox modifications (Fig. 15.1e and Sect. 15.3).

First, GAPDH oxidation is associated with a loss of catalytic activity, and in some cases alterations in its oligomerization and cellular localization (Arutyunova et al. 2003; Pierce et al. 2008). The effects of GAPDH oxidation by H_2O_2 have been shown to be either anti-apoptotic via interaction of oxidized GAPDH with PLD2 (Kim et al. 2003) or pro-apoptotic due to disulfide-mediated aggregation of GAPDH (Nakajima et al. 2007; Nakajima et al. 2009). In the latter however, oxidation of M46, rather than C152 was shown to initiate free radical-induced GAPDH aggregation (Samson et al. 2014).

Second, modification of GAPDH by H_2S (S-sulfhydration or S-sulfuration) affects GAPDH function (Filipovic 2015). S-sulfhydration of C152 has been shown to not only increase GAPDH oxidoreductase activity, but also promote actin polymerization in cells (Mustafa et al. 2009). In a recent report, however, S-sulfhydration of GAPDH by H_2S and polysulfides was shown to occur at C156 and C247, but not at C152, unless C156 was mutated into a Ser. In all cases, modified GAPDH showed decreased catalytic activity (Jarosz et al. 2015). While it is not clear how C247 modification inactivated GAPDH, it was proposed that C156 modification lead to alterations in C152 pKa and a decrease in reactivity. Regardless, the mechanism by which H_2S modifies GAPDH in cells is still unknown.

Under cellular oxidative conditions, GAPDH undergoes S-thiolation by glutathione and cysteine, leading to a reversible decrease in catalytic activity that could protect it from irreversible oxidation (Grant et al. 1999; Ravichandran et al. 1994; Schuppe-Koistinen et al. 1994). These modifications could play a key role in modulating metabolism during respiratory bursts.

Third, GAPDH is a sensor of NO via S-nitrosylation (or S-nitrosation) at C152 (Hara and Snyder 2006; Hara et al. 2005; Sen et al. 2008). This reversible modification inactivated GAPDH (Molina y Vedia et al. 1992) and blocked its function in heme insertion (Chakravarti et al. 2010). Recent studies have shown that GAPDH C247 is also a target for S-nitrosation, resulting in a loss of its chaperone function (Jia et al. 2012).

As discussed below, S-sulfhydration, S-thiolation, and S-nitrosylation are all important functional regulators of GAPDH and its role in cell death.

15.2.5 The Role of GAPDH in Cell Death

15.2.5.1 Apoptotic Pathway Involving GAPDH

Apoptosis can be triggered by a number of factors such as oxidative stress, hormonal stimulation, growth factor withdrawal, activation of cell-surface receptors, and cell cycle perturbation. GAPDH plays many roles in apoptosis as outlined below.

Extrinsic Apoptotic Pathway

The extrinsic pathway of apoptosis is triggered by death receptor activation and induction of caspases. Exposure of cells to transforming growth factor beta 1 (TGF-β1), or tumor necrosis factor related apoptosis inducing ligand (TRAIL) coincided with increases in GAPDH expression, nuclear accumulation, and activity (Barbini et al. 2007; Du et al. 2007), eventually leading to apoptosis. While the extrinsic pathway can in some cases carry out cell death directly, it often works in conjunction with the intrinsic pathway.

Intrinsic Apoptotic Pathway

GAPDH has been observed to play a role in many mitochondrial outer membrane permeabilization (MOMP)-inducing mechanisms (Ishitani et al. 2003), including cell stress (Ishitani and Chuang 1996; Ishitani et al. 1997; Ishitani et al. 1998; Saunders et al. 1997) and senescence (Ishitani and Chuang 1996). GAPDH has been observed to accumulate in the mitochondria (Huang et al. 2009; Tarze et al. 2007) and nucleus (Hara et al. 2005; Sawa et al. 1997) of multiple cell types during apoptosis. GAPDH overexpression was also observed to be associated with apoptosis (Ishitani and Chuang 1996; Ishitani et al. 1996; Ishitani et al. 1998). GAPDH has also been shown to induce MOMP via interactions with the mitochondrial voltage dependent anion channel VDAC1, an integral member of the permeability transition pore complex. This interaction was shown to be thiol dependent, implicating oxidative stress induced GAPDH action (Tarze et al. 2007). As well, an increase in GAPDH disulfide binding and protein aggregation in cells (Huang et al. 2009) and a decrease in oxidoreductase activity (Saunders et al. 1997) are often associated with apoptosis.

15.2.5.2 Energy Depletion

In astrocytes, MOMP could be delayed with glycolytically derived ATP, due to activation of AMPK, resulting in increased glycolytic output (Almeida et al. 2004). In this case, ATP depletion is in part thought to mediate cell death in response to MOMP. In cells treated with 3-bromopyruvate, GAPDH was shown to be specifically inhibited, coinciding with MOMP initiation, ATP depletion, and cell death (Ganapathy-Kanniappan et al. 2009).

While ATP depletion may induce cell death in some instances, other pathways that mediate cell death in lieu of MOMP have been recognized and are collectively referred to as caspase independent cell death (Colell et al. 2007). GAPDH plays a prominent role in these pathways.

15.2.5.3 GAPDH-Mediated Caspase-Independent Cell Death

During apoptotic cell death GAPDH has been observed to translocate to the nucleus in a number of cell systems (see above). However, GAPDH does not contain a nuclear localization signal, but does contain an export signal sequence (Brown et al. 2004) and is likely too large to readily enter the nucleus (Hara et al. 2005). Recently an intricate NO-mediated signaling cascade has been elucidated, triggered by GAPDH S-nitrosylation (Hara and Snyder 2006). First, cell stress activates nitric oxide synthase and p300/CBP associated factor (PCAF) leading to S-nitrosylation of GAPDH C152 (Hara et al. 2005), and acetylation of GAPDH at K117, K227, and K251 (Ventura et al. 2010). S-nitrosation of GAPDH triggers binding to the E3 ubiquitin ligase Siah1 (Hara et al. 2005) via GAPDH K227 (see Fig. 15.1b), and subsequent nuclear translocation of the complex. The GAPDH-Siah1 interaction stabilizes Siah-1, which then ubiquitinates and degrades a series of downstream targets, thus initiating cell death (Hara et al. 2005). Second, nuclear GAPDH is acetylated at K162 by PCAF, increasing autoacetylation of PCAF and its catalytic activity leading to downstream acetylation of several proteins (p53, p21, Bax, PUMA) and cell death (Sen et al. 2008). Third, SNO-GAPDH can transnitrosylate nuclear deacetylase proteins, like SIRT1 and HDAC2 following translocation (Kornberg et al. 2010), thus regulating nuclear activities. Conversely, GAPDH mediated nitrosylation of nucleophosmin (NPM) was shown to promote the interaction of NPM with Siah1 and thereby decrease its ubiquitination activity, eventually leading to cell survival (Lee et al. 2012). The GAPDH-Siah1 interaction is further inhibited via a competing interaction between GAPDH and GOSPEL (GAPDH's competitor of Siah Protein Enhances Life). S-nitrosylation of GOSPEL increases its binding to GAPDH (Fig. 15.1b), thus preventing Siah-1-mediated GAPDH nuclear translocation and NO-mediated apoptosis (Sen et al. 2009).

The GAPDH-Siah-1 cascade also plays key roles in other pathways, including hyperglycemia-induced cell death in retinal cells (Yego and Mohr 2010), and high dose cocaine-induced neuronal cell death (Harraz and Snyder 2015). Pharmacological disruption of this signaling cascade shows promising results for the treatment of neurodegenerative diseases (Hara et al. 2006).

15.2.6 Autophagy

GAPDH was found to associate with autophagosomes and lysosomes (Fengsrud et al. 2000; Sneve et al. 2005), and has since been shown to play multiple roles in autophagy that can act in either cytotoxic or cytoprotective capacities (Song and Finkel 2007). This differential outcome is dependent on nutrient conditions and regulatory protein-protein interactions. Cytotoxic autophagy was shown to occur in conditions of hyperglycemia that correlated with the upregulation of GAPDH and Atg12 (Adastra et al. 2011). Conversely, in hypoglycemic conditions, AMPK phosphorylated GAPDH at S122, inducing nuclear translocation and subsequent SIRT1 activation (Canto and Auwerx 2012), which promoted autophagy allowing for cell survival upon fuel depletion (Chang et al. 2015). In HeLa cells, GAPDH was shown to both increase ATP yields when the mitochondria was compromised and regulate autophagic gene expression, particularly of Atg12 (Colell et al. 2007).

Additionally, treatment of normal human fibroblasts and keratinocytes with nicotinamide (NAM) increased cell lifespan and replicative capacity (Kang et al. 2006). In these cells, ROS accumulation was delayed, telomere shortening decreased, and activation of p53 attenuated, all of which are processes in which GAPDH is involved. Interestingly, cellular mitochondrial potential increased and respiration decreased (Kang et al. 2006). These effects were found to be due to selective mitophagy, removing low potential mitochondria. GAPDH and Atg12 were found to be upregulated upon NAM treatment. These effects were mitigated by inhibition of NAM-NAD^+ conversion, further implying GAPDH involvement (Kang and Hwang 2009).

15.2.7 Iron Homeostasis

Iron is an essential element for prokaryotic and eukaryotic organisms. Maintenance of cellular iron concentrations is highly regulated and involves an orchestrated balance between uptake, trafficking, and efflux. Cellular uptake occurs via two main pathways: transferrin-bound (Tf) and non-transferrin-bound iron uptake (Lane et al. 2015; Silva and Faustino 2015). Recently, GAPDH was shown to be involved in iron homeostasis both in pathogens and in mammalian cells (Boradia et al. 2014b; Sheokand et al. 2014).

GAPDH, a known virulence factor (Pancholi and Fischetti 1992), was first identified as a potential Tf-binding surface-localized protein involved in Tf-iron acquisition in the staphylococci *S. aureus* and *S. epidermidis* (Modun and Williams1999), via interaction with the N-terminus of Tf (Modun et al. 2000). These results are controversial, as others found no evidence of interaction between GAPDH and Tf in *S. aureus* (Taylor and Heinrichs 2002). However, not all GAPDH molecules possess this function as it was not detected in either *S. saprophyticus* (Modun and Williams1999) or *B. stearothermophilus* (Modun et al. 2000), indicating that GAPDH-Tf binding is not a mechanism inherent to all gram-positive bacteria. Additionally, GAPDH-Tf mediated cellular uptake of iron was observed in *M. tuberculosis* and *M. smegmatis*, via binding to holo-Tf (iron bound) and not to apo-Tf (Boradia et al. 2014a). Similarly, cells utilize GAPDH for lactoferrin-mediated iron uptake, indicating a potentially general mechanism with all Tf family proteins (Rawat et al. 2012).

The role of GAPDH in regulating iron homeostasis appears to be conserved, as GAPDH is expressed on the surface of multiple eukaryotic cell types (Raje et al. 2007). GAPDH expression and surface localization have been shown to be increased either in high iron media (Polati et al. 2012), or low iron concentration (Quail and Yeoh 1995). The involvement of GAPDH in either iron influx or efflux may be related to the iron concentration and associated differential GAPDH post-translational modifications and localizations in the cell (reviewed in (Boradia et al. 2014b)). GAPDH acts not only as a membrane associated Tf receptor, but also as a soluble receptor in transferrin/lactoferrin iron uptake (Chauhan et al. 2015; Sheokand et al. 2014), correlating with its observed functions in staphylococci.

15.2.8 Heme Metabolism

GAPDH has recently been identified as a non-canonical heme binding protein. First, GAPDH, in conjunction with heat shock protein 90 (Hsp90), is involved in inducible nitric oxide synthase (NOS) heme insertion (Chakravarti et al. 2010). This activity is inhibited by S-nitrosylation of GAPDH C152. Additionally K227 was found to play a key role in heme insertion (Fig. 15.1b). GAPDH sub-stoichiometrically binds hemin with nM affinity to form a low-spin six-coordinate complex with an unknown Histidine ligand. GAPDH also binds ferrous heme, though to a lesser extent. These results support the hypothesis that GAPDH may act as a general transient heme carrier protein *in vivo*, whereby a redox-based change in heme affinity could facilitate the release of heme from GAPDH in cells (Hannibal et al. 2012). This process may be widely applicable to other heme proteins as it was recently described that heme insertion into soluble guanylate cyclase was dependent on Hsp90 and an unidentified heme carrier protein, which may turn out to be GAPDH (Ghosh and Stuehr 2016; Sarkar et al. 2015).

15.2.9 GAPDH Interactions Involving Nucleic Acids

15.2.9.1 DNA Binding Proteins

GAPDH plays a key role in mediating both gene transcription and DNA repair via interactions with additional regulatory proteins. GAPDH is involved in the cell cycle dependent gene regulation of Histone 2B. Histone 2B expression is activated by the transcription factor Oct-1 and the corresponding OCA-S (Oct-1 Co-Activator in S-phase) transcription complex. GAPDH interacts directly with Oct-1 and plays a central role in the formation of the OCA-S complex. Complex formation is partially redox regulated via the GAPDH-Oct-1 interaction (Dai et al. 2008; Zheng et al. 2003). GAPDH is also involved in receptor mediated gene expression via interactions with the androgen receptor (AR). GAPDH was identified as a co-activator protein that complexes with AR and facilitates its nuclear translocation allowing for AR activation (Harada et al. 2007). Additionally, GAPDH was observed to interact with APE-1 (AP Endonuclease-1) in the repair of DNA abasic sites as well as redox based transcriptional regulation. It was demonstrated that GAPDH could catalyze the reduction of oxidized, inactive APE-1, using both C152 and C156. This allowed for the maintenance of APE-1 activity during oxidative stress

(Azam et al. 2008). These functions show the inter-relatedness of additional GAPDH functions such as redox sensing and the dependence of GAPDH functions on subcellular localization.

15.2.9.2 Cellular DNA

GAPDH nucleic acid binding was first reported in 1977 (Perucho et al. 1977). Later studies showed that GAPDH possessed nucleic acid helix destabilizing activity (Karpel and Burchard 1981) and indicated its ability to bind both single-stranded and double-stranded DNA (Grosse et al. 1986). GAPDH-nucleic acid binding is redox sensitive (Arutyunova et al. 2003), and oxidative modification of GAPDH C152 reduces nucleic acid binding capacity as well as cofactor binding. The GAPDH-nucleic acid binding site is not known, but may involve the cofactor binding site (Arutyunova et al. 2003).

GAPDH was shown to play an active role in the DNA repair processes, as a uracil DNA glycosylase (Mansur et al. 1993). GAPDH has also been demonstrated to play a role in cell senescence. GAPDH could prevent telomere shortening by directly binding to telomeric DNA when cells were exposed to gemcitabine, doxorubicin (Demarse et al. 2009), or ceramide (Sundararaj et al. 2004). Conversely, other studies showed that GAPDH could induce cell senescence via GAPDH inhibition of telomerase (Nicholls et al. 2012). GAPDH binding to the 5′ and 3′-untranslated regions (UTRs) of the telomerase RNA component (TERC) via its Rossman fold did not directly cause inhibition, but promoted interaction of the GAPDH C-terminal domain with the telomerase reverse transcriptase (TERT) and its subsequent inhibition (Nicholls et al. 2012).

15.2.9.3 Cellular RNA

GAPDH was found to be one of the three major RNA binding proteins in rabbit reticulocytes (Ryazanov 1985). Probing fractionated HeLa nuclear extracts with wild-type $tRNA^{met}$ resulted in a single protein post-purification identified as GAPDH (Singh and Green 1993). GAPDH-tRNA interactions were shown to occur with high affinity (apparent K_D of 18 nM) and specificity and were influenced by the T stem, anticodon, and variable loops of the tRNA. These results were supported by the biophysical characterization of the interaction between $tRNA^{phe}$ and GAPDH using Raman spectroscopy and circular dichroism (Carmona et al. 1999).

The proclivity of GAPDH for adenine-uridine rich elements (AREs) in mRNAs was first discovered some 20 years ago (Nagy and Rigby 1995). Since then, GAPDH has been identified as a general ARE-binding protein (AUBP). GAPDH has been shown to bind AREs from the 3′-UTR of several clinically important genes such as: (i) important cancer regulatory genes, including colony stimulating factor-1 (CSF-1; (Bonafe et al. 2005)), granulocyte macrophage colony stimulating factor (CSF-2; (Nagy and Rigby 1995)), cyclooxygenase-2 (Cox-2; (Ikeda et al. 2012)), connective tissue growth factor (CTGF/Ccn-2; (Kondo et al. 2011)), and c-myc (Nagy and Rigby 1995); (ii) genes involved in cardiovascular homeostasis, including endothelin-1 (ET-1; (Rodriguez-Pascual et al. 2008)) and angiotensin II type I receptor (ATIR; (Backlund et al. 2009)); (iii) cytokine encoding genes, including interferon-γ (IFN-γ; (Chang et al. 2013; Nagy and Rigby 1995)), interleukin-2 (IL-2; (Chang et al. 2013; Nagy and Rigby 1995)), and tumor necrosis factor-α (TNF-α; (Millet et al. 2016; White et al. 2015)); and (iv) neuro-regulatory voltage gated sodium channel type 1 α subunit gene (SCN1A; (Zeng et al. 2014)). All of these interactions have been reviewed recently (White and Garcin 2016).

The consequences of several GAPDH-RNA interactions have not yet been elucidated, but those that have result in variable outcomes. GAPDH has been shown to stabilize CSF-1 (Bonafe et al. 2005) and CTGF/Ccn-2 (Kondo et al. 2011), both interactions having been linked to cancer progression. Conversely, GAPDH has been shown to decrease the stability of ET-1 (Rodriguez-Pascual et al. 2008), Cox-2 (Ikeda et al. 2012), and SCN1A (Zeng et al. 2014), dysregulation of which has been linked to cardiovascular disease, cancer, and Dravet syndrome, respectively. Alternatively, GAPDH can repress mRNA translation, as seen with IFN-γ and IL-2 (Chang et al. 2013), or TNF-α (Millet et al. 2016). The latter three interactions have been shown to be under metabolic control, wherein normal glucose levels prevent GAPDH from interacting with the RNAs, allowing for protein expression (Chang et al. 2013; Millet et al. 2016). GAPDH translational repression was also observed with AT1R (Backlund et al. 2009).

GAPDH interacts with additional cellular RNAs including myosin heavy chain (MyHC) mRNA (Kiri and Goldspink 2002), glucose transporter-1 (Glut1) mRNA (McGowan and Pekala 1996), and the TNF-α hammerhead ribozyme (Sioud and Jespersen 1996). While the roles of each of these interactions are not fully understood, it is postulated that GAPDH-MyHC mRNA interactions restrict mRNA diffusion allowing for increased concentration in regions that require rapid growth and repair (Kiri and Goldspink 2002), GAPDH-GLUT1 mRNA interactions regulate glucose transporter expression in response to cellular nutrient levels (McGowan and Pekala 1996), and GAPDH-ribozyme interactions allow for increased ribozyme activity and target degradation (Sioud and Jespersen 1996).

15.2.9.4 Viral Nucleic Acids

In addition to cellular RNAs, GAPDH also interacts with viral RNAs. The consequence of GAPDH-viral RNA interactions is either anti-viral or pro-viral. GAPDH antiviral functions occur via interactions with ss-positive sense (+) RNA viruses. GAPDH preferentially binds the 3′-UTR of (+)-progeny RNA of both bamboo mosaic virus (BaMV) and its satellite virus (satBaMV) skewing the asymmetric synthesis of (+)-RNA that the viruses rely on for replication (Buck 1996). Alternatively, GAPDH competes with the polypyrimidine tract-binding protein (PTB) for hepatitis A virus (HAV) RNA (Dollenmaier and Weitz 2003; Kusov et al. 1996; Nüesch et al. 1993), resulting in decreased gene expression (Schultz et al.

1996; Yi et al. 2000). GAPDH also serves an anti-viral function in the case of transmissible gastroenteritis virus (TGEV), though it is unknown by what mechanism this occurs (Galán et al. 2009).

GAPDH activity can conversely be co-opted to assist in viral propagation. GAPDH was shown to be important for the ss-(+)-RNA virus hepatitis C virus (HCV) propagation wherein GAPDH silencing was shown to decrease viral load (Petrik et al. 1999; Randall et al. 2007). Additionally, GAPDH was shown to preferentially bind to and sequester the (−)-progeny RNA of the ss-(+) tomato bushy stunt RNA virus (TBSV), thus promoting the release of (+)-progeny (Wang and Nagy 2008). *In vitro* replicase assays demonstrated the role of GAPDH in asymmetric synthesis where in the presence of GAPDH the amount of (+)-RNA was 10–100 times greater than that of negative, whereas in its absence the ratio was approximately 1:1 (Huang and Nagy 2011). GAPDH has also been shown to bind the (−)-progeny RNA of Japanese encephalitis virus (JEV; (Yang et al. 2009)). While the consequences of this interaction were not determined, it potentially acts in a manner similar to that of TBSV. GAPDH is also co-opted by the ss-(−)-RNA viruses hepatitis delta virus (HDV; (Lin et al. 2000)) and human parainfluenza virus (HPIV; (De BP et al. 1996)). Many of the GAPDH-viral RNA interactions are mediated by the formation of multiple ribonucleoprotein complexes. This phenomenon has also been observed *in vitro* with TNF-α mRNA (White et al. 2015), though the biological significance of these complexes has yet to be elucidated.

15.3 GAPDH Functional Regulation and Cellular Localization Diversity

The plethora of GAPDH functions are regulated by an intricate network of mediators including post-translational modifications, interacting partners, and subcellular localization (summarized in Tables 15.1, 15.2 and 15.3, respectively). These three facets overlap and cannot be examined separately from one another. The following section will give an overview of the interconnectedness of GAPDH functions as well as regulatory mechanisms, following closely with GAPDH localization. In addition, we have mapped GAPDH residues that have been shown to be important for its functions, post-translational modifications, cellular localizations, and interacting partners (Fig. 15.1).

15.3.1 Extracellular Localization

GAPDH has long been speculated to be secreted from cells and function in a paracrine/autocrine capacity. Indeed, GAPDH has been identified in the conditioned medium of several mammalian cell lines, and in rat serum (Yamaji et al. 2005), and in the culture supernatants of J774 and THP-1 cell lines (Raje et al. 2007). C152 was

Table 15.1 Post-translational modifications

PTM	Role/Effect	Residue	References
Phosphorylation			
	Membrane trafficking	Y42	Coussens et al. (1985), Sergienko et al. (1992) and Tisdale and Artalejo (2007)
		Ser (?)	Tisdale (2002)
	Sirt1 activation	S122	Chang et al. (2015)
	Apoptosis	T237	Huang et al. (2011)
S-sulfhydration			
	Actin polymerization	C152	Mustafa et al. (2009)
	Apoptosis	C156	Jarosz et al. (2015)
		C247	
	Tubulin oxidation	C152	(Landino et al. 2014)
	APE1 reduction	C152	(Azam et al. 2008)
		C156	
S-thiolation			
	Oxidative protection	C152	Schuppe-Koistinen et al. (1994),Ravichandran et al. (1994),Puder and Soberman (1997), Grant et al. (1999) and Kim et al. (2003)
S-nitrosylation			
		C152	Jaffrey et al. (2001)
	Apoptosis		Hara et al. (2005, 2006) and Baek et al. (2008)
	Heme metabolism		Chakravarti et al. (2010)
Oxidation			
	Aggregation	M46	Samson et al. (2014)
	Apoptosis	C152	Kim et al. (2003) and Baek et al. (2008)
Acetylation			
	Apoptosis	K162	Sen et al. (2008)
	Nuclear translocation	K117	Ventura et al. (2010)
		K227	Ventura et al. (2010)
		K251	Ventura et al. (2010)
ADP-ribosylation			
	Necrosis	C152	Du et al. (2003) and Devalaraja-Narashimha and Padanilam (2009)
Pyruvylation			
	Energy depletion	C152	Ganapathy-Kanniappan et al. (2009)
O-glycosylation			
	Nuclear translocation	T229	Park et al. (2009)

identified as a critical mediator of secreted GAPDH binding and uptake to cells (Chauhan et al. 2015; Yamaji et al. 2005). While at least one role of secreted GAPDH is known, that of orchestrating cellular iron homeostasis (Sect. 15.2.7), additional functions have yet to be identified.

Table 15.2 Protein-protein interactions

Function	Localization	Partner	References
Vesicular transport			
	Cytosol	Rab 2	Tisdale (2001) and Tisdale et al. (2004)
		Src	Coussens et al. (1985), Sergienko et al. (1992) and Tisdale and Artalejo (2007)
		PKCι/λ	Tisdale (2001, 2002) and Tisdale et al. (2004)
		Tubulin	Glaser et al. (2002) and Tisdale et al. (2009)
Microtubule-membrane interactions			
	Cytosol	p22	Andrade et al. (2004)
Triad junction formation			
	Cytosol	Calsequestrin	Caswell and Corbett (1985)
		MAP1B	Cueille et al. (2007)
Cytoskeletal polymerization			
	Cytosol	Actin	Méjean et al. (1989), Rogalski-Wilk and Cohen (1997), Yuan et al. (1999) and Mustafa et al. (2009)
		Tubulin	Kumagai and Sakai (1983)
			Huitorel and Pantaloni (1985), Durrieu et al. (1987a, b), Launay et al. (1989), Walsh et al. (1989), Somers et al. (1990), Volker and Knull (1993), Muronetz et al. (1994), Volker and Knull (1997) and Landino et al. (2014)
Cell death			
	Cytosol	PLD2	(Kim et al. 2003)
		Siah1	Hara et al. (2005) andHara and Snyder (2006)
		GOSPEL	Sen et al. (2009)
		Akt	Huang et al. (2011)
		PARP1	Du et al. (2003) and Devalaraja-Narashimha and Padanilam (2009)
	Nucleus	p300/CBP	Sen et al. (2008)
		NPM	Lee et al. (2012)
	Mitochondria	VDAC1	Tarze et al. (2007)
Autophagy			
	Cytosol	AMPK	Chang et al. (2015)
		Sirt1	Chang et al. (2015)
Iron homeostasis			
	Plasma membrane	Transferrin	Modun and Williams (1999, Modun et al. (2000), Taylor and Heinrichs (2002), Raje et al. (2007) and Boradia et al. (2014a)
		Lactoferrin	Rawat et al. (2012) and Chauhan et al. (2015)
		uPAR	Chauhan et al. (2015)
		Ferroportin	Sheokand et al. (2014)

(continued)

Table 15.2 (continued)

Function	Localization	Partner	References
Heme metabolism			
	Cytosol	iNOS	Chakravarti et al. (2010)
Gene regulation			
	Cytosol	Oct1	Zheng et al. (2003) and Dai et al. (2008)
		AR	Harada et al. (2007)
Nuclear shuttling			
	Cytosol	PCAF	Ventura et al. (2010)
	Nucleus	CRM1	Brown et al. (2004)
DNA maintenance			
	Nucleus	APE1	Azam et al. (2008)
		TERC	Nicholls et al. (2012)
Synaptic transmission			
	Plasma membrane	$GABA_AR$	Laschet (2004)
Unknown			
	Plasma membrane	AE1 (band 3)	Ercolani et al. (1992), Campanella et al. (2005) and Su et al. (2011)

15.3.2 Membrane Localization

As discussed previously, GAPDH has been shown to possess several plasma membrane associated functions. Beyond its fusogenic properties, GAPDH plays an important role in iron homeostasis via its interactions with transferrin family proteins (Boradia et al. 2014a; Chauhan et al. 2015; Modun and Williams 1999; Modun et al. 2000; Raje et al. 2007; Taylor and Heinrichs 2002). GAPDH functions in opposing roles where it is involved in iron intake via interactions with uPAR (Chauhan et al. 2015), or iron efflux via interactions with ferroportin (Sheokand et al. 2014). GAPDH also associates with the membrane transport protein anion exchanger 1 (AE1 or band 3) in both erythrocyte and intercalated kidney cells. The physiological significance of this association has not yet been elucidated but many hypotheses include providing local glycolytic flux, modulating redox environment via NAD^+/NADH ratios, or serving as a link between band 3 and the cytoskeleton, all of which GAPDH has been shown to do in other systems (Campanella et al. 2005; Ercolani et al. 1992; Su et al. 2011).

15.3.3 Cytosolic Localization

The majority of GAPDH functions occur within the cytosol. Here, GAPDH post-translational modifications linked to its function in stress response are intimately tied to its role in intermediary metabolism. Several known modifications, including

Table 15.3 Subcellular localization

Function	Start	End	References
Iron metabolism			
	Surface	Cytosol	Raje et al. (2007), Polati et al. (2012), Sheokand et al. (2014), Boradia et al. (2014b) and Chauhan et al. (2015)
	Endosome		
	Surface	Extracellular	Modun et al. (2000), Sheokand et al. (2014) and Chauhan et al. (2015)
Heme metabolism			
	Cytosol		Chakravarti et al. (2010)
Intermediary metabolism			
	Cytosol		Hildebrandt et al. (2015)
Membrane fusion			
	Cytosol		Robbins et al. (1995), Hessler et al. (1998) and Han et al. (1998)
Intracellular trafficking			
	ER	Golgi	Tisdale (2001), Tisdale et al. (2004) and Tisdale et al. (2009)
	VTC/ERGIC		
Nuclear shuttling			
	Cytosol	Nucleus	Park et al. (2009)
			Ventura et al. (2010)
	Nucleus	Cytosol	Brown et al. (2004)
Cell death			
Cell death	Cytosol	Nucleus	Sawa et al. (1997), Hara et al. (2005), Nakajima et al. (2007), Nakajima et al. (2009), Huang et al. (2011) and Samson et al. (2014)
CICD	Cytosol	Nucleus	Hara and Snyder (2006), Yego et al. (2009) and Yego and Mohr (2010)
Extrinsic apoptosis	Cytosol	Nucleus	Du et al. (2003) and Barbini et al. (2007)
Intrinsic apoptosis	Cytosol	Nucleus	Tarze et al. (2007) and Huang et al. (2011)
Autophagy			
	Cytosol	Autophagosome	Fengsrud et al. (2000) and Sneve et al. (2005)
	Cytosol	Nucleus	Colell et al. (2007), Canto and Auwerx (2012) and Chang et al. (2015)
Gene regulation			
	Cytosol	Nucleus	Zheng et al. (2003), Harada et al. (2007), Dai et al. (2008) and Nicholls et al. (2012)
DNA maintenance			
	Cytosol	Nucleus	Sundararaj et al. (2004) and Demarse et al. (2009)
RNA export			
	Nucleus	Cytosol	Singh and Green (1993)

oxidation (Nakajima et al. 2007, 2009; Samson et al. 2014), phosphorylation (Huang et al. 2011), acetylation (Sen et al. 2008; Ventura et al. 2010), S-sulfhydration (Jarosz et al. 2015), S-nitrosylation (Baek et al. 2008; Hara and Snyder 2006; Hara et al. 2005), pyruvylation (Ganapathy-Kanniappan et al. 2009), and NAD-attachment (Mohr et al. 1996) have been proposed to inhibit GAPDH oxidoreductase activity and/or disrupt oligomerization, frequently serving as indicators for cell death, in part due to cellular energy collapse (Almeida et al. 2004). Other reversible modifications such as S-sulfhydration (Jarosz et al. 2015; Mustafa et al. 2009) and S-thiolation (Grant et al. 1999; Ravichandran et al. 1994; Schuppe-Koistinen et al. 1994), have been shown to serve as protective modifications, mitigating irreversible oxidation or shifting glycolytic flux to favor the production of reducing equivalents (Ralser et al. 2007).

Many additional non-glycolytic functions of GAPDH occur in the cytosol, including microtubule and microfilament binding, heme insertion, and autophagy (see Sect. 15.2). The membrane fusion activity of GAPDH involved in endocytosis, exocytosis, and mediating intracellular trafficking is regulated via phosphorylation events. These events involve the translocation of GAPDH between multiple compartments and the plasma membrane. The well-established role of GAPDH in retrograde trafficking (Tisdale et al. 2009) from the Golgi back to the ER by way of VTC/ERGICs is mediated by phosphorylation events (Coussens et al. 1985; Sergienko et al. 1992; Tisdale 2002; Tisdale and Artalejo 2007) as well as interactions with Rab2 (Tisdale 2001; Tisdale et al. 2004), PKCι/λ (Tisdale 2001, 2002; Tisdale et al. 2004), Src (Coussens et al. 1985; Sergienko et al. 1992; Tisdale and Artalejo 2007), and tubulin (Tisdale et al. 2009). This process requires two phosphorylations, each by a different kinase at a different residue; the first catalyzed by Src at Tyr 41 facilitating macromolecular complex formation (Tisdale and Artalejo 2006) and the second occurring at an unknown Serine residue mediated by PKCι/λ modulating tubulin bundling and cross-linking as well as dynein recruitment (Tisdale 2002). GAPDH is also autophosphorylated and can act as a kinase to phosphorylate membrane bound microsomal proteins (Kawamoto and Caswell 1986).

15.3.4 Mitochondrial Localization

GAPDH has been shown to function in MOMP (Ishitani et al. 2003) in response to cell stress (Ishitani and Chuang 1996; Ishitani et al. 1997, 1998; Saunders et al. 1997) and as a function of cell senescence (Ishitani et al. 1996), and accumulates in the mitochondria during apoptosis (Huang et al. 2009; Tarze et al. 2007). The role of GAPDH in MOMP has not yet been fully elucidated, though its interactions with VDAC1 (Tarze et al. 2007) are a key mediator of the overall process (reviewed in Sect. 15.2.5).

15.3.5 Nuclear Localization

Several non-glycolytic functions of GAPDH involve translocation to the nucleus, usually in response to cell stress. Oxidation (Arutyunova et al. 2003; Pierce et al. 2008) and glycosylation (Park et al. 2009) have been shown to disrupt oligomerization and lead to nuclear translocation. GAPDH typically accumulates in the nucleus during apoptosis (Hara et al. 2005; Sawa et al. 1997). Indeed, GAPDH has multiple roles in the initiation and prevention of apoptosis, a critical aspect of which being nuclear translocation. The GAPDH-Siah1 pathway, mediated by S-nitrosylation and acetylation, represents the epitome of the interconnectedness of GAPDH modifications, protein-protein interactions, and subcellular localization (see Sect. 15.2.5.3).

GAPDH has additional functions with respect to nutrient sensing and control of cell death. GAPDH translocates to the nucleus of Muller cells under hyperglycemic conditions (Yego et al. 2009), via interaction with Siah1 (Yego and Mohr 2010) or o-linked n-acetyl glucosamine glycosylation of GAPDH T229 (Park et al. 2009). Conversely, serum depletion also induces GAPDH nuclear translocation (Kwon et al. 2010; Schmitz 2001). These processes are under the control of PI3K/Akt–AMPK interactions. AMPK-induced nuclear translocation was shown to occur via phosphorylation of GAPDH S122 (Canto and Auwerx 2012). Meanwhile, PI3K (Kwon et al. 2010; Schmitz 2001) and Akt (Kwon et al. 2010) function to promote GAPDH nuclear export and cytosolic retainment, where Akt phosphorylates T237 to prevent GAPDH nuclear translocation (Huang et al. 2011). Phosphorylation at T237 may hinder Siah1 binding as nearby K227 has been shown to be critical in the interaction.

Within the nucleus, GAPDH serves as a mediator of transcriptional regulation via interactions with Oct-1 (Dai et al. 2008; Zheng et al. 2003) and androgen receptor (Harada et al. 2007). GAPDH also serves to protect DNA integrity by thiol exchange to ensure APE1 is reduced and able to repair DNA (Azam et al. 2008) or it can induce cell senescence by interaction with and inhibition of telomerase reverse transcriptase (Nicholls et al. 2012). While GAPDH does not contain a NLS and most nuclear localization events are mediated by interactions with additional protein effectors, the protein does contain a CRM1 nuclear export domain (Brown et al. 2004). More studies are needed to elucidate the exact mechanisms by which GAPDH switches between its multiple roles and localizations, and how structural alterations, residue modifications, and changes in localization dictate functional outcomes.

15.4 Structural Studies of GAPDH

There are two GAPDH isozymes in higher eukaryotes: the spermatogenic GAPDS and the somatic GAPDH. GAPDS is critical to sperm mobility, thus making it an attractive target for contraception. The two isoforms share 55.7% sequence identity.

GAPDS possesses a testis-specific N-terminal extension of about 72 residues, which ensures its association with sperm flagellum. Human GAPDH is expressed as a 36 kDa polypeptide with 335 amino acids. Both isoforms share a similar domain organization: an NAD^+-binding domain, which adopts a classical Rossman fold and spans regions 1–151 and 315–335, and a catalytic domain, which includes residues 152–314. More than 100 x-ray structures of GAPDH from all three domains of life are currently available in the Protein DataBank (rcsb.org). It is intriguing that GAPDH has only been crystallized so far with NAD^+, $NADP^+$ and NAD^+ analogs, substrates, substrate analogs, intermediates, products, inhibitors targeting the active site C152, and small molecules present in crystallization or cryoprotectant solutions (glycerol, phosphate, sulfate, etc.). No crystal structure with nucleic acid, protein partner, or peptide is available. Regardless, this wealth of information offers opportunities to examine the special features of this protein, as well as similarities and differences among the GAPDH family members, outlined below.

15.4.1 Oligomerization and Interfaces

GAPDH is homotetrameric and is best described as a dimer of dimers, composed of four identical subunits, named O, P, Q, and R (Fig. 15.2). Three 2-fold axes (noted *P*, *Q*, *R*) exist between these subunits and delineate the dimer interface (*P*, between subunits O and P or Q and R), the tetramer interface (*R*, between dimers O/P and Q/R), and the NAD^+-binding grooves (*Q*, between subunits O and Q, or P and R). The dimer interface is the most extensive (burying ~1900 $Å^2$ per dimer) and involves the catalytic domains of adjacent subunits. Dimerization is mediated via hydrogen bonds, salt bridges, and hydrophobic interactions via a structural motif composed of a 4-stranded antiparallel β-sheet flanked by 2 α-helices. This motif is reminiscent of the RNA Recognition Motif (RRM) that has been implicated in RNA binding and protein-protein interactions (Maris et al. 2005). A structural alignment allowed us to identify structural neighbors to this region of GAPDH, including the RRM1 of heterogenous nuclear ribonucleoprotein L (hnRNP L, PDB entry 3R27). The two proteins display an rmsd of 3.8 Å for 78 atoms superimposed, suggesting similar folds, despite low sequence identity (White and Garcin 2016). The hnRNP L protein contains an RRM that plays a critical role in RNA binding, and therefore this interface has been proposed as a putative RNA binding site for GAPDH (Carmona et al. 1999; White et al. 2015). In addition, the dimer interface contains aromatic and positively-charged residues that could play a key role in RNA binding (White and Garcin 2016), however more studies are needed to determine its potential involvement in RNA binding

The tetrameric (*R*) interface is not as extensive as the dimer interface (buries ~3500 $Å^2$ total) and involves residues from loops from each subunit interacting via hydrogen bonds, salt bridges, and hydrophobic interactions. The *R* interface contains a central channel that is mostly filled with water molecules and has been proposed as the binding site for various molecules (see Sect. 15.4.5). Several reports

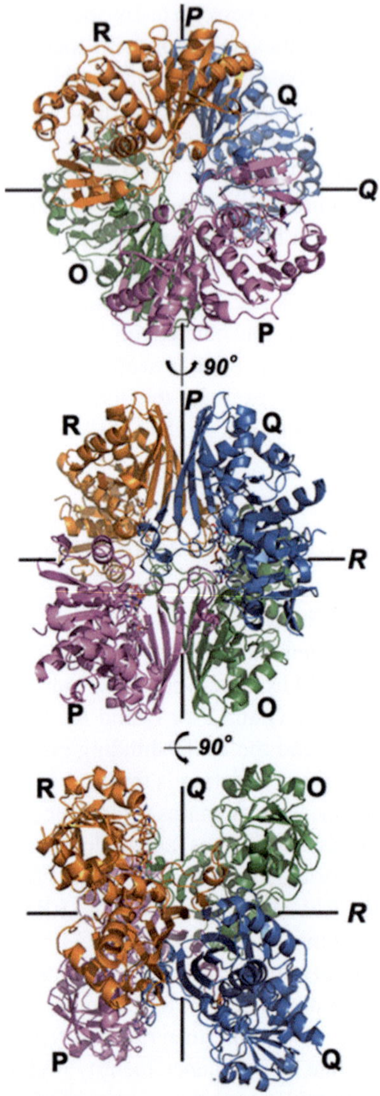

Fig. 15.2 Homotetrameric structure of human GAPDH (PDB entry 4WNC). The four molecules (O, P, Q, R) shown as cartoons relate to each other via the three 2-fold axes (*P*, *Q*, *R*). Bound NAD⁺ molecules are shown in *sticks*

claim that GAPDH can dissociate from tetramers to dimers, and even monomers (Arutyunova et al. 2003; Constantinides and Deal 1969; Sundararaj et al. 2004). How GAPDH would switch between these oligomeric states is not known. The NAD^+ cofactor has been suggested to promote tetramerization, while ATP would promote monomerization (Constantinides and Deal 1969). In other cases, post-translational modifications have been suggested to influence the oligomerization state of GAPDH. In particular, glycosylation of human GAPDH T229 by an unknown enzyme was proposed to induce monomerization of GAPDH and nuclear translocation (Park et al. 2009). The T229K mutation, which was proposed as a mimic for glycosylation, was also shown to disrupt tetramerization. We used a variety of biophysical characterizations, including size-exclusion chromatography, x-ray crystallography, small-angle x-ray scattering, and nano-ESI mass spectrometry and showed that this mutant remained tetrameric like wild-type GAPDH at concentrations as low as 0.2 μM (White et al. 2015). While we cannot explain the discrepancy between these results, we noted that only repeated freeze-thaw cycles seemed to disrupt GAPDH tetramers and lead to aggregation. This is in agreement with earlier studies that showed monomerization of GAPDH was induced at lower temperatures (Constantinides and Deal 1969). Regardless, these studies do not rule out that GAPDH could be present in cells as a mixture of oligomeric species, including tetramers, dimers, and monomers, though this remains to be determined unambiguously.

15.4.2 Cofactor Binding Site

The GAPDH tetramer contains four possible NAD^+-binding sites. Crystal structures with 2 (Cowan-Jacob et al. 2003), 3 (Jenkins and Tanner 2006; White et al. 2015), or 4 (Baker et al. 2014; Chaikuad et al. 2011; Ismail and Park 2005) NAD^+ molecules have been determined, but no crystal structure of the apo human GAPDH enzyme is available. In our hands, we find that purified GAPDH almost always contains NAD^+ even in the absence of added cofactor during bacterial growth or protein purification. Crystal structures of bacterial apo and holo GAPDH have been described and show that cofactor binding promotes conformational changes mostly in the N-terminal domain near the cofactor adenosine moiety (Skarżyński and Wonacott 1988; Yun et al. 2000). Comparison of the apo and holo human and mouse GAPDS isoform showed similar conformational changes in the N-terminal domain, as well as unique changes in the S-loop (residues 182–205) of the C-terminal domain (Danshina et al. 2016). The S-loop is centrally located at the tetramer interface, and accounts for 12 hydrogen bonds and 7 salt-bridges per monomer. This loop was proposed to be involved in the cooperativity observed between GAPDH subunits (Duée et al. 1996).

NAD^+ is bound in an extended conformation and makes numerous hydrogen bonds with water molecules and few with protein residues, including R13, I14, E79, and N316. Non-polar contacts of the aromatic rings with I14, Y320, P36, F37, T99, and F102 further contribute to binding. The NAD^+ binding pocket is well conserved among GAPDH enzymes and comprises residues 9–10, 12–14, 16, 34–38, 79–80, 98–103, 122–124, 151–152, 179, 182–183, 316–317, and 320. In addition, residues 188–191 of the S-loop from the subunit across the *Q* interface also contribute to the cofactor-binding site (Fig. 15.3a). This loop usually adopts a "closed" conformation close to the adjacent subunit (White et al. 2015), but has also been observed in an "open" conformation further from the adjacent subunit (Ismail and Park 2005). The significance of this difference is not clear, but could be due to crystallographic artifacts.

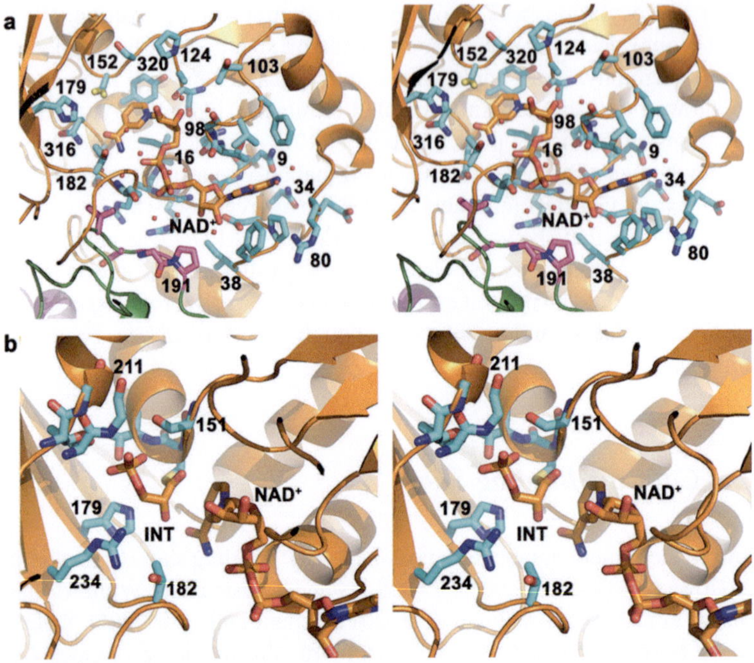

Fig. 15.3 Stereoview of the cofactor and substrate-binding sites. (**a**) The NAD^+ (*orange*) binding site is located in the N-terminal domain, and involves numerous water molecules (*red spheres*) and protein residues (*cyan, and magenta*; PDB entry 4WNC). (**b**) The substrate-binding site is located between the N-terminal and C-terminal domains and in close proximity to the NAD^+ binding site. The loop 211–212 shows two alternate conformations. In addition, the covalent thioacylglycerate 3-phosphate is shown (PDB entry 3CMC)

In protozoan GAPDHs, differences in residues surrounding the cofactor adenosyl moiety exist and have been exploited for the design of specific inhibitors against the parasitic enzymes (Aronov et al. 1999; Kim and Hol 1998; Ladame et al. 2003; Satchell et al. 2005; Suresh et al. 2001; Verlinde et al. 1994). A narrow hydrophobic cleft ("selectivity cleft") is present in these enzymes, but absent in human GAPDH, due to different conformations of regions 35–40 and 77–82. Inhibitors building on the adenosine scaffold have been designed and show excellent selectivity over the human enzyme (Bressi et al. 2001), thus possibly providing a promising route for the design of therapeutics against parasitic infections, including malaria, sleeping sickness, leishmaniasis, and Chagas' disease.

15.4.3 Active Site and Mechanism

GAPDH catalyzes the oxidative phosphorylation of D-glyceraldehyde 3-phosphate (GAP) into 1,3-bisphosphoglycerate (BPG). Extensive work has been done to elucidate the exact mechanism, and various structures of GAPDH with cofactor, substrate, analogs, products, and intermediate have been solved to answer this question. Recently, new light has been shed on this old mechanism with the structure of *Bacillus stearothermophilus* GAPDH bound to the thioacylglyceraldehyde 3-phosphate intermediate (Moniot et al. 2008). GAPDH utilizes a two-step mechanism. The first step is a nucleophilic attack of the active site cysteine C152 onto GAP followed by hydride transfer from the thiohemiacetal to the NAD^+ nicotinamide assisted by H179, which acts as a base catalyst (Talfournier et al. 1998). The second step is a nucleophilic attack of inorganic phosphate on the carbonyl group of the thioester resulting in its phosphorylation.

The active site is located at the border between the N-terminal NAD^+ binding site and the catalytic domain (Fig. 15.3b). Two anion sites were proposed (Castilho et al. 2003; Moras et al. 1975) to provide binding for inorganic phosphate (Pi site) and the phosphate group of GAP or BPG (Ps site). The Ps site is highly conserved in all GAPDH enzymes regardless of the presence of cofactor and/or ligands and is lined by residues T182, R234, and the 2'OH group of the nicotinamide ribose. The Pi site shows more variability depending on the species, identity of bound ligands, and conformation of the substrate loop 209–215 (Kim et al. 1995; Korndörfer et al. 1995; Moniot et al. 2008; Moras et al. 1975; Skarżyński and Wonacott 1988). The Pi site is lined by residues S151, T211, G212, and T153 (in the new Pi site). Based on the intermediate-bound crystal structure (Moniot et al. 2008), and that of the non-covalent Michaelis-Menten complex (Didierjean et al. 2003), a flip-flop mechanism was proposed:

1. The GAP substrate binds with its C3P group in the Ps site
2. Active site C152 (activated by H179) performs a nucleophilic attack on GAP C1 atom, followed by hydride transfer from GAP C1 to NAD^+ C4 via base catalysis of H179. At this point, C3P is most likely still bound in the Ps site
3. Conformational changes result in C3P flipping to the new Pi site and movement of O1 away from H179
4. NADH is released and NAD^+ binding induces conformational changes leading to flipping back of O1 closer to H179 and C3P returning to the Ps site
5. Inorganic phosphate bound in the Pi site attacks the intermediate and leads to BPG release, while the enzyme returns to its resting state for a new cycle

Surprisingly, the overall conformation of the active site varies little during catalysis, except for the 209–215 loop. Other structural studies of *Cryptosporidium parvum* GAPDH have shown that the C3P of GAP did not bind to the Ps site, but to the new Pi site and to another distinct new site, highlighting the flexibility of the substrate and its binding mode to the GAPDH pocket (Cook et al. 2009). More studies will be needed to determine exactly how and when flip-flopping occurs during catalysis.

15.4.4 Positive Groove

The GAPDH homotetramer possesses two 70-Å long grooves along the *P* interface that encompass the substrate- and NAD^+-binding regions of the two dimers (Fig. 15.4). These regions have been postulated as nucleic-acid binding regions, because of the large number of positively charged residues, and because they are wide enough to accommodate a nucleic-acid helix (Carmona et al. 1999; Demarse et al. 2009). This hypothesis was supported by experiments showing that GAP and NAD^+ could both compete with RNA binding. However, more studies are required to unambiguously identify the DNA and/or RNA binding site in GAPDH, critical to its function in gene regulation.

The positively charged groove is also an attractive binding site for the protein band 3 (anion transport protein), which locates to the membrane of human erythrocytes. GAPDH associates with several regions of band 3: N-terminal negatively charged peptides D_6DYED_{10}, $E_{19}YED_{23}$, and a region spanning residues 356–384 (Campanella et al. 2005; Chu and Low 2006; Chu et al. 2012; Ercolani et al. 1992; Su et al. 2011). In addition, binding of GAPDH to the N-terminal regions of band 3 inhibits its catalytic activity (Campanella et al. 2005; Tsai et al. 1982). The third binding site in band 3 is not inhibitory but increases binding affinity to GAPDH (Puchulu-Campanella et al. 2013).

The binding site of band 3 in GAPDH is not known, but some structural information on the band 3 N-terminal regions (residues 1–15) is available. NMR structure determination showed that residues 4–9 were well ordered (Schneider and Post 1995). This region has also been shown to bind and inhibit aldolase (Campanella et al. 2005). Molecular docking suggested that the negatively-charged band 3 peptide could potentially bind to or near the positively charged catalytic center of aldolase via electrostatic interactions. Other proteins known to interact with GAPDH contain a similar acidic motif: myocilin ($E_{230}YD_{232}$), tubulin ($E_{429}KDYEE_{434}$), GLUT4 ($E_{499}LEY_{502}$), F-actin ($E_{361}YDE_{364}$), and troponin T ($E_5EQEYEEE_{12}$), which also inhibits GAPDH (Chu and Low 2006). Further studies will be needed to determine whether the positively charged groove of GAPDH could serve as the docking site for band 3 and other partners.

15.4.5 The Central Channel

The central channel runs parallel to the *R* axis of the GAPDH homotetramer (Fig. 15.5). This channel has been proposed as the binding site for small molecules, including deprenyl-analogs CGP3466 and DES, that act as anti-apoptotic agents (Carlile et al. 2000). CGP3466 was shown to induce tetramer-to-dimer conversion of GAPDH *in vitro*, and its binding was inhibited by an anti-GAPDH antibody with a binding epitope located at the entrance of the channel. The channel is very narrow in the center of the homotetramer (4 Å by 10 Å), but expands towards the opening (7 Å by 12 Å), and can potentially accommodate small molecules. In addition, the

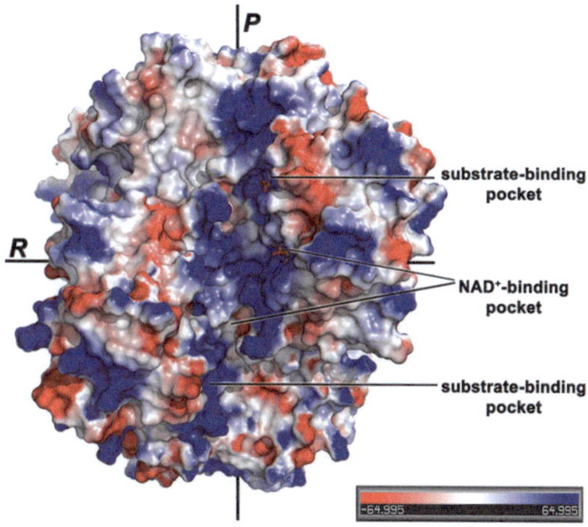

Fig. 15.4 Electrostatic potential of GAPDH mapped onto the solvent-accessible surface with positively-charged regions in *blue* and negatively charged regions in *red*. Tetramer axes *P* and *Q* are indicated for orientation

entrance of the channel is positively charged, suggesting a possible additional binding site for nucleic acids (Fig. 15.5). In the human GAPDH structure (White et al. 2015) and most GAPDH structures, the channel is occupied by ordered water molecules. Whether small molecules or nucleic acids can bind GAPDH in this channel remains to be determined.

15.4.6 The Connection Between the Different Structural Determinants

We have recently shown an unexpected connection between NAD^+ binding and GAPDH oligomeric interfaces (White et al. 2015). Using a combination of mass spectrometry techniques, we first showed that GAPDH is a mixture of NAD^+-free and NAD^+-bound species in solution. This result is somewhat inconsistent with the observation that most GAPDH structures contain NAD^+ (see Sect. 15.4.2). This may suggest that the NAD^+-bound GAPDH homotetramer shows a high tendency to crystallize, compared to the apo form. Second, using hydrogen-deuterium exchange mass spectrometry (HDXMS), we showed a connection between NAD^+ binding, the dimer interface, and the tetrameric interface. A single mutation at the dimer interface (T229K) significantly decreased the amount of NAD^+-bound GAPDH in solution. This mutation is located about 20 Å away from the NAD^+ binding site, therefore its effect on NAD^+ binding is puzzling. A comparison of the wild-type and mutant GAPDH crystal structures showed that the mutation promoted a cascade of subtle conformational changes at the dimer interface. Our HDXMS analysis of the proteins in solution confirmed these conformational changes, and showed additional short- and long-range conformational and dynamic changes in the mutant protein. First, three peptides that participate in the tetramer interface showed structural changes and increased dynamics in T229K GAPDH. Second, the mutation affected

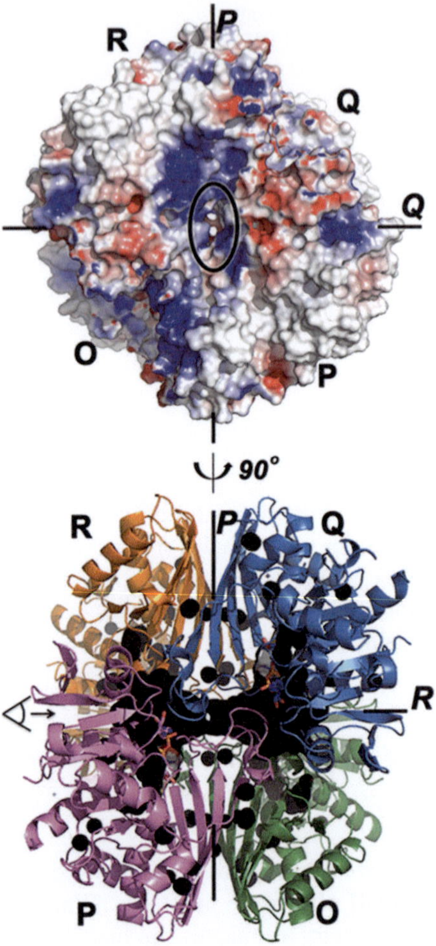

Fig. 15.5 The central channel in GAPDH. *Top*: view along the *R* axis showing the central channel (*circled*) with positively-charged (*blue*) regions lining the entrance of the channel. *Bottom*: View along the *Q* axis shows the channel and small cavities (*black*) spanning the entire width of the GAPDH tetramer along the *R* axis

several peptides participating in NAD^+ binding, confirming that the mutation allosterically perturbed NAD^+ binding. Overall, all observed structural and dynamic changes were confined to the homotetramer *P* interface. Collectively, these results strongly support a connection between NAD^+ binding, and the dimer and tetramer interfaces. These connections may explain some of the inhibitory effects of NAD^+ addition on GAPDH interactions with protein or nucleic acid partners.

15.4.7 Comparison of GAPDH from Different Domains of Life

Overall, GAPDH enzymes from bacteria and eukaryotes share high sequence similarity (>40%), and all use NAD^+ as a cofactor, except some plant GAPDHs that use $NADP^+$. As a result, the 3D structures of bacterial and eukaryotic GAPDHs are very similar.

On the other hand, archeal GAPDHs share low sequence identity with other GAPDHs (<20%) and they can use either NAD^+ or $NADP^+$ as cofactors. Because of the low sequence identity, these enzymes were originally assigned to a different protein family. However, structural information confirmed that archaeal GAPDH belonged to the same family as their eubacteria and eukaryotic homologs. The 3D structures from *Pyrococcus horikoshii* (PDB entry 2CZC), *Sulfolobus solfataricus* (PDB entry 1B7G (Isupov et al. 1999)), *Methanothermus fervidus* (PDB entry 1CF2 (Charron et al. 2000)), and *Methanocaldococcus jannaschii* (PDB entry 2YYY (Malay et al. 2009)) have been determined. These structures show a homotetrameric arrangement, as seen for other GAPDHs. Several features set the archeal GAPDHs apart however. First, additional secondary structures at the C-terminus provide extra contacts between the two domains. Second, disulfide bridging and several salt bridges are thought to provide additional stabilization for the thermophilic enzymes. Third, the residues important for catalysis and cofactor binding are relocated to different secondary structure elements. It was proposed that the archeal GAPDHs evolved independently to eventually converge on a similar 3D structure (Littlechild et al. 2004).

15.5 Conclusions and Future Perspectives

GAPDH is involved in a wide array of cellular functions beyond its role in glycolysis. In this, GAPDH likely serves as a link between cellular nutrient availability and alternative functions. These links have been speculated in the case of insulin secretion, owing to the role of GAPDH in islet plasma membrane and secretory granule fusion, and the control of Glut 1 expression, where GAPDH was shown to bind the 3′-UTR AU-rich region of GLUT-1 mRNA.

The link between GAPDH metabolic and auxiliary functions has been more clearly elucidated in several cases. One is the alternative phosphorylation by AMPK and Akt, where AMPK phosphorylation of S122 induces GAPDH nuclear translocation and Akt phosphorylation of T237 prevents GAPDH nuclear translocation. Another is the function of GAPDH under conditions of hyperglycemia. In high glucose environments, GAPDH mediates cell death by either interactions with Siah1 or induction of autophagy. Finally, the cytokines IFN-γ and IL-2, as well as TNF-α, have all been shown to be translationally repressed by GAPDH in low glucose conditions.

Further elucidation of structural determinants in GAPDH functions and the regulation thereof could serve to further the treatment of many pathologies associated with GAPDH dysfunction.

References

Adastra KL, Chi MM, Riley JK, Moley KH (2011) A differential autophagic response to hyperglycemia in the developing murine embryo. Reproduction 141:607–615. doi:10.1530/REP-10-0265

Almeida A, Moncada S, Bolaños JP (2004) Nitric oxide switches on glycolysis through the AMP protein kinase and 6-phosphofructo-2-kinase pathway. Nat Cell Biol 6:45–51. doi:10.1038/ncb1080

Andrade J, Pearce ST, Zhao H, Barroso M (2004) Interactions among p22, glyceraldehyde-3-phosphate dehydrogenase and microtubules. Biochem J 384:327–336. doi:10.1042/BJ20040622

Arnold H, Pette D (1968) Binding of glycolytic enzymes to structure proteins of the muscle. Eur J Biochem 6:163–171

Aronov AM, Suresh S, Buckner FS et al (1999) Structure-based design of submicromolar, biologically active inhibitors of trypanosomatid glyceraldehyde-3-phosphate dehydrogenase. Proc Natl Acad Sci 96:4273–4278. doi:10.1073/pnas.96.8.4273

Arutyunova EI, Danshina PV, Domnina LV et al (2003) Oxidation of glyceraldehyde-3-phosphate dehydrogenase enhances its binding to nucleic acids. Biochem Biophys Res Commun 307:547–552. doi:10.1016/S0006-291X(03)01222-1

Azam S, Jouvet N, Jilani A et al (2008) Human glyceraldehyde-3-phosphate dehydrogenase plays a direct role in reactivating oxidized forms of the DNA repair enzyme APE1. J Biol Chem 283:30632–30641

Backlund M, Paukku K, Daviet L et al (2009) Posttranscriptional regulation of angiotensin II type 1 receptor expression by glyceraldehyde 3-phosphate dehydrogenase. Nucleic Acids Res 37:2346–2358

Baek D, Jin Y, Jeong JC et al (2008) Suppression of reactive oxygen species by glyceraldehyde-3-phosphate dehydrogenase. Phytochemistry 69:333–338

Baker BY, Shi W, Wang B, Palczewski K (2014) High-resolution crystal structures of the photoreceptor glyceraldehyde 3-phosphate dehydrogenase (GAPDH) with three and four-bound NAD molecules: structures of the GAPDH Photoreceptor. Protein Sci 23:1629–1639. doi:10.1002/pro.2543

Barbini L, Rodríguez J, Dominguez F, Vega F (2007) Glyceraldehyde-3-phosphate dehydrogenase exerts different biologic activities in apoptotic and proliferating hepatocytes according to its subcellular localization. Mol Cell Biochem 300:19–28. doi:10.1007/s11010-006-9341-1

Bonafe N, Gilmore-Hebert M, Folk NL et al (2005) Glyceraldehyde-3-phosphate dehydrogenase binds to the AU-Rich 3′ untranslated region of colony-stimulating factor-1 (CSF-1) messenger RNA in human ovarian cancer cells: possible role in CSF-1 posttranscriptional regulation and tumor phenotype. Cancer Res 65:3762–3271

Boradia VM, Malhotra H, Thakkar JS et al (2014a) Mycobacterium tuberculosis acquires iron by cell-surface sequestration and internalization of human holo-transferrin. Nat Commun 5:4730. doi:10.1038/ncomms5730

Boradia VM, Raje M, Raje CI (2014b) Protein moonlighting in iron metabolism: glyceraldehyde-3-phosphate dehydrogenase (GAPDH). Biochem Soc Trans 42:1796–1801. doi:10.1042/BST20140220

Bressi JC, Verlinde CLMJ, Aronov AM et al (2001) Adenosine analogues as selective inhibitors of glyceraldehyde-3-phosphate dehydrogenase of trypanosomatidae via structure-based drug design. J Med Chem 44:2080–2093. doi:10.1021/jm000472o

Brown VM, Krynetski EY, Krynetskaia NF et al (2004) A novel CRM1-mediated nuclear export signal governs nuclear accumulation of glyceraldehyde-3-phosphate dehydrogenase following genotoxic stress. J Biol Chem 279:5984–5992. doi:10.1074/jbc.M307071200

Bruns G, Gerald P, Lalley P et al (1979) Gene mapping of the mouse by somatic cell hybridization. Cytogenet Cell Genet 25:139

Buck KW (1996) Comparison of the replication of positive-stranded RNA viruses of plants and animals. In: Maramorosch K, Murphy FA, Shatkin AJ (eds) Advances in virus research. Academic, Amsterdam, pp 159–251

Campanella ME, Chu H, Low PS (2005) Assembly and regulation of a glycolytic enzyme complex on the human erythrocyte membrane. Proc Natl Acad Sci U S A 102:2402–2407. doi:10.1073/pnas.0409741102

Canto C, Auwerx J (2012) Targeting sirtuin 1 to improve metabolism: all you need is NAD^{+}? Pharmacol Rev 64:166–187. doi:10.1124/pr.110.003905

Carlile GW, Chalmers-Redman RME, Tatton NA et al (2000) Reduced apoptosis after nerve growth factor and serum withdrawal: conversion of tetrameric glyceraldehyde-3-phosphate dehydrogenase to a dimer. Mol Pharmacol 57:2–12

Carmona P, Rodriguez-Casado A, Molina M (1999) Conformational structure and binding mode of glyceraldehyde-3-phosphate dehydrogenase to tRNA studied by Raman and CD spectroscopy. Biochim Biophys Acta 1432:222–233

Castilho MS, Pavão F, Oliva G et al (2003) Evidence for the two phosphate binding sites of an analogue of the thioacyl intermediate for the *Trypanosoma cruzi* glyceraldehyde-3-phosphate dehydrogenase-catalyzed reaction, from its crystal structure. Biochemistry 42:7143–7151. doi:10.1021/bi0206107

Caswell AH, Corbett AM (1985) Interaction of glyceraldehyde-3-phosphate dehydrogenase with isolated microsomal subfractions of skeletal muscle. J Biol Chem 260:6892–6898

Chaikuad A, Shafqat N, Al-Mokhtar R et al (2011) Structure and kinetic characterization of human sperm-specific glyceraldehyde-3-phosphate dehydrogenase, GAPDS. Biochem J 435:401–409. doi:10.1042/BJ20101442

Chakravarti R, Aulak KS, Fox PL, Stuehr DJ (2010) GAPDH regulates cellular heme insertion into inducible nitric oxide synthase. Proc Natl Acad Sci U S A 107:18004–18009

Chang C-H, Curtis JD, Maggi LB Jr et al (2013) Posttranscriptional control of T cell effector function by aerobic glycolysis. Cell 153:1239–1251. doi:10.1016/j.cell.2013.05.016

Chang C, Su H, Zhang D et al (2015) AMPK-dependent phosphorylation of GAPDH triggers sirt1 activation and is necessary for autophagy upon glucose starvation. Mol Cell 60:930–940. doi:10.1016/j.molcel.2015.10.037

Chao CC-K, Yam W-C, Lin-Chao S (1990) Coordinated induction of two unrelated glucose-regulated protein genes by a calcium ionophore: human BiPGRP78 and GAPDH. Biochem Biophys Res Commun 171:431–438. doi:10.1016/0006-291X(90)91411-K

Charron C, Talfournier F, Isupov M et al (2000) The crystal structure of d-glyceraldehyde-3-phosphate dehydrogenase from the hyperthermophilic archaeon Methanothermus fervidus in the presence of $NADP^{+}$ at 2.1 Å resolution. J Mol Biol 297:481–500. doi:10.1006/jmbi.2000.3565

Chauhan AS, Rawat P, Malhotra H et al (2015) Secreted multifunctional Glyceraldehyde-3-phosphate dehydrogenase sequesters lactoferrin and iron into cells via a non-canonical pathway. Sci Rep 5:18465

Chu H, Low PS (2006) Mapping of glycolytic enzyme-binding sites on human erythrocyte band 3. Biochem J 400:143–151. doi:10.1042/BJ20060792

Chu H, Puchulu-Campanella E, Galan JA et al (2012) Identification of cytoskeletal elements enclosing the ATP pools that fuel human red blood cell membrane cation pumps. Proc Natl Acad Sci U S A 109:12794–12799. doi:10.1073/pnas.1209014109

Colell A, Ricci J-E, Tait S et al (2007) GAPDH and autophagy preserve survival after apoptotic cytochrome c release in the absence of caspase activation. Cell 129:983–997. doi:10.1016/j.cell.2007.03.045

Colell A, Green DR, Ricci J-E (2009) Novel roles for GAPDH in cell death and carcinogenesis. Cell Death Differ 16:1573–1581. doi:10.1038/cdd.2009.137

Constantinides SM, Deal WC (1969) Reversible dissociation of tetrameric rabbit muscle glyceraldehyde 3-phosphate dehydrogenase into dimers or monomers by adenosine triphosphate. J Biol Chem 244:5695–5702

Cook WJ, Senkovich O, Chattopadhyay D (2009) An unexpected phosphate binding site in glyceraldehyde 3-phosphate dehydrogenase: crystal structures of apo, holo and ternary complex of *Cryptosporidium parvum* enzyme. BMC Struct Biol 9:9. doi:10.1186/1472-6807-9-9

Coussens PM, Cooper JA, Hunter T, Shalloway D (1985) Restriction of the in vitro and in vivo tyrosine protein kinase activities of pp60c-src relative to pp60v-src. Mol Cell Biol 5:2753–2763. doi:10.1128/MCB.5.10.2753

Cowan-Jacob SW, Kaufmann M, Anselmo AN et al (2003) Structure of rabbit-muscle glyceraldehyde-3-phosphate dehydrogenase. Acta Crystallogr Sect D 59:2218–2227. doi:10.1107/S0907444903020493

Cueille N, Blanc CT, Riederer IM, Riederer BM (2007) Microtubule-associated protein 1B binds glyceraldehyde-3-phosphate dehydrogenase. J Proteome Res 6:2640–2647. doi:10.1021/pr070081z

Dai R-P, Yu F-X, Goh S-R et al (2008) Histone 2B (H2B) expression is confined to a proper NAD+/NADH redox status. J Biol Chem 283:26894–26901. doi:10.1074/jbc.M804307200

Danshina PV, Qu W, Temple BR et al (2016) Structural analyses to identify selective inhibitors of glyceraldehyde 3-phosphate dehydrogenase-S, a sperm-specific glycolytic enzyme. Mol Hum Reprod 22:410. doi:10.1093/molehr/gaw016

Daubenberger CA, Tisdale EJ, Curcic M et al (2003) The N′-terminal domain of glyceraldehyde-3-phosphate dehydrogenase of the apicomplexan plasmodium falciparum mediates GTPase Rab2-dependent recruitment to membranes. Biol Chem 384:1227–1237

De BP, Gupta S, Zhao H et al (1996) Specific Interaction in vitro and in vivo of glyceraldehyde-3-phosphate dehydrogenase and LA protein with cis-acting RNAs of human parainfluenza virus type 3. J Biol Chem 271:24728–24735. doi:10.1074/jbc.271.40.24728

De Arcuri BF, Vechetti GF, Chehín RN et al (1999) Protein-induced fusion of phospholipid vesicles of heterogeneous sizes. Biochem Biophys Res Commun 262:586–590. doi:10.1006/bbrc.1999.1243

Demarse NA, Ponnusamy S, Spicer EK et al (2009) Direct binding of glyceraldehyde 3-phosphate dehydrogenase to telomeric DNA protects telomeres against chemotherapy-induced rapid degradation. J Mol Biol 394:789–803

Devalaraja-Narashimha K, Padanilam BJ (2009) PARP-1 inhibits glycolysis in ischemic idneys. J Am Soc Nephrol 20:95–103. doi:10.1681/ASN.2008030325

Didierjean C, Corbier C, Fatih M et al (2003) Crystal structure of two ternary complexes of phosphorylating glyceraldehyde-3-phosphate dehydrogenase from *Bacillus stearothermophilus* with NAD and D-Glyceraldehyde 3-Phosphate. J Biol Chem 278:12968–12976. doi:10.1074/jbc.M211040200

Dollenmaier G, Weitz M (2003) Interaction of glyceraldehyde-3-phosphate dehydrogenase with secondary and tertiary RNA structural elements of the hepatitis A virus 3′ translated and non-translated regions. J Gen Virol 84:403–414

Du X, Matsumura T, Edelstein D et al (2003) Inhibition of GAPDH activity by poly(ADP-ribose) polymerase activates three major pathways of hyperglycemic damage in endothelial cells. J Clin Invest 112:1049–1057. doi:10.1172/JCI18127

Du Z-X, Wang H-Q, Zhang H-Y, Gao D-X (2007) Involvement of glyceraldehyde-3-phosphate dehydrogenase in tumor necrosis factor-related apoptosis-inducing ligand-mediated death of thyroid cancer cells. Endocrinology 148:4352–4361. doi:10.1210/en.2006-1511

Duée E, Olivier-Deyris L, Fanchon E et al (1996) Comparison of the structures of wild-type and a N313T mutant of*Escherichia coli* Glyceraldehyde 3-phosphate dehydrogenases: implication for NAD binding and cooperativity. J Mol Biol 257:814–838. doi:10.1006/jmbi.1996.0204

Durrieu C, Bernier-Valentin F, Rousset B (1987a) Binding of glyceraldehyde 3-phosphate dehydrogenase to microtubules. Mol Cell Biochem 74:55–65. doi:10.1007/BF00221912

Durrieu C, Bernier-Valentin F, Rousset B (1987b) Microtubules bind glyceraldehyde 3-phosphate dehydrogenase and modulate its enzyme activity and quaternary structure. Arch Biochem Biophys 252:32–40. doi:10.1016/0003-9861(87)90005-1

Ercolani L, Brown D, Stuart-Tilley A, Alper SL (1992) Colocalization of GAPDH and band 3 (AE1) proteins in rat erythrocytes and kidney intercalated cell membranes. Am J Physiol Ren Physiol 262:F892–F896

Fengsrud M, Raiborg C, Berg TO et al (2000) Autophagosome-associated variant isoforms of cytosolic enzymes. Biochem J 352:773–781. doi:10.1042/bj3520773

Filipovic MR (2015) Persulfidation (S-sulfhydration) and H2S. In: Moore PK, Whiteman M (eds) Chemistry, biochemistry and pharmacology of hydrogen sulfide. Springer, Cham, pp 29–59

Galán C, Sola I, Nogales A et al (2009) Host cell proteins interacting with the 3′ end of TGEV coronavirus genome influence virus replication. Virology 391:304–314. doi:10.1016/j.virol.2009.06.006

Ganapathy-Kanniappan S, Geschwind J-FH, Kunjithapathham R et al (2009) Glyceraldehyde-3-phosphate dehydrogenase (GAPDH) is pyruvylated during 3-bromopyruvate mediated cancer cell death. Anticancer Res 29:4909–4918

Ghosh A, Stuehr D (2016) Regulation of sGC via hsp90, cellular heme, sGC agonists, and NO: new pathways and clinical perspectives. Antioxid Redox Signal 14:1039. doi:10.1089/ars.2016.6690

Glaser PE, Gross RW (1995) Rapid plasmenylethanolamine-selective fusion of membrane bilayers catalyzed by an isoform of glyceraldehyde-3-phosphate dehydrogenase: discrimination between glycolytic and fusogenic roles of individual isoforms. Biochemistry 34:12193–12203. doi:10.1021/bi00038a013

Glaser PE, Han X, Gross RW (2002) Tubulin is the endogenous inhibitor of the glyceraldehyde 3-phosphate dehydrogenase isoform that catalyzes membrane fusion: implications for the coordinated regulation of glycolysis and membrane fusion. Proc Natl Acad Sci 99:14104–14109. doi:10.1073/pnas.222542999

Grant CM, Quinn KA, Dawes IW (1999) Differential protein S-thiolation of glyceraldehyde-3-phosphate dehydrogenase isoenzymes influences sensitivity to oxidative stress. Mol Cell Biol 19:2650–2656. doi:10.1128/MCB.19.4.2650

Graven KK, Troxler RF, Kornfeld H et al (1994) Regulation of endothelial cell glyceraldehyde-3-phosphate dehydrogenase expression by hypoxia. J Biol Chem 269:24446–24453

Grosse F, Nasheuer H, Scholtissek S, Schomburg U (1986) Lactate dehydrogenase and glyceraldehyde-phosphate dehydrogenase are single-stranded DNA-binding proteins that affect the DNA-polymerase-alpha-primase comple15. Eur J Biochem 160:459–467

Han X, Ramanadham S, Turk J, Gross RW (1998) Reconstitution of membrane fusion between pancreatic islet secretory granules and plasma membranes: catalysis by a protein constituent recognized by monoclonal antibodies directed against glyceraldehyde-3-phosphate dehydrogenase. Biochim Biophys Acta Biomembr 1414:95–107. doi:10.1016/S0005-2736(98)00154-0

Hannibal L, Collins D, Brassard J et al (2012) Heme binding properties of glyceraldehyde-3-phosphate dehydrogenase. Biochemistry 51:8514–8529. doi:10.1021/bi300863a

Hara MR, Snyder SH (2006) Nitric oxide-GAPDH-Siah: a novel cell death cascade. Cell Mol Neurobiol 26:527–538

Hara MR, Agrawal N, Kim SF et al (2005) S-nitrosylated GAPDH initiates apoptotic cell death by nuclear translocation following Siah1 binding. Nat Cell Biol 7:665–674

Hara MR, Thomas B, Cascio MB et al (2006) Neuroprotection by pharmacologic blockade of the GAPDH death cascade. Proc Natl Acad Sci U S A 103:3887–3889. doi:10.1073/pnas.0511321103

Harada N, Yasunaga R, Higashimura Y et al (2007) Glyceraldehyde-3-phosphate dehydrogenase enhances transcriptional activity of androgen receptor in prostate cancer cells. J Biol Chem 282:22651–22661. doi:10.1074/jbc.M610724200

Harraz M, Snyder S (2015) Nitric oxide-GAPDH transcriptional signaling mediates behavioral actions of cocaine. Former Curr Drug Targets CNS Neurol Disord 14:757–763. doi:10.2174/1871527314666150529150143

Hessler RJ, Blackwood RA, Brock TG et al (1998) Identification of glyceraldehyde-3-phosphate dehydrogenase as a Ca2+-dependent fusogen in human neutrophil cytosol. J Leukoc Biol 63:331–336

Hildebrandt T, Knuesting J, Berndt C et al (2015) Cytosolic thiol switches regulating basic cellular functions: GAPDH as an information hub? Biol Chem 396:523. doi:10.1515/hsz-2014-0295

Huang T-S, Nagy PD (2011) Direct inhibition of tombusvirus plus-strand RNA synthesis by a dominant negative mutant of a host metabolic enzyme, glyceraldehyde-3-phosphate dehydrogenase, in yeast and plants. J Virol 85:9090–9102. doi:10.1128/JVI.00666-11

Huang J, Hao L, Xiong N et al (2009) Involvement of glyceraldehyde-3-phosphate dehydrogenase in rotenone-induced cell apoptosis: relevance to protein misfolding and aggregation. Brain Res 1279:1–8. doi:10.1016/j.brainres.2009.05.011

Huang Q, Lan F, Zheng Z et al (2011) Akt2 kinase suppresses Glyceraldehyde-3-phosphate Dehydrogenase (GAPDH)-mediated apoptosis in ovarian cancer cells via phosphorylating GAPDH at threonine 237 and decreasing its nuclear translocation. J Biol Chem 286:42211–42220. doi:10.1074/jbc.M111.296905

Huitorel P, Pantaloni D (1985) Bundling of microtubules by glyceraldehyde-3-phosphate dehydrogenase and its modulation by ATP. Eur J Biochem 150:265–269

Ikeda Y, Yamaji R, Irie K et al (2012) Glyceraldehyde-3-phosphate dehydrogenase regulates cyclooxygenase-2 expression by targeting mRNA stability. Arch Biochem Biophys 528:141–147. doi:10.1016/j.abb.2012.09.004

Ishitani R, Chuang DM (1996) Glyceraldehyde-3-phosphate dehydrogenase antisense oligodeoxynucleotides protect against cytosine arabinonucleoside-induced apoptosis in cultured cerebellar neurons. Proc Natl Acad Sci U S A 93:9937–9941

Ishitani R, Sunaga K, Hirano A et al (1996) Evidence that glyceraldehyde-3-phosphate dehydrogenase is involved in age-induced apoptosis in mature cerebellar neurons in culture. J Neurochem 66:928–935. doi:10.1046/j.1471-4159.1996.66030928.x

Ishitani R, Sunaga K, Tanaka M et al (1997) Overexpression of glyceraldehyde-3-phosphate dehydrogenase is involved in low K+-induced apoptosis but not necrosis of cultured cerebellar granule cells. Mol Pharmacol 51:542–550. doi:10.1124/mol.51.4.542

Ishitani R, Tanaka M, Sunaga K et al (1998) Nuclear localization of overexpressed glyceraldehyde-3-phosphate dehydrogenase in cultured cerebellar neurons undergoing apoptosis. Mol Pharmacol 53:701–707. doi:10.1124/mol.53.4.701

Ishitani R, Tajima H, Takata H et al (2003) Proapoptotic protein glyceraldehyde-3-phosphate dehydrogenase: a possible site of action of antiapoptotic drugs. Prog Neuro-Psychopharmacol Biol Psychiatry 27:291–301. doi:10.1016/S0278-5846(03)00024-1

Ismail SA, Park HW (2005) Structural analysis of human liver glyceraldehyde-3-phosphate dehydrogenase. Acta Crystallogr D Biol Crystallogr 61:1508–1513

Isupov MN, Fleming TM, Dalby AR et al (1999) Crystal structure of the glyceraldehyde-3-phosphate dehydrogenase from the hyperthermophilic archaeon Sulfolobus solfataricus. J Mol Biol 291:651–660. doi:10.1006/jmbi.1999.3003

Jacob F, Guertler R, Naim S et al (2013) Careful selection of reference genes is required for reliable performance of RT-qPCR in human normal and cancer cell lines. PLoS ONE 8:e59180. doi:10.1371/journal.pone.0059180

Jaffrey S, Erdjument-Bromage H, Ferris C et al (2001) Protein S-nitrosylation: a physiological signal for neuronal nitric oxide. Nat Cell Biol 3:193–197. doi:10.1038/35055104

Jarosz AP, Wei W, Gauld JW et al (2015) Glyceraldehyde 3-phosphate dehydrogenase (GAPDH) is inactivated by S-sulfuration in vitro. Free Radic Biol Med 89:512–521. doi:10.1016/j.freeradbiomed.2015.09.007

Jenkins JL, Tanner JJ (2006) High-resolution structure of human D-glyceraldehyde-3-phosphate dehydrogenase. Acta Crystallogr D Biol Crystallogr 62:290–301

Jia J, Arif A, Willard B et al (2012) Protection of extraribosomal RPL13a by GAPDH and dysregulation by S-nitrosylation. Mol Cell 47:656–663 doi: http://d15.doi.org/10.1016/j.molcel.2012.06.006

Kaneda M, Takeuchi K, Inoue K, Umeda M (1997) Localization of the phosphatidylserine-binding site of glyceraldehyde-3-phosphate dehydrogenase responsible for membrane fusion. J Biochem 122:1233–1240

Kang HT, Hwang ES (2009) Nicotinamide enhances mitochondria quality through autophagy activation in human cells. Aging Cell 8:426–438. doi:10.1111/j.1474-9726.2009.00487.x

Kang HT, Lee HI, Hwang ES (2006) Nicotinamide extends replicative lifespan of human cells. Aging Cell 5:423–436. doi:10.1111/j.1474-9726.2006.00234.x

Karpel RL, Burchard AC (1981) A basic isozyme of yeast glyceraldehyde-3-phosphate dehydrogenase with nucleic acid helix-destabilizing activity. Biochim Biophys Acta Nucl Acids Protein Synth 654:256–267 doi: http://d15.doi.org/10.1016/0005-2787(81)90180-5

Kawamoto R, Caswell A (1986) Autophosphorylation of glyceraldehydephosphate dehydrogenase and phosphorylation of protein from skeletal muscle microsome. Biochemistry 25:656–661

Kim H, Hol WGJ (1998) Crystal structure of Leishmania mexicana glycosomal glyceraldehyde-3-phosphate dehydrogenase in a new crystal form confirms the putative physiological active site structure. J Mol Biol 278:5–11. doi:10.1006/jmbi.1998.1661

Kim H, Feil IK, Verlinde CLMJ et al (1995) Crystal structure of glycosomal glyceraldehyde-3-phosphate dehydrogenase from Leishmania mexicana: implications for structure-based drug design and a new position for the inorganic phosphate binding site. Biochemistry 34:14975–14986. doi:10.1021/bi00046a004

Kim JH, Lee S, Park JB et al (2003) Hydrogen peroxide induces association between glyceraldehyde 3-phosphate dehydrogenase and phospholipase D2 to facilitate phospholipase D2 activation in PC12 cells. J Neurochem 85:1228–1236. doi:10.1046/j.1471-4159.2003.01755.x

Kiri A, Goldspink G (2002) RNA–protein interactions of the 3′ untranslated regions of myosin heavy chain transcripts. J Muscle Res Cell Motil 23:119–129. doi:10.1023/A:1020211729728

Kondo S, Kubota S, Mukudai Y et al (2011) Binding of glyceraldehyde-3-phosphate dehydrogenase to the cis-acting element of structure-anchored repression in ccn2 mRNA. Biochem Biophys Res Commun 405:382–387. doi:10.1016/j.bbrc.2011.01.034

Kornberg MD, Sen N, Hara MR et al (2010) GAPDH mediates nitrosylation of nuclear proteins. Nat Cell Biol 12:1094–1100. doi:10.1038/ncb2114

Korndörfer I, Steipe BS, Huber R et al (1995) The crystal structure of holo-glyceraldehyde-3-phosphate dehydrogenase from the hyperthermophilic bacterium thermotoga maritima at 2.5 Å resolution. J Mol Biol 246:511–521. doi:10.1006/jmbi.1994.0103

Kumagai H, Sakai H (1983) A porcine brain protein (35 K Protein) which bundles microtubules and its identification as glyceraldehyde 3-phosphate dehydrogenase. J Biochem 93:1259–1269

Kusov Y, Weitz M, Dollenmeier G et al (1996) RNA-protein interactions at the 3′ end of the hepatitis A virus RNA. J Virol 70:1890–1897

Kwon H, Rhim J, Jang I et al (2010) Activation of AMP-activated protein kinase stimulates the nuclear localization of glyceraldehyde 3-phosphate dehydrogenase in human diploid fibroblasts. Exp Mol Med 42:254–269. doi:10.3858/emm.2010.42.4.025

Ladame S, Castilho M, Silva C et al (2003) Crystal structure of trypanosoma cruzi glyceraldehyde-3-phosphate dehydrogenase complexed with an analogue of 1,3-bisphospho-d-glyceric acid. Eur J Biochem 270:4574–4586

Landino LM, Hagedorn TD, Kennett KL (2014) Evidence for thiol/disulfide exchange reactions between tubulin and glyceraldehyde-3-phosphate dehydrogenase: evidence for thiol/disulfide exchange reactions. Cytoskeleton 71:707–718. doi:10.1002/cm.21204

Lane DJR, Merlot AM, Huang ML-H et al (2015) Cellular iron uptake, trafficking and metabolism: key molecules and mechanisms and their roles in disease. Biochim Biophys Acta Mol Cell Res 1853:1130–1144. doi:10.1016/j.bbamcr.2015.01.021

Laschet JJ (2004) Glyceraldehyde-3-phosphate dehydrogenase is a GABAA receptor kinase linking glycolysis to neuronal inhibition. J Neurosci 24:7614–7622. doi:10.1523/JNEUROSCI.0868-04.2004

Launay JF, Jellali A, Vanier MT (1989) Glyceraldehyde-3-phosphate dehydrogenase is a microtubule binding protein in a human colon tumor cell line. Biochim Biophys Acta Protein Struct Mol Enzymol 996:103–109. doi:10.1016/0167-4838(89)90101-5

Lee SB, Kim CK, Lee K-H, Ahn J-Y (2012) S-nitrosylation of B23/nucleophosmin by GAPDH protects cells from the SIAH1–GAPDH death cascade. J Cell Biol 199:65–76. doi:10.1083/jcb.201205015

Lin S-S, Chang SC, Wang Y-H et al (2000) Specific interaction between the hepatitis delta virus RNA and glyceraldehyde 3-phosphate dehydrogenase: an enhancement on ribozyme catalysis. Virology 271:46–57. doi:10.1006/viro.2000.0302

Littlechild JA, Guy JE, Isupov MN (2004) Hyperthermophilic dehydrogenase enzymes. Biochem Soc Trans 32:255–258. doi:10.1042/bst0320255

Lowe SL, Adrian C, Ouporov IV et al (2003) Brownian dynamics simulations of glycolytic enzyme subsets with F-actin. Biopolymers 70:456–470. doi:10.1002/bip.10530

Malay AD, Bessho Y, Ellis MJ et al (2009) Structure of glyceraldehyde-3-phosphate dehydrogenase from the archaeal hyperthermophile *Methanocaldococcus jannaschii*. Acta Crystallogr Sect F Struct Biol Cryst Commun 65:1227–1233. doi:10.1107/S1744309109047046

Mansur NR, Meyer-Siegler K, Wurzer JC, Sirover MA (1993) Cell cycle regulation of the glyceraldehyde3phosphate dehydrogenaseluracil DNA glycosylase gene in normal human cells. Nucleic Acids Res 21:993–998. doi:10.1093/nar/21.4.993

Maris C, Dominguez C, Allain FH-T (2005) The RNA recognition motif, a plastic RNA-binding platform to regulate post-transcriptional gene expression: the RRM domain, a plastic RNA-binding platform. FEBS J 272:2118–2131. doi:10.1111/j.1742-4658.2005.04653.x

Mazzola JL, Sirover MA (2002) Alteration of nuclear glyceraldehyde-3-phosphate dehydrogenase structure in Huntington's disease fibroblasts. Mol Brain Res 100:95–101. doi:10.1016/S0169-328X(02)00160-2

Mazzola JL, Sirover MA (2003) Subcellular alteration of glyceraldehyde-3-phosphate dehydrogenase in Alzheimer's disease fibroblasts. J Neurosci Res 71:279–285. doi:10.1002/jnr.10484

McGowan K, Pekala PH (1996) Dehydrogenase binding to the 3′-untranslated region of GLUT1 mRNA. Biochem Biophys Res Commun 221:42–45. doi:10.1006/bbrc.1996.0541

Méjean C, Pons F, Benyamin Y, Roustan C (1989) Antigenic probes locate binding sites for the glycolytic enzymes glyceraldehyde-3-phosphate dehydrogenase, aldolase and phosphofructokinase on the actin monomer in microfilaments. Biochem J 264:671–677. doi:10.1042/bj2640671

Mezquita J, Pau M, Mezquita C (1998) Several novel transcripts of glyceraldehyde-3-phosphate dehydrogenase expressed in adult chicken testis. J Cell Biochem 71:127–139

Miki K, Qu W, Goulding E et al (2004) Glyceraldehyde 3-phosphate dehydrogenase-S, a sperm-specific glycolytic enzyme, is required for sperm motility and male fertility. Proc Natl Acad Sci 101:16501–16506. doi:10.1073/pnas.0407708101

Millet P, Vachharajani V, McPhail L et al (2016) GAPDH binding to TNF-alpha mRNA contributes to posttranscriptional repression in monocytes: a novel mechanism of communication between inflammation and metabolism. J Immunol 196:2541–2551. doi:10.4049/jimmunol.1501345

Modun B, Williams P (1999) The staphylococcal transferrin-binding protein is a cell wall glyceraldehyde-3-phosphate dehydrogenase. Infect Immun 67:1086–1092

Modun B, Morrissey J, Williams P (2000) The staphylococcal transferrin receptor: a glycolytic enzyme with novel functions. Trends Microbiol 8:231–237. doi:10.1016/S0966-842X(00)01728-5

Mohr S, Stamler JS, Brune B (1996) Posttranslational modification of glyceraldehyde-3-phosphate dehydrogenase by S-nitrosylation and subsequent NADH attachment. J Biol Chem 271:4209–4214

Molina y Vedia L, McDonald B, Reep B et al (1992) Nitric oxide-induced S-nitrosylation of glyceraldehyde-3-phosphate dehydrogenase inhibits enzymatic activity and increases endogenous ADP-ribosylation. J Biol Chem 267:24929–24932

Moniot S, Bruno S, Vonrhein C et al (2008) Trapping of the thioacylglyceraldehyde-3-phosphate dehydrogenase intermediate from *Bacillus stearothermophilus*: direct evidence for a flip-flop mechanism. J Biol Chem 283:21693–21702. doi:10.1074/jbc.M802286200

Moras D, Olsen KW, Sabesan MN et al (1975) Studies of asymmetry in the three-dimensional structure of lobster D-glyceraldehyde-3-phosphate dehydrogenase. J Biol Chem 250:9137–9162

Muronetz VI, Wang ZX, Keith TJ et al (1994) Binding constants and stoichiometries of glyceraldehyde 3-phosphate dehydrogenase-tubulin complexes. Arch Biochem Biophys 313:253–260. doi:10.1006/abbi.1994.1385

Mustafa AK, Gadalla MM, Sen N et al (2009) H2S signals through protein S-sulfhydration. Sci Signal 2:ra72. doi:10.1126/scisignal.2000464

Nagy E, Rigby WFC (1995) Glyceraldehyde-3-phosphate dehydrogenase selectively binds AU-rich RNA in the NAD+-binding region (Rossmann fold). J Biol Chem 270:2755–2763

Nakagawa T, Hirano Y, Inomata A et al (2003) Participation of a fusogenic protein, glyceraldehyde-3-phosphate dehydrogenase, in nuclear membrane assembly. J Biol Chem 278:20395–20404. doi:10.1074/jbc.M210824200

Nakajima H, Amano W, Fujita A et al (2007) The active site cysteine of the proapoptotic protein glyceraldehyde-3-phosphate dehydrogenase is essential in oxidative stress-induced aggregation and cell death. J Biol Chem 282:26562–26574

Nakajima H, Amano W, Kubo T et al (2009) Glyceraldehyde-3-phosphate dehydrogenase aggregate formation participates in oxidative stress-induced cell death. J Biol Chem 284:34331–34341. doi:10.1074/jbc.M109.027698

Nicholls C, Pinto AR, Li H et al (2012) Glyceraldehyde-3-phosphate dehydrogenase (GAPDH) induces cancer cell senescence by interacting with telomerase RNA component. Proc Natl Acad Sci 109:13308–13313. doi:10.1073/pnas.1206672109

Nüesch JPF, Weitz M, Siegl G (1993) Proteins specifically binding to the 3′ untranslated region of hepatitis A virus RNA in persistently infected cells. Arch Virol 128:65–79. doi:10.1007/BF01309789

Ouporov IV, Keith TJ, Knull HR, Thomasson KA (2000) Computer simulations of glycolytic enzyme interactions with F-actin. J Biomol Struct Dyn 18:311–323. doi:10.1080/07391102.2000.10506668

Ouporov IV, Knull HR, Lowe SL, Thomasson KA (2001) Interactions of glyceraldehyde-3-phosphate dehydrogenase with G- and F-actin predicted by Brownian dynamics. J Mol Recognit 14:29–41. doi:10.1002/1099-1352(200101/02)14:1<29::AID-JMR517>3.0.CO;2-T

Pancholi V, Fischetti VA (1992) A major surface protein on group A streptococci is a glyceraldehyde-3-phosphate-dehydrogenase with multiple binding activity. J Exp Med 176:415–426. doi:10.1084/jem.176.2.415

Park J, Han D, Kim K et al (2009) O-GlcNAcylation disrupts glyceraldehyde-3-phosphate dehydrogenase homo-tetramer formation and mediates its nuclear translocation. Biochim Biophys Acta Proteins Proteomics 1794:254–262. doi:10.1016/j.bbapap.2008.10.003

Persons DA, Schek N, Hall BL, Finn OJ (1989) Increased expression of glycolysis-associated genes in oncogene-transformed and growth-accelerated states. Mol Carcinog 2:88–94

Perucho M, Salas J, Salas M (1977) Identification of mammalian DNA-binding protein P8 as glyceraldehyde-3-phosphate dehydrogenase. Eur J Bochem 81:557–562

Petrik J, Parker H, Alexander GJM (1999) Human hepatic glyceraldehyde-3-phosphate dehydrogenase binds to the poly(U) tract of the 3′ non-coding region of hepatitis C virus genomic RNA. J Gen Virol 80:3109–3113. doi:10.1099/0022-1317-80-12-3109

Piechaczyk M, Blanchard J, Sabouty S et al (1984) Unusual abundance of vertebrate 3-phosphate dehydrogenase pseudogenes. Nature 312:469–471

Pierce A, Mirzaei H, Muller F et al (2008) GAPDH is conformationally and functionally altered in association with oxidative stress in mouse models of amyotrophic lateral sclerosis. J Mol Biol 382:1195–1210

Polati R, Castagna A, Bossi AM et al (2012) Murine macrophages response to iron. J Proteome 76:10–27. doi:10.1016/j.jprot.2012.07.018

Puchulu-Campanella E, Chu H, Anstee DJ et al (2013) Identification of the components of a glycolytic enzyme metabolon on the human red blood cell membrane. J Biol Chem 288:848–858. doi:10.1074/jbc.M112.428573

Puder M, Soberman RJ (1997) Glutathione conjugates recognize the Rossmann fold of glyceraldehyde-3-phosphate dehydrogenase. J Biol Chem 272:10936–10940. doi:10.1074/jbc.272.16.10936

Quail EA, Yeoh GCT (1995) The effect of iron status on glyceraldehyde 3-phosphate dehydrogenase expression in rat liver. FEBS Lett 359:126–128. doi:10.1016/0014-5793(95)00023-3

Raje CI, Kumar S, Harle A et al (2007) The macrophage cell surface glyceraldehyde-3-phosphate dehydrogenase is a novel transferrin receptor. J Biol Chem 282:3252–3261. doi:10.1074/jbc.M608328200

Ralser M, Wamelink MM, Kowald A et al (2007) Dynamic rerouting of the carbohydrate flux is key to counteracting oxidative stress. J Biol 6:10. doi:10.1186/jbiol61

Randall G, Panis M, Cooper JD et al (2007) Cellular cofactors affecting hepatitis C virus infection and replication. Proc Natl Acad Sci USA 104:12884–12889

Ravichandran V, Seres T, Moriguchi T et al (1994) S-thiolation of glyceraldehyde-3-phosphate dehydrogenase induced by the phagocytosis-associated respiratory burst in blood monocytes. J Biol Chem 269:25010–25015

Rawat P, Kumar S, Sheokand N et al (2012) The multifunctional glycolytic protein glyceraldehyde-3-phosphate dehydrogenase (GAPDH) is a novel macrophage lactoferrin receptor. Biochem Cell Biol 90:329–338. doi:10.1139/o11-058

Robbins AR, Ward RD, Oliver C (1995) A mutation in glyceraldehyde 3-phosphate dehydrogenase alters endocytosis in CHO cells. J Cell Biol 130:1093–1104. doi:10.1083/jcb.130.5.1093

Rodriguez-Pascual F, Redondo-Horcajo M, Magan-Marchal N et al (2008) Glyceraldehyde-3-phosphate dehydrogenase regulates endothelin-1 expression by a novel, redox-sensitive mechanism involving mRNA stability. Mol Cell Biol 28:7139–7155

Rogalski-Wilk AA, Cohen RS (1997) Glyceraldehyde-3-phosphate dehydrogenase activity and F-actin associations in synaptosomes and postsynaptic densities of porcine cerebral corte15. Cell Mol Neurobiol 17:51–70. doi:10.1023/A:1026377004261

Ryazanov AG (1985) Glyceraldehyde-3-phosphate dehydrogenase is one of the three major RNA-binding proteins of rabbit reticulocytes. FEBS Lett 192:131–134. doi:10.1016/0014-5793(85)80058-2

Samson AL, Knaupp AS, Kass I et al (2014) Oxidation of an exposed methionine instigates the aggregation of glyceraldehyde-3-phosphate dehydrogenase. J Biol Chem 289:26922–26936. doi:10.1074/jbc.M114.570275

Sarkar A, Dai Y, Haque MM et al (2015) Heat shock protein 90 associates with the Per-Arnt-Sim domain of Heme-free soluble guanylate cyclase: implications for enzyme maturation. J Biol Chem 290:21615–21628. doi:10.1074/jbc.M115.645515

Satchell JF, Malby RL, Luo CS et al (2005) Structure of glyceraldehyde-3-phosphate dehydrogenase from *Plasmodium falciparum*. Acta Crystallogr D Biol Crystallogr 61:1213–1221. doi:10.1107/S0907444905018317

Saunders PA, Chalecka-Franaszek E, Chuang D-M (1997) Subcellular distribution of glyceraldehyde-3-phosphate dehydrogenase in cerebellar granule cells undergoing cytosine arabinoside-induced apoptosis. J Neurochem 69:1820–1828. doi:10.1046/j.1471-4159.1997.69051820.x

Sawa A, Khan AA, Hester LD, Snyder SH (1997) Glyceraldehyde-3-phosphate dehydrogenase: nuclear translocation participates in neuronal and nonneuronal cell death. Proc Natl Acad Sci 94:11669–11674

Schmitz H (2001) Reversible nuclear translocation of glyceraldehyde-3-phosphate dehydrogenase upon serum depletion. Eur J Cell Biol 80:419–427

Schmitz H, Bereiter-Hahn J (2002) Glyceraldehyde-3-phosphate dehydrogenase associates with actin filaments in serum deprived NIH 3 T3 cells only. Cell Biol Int 26:155–164. doi:10.1006/cbir.2001.0819

Schneider ML, Post CB (1995) Solution structure of a band 3 peptide inhibitor bound to aldolase: a proposed mechanism for regulating binding by tyrosine phosphorylation. Biochemistry 34:16574–16584. doi:10.1021/bi00051a005

Schultz D, Hardin C, Lemon S (1996) Specific interaction of glyceraldehyde 3-phopshate dehydrogenase with the 5′-nontranslated RNA of hepatitis A virus. J Biol Chem 271:14134–14142

Schuppe-Koistinen I, Moldéus P, Bergman T, Cotgreave I (1994) S-thiolation of human endothelial cell glyceraldehyde-3-phosphate dehydrogenase after hydrogen peroxide treatment. Eur J Biochem 221:1033–1037

Sen N, Hara MR, Kornberg MD et al (2008) Nitric oxide-induced nuclear GAPDH activates p300/CBP and mediates apoptosis. Nat Cell Biol 10:866–873

Sen N, Hara MR, Ahmad AS et al (2009) GOSPEL: a neuroprotective protein that binds to GAPDH upon S-nitrosylation. Neuron 63:81–91. doi:10.1016/j.neuron.2009.05.024

Sergienko EA, Kharitonenkov AI, Bulargina TV et al (1992) d-Glyceraldehyde-3-phosphate dehydrogenase purified from rabbit muscle contains phosphotyrosine. FEBS Lett 304:21–23. doi:10.1016/0014-5793(92)80580-A

Sheedy R, Clarke F (2001) Predicting interaction sites between glycolytic enzymes and cytoskeletal proteins employing the concepts of the molecular recognition theory. In: Thomas DD, dos Remedios CG (eds) Molecular interactions of actin. Actin structure and actin-binding proteins. Springer-Verlag, Heidelberg, pp 155–164

Sheokand N, Malhotra H, Kumar S et al (2014) Moonlighting cell-surface GAPDH recruits apotransferrin to effect iron egress from mammalian cells. J Cell Sci 127:4279–4291. doi:10.1242/jcs.154005

Shiozawa M, Fukutani Y, Arai N et al (2003) Glyceraldehyde 3-phosphate dehydrogenase and endothelin-1 immunoreactivity is associated with cerebral white matter damage in dentatorubral–pallidoluysian atrophy. Neuropathology 23:36–43. doi:10.1046/j.1440-1789.2003.00480.x

Silva B, Faustino P (2015) An overview of molecular basis of iron metabolism regulation and the associated pathologies. Biochim Biophys Acta (BBA) - Mol Basis Dis 1852:1347–1359. doi:10.1016/j.bbadis.2015.03.011

Singh R, Green M (1993) Sequence-specific binding of transfer RNA by glyceraldehyde-3-phosphate dehydrogenase. Science 259:365–368. doi:10.1126/science.8420004

Sioud M, Jespersen L (1996) Enhancement of hammerhead ribozyme catalysis by glyceraldehyde-3-phosphate dehydrogenase. J Mol Biol 257:775–789. doi:10.1006/jmbi.1996.0201

Sirover MA (2012) Subcellular dynamics of multifunctional protein regulation: mechanisms of GAPDH intracellular translocation. J Cell Biochem 113:2193–2200. doi:10.1002/jcb.24113

Sirover MA (2014) Structural analysis of glyceraldehyde-3-phosphate dehydrogenase functional diversity. Int J Biochem Cell Biol 57:20–26. doi:10.1016/j.biocel.2014.09.026

Skarżyński T, Wonacott AJ (1988) Coenzyme-induced conformational changes in glyceraldehyde-3-phosphate dehydrogenase from *Bacillus stearothermophilus*. J Mol Biol 203:1097–1118. doi:10.1016/0022-2836(88)90130-1

Sneve ML, Øverbye A, Fengsrud M, Seglen PO (2005) Comigration of two autophagosome-associated dehydrogenases on two-dimensional polyacrylamide gels. Autophagy 1:157–162. doi:10.4161/auto.1.3.2037

Somers M, Engelborghs Y, Baert J (1990) Analysis of the binding of glyceraldehyde-3-phosphate dehydrogenase to microtubules, the mechanism of bundle formation and the linkage effect. Eur J Biochem 193:437–444

Song S, Finkel T (2007) GAPDH and the search for alternative energy. Nat Cell Biol 9:869–870. doi:10.1038/ncb0807-869

Stone E, Rothblum K, Alevy M et al (1985) Complete sequence of the chicken glyceraldehyde-3-phosphate dehydrogenase gene. Proc Natl Acad Sci 82:1628–1632. doi:10.1073/pnas.82.6.1628

Su Y, Blake-Palmer KG, Fry AC et al (2011) Glyceraldehyde 3-phosphate dehydrogenase is required for band 3 (anion exchanger 1) membrane residency in the mammalian kidney. Am J Physiol Ren Physiol 300:F157–F166. doi:10.1152/ajprenal.00228.2010

Sundararaj KP, Wood RE, Ponnusamy S et al (2004) Rapid shortening of telomere length in response to ceramide involves the inhibition of telomere binding activity of nuclear glyceraldehyde-3-phosphate dehydrogenase. J Biol Chem 279:6152–6162

Suresh S, Bressi JC, Kennedy KJ et al (2001) Conformational changes in Leishmania mexicana glyceraldehyde-3-phosphate dehydrogenase induced by designed inhibitors. J Mol Biol 309:423–435. doi:10.1006/jmbi.2001.4588

Talfournier F, Colloc'h N, Mornon JP, Branlant G (1998) Comparative study of the catalytic domain of phosphorylating glyceraldehyde-3-phosphate dehydrogenases from bacteria and archaea via essential cysteine probes and site-directed mutagenesis. Eur J Biochem 252:447–457

Tarze A, Deniaud A, Le Bras M et al (2007) GAPDH, a novel regulator of the pro-apoptotic mitochondrial membrane permeabilization. Oncogene 26:2606–2620. doi:10.1038/sj.onc.1210074

Tatton NA (2000) Increased caspase 3 and bax immunoreactivity accompany nuclear GAPDH translocation and neuronal apoptosis in Parkinson's disease. Exp Neurol 166:29–43. doi:10.1006/exnr.2000.7489

Taylor J, Heinrichs D (2002) Transferrin binding in *Staphylococcus aureus*: involvement of a cell wall-anchored protein. Mol Microbiol 43:1603–1614

Tisdale EJ (2001) Glyceraldehyde-3-phosphate dehydrogenase is required for vesicular transport in the early secretory pathway. J Biol Chem 276:2480–2486

Tisdale EJ (2002) Glyceraldehyde-3-phosphate dehydrogenase is phosphorylated by protein kinase C/and plays a role in microtubule dynamics in the early secretory pathway. J Biol Chem 277:3334–3341. doi:10.1074/jbc.M109744200

Tisdale EJ, Artalejo CR (2006) Src-dependent a protein kinase C/(aPKC /) tyrosine phosphorylation is required for aPKC/association with Rab2 and glyceraldehyde-3-phosphate dehydrogenase on pre-golgi intermediates. J Biol Chem 281:8436–8442. doi:10.1074/jbc.M513031200

Tisdale EJ, Artalejo CR (2007) A GAPDH mutant defective in Src-dependent tyrosine phosphorylationimpedesRab2-mediatedevents.Traffic8:733–741.doi:10.1111/j.1600-0854.2007.00569.x

Tisdale EJ, Kelly C, Artalejo CR (2004) Glyceraldehyde-3-phosphate dehydrogenase Interacts with Rab2 and plays an essential role in endoplasmic reticulum to golgi transport exclusive of its glycolytic activity. J Biol Chem 279:54046–54052. doi:10.1074/jbc.M409472200

Tisdale EJ, Azizi F, Artalejo CR (2009) Rab2 utilizes glyceraldehyde-3-phosphate dehydrogenase and protein kinase C to associate with microtubules and to recruit dynein. J Biol Chem 284:5876–5884. doi:10.1074/jbc.M807756200

Tokunaga K, Nakamura Y, Sakata K et al (1987) Enhanced expression of a glyceraldehyde-3-phosphate dehydrogenase gene in human lung cancers. Cancer Res 47:5616–5619

Tristan C, Shahani N, Sedlak TW, Sawa A (2011) The diverse functions of GAPDH: views from different subcellular compartments. Cell Signal 23:317–323. doi:10.1016/j.cellsig.2010.08.003

Tsai IH, Murthy SN, Steck TL (1982) Effect of red cell membrane binding on the catalytic activity of glyceraldehyde-3-phosphate dehydrogenase. J Biol Chem 257:1438–1442

Tsuchiya K, Tajima H, Yamada M et al (2004) Disclosure of a pro-apoptotic glyceraldehyde-3-phosphate dehydrogenase promoter: anti-dementia drugs depress its activation in apoptosis. Life Sci 74:3245–3258. doi:10.1016/j.lfs.2003.11.029

Ventura M, Mateo F, Serratosa J et al (2010) Nuclear translocation of glyceraldehyde-3-phosphate dehydrogenase is regulated by acetylation. Int J Biochem Cell Biol 42:1672–1680 doi: http://d15.doi.org/10.1016/j.biocel.2010.06.014

Verlinde CLMJ, Callens M, Van Calenbergh S et al (1994) Selective inhibition of trypanosomal glyceraldehyde-3-phosphate dehydrogenase by protein structure-based design: toward new drugs for the treatment of sleeping sickness. J Med Chem 37:3605–3613. doi:10.1021/jm00047a017

Volker KW, Knull HR (1993) Glycolytic enzyme-tubulin interactions: role of tubulin carboxy terminals. J Mol Recognit 6:167–177. doi:10.1002/jmr.300060405

Volker KW, Knull HR (1997) A glycolytic enzyme binding domain on tubulin. Arch Biochem Biophys 338:237–243. doi:10.1006/abbi.1996.9819

Waingeh VF, Lowe SL, Thomasson KA (2004) Brownian dynamics of interactions between glyceraldehyde-3-phosphate dehydrogenase (GAPDH) mutants and F-actin. Biopolymers 73:533–541. doi:10.1002/bip.10560

Walsh JL, Keith TJ, Knull HR (1989) Glycolytic enzyme interactions with tubulin and microtubules. Biochim Biophys Acta Protein Struct Mol Enzymol 999:64–70. doi:10.1016/0167-4838(89)90031-9

Wang RY, Nagy PD (2008) Tomato bushy stunt virus co-opts the RNA-binding function of a host metabolic enzyme for viral genomic RNA synthesis. Cell Host Microbe 3:178–187

White MR, Garcin ED (2016) The sweet side of RNA regulation: glyceraldehyde-3-phosphate dehydrogenase as a noncanonical RNA-binding protein. Wiley Interdiscip Rev RNA 7:53–70. doi:10.1002/wrna.1315

White MR, Khan MM, Deredge D et al (2015) A dimer interface mutation in glyceraldehyde-3-phosphate dehydrogenase regulates its binding to AU-rich RNA. J Biol Chem 290:1770–1785. doi:10.1074/jbc.M114.618165

Yamaji R, Chatani E, Harada N et al (2005) Glyceraldehyde-3-phosphate dehydrogenase in the extracellular space inhibits cell spreading. Biochim Biophys Acta Gen Subj 1726:261–271. doi:10.1016/j.bbagen.2005.07.013

Yang S-H, Liu M-L, Tien C-F et al (2009) Glyceraldehyde-3-phosphate dehydrogenase (GAPDH) interaction with 3′ ends of Japanese encephalitis virus RNA and colocalization with the viral NS5 protein. J Biomed Sci 16:40. doi:10.1186/1423-0127-16-40

Yego ECK, Mohr S (2010) siah-1 protein is necessary for high glucose-induced glyceraldehyde-3-phosphate dehydrogenase nuclear accumulation and cell death in Muller cells. J Biol Chem 285:3181–3190. doi:10.1074/jbc.M109.083907

Yego ECK, Vincent JA, Sarthy V et al (2009) Differential regulation of high glucose–induced glyceraldehyde-3-phosphate dehydrogenase nuclear accumulation in Müller cells by IL-1β and IL-6. Invest Opthalmol Vis Sci 50:1920. doi:10.1167/iovs.08-2082

Yi M, Schultz DE, Lemon SM (2000) Functional significance of the interaction of Hepatitis A virus RNA with Glyceraldehyde 3-Phosphate Dehydrogenase (GAPDH): opposing effects of GAPDH and polypyrimidine tract binding protein on internal ribosome entry site function. J Virol 74:6459–6468. doi:10.1128/JVI.74.14.6459-6468.2000

Yuan A, Mills RG, Bamburg JR, Bray JJ (1999) Cotransport of glyceraldehyde-3-phosphate dehydrogenase and actin in axons of chicken motoneurons. Cell Mol Neurobiol 19:733–744. doi:10.1023/A:1006953022763

Yun M, Park C-G, Kim J-Y, Park H-W (2000) Structural analysis of glyceraldehyde 3-phosphate dehydrogenase from *Escherichia coli*: direct evidence of substrate binding and cofactor-induced conformational changes. Biochemistry 39:10702–10710. doi:10.1021/bi9927080

Zeng T, Dong Z-F, Liu S-J et al (2014) A novel variant in the 3′ UTR of human SCN1A gene from a patient with Dravet syndrome decreases mRNA stability mediated by GAPDH's binding. Hum Genet 133:801–811. doi:10.1007/s00439-014-1422-8

Zheng L, Roeder RG, Luo Y (2003) S phase activation of the histone H2B promoter by OCA-S, a coactivator complex that contains GAPDH as a key component. Cell 114:255–266

Zhou Y, Yi X, Stoffer JB et al (2008) The multifunctional protein glyceraldehyde-3-phosphate dehydrogenase is both regulated and controls colony-stimulating factor-1 messenger RNA stability in ovarian cancer. Mol Cancer Res 6:1375–1384

Chapter 16
Protein Complexes in the Nucleus: The Control of Chromosome Segregation

Victor M. Bolanos-Garcia

Abstract Mistakes in the process of cell division can lead to the loss, gain or rearrangement of chromosomes. Significant chromosomal abnormalities are usually lethal to the cells and cause spontaneous miscarriages. However, in some cases, defects in the spindle assembly checkpoint lead to severe diseases, such as cancer and birth and development defects, including Down's syndrome. The timely and accurate control of chromosome segregation in mitosis relies on the spindle assembly checkpoint (SAC), an evolutionary conserved, self-regulated signalling system present in higher organisms. The spindle assembly checkpoint is orchestrated by dynamic interactions between spindle microtubules and the kinetochore, a multiprotein complex that constitutes the site for attachment of chromosomes to microtubule polymers to pull sister chromatids apart during cell division. This chapter discusses the current molecular understanding of the essential, highly dynamic molecular interactions underpinning spindle assembly checkpoint signalling and how the complex choreography of interactions can be coordinated in time and space to finely regulate the process. The potential of targeting this signalling pathway to interfere with the abnormal segregation of chromosomes, which occurs in diverse malignancies and the new opportunities that recent technological developments are opening up for a deeper understanding of the spindle assembly checkpoint are also discussed.

Keywords Spindle assembly checkpoint (SAC) • Kinetochore • Genome instability • Chromosome segregation • Cancer • Protein complexes • Folding upon binding • Disorder-to-order transitions • KMN network

---

V.M. Bolanos-Garcia (✉)
Faculty of Health and Life Sciences, Department of Biological and Medical Sciences, Oxford Brookes University, Oxford OX3 0BP, UK
e-mail: vbolanos-garcia@brookes.ac.uk

J.R. Harris, J. Marles-Wright (eds.), *Macromolecular Protein Complexes*, Subcellular Biochemistry 83, DOI 10.1007/978-3-319-46503-6_16

Abbreviations

APC/C	Anaphase Promoting Complex/Cyclosome
ATP	Adenosine triphosphate
Bub1	Budding uninhibited by benzimidazoles 1
BubR1	Budding uninhibited by benzimidazoles related 1
CCAN	Centromeric nucleosome-associated network
Cdc20	Cell division cycle protein 20
CENP-E	Centromere-associated protein E
CENP-F	Centromere-associated protein F
CIN	Chromosomal instability
CLASP-1	CLIP-associating protein 1
CLASP-2	CLIP-associating protein 2
FRET	Förster resonance energy transfer
KMN	KNL1/Mis12/Ndc80 network
KNL1	Kinetochore-null phenotype 1
Mad1	Mitotic arrest deficient 1
Mad2	Mitotic arrest deficient 2
MIND complex	Mis12 complex
Mps1	Monopolar spindle 1
NMR	Nuclear magnetic resonance
ROD	Rough deal
RZZ-complex	Rod, Zwilch and ZW10 complex
SAC	Spindle assembly checkpoint
SAXS	Small angle x-ray scattering
Spc105	Spindle pole body 105
Spc105-related	Spc105R
TEM	cryo-transmission electron microscopy
XFEL	Ultrafast X-ray free-electron laser
ZW10	Zeste-white 10

16.1 Introduction

16.1.1 The SAC-KMN axis

The accurate and timely segregation of chromosomes during mitosis requires the formation of a bipolar mitotic spindle with stably attached chromosomes. Once all of the chromosomes are aligned properly, the connection between the sister chromatids is severed by the action of separase, a cysteine protease. Separase also contributes to centriole disengagement at the end of mitosis. Temporal and spatial coordination of these two activities with the rest of the cell cycle is required for the successful completion of mitosis. The accurate segregation of chromosomes when cells divide is ensured by the spindle assembly checkpoint (SAC), a highly intricate regulatory mechanism that monitors and corrects defects in chromosome attachment to the metaphase plate. The SAC is controlled by a panel of protein kinases that includes Bub1, BubR1, Mps1 and Aurora B and the non-kinase proteins Mad1, Mad2, Bub3, BuGZ and Cdc20.

16.1.2 Multidomain Protein Kinases Regulate the SAC

Budding uninhibited by benzimidazoles 1 (Bub1), Budding uninhibited by benzimidazoles related 1 (BubR1), dual-specificity kinase Monopolar spindle 1 (Mps1) and Aurora B are multidomain serine/threonine protein kinases with essential roles in the SAC signalling pathway in higher organisms (Krenn and Musacchio 2015; Elowe 2011; Musacchio 2011; Boyarchuk et al. 2007; Abrieu et al. 2001). These SAC proteins have evolutionarily conserved catalytic domains in organisms ranging from budding and fission yeasts to worms to humans (Bavetsias and Linardopoulos 2015; Bolanos-Garcia and Blundell 2011). For instance, Bub1 is required for the proper assembly of the inner centromere (Boyarchuk et al. 2007). Phosphorylation of the kinetochore organiser protein KNL1 by Mps1 is required for the recruitment of Bub1, BubR1 and Bub3 to the kinetochore while Mps1 from fission yeast (known as Mph1 in this specie) phosphorylates Mad3 to inhibit Cdc20 (known as Slp1 in fission yeast) and this post-translational modification appears important to maintain SAC arrest (Zich et al. 2016). It has been proposed that Mps1 compete with microtubules to bind at kinetochores and that such competitive binding contributes to regulate SAC signalling (Hiruma et al. 2015; Ji et al. 2015). Whether additional interactions mediated by Mps1, microtubules and/or the Ndc80 complex are involved in this process is an aspect that remains to be clarified (Aravamudhan et al. 2015; Krenn and Musacchio 2015; Nilsson 2015).

The proteins Mitotic arrest deficient 1 and 2 (Mad1 and Mad2, respectively); Budding uninhibited by benzimidazoles 3 (Bub3); Cell division cycle protein 20 (Cdc20); Bub3-interacting GLEBS-motif-containing ZNF207 (BuGZ); and Rod, Zwilch and ZW10, which define the RZZ-complex (Lara-Gonzalez et al. 2012), are also central components of the SAC. BubR1 (known as Mad3 in yeasts) interacts with Bub3, Mad2 and Cdc20 to form the Mitotic Checkpoint Complex (MCC; *see* Fig. 16.1), which inhibits the Anaphase Promoting Complex/Cyclosome (APC/C) to prevent metaphase-anaphase transition (Zhang et al. 2016; Musacchio 2011; Chao et al. 2012). Some of the key interactions underpinning SAC signalling are presented in Table 16.1. It has been shown that in both mammalian cells and in the fission yeast *Schizosaccharomyces pombe*, kinetochores shortened after microtubule severing. However, whereas in fission yeast all kinetochores could relax to a similar length, in human cells the more stretched kinetochores remained more stretched suggesting that the differences are due to the increased structural complexity of the mammalian kinetochore (Cojoc et al. 2016).

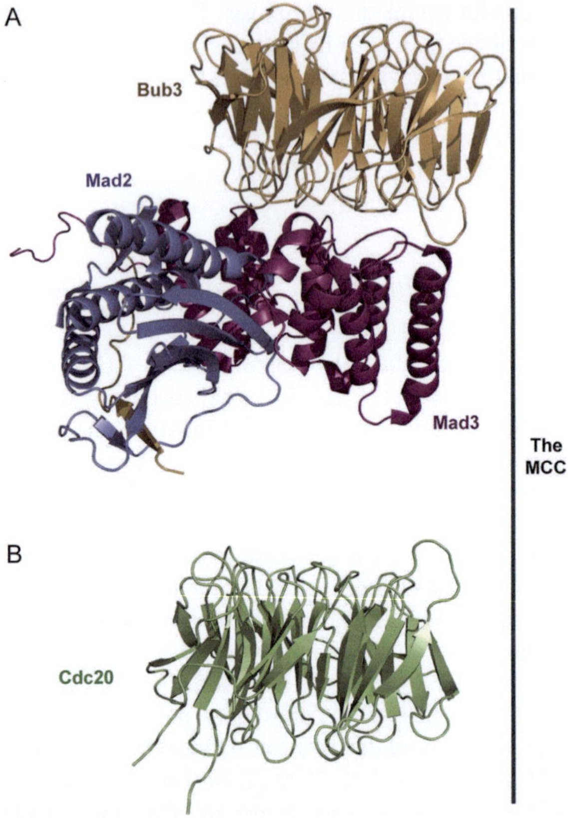

Fig. 16.1 A, crystal structure of a ternary heterotrimer of the Mitotic Checkpoint Complex (MCC) from *Schizosaccharomyces pombe* (pdb id 4AEZ). B, crystal structure of Cdc20 (pdb id 4GGA), another WD 40 fold protein that forms part of the MCC. Figures generated with PyMOL (DeLano 2002)

Table 16.1 Summary of key SAC protein interactions that regulate cell cycle progression

SAC protein	Interacting partner	Function
Aurora B	INCEP, H2A, Nd80	Bipolar attachment of microtubules and controls levels of Mps1 protein
Bub1	Bub3, BuGZ, KNL1	Binds to and phosphorylates Mad1
BubR1	Mad1, Cdc20, BuGZ, KNL1, Mad2	Recruiting Mad1 and Mad2 to the kinetochore and inhibiting APC/C
Bub3	Bub1, BubR1 (Mad3), Bub3	Facilitates kinetochore localisation of Bub1 and BubR1
BuGZ	Mad2, BubR1	Promotes kinetochore localisation of Bub1-Bub1 and BubR1-Bub3 complexes
Cdc20	Mad2, BubR1 (Mad3)	Activates APC/C when SAC is satisfied to promotes cell cycle progression
Mad1	Mad2, Bub1/Bub3 complex	Recruitment of Mad2 to the kinetochore; when SAC is unsatisfied, it is phosphorylated by Bub1 and Mps1
Mad2	Mad1, Cdc20, Bub3	Binds to Cdc20 to inhibit APC/C activity
Mps1	Mad2 and BubR1	Phosphorylates MELT motifs in KNL1

The protein BuGZ was identified recently as a critical component of the SAC that contributes to the targeting of the Bub1-Bub3 and BubR1-Bub3 heterodimer complexes to the kinetochore. BuGZ is predicted to contain a N-terminal zinc finger domain; a GLEBS motif that is conserved from yeast to human and essential to bind Bub3, and a region of low structural complexity (Jiang et al. 2014; Toledo et al. 2014). BuGZ binds to microtubules and tubulin to regulate the SAC in a process that involves BuGZ phase transition (coacervation) (Jiang et al. 2015). It will be important to clarify if coacervation is a physicochemical feature shared by other proteins that regulate spindle assembly. Ultimately, this knowledge can be used to define new cancer therapies that rely on mitosis inhibition (Herman et al. 2015).

Zeste-white 10 (ZW10) and rough deal (ROD) were initially identified in *D. melanogaster*. Both proteins are highly conserved among multicellular eukaryotes (Karess 2005; Scaërou et al. 1999, 2001; Williams et al. 1992; Karess and Glover 1989). Null mutations in the *zw10* and *rod* genes and depletion of these proteins in *C. elegans* and vertebrate cells result in chromosome segregation defects and extensive aneuploidy in mitotic and meiotic cells. Co-immunoprecipitation of ZW10 and ROD in fly and human cell extracts and their immunostaining in mitotic cells suggest an interdependent recruitment of the proteins to the kinetochore (Williams et al. 1992). Zwilch mutations cause a similar mitotic phenotype as ROD and ZW10, thus confirming a role of this protein in the SAC. Zwilch, ROD and ZW10 from human extracts isolated by affinity chromatography methods using ZW10 as bait show the three proteins associate to form a stable complex that seems to contain two copies of each protein (Civril et al., 2010; Kops et al. 2005a; Williams et al. 2003). The recent crystallisation and preliminary X-ray crystallographic analysis of a human ROD-ZW10-Zwilch complex (Altenfeld et al. 2015) further supports the notion of a complex of the three constitutive subunits with a 2:2:2 stoichiometry.

The control of chromosome segregation in higher organisms requires communication of SAC sub-complexes with the KMN (KNL1/Mis12/Ndc80) network, a multiprotein macromolecular assembly that constitutes the structural core of the kinetochore and is essential for the establishment of proper kinetochore-microtubule attachments (Liu et al. 2016; Aravamudhan et al. 2015; Ghongane et al. 2014; Santaguida and Musacchio 2009). Recently considerable progress has been made in understanding the composition of the kinetochore, the recruitment hierarchy of its components, and the principles of its regulation (revised in Agarwal and Varma 2014; Przewloka and Glover 2009). However, structural details of the molecular interactions of the KMN network have proved elusive even though they are clearly indispensable for the mechanism of kinetochore assembly and SAC signalling. The understanding of the function of the individual subunits and the different sub-complexes of the KMN network requires a description of the functional and structural features of its central components, and this is presented below.

16.2 The Kinetochore Null Mutant 1 (KNL1)

The kinetochore protein KNL1 (also known as CASC5, AF15q14 and Blinkin in humans; Spc105 in budding yeast, Spc7 in fission yeast, and KNL1 in humans and *C. elegans*) acts as a multi-substrate docking platform of the KMN network. KNL1 was initially identified in *Saccharomyces cerevisiae* as component of the spindle pole body (hence the acronym Spc105) (Nekrasov et al. 2003), which in *C. elegans* is commonly referred to as KNL1 (kinetochore-null phenotype 1) (Cheeseman et al. 2004) and Spc105R (Spc105-related) in *Drosophila* (revised in Przewloka and Glover 2009). Spc105R from Drosophila shows considerable sequence divergence compared to other species. This feature of the KMN contributes to define a distinctive structural organisation to the KMN network of the fruit fly. Depletion of KNL1 of higher organisms by RNAi causes severe chromosomal segregation defects that resemble phenotypes observed after depletion of the SAC kinases Bub1 and BubR1, including premature exit from mitosis and early onset of anaphase (Przewloka and Glover 2009). KNL1 is the largest subunit of the KMN network and is required for accurate chromosome segregation during mitosis (Desai et al. 2003). KNL1 integrates SAC kinase and phosphatase activities and contributes to the formation of kinetochore-microtubule attachments (Przewloka and Glover 2009; Santaguida and Musacchio 2009). Evidence of the precise role of functional regions in KNL1, including the motifs SILK, RVSF, MELT, KI, and a domain that adopts the RWD fold is providing new insights how KNL1 coordinates SAC activity. This is an important aspect of the SAC that has been revised recently (Caldas and DeLuca 2014; Ghongane et al. 2014).

16.3 The Ndc80 Complex

Microtubules contribute to key biological processes controlled by cell motility, including the maintenance of cell orientation and the regulation of focal adhesion turnover (Alushin et al. 2010). In human cells the Ndc80 complex is linked to centromeric chromatin to mediate end-on attachment of spindle microtubules in a process that requires Ndc80 binding to the kinetochore proteins CENP-T and CENP-C (Suzuki et al. 2015; Tanaka 2013). In humans, the Ndc80 complex is composed of the proteins Hec1 (a subunit that is also commonly referred to as Ndc80), Nuf2, Spc24 and Spc25 (DeLuca and Musacchio 2012; Tooley and Stukenberg 2011; Varma and Salmon 2012). The Ndc80 complex adopts a dumbbell-shape architecture with the subcomplexes Nuf2-Ndc80 and Spc24-Spc25 located in opposite ends of the molecule (*see* Fig. 16.2) (Ciferri et al. 2005; Wei et al. 2005). In one hand, Nuf2 binding to Ndc80 is required for the localisation of the Ndc80 complex to microtubules. On the other hand, Spc24-Spc25 heterodimer complex formation appears to be a pre-requisite for the binding of Spc24-Spc25 to KNL1 and the Mis12 complex (Cheeseman et al. 2006; Kiyomitsu et al. 2007; Wei et al. 2007; Ciferri

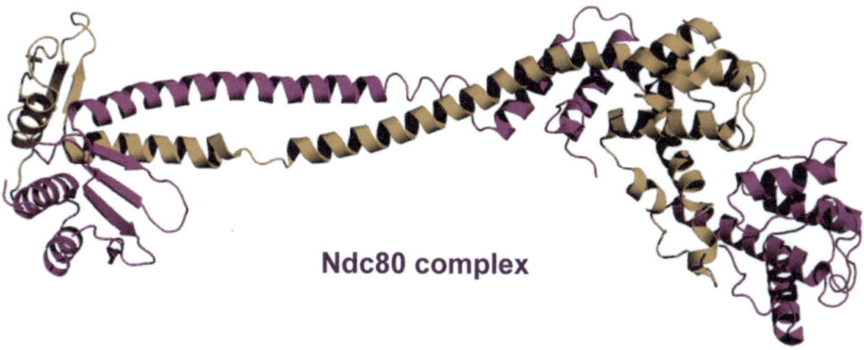

Fig. 16.2 Crystal structure of a chimeric (bonsai) Ndc80 complex (pdb id 2VE7) revealed a dumbbell-shape topology. The sub-complexes Nuf2-Ndc80 and Spc24-Spc25 are located in opposite ends of the molecule

et al. 2008; Wan et al. 2009; Joglekar and DeLuca 2009). The hairpin region of Ndc80 seems to be an important structural requirement for the effective kinetochore recruitment of Mps1 (Mph1), at least in fission yeast (Chmielewska et al. 2016).

Protein depletion assays coupled to quantification of kinetochore protein copy numbers in human cells have provided clues of the extent of the interactions between centromeric chromatin and the microtubule-binding Ndc80 complex. Such studies revealed about 244 Ndc80 complexes per human kinetochore (i.e., approximately 14 per kinetochore microtubule) and 151 Ndc80 complexes are associated to the KMN network. These studies also showed that each CENP-T molecule recruited approximately two copies of the Ndc80 complex (one as part of a KMN network) with nearly 40% of CENP-C recruited exclusively to the KMN network (Suzuki et al. 2015). Undoubtedly, the quantification of kinetochore protein copy numbers in other species should provide important new insights into the evolution and the subtle differences in the mode of regulation of the KMN network.

16.4 Mis12 Complex

In humans, the Mis12 complex (also known as the MIND complex) consists of the proteins Mis12, Dsn1, Nsl1 and Nnf1. The complex is a central component of the KMN network. Nsl1 has been identified as a link between the human Mis12 and Ndc80 complexes (Petrovic et al. 2010) and this protein seems to play a similar important role in yeast kinetochores (Kudalkar et al. 2015). Indeed, recent studies in yeast cells have shown that the interaction between the Mis12 complex (which in yeast is made up of the proteins Mtw1, Nsl1, Nnf1, Dsn1) and the Ndc80 complex (constituted by the proteins Ndc80, Nuf2, Spc24, Spc25) (Biggins 2013) is mediated by an extensive number of contacts (Kudalkar et al. 2015). Mis12 alone does not bind to microtubules but it does bind to the Ndc80 complex, an interaction that enhances Ndc80 binding to microtubules (Kudalkar et al. 2015). In Drosophila, the Mis12 complex localises to the mitotic centromere in a process that implicates the binding of the N-terminal regions of Mis12 and Nnf1 to CENP-C (Richter et al. 2016). Reconstitution of the yeast Mis12-Ndc80 assembly coupled to cross-linking analysis revealed an intricate set of interactions involving five of the eight proteins within the Mis12 and Ndc80 complexes and a direct interaction between the proteins Nsl1 and Spc24/Spc25. The latter interaction defines a unique interface thus suggesting that in different organisms the regulation of Ndc80 functions may be achieved by a distinctive mode of interactions (Kudalkar et al. 2015).

16.5 Centromere-Associated Protein E (CENP-E)

CENP-E is required to maintain a stable genome through the stabilisation of microtubule capture in the kinetochore. CENP-E functions as a highly processive plus end-directed motor that couples chromosome position with microtubule depolymerisation thus linking kinetochores to dynamic spindle microtubules. CENP-E participates in the recruitment of BubR1, Mad1 and Mad2 to attached and newly unattached kinetochores and its binding to the SKAP protein is required for accurate chromosome segregation in mitotic cells (Huang et al. 2012). SKAP seems to form part of the kinetochore corona fibres of mammalian centromeres as judged by immunoelectron microscope imaging. The interaction between CENP-E and SKAP, which involves the C-terminal tail of the former protein, is thought to be essential for kinetochore-microtubule attachment *in vivo*. Depletion of SKAP or CENP-E by RNA interference drastically reduces inter-kinetochore tension, thus leading to chromosome segregation defects and a prolonged delay to fulfill metaphase alignment (Huang et al. 2012).

In human cells CENP-E kinetochore localisation depends on its binding to Nuf2, an interaction that is mediated by C-terminal regions of both CENP-E and Nuf2 as determined with the yeast two-hybrid system and pulldown assays (Liu et al. 2007). Moreover, depletion of human Nuf2 by small interfering RNA abolished CENP-E kinetochore localisation and resulted in chromosome segregation defects, thus confirming the requirement of Nuf2 for CENP-E localisation to the kinetochore and the essential role of the interaction for the correct segregation of chromosomes in mitosis (Liu et al. 2007).

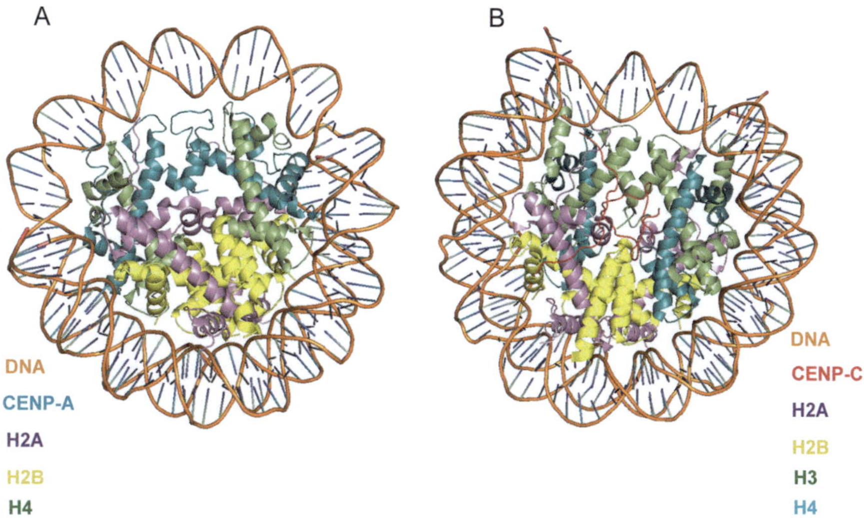

Fig. 16.3 (**a**) Crystal structure of a centromeric nucleosome in complex with CENP-A; (**b**) crystal structure of a centromeric nucleosome in complex with CENP-C. In both cartoons the view is in the axis of the DNA supercoil

16.6 The Centromeric Nucleosome-Associated Network (CCAN)

Centromeres are differentiated chromatin domains, present once per chromosome, that direct segregation of the genome in mitosis and meiosis by specifying assembly of the kinetochore. The latter provides an essential link that brings together chromosomes and spindle microtubules (Pesenti et al. 2016; Rago et al. 2015). The specific spatial configuration of the centromere is likely to contribute to the tight regulation of mitosis and the dynamics of kinetochore-microtubule attachments (George and Walworth 2016). The centromeric nucleosome-associated network (CCAN) is a constitutive complex that is assembled onto centromeric CENP-A chromatin and widely considered as the prime candidate for specifying centromere identity (Foltz et al. 2006). The CCAN is composed by the proteins CENP-C, CENP-H/ CENP-I/ CENP-K, CENP-L/ CENP-M/ CENP-N, CENP-O/CENP-P/CENP-Q/CENP-R/ CENP-U, CENP-T/CENP-W, and CENP-S/CENP-X (Foltz et al. 2006; revised by Perpelescu and Fukagawa 2011). The CCAN recruits the outer kinetochore components of the KMN network KNL1, the Mis12 complex, and the Ndc80 complex thus bringing together kinetochore proteins and spindle microtubules. Disruption of the interaction between CENP-A and CCAN causes errors of chromosome alignment and segregation that prevent cell survival (Foltz et al. 2006).

CENP-A is a centromere-specific isoform of histone H3 (Perspelescu and Fukugawa 2011; Stoler et al. 1995; Palmer et al. 1991) that contributes to kinetochore formation and centromere-kinetochore assembly thus guiding the movement of chromosomes and cell cycle progression throughout mitosis (Fachinetti et al. 2013; Mendiburo et al. 2011; Barnhart et al. 2011; Wan et al. 2009). The crystal structures of two human centromeric nucleosomes, one containing CENP-A (pdb id 3AN2) and one containing CENP-C in complex with the cognate α-satellite DNA derivative (pdb id 4X23) revealed that the latter molecule wraps around a histone octamer (*see* Fig. 16.3). Such octameric complex is defined by the assembly of two copies of histones H2A, H2B, H4 and CENP-A (Tachinawa et al. 2011).

In addition to CENP-A, CENP-C and CENP-T contribute to kinetochore assembly in vertebrates as shown by studies in which the DNA-binding regions of CENP-C and CENP-T were replaced with alternate chromosome-targeting domains, thus resulting in the localisation of functional CENP-C and CENP-T to ectopic loci and a CENP-A-independent assembly of the kinetochore (Gascoigne et al. 2011). Furthermore, phosphorylation of CENP-T appears as an important requirement for proper mitotic assembly of both endogenous and ectopic kinetochores (Gascoigne et al. 2011).

CENP-H is an inner kinetochore protein that is highly conserved amongst eukaryotes (Orthaus et al. 2006). CENP-H directly interacts with CENP-K through multiple contacts to form a stable heterodimeric complex. CENP-H and CENP-K are predicted to contain extensive coiled-coil regions that seem to play an important role in the stabilisation of the CENP-H-CENP-K heterocomplex (Qiu et al. 2009). Depletion of CENP-H in human cells led to severe mitotic phenotypes including misaligned chromosomes and multipolar spindles but not mitotic arrest (Orthaus et al. 2006). CENP-H depletion results in reduced levels of CENP-E but only slightly affects the levels of CENP-C bound to the kinetochore while suppression of CENP-H expression has not effect on BubR1 kinetochore localisation and a SAC response (Orthaus et al. 2006).

The CENP-T/W complex assembles in late S and G2 phases of the cell cycle and is required for mitosis. The CENP-T/W complex is integrated with centromeric chromatin in association with Histone H3 nucleosomes (Prendergast et al. 2011) but it does not persist across cell generations. Instead, association of H3 with the CENP--T/W complex seems to be specific for the regulation of kinetochore activity (Prendergast et al. 2011). CENP-T centromere localisation is restricted to the S-phase of the cell cycle. CENP-T directly associates with CENP-A and CENP-B as shown by Förster resonance energy transfer (FRET) studies. Taken together these studies indicate that CENP-T is required for the recruitment of other proteins to the kinetochore (Hellwig et al., 2008). Furthermore, centromeric-bound CENP-T-W and CENP-S-X subcomplexes associate to form a stable CENP-T-W-S-X heterotetramer that binds to DNA to form supercoil structures (Takeuchi et al. 2014; Nishino et al. 2012). High-resolution structural analyses of the individual subcomplexes and the tetramer have revealed important structural similarities with the nucleosome and certain histone fold-containing complexes (Nishino et al. 2012).

In human cells the inner kinetochore components CENP-C and CENP-T function in parallel pathways to recruit the KMN network to the kinetochore (Nishino et al. 2013; Schleiffer et al. 2012) as shown by independent ectopic targeting of these proteins to a chromosomal locus (Rago et al. 2015). For instance, the physical interaction of CENP-C with KNL1 and the Mis12 complex is required for the recruitment of the Ndc80 complex to the kinetochore whereas CENP-T kinetochore recruitment is only dependent of CENP-T binding to the Ndc80 complex. Furthermore, the CENP-T-Ndc80 complex assembly in turn promotes KNL1/Mis12 complex recruitment in a process that implicates a separate region on CENP-T (Rago et al. 2015). The formation of the CENP-C and CENP-T sub-complexes seems to obey different regulatory controls: the recruitment of the KMN network to CENP-C is stimulated by Aurora B kinase while that of CENP-T is regulated by cyclin-dependent kinase (Cdk) (Rago et al. 2015).

A number of additional microtubule plus-end binding proteins that have been associated with the kinetochore include the CLIP-associating protein 1 (CLASP-1) and 2 (CLASP-2), Astrin, Kinastrin, KIF2B, Kif18A and SKAP. The general structural and functional features of these proteins and their roles in SAC signalling are described below.

16.7 CLASP-1 and CLASP-2

The microtubule plus-end binding proteins CLASP-1 and CLASP-2 play important roles in the regulation of the density, length distribution and stability of interphase microtubules thus integrating spindle and kinetochore functions (Pereira et al. 2006; Maiato et al. 2003). In yeast, Drosophila, and Xenopus, one CLASP orthologue is present, whereas in human two proteins have been identified: CLASP-1 and CLASP-2. In all these organisms CLASP proteins are required for mitotic spindle assembly through the regulation of microtubule dynamics at the kinetochore. In mitotic cells both proteins associate with the ends of growing microtubules and with kinetochores in a process that requires the binding of these proteins to EB1 (Mimori-Kiyosue et al. 2005). The interaction of CLASP-1 and CLASP-2 with EB1 implicates the middle region of both CLASPs (Mimori-Kiyosue et al. 2005). At least in HeLa cells CLASP-1 and CLASP-2 show similar and at least partially redundant roles in organising the mitotic apparatus (Pereira et al. 2006). Their simultaneous depletion results in extensive mitotic spindle defects and an abnormal exit from mitosis. Targeting CLASP-1 with specific anti-CLASP-1 antibodies impairs microtubule dynamics in the kinetochore and the mitotic spindle, leading to the formation of abnormal monopolar asters in which the chromosomes are found buried in the interior. Similarly, the expression of a truncated form of CLASP-1 lacking the kinetochore binding domain results in the formation of depolymerisation-resistant microtubule bundles with a radial array (Maiato et al. 2003). Inhibition of glycogen synthase kinase-3 (GSK3) activity by the tyrosine kinase receptor ErbB2 regulates microtubule capture and stabilisation. Inhibition of Glycogen synthase kinase 3 beta (GSK3b) causes relocalisation of CLASP-2 to the plasma membrane and ruffles (Zaoui et al. 2010). All these observations strongly support an important role for CLASP-1 and CLASP-2 in the organisation of the mitotic spindle and the control of microtubules attachments.

It has been suggested that microtubules in vertebrate somatic cells are not only formed by the centrosome but that a significant number of them originate from the Golgi apparatus in a centrosome-independent manner. The process requires CLASPs recruitment to the trans-Golgi network by the protein GCC185 (Efimov et al. 2007; Zhonghua et al. 2007). However, mechanistic details of the regulation of spindle assembly in mitosis by membrane systems remain largely obscure.

16.8 Astrin and Kinastrin

Astrin is a mitotic spindle-associated protein found in most human cell lines and tissues that is required for proper chromosome alignment at the metaphase plate; is essential for progression through mitosis and contributes to the regulation of separase activity (Dunsch et al. 2011; Thein et al. 2007; Gruber et al. 2002). Depletion of this protein by RNA interference delays chromosome alignment, leads to the loss of spindle architecture and sister chromatid cohesion before the onset of anaphase, and ultimately results in apoptosis (Gruber et al. 2002). Amino acid sequence analysis and fold recognition bioinformatics tools suggest that Astrin has an N-terminal globular domain and an extended coiled-coil domain. Electron microscopy studies of recombinant Astrin showed that this protein self-associates to form parallel dimers with head-stalk structures reminiscent of motor proteins. However, the low amino acid sequence identity and structural similarity to known motor proteins requires further investigations to establish to what extent there is a functional correspondence between Astrin and kinesins.

Kinastrin is the major interacting partner of Astrin in mitotic cells and the interaction is required for Astrin targeting to microtubule plus ends. Overexpression or depletion of Kinastrin mislocalise Astrin and causes mitotic defects that resemble those observed in Astrin-depleted cells. Astrin and Kinastrin can form a complex with SKAP, which also co-localises to microtubule plus ends to facilitate chromosome alignment (Dunsch et al. 2011). These observations support the notion that the microtubule plus end targeting activity of Astrin is required to sustain spindle architecture and to ensure chromosome alignment and that perturbation of these interactions delay mitosis and cause the premature activation of separase (Dunsch et al. 2011). Interestingly, Astrin acts as a negative regulator of mTORC1, which seems to be essential to elicit a cellular stress response. Under stress conditions, Astrin blocks mTORC1 self-association and recruits Raptor, a protein component of mTORC1, to stress granules, thus preventing apoptosis caused by the induction of mTORC1 hyperactivation (Thedieck et al. 2013). This is an exciting finding that suggests a potential link between cellular stress response and the control of chromosome segregation. Further studies should aim to clarify this aspect of SAC signaling.

16.9 KIF2B and Kif18A

The human genome has three genes (Kif2a, Kif2b, and MCAK [also known as Kif2c]) that encode for kinesin-13 proteins. Kif2a, Kif2b, and MCAK fulfill distinct functions during mitosis in human cells (Hood et al. 2012; Manning et al. 2007). Human kinesin Kif18A is a kinesin-8 protein and microtubule-depolymerising protein that contributes to stabilise the CENP-E-Bub1 complex at the kinetochores during early mitosis (Mayr et al. 2007). *In vitro*, Kif18A shows a slow plus-end-directed microtubule depolymerising activity whereas in mitotic cells *in vivo* Kif18A localises close to the plus ends of kinetochore microtubules. Depletion of Kif18A induces aberrant mitotic spindles and loss of tension across sister kinetochores and activates the SAC (Mayr et al. 2007). During vertebrate cell division, chromosomes oscillate with periods of smooth motion and rapid reversals in direction. These fluctuations must be spatially constrained to ensure the proper alignment and high fidelity segregation of chromosomes. In humans, Kif18A plays an essential role in the control of chromosome oscillations by reducing the amplitude of pre-anaphase oscillations and slowing down poleward movements during anaphase. This manner, Kif18A contributes to the control of kinetochore microtubule dynamics underlying chromosome positioning in mitosis (Gardner et al. 2008; Stumpff et al. 2008). Moreover, Kif18A physically interact with CENP-E and BubR1 during mitosis as revealed by co-immunoprecipitation studies. Kif18A depletion results in mitotic arrest and chromosome missalignment and stimulates CENP-E degradation indicating that chromosome congression defects due to Kif18A depletion are at least in part mediated through destabilisation of CENP-E (Huang et al. 2009).

16.10 SKAP

SKAP is an essential component of the mitotic spindle that associates with kinetochores and is required for chromosome alignment, normal timing of sister chromatid segregation and maintenance of spindle pole architecture (Fang et al. 2009). SKAP also plays a role in the control of kinetochore oscillations and the regulation of microtubule plus-ends dynamics during mitosis (Wang et al. 2012a). Although suppression of SKAP expression does not stimulate the SAC, it substantially increases the duration of metaphase, delays the activation of separase and decreases the fidelity of chromosome segregation (Fang et al. 2009).

SKAP binds to microtubules *in vitro*, an interaction that is synergised by CENP-E. Thus, CENP-E and SKAP work together to control dynamic kinetochore-microtubule interactions (Huang et al. 2012). SKAP binds to the C-terminal tail of CENP-E *in vitro* and is essential for an accurate kinetochore-microtubule attachment *in vivo*. Depletion of SKAP or CENP-E by RNA interference drastically impairs inter-kinetochore tension and causes chromosome missegregation (Wang et al. 2012b; Huang et al. 2012). SKAP also interacts with Mis13, which seems

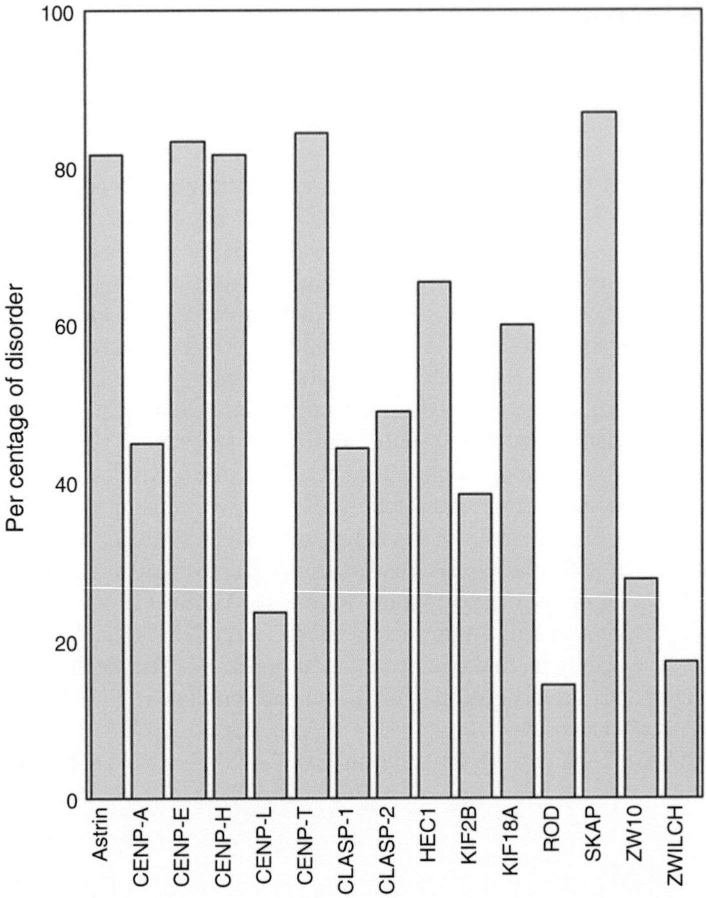

Fig. 16.4 Different KNL1 complexes revealed a similar mode of binding underlying disorder-to-order transitions. A, crystal structure of human TPR BubR1 in complex with the KNL1 KI motif (pdb id 3SI5); crystal structure of a Bub3-Bub1 GLEBS motif-KNL1 MELT motif complex (pdb id 4BL0). C, KNL1 RWD domain in complex with a Nsl1 fragment (pdb id 4NF9).

important for the accurate interaction between kinetochore and dynamic spindle microtubules. SKAP directly binds Mis13 and the interaction specifies the kinetochore localisation of the former protein, an observation that has been confirmed by small interfering RNA studies to suppress Mis13 expression (Wang et al. 2012b). A complex formed between SKAP and Astrin-Kinastrin localises to microtubule plus ends to facilitate proper chromosome alignment (Dunsch et al. 2011). Further studies should aim to clarify the role of these interactions in the control of SAC signalling.

Fig. 16.5 The presence of large segments of disordered regions in multiple kinetochore binding proteins is a common structural feature. The plot shows the disorder predictions based on PONDR-FIT (Xue et al. 2010) analyses.

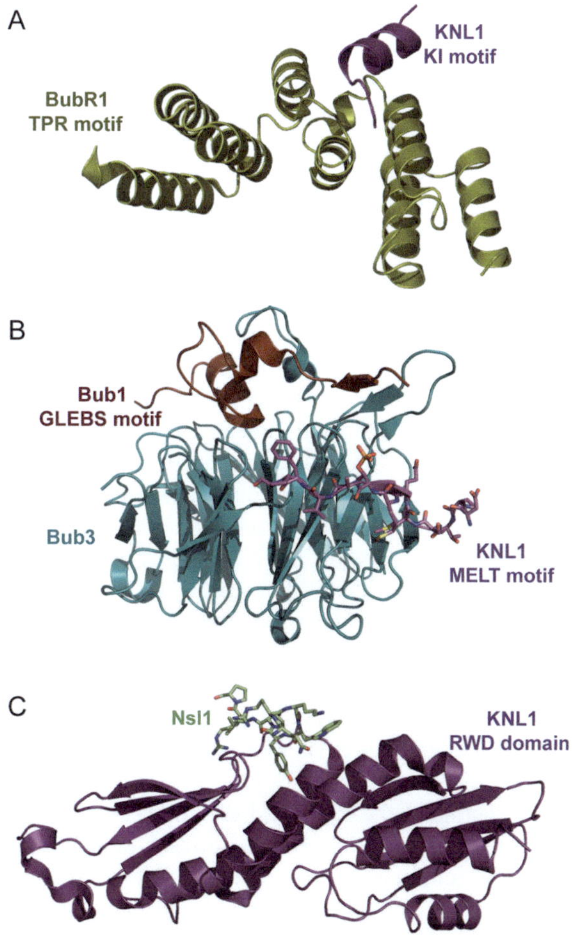

16.11 Disorder-to-Order Transitions in the KMN

It is worth noting that many of the kinetochore proteins described above are predicted to contain large regions of low structural complexity (*see* Fig. 16.4). A pattern of disorder-to-order transitions in SAC signalling has emerged from the structures of diverse complexes involving the kinetochore organiser protein KNL1, including the N-terminal TPR-containing domains of Bub1 and BubR1 in complex with the KNL1 N-terminal KI motifs; Bub3 bound to KNL1 MELT motifs and the KNL1 RWD domain in complex with a synthetic peptide that mimics the protein Nls1 (Bolanos-Garcia et al. 2011; Krenn et al. 2012; Primorac et al. 2013; Petrovic et al. 2014) (*see* Fig. 16.5). One distinctive feature that emerges from the analysis of the above mentioned complexes is the predominance of cooperative hydrophobic interactions that stabilise the complexes. With the exception of the C-terminal region which contains a globular RWD domain, multiple regions of low structure complexity that span most of the polypeptide chain occur in KNL1. This is not surprising because multiple regions of low structure complexity occur often in hub proteins that define interactome networks (Babu et al. 2012; Kim et al. 2006; Dosztányi et al. 2006; Dunker et al. 2005, 2008; Haynes et al. 2006). Indeed, in multiple biological systems cooperative interactions involve the recognition of a flexible protein by a globular one, leading to concerted folding and binding (Blundell et al. 2002). This is particularly evident in hub proteins that define interactome networks because such proteins contain intrinsic local disordered regions (Dunker et al., 1998; Gsponer and Babu, 2009) that often associate with interacting partners through concerted binding and folding (Uversky 2015; Dosztányi et al. 2006; Dunker et al. 2005). The general model for concerted folding upon binding appears to be initial binding of a large side chain into a deep pocket, usually followed by interaction at a second and sometimes third pocket, forming a cluster of small pockets (Fuller et al. 2009). Less conserved interactions involving regions N- or C-terminal to the conserved motif then fold cooperatively onto the surface of the globular partner. There are examples of this type of interactions in SAC signalling, including the binding of the KI motif of KNL1 to BubR1 (Bolanos-Garcia et al. 2011) and possibly that of an equivalent KI motif in KNL1 specific to Bub1 (Krenn et al. 2012) and the binding of Bub3 to the MELT motifs of KNL1, a sequential, multisite interaction that is subjected to phospho-regulation (Vleugel et al. 2015). Indeed, the interplay of phosphorylation and dephosphorylation cascades rises as an important mechanism to regulate the SAC (Manic et al. 2017; Funabiki and Wynne 2013; London et al. 2012; Shepperd et al. 2012; Rosenberg et al. 2011; Liu et al. 2010).

Furthermore, the reciprocal communication of disorder-to-order transitions on two or more distant functional surfaces of high intrinsic disorder can maximise allosteric coupling between proteins. This mode of molecular recognition and signal amplification may obey the same mechano-chemical principles underlying the interaction of simpler systems such as binding of a biotin repressor to biotin protein ligase (Egington et al. 2015). Also, macromolecular crowding effects (Mourão et al. 2014; Cino et al. 2012; Babu et al. 2012; Wang et al. 2012a) can be anticipated to play a major role in the regulation of the SAC given the prominent role of proteins with multiple regions of low structural complexity in the process, including kinesin motors (Leduc et al. 2012).

In summary, regulation of the rate in which spindle microtubules attach/detach to/from kinetochores plays a central role in the control of chromosome segregation. Multiple mechanisms of assembly and holistic models that take into account the role of protein receptors, signalling networks and regulatory feedback mechanisms have been proposed in an attempt to describe more precisely the role of kinetochore-microtubules interactions for the control mitotic progression in higher organisms (Kim and Yu 2015; Godek et al. 2015). As discussed below, disruption of this balance quickly results in aneuploidy, genome instability, cancer and diverse birth and development defects.

16.12 The SAC-KMN Axis in Disease

Mistakes in the process of cell division can lead to the rearrangement, the loss or gain of chromosomes (aneuploidy). Solid tumors are frequently aneuploid, and many display high rates of chromosome missegregation and chromosomal instability (CIN). The most common cause of CIN is the persistence of aberrant kinetochore-microtubule attachments, which manifest as lagging chromosomes in anaphase. Errors in kinetochore-microtubule attachments during prometaphase can be due to stochastic interactions between kinetochores and microtubules.

Mps1 has been identified in the signature of the top 25 genes overexpressed in CIN and aneuploid tumours (Kops et al. 2005b; Carter et al. 2006) and found to be upregulated in a number of tumours of different origins including bladder, anaplastic thyroid, breast, lung, esophagus, and prostate. In the absence of a functional mitotic checkpoint, as occurs when Mps1 function is lost, cells become rapidly aneuploid and subsequently die (Kops et al. 2005b; Janssen et al. 2009). This feature, together with the observations that inhibiting Mps1 with chemical inhibitors kills cultured tumour cells (Kwiatkowski et al. 2010) and that even its partial inhibition creates tumour cells more sensitive to clinical doses of taxol (Janssen et al. 2009), show that targeting Mps1 with drugs may be beneficial to arrest proliferation of tumour cells. Significant chromosomal abnormalities are the cause of severe diseases such as breast cancer, the most common cancer in the UK. This year alone, 50,000 people in the UK will find out they have breast cancer and 12,000 people will die from it (Cancer Research UK organisation). The fact that Mps1 inhibition in tumour xenograft models significantly reduces tumour growth rates while leaving normal cell growth unaffected (Daniel et al. 2011) makes Mps1 an attractive target for cancer therapy (Kapanidou and Bolanos-Garcia 2014). In addition to Mps1, Aurora B kinase is the cellular target of diverse Medicinal Chemistry campaigns to develop inhibitors that function as adenosine triphosphate (ATP) competitors of these protein kinases. Also important is the search of ubiquitin ligase inhibitors that target the E3 ubiquitin ligase activity of the APC/C complex and APC/C regulators (Zhang et al. 2014, 2016; Zhou et al. 2013, 2016; Fujimitsu et al. 2016).It can be anticipated a steady increase of activity in this area in the coming years.

Defects in centrosome and spindle-associated functionsare the most frequent cause of primary microcephaly syndromes in humans. For example, mutations in CENP-E have defined a novel kinetochore-centromeric mechanism for microcephalic primordial dwarfism (Mirzaa et al. 2014) while centromere protein F (CENP-F) has been implicated in Hutchinson-Gilford progeria syndrome, a rare disorder that leads to premature ageing and death due to myocardial infarction or stroke. The disease is caused by expression of the protein Progerin, which is a truncated version of the protein prelamin A (Eisch et al. 2016). Progerin displaces CENP-F from metaphase chromosome kinetochores, thus increasing chromatin lagging and causing genome instability (Eisch et al. 2016).

The range of malignancies described above indicate that the development of new drugs to interfere with abnormal cell proliferation is urgently required. The development of new drugs to interfere with defective SAC signalling and its communication with the KMN network in human tumours appears as an attractive alternative to prevent the proliferation of cells carrying abnormalities in chromosome structure and number. Multiple protein-protein interactions in regulatory hubs that control chromosome segregation and mitosis progression may constitute an important pool of novel drug targets. Structural insight into the molecular architecture of key interactions that regulate the SAC should pave the way for drug target identification and validation. In the absence of high resolution structural data, definition of the relationship between hub proteins and drug targets based on the combinatorial analysis of intrinsic structural disorder and gene onthology seems particularly attractive (Fu et al. 2015; The Gene Ontology Consortium 2010).

16.13 Emerging Methods in Structural Biology

The dynamic and coordinated assembly and disassembly of protein complexes in time and space follows sequential obligate stages that result in an enhanced selectivity with a low margin for errors in the process. At the same time, the dynamics of protein complex assembly and disassembly represents a great challenge for their structural and functional characterisation and often requires a combinatorial multidisciplinary approach involving a range of biochemical, biophysical, molecular and cellular approaches. Recent advances in Förster resonance energy transfer by fluorescence lifetime imaging microscopy; laser ablation; small angle x-ray scattering in structural biology (SAXS); nuclear magnetic resonance (NMR); serial femtosecond crystallography; and cryo-transmission electron microscopy (TEM) represent new exciting opportunities to understand the complex dynamics and mode of regulation of the SAC-KMN-microtubule signaling axis through the combinatorial use of the techniques.

Advances in SAXS methods allow the study of macromolecular complexes in solution that provide information about the shapes, conformations; oligomeric states of globular, non-globular and disordered macromolecules (Chaudhuri 2015) whereas multinuclear relaxation dispersion NMR methods permit to follow molecular recognition events of intrinsically disordered proteins in solution (Schneider et al. 2015; Parigi et al. 2014). Time-resolved protein crystallography using ultrafast X-ray free-electron lasers (XFELs) make it possible to follow rapid structural changes resulting from photolysis in the crystalline state and to resolve reaction intermediates at impressive high resolution (Barends et al. 2015; Tenboer et al. 2014). More recent improvements in serial femtosecond crystallography allowed the collection of X-ray diffraction patterns using X-ray pulses of 50 femtosecond duration that contained approximately 2×10^{12} photons per pulse to achieve a high-resolution XFEL structure of 1.75 Å (Ginn et al. 2015). Laser ablation has been used recently to separate microtubules attached to a merotelic kinetochore to study the mechanical response of the kinetochore resulting from changes of its length (Cojoc et al. 2016). At the same time, the study shows that the use of merotelic kinetochores emerges as an attractive experimental model for studying the mechanical properties of the kinetochore in live cells (Cojoc et al. 2016).

Equally impressive is the pace of instrumental and computational improvements in TEM where modern electron microscopes can produce images at a resolution higher than 2.0 Å (Glaeser 2016; Nogales 2016). A big gain that TEM offers is the possibility of studying samples of a relatively heterogeneous nature, thus allowing the analysis of multiple structural states that recapitulate the dynamics of complex protein-protein interactions including their mode of regulation and assembly/disassembly under diverse conditions (Weis et al. 2015; Louder et al. 2016). Parallel advances in cryo-electron tomography now allow the visualisation of macromolecular assemblies of irregular shapes; of organelles and even entire cells at the subnanometre resolution scale. An excellent review on this topic has been reported recently by Helen Saibil's group (2016).

16.14 Closing Remarks

The study of cell division, the mechanism of transmission of the genetic material to descendants and the molecular basis of premature aging and cancer are areas of great interest in the Biomedical Sciences. Spindle assembly checkpoint (SAC) signalling is a truly fundamental cellular process of higher organisms that ensures the faithful segregation of chromosomes each time a cell divides. Undoubtedly, the inhibition of aberrant SAC signalling will benefit a wide range of disciplines, ranging from the cellular and molecular understanding of cell division in health and disease to the study of cell development, genome stability, ageing and comparative genomics.

The synergistic combination of biochemical, biophysical and structural biology methods for the characterisation of dynamic macromolecular complexes together with cellular and systems biology approaches should lead to a more comprehensive understanding of the cell and provide insights into how defects in molecular interactions can lead to the impairment of cellular regulation and function.

As the use of these experimental techniques alongside with molecular and computational methods begin to give insights into the dynamics of protein assembly/disassembly and their architecture, we will learn more mechanistic details of the remarkable complexity of the network of interactions between thousands of protein components that regulate metabolic and signalling pathways essential to all eukaryotes.

Because large multi-protein complexes play critical roles in cell regulation, interfering with the dynamics of their assembly and/or dissociation rises as an attractive strategy for the treatment of diseases. Extending the study of the structure and the dynamics of isolated SAC-KMN-microtubule sub-complexes to the molecular understanding of the mode of organisation of larger assemblies that ensure signal generation and amplification in a narrow spatial-temporal framework continues to represent a major challenge. Recent advances in TEM, electron-free lasers and a range of biophysical methods herald a new and exciting era for the molecular understanding of nuclear complexes that ensure genome stability to an unprecedented level of detail.

References

Abrieu A, Magnaghi-Jaulin L, Kahana JA, Peter M, Castro A, Vigneron S, Lorca T, Cleveland DW, Labbé JC (2001) Mps1 is a kinetochore-associated kinase essential for the vertebrate mitotic checkpoint. Cell 106:83–93

Agarwal S, Varma D (2014) How the SAC gets the axe: Integrating kinetochore microtubule attachments with spindle assembly checkpoint signaling. BioArchitecture 5:1–12

Altenfeld A, Wohlgemuth S, Wehenkel A, Vetter IR, Musacchio A (2015) Complex assembly, crystallization and preliminary X-ray crystallographic analysis of the human Rod-Zwilch-ZW10 (RZZ) complex. Acta Crystallogr F Struct Biol Commun 71:438–442

Alushin GM, Ramey VH, Pasqualato S, Ball DA, Grigorieff N, Musacchio A, Nogales E (2010) The Ndc80 kinetochore complex forms oligomeric arrays along microtubules. Nature 467:805–810

Aravamudhan P, Goldfarb AA, Joglekar AP (2015) The kinetochore encodes a mechanical switch to disrupt spindle assembly checkpoint signalling. Nat Cell Biol 17:868–879

Babu MM, Kriwacki RW, Pappu RV (2012) Structural biology. Versatility from protein disorder Science 337:1460–1461

Barends TR, Foucar L, Ardevol A, Nass K, Aquila A, Botha S, Doak RB, Falahati K, Hartmann E, Hilpert M, Heinz M, Hoffmann MC, Köfinger J, Koglin JE, Kovacsova G, Liang M, Milathianaki D, Lemke HT, Reinstein J, Roome CM, Shoeman RL, Williams GJ, Burghardt I, Hummer G, Boutet S, Schlichting I (2015) Direct observation of ultrafast collective motions in CO myoglobin upon ligand dissociation. Science 350:445–450

Barnhart MC, Kuich PH, Stellfox ME, Ward JA, Bassett EA, Black BE, Foltz DR (2011) HJURP is a CENP-A chromatin assembly factor sufficient to form a functional de novo kinetochore. J Cell Biol 194:229–243

Bavetsias V, Linardopoulos S (2015) Aurora Kinase Inhibitors: Current Status and Outlook. Front Oncol 5:278

Biggins S (2013) The composition, functions, and regulation of the budding yeast kinetochore. Genetics 194:817–846

Blundell TL, Bolanos-Garcia VM, Chirgadze DY, Harmer NJ, Lo T, Pellegrini L, Sibanda BL (2002) Asymmetry in the multiprotein systems of molecular biology. Struct Chem 13:405–412

Bolanos-Garcia VM, Blundell TL (2011) BUB1 and BUBR1: multifaceted kinases of the cell cycle. Trends Biochem Sci 36:141–150

Bolanos-Garcia VM, Lischetti T, Matak-Vinkovic D, Cota E, Simpson PJ, Chirgadze DY, Spring DR, Robinson CV, Nilsson J, Blundell TL (2011) Structure of a Blinkin-BUBR1 complex reveals an interaction crucial for kinetochore-mitotic checkpoint regulation via an unanticipated binding Site. Structure 19:1691–1700

Bolanos-Garcia VM, Wu Q, Ochi T, Chirgadze DY, Sibanda BL, Blundell TL (2012) Spatial and temporal organisation of multiprotein assemblies: achieving sensitive control in information-rich cell regulatory systems. Philos Transact A Math Phys Eng Sci 370:3023–3039

Boyarchuk Y, Salic A, Dasso M, Arnaoutov A (2007) Bub1 is essential for assembly of the functional inner centromere. J Cell Biol 176:919–928

Caldas GV, DeLuca JG (2014) KNL1: bringing order to the kinetochore. Chromosoma 123:169–181

Carroni M, Saibil HR (2016) Cryo electron microscopy to determine the structure of macromolecular complexes. Methods 95:78–85

Carter SL, Eklund AC, Kohane IS, Harris LN. Szallasi Z (2006) A signature of chromosomal instability inferred from gene expression profiles predicts clinical outcome in multiple human cancers. Nat Genet 38:1043–1048.

Chao W, Kulkarni K, Zhang Z, Kong E, Barford D (2012) Structure of the mitotic checkpoint complex. Nature 484:208–213

Chaudhuri BN (2015) Emerging applications of small angle solution scattering in structural biology. Protein Sci 24:267–276

Cheeseman IM, Niessen S, Anderson S, Hyndman F, Yates JR 3rd, Oegema K, Desai A (2004) A conserved protein network controls assembly of the outer kinetochore and its ability to sustain tension. Genes Dev 18:2255–2268

Cheeseman IM, Chappie JS, Wilson-Kubalek EM, Desai A (2006) The conserved KMN network constitutes the core microtubule-binding site of the kinetochore. Cell 127:983–997

Chmielewska AE, Tang NH, Toda T (2016) The hairpin region of Ndc80 is important for the kinetochore recruitment of Mph1/MPS1 in fission yeast. Cell Cycle 15:740–747

Ciferri C, De Luca J, Monzani S, Ferrari KJ, Ristic D, Wyman C, Stark H, Kilmartin J, Salmon ED, Musacchio A (2005) Architecture of the human ndc80-hec1 complex, a critical constituent of the outer kinetochore. J Biol Chem 280:29088–29095

Ciferri C, Pasqualato S, Screpanti E, Varetti G, Santaguida S, Dos Reis G, Maiolica A, Polka J, De Luca JG, De Wulf P, Salek M, Rappsilber J, Moores CA, Salmon ED, Musacchio A (2008) Implications for kinetochore-microtubule attachment from the structure of an engineered Ndc80 complex. Cell 133:427–439

Cino EA, Karttunen M, Choy W-Y (2012) Effects of Molecular Crowding on the Dynamics of Intrinsically Disordered Proteins. PLoS One 7:e49876

Civril F, Wehenkel A, Giorgi FM, Santaguida S, Di Fonzo A, Grigorean G, Ciccarelli FD, Musacchio A (2010) Structural analysis of the RZZ complex reveals common ancestry with multisubunit vesicle tethering machinery. Structure 18:616–626

Cojoc G, Roscioli E, Zhang L, García-Ulloa A, Shah JV, Berns MW, Pavin N, Cimini D, Tolić IM, Gregan J (2016) Laser microsurgery reveals conserved viscoelastic behavior of the kinetochore. J Cell Biol 212:767–776

Daniel J, Coulter J, Woo JH, Wilsbach K, Gabrielson E (2011) High levels of the Mps1 checkpoint protein are protective of aneuploidy in breast cancer cells. Proc Natl Acad Sci U S A 108:5384–5389

DeLano WL (2002) The PyMOL molecular graphics system. DeLano Scientific, San Carlos

DeLuca JG, Musacchio A (2012) Structural organization of the kinetochore-microtubule interface. Curr Opin Cell Biol 24:48–56

Desai A, Rybina S, Müller-Reichert T, Shevchenko A, Shevchenko A, Hyman A, Oegema K (2003) KNL-1 directs assembly of the microtubule-binding interface of the kinetochore in *C. elegans*. Genes Dev 17:2421–2435

Dosztányi Z, Chen J, Dunker AK, Simon I, Tompa P (2006) Disorder and sequence repeats in hub proteins and their implications for network evolution. J Proteome Res 5:2985–2995

Dunker AK, Garner E, Guilliot S, Romero P, Albrecht K, Hart J, Obradovic Z, Kissinger C, Villafranca JE (1998) Protein disorder and the evolution of molecular recognition: theory, predictions and observations. Pac Symp Biocomput:473–484

Dunker AK, Cortese MS, Romero P, Iakoucheva LM, Uversky VN (2005) Flexible nets: the roles of intrinsic disorder in protein interaction networks. FEBS J 272:5129–5148

Dunker AK, Oldfield CJ, Meng J, Romero P, Yang JY, Chen JW, Vacic V, Obradovic Z, Uversky VN (2008) The unfoldomics decade: an update on intrinsically disordered proteins. BMC Genomics 9:S1

Dunsch AK, Linnane E, Barr FA, Gruneberg U (2011) The astrin-kinastrin/SKAP complex localizes to microtubule plus ends and facilitates chromosome alignment. J Cell Biol 192:959–968

Efimov A, Kharitonov A, Efimova N, Loncarek J, Miller PM, Andreyeva N, Gleeson P, Galjart N, Maia AR, McLeod IX, Yates JR 3rd, Maiato H, Khodjakov A, Akhmanova A, Kaverina I (2007) Asymmetric CLASP-dependent nucleation of noncentrosomal microtubules at the trans-Golgi network. Dev Cell 12:917–930

Eginton C, Cressman WJ, Bachas S, Wade H, Beckett D (2015) Allosteric Coupling via Distant Disorder-to-Order Transitions. J Mol Biol 427:1695–1704

Eisch V, Lu X, Gabriel D, Djaali K (2016) Progerin impairs chromosome maintenance by depleting CENP-F from metaphase kinetochores in Hutchinson-Gilford progeria fibroblasts. Oncotarget Mar 22. 7(17):24700–24718 doi: 10.18632/oncotarget.8267.

Elowe S (2011) Bub1 and BubR1: at the Interface between Chromosome Attachment and the Spindle Checkpoint. Mol Cell Biol 31:3085–3093

Fachinetti D, Folco HD, Nechemia-Arbely Y, Valente LP, Nguyen K, Wong AJ, Zhu Q, Holland AJ, Desai A, Jansen LE, Cleveland DW (2013) A two-step mechanism for epigenetic specification of centromere identity and function. Nat Cell Biol 15:1056–1066

Fang L, Seki A, Fang G (2009) SKAP associates with kinetochores and promotes the metaphase-to-anaphase transition. Cell Cycle 8:2819–2827

Foltz DR, Jansen LE, Black BE, Bailey AO, Yates JR 3rd, Cleveland DW (2006) The human CENP-A centromeric nucleosome-associated complex. Nat Cell Biol 8:458–469

Fu Y, Guo Y, Wang Y, Luo J, Pu X, Li M, Zhang Z (2015) Exploring the relationship between hub proteins and drug targets based on GO and intrinsic disorder. Comput Biol Chem 56:41–48

Fujimitsu K, Grimaldi M, Yamano H (2016) Cyclin-dependent kinase 1-dependent activation of APC/C ubiquitin ligase. Science 352:1121–1124

Fuller AA, Du D, Liu F, Davoren JE, Bhabha G, Kroon G, Case DA, Dyson HJ, Powers ET, Wipf P, Gruebele M, Kelly JW (2009) Evaluating beta-turn mimics as beta-sheet folding nucleators. Proc Natl Acad Sci U S A 106:11067–11072

Funabiki H, Wynne DJ (2013) Making an effective switch at the kinetochore by phosphorylation and dephosphorylation. Chromosoma 122:135–158

Gardner MK, Odde DJ, Bloom K (2008) Kinesin-8 molecular motors: putting the brakes on chromosome oscillations. Trends Cell Biol 18:307–310

Gascoigne KE, Takeuchi K, Suzuki A, Hori T, Fukagawa T, Cheeseman IM (2011) Induced ectopic kinetochore assembly bypasses the requirement for CENP-A nucleosomes. Cell 145:410–422

George AA, Walworth NC (2016) Microtubule dynamics decoded by the epigenetic state of centromeric chromatin. Curr Genet Mar 14. 62(4):691–695 PMID: 26976145.

Ghongane P, Kapanidou M, Asghar A, Elowe S, Bolanos-Garcia VM (2014) The dynamic protein Knl1- a kinetochore rendezvous. J Cell Sci 127:3415–3423

Ginn HM, Messerschmidt M, Ji X, Zhang H, Axford D, Gildea RJ, Winter G, Brewster AS, Hattne J, Wagner A, Grimes JM, Evans G, Sauter NK, Sutton G, Stuart DI (2015) Structure of CPV17 polyhedrin determined by the improved analysis of serial femtosecond crystallographic data. Nat Commun 6:6435

Glaeser RM (2016) How good can cryo-EM become? Nat Methods 13:28–32

Godek KM, Kabeche L, Compton DA (2015) Regulation of kinetochore-microtubule attachments through homeostatic control during mitosis. Nat Rev Mol Cell Biol 16:57–64

Gruber J, Harborth J, Schnabel J, Weber K, Hatzfeld M (2002) The mitotic-spindle-associated protein astrin is essential for progression through mitosis. J Cell Sci 115:4053–4059

Gsponer J, Babu MM (2009) The rules of disorder or why disorder rules. Prog Biophys Mol Biol 99:94–103

Haynes C, Oldfield CJ, Ji F, Klitgord N, Cusick ME, Radivojac P, Uversky VN, Vidal M, Iakoucheva LM (2006) Intrinsic disorder is a common feature of hub proteins from four eukaryotic interactomes. PLoS Comput Biol 2:e100

Hellwig D, Münch S, Orthaus S, Hoischen C, Hemmerich P, Diekmann S (2008) Live-cell imaging reveals sustained centromere binding of CENP-T via CENP-A and CENP-B. J Biophotonics 1:245–254

Herman JA, Toledo CM, Olson JM, DeLuca JG, Paddison PJ (2015) Molecular pathways: regulation and targeting of kinetochore-microtubule attachment in cancer. Clin Cancer Res 21:233–239

Hiruma Y, Sacristan C, Pachis ST, Adamopoulos A, Kuijt T, Ubbink M, von Castelmur E, Perrakis A, Kops GJ (2015) Competition between MPS1 and microtubules at kinetochores regulates spindle checkpoint signaling. Science 348:1264–1267

Hood EA, Kettenbach AN, Gerber SA, Compton DA (2012) Plk1 regulates the kinesin-13 protein Kif2b to promote faithful chromosome segregation. Mol Biol Cell 23:2264–2274

Huang Y, Yao Y, Xu HZ, Wang ZG, Lu L, Dai W (2009) Defects in chromosome congression and mitotic progression in KIF18A-deficient cells are partly mediated through impaired functions of CENP-E. Cell Cycle 8:2643–2649

Huang Y, Wang W, Yao P, Wang X, Liu X, Zhuang X, Yan F, Zhou J, Du J, Ward T, Zou H, Zhang J, Fang G, Ding X, Dou Z, Yao X (2012) CENP-E kinesin interacts with SKAP protein to orchestrate accurate chromosome segregation in mitosis. J Biol Chem 287:1500–1509

Janssen A, Kops GJ, Medema RH (2009) Elevating the frequency of chromosome mis-segregation as a strategy to kill tumor cells. Proc Natl Acad Sci U S A 106:19108–19113

Ji Z, Gao H, Yu H (2015) Kinetochore attachment sensed by competitive Mps1 and microtubule binding to Ndc80C. Science 48:1260–1264

Jiang H, He X, Wang S, Jia J, Wan Y, Wang Y, Zeng R, Yates J 3rd, Zhu X, Zheng Y (2014) A Microtubule-Associated Zinc Finger Protein, BuGZ, Regulates Mitotic Chromosome Alignment by Ensuring Bub3 Stability and Kinetochore Targeting. Dev Cell 28:268–281

Jiang H, Wang S, Huang Y, He X, Cui H, Zhu X, Zheng Y (2015) Phase transition of spindle-associated protein regulate spindle apparatus assembly. Cell 163:108–122

Joglekar AP, DeLuca JG (2009) Chromosome segregation: Ndc80 can carry the load. Curr Biol 19:R404–R407

Kapanidou M, Bolanos-Garcia VM (2014) Spindle Assembly Checkpoint (SAC): More New Targets for Anti-Cancer Drug Therapies. Adv Cancer Drug Targ 2:54–79

Karess R (2005) Rod-ZW10-ZWILCH: a key player in the spindle checkpoint. Trends Cell Biol 15:386–392

Karess RE, Glover DM (1989) Rough deal: a gene required for proper mitotic segregation in Drosophila. J Cell Biol 109:2951–2961

Kim S, Yu H (2015) Multiple assembly mechanisms anchor the KMN spindle checkpoint platform at human mitotic kinetochores. J Cell Biol 208:181–196

Kim PM, Lu LJ, Xia Y, Gerstein MB (2006) Relating three-dimensional structures to protein networks provides evolutionary insights. Science 314:1938–1941

Kiyomitsu T, Obuse C, Yanagida M (2007) Human Blinkin/AF15q14 is required for chromosome alignment and the mitotic checkpoint through direct interaction with Bub1 and BubR1. Dev Cell 13:663–676

Kops GJ, Kim Y, Weaver BA, Mao Y, McLeod I, Yates JR 3rd, Tagaya M, Cleveland DW (2005a) ZW10 links mitotic checkpoint signaling to the structural kinetochore. J Cell Biol 169:49–60

Kops GJ, Weaver BA, Cleveland DW (2005b) On the road to cancer: aneuploidy and the mitotic checkpoint. Nat Rev Cancer 5:773–785

Krenn V, Musacchio A (2015) The Aurora B Kinase in Chromosome Bi-Orientation and Spindle Checkpoint Signaling. Front Oncol 5:225

Krenn V, Wehenkel A, Li X, Santaguida S, Musacchio A (2012) Structural analysis reveals features of the spindle checkpoint kinase Bub1-kinetochore subunit Knl1 interaction. J Cell Biol 196:451–467

Kudalkar EM, Scarborough EA, Umbreit NT, Zelter A, Gestaut DR, Riffle M, Johnson RS, MacCoss MJ, Asbury CL, Davis TN (2015) Regulation of outer kinetochore Ndc80 complex-based microtubule attachments by the central kinetochore Mis12/MIND complex. Proc Natl Acad Sci U S A 112:E5583–E5589

Kwiatkowski N, Jelluma N, Filippakopoulos P, Soundararajan M, Manak MS, Kwon M, Choi HG, Sim T, Deveraux QL, Rottmann S, Pellman D, Shah JV, Kops GJ, Knapp S, Gray NS (2010)

Small-molecule kinase inhibitors provide insight into Mps1 cell cycle function. Nat Chem Biol 6:359–368

Lara-Gonzalez P, Westhorpe F, Taylor S (2012) The Spindle Assembly Checkpoint. Curr Biol 22:966–980

Leduc C, Padberg-Gehle K, Varga V, Helbing D, Diez S, Howard J (2012) Molecular crowding creates traffic jams of kinesin motors on microtubules. Proc Natl Acad Sci U S A 109:6100–6105

Liu D, Ding X, Du J, Cai X, Huang Y, Ward T, Shaw A, Yang Y, Hu R, Jin C, Yao X (2007) Human NUF2 interacts with centromere-associated protein E and is essential for a stable spindle microtubule-kinetochore attachment. J Biol Chem 282:21415–21424

Liu D, Vleugel M, Backer CB, Hori T, Fukagawa T, Cheeseman IM, Lampson LA (2010) Regulated targeting of protein phosphatase 1 to the outer kinetochore by KNL1 opposes Aurora B kinase. J Cell Biol 188:809–920

Liu Y, Petrovic A, Rombaut P, Mosalaganti S, Keller J, Raunser S, Herzog F, Musacchio A (2016) Insights from the reconstitution of the divergent outer kinetochore of *Drosophila melanogaster*. Open Biol 6(2) pii: 150236.

London N, Ceto S, Ranish JA, Biggins S (2012) Phosphoregulation of Spc105 by Mps1 and PP1 regulates Bub1 localization to kinetochores. Curr Biol 22:900–906

Louder RK, He Y, López-Blanco JR, Fang J, Chacón P, Nogales E (2016) Structure of promoter-bound TFIID and model of human pre-initiation complex assembly. Nature 531:604–609

Maiato H, Fairley EA, Rieder CL, Swedlow JR, Sunkel CE, Earnshaw WC (2003) Human CLASP1 is an outer kinetochore component that regulates spindle microtubule dynamics. Cell 113:891–904

Manic G, Corradi F, Sistigu A, Siteni S, Vitale I (2017) Molecular regulation of the spindle assembly checkpoint by kinases and phosphatases. Int Rev Cell Mol Biol 328:105–161

Manning AL, Ganem NJ, Bakhoum SF, Wagenbach M, Wordeman L, Compton DA (2007) The kinesin-13 proteins Kif2a, Kif2b, and Kif2c/MCAK have distinct roles during mitosis in human cells. Mol Biol Cell 18:2970–2979

Mayr MI, Hümmer S, Bormann J, Grüner T, Adio S, Woehlke G, Mayer TU (2007) The human kinesin Kif18A is a motile microtubule depolymerase essential for chromosome congression. Curr Biol 17:488–498

Mendiburo MJ, Padeken J, Fulop S, Schepers A, Heun P (2011) Drosophila CENH3 is sufficient for centromere formation. Science 334:686–690

Mimori-Kiyosue Y, Grigoriev I, Lansbergen G, Sasaki H, Matsui C, Severin F, Galjart N, Grosveld F, Vorobjev I, Tsukita S, Akhmanova A (2005) CLASP1 and CLASP2 bind to EB1 and regulate microtubule plus-end dynamics at the cell cortex. J Cell Biol 168:141–153

Mirzaa GM, Vitre B, Carpenter G, Abramowicz I, Gleeson JG, Paciorkowski AR, Cleveland DW, Dobyns WB, O'Driscoll M (2014) Mutations in CENPE define a novel kinetochore-centromeric mechanism for microcephalic primordial dwarfism. Hum Genet 133:1023–1039

Mourão MA, Hakim JB, Schnell S (2014) Connecting the dots: the effects of macromolecular crowding on cell physiology. Biophys J 107:2761–2766

Musacchio A (2011) Spindle assembly checkpoint: the third decade. Philos Trans R Soc Lond Ser B Biol Sci 366:3595–3604

Nekrasov VS, Smith MA, Peak-Chew S, Kilmartin JV (2003) Interactions between centromere complexes in *Saccharomyces cerevisiae*. Mol Biol Cell 14:4931–4946

Nilsson J (2015) Mps1-Ndc80: one interaction to rule them all. Oncotarget 6:16822–16823

Nishino T, Takeuchi K, Gascoigne KE, Suzuki A, Hori T, Oyama T, Morikawa K, Cheeseman IM, Fukagawa T (2012) CENP-T-W-S-X forms a unique centromeric chromatin structure with a histone-like fold. Cell 148:487–501

Nishino T, Rago F, Hori T, Tomii K, Cheeseman IM, Fukagawa T (2013) CENP-T provides a structural platform for outer kinetochore assembly. EMBO J 32:424–436

Nogales E (2016) The development of cryo-EM into a mainstream structural biology technique. Nat Methods 13:24–27

Orthaus S, Ohndorf S, Diekmann S (2006) RNAi knockdown of human kinetochore protein CENP-H. Biochem Biophys Res Commun 348:36–46

Palmer DK, O'Day K, Trong HL, Charbonneau H, Margolis RL (1991) Purification of the centromerespecific protein CENP-A and demonstration that it is a distinctive histone. Proc Natl Acad Sci U S A 88:3734–3738

Parigi G, Rezaei-Ghaleh N, Giachetti A, Becker S, Fernandez C, Blackledge M, Griesinger C, Zweckstetter M, Luchinat C (2014) Long-range correlated dynamics in intrinsically disordered proteins. J Am Chem Soc 136:16201–16209

Pereira AL, Pereira AJ, Maia AR, Drabek K, Sayas CL, Hergert PJ, Lince-Faria M, Matos I, Duque C, Stepanova T, Rieder CL, Earnshaw WC, Galjart N, Maiato H (2006) Mammalian CLASP1 and CLASP2 cooperate to ensure mitotic fidelity by regulating spindle and kinetochore function. Mol Biol Cell 17:4526–4542

Perpelescu M, Fukagawa T (2011) The ABCs of CENPs. Chromosoma 120:425–446

Pesenti ME, Weir JR, Musacchio A (2016) Progress in the structural and functional characterization of kinetochores. Curr Opin Struct Biol 37:152–163

Petrovic A, Pasqualato S, Dube P, Krenn V, Santaguida S, Cittaro D, Monzani S, Massimiliano L, Keller J, Tarricone A, Maiolica A, Stark H, Musacchio A (2010) The MIS12 complex is a protein interaction hub for outer kinetochore assembly. J Cell Biol 190:835–852

Petrovic A, Mosalaganti S, Keller J, Mattiuzzo M, Overlack K, Krenn V, De Antoni A, Wohlgemuth S, Cecatiello V, Pasqualato S, Raunser S, Musacchio A (2014) Modular assembly of RWD domains on the Mis12 complex underlies outer kinetochore organization. Mol Cell 53:591–605

Prendergast L, van Vuuren C, Kaczmarczyk A, Doering V, Hellwig D, Quinn N, Hoischen C, Diekmann S, Sullivan KF (2011) Premitotic assembly of human CENPs -T and -W switches centromeric chromatin to a mitotic state. PLoS Biol 9:e1001082

Primorac I, Weir JR, Chiroli E, Gross F, Hoffmann I, van Gerwen S, Ciliberto A. Musacchio A (2013) Bub3 reads phosphorylated MELT repeats to promote spindle assembly checkpoint signaling. Elife 2:01030.

Przewloka MR, Glover DM (2009) The kinetochore and the centromere: a working long distance relationship. Annu Rev Genet 43:439–465

Qiu S, Wang J, Yu C, He D (2009) CENP-K and CENP-H may form coiled-coils in the kinetochores. Sci China C Life Sci 52:352–359

Rago F, Gascoigne KE, Cheeseman IM (2015) Distinct Organization and Regulation of the Outer Kinetochore KMN Network Downstream of CENP-C and CENP-T. Curr Biol 25:671–677

Richter MM, Poznanski J, Zdziarska A, Czarnocki-Cieciura M, Lipinszki Z, Dadlez M, Glover DM, Przewloka MR (2016) Network of protein interactions within the Drosophila inner kinetochore. Open Biol 6:150238

Rosenberg JS, Cross FR, Funabiki H (2011) KNL1/Spc105 recruits PP1 to silence the spindle assembly checkpoint. Curr Biol 21:942–947

Santaguida S, Musacchio A (2009) The life and miracles of kinetochores. EMBO J 28:2511–2531

Scaërou F, Aguilera I, Saunders R, Kane N, Blottière L, Karess R (1999) The rough deal protein is a new kinetochore component required for accurate chromosome segregation in Drosophila. J Cell Sci 112:3757–3768

Scaërou F, Starr DA, Piano F, Papoulas O, Karess RE, Goldberg ML (2001) The ZW10 and Rough Deal checkpoint proteins function together in a large, evolutionarily conserved complex targeted to the kinetochore. J Cell Sci 114:3103–3114

Schleiffer A, Maier M, Litos G, Lampert F, Hornung P, Mechtler K, Westermann S (2012) CENP-T proteins are conserved centromere receptors of the Ndc80 complex. Nat Cell Biol 14:604–613

Schneider R, Maurin D, Communie G, Kragelj J, Hansen DF, Ruigrok RW, Jensen MR, Blackledge M (2015) Visualizing the molecular recognition trajectory of an intrinsically disordered protein using multinuclear relaxation dispersion NMR. J Am Chem Soc 137:1220–1229

Shepperd LA, Meadows JC, Sochaj AM, Lancaster TC, Zou J, Buttrick GJ, Rappsilber J, Hardwick KG, Millar JB (2012) Phosphodependent recruitment of Bub1 and Bub3 to Spc7/KNL1 by Mph1 kinase maintains the spindle checkpoint. Curr Biol 22:891–899
Stoler S, Keith KC, Curnick KE, Fitzgerald-Hayes M (1995) A mutation in CSE4, an essential gene encoding a novel chromatin-associated protein in yeast, causes chromosome nondisjunction and cell cycle arrest at mitosis. Genes Dev 9:573–586
Stumpff J, von Dassow G, Wagenbach M, Asbury C, Wordeman L (2008) The kinesin-8 motor Kif18A suppresses kinetochore movements to control mitotic chromosome alignment. Dev Cell 14:252–262
Suzuki A, Badger BL, Salmon ED (2015) A quantitative description of Ndc80 complex linkage to human kinetochores. Nat Commun 6:8161
Tachiwana H, Kagawa W, Shiga T, Osakabe A, Miya Y, Saito K, Hayashi-Takanaka Y, Oda T, Sato M, Park SY, Kimura H, Kurumizaka H (2011) Crystal structure of the human centromeric nucleosome containing CENP-A. Nature 476:232–235
Takeuchi K, Nishino T, Mayanagi K, Horikoshi N, Osakabe A, Tachiwana H, Hori T, Kurumizaka H, Fukagawa T (2014) The centromeric nucleosome-like CENP-T-W-S-X complex induces positive supercoils into DNA. Nucleic Acids Res 42:1644–1655
Tanaka K (2013) Regulatory mechanisms of kinetochore-microtubule interaction in mitosis. Cell Mol Life Sci 70:559–579
Tenboer J, Basu S, Zatsepin N, Pande K, Milathianaki D, Frank M, Hunter M, Boutet S, Williams GJ, Koglin JE, Oberthuer D, Heymann M, Kupitz C, Conrad C, Coe J, Roy-Chowdhury S, Weierstall U, James D, Wang D, Grant T, Barty A, Yefanov O, Scales J, Gati C, Seuring C, Srajer V, Henning R, Schwander P, Fromme R, Ourmazd A, Moffat K, Van Thor JJ, Spence JC, Fromme P, Chapman HN, Schmidt M (2014) Time-resolved serial crystallography captures high-resolution intermediates of photoactive yellow protein. Science 346:1242–1246
The Gene Ontology Consortium (2010) The gene ontology in 2010: extensions and refinements. Nucleic Acids Res 38:D331–D335
Thedieck K, Holzwarth B, Prentzell MT, Boehlke C, Kläsener K, Ruf S, Sonntag AG, Maerz L, Grellscheid SN, Kremmer E, Nitschke R, Kuehn EW, Jonker JW, Groen AK, Reth M, Hall MN, Baumeister R (2013) Inhibition of mTORC1 by astrin and stress granules prevents apoptosis in cancer cells. Cell 154:859–874
Thein KH, Kleylein-Sohn J, Nigg EA, Gruneberg U (2007) Astrin is required for the maintenance of sister chromatid cohesion and centrosome integrity. J Cell Biol 178:345–354
Toledo CM, Herman JA, Olsen JB, Ding Y, Corrin P, Girard EJ, Olson JM, Emili A, DeLuca JG, Paddison PJ (2014) BuGZ Is Required for Bub3 Stability, Bub1 Kinetochore Function, and Chromosome Alignment. Dev Cell 28:282–294
Tooley J, Stukenberg PT (2011) The Ndc80 complex: integrating the kinetochore's many movements. Chromosom Res 19:377–391
Uversky VN (2015) The multifaceted roles of intrinsic disorder in protein complexes. FEBS Lett 589:2498–2506
Varma D, Salmon ED (2012) The KMN protein network--chief conductors of the kinetochore orchestra. J Cell Sci 125:5927–5936
Vleugel M, Omerzu M, Groenewold V, Hadders MA, Lens SM, Kops GJ (2015) Sequential Multisite Phospho-Regulation of KNL1-BUB3 Interfaces at Mitotic Kinetochores. Mol Cell 57:824–835
Wan X, O'Quinn RP, Pierce HL, Joglekar AP, Gall WE, DeLuca JG, Carroll CW, Liu ST, Yen TJ, McEwen BF, Stukenberg PT, Desai A, Salmon ED (2009) Protein architecture of the human kinetochore microtubule attachment site. Cell 137:672–684
Wang X, Zhuang X, Cao D, Chu Y, Yao P, Liu W, Liu L, Adams G, Fang G, Dou Z, Ding X, Huang Y, Wang D, Yao X (2012a) Mitotic regulator SKAP forms a link between kinetochore core complex KMN and dynamic spindle microtubules. J Biol Chem 287:39380–39390
Wang Y, Sarkar M, Smith AE, Krois AS, Pielak GJ (2012b) Macromolecular Crowding and Protein Stability. J Am Chem Soc 134:16614–16618

Wei RR, Sorger PK, Harrison SC (2005) Molecular organization of the Ndc80 complex, an essential kinetochore component. Proc Natl Acad Sci U S A 102:5363–5367

Wei RR, Al-Bassam J, Harrison SC (2007) The Ndc80/HEC1 complex is a contact point for kinetochore-microtubule attachment. Nat Struct Mol Biol 14:54–59

Weis F, Giudice E, Churcher M, Jin L, Hilcenko C, Wong CC, Traynor D, Kay RR, Warren AJ (2015) Mechanism of eIF6 release from the nascent 60S ribosomal subunit. Nat Struct Mol Biol 22:914–919

Williams BC, Karr TL, Montgomery JM, Goldberg ML (1992) The Drosophila l(1)ZW10 gene product, required for accurate mitotic chromosome segregation, is redistributed at anaphase onset. J Cell Biol 118:759–773

Williams BC, Li Z, Liu S, Williams EV, Leung G, Yen TJ, Goldberg ML (2003) ZWILCH, a New Component of the ZW10/ROD Complex Required for Kinetochore Functions. Mol Biol Cell 14:1379–1391

Xue B, Dunbrack RL, Williams RW, Dunker AK, Uversky VN (2010) PONDR-FIT: a meta-predictor of intrinsically disordered amino acids. Biochim Biophys Acta 180:996–1010

Zaoui K, Benseddik K, Daou P, Salaün D, Badache A (2010) ErbB2 receptor controls microtubule capture by recruiting ACF7 to the plasma membrane of migrating cells. Proc Natl Acad Sci U S A 107:18517–18522

Zhang J, Wan L, Dai X, Sun Y, Wei W (2014) Functional characterization of Anaphase Promoting Complex/Cyclosome (APC/C) E3 ubiquitin ligases in tumorigenesis. Biochim Biophys Acta 1845:277–293

Zhang S, Chang L, Alfieri C, Zhang Z, Yang J, Maslen S, Skehel M, Barford D (2016) Molecular mechanism of APC/C activation by mitotic phosphorylation. Nature 533:260–264

Zhonghua Liu, Vong QP, Yixian Zheng (2007) CLASPing Microtubules at the trans-Golgi Network. Dev Cell 12:839–840

Zhou W, Wei W, Sun Y (2013) Genetically engineered mouse models for functional studies of SKP1-CUL1-F-box-protein (SCF) E3 ubiquitin ligases. Cell Res 23:599–619

Zhou Z, He M, Shah AA, Wan Y (2016) Insights into APC/C: from cellular function to diseases and therapeutics. Cell Div 11:9

Zich J, May K, Paraskevopoulos K, Sen O, Syred HM, van der Sar S, Patel H, Moresco JJ, Sarkeshik A, Yates JR 3rd, Rappsilber J, Hardwick KG (2016) Mps1Mph1 Kinase Phosphorylates Mad3 to Inhibit Cdc20Slp1-APC/C and Maintain Spindle Checkpoint Arrests. PLoS Genet 12:e1005834

Chapter 17
GroEL and the GroEL-GroES Complex

Noriyuki Ishii

Abstract Chaperonin is categorized as a molecular chaperone and mediates the formation of the native conformation of proteins by first preventing folding during synthesis or membrane translocation and subsequently by mediating the step-wise ATP-dependent release that result in proper folding. In the GroEL-GroES complex, a single heptameric GroEL ring binds one GroES ring in the presence of ATP/ADP, in this vein, the double ring GroEL tetradecamer is present in two distinct types of GroEL-GroES complexes: asymmetric 1:1 "bullet"-shaped GroEL:GroES and symmetric 1:2 "football" (American football)-shaped GroEL:GroES$_2$. There have been debates as to which complex is critical to the productive protein folding mediated by the GroEL-GroES complex, and how GroES coordinates with GroEL in the chaperonin reaction cycle in association with regulation by adenine nucleotides and through the interplay of substrate proteins. A lot of knowledge on chaperonins has been accumulating as if expanding as ripples spread around the GroEL-GroES from *Escherichia coli*. In this article, an overview is presented on GroEL and the GroEL-GroES complex, with emphasis on their morphological variations, and some potential applications to the fabrication of nanocomposites using GroEL as a nano-block. In parallel, a guideline is presented that supports the recognition that the *E. coli* and its GroEL-GroES complex do not always receive in standard literature because the biochemical features of chaperonins derived from others special, such as mammals, are not always the same as those confirmed using GroEL-GroES derived from *E. coli*.

Keywords Chaperonin • Electron microscopy • Equatorial split • GroEL • GroES • Hsp60 • Nanocomposite • Self-assembly • Chaperone • Protein folding

---

N. Ishii (✉)
Biomedical Research Institute, Department of Life Science and Biotechnology, National Institute of Advanced Industrial Science and Technology (AIST),
Tsukuba Central-6, 1-1-1 Higashi Tsukuba Ibaraki, 305-8566 Tsukuba, Japan
e-mail: n.ishii@aist.go.jp

J.R. Harris, J. Marles-Wright (eds.), *Macromolecular Protein Complexes*,
Subcellular Biochemistry 83, DOI 10.1007/978-3-319-46503-6_17

17.1 Introduction

Synthesis of proteins in the cell is promoted by many ribosomes distributed in the cytoplasm surrounding the nucleus. The amino acids are linked sequentially by peptide bonds and form a chain of polypeptide according to the translation of genetic information, leading to the autonomous folding of protein into its own active three-dimensional structure. More strictly, protein folding into its native tertiary structure is achieved automatically by each protein's amino acid sequence. This is the well known Anfinsen's dogma in molecular biology which implies in this context that the full information for both the native conformation and folding pathway is encoded in its primary structure (Anfinsen 1973). Anfinsen shared the Nobel Prize for chemistry in 1972 for their historic work with ribonuclease A. Since then, many purified proteins, which once had been denatured to random coil-like structures, have been confirmed to refold spontaneously *in vitro*. The folding is explained as a process driven by small differences in the Gibbs free energy between the unfolded and native states. Many exotic and exciting results relevant to protein folding and manners of self-assembly have been accumulating on the basis of the dogma. Consequently, scholars and researchers have been accepting the dogma. It may be said that the study of protein folding especially *in vitro* folding from denaturation is indeed itself the history of protein science.

How polypeptide chains of amino acids routinely and correctly fold into an active form in three-dimension is one of the most outstanding questions in structural biology. The entropy driven collapse of random coil-like conformation converges towards the so-called pre-fold 'molten globule' state. Some specific regions in the string of the amino acid sequence defined by genetic translation achieve secondary structures such as α-helices or β-strands, which successively makes an acquisition of the peculiar tertiary structures in an intermediate structure of 'molten globule' and in some cases, further followed by assembling into quaternary structures. However, how each sequence finds and reaches the correct structure among the number of possible conformations available within surprisingly short timescales is still the large challenging subject. Protein folding research has been promoted as one of the important issues in life sciences requiring further clarification and understanding despite having been studied by many scholars and researchers worldwide for many years. Currently, genomic information of many organisms has been deciphered, and amino acid sequence information of tens of thousands of proteins is available, yet it remains difficult even now to predict their three-dimensional structures.

On the other hand, although it is not the direct purpose of this article, the author will only briefly mention that there are exceptions to the rule that sequence uniquely determines structure, and that counterexamples have been discovered in the early 1980s. Prions are proteins which can have the same sequence but different tertiary structures. For example, a prion protein can exist as an aberrant β-sheet form or the normal α-helix form; the former causes a prion disease while the latter is expressed in a healthy human body without harm. In some cases, native proteins refold into a different stable conformation with inter-molecular interactions, which causes fatal amyloid formation. There are other structural states where the Anfinsen's dogma falls short. Proteins involved in Alzheimer's and Parkinson's diseases are counter-examples to the dogmas. It may be said that amyloid and related diseases are caused by changes in protein folding (misfolding). Furthermore, the existence of intrinsically disordered proteins is another interesting issue in connection with protein folding, however, the author shall not attempt to deeply cover this very important area at this time (Chouard 2011).

Getting back to the main subject, through billions of years of evolution, cells have developed the protein homeostasis (or proteostasis) network in order to cope with environmental stresses and to facilitate the correct folding for complicated larger protein molecules. Several proteinaceous components have been discovered that mediates the process. These components under the network form a family called "molecular chaperones". Ellis introduced the concept of molecular chaperone, and the current definition remains very similar to that given in 1987 (Ellis 1987). The component which differs from the original is associated with the final oligomeric structure, defined as a molecular chaperone. They are expressed ubiquitously in prokaryotes, eukaryotes, and in the cytosol as well as within organelles. In spite of the extremely crowded environment within the cell, thanks to the conserved systems of chaperones such as GroE (GroEL/GroES) or DnaK (DnaK/DnaJ/GrpE) systems (Ellis 1996; Horwich et al. 2006; Yamakoshi et al. 1998; Motohashi et al. 1994), newly synthesized proteins and possible refolding proteins, after having been post-translationally imported across biological membranes, are able to achieve their functionally active form by avoiding inter-molecular aggregation formation. Nearly 30% of *E. coli* proteins are reported to be aggregation prone without the assistance by chaperones (Niwa et al. 2009).

The most representative of the chaperone class of molecules is chaperonins, namely GroEL and GroES, which is the focus of this article. Chaperonins are distributed in all three kingdoms, and are classified into Group I and Group II according to the presence or absence of co-chaperonin, their amino acid sequences, and oligomeric structures. Group I chaperonins are found in the bacterial cytosol (GroEL) (Fig. 17.1), eukaryotic organelles (Hsp60) such as mitochondria (mtHsp60) and chloroplast (Rubisco binding protein), and some archaea cytosol (Klunker et al. 2003). Group II consists of the archaeal (thermosomes) (Trent et al. 1991) and eukaryotic cytosolic variants (TCP1, TCP1 ring complex (TRiC), or chaperonin containing TCP1 (CCT)) (Frydman et al. 1992; Kubota et al. 1994). Although some of those members can be heat-induced, chaperonins are essential in protein folding at almost all temperature ranges. Furthermore, chaperonins are constitutive proteins and are induced under a large variety of cellular stresses.

Proteins are the central building blocks to life. It is a polymer consisting of the amino acids which determine its three-dimensional shape. In a sense, it is the chaperonin that mediates the process by providing favorable conditions. Chaperonin assists in oligomeric assembly by preventing the formation of improper protein aggregates. The chaperonin-mediated folding reaction remains an interesting feature, for example, allosteric mechanism of the chaperonin coupled to ATP binding.

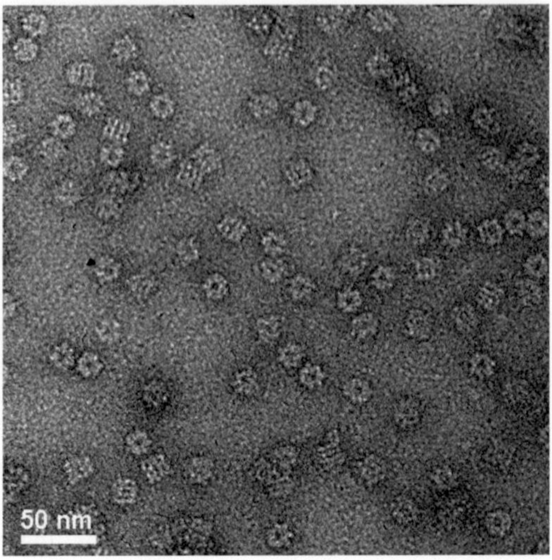

Fig. 17.1 An electron microscopic image of GroEL tetradecamer from *E. coli*. Seven-membered rings as viewed from top and rectangular (*square*) shapes with stripes from the side are seen

Although there are numerous superb publications reviewing chaperonins (Ellis et al. 1993; Ellis 1996; Lorimer and Todd 1996; Fink and Goto 1998; Dobson 2003; Horwich et al. 2006; Zanin-Zhorov and Cohen 2007; Richter et al. 2010; Hartl et al. 2011; Saibil et al. 2013), especially the GroEL-GroES complexes, one of the greatest unsolved problem thus far in the chaperonin field is how GroES coordinates with GroEL in the chaperonin function reactions regulated by ATP (both binding and hydrolysis reaction) and with the existence or nonexistence of substrate protein folding intermediates. In this article, the author will present an overview on chaperonins of biological macromolecular machines, GroEL and the GroEL-GroES complex, and will introduce some potential applications to the fabrication of stimuli-responsive, mechanical movable nanocomposites using GroEL as a nano-block.

17.2 Molecular Structures

It is now generally accepted that protein folding in the cell needs mediation by molecular chaperone components. Chaperonins are member of a subcategory under molecular chaperones. Although the term "GroEL" has been used synonymously with chaperonin, strictly speaking, GroEL is a product of *groEL* gene (growth essential large) from *E. coli* (Fig. 17.1). The *groEL* gene encodes a polypeptide of 548 amino acids, and is part of *groE* operon that also contains *groES* gene (growth essential small) for GroES. GroES is known as cochaperonin and consists of 97 amino acids (Chandrasekhar et al. 1986). The primary structures for both, GroEL and GroES, as well as their homologs from different organismic species are available in internet databases (Hill et al. 2004, http://www.cpndb.ca).

When talking about structure-function relationship, there always seems to be a compromise between the best achievable resolution of the analytical technique versus the depth we desire to know about a molecular functional mechanism on the basis of the protein structure. In the early stage of the investigation on GroEL and of course thus far, transmission electron microscopy (TEM) has played an important role providing precious morphological information on its structures to delineate as a molecular model (Hohn et al. 1979; Hendrix 1979). While X-ray crystallography needs three-dimensional single crystals with appropriate sizes in which each protein molecule in a unique conformation is regularly arranged, electron microscopy is suitable for structural analysis of huge macromolecules with larger dynamical structural changes. Such characteristics of TEM have been, in later years, combined with the cryogenic technology, which has brought a new horizon in protein structural biology (Fujiyoshi 1998; Glaeser 1971, 1985; Roseman et al. 1996). Due to the limitation on the resolution available at the early days of investigation, over-speculation regarding molecular model reconstruction based only on projection image data by electron microscopy under the limited resolution had frequently occurred. We have presented another model, correcting the previously prevailing model, as to the manner of subunit arrangements in the GroEL tetradecamer by using the electron microscopic projection image of GroEL homolog isolated from a thermophilic eubacteria, *Thermus thermophiles* (Ishii et al. 1992). We have reported the functionally active form as holo-chaperonin, which is now known to be an asymmetric 1:1 "bullet"-shaped GroEL:GroES complex, although the origin is different from *E. coli* (Ishii et al. 1992; Taguchi et al. 1991). Later, the structure of GroEL tetradecamer determined by X-ray crystallography has verified the correctness in the subunits' arrangement of our molecular structure model for GroEL (Braig et al. 1994).

GroEL generally exists as double layered homo-heptameric rings having totally about 800 kDa, in which two rings with seven 57 kDa subunits are stacked back to back (isologously), subsequently adopting cylindrical and hollow shaped appearance and enclosing two large but non-contiguous central cavities. The cochaperonin, GroES is a single layered heptameric ring consisting of seven 10 kDa subunits with mobile loops extending from the rim, which attaches coaxially to the ends of the GroEL cylinder. The X-ray structure of the cylindrical GroEL tetradecamer has been determined at 2.8 Å resolution (Braig et al. 1994) (PDB entry ID: 1GRL), and that of the asymmetric 1:1 bullet-shaped GroEL:GroES complex with ADP_7 has been determined at 3.0 Å resolution (Xu et al. 1997) (PDB entry ID: 1AON). According to the crystal structures determined by X-ray crystallography at atomic resolution, 18 α-helices and 19 β-strands are assigned for the secondary structural elements in one GroEL subunit. Meanwhile, 9 β-strands and no α-helices are assigned in GroES subunit. In GroEL, the subunit consists of three structurally distinguishable domains as follows (Fig. 17.2): the apical domain that plays an important role in binding to GroES or substrate proteins using its hydrophobic surface region, which constructs the end-on surface of GroEL; the equatorial domain that has an adenine nucleotide (ATP/ADP) binding pocket, is located at the middle region in the cylindrical GroEL tetradecamer so as to form an equatorial interface

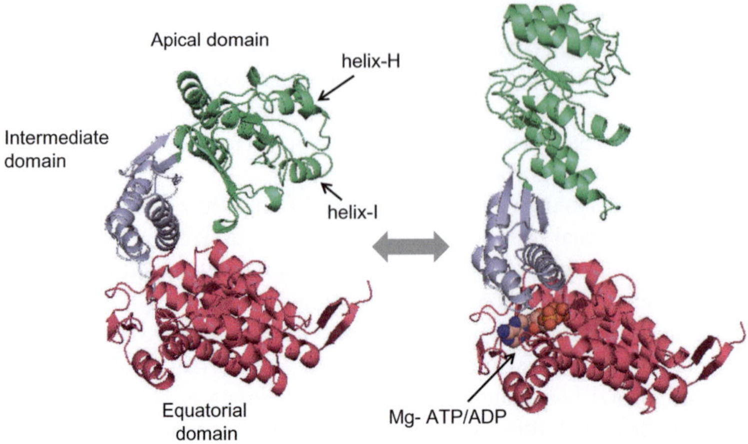

Fig. 17.2 The crystal structures of GroEL subunits, with and without a bound nucleotide. Helices H and I, are the binding site for the substrate protein. Upon binding of nucleotide, the upward movement of apical- and intermediate- domains creates a large expanded cavity on the *cis*-side, which facilitates the substrate protein encapsulation to the Anfinsen's cage

between the two heptameric GroEL rings (*cis*- and *trans*-rings); and the intermediate domain with a hinge region is located between the apical and the equatorial domains connecting both, and that also transmits allosteric signals between them. Upon binding of ATP to the GroEL ring, it induces intra-ring positive cooperative upward movement of the intermediate and apical domains, which prompts the formation of GroEL that binds GroES (*cis*-GroEL ring). The formation of a metastable GroEL-GroES complex with a folding intermediate appears to be taking place at the common site for substrates that interact with GroEL (Horowitz et al. 1999) and yet, there is still some debate regarding this.

17.3 Various Structural Forms of the GroEL-GroES Complex

GroEL from *E. coli* is usually isolated and purified as GroEL tetradecamer alone (without GroES bound), and its crystal structure has been determined by X-ray crystallography (Braig et al. 1994). Although the GroEL tetradecamer has been considered as a constituent unit, which is the general consensus nowadays among researchers, a single-ring form of GroEL heptamer has been detected in some Group I chaperonins such as mitochondrial Hsp60 (Viitanen et al. 1992), *Paracoccus denitrificans* GroEL (Ishii et al. 1992; Sumi et al. 1992) and *Thermoanaerobacter brockii* GroEL (Todd et al. 1995). The mammalian mitochondrial chaperonin, mtHsp60 has been reported to function predominantly as a single-ring rather than a double-ring complex (Viitanen et al. 1992; Nielsen and Cowan 1998). Chen et al. have reported on the reconstruction from a single particle analysis applied to the cryogenic electron microscopic images of the single-ring GroEL-GropES complex using an ATPase-deficient single ring mutant (SR398) of GroEL and has a discussion on the dynamic conformational change (Chen et al. 2006). Interestingly, the C-terminal region of GroEL subunit that was unresolved in the crystal structure (Braig et al. 1994) emerged from the density map, and the truncated cone-shaped cavity appears to transform to an ellipsoidal cavity when encapsulating the larger substrate protein (Chen et al. 2006). The dynamic nature of conformational changes and its continuous structural variation would not be easily determined by crystallographic approach. There may be more conformational variability that has not yet been previously observed. It should be important to investigate further the aspect of single-ring chaperonin behavior.

Since the single GroEL ring consisting of seven 57 kDa subunits binds one GroES heptamer ring in the presence of ATP/ADP, in this vein, the double ring form of GroEL tetradecamer has allowed to take two distinct types of GroEL-GroES complex formation, that is, asymmetric 1:1 bullet-shaped GroEL:GroES and symmetric 1:2 "football" (American football) -shaped GroEL:GroES$_2$ complexes. The bullet-shaped GroEL-GroES complex is the form with GroES bound to only one end of the GroEL tetradecamer, and the football-shaped GroEL-GroES$_2$ complex is with two GroES bound to both ends of the GroEL tetradecamer (Fig. 17.3). The X-ray crystal structures have been determined for both, the bullet-shaped GroEL-GroES (Xu et al. 1997) and the football-shaped GroEL-GroES$_2$ complexes (Koike-Takeshita et al. 2014) (PDB entry ID: 3WVL).

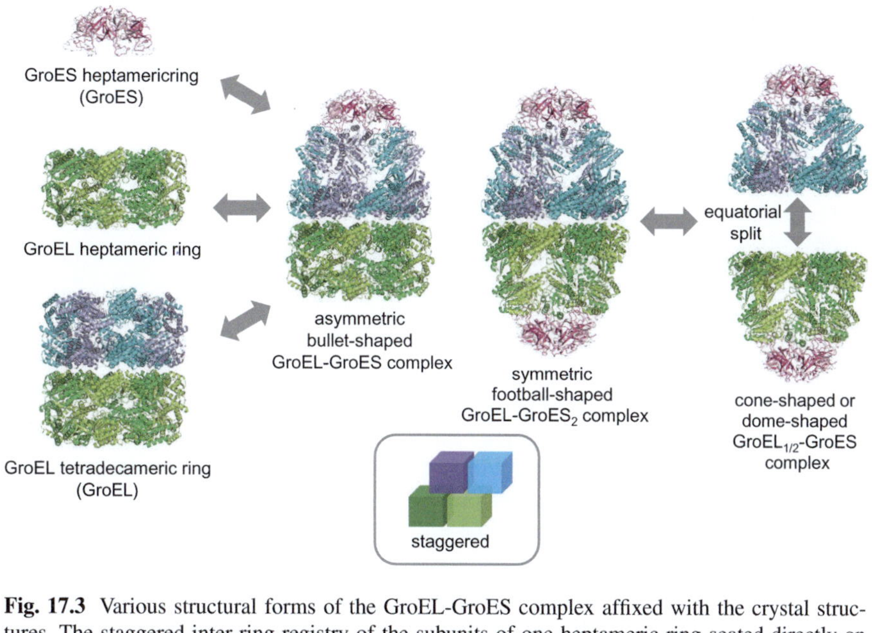

Fig. 17.3 Various structural forms of the GroEL-GroES complex affixed with the crystal structures. The staggered inter-ring registry of the subunits of one heptameric ring seated directly on subunits in the second ring is illustrated (PDB entry IDs: 1GRL, 1AON, 3WVL)

Recently the bullet-shaped complex formation is understood to occur only in the presence of ADP, meanwhile formation of both the bullet-shaped and the football-shaped complexes are reconciled with the presence of adenine nucleotides having ATP-configuration including the non-*hydrolyzable* ATP analogues. The football-shaped GroEL-GroES$_2$ complex formation appears not to be controlled by a single parameter, because the football-shaped complex form has been detected under a variety of combined conditions such as [ATP]/[ADP] ratio, [K^+] concentration, and further the presence or the absence of non-native substrate protein or folding intermediate (Azem et al. 1994; Schmidt et al. 1994).

Taking a closer look at the crystal structure of GroEL, the Group I chaperonin, it should be important to describe the manner in which the subunits of one heptameric ring are seated directly on a subunit in the second ring, which is the staggered inter-ring registry as illustrated in Fig. 17.3 (Lopez et al. 2015). This is an important feature in understanding the dynamic morphological transition of the GroEL-GroES complex. It is now believed that the symmetric football-shaped complex is a precursor to the equatorial split, which leads to a single-ring GroEL-GropES complex (cone-shaped or dome-shaped complex) formation (Ishii et al. 1995). The split at the equatorial plane of the symmetric GroEL-GroES complexes would be attributed to the weakened inter-ring interactions between the *cis*- and *trans*- heptamer rings in the complex, which has smaller contact interaction energy due to lack of a salt bridge at the inter-ring interface, as compared with the bullet-shaped complex (Ishii and Sato 2013; Taguchi 2015).

On the other hand, the filamentous complex (one-dimensional assembly) of*E. coli* GroEL and GroES has been reported to form when GroEL and GroES at relatively low molecular rate (1:1–1:3) are incubated at room temperature in the presence of ATP and magnesium (Harris et al. 1995). Such filamentous cytoskeleton-like assemblies consisting of chaperonin molecules has attracted keen interests from nano-biotechnology and nano-biomedical research fields (Biswas et al. 2009; Muramatsu et al. 2006). The author will introduce examples that may lead to further insight into bridging protein (bio-macromolecular) science with chemical nano-biotechnological research.

17.4 Function and Chaperonin Reaction ATPase Cycle

A cylindrical GroEL tetradecamer and its lid-like cofactor GroES heptamer form a nano-cage in which a single polypeptide chain is transiently enclosed and prompted to fold properly. Apparently during folding, the substrate proteins or polypeptides are isolated from the bulk phase of the solution (or cytosol in the cell). Interestingly, GroEL tetradecamer is also required for its own assembly (Cheng et al. 1990; Motojima and Yoshida 2010). Urea-denatured GroEL appears to first reassemble spontaneously into tetradecamer and then catalyze further assembly in Mg-ATP dependent chaperonin reaction (Lissin et al. 1990). On the other hand, GroES has been reported to self-assemble under appropriate condition *in vitro* (Mascagni et al. 1991).

The function and the GroEL-GroES chaperonin reaction mechanism by which there is general consensus is as follows (Horwich et al. 2006; Walter 2002); First, a non-native protein or a polypeptide chain binds as a molten globule-like folding intermediate to hydrophobic sites located on one of the apical domain of seven candidate subunits of GroEL and then followed by intra-ring positive cooperativity. Subsequently, upon binding of ATP to GroEL, *cis*-GroEL subunits change to the elongated conformation (Fig. 17.2) and GroES caps the GroEL ring that holds the substrate (termed *cis*-ring of GroEL), then the unfolded protein is encapsulated in the cavity underneath GroES. The enclosed chamber known as Anfinsen's cage is large enough to accommodate proteins up to ~60 kDa. The second GroES and another ATP cannot bind to the opposite GroEL ring (termed *trans*-ring) until ATP in the *cis*-ring is hydrolyzed, due to the negative cooperativity between the two rings of GroEL (Horovitz and Willison 2005). Thus the encapsulated protein is rendered to fold in the hydrophobic environment of the elongated cavity of GroEL caped with GroES. The grace time for the protein substrate to (re)fold is equal to a period required for the hydrolysis of the seven ATP molecules bound in the *cis*-ring. After ATP in the *cis*-ring is hydrolyzed, GroES, ADP, and the committed protein are released in a step-wise manner from the *cis*-ring by the binding of ATP and another unfolded protein to the *trans*-ring. GroES is thought to regulate the step-wise process. This occurs irrespective of whether the protein has fully folded, meaning most of proteins released are still in non-native form that has to be rebound by GroEL for a further trial at folding. The second GroES associates with the *trans*-ring of GroEL, forming another *cis*-ring to succeed the chaperonin reaction cycle (Hayer-Hartl et al. 1995; Engel et al. 1995). Therefore, which GroEL heptamer ring may be active at a time for binding to the substrate protein should be dependent on the regulation by GroES. The process can be interpreted as a two-stroke engine, and the current mechanistic model appears to well explain the relationship between the chaperonin-assisted protein folding and the molecular morphology of the asymmetric bullet-shaped GroEL-GroES complex.

Electron microscopy has often played a critical role in structural biology, and continues to open new horizons. Analyses by electron microscopy revealed the existence of football-shaped complex during the chaperonin reaction cycle (Grallert and Buchner 2001). There had been an agreement that football-shaped complex might occur as a transient intermediate state in the reaction cycle, when GroES bound to the *trans*-ring of GroEL before the release of GroES from *cis*-ring was completed. Koike-Takeshita et al. demonstrated the formation of a symmetric GroEL-GroES$_2$ complex using a slow ATP-hydrolyzing GroEL mutant (D398A) in the presence of ATP (Koike-Takeshita et al. 2008). The interaction between GroEL and GroES was investigated using fluorescence resonance energy transfer (FRET) by Sameshima et al., and they found that nearly equivalent amounts of the asymmetric GroEL-GroES and the symmetric GroEL-GroES$_2$ complexes coexist during the reaction cycle (Sameshima et al. 2008). They further characterized the emergence of the football-shaped complex and concluded that denatured proteins facilitate the dissociation of ADP from the *trans*-ring of GroEL and the concomitant association of ATP and the second GroES, subsequently facilitating the formation of the symmetric football-shaped complex (Sameshima et al. 2010).

Recent reports suggest that the presence of non-native substrate protein affects the GroEL-GroES reaction by shifting its morphology from the asymmetric bullet-shaped to symmetric football-shaped complexes. Haldar et al. have reported on the relative population in the formation between the bullet-shaped and the football-shaped GroEL-GroES complexes recently (Haldar et al. 2015). They have characterized the symmetric GroEL-GroES$_2$ complex as substantially populated by the association of non-foldable proteins with asymmetric GroEL-GroES complex. Such non-foldable proteins like β-casein, α-lactalbumin, and so on appear to overstimulate the GroEL ATPase activity and uncouple the negative GroEL inter-ring allostery. It is also worthwhile to investigate the biochemical role of intrinsically disordered or unfolded proteins in the cell. Thus in the symmetric GroEL-GroES$_2$ complex mode, both GroEL rings bind to GroES simultaneously, realizing the folding active form. They also mention that the asymmetric GroEL-GroES complex is dominant both in the absence of substrate proteins and in the presence of foldable substrate proteins. Furthermore, uncoupling of the GroEL ring and formation of symmetric GroEL-GroES$_2$ complex is suppressed at physiological [ATP]/[ADP] concentration. The interaction with non-foldable substrate proteins appears to weaken the negative cooperativity between the two GroEL rings, extending the time to bind and hydrolyze ATP, but it does allow binding to GroES. However, their conclusion is that the asymmetric GroEL-GroES complex represents the main folding active form. As they suggested, whether the presence of non-foldable or foldable substrate proteins makes a difference on the GroEL-GroES complex formation to either asymmetric bullet-shaped or symmetric football-shaped complexes remains to be clarified.

Taguchi has recently studied the GroEL-GroES chaperonin reaction cycle from the viewpoint of the symmetric football-shaped GroEL-GroES$_2$ complex (Taguchi 2015). The ATPase kinetics of GroEL appears to change when substrate proteins exist, and the nucleotide exchange should be accelerated by the substrate protein, leading to the symmetric football-shaped formation. Although there have been debates as to which complex, either the asymmetric bullet-shaped or symmetric football-shaped, is critical to the productive folding assisted by GroEL-GroES chaperonin, both complexes appear not to be mutually exclusive. The substrate protein itself configures the shape of the folding cradle that suits each protein, switching GroEL-GroES chaperonin reaction cycling from the asymmetric bullet-shaped to the symmetric football-shaped complexes, and *vice versa*. Fundamentally different mechanisms might underlie the GroEL-GroES reaction cycle in the absence and in the presence of substrate protein, and depend on whether the substrate protein is foldable or non-foldable.

Our understanding of how the substrate protein interacts with the Anfinsen's cage of *cis*-GroEL ring capped with GroES has been undergoing revision. Although the polypeptide of unfolded substrate protein is completely contained in the Anfinsen's cage, which has become a current prevailing dogma described in textbooks, recent stimulating results reported by Motojima and Yoshida (2010) seem to support that the polypeptide is not completely confined to the cavity, and that the folding proceeds while the elongated polypeptide partly protrudes out of the cavity (Motojima and Yoshida 2010; Ishii et al. 1994).

Furthermore, taking together the fact that the equatorial split happens at the equator plane of the complex, resulting in two-cone (or dome-shaped) $GroEL_{1/2}$-GroES complexes (here, $GroEL_{1/2}$ stands for GroEL heptameric ring), investigation and analysis to clarify mutual relationships between morphological transitions is the next challenge. In such a challenge, characteristic figures such as asymmetric bullet-shaped, symmetric football-shaped and cone- or dome-shaped GroEL-GroES complexes as well as single heptameric rings of GroEL and GroES are positioned in the chaperonin-mediate reaction cycles. Clarifying the connections and positioning of these complexes to appropriate sites in the reaction cycles should be a primary goal leading to better understanding of the chaperonin reaction cycle.

17.5 Applications as the Nanomolecular Machine

A protein has a peculiar higher-order structure, and each has a physiological function based on the original structure. The sophisticated structure and function are products of a long process of protein evolution. Some architecture possesses such sophisticated and minute functionality at the nanometer-scale which can hardly be achieved by human engineering (Ishii et al. 1991, 2010; Zahn et al. 1993; Ishii 2014). In this section, the author will introduce the development of exotic functional materials by utilizing the cylindrical structure of GroEL.

17.5.1 As a Carrier of the Artificial Substances

In order to utilize the GroEL macromolecule as a functional nano-material, one of the important criteria should be the possibility of introducing the different artificial substances from the original substrate of protein folding intermediates into the cavity of the molecular complex. Another criterion is whether it can be released by intra-molecular structural changes using energy from the hydrolysis of Mg-ATP, which is exactly the feature of GroEL. As the details have been reported by Ishii et al. (2003), here we briefly outline the procedures (Fig. 17.4). First, we examined experimental conditions for the preparation of the nanoparticle-GroEL conjugates, by introducing the semiconductor nanoparticles (CdS) with proper size (2.2 nm in a diameter), for the stabilization of the nanoparticles, and for the releasing of nanoparticles in response to the hydrolysis of Mg-ATP. The semiconductor nanoparticles such as CdS have fluorescent properties, which aid in detecting the uptake and the release of CdS by GroEL. The CdS nanoparticles prepared in the water-soluble solvent, dimethylformamide was mixed and gently stirred with GroEL in 25 mM Tris-HCl buffer (pH 7.5) containing 100 mM KCl for ~100 h at 4 °C, then purified

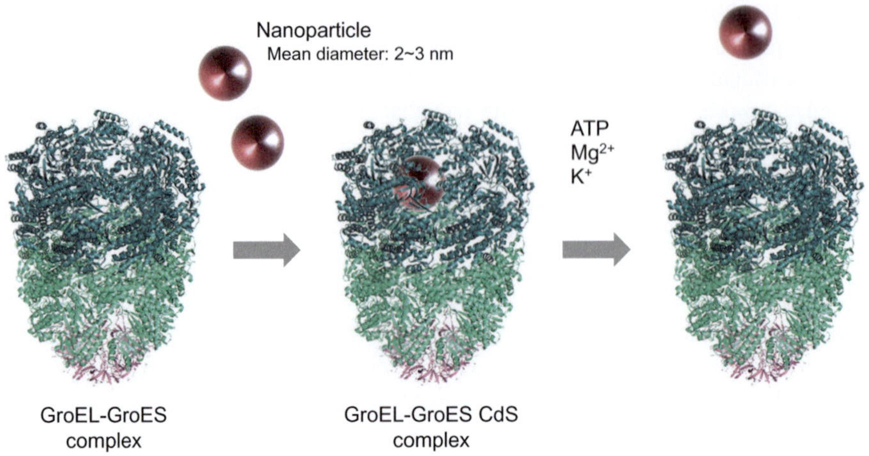

Fig. 17.4 Schematic representation of the formation of GroEL-GroES plus CdS nanoparticle complexes by inclusion of CdS nanoparticles into the cylindrical cavity of the asymmetric bullet-shaped GroEL-GroES, and its ATP-triggered guest release (Ishii et al. 2003)

and fractionated by gel filtration column chromatography. The formation of the GroEL, which encapsulated a CdS nanoparticle in the central cavity, was confirmed visually by TEM, as well as by fluorescence spectroscopy. The resultant complex was thermally stable against any denaturation, decomposition, and the release of nanoparticles up to 50 °C. The heat-resistant temperature of the complex appeared to be consistent with that of the species habitat of GroEL origin. Although the GroEL-CdS nanoparticle composite body appears stable, CdS nanoparticles which have been encapsulated is released immediately after the addition of ATP, MgCl, and KCl into the buffer solution. It is remarkable that even after the release of the nanoparticles, GroEL maintains its higher-order structure. The condition for such release is realized only when the three components of ATP, Mg^{2+}, K^+ ions are present, and it is similar to the conditions required for functional expression of GroEL chaperonin *in vivo*. From the above, it has been found that GroEL chaperonin uptakes guest compounds even for the artificial substances employing the same molecular mechanism as the native protein folding intermediates, and then release them under the control of ATP binding and hydrolysis.

17.5.2 As an "AND" Logic Gate

As mentioned in the previous section, GroEL chaperonin can uptake (even artificial) substances and release under ATP control. When considering its use as a carrier for functional materials, it would be more advantageous to be controlled by another physical stimulus. In addition to the chemical control responding to ATP, Muramatsu et al. (2006) have reported on the creation of a photo-responsive functional GroEL that would be controllable by light. Although there are three cysteine residues in the 57 kDa subunit constituting GroEL, all of them were replaced by genetic engineering to alanine, then, the mutant subunit was prepared so that lysine-231 which is located at the entrance of the central cavity was substituted with cysteine. The modified GroEL mutant introduces an azobenzene moiety as a photo-responsive gate around the entrance of the central cavity. This was prepared by chemical binding the azobenzene derivative having a maleimide site to K231C GroEL (Fig. 17.5). The modified GroEL chaperonin was confirmed to have similar architecture to wild-type GroEL, and is capable of incorporating the denatured protein (green fluorescent protein) into the cavity. It is well known that the azobenzene moiety can change its form from *trans*- to *cis*-form by ultraviolet irradiation, and from *cis*- to *trans*-form by visible light irradiation, meaning that its isomerization can be controlled reversibly by the wavelength of irradiation. In *trans*-form, the molecule takes the extended structure, and the transition has resulted in *cis*-form in a relatively short structure, therefore, it is possible to control the size of the opening of the gate by selecting irradiation as either ultraviolet or visible light (Fig. 17.5). The effect of ATP on modified GroEL appears to be similar to that of the wild-type GroEL, that is, the guest emission is very slow in the absence of ATP. In contrast, in the presence of ATP, when azobenzene site is isomerized to *cis*-form under the irradiation with ultraviolet light, the guest protein is released rapidly. On the other hand, when isomerized to *trans*-azobenzene form by the irradiation with visible light, the release rate has been significantly reduced. These results can be interpreted as follows; taking each of the ATP addition and the ultraviolet light irradiation as an independent input signal, only when both are loaded, the output (release of denatured protein or folding intermediate) to respond occurs, and this is considered as an "AND" logic circuit. The above is one example of the exotic nano-scaled applications using the protein function of GroEL in the construction of the logical operator in chemical biotechnology.

17.5.3 As a GroEL Nanotube

Thus far, even in the wild-type GroEL-GroES system, formation of the one-dimensional tubular-like assembly has been reported (Harris et al. 1995). However, its formation is difficult to control, and such tubular-like assemblies appear to be by-chance products associated with specimen grid preparation for electron microscopy.

The addition of ATP binding and light control to the functions of the GroEL chaperonin, the chemically modified mutant GroEL can form a tubular assembly aligned one-dimensionally under certain conditions. The author would like to introduce this achievement by Biswas et al. briefly (Biswas et al. 2009, 2013).

The spiropyran known as photochromic molecules of which polarity is greatly changed by light irradiation has been introduced site-specifically at around the entrance of the central cavity of GroEL. Thereby the spiropyran / merocyanine units

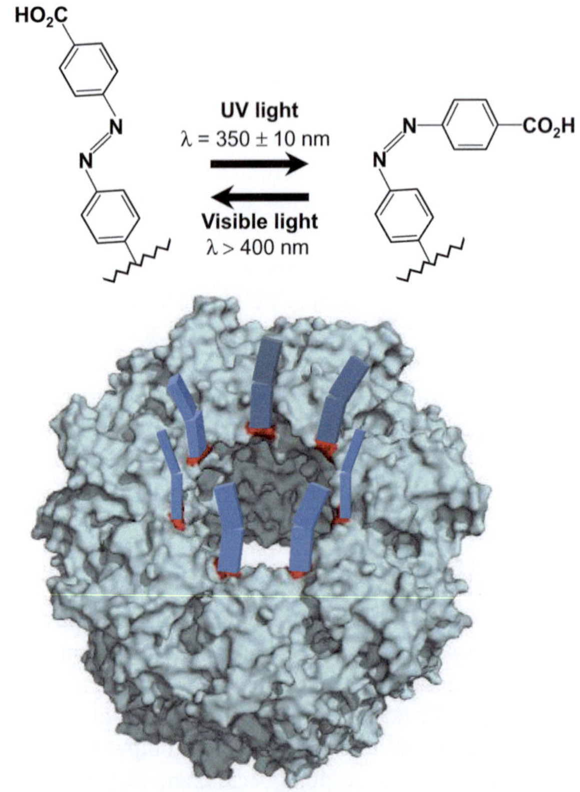

Fig. 17.5 Schematic representation of structural changes of the azobenzene-based photomechanical gates of azo-GroEL upon exposure to UV and visible lights (Muramatsu et al. 2006)

contribute to the gate switching between the open and the close motions of GroEL by ATP and also light. Construction of GroEL$_{SP/MC}$ was as follows; Briefly, all cysteine residues of each subunit of the native GroEL have been replaced with alanines, and then lysine-311 and leucine-314, and those residing on the end-on surface have been each substituted with a cysteine, respectively. Finally, the GroEL mutant having 14 cysteine residues in each entrance of the cavity was prepared, which means the resultant GroEL mutant has 28 cysteine residues that can be modified on the both end-on surfaces. The introduction of spiropyran to the cysteine sites has been achieved by applying a spiropyran derivative having a maleimide site to the GroEL mutant. After incubation, allowing the reaction of cysteines with spirobenzopyran-appended maleimide for 12 h at 4 °C, the color of the mixture turned light-purple indicating that the spontaneous reaction of partial isomerization of spiropyran to merocyanine had occurred. Therefore, spiropyran and merocyanine coexist in the buffer (Fig. 17.6). Using the gel filtration chromatography to remove the unreacted substances, the purified protein was used as GroEL$_{SP/MC}$. After treatment with the modification reaction, and when magnesium chloride was supplemented to the GroEL$_{SP/MC}$ mutant, the ratio of the polymeric regions of the elution profile from the gel permeation chromatography (GPC) significantly increased.

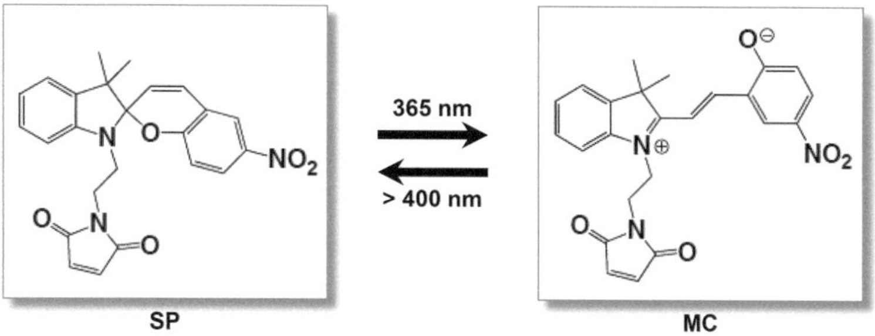

Fig. 17.6 Isomerizaton behavior of a spiropyran (SP)/melocyanine (MC) derivative (Biswas et al. 2013)

In the TEM observation of the eluate fraction corresponding to the polymeric region, unexpectedly, the modified $GroEL_{SP/MC}$ chaperonins self-assembled into a long tubular architecture (Fig. 17.7). The conditions for $GroEL_{SP/MC}$ to form self-linked one-dimensional assemblies, that is, $GroEL_{SP/MC}$ chaperonin nanotubes, were optimized assessing the buffer conditions, incubation temperature and periods and so on. Careful inspection confirmed that the chemical modification with spiropyran / merocyanine played critical role in contributing to its formation. The influence of coexistence of monovalent, or divalent cations, and with/without addition of nucleotides, on the nanotube formation was investigated by TEM (Ishii 2013). From a sequential study of the addition of metal salt, not only magnesium chloride but divalent metal ions such as calcium and zinc have been found to be effective in facilitating nanotube formation. From the above considerations, the locations of the spiropyran / merocyanine sites with the coordination of the metal ions are the most important factors to contribute the $GroEL_{SP/MC}$ nanotube formation. Moreover, the addition of ethylenediaminetetraacetic acid (EDTA) to the formed nanotubes resulted in the nanotube dissociation into the components of the building blocks. The divalent metal ions are chelated by EDTA and removed, therefore, it has been confirmed that the coordination plays an important role in the $GroEL_{SP/MC}$ nanotube forming.

Next, when the bovine α-lactalbumin, which had been denatured and fluorescently labeled, was applied to the $GroEL_{SP/MC}$ in the absence of any divalent metal ions, the unfolded α-lactalbumin was found to be incorporated by the $GroEL_{SP/MC}$ in the same manner as the wild-type GroEL. When magnesium chloride was added to this condition, it was found that the nanotube formation occurs while $GroEL_{SP/MC}$ encapsulates the modified α-lactalbumin in the cavities. Thus, by taking advantage of the chaperonin function of capturing the guest compound within the central cavity, Biswas et al. have succeeded in the $GroEL_{SP/MC}$ nanotube formation keeping the guest molecules within the nanotube (Biswas et al. 2013). Surprisingly, the chaperonin nanotube encapsulating the substances in the cavities can penetrate across the bio-membrane into the living cell, and then dissociate into compartments releasing the guest substances in the cellular ATP-responsive manner (Biswas et al. 2013). Furthermore, Sim et al. recently reported the construction of magnetic field

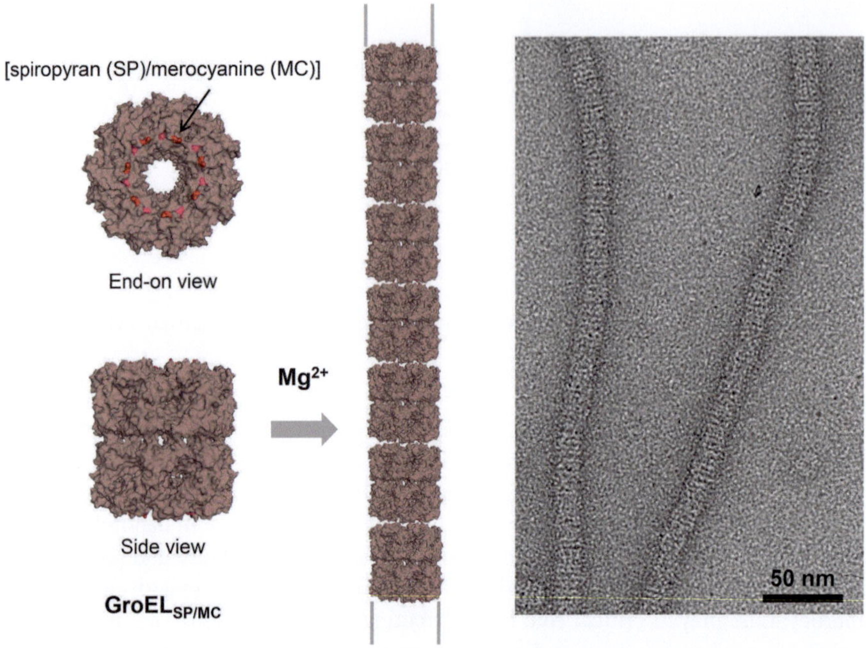

Fig. 17.7 Schematic representations of the $GroEL_{SP/MC}$ forming into a nanotube from mutant $GroEL_{SP/MC}$ and Mg^{2+}-mediated supramolecular polymerization, affixed with TEM image

sensitive GroEL nanotube (Sim et al. 2015). These nano-machines are potential components towards the fabrication of intelligent nano-devices.

17.6 Closing Remarks

The focus of modern protein science research has moved from the study of individual proteins to protein systems research, targeting the dynamics of the cellular process and interactions on the set of molecules involved using a wide range of techniques from genomics to molecular probe and imaging science. Although the fundamental mechanisms adopted in any biological processes in different species appear to be the same, there have been variations in the regulation mechanisms. For example, proteins sharing high sequence similarities over species which were evolutionarily derived from a common ancestor sometimes show different characteristics with distinguishable mechanisms. Proteins found in different species yet classified within the same nomenclature can exhibit differences in mechanism, since they adapt to changing physiological states (homeostasis) on its evolutionary path to diversity. During the process, proteins must evolve and cope with the host cell to build compensation mechanisms and networks entangled intricately by both intracellular and inter-cell interactions. Therefore, even if one protein derived from a certain species is characterized well with clear-cut results, its characteristics do not always agree with its counterpart proteins of the same family but from a different species.

By X-ray crystal structure analysis which began in 1960, the three-dimensional protein structures, such as myoglobin, hemoglobin, and lysozyme were determined one after the other. That of the huge protein complex of GroEL was on the cover of the journal *Nature* in 1994. The era has arrived when one can understand the function of a certain protein on the basis of structure at the atomic resolution. At the same time, a concern about whether a protein (or protein complex) coming from a particular species can constitute a representative model for all species has emerged in the author's mind as follows; lysozyme for example, has been investigated thoroughly and is thus frequently featured as a typical model protein in textbooks of protein science. Lysozyme crystals are obtained using NaCl as a precipitant. However, treating lysozyme as the archetypal protein for crystallization trial procedures to be mimicked with other proteins can lead to failure.

It is inevitable that a specific target protein from a particular organism will gain more exposure and significance with time, especially if it was discovered at the early stages of research. Therefore, finding counterexamples to the prevailing theory can at times be misinterpreted as being exceptions to the prevailing model. The Group I chaperonin research has been expanding around the GroEL-GroES derived from *E. coli*, and it is certain that a lot of valuable knowledge has been obtained. It is about time to that we have to recognize that the *E. coli* and the GroEL-GroES complex derived from *E. coli* should not be treated exclusively as the archetype model for the chaperonin complex. As Okamoto et al. have recently reported, the mammalian Group I chaperonin homolog (Hsp60-Hsp10) is poorly understood while GroEL-GroES system of bacterial origin has been analyzed in detail. In mammalian cells, many physiological functions such as protein folding, assembly, transports across membranes are assumed to be mediated by Hsp60, and there is a report that mammalian Hsp60 exists in heptameric single ring structure (Okamoto et al. 2015) and that the Hsp60-Hsp10 reaction cycle is different from the GroEL-GroES reaction cycle (Parnas et al. 2012; Vilasi et al. 2014). Biochemical features of the chaperonins derived from mammals and culture cells have been accumulating recently, and those are not always the same as those confirmed with *E. coli* GroEL-GroES.

Recently, due to the more and more progress in information technology, and the availability of easy crystallization kits, the difficulty threshold of protein structural analysis has been lowering. If one wants to determine a structure of a certain protein by electron microscopic and/or X-ray crystal structure analyses, the structure obtained at the initial stage is sometimes slightly crude with inadequate resolution even after the refinement routine. Because the field of chaperonin continues to be replete with mysteries such as the robustness of the inter-subunit interactions in the GroEL-GroES complexes and the chaperonin function and structure relationship, deeper understanding may require bold feats of imagination. Anyway, it should be important to realize what is measured in the structure-determining method employed. In the protein structure, usually the electrostatic potential at any point varies and can be represented by a function Ψ(x, y, z) or in polar coordinates Ψ(r, θ, φ). While the values of potential are in the form of a scalar field, for an arbitrary potential distribution it is possible to imagine equipotential surfaces. Considering one equipotential surface to a nearby one along the normal to the equipotential surface, the gradient of the potential can be defined as the interaction forces between such subunits, those form a vector field (Ishii and Sato 2013). Therefore, thermodynamic evaluation of interaction between the constituent subunits in the asymmetric bullet-shaped, symmetric football-shaped complexes and so on should be the next challenging issues. Needless to say, unfounded speculations must be avoided and it is instructive not overstate any conclusions regarding details finer than the real resolution where the structural model is built.

Although the progress in this subcellular biochemistry area for macromolecular chaperonin complexes has been very rapid in recent years, much remains to be clarified. In this article, the author would have tried to summarize the recent progresses on the quaternary structures reported for the GroEL-GroES complexes while taking the chaperonin function into consideration. It would be very satisfying for the chapter author, if this article has successfully conveyed the growing progress in elucidating Group I chaperonin GroEL-GroES complex through its nearly 30-year history. Looking back at its consequences today, Anfinsen's dogma can be read as a suggestion that spontaneous folding occurs if it meets the following conditions; "uniqueness", which means the sequence does not have any other configuration with a comparable free energy; "stability", which means small changes in the surrounding environment cannot give rise to the minimum configuration changes; "kinetical accessibility", which means the final shape can be formed without going through any highly complex changes in the shape. At the same time we have to be aware how far the hypothesis and/or the structure model can be applied, namely whether it is applicable for events in cells, *in vivo* or just for those in a test tube, *in vitro*. Certain protein structure determined by X-ray crystallography only reflects the structure under the crystallization buffer condition therefore it does not guarantee the same in different solution conditions.

This article serves as an introductory exploration into the study of the Group I chaperonin GroEL-GroES complex. It is hoped that this article can serve as a background reference that will motivate the reading of more specialized or advanced accounts in literature, inspiring future generation of researchers in this field.

Acknowledgements The author would like to express thanks to the book editor, Professor Dr. Robin Harris for giving him this precious opportunity to review the progress thus far made on the Group I chaperonins, GroEL and the GroEL-GroES complexes research and to introduce some interdisciplinary applications using GroEL as a building nano-block, which forms the basis of a collaboration with Professor Dr. Takuzo Aida. The author's appreciation is directed to him and his group members. He also appreciates Dr. Kenneth S. Kim for critical reading of the draft stage of the article.

References

Anfinsen CB (1973) Principles that govern the folding of protein chains. Science 181:223–230

Azem A, Kessel M, Goloubinoff P (1994) Characterization of a functional $GroEL_{14}(GroES_7)_2$ chaperonin hetero-oligomer. Science 265:653–656

Biswas S, Kinbara K, Oya N, Ishii N, Taguchi H, Aida T (2009) A tubular biocontainer: metal ion-induced 1D assembly of a molecularly engineered chaperonin. J Am Chem Soc 131:7556–7557

Biswas S, Kinbara K, Niwa T, Taguchi H, Ishii N, Watanabe S, Miyata K, Kataoka K, Aida T (2013) Biomolecular robotics for chemomechanically driven guest delivery fuelled by intracellular ATP. Nat Chem 5:613–620

Braig K, Otwinowski Z, Hegde R, Boisvert DC, Joachimiak A, Horwich AL, Sigler PB (1994) The crystal structure of the bacterial chaperonin GroEL at 2.8 Å. Nature 371:578–586

Chandrasekhar GN, Tilly K, Woolford C, Hendrix R, Georgopoulos C (1986) Purification and properties of the groES morphogenetic protein of *Escherichia coli.* J Biol Chem 261:12414–12419

Chen DH, Song JL, Chuang DT, Chiu W, Ludtke SJ (2006) An expanded conformation of single-ring GroEL-GroES complex encapsulates an 86 kDa substrate. Structure 14:1711–1722

Cheng M, Hartl Y, Horwich FU (1990) The mitochondrial chaperonin hsp60 is required for its own assembly. Nature 348:455–458

Chouard T (2011) Breaking the protein rules. Nature 471:151–153

Dobson CM (2003) Protein folding and misfolding. Nature 426:884–890

Ellis RJ (1987) Proteins as molecular chaperones. Nature 328:378–379

Ellis RJ (ed) (1996) The Chaperonins. Academic, San Diego

Ellis RJ, Laskey RA, Lorimer GH (eds) (1993) Molecular chaperones. Chapman and Hall, London

Engel A, Hayer-Hartl MK, Goldie KN, Pfeifer G, Hegerl R, Muller S, da Silva AC, Baumeister W, Hartl FU (1995) Functional significance of symmetrical versus asymmetrical GroEL-GroES chaperonin complexes. Science 269:832–836

Fink AL, Goto Y (eds) (1998) Molecular chaperons in the life cycle of proteins. Structure, function, and mode of action. Marcell Dekker, Inc., New York

Frydman J, Nimmesgern E, Erdjument-Bromage H, Wall JS, Tempst P, Hartl FU (1992) Function in protein folding of TRiC, a cytosolic ring complex containing TCP-1 and structurally related subunits. EMBO J 11(13):4767–4778

Fujiyoshi Y (1998) The structural study of membrane proteins by electron crystallography. Adv Biophys 35:25–80

Glaeser RM (1971) Limitation to significant information in biological electron microscopy as a result of radiation damage. J Ultrastruct Res 36:466–482

Glaeser RM (1985) Electron crystallography of biological macromolecules. Annu Rev Phys Chem 36:243–275

Grallert H, Buchner J (2001) A structural view of the GroE chaperone cycle. J Struct Biol 135(2):95–103

Haldar S, Gupta AJ, Yan X, Miličić G, Hartl FU, Hayer-Hartl M (2015) Chaperonin-assisted protein folding: relative population of asymmetric and symmetric GroEL:GroES complexes. J Mol Biol 427(12):2244–2255

Harris JR, Zahn R, Pluckthun A (1995) Electron microscopy of the GroEL-GroES filament. J Struct Biol 115:68–77

Hartl FU, Bracher A, Hayer-Hartl M (2011) Molecular chaperones in protein folding and proteostasis. Nature 475:324–332

Hayer-Hartl MK, Martin J, Hartl FU (1995) Asymmetrical interaction of GroEL and GroES in the ATPase cycle of assisted protein folding. Science 269:836–841

Hendrix RW (1979) Purification and properties of GroE, a host protein involved in bacteriophage assembly. J Mol Biol 129:375–392

Hill JE, Penny SL, Crowell KG, Goh SH, Hemmingsen SM (2004) cpnDB: a chaperonin sequence database. Genome Res 14:1669–1675 http://www.cpndb.ca

Hohn T, Hohn B, Engel A, Wurtz M, Smith PR (1979) Isolation and characterization of the host protein groE involved in bacteriophage lambda assembly. J Mol Biol 129:359–373

Horovitz A, Willison KR (2005) Allosteric regulation of chaperonins. Curr Opin Struct Biol 15:646–651

Horowitz PM, Lorimer GH, Ybarra J (1999) GroES in the asymmetric $GroEL_{14}$-$GroES_7$ complex exchanges via an associative mechanism. Proc Natl Acad Sci U S A 96:2682–2686

Horwich AL, Farr GW, Fenton WA (2006) GroEL-GroES-mediated protein folding. Chem Rev 106:1917–1930

Ishii N (2013) Observation by transmission electron microscopy of organic nano-tubular architectures. In: Mendez-Vilas A (ed) Current microscopy contributions to advances in science and technology. Formatex Research Center, Badajoz, pp 1225–1233

Ishii N (2014) Image analyses of two-dimensional crystalline arrays of membrane proteins and protein supramolecular complexes. In: Echon RM (ed) Advances in image analysis research. Nova Science Publishers Inc., New York, pp 189–216

Ishii N, Sato T (2013) Anisotropic intersubunit and inter-ring interactions revealed in the native bullet-shaped chaperonin complex from *Thermus thermophilus*. Biochim Biophys Acta 1830:2907–2916

Ishii N, Taguchi H, Yoshida M, Yoshimura H, Nagayama K (1991) Image analysis by electron microscopy of two-dimensional crystals developed on a mercury surface of chaperonin from *Thermus thermophilus*. J Biochem 110:905–908

Ishii N, Taguchi H, Sumi M, Yoshida M (1992) Structure of holo-chaperonin studied with electron microscopy: oligomeric cpn10 on top of two layers of cpn60 rings with two stripes each. FEBS Lett 299:169–174

Ishii N, Taguchi H, Sasabe H, Yoshida M (1994) Folding intermediate binds to the bottom of bullet-shaped holo-chaperonin and is readily accessible to antibody. J Mol Biol 236:691–696

Ishii N, Taguchi H, Sasabe H, Yoshida M (1995) Equatorial split of holo-chaperonin from *Thermus thermophilus* by ATP and K^+. FEBS Lett 362(2):121–125

Ishii D, Kinbara K, Ishida Y, Ishii N, Okochi M, Yohda M, Aida T (2003) Chaperonin-mediated stabilization and ATP-triggered release of semiconductor nanoparticles. Nature 423(6940):628–632

Ishii N, Okuro K, Kinbara K, Aida T (2010) Image analysis of alpha/beta-tubulin rings in two-dimensional crystalline arrays of periodic mesoporous nanostructures. J Biochem 147(4):555–563

Klunker D, Haas B, Hirtreiter A, Figueiredo L, Naylor DJ, Pfeifer G, Müller V, Deppenmeier U, Gottschalk G, Hartl FU, Hayer-Hartl M (2003) Coexistence of group I and group II chaperonins in the archaeon *Methanosarcina mazei*. J Biol Chem 278(35):33256–33267

Koike-Takeshita A, Yoshida M, Taguchi H (2008) Revisiting the GroEL-GroES reaction cycle via the symmetric intermediate implied by novel aspects of the GroEL(D398A) mutant. J Biol Chem 283(35):23774–23781

Koike-Takeshita A, Arakawa T, Taguchi H, Shimamura T (2014) Crystal structure of a symmetric football-shaped GroEL:$GroES_2$-ATP_{14} complex determined at 3.8 Å reveals rearrangement between two GroEL rings. J Mol Biol 426(21):3634–3641

Kubota H, Hynes G, Carne A, Ashworth A, Willison K (1994) Identification of six Tcp-1-related genes encoding divergent subunits of the TCP-1-containing chaperonin. Curr Biol 4(2):89–99

Lissin NM, Venyaminov SY, Girshovich AS (1990) (Mg-ATP)-dependent self-assembly of molecular chaperone GroEL. Nature 348:339–342

Lopez T, Dalton K, Frydman J (2015) The mechanism and function of Group II chaperonins. J Mol Biol 427:2919–2930

Lorimer GH, Todd MJ (1996) GroE structures galore. Nat Struct Biol 3:116–121

Mascagni P, Tonolo M, Ball H, Lim M, Ellis RJ, Coates A (1991) Chemical synthesis of 10 kDa chaperonin. Biological activity suggests chaperonins do not require other molecular chaperones. FEBS Lett 286:201–203

Motohashi K, Taguchi H, Ishii N, Yoshida M (1994) Isolation of the stable trimeric DnaK/DnaJ complex from *Thermus thermophiles*. J Biol Chem 269:27074–27079

Motojima F, Yoshida M (2010) Polypeptide in the chaperonin cage partly protrudes out and then folds inside or escapes outside. EMBO J 29(23):4008–4019

Muramatsu S, Kinbara K, Taguchi H, Ishii N, Aida T (2006) Semibiological molecular machine with an implemented "AND" logic gate for regulation of protein folding. J Am Chem Soc 128:3764–3769

Nielsen KL, Cowan NJ (1998) A single ring is sufficient for productive chaperonin-mediated folding *in vivo*. Mol Cell 2(1):93–99

Niwa T, Ying BW, Saito K, Jin W, Takada S, Ueda T, Taguchi H (2009) Bimodal protein solubility distribution revealed by an aggregation analysis of the entire ensemble of *Escherichia coli* proteins. Proc Natl Acad Sci U S A 106(11):4201–4206

Okamoto T, Ishida R, Yamamoto H, Tanabe-Ishida M, Haga A, Takahashi H, Takahashi K, Goto D, Grave E, Itoh H (2015) Functional structure and physiological functions of mammalian wild-type HSP60. Arch Biochem Biophys 586:10–19

Parnas A, Nisemblat S, Weiss C, Levy-Rimler G, Pri-Or A, Zor T, Lund PA, Bross P, Azem A (2012) Identification of elements that dictate the specificity of mitochondrial Hsp60 for its co-chaperonin. PLoS One 7:e50318

Richter K, Haslbeck M, Buchner J (2010) The heat shock response: life on the verge of death. Mol Cell 40:253–266

Roseman AM, Chen S, White H, Braig K, Saibil HR (1996) The chaperonin ATPase cycle: mechanism of allosteric switching and movements of substrate-binding domains in GroEL. Cell 87:241–251

Sameshima T, Iizuka R, Ueno T, Funatsu T (2010) Denatured proteins facilitate the formation of the football-shaped GroEL- $(GroES)_2$ complex. Biochem J 427:247–254

Saibil HR, Fenton WA, Clare DK, Horwich AL (2013) Structure and allostery of the chaperonin GroEL. J Mol Biol 425:1476–1487

Sameshima T, Ueno T, Iizuka R, Ishii N, Terada N, Okabe K, Funatsu T (2008) Football- and bullet-shaped GroEL-GroES complexes coexist during the reaction cycle. J Biol Chem 283(35):23765–23773

Schmidt M, Rutkat K, Rachel R, Pfeifer G, Jaenicke R, Viitanen P, Lorimer G, Buchner J (1994) Symmetric complexes of GroE chaperonins as part of the functional cycle. Science 265:656–689

Sim S, Miyajima D, Niwa T, Taguchi H, Aida T (2015) Tailoring micrometer-long high-integrity 1D array of superparamagnetic nanoparticles in a nanotubular protein jacket and its lateral magnetic assembling behavior. J Am Chem Soc 137:4658–4661

Sumi M, Taguchi H, Yokoyama K, Ishii N, Yoshida M (1992) Identification and characterization of a chaperonin from *Paracoccus denitrificans*. Life Sci Adv 11:179–182

Taguchi H (2015) Reaction cycle of chaperonin GroEL via symmetric "football" intermediate. J Mol Biol 427:2912–2918

Taguchi H, Konishi J, Ishii N, Yoshida M (1991) A chaperonin from a thermophilic bacterium, *Thermus thermophilus*, that controls refolding of several thermophilic enzymes. J Biol Chem 226:22411–22418

Todd MJ, Walke S, Lorimer G, Truscott K, Scopes RK (1995) The single-ring *Thermoanaerobacter brockii* chaperonin 60 (Tbr-EL_7) dimerizes to Tbr-EL_{14}·Tbr-ES_7 under protein folding conditions. Biochemistry 34:14932–14941

Trent JD, Nimmesgern E, Wall JS, Hartl FU, Horwich AL (1991) A molecular chaperone from a thermophilic archaebacterium is related to the eukaryotic protein t-complex polypeptide-1. Nature 354:490–493

Viitanen PV, Lorimer GH, Seetharam R, Gupta RS, Oppenheim J, Thomas JO, Cowan NJ (1992) Mammalian mitochondrial chaperonin 60 functions as a single toroidal ring. J Biol Chem 267:695–698

Vilasi S, Carrotta R, Mangione MR, Campanella C, Librizzi F, Randazzo L, Martorana V, Marino Gammazza A, Ortore MG, Vilasi A, Pocsfalvi G, Burgio G, Corona D, Palumbo Piccionello A, Zummo G, Bulone D, Conway de Macario E, Macario AJL, San Biagio PL, Cappello F (2014) Human Hsp60 with its mitochondrial import signal occurs in solution as heptamers and tetradecamers remarkably stable over a wide range of concentrations. PLoS One 9:e97657

Walter S (2002) Structure and function of the GroE chaperone. Cell Mol Life Sci 59:1589–1597

Xu Z, Horwich AL, Sigler PB (1997) The crystal structure of the asymmetric GroEL-GroES-$(ADP)_7$ chaperonin complex. Nature 388:741–750

Yamakoshi M, Taguchi H, Yoshida M, Ishii N (1998) A chaperonin from a thermophilic bacterium, *Thermus thermophiles*. In: Fink AL, Goto Y (eds) Molecular chaperons in the life cycle of proteins. Structure, function, and mode of action. Marcell Dekker, Inc, New York, pp 301–330

Zahn R, Harris JR, Pfeifer G, Plückthun A, Baumeister W (1993) Two-dimensional crystals of the molecular chaperone GroEL reveal structural plasticity. J Mol Biol 229:579–584

Zanin-Zhorov A, Cohen IR (2007) HSP60: a pleiotropic immune signal. In: Asea AAA, De Maio A (eds) Heat shock proteins: potent mediators of inflammation and immunity. Springer, Dordrecht, pp 265–272

Chapter 18
The Aminoacyl-tRNA Synthetase Complex

Marc Mirande

Abstract Aminoacyl-tRNA synthetases (AARSs) are essential enzymes that specifically aminoacylate one tRNA molecule by the cognate amino acid. They are a family of twenty enzymes, one for each amino acid. By coupling an amino acid to a specific RNA triplet, the anticodon, they are responsible for interpretation of the genetic code. In addition to this translational, canonical role, several aminoacyl-tRNA synthetases also fulfill nontranslational, moonlighting functions. In mammals, nine synthetases, those specific for amino acids Arg, Asp, Gln, Glu, Ile, Leu, Lys, Met and Pro, associate into a multi-aminoacyl-tRNA synthetase complex, an association which is believed to play a key role in the cellular organization of translation, but also in the regulation of the translational and nontranslational functions of these enzymes. Because the balance between their alternative functions rests on the assembly and disassembly of this supramolecular entity, it is essential to get precise insight into the structural organization of this complex. The high-resolution 3D-structure of the native particle, with a molecular weight of about 1.5 MDa, is not yet known. Low-resolution structures of the multi-aminoacyl-tRNA synthetase complex, as determined by cryo-EM or SAXS, have been reported. High-resolution data have been reported for individual enzymes of the complex, or for small subcomplexes. This review aims to present a critical view of our present knowledge of the aminoacyl-tRNA synthetase complex in 3D. These preliminary data shed some light on the mechanisms responsible for the balance between the translational and nontranslational functions of some of its components.

Keywords Aminoacyl-tRNA synthetase (AARS) • Multi-aminacyl-tRNA synthetase complex (MSC) • MSC assembly • Core synthetases • Crystal structure

18.1 Introduction

Aminoacyl-tRNA synthetases (AARSs) are a family of twenty enzymes, one for each amino acid. These ubiquitous enzymes are responsible for interpreting the genetic code, by pairing a specific amino acid to its cognate tRNA (Ibba and Söll 2000).

M. Mirande (✉)
Institute for Integrative Biology of the Cell (I2BC), CEA, CNRS, Univ. Paris-Sud, Université Paris-Saclay, 1 avenue de la Terrasse, 91190 Gif-sur-Yvette, Paris, France
e-mail: Marc.Mirande@i2bc.paris-saclay.fr

J.R. Harris, J. Marles-Wright (eds.), *Macromolecular Protein Complexes*, Subcellular Biochemistry 83, DOI 10.1007/978-3-319-46503-6_18

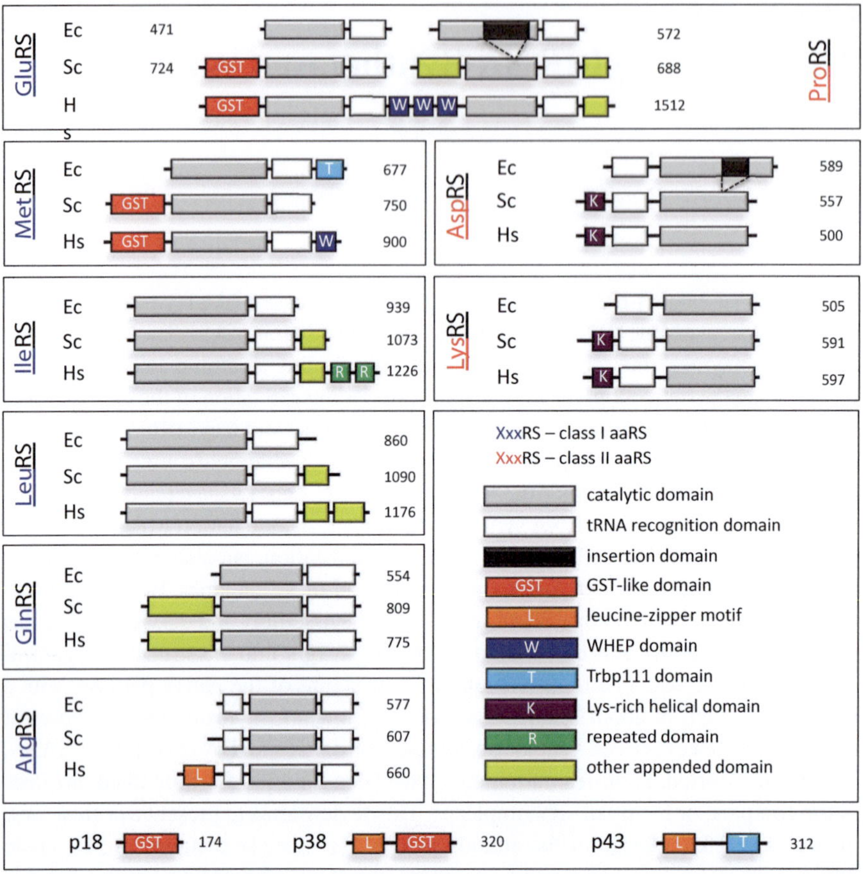

Fig. 18.1 Additional structural domains of AARSs from the MSC. The catalytic and tRNA recognition domains of AARSs from *Escherichia coli* (EC), *Saccharomyces cerevisiae* (Sc), and *Homo sapiens* (Hs) are conserved between species. In human, structural domains are grafted at the N- or C-terminus of the core enzymes, or link the two synthetases in the bifunctional glutamyl-prolyl-tRNA synthetase (GluProRS). The 3 auxiliary proteins p18, p38 and p43 are essential for the assembly of the MSC

The 20 AARSs are divided into two subfamilies of 10-members each, according to the structural organization of their catalytic domain, which is built around a parallel (class I) or anti-parallel (class II) β-sheet (Eriani et al. 1990). A tRNA-binding domain is appended at the N- or C-terminus of the catalytic domain. As a general rule, the structural organization of these enzymes is well conserved throughout evolution, including the catalytic and tRNA-binding domains that form the fundamental core of the AARSs (Fig. 18.1). The main structural difference between AARSs from bacteria and vertebrates lies in the addition of specific domains, which are grafted at the N- or C-terminus of the core synthetases (Fig. 18.1). Another specific feature of vertebrate AARSs is the assembly of nine of the family members within a multi-synthetase complex (MSC), which contains an additional three structural auxiliary proteins as well (Mirande 1991; Han et al. 2003; Guo et al. 2010; Havrylenko and Mirande 2015). It contains the nine AARSs specific for the amino acids Arg, Asp, Gln, Glu, Ile, Leu, Lys, Met and Pro, the 3 auxiliary proteins are p43, p38 and p18 (also known as AIMP1, AIMP2 and AIMP3, respectively). Two AARSs – GluRS and ProRS – are fused into a single polypeptide, with the two core enzymes linked by three WHEP domains (Fig. 18.1) (Cérini et al. 1991; Ray et al. 2011).

The N-terminal appended domain of MRS mediates its association within the MSC (Mirande et al. 1982; Kaminska et al. 2001). It is a GST (Glutathione S-transferase)-like domain that can also be found at the N-terminus of bifunctional GluProRS, at the C-terminus of p38 (Quevillon et al. 1999), the scaffold protein of the MSC, and in p18 (Quevillon and Mirande 1996), the auxiliary protein that links MetRS to the MSC (Fig. 18.1). This structural module is essential to the assembly of the MSC (Kaminska et al. 2009a). In addition to this protein-protein interaction module, MetRS also has a C-terminal appended domain, referred to as a WHEP domain, which is also found in triplicate in the linker region of GluProRS (Fig. 18.1). This WHEP domain displays RNA-binding properties (Cahuzac et al. 2000; Kaminska et al. 2001). The lysine-rich N-terminal domains of AspRS and LysRS (Fig. 18.1), as well as the C-terminal Trbp111-like domain of p43, are also known as RNA-binding domains (Francin et al. 2002; Francin and Mirande 2003; Guigou et al. 2004; Shalak et al. 2007).

Another type of protein-protein interaction module, the leucine-zipper motif, is found in the N-terminal region of ArgRS, and in the two auxiliary proteins, p38 and p43 (Fig.18.1). The N-terminal appended domain of ArgRS mediates its association within the MSC (Lazard and Mirande 1993; Robinson et al. 2000), through the interaction with the N-terminal domain of p43 (Robinson et al. 2000; Guigou et al. 2004). The leucine-zipper motif of p43 associates with the leucine-zipper motif of p38, which form a coiled-coil (Ahn et al. 2003). Other types of appended domains are also observed at the N- or C-termini of IleRS, LeuRS, or GlnRS, but their actual role has not yet been deciphered.

The assembly of the MSC is crucial for translation. *In cellulo*, after stable translational silencing of p18, p38 or p43 with shRNAs in HeLa cells, the MSC was in part dismantled, and a growth retardation phenotype was observed (Kaminska et al. 2009a). Because the nine AARSs associated within the MSC are the sole source of these AARSs in the cell, subcellular organization of the MSC within the translation apparatus, via interaction with polysomes (Kaminska et al. 2009b; David et al. 2011), is believed to be essential for the translational function of its components. Two forms of ArgRS are produced in mammalian cells, for two distinct in frame AUG start codons on the same mRNA (Lazard and Mirande 1993; Yang et al. 2014). The full-length ArgRS species is a component of the MSC and the shorter form, which does not contain the 72 N-terminal residues, is a free enzyme. Arg-tRNA produced by MSC-ArgRS is a better substrate for protein synthesis than Arg-tRNA generated by free-ArgRS (Kyriacou and Deutscher 2008). The MSC is also proposed to be a depot for releasable regulatory proteins involved in nontranslational cellular processes (Ray et al. 2007). For instance, in stimulated mast cells, LysRS from the MSC is translocated to the nucleus, where it activates the microphtalmia-associated transcription factor – MITF (Yannay-Cohen et al. 2009). GluProRS switches from a translational function to a translational silencing function after dissociation from the MSC and association with alternative partners within the GAIT (γ-interferon activated inhibitor of translation) complex (Sampath et al. 2004). The fine-tuning between their translational and nontranslational functions certainly requires a tight control of their spatio-temporal organization in living cells.

The finding that the MSC could be endowed with cellular functions well beyond its role in translation stimulated a particular interest to understand its structural organization at the atomic level, and to provide interesting clues concerning the mechanisms of the controlled release of its components to fulfill the observed non-conventional functions.

18.2 Crystal Structure of Aminoacyl-tRNA Synthetase Components of the MSC

In an attempt to understand the molecular features responsible for the assembly of the synthetases within the MSC, a crystallographic study of individual enzymes was undertaken.

18.2.1 Class 2 AARS

The crystal structures of AspRS (Kim et al. 2013), LysRS (Guo et al. 2008) and of the ProRS moiety of the bifunctional GluProRS (Zhou et al. 2013), the three class 2 AARSs of the MSC, have been reported (Fig. 18.2). Crystals of LysRS were obtained after deletion of the 70 N-terminal residues that form its eukaryote-specific lysine-rich helical domain, crystals of ProRS were limited to the domains of the core enzyme and, even if full-length AspRS was present in the crystal, its N-terminal specific domain could not be modeled. Thus, the crystal structures of the core enzymes were very similar to those already described for the homologous synthetases from bacteria or yeast, and did not provide further insight into the roles of the appended domains. They contain the classical catalytic domain of class 2 AARSs, built around a β-sheet made of anti-parallel β-strands. The dimer interface of these three synthetases is mediated by motif 1 of class 2 AARSs (Eriani et al. 1990). The tRNA-recognition domain of AspRS and LysRS form an oligonucleotide-binding (OB) fold. This β-barrel domain is common to many RNA-binding proteins. The tRNA-binding domain of ProRS has an α/β fold characteristic of class 2a synthetases, including ProRS, HisRS, GlyRS and ThrRS (Cusack 1995; Yaremchuk et al. 2000). ProRS from the MSC also shares the eukaryote/archeon-specific C-terminal zinc-binding domain that folds back toward the active site.

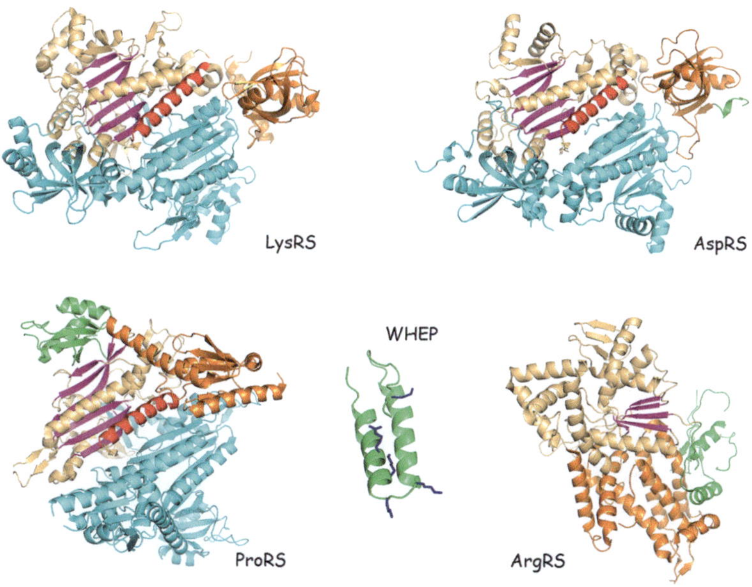

Fig. 18.2 Individual fragments of the AARS components of the MSC. The crystal structures of the LysRS (PDB 3BJU) and AspRS (PDB 4 J15) components minus their lysine-rich N-terminal domains, of the C-terminal ProRS (PDB 4HVC) fragment of bifunctionnal GluProRS, and of the ArgRS (PDB 4Q2Y) component minus its leucine-zipper N-terminal domain, are shown. The solution structure of one WHEP domain (PDB 1R1B and 1FYJ) of GluProRS is shown. The catalytic domains of the 4 AARSs are shown in *light orange*, the tRNA recognition domains in *orange*. Additional domains are in *green*. The β-strands that form the platforms of the catalytic domains are in *magenta*. The long α-helix from motif 1 of class 2 AARSs is in *red*. The second monomer of dimeric class 2 LysRS, AspRS and ProRS is in *cyan*. The conserved lysine and arginine residues of the WHEP repeat are indicated

18.2.2 WHEP Domains

The three WHEP domains (named after Trp(W)RS, His(H)RS and GluPro(EP)RS where these domains were first identified), which link the GluRS and ProRS domains of GluProRS, form a helix-turn-helix motif (Fig. 18.2) (Cahuzac et al. 2000; Jeong et al. 2000). Five conserved lysine and arginine residues form a basic patch on one side of this domain that has been identified as a tRNA- (Cahuzac et al. 2000) or RNA- (Jia et al. 2008) binding domain. A single WHEP element is also appended at the C-terminus of MetRS in the MSC and plays an important functional role in tRNA binding (Kaminska et al. 2001).

18.2.3 Class 1 AARS

The crystal structure of ArgRS, one of the six class 1 synthetases of the MSC, was determined after crystallization of the free form of ArgRS, which does not possess the 72 N-terminal residues carrying the leucine-zipper motif necessary for its association with p43. The overall structure is very similar to that described for the yeast enzyme. Its catalytic domain is arranged around a Rossmann fold containing five parallel β-strands, its bipartite tRNA-recognition domain is composed of the C-terminal α-helical domain end of its N-terminal subdomain (Fig. 18.2) (Kim et al. 2014).

18.3 Crystal Structures of Auxiliary Components of the MSC

Assembly of the MSC relies on the presence of the three auxiliary proteins, p18, p38 and p43, which fulfill essential structural functions. Early experiments suggested that the p38 protein represents the scaffold protein of the complex (Quevillon et al. 1999). Its complete knock-down in cultured cells led to the release of LysRS and AspRS from the MSC, which was dissociated in two stable subcomplexes: subcomplex I contained GluProRS, IleRS, LeuRS, p18 and MetRS, where subcomplex II contained ArgRS, GlnRS and p43 (Kaminska et al. 2009a). Knockdown of p18 was just accompanied by the release of MetRS from the MSC. Knockdown of p43 led to a MSC deprived of ArgRS and GlnRS.

18.3.1 p18: Two Potential Protein-Binding Sites

The crystal structure of the p18 component of the MSC revealed a structural fold similar to that described for the large family of glutathione S-transferases (GST) (Kim et al. 2008), as expected from sequence similarities previously observed. Two p18 subunits were recovered in the asymmetric crystal unit, related by a non-crystallographic two-fold axis. This putative dimer interface involves helix α7 and the loop between helices α4 and α5 (Fig. 18.3a). Another putative dimer interface was observed in the crystal. Two molecules of p18 were related by a crystallographic two-fold axis, with a new interface involving helices α2 and α3 from the two monomers that form a 4-helix bundle (Fig. 18.3b). This structural arrangement is similar to that found in a typical GST dimer. These observations suggest that p18 possesses two different protein interaction interfaces, and that it is able to associate with two different proteins within the MSC that carry GST-like domains.

This observation is reminiscent of the structural arrangement described in yeast between Arc1p, a p43-like protein, and the two synthetases MetRS and GluRS

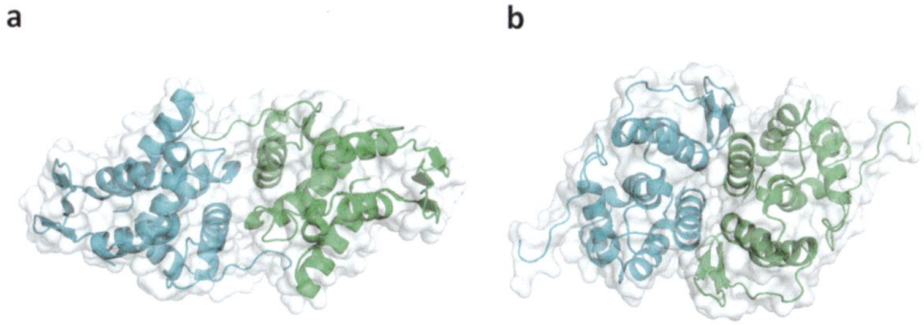

Fig. 18.3 The two dimer interfaces of the p18 protein. (**a**) interaction of two p18 subunits in the asymmetric crystal unit according to a non-crystallographic two-fold axis, also known as interface 2. (**b**) interaction of two p18 subunits according to the crystallographic two-fold axis, also known as interface 1 (PDB 2UZ8)

(Simader et al. 2006). In the yeast *Saccharomyces cerevisiae*, MetRS and GluRS form a multisynthetase complex that dissociates upon switching from fermentation to respiration, which induces translocation of MetRS into the nucleus, and of GluRS into the mitochondria, to synchronize expression of ATP synthase genes (Frechin et al. 2014). The N-terminal domains of Arc1p, MetRS and GluRS all have a GST-like fold that form a trimeric assembly (Simader et al. 2006). Association of MetRS and Arc1p resembles a classical GST homo dimer, and is similar to the interaction observed between two p18 molecules related by a crystallographic two-fold symmetry (Fig. 18.3b). Association of GluRS with Arc1p involves a non-classical GST interface similar to that observed for two molecules of p18 in the asymmetric crystal unit (Fig. 18.3a). This suggested that p18 in the MSC could use these two interfaces to interact with two AARSs (Kim et al. 2008).

The p18 protein has also been described as an activator of p53 through its interaction with the ATM and ATR kinases in the nucleus (Park et al. 2005). In response to DNA damage, p18 would be dissociated from the MSC, to be subsequently translocated into the nucleus. Mutations that affect the two potential protein interfaces in p18 have been isolated in patients with chronic myeloid leukemia (Kim et al. 2008). The p18 binding interface that involves helix $\alpha 7$ and the loop between helices $\alpha 4$ and $\alpha 5$ (Fig. 18.3a) is used for the interaction with ATM (Kim et al. 2008). The mechanism of release of p18 from the cytosolic MSC to act as a potential tumor suppressor in the nucleus is not known.

18.3.2 p43: a Two-Domain Protein

The p43 component of MARS is essential for the association of GlnRS and ArgRS with the MSC (Kaminska et al. 2009a). It is made up of two structural domains. Its 150 residue long N-terminal domain can replace the full-length protein to build the MSC and its C-terminal domain does not associate with the MSC (Kaminska et al. 2009a). The C-domain of p43 can be released from the MSC after cleavage by

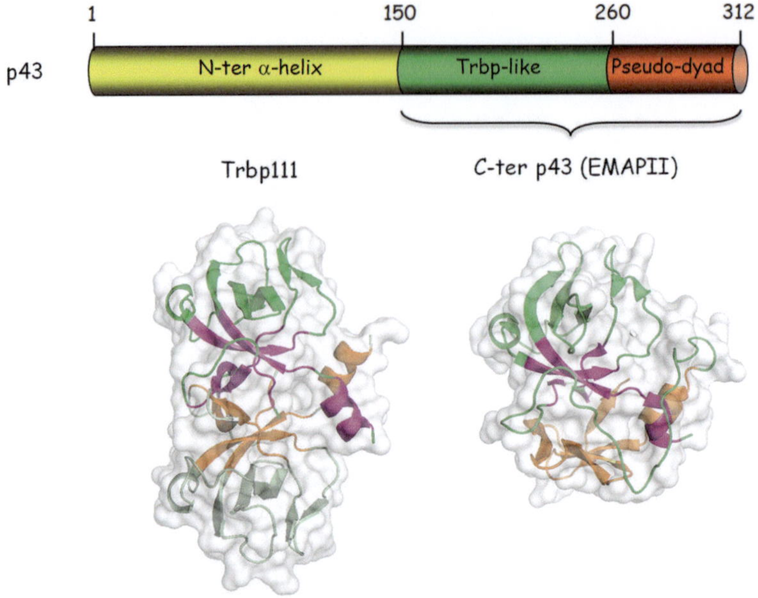

Fig. 18.4 The internal pseudo-dyad of monomeric p43. The p43 component of the MSC is made of a long N-terminal α-helix that contains a leucine-zipper motif, and an EMAPII C-terminal domain made of a Trbp111-like subunit and an internal pseudo-dyad. The crystal structures of *Thermus thermophilus* Trbp111 (PDB 1GD7) and human EMAPII (PDB 1FLO) are shown. One monomer of Trbp111 is in *green* and *magenta*, the other in *light green* and *orange*. The corresponding structural domains in EMAPII are drawn with the same colors

caspase 7 (Shalak et al. 2001). It has been identified as the endothelial-monocyte-activating polypeptide II (EMAPII), a tumor-derived inflammatory cytokine (Kao et al. 1992).

Its N-terminal domain contains a leucine-zipper motif and is believed to form a long α-helix that builds a coiled-coil with the N-terminal domain of p38 (Ahn et al. 2003). Its C-terminal domain is similar to bacterial Trbp111, a protein that binds nucleic acids. Trbp111 has a β-barrel architecture, which is referred to as an OB-fold (Fig. 18.4) (Swairjo et al. 2000). However, Trbp111 is a dimer and p43 is a monomeric protein. The crystal structure of EMAPII showed that the Trbp-like domain in p43 also forms an OB-fold, but it also revealed that it contains an extra C-terminal domain related to the OB-fold by an internal two-fold symmetry (Renault et al. 2001). This extra-domain forms a pseudo-dyad and mimics the dimer-interface of the second monomer in Trbp111 (Fig. 18.4). The crystal structure of EMAPII provides a rational explanation for the monomeric state of p43.

The peptide ^{158}RIGRIVT^{164} in EMAPII, which was believed to be the shortest peptide able to recapitulate the biological activity of EMAPII (Kao et al. 1994), corresponds to strand β1 of the OB-fold and is buried in the protein core (Renault et al. 2001). This suggests that the biological effects of the peptide are serendipitous.

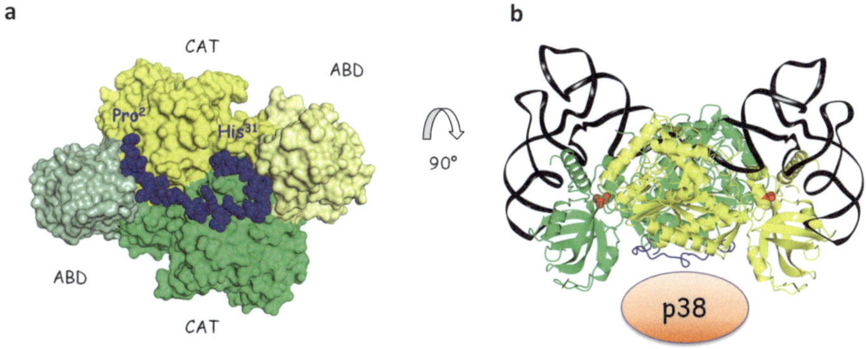

Fig. 18.5 Association of LysRS with the scaffold protein of the MSC. (**a**) The catalytic (CAT) and anticodon binding (ABD) domains of the 2 subunits of LysRS are shown in *green* and *yellow*, the N-terminal peptide of p38, from Pro^2 to His^{31} is shown in *blue* (PDB 4DPG). (**b**) Location of the tRNA- and p38-binding sites on LysRS. Docking of tRNA is based on the co-crystal structure of yeast AspRS with $tRNA^{Asp}$ (1ASY). Ser^{207} is in *red*

18.4 Structural Organization of Subcomplexes

18.4.1 The Binary Complex LysRS:p38

Association of LysRS to the MSC involves an interaction between its catalytic domain and the very N-terminal 42-aa peptide of p38 (Quevillon et al. 1999; Robinson et al. 2000). LysRS, with a deletion of its N-terminal 70 residues, and a peptide containing the N-terminal 48 residues of p38 were co-crystalized, and the structure of the complex was solved at 2.86 Å resolution (Ofir-Birin et al. 2013).

In the crystal, only the N-terminal 30 residues of p38 were visible, the remaining part was disordered (Fig. 18.5a). Only one p38 peptide interacts per dimer of LysRS. Two interaction areas are observed. Residues 10 to 23 of p38 interact with the catalytic domain of one monomer of LysRS, while the two peptide segments from residues 2–9 to 24–31, which form two related motifs, bind in the two symmetrical grooves formed between the catalytic domain of one LysRS subunit and the tRNA-binding domain of the other subunit (Ofir-Birin et al. 2013). These two grooves are mainly hydrophobic, and the residues in LysRS that from direct contacts with the two related motifs in p38 are conserved in yeast LysRS, an enzyme that cannot associate in the MSC (Agou et al. 1996). This suggests that the N- and C-terminal extremities of p38 that are observed in the hydrophobic pocket of LysRS do not recapitulate the folding of these two regions in native p38. Probing of the protein interfaces between LysRS and full-length, native p38 confirms that the binding interface involves residues 8-26 of p38 (Rémion et al. 2016).

Interestingly, the binding site of p38 on LysRS is located at the side of the dimer that is underneath the area of LysRS that is involved in tRNA binding (Fig. 18.5b). This is consistent with the finding that LysRS in the MSC and a free form of LysRS aminoacylate tRNA with similar efficiencies (Mirande et al. 1983).

Cytosolic LysRS in the MSC has an essential translational function and serves as a regulator of the immune response after dissociation from the MSC and import into the nucleus (Lee and Razin 2005). LysRS is phosphorylated on Ser207 in a MAPK-dependent manner upon immune stimulation of mast cells (Yannay-Cohen et al. 2009). The co-crystal structure of LysRS with a N-terminal peptide of p38, together with small-angle X-ray scattering (SAXS) analysis of LysRS and of a phosphomimetic mutant, LysRS-S207D, revealed that phosphorylation at Ser207 results in an open conformer of LysRS that loses its ability to associate in the MSC (Ofir-Birin et al. 2013). Ser207 is located in the anticodon-binding domain of LysRS at the interface between the anticodon-binding domain of one subunit and the catalytic domain of the other subunit in the dimer (Fig. 18.5b). Mutations at the subunit interface, that should also destabilize the dimeric organization of LysRS, result in its inactivation in the tRNA-aminoacylation reaction. In the meantime, LysRS activity in the synthesis of diadenosine tetraphosphate (Ap4A), a molecule described a long time ago as a second messenger (Zamecnik 1983), is stimulated. This side-product, which is synthesized after a reaction of aminoacyl-adenylate with a second ATP molecule, activates the transcription factor MITF.

18.4.2 The ArgRS:GlnRS:p43 Complex, the Subcomplex II of the MSC

The crystal structure of an N-terminally truncated derivative of ArgRS (Kim et al. 2014) was not informative to understand how it is assembled into the MSC, since its interaction with p43 in the MSC requires the N-terminal domain (Lazard and Mirande 1993; Robinson et al. 2000; Guigou et al. 2004). Based on the observation that subcomplex II of the MSC, containing ArgRS, GlnRS and p43, can be reconstituted *in vitro* starting from the three purified individual components (Robinson et al. 2000), the crystal structure of this MSC subcomplex was determined at 4 Å resolution (Fu et al. 2014).

The crystal contained one molecule of each of the 3 components in the asymmetric unit (Fig. 18.6). The 72-aa long N-terminal domain of MSC-ArgRS forms two long α-helices, Ha from residues 7 to 29, and Hb from residues 40 to 65. In the crystal structure, only the 80-aa N-terminal region of p43 is visible. It also forms a long α-helix. The helical region of p43 from Lys46 to Leu66, which contains a leucine-zipper motif, forms a coiled-coil with helix Hb of ArgRS. Helix Ha of ArgRS could interact with residues from the loop between strands S21 and S22 of the anticodon-binding domain of GlnRS, but the main interaction area of GlnRS with ArgRS and p43 seems to involve its N-terminal catalytic domain, in agreement with previous biochemical data (Kim et al. 2000). Mutation Arg403Trp in the catalytic domain of GlnRS, which is the causative mutation in progressive microcephaly (Zhang et al. 2014), disrupts the interaction between GlnRS and ArgRS.

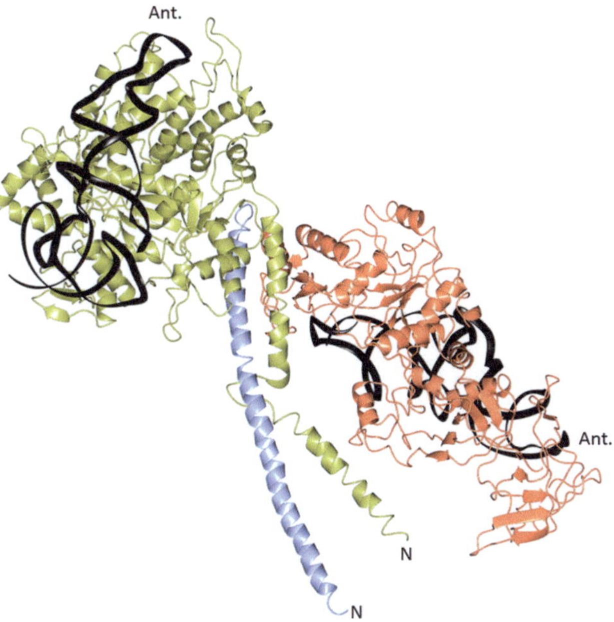

Fig. 18.6 The trimeric ArgRS:GlnRS:p43 subcomplex of the MSC. The trimer contained in the asymmetric unit is shown (PDB 4R3Z).The crystal structure of full-length ArgRS is in *yellow*, of N-terminally truncated GlnRS in *orange*, of the N-terminal domain of p43 in *blue*. The N-termini of the long helical domain of p43 and of ArgRS are indicated (N). Docking of $tRNA^{Arg}$ and $tRNA^{Gln}$ is based on the co-crystal structures of yeast ArgRS (1F7U) and *E. coli* GlnRS (1QTQ) in presence of their cognate tRNAs. The positions of the anticodon loops of the 2 tRNAs is indicated (Ant.)

The helical segment of p43, from Val7 to Glu44, interacts with the N-terminal α-helical fragment of p38 (Ahn et al. 2003; Fu et al. 2014). In the crystal structure of the ArgRS:GlnRS:p43 complex, which does not contain p38, the N-terminal helix of p43 also interacts with helix Ha of ArgRS (Fu et al. 2014). However, this interaction involves two ternary complexes related by a crystallographic two-fold axis. Thus, the N-terminal helix of p43 from one trimer interacts with helix Ha of ArgRS from the second trimer, and *vice versa*. However, in the symmetry-related hexameric structure containing two subunits of each ArgRS, GlnRS and p43, the active site of ArgRS is not directly accessible to the tRNA substrate, but would require a large conformational change. Alternatively, the hexameric complex could be the result of crystal packing effects that are only allowed in the absence of p38.

The C-terminal EMAPII domain of p43 (Fig. 18.4) was not visible in the crystal, suggesting its flexibility in the MSC, which is consistent with the finding that it can be removed without affecting the stability of the complex (Shalak et al. 2001). The N-terminal 200-aa long N-terminal domain of GlnRS was also not visible in the crystal. In yeast, GlnRS possesses a similar domain, which was shown to be important for tRNA binding (Wang et al. 2000). This is consistent with the observation that, in the absence of tRNA in the crystal, the N-domain of human GlnRS could be very flexible.

18.4.3 Role of GST-Homology Domains in the Assembly of the MSC

GST-homology domains are recovered at the N-terminus of GluProRS and MetRS, at the C-terminus of p38 and in p18 (Fig. 18.1). Taking into account that GST-like domains are structural domains commonly used for protein assembly, association between the four domains identified in the MSC was analyzed *in vitro*, and the crystal structures of the observed binary complexes were determined (Cho et al. 2015).

Three different heterodimers between these GST-like domains were reported. The p18 component of the MSC, which has the potential to form two types of homodimers (Fig. 18.3), is able to form two types of complexes, one with the GST-like domain of MetRS ($MetRS^{GST}$), and one with $EPRS^{GST}$. The complex between $MetRS^{GST}$ and $p18^{GST}$ resembles a classical GST homo dimer, and involves interface 1 of $p18^{GST}$, corresponding to two p18 subunits related by a crystallographic two-fold symmetry (Kim et al. 2008), and similar to the interaction observed between Arc1p and MetRS in yeast (Simader et al. 2006). Association between $p18^{GST}$ and $GluProRS^{GST}$ uses interface 2 of p18, corresponding to the mode of interaction of two molecules of p18 in the asymmetric crystal unit, which is similar to the interaction described between Arc1p and GluRS in yeast. Thus, human p18 in the MSC replaces the C-terminal domain of yeast Arc1p, and the modes of interaction with MetRS and GluRS are conserved between yeast and human. The potential interfaces 1 of $GluProRS^{GST}$ and of $p38^{GST}$ are able to associate to form a hetero dimer. MRS^{GST} and $EPRS^{GST}$ can bind p18 simultaneously (Cho et al. 2015).

Based on these observations, and on the other published data summarized above, a model of the assembly of nine components of the MSC was proposed (Fig. 18.7) (Cho et al. 2015). According to this model, interface 1 of the GST-like domain of p38 would interact with interface 1 of $GluProRS^{GST}$, which would interact via its interface 2 with interface 2 of $p18^{GST}$, which would in turn interact via its interface 1 with interface 1 of $MetRS^{GST}$. This assembly would recapitulate subcomplex I of the MSC, which also contains LeuRS and IleRS (Kaminska et al. 2009a). The other complex-forming domain of the scaffold protein p38, corresponding to its N-terminal leucine-zipper domain, would associate subcomplex I into the MSC (Cho et al. 2015). The leucine-zipper motif of p38 would interact with the N-terminal region of the long helical domain of p43, that forms a coiled-coil with the N-terminal Ha helix of ArgRS (Fig. 18.6), which forms the platform for interaction with GlnRS (Fu et al. 2014). The very N-terminal peptide of p38 will also associate one dimer of LysRS into the MSC (Ofir-Birin et al. 2013).

The assembly described above should correspond to only one half of the MSC (right or left part of Fig. 18.7). Since the ProRS domain of GluProRS requires to be

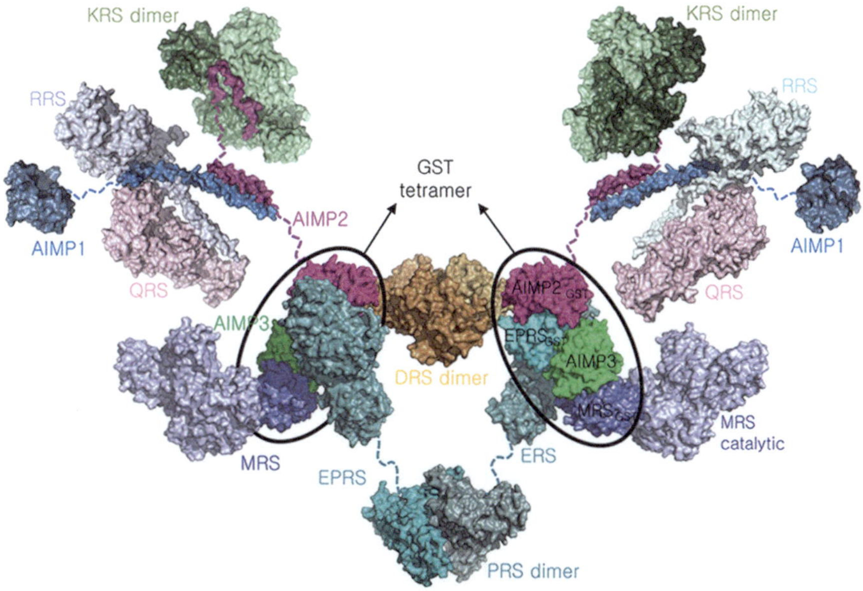

Fig. 18.7 General model of organization of the components of the MSC. The model of the assembly of the MSC is based on the co-crystal structure of GluProRS^{GST} with p38^{GST} (AIMP2) (PDB 5A34), of GluProRS^{GST} with p18^{GST} (AIMP3) (PDB 5BMU), and of MetRS^{GST} with p18^{GST} (AIMP3) (PDB 4BVX) (Reproduced from Cho et al. 2015)

in a dimer to be functional (Kerjan et al. 1992), two GluProRS subunits have to be present into the MSC. In the proposed model, AspRS could be the link between the two half-MSCs (Cho et al. 2015). However, it is known that AspRS is readily exchangeable within the MSC (Agou and Mirande 1997), which is not consistent with its central position in the model. Alternatively, the scaffold protein p38 itself, which has the potential to form a dimer in solution (Robinson et al. 2000), could be responsible for the assembly of two half-MSCs. Interface 2 of its p38^{GST} might be involved, at least in part, in this process. The possible location and stoichiometry of LeuRS and IleRS within the MSC remains unclear, but biochemical data suggest that LeuRS, IleRS, and MetRS are present in equal amounts in the complex (Johnson and Yang 1981; Mirande et al. 1982). Based on the proposed model (Fig. 18.7) and on the expected stoichiometries for LeuRS and IleRS, the MSC would contain two subunits of each of its components, except for LysRS that would be present with a stoichiometry of two dimers per MSC complex. The molecular weight of the MSC would be close to 2 MDa, a value larger than the value of 1.5 MDa previously anticipated (Dias et al. 2013).

This model of the assembly of the MSC is consistent with the observations that the EMAPII domain of p43 can be easily released after cellular proteolysis during apoptosis (Shalak et al. 2001), or that LysRS (Ofir-Birin et al. 2013) or MetRS (Ko et al. 2000), two peripheral components, may leave the MSC without impairing its overall organization. Concerning p18, which has been described as a tumor suppres-

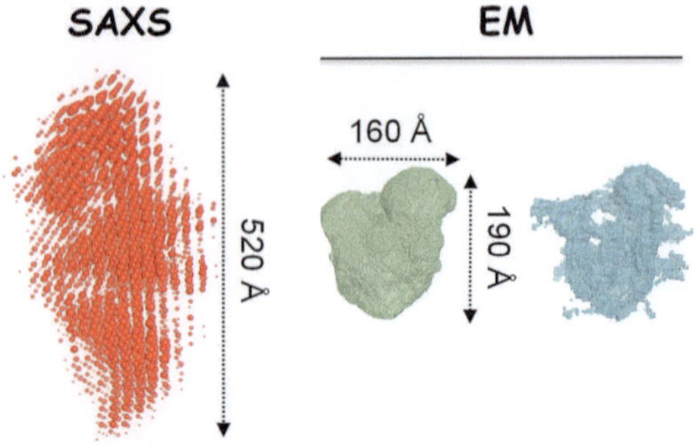

Fig. 18.8 Low-resolution structure of the MSC. The low resolution envelopes of the MSC obtained by SAXS (Dias et al. 2013) or from the negative stain (*green*) and cryo (*cyan*) EM three-dimensional reconstructions (Norcum and Boisset 2002) are shown

sor in the nucleus (Park et al. 2005), its release from the other components of the cytosolic MSC would require two independent events, one that would disrupt its interaction with MetRS, and one that would prevent its association with p38. In the case of GluProRS, which forms an inflammation-responsive mRNP after dissociation from the MSC (Sampath et al. 2004), the present model does not provide clues to understand the mechanism of its controlled release from the MSC, without impairing the overall architecture of the complex. It has been reported that phosphorylation of Ser886 and Ser999, located in the linker domain of GluProRS made of three WHEP elements (Fig.18.1), is the key factor that induces its release from the MSC (Arif et al. 2009). The model shown in Fig. 18.7 suggests that anchoring of GluProRS to the complex is mainly mediated by its GST-like N-terminal domain and does not involve the WHEP-linker. As a consequence of the release of GluProRS from the MSC, p18 and MetRS would lose their binding interface in the MSC, the GluProRS^{GST} domain, and should be concomitantly released from the MSC, which has not been reported.

18.5 Low-Resolution Structures of the Native MSC

Although the data reported above certainly recapitulate many of the structural characteristics responsible for the assembly of the 11 components of the MSC, structural analysis of the native particle is needed to get a more functional view of the MSC. Structural analysis of the native complex isolated from natural sources, from mammalian tissues or cells, has been limited by the availability of sufficient amounts of the MSC purified to homogeneity.

The 3-dimensional architecture of the eukaryotic MSC, isolated from rabbit or human sources, was determined from negatively stained- and cryo-electron micrographs (Norcum 1989; Norcum and Dignam 1999; Norcum and Boisset 2002; Wolfe et al. 2005), and from small-angle X-ray scattering (SAXS) analysis (Dias et al. 2013) (Fig. 18.8). Results obtained at a resolution not better than 30 Å do not provide structural details, but allow us to determine the global size of the envelope of the particle. Particle dimensions were estimated to be 25 × 30 × 23 nm (Norcum

1999) or 19 × 16 × 10 nm (Norcum and Boisset 2002) according to EM measurements, or 50 nm in the largest intramolecular distance as determined by SAXS (Dias et al. 2013), suggesting that the MSC has an elongated shape in solution. The large divergence in the estimated size of the particle could be due to the different methods of sample preparation for EM and SAXS, or to the quality of the complexes that might be heterogeneous in size and composition due to uncontrolled proteolysis that may especially occur in the peptides that link the different domains of its components (Fig. 18.1) during isolation of the particles. Several attempts were conducted to visualize the various components of the MSC using antibodies and immunogold labeling (Norcum and Dignam 1999; Wolfe et al. 2005) but higher resolution studies are needed to get a precise structural view of the MSC.

The non-compact and probably flexible shape of the MSC observed in solution (Dias et al. 2013) is consistent with the idea according to which several component of the MSC have the ability to dissociate from the complex to fulfill non-translational functions. The large overall dimension of the MSC (Fig. 18.8) and the organization of the AARS components at the periphery of the particle (Fig. 18.7) seem to be especially well designed to allow the concomitant functioning of all the synthetase components of the complex in the tRNA-aminoacylation reaction, to accomplish their primary and essential translational function.

References

Agou F, Mirande M (1997) Aspartyl-tRNA synthetase from rat – *In vitro* functional analysis of its assembly into the multisynthetase complex. Eur J Biochem 243:259–267

Agou F, Quevillon S, Kerjan P, Latreille MT, Mirande M (1996) Functional replacement of hamster lysyl-tRNA synthetase by the yeast enzyme requires cognate amino acid sequences for proper tRNA recognition. Biochemistry 35:15322–15331

Ahn HC, Kim S, Lee BJ (2003) Solution structure and p43 binding of the p38 leucine zipper motif: coiled-coil interactions mediate the association between p38 and p43. FEBS Lett 542:119–124

Arif A, Jia J, Mukhopadhyay R, Willard B, Kinter M, Fox PL (2009) Two-site phosphorylation of EPRS coordinates multimodal regulation of noncanonical translational control activity. Mol Cell 35:164–180

Cahuzac B, Berthonneau E, Birlirakis N, Guittet E, Mirande M (2000) A recurrent RNA-binding domain is appended to eukaryotic aminoacyl-tRNA synthetases. EMBO J 19:445–452

Cérini C, Kerjan P, Astier M, Gratecos D, Mirande M, Semeriva M (1991) A component of the multisynthetase complex is a multifunctional aminoacyl-tRNA synthetase. EMBO J 10:4267–4277

Cho HY, Maeng SJ, Cho HJ, Choi YS, Chung JM, Lee S, Kim HK, Kim JH, Eom C-Y, Kim Y-G, Guo M, Jung HS, Kang BS, Kim S (2015) Assembly of multi-tRNA synthetase complex via heterotetrameric glutathione transferase-homology domains. J Biol Chem 290:29313–29328

Cusack S (1995) Eleven down and nine to go. Nat Struct Biol 2:824–831

David A, Netzer N, Strader MB, Das SR, Chen CY, Gibbs J, Pierre P, Bennink JR, Yewdell JW (2011) RNA binding targets aminoacyl-tRNA synthetases to translating ribosomes. J Biol Chem 286:20688–20700

Dias J, Renault L, Perez J, Mirande M (2013) Small-angle X-ray solution scattering study of the multi-aminoacyl-tRNA synthetase complex reveals an elongated and multi-armed particle. J Biol Chem 288:23979–23989
Eriani G, Delarue M, Poch O, Gangloff J, Moras D (1990) Partition of tRNA synthetases into two classes based on mutually exclusive sets of sequence motifs. Nature 347:203–206
Francin M, Mirande M (2003) Functional dissection of the eukaryotic-specific tRNA-interacting factor of lysyl-tRNA synthetase. J Biol Chem 278:1472–1479
Francin M, Kaminska M, Kerjan P, Mirande M (2002) The N-terminal domain of mammalian lysyl-tRNA synthetase is a functional tRNA-binding domain. J Biol Chem 277:1762–1769
Frechin M, Enkler L, Tetaud E, Laporte D, Senger B, Blancard C, Hammann P, Bader G, Clauder-Münster S, Steinmetz LM, Martin RP, di Rago J-P, Becker HD (2014) Expression of nuclear and mitochondrial genes encoding ATP synthase is synchronized by disassembly of a multisynthetase complex. Mol Cell 56:763–776
Fu Y, Kim Y, Jin KS, Kim HS, Kim JH, Wang D, Park M, Jo CH, Kwon NH, Kim D, Kim MH, Jeon YH, Hwang KY, Kim S, Cho Y (2014) Structure of the ArgRS-GlnRS-AIMP1 complex and its implications for mammalian translation. Proc Natl Acad Sci U S A 111:15084–15089
Guigou L, Shalak V, Mirande M (2004) The tRNA-interacting factor p43 associates with mammalian arginyl-tRNA synthetase but does not modify its tRNA aminoacylation properties. Biochemistry 43:4592–4600
Guo M, Ignatov M, Musier-Forsyth K, Schimmel P, Yang XL (2008) Crystal structure of tetrameric form of human lysyl-tRNA synthetase: implications for multisynthetase complex formation. Proc Natl Acad Sci U S A 105:2331–2336
Guo M, Schimmel P, Yang XL (2010) Functional expansion of human tRNA synthetases achieved by structural inventions. FEBS Lett 584:434–442
Han JM, Kim JY, Kim S (2003) Molecular network and functional implications of macromolecular tRNA synthetase complex. Biochem Biophys Res Commun 303:985–993
Havrylenko S, Mirande M (2015) Aminoacyl-tRNA synthetase complexes in evolution. Int J Mol Sci 16:6571–6594
Ibba M, Söll D (2000) Aminoacyl-tRNA synthesis. Annu Rev Biochem 69:617–650
Jeong EJ, Hwang GS, Kim KH, Kim MJ, Kim S, Kim KS (2000) Structural analysis of multifunctional peptide motifs in human bifunctional tRNA synthetase: identification of RNA-binding residues and functional implications for tandem repeats. Biochemistry 39:15775–15782
Jia J, Arif A, Ray PS, Fox PL (2008) WHEP domains direct noncanonical function of glutamyl-prolyl tRNA synthetase in translational control of gene expression. Mol Cell 29:679–690
Johnson DL, Yang DCH (1981) Stoichiometry and composition of an aminoacyl-tRNA synthetase complex from rat liver. Proc Natl Acad Sci U S A 78:4059–4062
Kaminska M, Shalak V, Mirande M (2001) The appended C-domain of human methionyl-tRNA synthetase has a tRNA-sequestering function. Biochemistry 40:14309–14316
Kaminska M, Havrylenko S, Decottignies P, Gillet S, Marechal PL, Negrutskii B, Mirande M (2009a) Dissection of the structural organization of the aminoacyl-tRNA synthetase complex. J Biol Chem 284:6053–6060
Kaminska M, Havrylenko S, Decottignies P, Le Marechal P, Negrutskii B, Mirande M (2009b) Dynamic organization of aminoacyl-tRNA synthetase complexes in the cytoplasm of human cells. J Biol Chem 284:13746–13754
Kao J, Ryan J, Brett G, Chen J, Shen H, Fan YG, Godman G, Familletti PC, Wang F, Pan YC, Stern D, Clauss M (1992) Endothelial monocyte-activating polypeptide II. A novel tumor-derived polypeptide that activates host-response mechanisms. J Biol Chem 267:20239–20247
Kao J, Fan YG, Haehnel I, Brett J, Greenberg S, Clauss M, Kayton M, Houck K, Kisiel W, Seljelid R, Burnier J, Stern D (1994) A peptide derived from the amino terminus of endothelial-monocyte-activating polypeptide II modulates mononuclear and polymorphonuclear leukocyte functions, defines an apparently novel cellular interaction site, and induces an acute inflammatory response. J Biol Chem 269:9774–9782

Kerjan P, Triconnet M, Waller JP (1992) Mammalian prolyl-tRNA synthetase corresponds to the 150 kDa subunit of the high-Mr aminoacyl-tRNA synthetase complex. Biochimie 74:195–205

Kim T, Park SG, Kim JE, Seol W, Ko YG, Kim S (2000) Catalytic peptide of human glutaminyl-tRNA synthetase is essential for its assembly to the aminoacyl-tRNA synthetase complex. J Biol Chem 275:21768–21772

Kim KJ, Park MC, Choi SJ, Oh YS, Choi EC, Cho HJ, Kim MH, Kim SH, Kim DW, Kim S, Kang BS (2008) Determination of three-dimensional structure and residues of the novel tumor suppressor AIMP3/p18 required for the interaction with ATM. J Biol Chem 283:14032–14040

Kim KR, Park SH, Kim HS, Rhee KH, Kim BG, Kim DG, Park MS, Kim HJ, Kim S, Han BW (2013) Crystal structure of human cytosolic aspartyl-tRNA synthetase, a component of multi-tRNA synthetase complex. Proteins 81:1840–1846

Kim HS, Cha SY, Jo CH, Han A, Hwang KY (2014) The crystal structure of arginyl-tRNA synthetase from *Homo sapiens*. FEBS Lett 588:2328–2334

Ko YG, Kang YS, Kim EK, Park SG, Kim S (2000) Nucleolar localization of human methionyl-tRNA synthetase and its role in ribosomal RNA synthesis. J Cell Biol 149:567–574

Kyriacou SV, Deutscher MP (2008) An important role for the multienzyme aminoacyl-tRNA synthetase complex in mammalian translation and cell growth. Mol Cell 29:419–427

Lazard M, Mirande M (1993) Cloning and analysis of a cDNA encoding mammalian arginyl-tRNA synthetase, a component of the multisynthetase complex with a hydrophobic N-terminal extension. Gene 132:237–245

Lee YN, Razin E (2005) Nonconventional involvement of LysRS in the molecular mechanism of USF2 transcriptional activity in FcepsilonRI-activated mast cells. Mol Cell Biol 25:8904–8912

Mirande M (1991) Aminoacyl-tRNA synthetase family from prokaryotes and eukaryotes: structural domains and their implications. Prog Nucleic Acid Res Mol Biol 40:95–142

Mirande M, Kellermann O, Waller JP (1982) Macromolecular complexes from sheep and rabbit containing seven aminoacyl-tRNA synthetases. II Structural characterization of the polypeptide components and immunological identification of the methionyl-tRNA synthetase subunit. J Biol Chem 257:11049–11055

Mirande M, Cirakoglu B, Waller JP (1983) Seven mammalian aminoacyl-tRNA synthetases associated within the same complex are functionally independent. Eur J Biochem 131:163–170

Norcum MT (1989) Isolation and electron microscopic characterization of the high molecular mass aminoacyl-tRNA synthetase complex from murine erythroleukemia cells. J Biol Chem 264:15043–15051

Norcum MT (1999) Ultrastructure of the eukaryotic aminoacyl-tRNA synthetase complex derived from two dimensional averaging and classification of negatively stained electron microscopic images. FEBS Lett 447:217–222

Norcum MT, Boisset N (2002) Three-dimensional architecture of the eukaryotic multisynthetase complex determined from negatively stained and cryoelectron micrographs. FEBS Lett 512:298–302

Norcum MT, Dignam JD (1999) Immunoelectron microscopic localization of glutamyl–/prolyl-tRNA synthetase within the eukaryotic multisynthetase complex. J Biol Chem 274:12205–12208

Ofir-Birin Y, Fang P, Bennett SP, Zhang HM, Wang J, Rachmin I, Shapiro R, Song J, Dagan A, Pozo J, Kim S, Marshall AG, Schimmel P, Yang XL, Nechushtan H, Razin E, Guo M (2013) Structural switch of lysyl-tRNA synthetase between translation and transcription. Mol Cell 49:30–42

Park BJ, Kang JW, Lee SW, Choi SJ, Shin YK, Ahn YH, Choi YH, Choi D, Lee KS, Kim S (2005) The haploinsufficient tumor suppressor p18 upregulates p53 via interactions with ATM/ATR. Cell 120:209–221

Quevillon S, Mirande M (1996) The p18 component of the multisynthetase complex shares a protein motif with the β and γ subunits of eukaryotic elongation factor 1. FEBS Lett 395:63–67

Quevillon S, Robinson JC, Berthonneau E, Siatecka M, Mirande M (1999) Macromolecular assemblage of aminoacyl-tRNA synthetases: identification of protein-protein interactions and characterization of a core protein. J Mol Biol 285:183–195
Ray PS, Arif A, Fox PL (2007) Macromolecular complexes as depots for releasable regulatory proteins. Trends Biochem Sci 32:158–164
Ray PS, Sullivan JC, Jia J, Francis J, Finnerty JR, Fox PL (2011) Evolution of function of a fused metazoan tRNA synthetase. Mol Biol Evol 28:437–447
Rémion A, Khoder-Agha F, Cornu D, Argentini M, Redeker V, Mirande M (2016) Identification of protein interfaces within the multi-aminoacyl-tRNA synthetase complex: the case of lysyl-tRNA synthetase and the scaffold protein p38. FEBS Open Biol 6:696–706
Renault L, Kerjan P, Pasqualato S, Menetrey J, Robinson JC, Kawaguchi S, Vassylyev DG, Yokoyama S, Mirande M, Cherfils J (2001) Structure of the EMAPII domain of human aminoacyl-tRNA synthetase complex reveals evolutionary dimer mimicry. EMBO J 20:570–578
Robinson JC, Kerjan P, Mirande M (2000) Macromolecular assemblage of aminoacyl-tRNA synthetases: quantitative analysis of protein-protein interactions and mechanism of complex assembly. J Mol Biol 304:983–994
Sampath P, Mazumder B, Seshadri V, Gerber CA, Chavatte L, Kinter M, Ting SM, Dignam JD, Kim S, Driscoll DM, Fox PL (2004) Noncanonical function of glutamyl-prolyl-tRNA synthetase: gene-specific silencing of translation. Cell 119:195–208
Shalak V, Kaminska M, Mitnacht-Kraus R, Vandenabeele P, Clauss M, Mirande M (2001) The EMAPII cytokine is released from the mammalian multisynthetase complex after cleavage of its p43/proEMAPII component. J Biol Chem 276:23769–23776
Shalak V, Guigou L, Kaminska M, Wautier MP, Wautier JL, Mirande M (2007) Characterization of p43(ARF), a derivative of the p43 component of multiaminoacyl-tRNA synthetase complex released during apoptosis. J Biol Chem 282:10935–10943
Simader H, Hothorn M, Kohler C, Basquin J, Simos G, Suck D (2006) Structural basis of yeast aminoacyl-tRNA synthetase complex formation revealed by crystal structures of two binary sub-complexes. Nucleic Acids Res 34:3968–3979
Swairjo MA, Morales AJ, Wang CC, Ortiz AR, Schimmel P (2000) Crystal structure of Trbp111: a structure-specific tRNA-binding protein. EMBO J 19:6287–6298
Wang CC, Morales AJ, Schimmel P (2000) Functional redundancy in the nonspecific RNA binding domain of a class I tRNA synthetase. J Biol Chem 275:17180–17186
Wolfe CL, Warrington JA, Treadwell L, Norcum MT (2005) A three-dimensional working model of the multienzyme complex of aminoacyl-tRNA synthetases based on electron microscopic placements of tRNA and proteins. J Biol Chem 280:38870–38878
Yang F, Ji QQ, Ruan LL, Ye Q, Wang ED (2014) The mRNA of human cytoplasmic arginyl-tRNA synthetase recruits prokaryotic ribosomes independently. J Biol Chem 289:20953–20959
Yannay-Cohen N, Carmi-Levy I, Kay G, Yang CM, Han JM, Kemeny DM, Kim S, Nechushtan H, Razin E (2009) LysRS serves as a key signaling molecule in the immune response by regulating gene expression. Mol Cell 34:603–611
Yaremchuk A, Cusack S, Tukalo M (2000) Crystal structure of a eukaryote/archaeon-like prolyl-tRNA synthetase and its complex with $tRNA^{Pro}$(CGG). EMBO J 19:4745–4758
Zamecnik P (1983) Diadenosine 5′,5″′-P1,P4-tetraphosphate (Ap4A): its role in cellular metabolism. Anal Biochem 134:1–10
Zhang X, Ling J, Barcia G, Jing L, Wu J, Barry BJ, Mochida GH, Hill RS, Weimer JM, Stein Q, Poduri A, Partlow JN, Ville D, Dulac O, Yu TW, Lam AT, Servattalab S, Rodriguez J, Boddaert N, Munnich A, Colleaux L, Zon LI, Soll D, Walsh CA, Nabbout R (2014) Mutations in QARS, encoding glutaminyl-tRNA synthetase, cause progressive microcephaly, cerebral-cerebellar atrophy, and intractable seizures. Am J Hum Genet 94:547–558
Zhou H, Sun L, Yang X, Schimmel P (2013) ATP-directed capture of bioactive herbal-based medicine on human tRNA synthetase. Nature 494:121–124

Chapter 19
The Pyruvate Dehydrogenase Complex and Related Assemblies in Health and Disease

Olwyn Byron and John Gordon Lindsay

Abstract The family of 2-oxoacid dehydrogenase complexes (2-OADC), typified by the pyruvate dehydrogenase multi-enzyme complex (PDC) as its most prominent member, are massive molecular machines (M_r, 4–10 million) controlling key steps in glucose homeostasis (PDC), citric acid cycle flux (OGDC, 2-oxoglutarate dehydrogenase) and the metabolism of the branched-chain amino acids, leucine, isoleucine and valine (BCOADC, branched-chain 2-OADC). These highly organised mitochondrial arrays, composed of multiple copies of three separate enzymes, have been widely studied as paradigms for the analysis of enzyme cooperativity, substrate channelling, protein-protein interactions and the regulation of activity by phosphorylation. This chapter will highlight recent advances in our understanding of the structure-function relationships, the overall organisation and the transport and assembly of PDC in particular, focussing on both native and recombinant forms of the complex and their individual components or constituent domains. Biophysical approaches, including X-ray crystallography (MX), nuclear magnetic resonance spectroscopy (NMR), cryo-EM imaging, analytical ultracentrifugation (AUC) and small angle X-ray and neutron scattering (SAXS and SANS), have all contributed significant new information on PDC subunit organisation, stoichiometry, regulatory mechanisms and mode of assembly. Moreover, the recognition of specific genetic defects linked to PDC deficiency, in combination with the ability to analyse recombinant PDCs housing both novel naturally-occurring and engineered mutations, have all stimulated renewed interest in these classical metabolic assemblies. In addition, the role played by PDC, and its constituent proteins, in certain disease states will be briefly reviewed, focussing on the development of new and exciting areas of medical and immunological research.

O. Byron
School of Life Sciences, University of Glasgow, Glasgow G12 8QQ, UK
e-mail: Olwyn.Byron@glasgow.ac.uk

J.G. Lindsay (✉)
Institute of Molecular, Cell and Systems Biology, Davidson Building, College of Medicine, Veterinary and Life Sciences, University of Glasgow, Glasgow G12 8QQ, UK
e-mail: Gordon.Lindsay@glasgow.ac.uk

J.R. Harris, J. Marles-Wright (eds.), *Macromolecular Protein Complexes*, Subcellular Biochemistry 83, DOI 10.1007/978-3-319-46503-6_19

Keywords Pyruvate dehydrogenase complex • Reaction cycle • Supramolecular architecture • Structure-function relationship • Biophysical characterisation • Phosphorylation • Intracellular trafficking • PDC deficiency • Genetic defects • Potential drug target

19.1 Introduction to Overall Complex Architecture, Macromolecular Organisation and the Reaction Cycle

Elucidation of the general architecture, subunit organisation and enzymology of bacterial and mammalian pyruvate dehydrogenase multi-enzyme complexes (PDCs) over the past 50 years has been achieved by a number of laboratories, including pioneering contributions from the groups of Lester Reed in Austin (Reed and Oliver 1982; Behal et al. 1989a, b; Reed et al. 1992); Richard Perham in Cambridge (Perham and Reche 1998; Perham et al. 1987); John Guest in Sheffield (Guest et al. 1989; Jordan et al. 1998) and Robert Harris in Indianapolis (Harris et al. 1997b, 2002; Popov et al. 1997). Each member of the 2-oxoacid dehydrogenase complex (2-OADC) family is composed of multiple copies of three distinct enzymes, termed E1, E2 and E3. Their overall morphology is determined by the central E2 enzyme that forms a pentagonal dodecahedral (60-meric) or cubic (24-meric) core acting as a framework to which multiple dimeric or hetero-tetrameric E1 and homo-dimeric E3 enzymes are tethered tightly, but non-covalently (Reed and Oliver 1982; Wagenknecht et al. 1990). In each case, these large central E2 assemblies are constructed from basic trimeric units that coalesce to form the 8 or 20 vertices of the E2 cube or dodecahedron, respectively. The 60-meric core is confined exclusively to PDCs from eukaryotic sources and Gram-positive bacteria. Interestingly, the E2 enzymes are all acyltransferases that are distantly related to chloramphenicol acetyltransferase, a naturally-occurring homo-trimer.

As shown in Fig. 19.1, the mammalian PDC reaction cycle involves the initial oxidative decarboxylation of pyruvate by its thiamine-diphosphate (ThDP)-requiring $\alpha_2\beta_2$ E1 enzyme, which also mediates transfer of the residual 2-carbon unit via an ethylidene-ThDP intermediate to reductively acetylate multiple lipoamide cofactors that are covalently linked to E2. Thereafter, E2 catalyses the transfer of the acetyl group to CoA, allowing the homodimeric dihydrolipoamide dehydrogenase (E3) to restore the reduced S_6, S_8 lipoamide thiols to their oxidised (disulphide) state, yielding acetyl CoA and NADH as final products.

19.1.1 E1, Pyruvate Dehydrogenase/Decarboxylase

E1 is a ThDP-dependent decarboxylase that catalyses the first, rate-limiting, irreversible step in the catalytic cycle (Cate et al. 1980; Berg et al. 1998). Initial decarboxylation of pyruvate is followed by the transfer of the ThDP-bound ethylidene

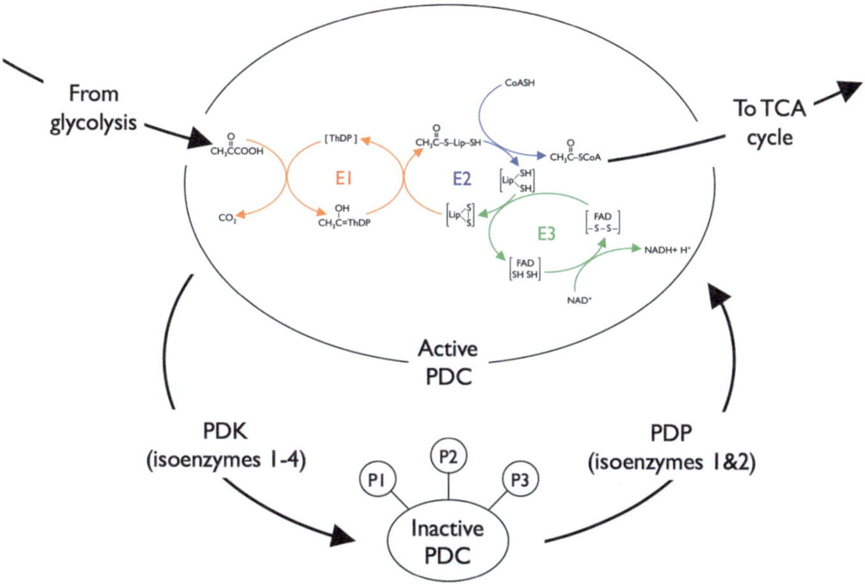

Fig. 19.1 PDC reaction cycle showing the active and inactive states of the multi-enzyme complex and the role of its associated kinases (PDK) and phosphatases (PDP) in its regulation by phosphorylation (P1-P3) and dephosphorylation of three specific serines in the E1α subunit

intermediate onto E2-linked lipoamide moieties in a reductive acetylation reaction. Both α_2 dimeric and $\alpha_2\beta_2$ tetrameric E1s house two active sites, each requiring Mg^{2+} ions and ThDP for activity. Several E1 crystal structures have been solved in recent years, including from *Escherichia coli* (Arjunan et al. 2002) and humans (Ciszak et al. 2003).

In the case of hetero-tetrameric E1s, the α and β subunits are related by a crystallographic two-fold axis of symmetry and arranged in a tetrahedral fashion. The active sites that bind ThDP are located in a 20 Å funnel-shaped tunnel formed at the α-β subunit interface, thereby facilitating access to the lipoyl-lysine moiety during the catalytic cycle. Hydrophobic contacts are primarily responsible for the stability of α/β subunit interactions. Biochemical, kinetic and spectral studies have demonstrated that E1 acts by a 'flip-flop' or alternating active-site mechanism. Thus, while one active site is participating in the pyruvate decarboxylation step, the other is simultaneously catalysing the reductive acetylation of an E2-lipoamide moiety (Yi et al. 1996; Sergienko et al. 2000; Sergienko and Jordan 2002; Kovina and Kochetov 1998; Khailova et al. 1990). X-ray structural studies have revealed that the ThDPs in the two E1 active sites communicate over a distance of 20 Å by reversibly shuttling a proton through an acidic tunnel in the protein. This proton 'wire' permits the cofactors to serve reciprocally as general acid/base catalysts and to switch the conformation of crucial active site peptide loops in a concerted manner (Frank et al. 2004).

Recently, an X-ray structure for a homo-dimeric E1:E2 di-domain (lipoyl domain (LD) plus subunit-binding domain (SBD), see below) sub-complex from *E. coli* has revealed a new folding motif located in the N-terminal region of E1 that is responsible for promoting a flexible and dynamic interaction with the E2 di-domain (Arjunan et al. 2014). This mode of interaction differs substantially in both location and nature from a previously-reported sub-complex of a SBD with an $\alpha_2\beta_2$ E1 from *Bacillus stearothermophilus* (Frank et al. 2005).

19.1.2 E2, Dihydrolipoamide Acetyltransferase

No high-resolution X-ray structures for any full-length E2-core assembly are available at present; however, crystal structures have been determined for both truncated 24- and 60-meric E2 constructs lacking their highly-mobile, modular N-terminal LDs and SBDs, as these are still capable of self-assembly (*Azotobacter vinelandii* (Mattevi et al. 1992); *B. stearothermophilus* (Milne et al. 2002); yeast (Stoops et al. 1992, 1997) and bovine PDC (Zhou et al. 2001b; Behal et al. 1994)). In both types, the fundamental building block is a basic trimeric unit, with 8 or 20 trimers associating to form a cube or pentagonal dodecahedron. Interestingly the inter-trimer contacts are much weaker than the intra-trimer ones, giving rise to a flexible, but well-connected core that is capable of 'breathing', a characteristic that may be essential for optimal PDC function by facilitating substrate channelling and active-site coupling (Zhou et al. 2001a; Kong et al. 2003).

High resolution cryo-EM reconstructions of human (Yu et al. 2008), bovine (Zhou et al. 2001b) and *Saccharomyces cerevisiae* (Stoops et al. 1992, 1997) cores as well as X-ray crystallographic structures of those from *Azotobacter vinelandii* (Mattevi et al. 1992, 1993) and *B. stearothermophilus* (Izard et al. 1999) have been determined. The C-terminal domains (CTDs) contain the conserved, active-site sequence motif, DHRXXDG, with the histidine and aspartate necessary for catalysis. The active-site is located in a 29 Å long channel that runs across the trimer at the CTD subunit interface. Interestingly, the two substrates that participate in the trans-acetylation reaction enter the active site from opposite directions. Coenzyme A enters from the interior of the hollow, highly-solvated core whereas the acetylated lipoamide moiety enters from the outside.

In recent years, the crystal structure of a unique 42-meric catalytic core of a 2-OADC from the thermophile *Thermoplasma acidophilum* has been solved at 4 Å resolution. In this structure the basic homo-trimeric units assemble into an oblate spheroid combining cubic and dodecahedral geometric elements (Marrott et al. 2012). Remarkably, the identical homo-trimeric building blocks of the 42-mer display differential modes of assembly, forming a polyhedral sphere with six pentagonal and three quadrilateral faces. In an extension of this study, removal of the five C-terminal amino acids of the E2 polypeptide (IIYEI) containing a key tyrosine involved in inter-trimer contacts results in the generation of a trimeric enzyme (Marrott et al. 2014). This minimal complex can bind both E1 and E3 components to generate a branched-chain 2-oxoacid dehydrogenase complex with catalytic activity similar to the native multi-enzyme complex, raising the question as to why these assemblies are so large in nature. In this regard, it seems probable that the catalytic advantages conferred by their general ability to perform active-site coupling accounts for this phenomenon. Active-site coupling, first observed in *E. coli* PDC (Bates et al. 1977), involves the rapid transfer of acetyl groups from one lipoyl group to another around the entire surface of the core. Thus, the full complement of LDs can be acetylated from pyruvate entering via a single E1 enzyme. This remarkable cooperation of active sites provides a means of linking pyruvate and CoA, regardless of where they bind to the complex, ensuring enhanced rates of reaction even at low substrate concentrations (Marrott et al. 2014).

Central to the catalytic mechanism is the mobility of the N-terminally-located, E2-linked LDs that act as 'swinging arms', enabling their prosthetic groups to visit the active sites of all three enzymes during the catalytic cycle. This allows the lipoamide moieties to:

I. Be reductively acetylated by E1 (in the case of PDC)
II. Participate in E2-directed acetyl group transfer to CoA
III. Be returned to their oxidised state by reaction with E3 (Fig. 19.1)

The structural basis for these large-scale domain movements is the modular organisation of E2 that contains between one and three small N-terminal LDs (approximately 80 amino acids) separated by pro-ala-rich flexible linker regions of approximately 30 amino acids in length. These N-terminal domains extend outwards from the 24/60-meric core comprising the CTDs that are responsible for self-assembly and also house the acyltransferase active site. The overall LD structure is that of two four-stranded β-sheets forming a flattened β-barrel stabilised by a core of hydrophobic residues. The lipoamide prosthetic groups are bound to the ε-amino terminal of a key lysine residue situated at the exposed tip of a type 1 β-turn at one end of the domain with the N and C-termini close together at the opposite end on adjacent β-strands (Fig. 19.3a) (Howard et al. 1998). In this location, the dithiolane ring of the prosthetic group is itself situated approximately 14 Å above the surface of each LD as a result of its octanoate 'tail', allowing further extension of its reach and conformational flexibility (Perham 2000). A prominent conserved feature of LDs is a two-fold axis of quasi-symmetry relating the N and C-terminal halves of the domain (Dardel et al. 1993; Green et al. 1995; Ricaud et al. 1996).

NMR spectroscopy approaches have yielded vital insights into LD structure, dynamics and substrate specificity for their cognate E1/E3 enzymes and regulatory kinases (Kumaran et al. 2013). This approach has been aided significantly by the modular domain organisation of E2, the compact size of the LDs and SBDs and their ability to fold independently into functional entities. Moreover, the properties of linker regions that are rich in ala, pro and acidic amino acids have also been defined by NMR indicating that they have sufficient flexibility to account for the large-scale movements of the LDs during catalysis while retaining sufficient rigidity to ensure that they are held above the surface of the E2 core (Green et al. 1992). Mutational analysis has shown that the linker regions in *E. coli* PDC (approximately

30 amino acids) can be truncated by 50% before there is a significant effect on overall complex activity (Miles et al. 1988). In addition, peptide-specific hydrogen/deuterium mass spectrometry and NMR studies have been used to identify interaction loci on the E2/E3BP core with associated kinases (PDKs) 1 and 2 (Wang et al. 2015).

Enzymatic studies, using a series of E2 constructs containing multiple LDs, have also highlighted that the presence of 3 LDs per E2 chain is optimal in *E. coli* but that only the outermost domain need be lipoylated for maximal activity (Guest et al. 1997). Thus, the retention of 3 LDs in tandem repeat appears to be a mechanism for extending the reach of the outer domain prosthetic group rather than providing extra cofactors for catalysis. Interestingly, only a single LD is present on all members of the OGDC and BCOADC family that have an underlying 24-meric cubic core geometry in common with *E. coli* PDC.

19.1.3 E3, Dihydrolipoamide Dehydrogenase (E3)

Crystal structures of the FAD-containing E3 dimer have been solved from several prokaryotic and eukaryotic sources, namely *A. vinelandii* (Mattevi et al. 1991); *Pseudomonas fluorescens* (Mattevi et al. 1993); *B. stearothermophilus* (Mande et al. 1996); *S. cerevisiae* (Toyoda et al. 1998) and human (Brautigam et al. 2006). In all cases, E3 operates by a similar catalytic mechanism. In mammalian systems, E3 is a common component of all members of the 2-OADC family although complex-specific forms of E3 are found in some bacteria (Carothers et al. 1989; Burns et al. 1989). E3 catalyses the oxidation of the dihydrolipoamide prosthetic group on the LD of E2 via a 2-step ‘ping-pong’ mechanism. In the first step, electrons are transferred from reduced lipoamide via a reactive disulphide bridge to FAD, forming a stable intermediate complex. Subsequently, reducing equivalents are transferred to NAD^+ as the final electron acceptor. Despite a low sequence identity, the tertiary structures of prokaryotic and eukaryotic E3s are very similar. The E3 dimer is composed of identical subunits, each comprising 4 domains: an FAD domain, an NAD domain, a central domain and an interface domain. In the absence of FAD *in vitro*, the mammalian E3 polypeptide can fold to a significant extent, attaining a structure similar to its native state. However, it requires the presence of its FAD cofactor to promote the final conformational changes necessary for dimerisation and activation (Lindsay et al. 2000).

19.1.4 The Interaction of E1 and E3 with the Core Complex

Located between the CTD and LDs of E2 is a small and compact SBD, which, in 60-meric Gram-positive bacterial cores (e.g. *B. stearothermophilus*), serves to position E1 and E3 tightly with a 1:1 stoichiometry in a mutually exclusive manner, with affinities in the nanomolar range (Mande et al. 1996). The E2-SBD binding sites on E1 and E3 are located across and close to the two-fold axis of symmetry, respectively (Frank et al. 2005). Occupation of both binding sites is prevented by steric clashes in one of the loop regions. In mammalian and yeast PDC, the equivalent domain has sole responsibility for promoting tight E1 interaction with the core. In these complexes, E3 is able to bind independently via an additional E2-related subunit referred to as E3 binding protein (E3BP). For many years E3BP was thought to be an E2 degradation product as it also contains an active LD; however, it was confirmed as an immunologically-distinct polypeptide in 1985 (De Marcucci and Lindsay 1985) and subsequently shown to be an integral, structural component of mammalian and yeast PDC responsible for E3 binding (Rahmatullah et al. 1989; Neagle and Lindsay 1991; Lawson et al. 1991). *In vitro* reconstitution of the E2 core lacking its E3BP partner confirmed the role of the latter subunit in high affinity E3 binding (McCartney et al. 1997).

The exact stoichiometry of E1 and E3 association with the SBDs of E2 and E3 remains unclear. While the crystal structure of the E3BP-SBD:E3 sub-complex suggests a 1:1 stoichiometry (Brautigam et al. 2006; Ciszak et al. 2006), solution structural studies have provided evidence for a 2:1 binding relationship (Smolle et al. 2006; Vijayakrishnan et al. 2011). In this model, E3 dimers (and possibly also E1 hetero-tetramers) have the potential to form 'cross-bridges' between adjacent E3BPs (and E2s) around the core surface (Vijayakrishnan et al. 2010). The presence of such 'cross-bridges' would provide a well-connected and robust network that may in turn promote enhanced movement and transfer of intermediates around the surface, thereby facilitating the efficiency of substrate channelling and active-site coupling.

19.1.5 Core Complex Heterogeneity

To date, it has not proved possible to crystallise any member of the 2-OADC family in its intact state. This is normally attributed to the size and complexity of these assemblies and in particular the high mobility of the E2-linked N-terminal lipoyl 'swinging arms' that are central to the multi-step catalytic mechanism. However, it is now apparent that the complexes also display intrinsic heterogeneity, so are not unique structures possessing a defined stoichiometry and subunit organisation. For example, in the PDC from *B. stearothermophilus*, the 60-meric E2 assembly can accommodate up to 60 E1 or 60 E3 enzymes that bind with similar affinities to overlapping, mutually-exclusive sites on E2 (Jung et al. 2002). Thus, the numbers and distribution of E1s and E3s around the core surface will vary from complex to complex and probably also in a time-dependent manner. In addition, there is the question of occupancy as there is no requirement to maintain saturation of all E1/E3 binding sites.

In eukaryotes, the structure and subunit composition of the E2/E3BP core is crucial for the efficient functioning of PDC. Initially, it was reported that 12E3BPs were associated with the E2 core prompting the suggestion that individual E3 subunits were situated on the twelve pentagonal faces of the E2 icosahedral assembly-the so-called 'addition' model of core organisation (60E2 + 12E3BP) (Maeng et al. 1994, 1996; Sanderson et al. 1996). However, in subsequent 48E2+12E3BP and 40E2+20E3BP 'substitution' models, twelve or twenty E3BPs were proposed to replace an equivalent number of E2s within the 60-meric core assembly (Brautigam et al. 2009; Hiromasa et al. 2004). SAXS and SANS solution studies combined with cryo-EM reconstructions of recombinant human E2/E3BP cores reveal the presence of open pentagonal faces, strongly supporting the 'substitution' model (Vijayakrishnan et al. 2010) (Fig. 19.2). Variations in the reported subunit composition of the E2/E3BP core in the substitution model have been attributed to genuine differences between native and recombinant assemblies. In the latter case, a 40:20 E2/E3BP core is produced under defined conditions where E3BP is present in excess, suggesting that a single E3BP can be accommodated per trimer. In contrast, the native bovine E2/E3BP assembly has a 48:12 subunit stoichiometry. Mathematical modelling has predicted that an 'average' 48E2:12E3BP arrangement allows maximum flexibility in assembly while providing the appropriate balance of bound E1 and E3 enzymes for optimal catalytic efficiency and regulatory fine tuning (Vijayakrishnan et al. 2011).

In the case of eukaryotic PDCs where the central core 'scaffold' consists of two types of subunit, its overall composition is approximately 48E2:12E3BP (Vijayakrishnan et al. 2011); however, as discussed below, studies on recombinant E2/E3BP cores have shown that up to 20 E3BPs can substitute for 20 E2 subunits within the 60-meric structure suggesting that the composition of the native assembly may represent that of an 'average structure' derived from a heterogeneous population (Vijayakrishnan et al. 2011). It is also possible that the average subunit composition of the E2/E3BP 60-mer may vary from tissue to tissue as an additional means of fine tuning its activity to the prevailing metabolic conditions.

Despite these limitations, a variety of structural approaches have been employed successfully to gain new insights into the overall architecture and distinctive geometry of PDC and the molecular dynamics underlying its enzymatic processes. In most cases, high resolution X-ray structural data have been obtained for individual enzymes, e.g. the homo-dimeric E1 of *E. coli* (Arjunan et al. 2002); the human $\alpha_2\beta_2$ E1 tetramer (Ciszak et al. 2003) and mammalian E3 (Brautigam et al. 2006). Truncated versions of the cubic and icosahedral E2 cores lacking their N-terminal LDs and SBDs have also proved amenable to crystallisation, providing detailed information on the E2 active-site geometry and the subunit contacts involved in oligomerisation in *Azotobacter vinelandii* (Mattevi et al. 1992, 1993); *B. stearothermophilus* (Milne et al. 2002; Izard et al. 1999); yeast (Stoops et al. 1992, 1997) and bovine PDC (Zhou et al. 2001b; Behal et al. 1994). SAXS, SANS and AUC studies have further contributed to resolution between differing models of E2/E3BP core organisation and the modes of attachment of the peripheral E1 and E3 enzymes (Vijayakrishnan et al. 2010, 2011) (Fig. 19.2).

Over-expression in *E. coli*, purification and reconstitution of the component enzymes of human PDC has been successfully achieved in recent years although the

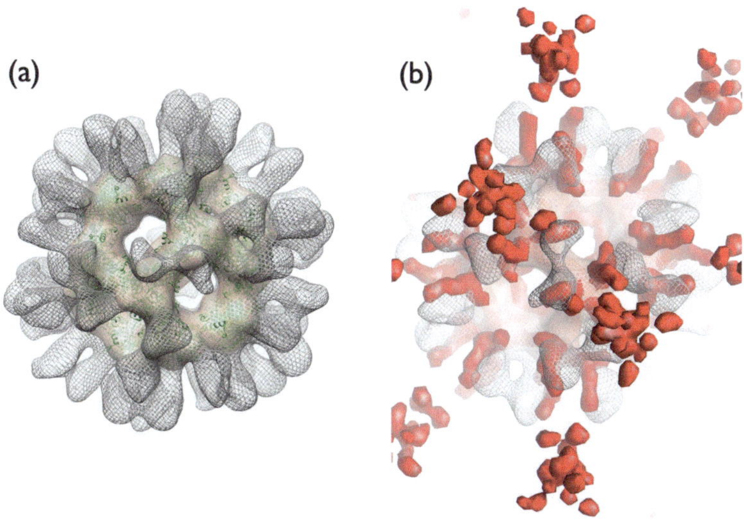

Fig. 19.2 Cryo-EM and SAXS models of recombinant human E2/E3BP and E2/E3BP:E3 complexes. (**a**) Cryo-EM reconstructions of the E2/E3BP (semi-transparent salmon-coloured surface) and E2/E3BP:E3 core (*grey mesh*), with the crystal structure of the *B. stearothermophilus* E2 core fitted within (*green ribbon*). The open faces of the core complex are easily visualised, consistent with the substitution model. (**b**) The E2/E3BP:E3 structure from (**a**) superimposed on the *ab initio* SAXS-derived E2/E3BP model (*red*). Extensions from the core surface consistent with SAXS data represent the LDs and SBDs not visualised by cryo-EM (Adapted with permission from Vijayakrishnan et al. 2010)

constituent subunits of both the $\alpha_2\beta_2$ E1 and E2/E3BP core enzymes must be co-expressed to promote their ordered integration and achieve the recovery of active enzyme (Korotchkina et al. 1995; Harris et al. 1997a). Reconstitution of fully-active recombinant human PDC is now a routine operation, as its three enzymes have an intrinsic capacity for self-association. Access to recombinant versions of the human complex is proving to be invaluable in the analysis of naturally-occurring and engineered mutations that provide new insights into the fundamental mechanisms of PDC function in health and disease (see below).

19.2 Biosynthesis and Assembly of Mammalian PDC

In common with the vast majority of mitochondrial polypeptides that are nuclear-encoded, the constituent subunits of mammalian PDC are synthesised on cytosolic ribosomes as individual precursor polypeptides containing an N-terminal targeting sequence that is recognised by specific receptors on the mitochondrial outer surface (Neupert 1997; Dudek et al. 2013; Becker et al. 2012). For matrix-located polypeptides, nascent chains are maintained in a loosely-folded, translocation competent state by cytoplasmic Hsp70 and Hsp90 chaperone complexes *en route* to the organelle. They are then delivered to the mitochondrial matrix via the TOM 40 and TIM 23 (translocases of the outer and inner membranes) complexes located at specialised contact sites directing passage through the two membranes of the organelle. Translocation is facilitated by a matrix-located mitochondrial Hsp70 that transfers the nascent chains to the interior of the mitochondrial compartment. During this passage, their N-terminal extensions are removed by a pre-sequence processing protease situated on the interior aspect of the inner membrane. A matrix-located Hsp60/10 chaperone system, analogous to that found in bacteria, promotes the final maturation and assembly of the constituent enzymes of PDC prior to their rapid assembly into a mature, intact complex of over 200 polypeptides (Fox 2012; De Marcucci et al. 1988; Chacinska et al. 2009).

Remarkably, recent research has highlighted the possibility that the intact PDC can exit the mitochondrion and be transported *en bloc* to the nucleus in a cell cycle-dependent manner to provide a source of acetyl CoA for histone modification during S-phase (Sutendra et al. 2014; de Boer and Houten 2014). Although the validity of this research awaits independent confirmation, it raises the intriguing possibility that a novel pathway exists for nuclear-organelle communication and the export of mature, folded proteins and complexes. Potential candidates for this novel trafficking pathway are mitochondrial-derived vesicles (MDVs) for which a role in intracellular communication has been identified in recent years (Sugiura et al. (2014) and references therein). However, the size of these vesicles (70–100 Å) would appear to preclude their involvement in the transport of intact PDC (~450 Å), although partial disassembly of PDC into its component enzymes prior to transit is also a possibility (Sun et al. 2015; Soubannier et al. 2012; McLelland et al. 2014). Further work is essential to establish the authenticity of this novel mitochondrial export pathway or provide further insights into alternative mechanisms for translocation of mitochondrial polypeptides/complexes to various intracellular locations (Ng and Tang 2014).

19.3 Regulation of PDC by End-Product Inhibition and Phosphorylation

Control of PDC activity plays an important role in dictating the fuel used by various tissues in different nutritional and hormonal states. Thus, mammalian PDC is unusual amongst mitochondrial enzymes/complexes as, in addition to being subject to end-product inhibition by acetyl CoA and NADH, it is subject to control via its state of phosphorylation on a minute-by-minute basis. Fine control of PDC activity is of crucial importance in that it catalyses the key committed step in glucose utilisation, ensuring the appropriate balance between glucose or fatty acid and ketone body production. For example, it is down-regulated during starvation or diabetes in order to preserve carbohydrate reserves for tissues that are primarily dependent on glucose for energy, whereas it is up-regulated after a meal to maximise energy production and fatty acid biosynthesis.

This acute or short-term regulation is primarily accomplished by four tightly-bound kinases (PDK1-4, related to bacterial histidine kinases (Fig. 19.1)) that are expressed in a tissue-specific manner at the level of one or two molecules per complex. Thus the phosphorylation state of PDC determines the flux of carbon through this vital committed step in carbohydrate metabolism since no net formation of glucose is possible from acetyl CoA. Collectively, this group of kinases targets one or more of three specific serine residues (site 1, Ser264; site 2, Ser271 and site 3, Ser203) on the E1α subunit with phosphorylation leading to complete inactivation in each case (Korotchkina and Patel 1995; Kolobova et al. 2001; Kato et al. 2008; Bao et al. 2004). PDK1 and 4 are highly expressed in heart, liver and muscle while PDK3 is prevalent in kidney, brain and testis and PDK2 is present in most tissues (Sugden and Holness 2003; Gudi et al. 1995; Bowker-Kinley et al. 1998). PDK1 phosphorylates all 3 serine residues in contrast to PDKs2–4 that can modify only sites 1 and 2 (Patel and Korotchkina 2001). Interestingly all four kinases modify these three serine residues at different rates and with different efficiencies, although phosphorylation at a single site is sufficient to cause inactivation, apparently by inhibiting substrate and/or ThDP binding in a site-specific manner (Korotchkina and Patel 2001). More specifically, an X-ray structure for a phosphorylated version of human E1 (site 1, Ser264) indicates that the phosphoryl group prevents ThDP-induced ordering of the two loops housing the three phosphorylation sites. These disordered loops impede binding of the LDs to E1, blocking their reductive acetylation and leading to disruption of substrate channelling (Kato et al. 2008).

PDKs1–3 are reported to interact with the E2 inner LD (ILD) while PDK4 binds to the E3BP-LD (Roche et al. 2003). PDK1-3 activity, in particular, is stimulated by their interaction with the E2-ILD and the lipoamide cofactor. Crystal structures of the ADP-bound form of the PDK2 and PDK3 dimers in association with the lipoylated E2-ILD have revealed in fine structural detail how their close association stimulates kinase activity. Conformational changes in the active-site clefts of both subunits result in disordering of the so-called ATP 'lids' in the ADP-bound state, thereby promoting rapid nucleotide release. Thus, structural and biochemical data suggest that E2-LD binding stimulates kinase activity by relieving end product inhibition (Kato et al. 2005; Green et al. 2008). A 2 Å resolution X-ray structure for PDK4 that normally binds the E3BP-LD, but is not significantly activated by this association, confirms this hypothesis. Its structure with bound ADP reveals an open conformation (in contrast to the closed conformation observed in the PDK2 and PDK3-ADP structures) in which there is a wide active-site cleft containing partially disordered C-terminal tails that facilitate the dissociation of bound nucleotide.

It is proposed that PDK4 with bound ADP exists in equilibrium between open and closed states with the favoured metastable open conformation being responsible for the robust basal activity of free PDK4 (Wynn et al. 2008). NADH, in addition to promoting end-product inhibition, is also able to enhance complex-bound kinase activity indirectly by altering the reduction/acetylation status of the lipoic acid prosthetic group on the E2/E3BP LDs. Recent H/D exchange mass spectrometry and

NMR studies have defined differences in the binding specificities and interaction loci of PDKs 1 and 2 for the two E2-linked LDs (Wang et al. 2015).

Reactivation of E1 is achieved by two specific heterodimeric phosphatases, PDPI and PDP2, comprising a catalytic and regulatory subunit allowing rapid fine tuning of PDC activity in response to prevailing metabolic conditions. PDP expression is also tissue-specific; PDP1 is mainly expressed in muscle whereas is PDP2 is abundant in liver and adipose tissue (Huang et al. 1998). PDP1, the dominant isoform, requires Mg^{2+} and is stimulated by micromolar levels of Ca^{2+} that are necessary to mediate binding to the E2-ILD (Damuni and Reed 1987). Pyruvate, acetyl CoA/CoA and NADH/NAD^+ ratios act as key effectors in feedback inhibition of PDC by influencing the redox and acetylation status of the lipoamide cofactors. While pyruvate inhibits PDK and acts as a positive inhibitor, NADH and acetyl CoA activate PDK and thereby act as negative regulators. A notable feature of the PDP1 crystal structure is a long insertion in the catalytic subunit creating a unique hydrophobic pocket capable of housing the lipoamide cofactor in a manner similar to that of PDKs, providing a structural basis for the indirect action of these effectors (Vassylyev and Symersky 2007).

Recent reports have also highlighted a novel mode of regulation of PDC activity, namely by control of its degree of lipoylation by SIRT4, a member of a family of seven mammalian NAD^+-dependent enzymes that govern diverse biological processes including genome regulation, stress responses and aging (Mathias et al. 2014). This ubiquitous group of enzymes, primarily known as lysine deacetylases, are involved in the removal of acetyl groups from key lysine residues, e.g. from histones during chromatin remodelling. However, evidence now suggests that several SIRTs can remove alternative lysine modifications such as succinyl and malonyl more efficiently than acetyl groups. Moreover, SIRTs 3–5 are mitochondrially-located while SIRTs 4 and 5 display little or no deacetylase activity. SIRT4 has now been shown to interact with E2-PDC where it promotes the efficient removal of lipoyl groups from the key lipoyl-lysines of the E2-LDs, diminishing PDC activity both *in vitro* and *in vivo* under appropriate metabolic conditions. The lipoamidase activity of SIRT4 may be a key factor in controlling the extent of lipoylation of all three members of the 2-OADC family, highlighting its central role as a guardian of cellular metabolism. Changes in lipoylation status will also impact on kinase and phosphatase regulation indirectly, given the importance of the lipoamide cofactor in mediating both kinase and phosphatase interactions with the complex.

19.4 Clinical Consequences of PDC Deficiency

The critical nature of the key metabolic step catalysed by the PDC implies that any loss-of-function mutations will be embryonic lethal. Moreover, specific natural mutations causing alterations in PDC activity were not readily identifiable until relatively recently in that the major indicator of PDC deficiency at the biochemical level, metabolic acidosis, is a general hallmark of mitochondrial dysfunction (Brown et al. 1994). Neurodevelopmental delay, hypotonia and structural brain abnormalities typical of Leigh syndrome are the commonest clinical signs of PDC deficiency. Genetic lesions in all the constituent enzymes of PDC have now been identified (see Imbard et al. (2011) and Patel et al. (2012) for comprehensive reviews). In contrast, defects in the BCOADC give rise to maple syrup urine disease (MSUD) accompanied by elevated blood and urine levels of the branched-chain 2-oxoacids derived from leucine, isoleucine and valine that are also excreted giving rise to a characteristic sweet-smelling urine (Harper et al. 1990). All new-born infants are tested at birth for MSUD as exposure to high levels of these metabolites rapidly results in severe neurological and developmental problems. At present strict dietary control of branched-chain amino acid intake is the sole effective treatment for this genetic condition. Reports of genetic defects in OGDC are extremely rare, reflecting the central role of this complex in maintaining the integrity of the citric acid cycle (Gibson et al. 2012; Dumont et al. 2009).

Although complexes I and III of the mitochondrial respiratory chain are considered to be the primary generators of ROS in this organelle, PDC, OGDC and BCOADC are now recognised to be significant sources of superoxide/hydrogen peroxide production that can generate ROS at much higher rates than complex I (Quinlan et al. 2013, 2014; Tretter and Adam-Vizi 2005). Deficiencies in these complexes can also lead to increased oxidative stress via enhanced ROS generation. For example, induced mitochondrial dihydrolipoyl succinyltransferase (E2-OGDC) deficiency in a transgenic mouse model increases nitrotyrosine levels, amyloid plaque burden and results in the occurrence of spatial learning and memory defects. Thus, current evidence suggests that OGDC may be involved in the pathogenesis of Alzheimer's disease through increased oxidative stress (Dumont et al. 2009). Similarly, differences in phenotype and severity of symptoms in patients carrying specific mutations in these complexes may partly be accounted for by enhanced ROS generation in individual cases. A further complication is that PDKs (especially PDK2) are susceptible to inactivation by mitochondrial ROS, undermining the normal regulatory mechanisms governing PDC and BCOADC activity (Hurd et al. 2012).

19.5 Genetic Defects in PDC Deficient Patients: *In Vitro* Evaluation Using Recombinant Human PDC

As stated previously, a major advance in our understanding of PDC structure and function occurred in the 1980s and 1990s, initially from selective proteolysis and immunological studies on bovine PDC demonstrating that E3BP (originally called protein X) was an integral subunit of the complex that contributed to its overall function. Subsequent cloning and sequencing of genes encoding yeast and human E3BP revealed a high degree of sequence similarity to E2 although they lacked the active site motif DHRXXDG essential for acetyltransferase activity (Harris et al. 1997a; Behal et al. 1989a). Disruption and mutagenesis of the *PDX1* gene encoding E3BP in yeast and studies of E3 binding to reassembled bovine E2 cores lacking E3BP confirmed the primary role of this subunit in high-affinity E3BP binding (McCartney et al. 1997; Lawson et al. 1991). In addition, immunoblotting of PDC subunits in cell and tissue samples from several patients with putative PDC deficiencies revealed a complete lack of the E3BP subunit while retaining partial PDC activity estimated at 10–20% of controls (Marsac et al. 1993). This residual activity has been attributed to the low-affinity binding of E3 to the E2 core even in the absence of E3BP (Singh 2008). Patients lacking any detectable E3BP protein are generally found to have deletions, nonsense mutations or point mutations at intron-exon boundaries that lead to aberrant splicing and prevent the production of a viable protein product (Brown et al. 2006; Aral et al. 1997).

In vitro reconstitution of the recombinant complex with recombinant E2/E3BP or E2 only core assemblies demonstrate that the native assembly becomes fully-activated with stoichiometric amounts of E3. Mutant PDC proved to be virtually inactive (3–5%) under these conditions; however, activity increased gradually to 40–50% of controls with a substantial excess of E3 (100-fold or greater (Singh 2008)). In contrast, a gradual decline in wild-type activity was observed at high E3 levels suggesting that E3 also interacts weakly with the E2-SBD and, at high concentrations, partially displaces the rate-limiting E1 enzyme. Surface plasmon resonance (SPR) data support this conclusion indicating that E3 binds to an E3BP-di-domain construct with a K_d of 5.3 ± 1.8 nM, whereas binding to the equivalent E2-di-domain was correspondingly weaker (80–100 fold) with a K_d of 0.4 ± 0.11 μM. This mutually exclusive binding is reminiscent of the situation in *B. stearothermophilus* PDC where E1 and E3 compete for an overlapping site on the E2 core.

Two unrelated cases of PDC deficiency linked to mutations in E2 have been described to date (Head et al. 2005). Of particular interest was the discovery of a patient with an unusual 'in-frame' 3-bp deletion encoding Glu121 of the mitochondrial precursor (Glu35 of the mature protein after removal of its 86 amino acid presequence prior to assembly). In this patient, who displayed relatively mild symptoms of PDC deficiency, activity was estimated at 25–35% of controls. As shown in Fig. 19.3c, Glu35 is situated in the central position of a small three-residue loop in the outer LD (OLD) joining β-strands 3 and 4. *In vitro* studies, employing recombinant wild-type and mutant LDs expressed as GST fusion proteins indicate that the ΔE121 deletion prevents its lipoylation and induces formation of LD dimers (Singh 2008). Reconstituted human PDC carrying the E35 deletion on the E2 OLD has 10–20% of overall complex activity, similar to patient values. In contrast, substitution of E35 with asp or gln, has no significant effect on the lipoylation status of the domain, does not induce dimerisation and has only minimal effects on the activity of reconstituted PDC (80–100% of controls). Biophysical studies (CD, tryptophan fluorescence spectroscopy and AUC) indicate that the presence of the ΔE121 OLD promotes large structural perturbations by interfering with the ordered, C-terminally-directed assembly of the E2/E3BP core as evidenced by the presence of multiple non-specific aggregates. The increased tendency of these mutant OLDs to interact in an inappropriate fashion promoting dimer formation has a profound negative impact on normal E2/E3BP core formation (Fig. 19.3d) resulting in a disordered E2/E3BP core that is responsible for the symptoms of PDC deficiency in this patient.

Interestingly the removal of 2–4 residues from a prominent surface loop joining β-strands 1 and 2 of an *E. coli* LD (Fig. 19.3a) is also accompanied by an increasing tendency to form a novel misfolded dimer. Detailed NMR structural studies of this misfolded structure reveal that their N-terminal halves are unfolded and dynamic, causing the C-terminal halves of two monomers to associate forming a structure with twofold symmetry and a topology mimicking that of the folded monomer (Fig. 19.3b). In summary, surface loops, in particular their length, appear to be key determinants in the folding process that leads to normal LD formation (Stott et al. 2009).

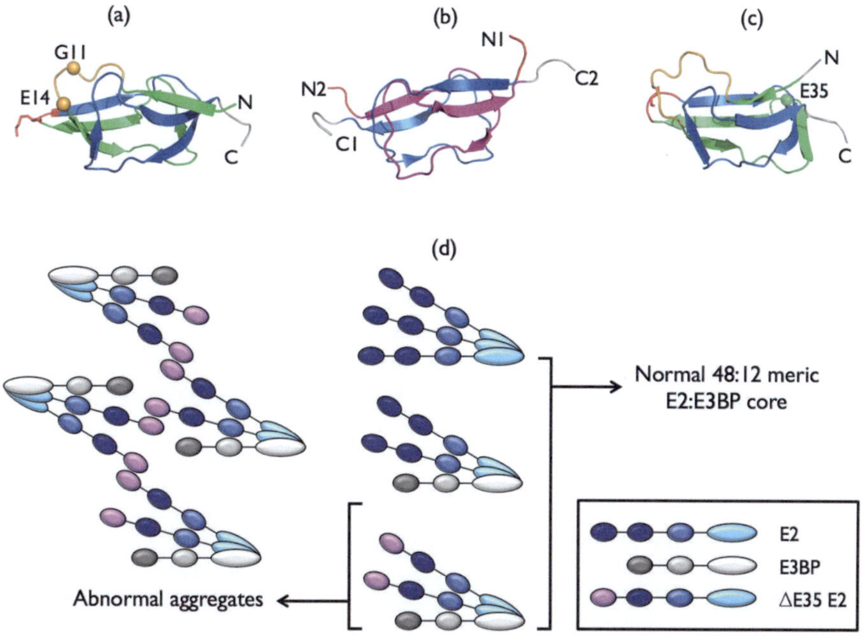

Fig. 19.3 Dimerisation of lipoyl domains induced by deletions in two key loop regions. (**a**) The NMR-derived structure of the ILD from *E. coli* (PDB ID: 1QJO, model 1) showing the position of selected residues (G11 to E14) within a prominent surface loop (*orange*) that when deleted result in the formation of dimers (**b**) as the result of mis-folding. At the core of the dimer (also observed via NMR spectroscopy (PDB ID: 2K7V)) are two β-sheets composed of the C-terminal halves (*blue* and *pink*) of two separate LDs. The unstructured N-terminal halves of the domains are not visualised here, for clarity (Stott et al. 2009). In (*c*) the position of E35 of the human ILD structure (PDB ID: 1FYC (Howard et al. 1998) in a different surface loop, is indicated since there is no structure for a human OLD. Deletion of this residue in recombinant human OLD also induces dimerisation of the domain (Singh 2008) as well as the formation of abnormal aggregates of the E2/E3BP core. The three residues that contain the lipoly-lysine are in red (*d*) trimers of E2 alone or which include one molecule of E3BP can then self-assemble to form normal (e.g. 48:12-meric) E2:E3BP core complex via interaction of the C-terminal domains. However, the ΔE35 (ΔE121 in the mitochondrial precursor) mutant tends to dimerise through its N-terminal domain, leading to the formation of abnormal aggregates

There is an old saying amongst biochemists and molecular biologists that 'dead proteins tell no tales' and indeed embarking on a site-directed mutagenesis programme with little or no knowledge of protein structure and mechanism of action can lead to difficulties in interpretation, often proving to be a rather fruitless and uninformative approach. However, recombinant PDC analysis of natural mutations in cases of PDC deficiency where there is always residual complex activity can provide detailed insights into the molecular basis of the disease (as in the above cases) while also contributing novel insights into enzyme mechanisms and mode of assembly.

19.6 PDC and the Major Diseases

In recent years, aberrant pyruvate metabolism linked to alterations in PDC activity, regulation and gene expression have been implicated in several of the major diseases currently affecting the human population (see Gray et al. (2014) for review and references therein). These include general neurodegenerative disorders such as Alzheimer's disease and Parkinson's disease, cancer (Saunier et al. 2016; Jeoung 2015), diabetes (Lee 2014; Jeoung 2015) and cardiovascular disease (Sun et al. 2015). In most cases, however, the aetiology of these progressive, long-term conditions and the direct involvement of PDC in particular is not well characterised so will not be discussed further at this stage. Instead the final section of this chapter will focus on three areas where there have been recent advances at the molecular level in highlighting a central role for PDC and its subunits, namely in the autoimmune condition, primary biliary cirrhosis (PBC), malarial parasite infection and multidrug resistance in bacteria.

19.6.1 Primary Biliary Cirrhosis (PBC)

Primary biliary cirrhosis (PBC) is a chronic, immune-mediated liver disease for which there is no cure and which can be halted only by liver transplant (itself no guarantee of escape from the disease). The bile duct damage associated with PBC is thought to arise from unique characteristics of biliary epithelial cell (BEC) apoptosis, wherein exposure of autoantigen to immune system effectors (Wang et al. 2014) causes lesions (Fig. 19.4). The autoantigens of PBC are E2 subunits of members of the 2-OADC family, chiefly PDC-E2. An anomaly in BEC apoptosis prevents S-glutathionylation-mediated degradation of PDC-E2 which is then exposed to autoimmune effector agents and itself becomes an autoantigen (Yamagiwa et al. 2014; Selmi et al. 2011). Eric Gershwin and colleagues have worked for many years to elucidate the etiology of this complex disease (Wang et al. 2014; Gershwin et al. 2000; Leung et al. 1996) and their findings, together with the work of others had strongly indicated that the major E2 epitope was the lipoamide moiety (including

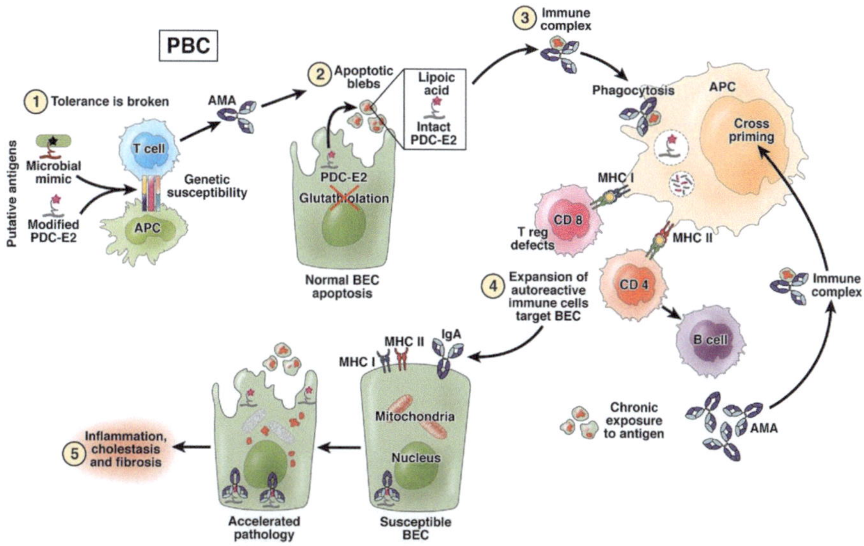

Fig. 19.4 The role of PDC in primary biliary cirrhosis (PBC). PBC is a complex, multi-factorial disease characterised by the loss of immune tolerance (1) to PDC E2, found in biliary epithelial cell (BEC) apoptotic blebs (2). Upon presentation of E2 to immune cells, immune complexes form (3), leading to the expansion of autoreactive immune cells (4) that target the BECs and cause chronic inflammation, cholestasis, and fibrosis (5). AMA, anti-mitochondrial antibody; APC, antigen-presenting cell; MHC, major histocompatibility complex. (Reproduced with permission from Hirschfield et al. 2010)

the lipoylated key lysine residue situated at the exposed type 1 β-turn tip) of the ILD.

The diagnosis of PBC relies on the indirect immunofluorescence detection of serum anti-mitochondrial antibodies (AMA, directed against PDC-E2) that are found in 95% of PBC cases. The AMA are tested using whole rat kidney tissue as the substrate but diagnosis could be better standardised if, instead, a panel of well-defined specific molecular targets was available. To this end Pacini et al. (2015) have developed a number of simple synthetic peptide antigens mimicking the main epitope on PDC-E2. In their work, solid-phase ELISA determination of the recognition of PBC patient sera by a panel of synthetic lipoylated and unlipoylated peptides, centred around the ILD key lysine (K173), revealed that lipoamide plays only a marginal role in IgM and IgG recognition. In order to understand the structural basis for this surprising finding, the structures of the synthetic peptides were determined by NMR spectroscopy, revealing a switch from random coil to helical conformation upon lipoylation. This modified, helical segment was then built into the existing NMR-derived structure for human unlipoylated (apo) PDC-E2 (PDB ID: 1FYC (Fig. 19.3c) (Howard et al. 1998)). After molecular dynamics (MD) simulations it was found that this helical region persisted, and owing to interactions of the lipoyl chain with proximal hydrophobic Leu and Ile side-chains in a groove formed by strands 1 and 5 of the ILD, the lipoylated lysine becomes obscured and is thus less likely to form the PDC E2 ILD epitope (Selmi et al. 2011), the residual antigenicity of the lipoylated lysine observed arising from limited remaining conformational flexibility that may allow it to be occasionally accessible to antibody binding. Until an atomic resolution structure is determined for a lipoylated PDC-E2 ILD, it will not be possible to conclude whether this model does truly represent the positioning of the lipoyl group and whether its role in PBC requires re-evaluation.

19.6.2 Malaria

In the human malaria parasite *Plasmodium falciparum*, PDC is located exclusively in the apicoplast, a plastid-like organelle found in most apicomplexan parasites, where it provides acetyl-CoA for fatty acid biosynthesis and has no apparent function in providing acetyl-CoA for TCA activity (Jacot et al. 2016). Instead, *Plasmodium* uses its branched-chain 2-OADC complex to convert pyruvate to acetyl-CoA. *Plasmodium* PDC is essential for parasite survival in the mosquito vector and for late liver-stage development in the human host, suggesting its suitability as a target for intervention strategies against malaria. A study of *P. falciparum* apicoplast E3 (*Pfa*E3) used AUC, SAXS and discrete molecular dynamics (DMD) to demonstrate that two parasite-specific regions of the bacterially overexpressed recombinant protein extend from a dimeric compact core (Laine et al. 2015). In addition to the two extra loops, the high-resolution model consistent with the experimental AUC and SAXS data includes a two-stranded anti-parallel β motif in the space that, in the human E3 structure, is occupied by the E3BP SBD (Brautigam et al. 2006). Since the *P. falciparum* genome does not encode an *e3bp* homologue, *Pfa*E3 does not interact with E3BP SBD, but rather the SBD of *Pf*E2. If the presence of this hypothetical additional β motif could be experimentally confirmed (e.g. via MX) it would confirm the difference in this interaction from that observed for human E3-E3BP, supporting the suitability of *Pfa*E3 as a target for the development of intervention strategies. In addition to this, *Plasmodium* aE1α, aE1β and aE2 have been identified, through gene (protein and transcript) expression analysis studies, as three of several candidate genes for the generation of genetically attenuated parasites (GAPs) for use as live vaccines (Kumar et al. 2015). Because *Pfa*E2 lacks a human orthologue, it, like *Pfa*E3, might also make a good anti-malarial drug target.

19.6.3 Bacteria and Antimicrobials

It is known that certain bacteria, especially those with small genomes, double-up the function of some of their proteins such that glycolytic enzymes can be found on the bacterial surface interacting with host proteins. This has recently been demonstrated for two members of the genus *Mycoplasma*, namely *M. pneumoniae* and *M.*

agalactiae. Three PDC components (E1α, E1β and E2) have been found on the surface of *M. pneumoniae*. Additionally, in the presence of the serine kinase activator urokinase, *M. pneumoniae* PDC E1β was found to be capable of activating host plasminogen to plasmin and degrading fibrinogen (Gründel et al. 2015). The interaction between the PDC components and plasminogen is ascribed to the propensity of its five kringle domains to bind to lysine residues, in which bacterial proteins are especially rich. This work indicates that these surface-associated PDC components may play a role in *M. pneumoniae* virulence and is important in expanding our understanding of the process of human respiratory epithelium colonization by this pathogen. A mutant *M. agalactiae* defective in growth *in vitro* was generated by Hegde et al. (2015) via transposon mutagenesis insertion in the *pdhB* gene (which encodes PDC E1β). The mutant was observed to be far less effective in the invasion of HeLa cells but no less effective in cell adhesion (compared with the wild type strain) suggesting that although PDC E1β acts at a stage secondary to initial contact for internalisation into eukaryotic cells, it nonethless plays a significant role in infection by *M. agalactiae*. In neither report is the mechanism of PDC protein delivery to the bacterial cell surface investigated, nor the form in which the proteins are "displayed". Therefore it remains to be determined whether in *M. pneumoniae* E2 is present there as an oligomeric complex or as trimers or even monomers or whether E1β is within the E1 heterodimer, for instance.

Together with PDC E2 and E3, E1 is additionally reported to play a role in chemokine CXCL10-mediated antimicrobial activity in multidrug-resistant *E. coli* (Schutte et al. 2016). Sub-lethal concentrations of CXCL10 disrupt the ability of the *E. coli* to form long chains of cells. Transposon mutagenesis was once again used to identify the gene encoding *E. coli* PDC E1 as has being critical to the action of CXCL10, since the mutants were resistant to its antibacterial effects. Interestingly, generation of further mutants in E2 and E3 suggests that it is disruption of the PDC in its entirety that confers increased resistance to CXCL10-mediated killing. Immunogold-labelled CXCL10 located to the surface of *E. coli*, regardless of whether the gene for E1 had been disrupted or not, suggesting that the chemokine binds initially to something other than E1. Instead, the effect may be indirect and may arise from a subtle change in bacterial metabolism. For instance, the authors speculate that the mutant bacteria may resist CXCL10-mediated killing by directing energy production away from aerobic respiration and the TCA cycle and instead forcing it to proceed through other, less efficient, metabolic pathways. Since we face an era of increasing resistance to existing antibiotics, a better understanding of the mechanism by which CXCL10 exerts its antimicrobial effect may be an important step in the pathway to new drug targets for *E. coli* or other Gram-negative pathogens.

19.7 Outlook

Where next in our further understanding of the structural basis for the function of PDC and its related assemblies? Owing to its probable inherent heterogeneity, it is unlikely that an X-ray crystallographic structure for intact PDC will be determined. Instead, recent breath-taking advances in cryo-EM, delivering structures to 2.2 Å resolution (Doerr 2015), probably hold the key since this method can accommodate heterogeneity in structure and can capture the conformations of molecules and complexes whose flexibility evades structural study by MX. It has even become possible to determine (albeit lower-resolution) complex structures *in situ* (Engel et al. 2015), offering the tantalising possibility of direct observation of variations in PDC architecture from tissue to tissue and as a function of metabolic or disease state.

Acknowledgments The work of Swetha Vijayakrishnan and Geetanjali Singh that contributed to Figs. 19.2 and 19.3 of this chapter is gratefully acknowledged.

References

Aral B, Benelli C, Ait-Ghezala G, Amessou M, Fouque F, Maunoury C, Créau N, Kamoun P, Marsac C (1997) Mutations in *PDX1*, the human lipoyl-containing component X of the pyruvate dehydrogenase-complex gene on chromosome 11p1, in congenital lactic acidosis. Am J Hum Genet 61(6):1318–1326. doi:10.1086/301653

Arjunan P, Nemeria N, Brunskill A, Chandrasekhar K, Sax M, Yan Y, Jordan F, Guest JR, Furey W (2002) Structure of the pyruvate dehydrogenase multienzyme complex E1 component from *Escherichia coli* at 1.85 Å resolution. Biochemistry 41(16):5213–5221. doi:10.1021/bi0118557

Arjunan P, Wang J, Nemeria NS, Reynolds S, Brown I, Chandrasekhar K, Calero G, Jordan F, Furey W (2014) Novel binding motif and new flexibility revealed by structural analyses of a pyruvate dehydrogenase-dihydrolipoyl acetyltransferase subcomplex from the *Escherichia coli* pyruvate dehydrogenase multienzyme complex. J Biol Chem 289(43):30161–30176. doi:10.1074/jbc.M114.592915

Bao HY, Kasten SA, Yan XH, Roche TE (2004) Pyruvate dehydrogenase kinase isoform 2 activity limited and further inhibited by slowing down the rate of dissociation of ADP. Biochemistry 43(42):13432–13441. doi:10.1021/bi049488x

Bates DL, Danson MJ, Hale G, Hooper EA, Perham RN (1977) Self-assembly and catalytic activity of pyruvate dehydrogenase multienzyme complex of *Escherichia coli*. Nature 268(5618):313–316. doi:10.1038/268313a0

Becker T, Böttinger L, Pfanner N (2012) Mitochondrial protein import: from transport pathways to an integrated network. Trends Biochem Sci 37(3):85–91. doi:10.1016/j.tibs.2011.11.004

Behal RH, Browning KS, Hall TB, Reed LJ (1989a) Cloning and nucleotide-sequence of the gene for protein X from *Saccharomyces cerevisiae*. Proc Natl Acad Sci U S A 86(22):8732–8736. doi:10.1073/pnas.86.22.8732

Behal RH, Browning KS, Reed LJ (1989b) Nucleotide and deduced amino-acid sequence of the alpha-subunit of yeast pyruvate-dehydrogenase. Biochem Biophys Res Commun 164(2):941–946. doi:10.1016/0006-291x(89)91549-0

Behal RH, DeBuysere MS, Demeler B, Hansen JC, Olson MS (1994) Pyruvate dehydrogenase multienzyme complex – characterization of assembly intermediates by sedimentation velocity analysis. J Biol Chem 269(50):31372–31377

Berg A, Westphal AH, Bosma HJ, de Kok A (1998) Kinetics and specificity of reductive acylation of mild-type and mutated lipoyl domains of 2-oxo-acid dehydrogenase complexes from *Azotobacter vinelandii*. Eur J Biochem 252(1):45–50. doi:10.1046/j.1432-1327.1998.2520045.x

Bowker-Kinley MM, Davis WI, Wu PF, Harris RA, Popov KM (1998) Evidence for existence of tissue-specific regulation of the mammalian pyruvate dehydrogenase complex. Biochem J 329(1):191–196

Brautigam CA, Wynn RM, Chuang JL, Machius M, Tomchick DR, Chuang DT (2006) Structural insight into interactions between dihydrolipoamide dehydrogenase (E3) and E3 binding protein of human pyruvate dehydrogenase complex. Structure 14:611–621

Brautigam CA, Wynn RM, Chuang JL, Chuang DT (2009) Subunit and catalytic component stoichiometries of an *in vitro* reconstituted human pyruvate dehydrogenase complex. J Biol Chem 284(19):13086–13098. doi:10.1074/jbc.M806563200

Brown GK, Otero LJ, Le Gris M, Brown RM (1994) Pyruvate dehydrogenase deficiency. J Med Genet 31(11):875–879. doi:10.1136/jmg.31.11.875

Brown RM, Head RA, Morris AAM, Raiman JAJ, Walter JH, Whitehouse WP, Brown GK (2006) Pyruvate dehydrogenase E3 binding protein (protein X) deficiency. Dev Med Child Neurol 48(9):756–760. doi:10.1017/s0012162206001617

Burns G, Sykes PJ, Hatter K, Sokatch JR (1989) Isolation of a 3rd lipoamide dehydrogenase from *Pseudomonas putida*. J Bacteriol 171(2):665–668

Carothers DJ, Pons G, Patel MS (1989) Dihydrolipoamide dehydrogenase: functional similarities and divergent evolution of the pyridine nucleotide-disulfide oxidoreductases. Arch Biochem Biophys 268(2):409–425. doi:10.1016/0003-9861(89)90309-3

Cate RL, Roche TE, Davis LC (1980) Rapid intersite transfer of acetyl groups and movement of pyruvate dehydrogenase component in the kidney pyruvate dehydrogenase complex. J Biol Chem 255(16):7556–7562

Chacinska A, Koehler CM, Milenkovic D, Lithgow T, Pfanner N (2009) Importing mitochondrial proteins: machineries and mechanisms. Cell 138(4):628–644. doi:10.1016/j.cell.2009.08.005

Ciszak EM, Korotchkina LG, Dominiak PM, Sidhu S, Patel MS (2003) Structural basis for flip-flop action of thiamin pyrophosphate-dependent enzymes revealed by human pyruvate dehydrogenase. J Biol Chem 278(23):21240–21246. doi:10.1074/jbc.M300339200

Ciszak EM, Makal A, Hong YS, Vettaikkorumakankauv AK, Korochkina LG, Patel MS (2006) How dihydrolipoamide dehydrogenase-binding protein binds dihydrolipoamide dehydrogenase in the human pyruvate dehydrogenase complex. J Biol Chem 281:648–655

Damuni Z, Reed LJ (1987) Purification and characterization of a divalent cation-independent, spermine-stimulated protein phosphatase from bovine kidney mitochondria. J Biol Chem 262(11):5133–5138

Dardel F, Davis AL, Laue ED, Perham RN (1993) 3-dimensional structure of the lipoyl domain from *Bacillus stearothermophilus* pyruvate dehydrogenase multienzyme complex. J Mol Biol 229(4):1037–1048. doi:10.1006/jmbi.1993.1103

de Boer VCJ, Houten SM (2014) A mitochondrial expatriate: nuclear pyruvate dehydrogenase. Cell 158(1):9–10. doi:10.1016/j.cell.2014.06.018

De Marcucci O, Lindsay JG (1985) Component X - an immunologically distinct polypeptide associated with mammalian pyruvate-dehydrogenase multi-enzyme complex. Eur J Biochem 149(3):641–648. doi:10.1111/j.1432-1033.1985.tb08972.x

De Marcucci OGL, Gibb GM, Dick J, Lindsay JG (1988) Biosynthesis, import and processing of precursor polypeptides of mammalian mitochondrial pyruvate dehydrogenase complex. Biochem J 251(3):817–823

Doerr A (2015) Cryo-EM goes high-resolution. Nat Methods 12(7):598–599. doi:10.1038/nmeth.3469

Dudek J, Rehling P, van der Laan M (2013) Mitochondrial protein import: common principles and physiological networks. Biochimica Et Biophysica Acta-Mol Cell Res 1833(2):274–285. doi:10.1016/j.bbamcr.2012.05.028

Dumont M, Ho DJ, Calingasan NY, Xu H, Gibson G, Beal MF (2009) Mitochondrial dihydrolipoyl succinyltransferase deficiency accelerates amyloid pathology and memory deficit in a transgenic mouse model of amyloid deposition. Free Radic Biol Med 47(7):1019–1027. doi:10.1016/j.freeradbiomed.2009.07.008

Engel BD, Schaffer M, Albert S, Asano S, Plitzko JM, Baumeister W (2015) *In situ* structural analysis of Golgi intracisternal protein arrays. Proc Natl Acad Sci U S A 112(36):11264–11269. doi:10.1073/pnas.1515337112

Fox TD (2012) Mitochondrial protein synthesis, import and assembly. Genetics 192(4):1203–1234. doi:10.1534/genetics.112.141267

Frank RAW, Titman CM, Pratap JV, Luisi BF, Perham RN (2004) A molecular switch and proton wire synchronize the active sites in thiamine enzymes. Science 306(5697):872–876. doi:10.1126/science.1101030

Frank RAW, Pratap JV, Pei XY, Perham RN, Luisi BF (2005) The molecular origins of specificity in the assembly of a multienzyme complex. Structure 13(8):1119–1130. doi:10.1016/j.str.2005.04.021

Gershwin ME, Ansari AA, Mackay IR, Nakanuma Y, Nishio A, Rowley MJ, Coppel RL (2000) Primary biliary cirrhosis: an orchestrated immune response against epithelial cells. Immunol Rev 174:210–225

Gibson GE, Chen H-L, Xu H, Qiu L, Xu Z, Denton TT, Shi Q (2012) Deficits in the mitochondrial enzyme alpha-ketoglutarate dehydrogenase lead to Alzheimer's disease-like calcium dysregulation. Neurobiol Aging 33(6). doi:10.1016/j.neurobiolaging.2011.11.003

Gray LR, Tompkins SC, Taylor EB (2014) Regulation of pyruvate metabolism and human disease. Cell Mol Life Sci 71(14):2577–2604. doi:10.1007/s00018-013-1539-2

Green JDF, Perham RN, Ullrich SJ, Appella E (1992) Conformational studies of the interdomain linker peptides in the dihydrolipoyl acetyltransferase component of the pyruvate dehydrogenase multienzyme complex of *Escherichia coli*. J Biol Chem 267(33):23484–23488

Green JDF, Laue ED, Perham RN, Ali ST, Guest JR (1995) 3-dimensional structure of a lipoyl domain from the dihydrolipoyl acetyltransferase component of the pyruvate dehydrogenase multienzyme complex of *Escherichia coli*. J Mol Biol 248(2):328–343. doi:10.1006/jmbi.1995.0225

Green T, Grigorian A, Klyuyeva A, Tuganova A, Luo M, Popov KM (2008) Structural and functional insights into the molecular mechanisms responsible for the regulation of pyruvate dehydrogenase kinase 2. J Biol Chem 283(23):15789–15798. doi:10.1074/jbc.M800311200

Gründel A, Friedrich K, Pfeiffer M, Jacobs E, Dumke R (2015) Subunits of the pyruvate dehydrogenase cluster of *Mycoplasma pneumoniae* are surface-displayed proteins that bind and activate human plasminogen. PLoS One 10(5):e0126600. doi:10.1371/journal.pone.0126600

Gudi R, Bowker-Kinley MM, Kedishvili NY, Zhao Y, Popov KM (1995) Diversity of the pyruvate dehydrogenase kinase gene family in humans. J Biol Chem 270(48):28989–28994

Guest JR, Angier SJ, Russell GC (1989) Structure, expression, and protein engineering of the pyruvate dehydrogenase complex of *Escherichia coli*. Ann N Y Acad Sci 573:76–99. doi:10.1111/j.1749-6632.1989.tb14988.x

Guest JR, Attwood MM, Machado RS, Matqi KY, Shaw JE, Turner SL (1997) Enzymological and physiological consequences of restructuring the lipoyl domain content of the pyruvate dehydrogenase complex of *Escherichia coli*. Microbiology 143(2):457–466

Harper PAW, Healy PJ, Dennis JA (1990) Animal model of humandisease - maple syrup urine disease (branched-chain ketoaciduria). Am J Pathol 136(6):1445–1447

Harris RA, Bowker-Kinley MM, Wu PF, Jeng JJ, Popov KM (1997a) Dihydrolipoamide dehydrogenase-binding protein of the human pyruvate dehydrogenase complex – DNA-derived amino acid sequence, expression, and reconstitution of the pyruvate dehydrogenase complex. J Biol Chem 272(32):19746–19751. doi:10.1074/jbc.272.32.19746

Harris RA, Hawes JW, Popov KM, Zhao Y, Shimomura Y, Sato J, Jaskiewicz J, Hurley TD (1997b) Studies on the regulation of the mitochondrial alpha-ketoacid dehydrogenase complexes and their kinases. In: Weber G (ed) Advances in Enzyme Regulation, vol 37, pp 271–293. doi:10.1016/s0065-2571(96)00009-x

Harris RA, Bowker-Kinley MM, Huang BL, Wu PF (2002) Regulation of the activity of the pyruvate dehydrogenase complex. In: Weber G (ed) Advances in Enzyme Regulation, vol 42, pp 249–259. doi:10.1016/s0065-2571(01)00061-9

Head RA, Brown RM, Zolkipli Z, Shahdadpuri R, King MD, Clayton PT, Brown GK (2005) Clinical and genetic spectrum of pyruvate dehydrogenase deficiency: dihydrolipoamide acetyltransferase (E2) deficiency. Ann Neurol 58(2):234–241. doi:10.1002/ana.20550

Hegde S, Rosengarten R, Chopra-Dewasthaly R (2015) Disruption of the *pdhB* pyruvate dehydrogenase gene affects colony morphology, *in vitro* growth and cell invasiveness of *Mycoplasma agalactiae*. PLOS One 10(6). doi:10.1371/journal.pone.0131134

Hiromasa Y, Fujisawa T, Aso Y, Roche E (2004) Organization of the cores of the mammalian pyruvate dehydrogenase complex formed by E2 and E2 plus the E3-binding protein and their capacities to bind the E1 and E3 components. J Biol Chem 279:6921–6933

Hirschfield GM, Heathcote EJ, Gershwin ME (2010) Pathogenesis of cholestatic liver disease and therapeutic approaches. Gastroenterology 139(5):1481–1496. doi:10.1053/j.gastro.2010.09.004

Howard MJ, Fuller C, Broadhurst RW, Perham RN, Tang JG, Quinn J, Diamond AG, Yeaman SJ (1998) Three-dimensional structure of the major autoantigen in primary biliary cirrhosis. Gastroenterology 115(1):139–146. doi:10.1016/s0016-5085(98)70375-0

Huang BL, Gudi R, Wu PF, Harris RA, Hamilton J, Popov KM (1998) Isoenzymes of pyruvate dehydrogenase phosphatase – DNA-derived amino acid sequences, expression, and regulation. J Biol Chem 273(28):17680–17688. doi:10.1074/jbc.273.28.17680

Hurd TR, Collins Y, Abakumova I, Chouchani ET, Baranowski B, Fearnley IM, Prime TA, Murphy MP, James AM (2012) Inactivation of pyruvate dehydrogenase kinase 2 by mitochondrial reactive oxygen species. J Biol Chem 287(42):35153–35160. doi:10.1074/jbc.M112.400002

Imbard A, Boutron A, Vequaud C, Zater M, de Lonlay P, Ogier de Baulny H, Barnerias C, Miné M, Marsac C, Saudubray J-M, Brivet M (2011) Molecular characterization of 82 patients with pyruvate dehydrogenase complex deficiency. Structural implications of novel amino acid substitutions in E1 protein. Mol Genet Metab 104(4):507–516. doi:10.1016/j.ymgme.2011.08.008

Izard T, Ævarsson A, Allen MD, Westphal AH, Perham RN, de Kok A, Hol WGJ (1999) Principles of quasi-equivalence and Euclidean geometry govern the assembly of cubic and dodecahedral cores of pyruvate dehydrogenase complexes. Proc Natl Acad Sci U S A 96(4):1240–1245. doi:10.1073/pnas.96.4.1240

Jacot D, Waller RF, Soldati-Favre D, MacPherson DA, MacRae JI (2016) Apicomplexan energy metabolism: carbon source promiscuity and the quiescence hyperbole. Trend Parasitol 32(1):56–70. doi:10.1016/j.pt.2015.09.001

Jeoung NH (2015) Pyruvate dehydrogenase kinases: therapeutic targets for diabetes and cancers. Diabetes Metab J 39(3):188–197. doi:10.4093/dmj.2015.39.3.188

Jordan F, Nemeria N, Guo FS, Baburina I, Gao YH, Kahyaoglu A, Li HJ, Wang J, Yi JZ, Guest JR, Furey W (1998) Regulation of thiamin diphosphate-dependent 2-oxo acid decarboxylases by substrate and thiamin diphosphate.Mg(II) - evidence for tertiary and quaternary interactions. Biochim Biophys Acta Protein Struct Mol Enzymol 1385(2):287–306. doi:10.1016/s0167-4838(98)00075-2

Jung HI, Bowden SJ, Cooper A, Perham RN (2002) Thermodynamic analysis of the binding of component enzymes in the assembly of the pyruvate dehydrogenase multienzyme complex of *Bacillus stearothermophilus*. Protein Sci 11(5):1091–1100. doi:10.1110/ps.4970102

Kato M, Chuang JL, Tso SC, Wynn RM, Chuang DT (2005) Crystal structure of pyruvate dehydrogenase kinase 3 bound to lipoyl domain 2 of human pyruvate dehydrogenase complex. EMBO J 24(10):1763–1774. doi:10.1038/sj.emboj.7600663

Kato M, Wynn RM, Chuang JL, Tso S-C, Machius M, Li J, Chuang DT (2008) Structural basis for inactivation of the human pyruvate dehydrogenase complex by phosphorylation: role of disordered phosphorylation loops. Structure 16(12):1849–1859. doi:10.1016/j.str.2008.10.010

Khailova L, Korochkina L, Severin S (1990) Intersite cooperativity in enzyme action of pyruvate dehydrogenase. In: Ju HB (ed) Biochemistry and Physiology of TDP Enzymes. VCH Weinheim, Blaubeuren, pp 251–261

Kolobova E, Tuganova A, Boulatnikov I, Popov KM (2001) Regulation of pyruvate dehydrogenase activity through phosphorylation at multiple sites. Biochem J 358:69–77. doi:10.1042/0264-6021:3580069

Kong YF, Ming DM, Wu YH, Stoops JK, Zhou ZH, Ma JP (2003) Conformational flexibility of pyruvate dehydrogenase complexes: a computational analysis by quantized elastic deformational model. J Mol Biol 330(1):129–135. doi:10.1016/s0022-2836(03)00555-2

Korotchkina LG, Patel MS (1995) Mutagenesis studies of the phosphorylation sites of recombinant human pyruvate dehydrogenase: site-specific regulation. J Biol Chem 270(24):14297–14304

Korotchkina LG, Patel MS (2001) Site specificity of four pyruvate dehydrogenase kinase isoenzymes toward the three phosphorylation sites of human pyruvate dehydrogenase. J Biol Chem 276(40):37223–37229. doi:10.1074/jbc.M103069200

Korotchkina LG, Tucker MM, Thekkumkara TJ, Madhusudhan KT, Pons G, Kim HJ, Patel MS (1995) Overexpression and characterization of human tetrameric pyruvate dehydrogenase and its individual subunits. Protein Expr Purif 6(1):79–90. doi:10.1006/prep.1995.1011

Kovina MV, Kochetov GA (1998) Cooperativity and flexibility of active sites in homodimeric transketolase. FEBS Lett 440(1–2):81–84. doi:10.1016/s0014-5793(98)01423-9

Kumar H, Frischknecht F, Mair GR, Gomes J (2015) *In silico* identification of genetically attenuated vaccine candidate genes for *Plasmodium* liver stage. Infect Genet Evol 36:72–81. doi:10.1016/j.meegid.2015.09.002

Kumaran S, Patel MS, Jordan F (2013) Nuclear magnetic resonance approaches in the study of 2-oxo acid dehydrogenase multienzyme complexes – a literature review. Molecules 18(10):11873–11903. doi:10.3390/molecules181011873

Laine M, Biddau M, Byron O, Müller S (2015) Biochemical and structural characterization of the apicoplast dihydrolipoamide dehydrogenase of *Plasmodium falciparum*. Biosci Rep 35(1). doi:10.1042/bsr20140150

Lawson JE, Behal RH, Reed LJ (1991) Disruption and mutagenesis of the *Saccharomyces cerevisiae* PDX1 gene encoding the protein X component of the pyruvate dehydrogenase complex. Biochemistry 30(11):2834–2839. doi:10.1021/bi00225a015

Lee I-K (2014) The role of pyruvate dehydrogenase kinase in diabetes and obesity. Diabetes Metab J 38(3):181–186. doi:10.4093/dmj.2014.38.3.181

Leung PSC, VandeWater J, Coppel RL, Nakanuma Y, Munoz S, Gershwin ME (1996) Molecular aspects and the pathological basis of primary biliary cirrhosis. J Autoimmun 9:119–128

Lindsay H, Beaumont E, Richards SD, Kelly SM, Sanderson SJ, Price NC, Lindsay JG (2000) FAD insertion is essential for attaining the assembly competence of the dihydrolipoamide dehydrogenase (E3) monomer from *Escherichia coli*. J Biol Chem 275(47):36665–36670. doi:10.1074/jbc.M004777200

Maeng CY, Yazdi MA, Niu XD, Lee HY, Reed LJ (1994) Expression, purification, and characterization of the dihydrolipoamide dehydrogenase-binding protein of the pyruvate dehydrogenase complex from *Saccharomyces cerevisiae*. Biochemistry 33(46):13801–13807. doi:10.1021/bi00250a034

Maeng CY, Yazdi MA, Reed LJ (1996) Stoichiometry of binding of mature and truncated forms of the dihydrolipoamide dehydrogenase-binding protein to the dihydrolipoamide acetyltransferase core of the pyruvate dehydrogenase complex from *Saccharomyces cerevisiae*. Biochemistry 35(18):5879–5882. doi:10.1021/bi9600254

Mande SS, Sarfaty S, Allen MD, Perham RN, Hol WGJ (1996) Protein-protein interactions in the pyruvate dehydrogenase multienzyme complex: dihydrolipoamide dehydrogenase complexed with the binding domain of dihydrolipoamide acetyltransferase. Structure 4(3):277–286. doi:10.1016/s0969-2126(96)00032-9

Marrott NL, Marshall JJT, Svergun DI, Crennell SJ, Hough DW, Danson MJ, van den Elsen JMH (2012) The catalytic core of an archaeal 2-oxoacid dehydrogenase multienzyme complex is a 42-mer protein assembly. FEBS J 279(5):713–723. doi:10.1111/j.1742-4658.2011.08461.x

Marrott NL, Marshall JJT, Svergun DI, Crennell SJ, Hough DW, van den Elsen JMH, Danson MJ (2014) Why are the 2-oxoacid dehydrogenase complexes so large? Generation of an active trimeric complex. Biochem J 463:405–412. doi:10.1042/bj20140359

Marsac C, Stansbie D, Bonne G, Cousin J, Jehenson P, Benelli C, Leroux J-P, Lindsay G (1993) Defect in the lipoyl-bearing protein-X subunit of the pyruvate-dehydrogenase complex in 2 patients with encephalomyelopathy. J Pediatr 123(6):915–920. doi:10.1016/s0022-3476(05)80387-7

Mathias RA, Greco TM, Oberstein A, Budayeva HG, Chakrabarti R, Rowland EA, Kang Y, Shenk T, Cristea IM (2014) Sirtuin 4 is a lipoamidase regulating pyruvate dehydrogenase complex activity. Cell 159(7):1615–1625. doi:10.1016/j.cell.2014.11.046

Mattevi A, Schierbeek AJ, Hol WGJ (1991) Refined crystal structure of lipoamide dehydrogenase from *Azotobacter vinelandii* at 2.2 Å resolution – a comparison with the structure of glutathione reductase. J Mol Biol 220(4):975–994. doi:10.1016/0022-2836(91)90367-f

Mattevi A, Obmolova G, Schulze E, Kalk KH, Westphal AH, de Kok A, Hol WGJ (1992) Atomic structure of the cubic core of the pyruvate dehydrogenase multienzyme complex. Science 255(5051):1544–1550. doi:10.1126/science.1549782

Mattevi A, Obmolova G, Kalk KH, Westphal AH, de Kok A, Hol WGJ (1993) Refined crystal structure of the catalytic domain of dihydrolipoyl transacetylase (E2p) from *Azotobacter vinelandii* at 2.6 Å resolution. J Mol Biol 230(4):1183–1199. doi:10.1006/jmbi.1993.1235

McCartney RG, Sanderson SJ, Lindsay JG (1997) Refolding and reconstitution studies on the transacetylase protein X (E2/X) subcomplex of the mammalian pyruvate dehydrogenase complex: evidence for specific binding of the dihydrolipoamide dehydrogenase component to sites on reassembled E2. Biochemistry 36(22):6819–6826. doi:10.1021/bi9630016

McLelland G-L, Soubannier V, Chen CX, McBride HM, Fon EA (2014) Parkin and PINK1 function in a vesicular trafficking pathway regulating mitochondrial quality control. EMBO J 33(4):282–295. doi:10.1002/embj.201385902

Miles JS, Guest JR, Radford SE, Perham RN (1988) Investigation of the mechanism of active site coupling in the pyruvate dehydrogenase multienzyme complex of *Escherichia coli* by protein engineering. J Mol Biol 202(1):97–106. doi:10.1016/0022-2836(88)90522-0

Milne JLS, Shi D, Rosenthal PB, Sunshine JS, Domingo GJ, Wu XW, Brooks BR, Perham RN, Henderson R, Subramaniam S (2002) Molecular architecture and mechanism of an icosahedral pyruvate dehydrogenase complex: a multifunctional catalytic machine. EMBO J 21(21):5587–5598. doi:10.1093/emboj/cdf574

Neagle JC, Lindsay JG (1991) Selective proteolysis of the protein X subunit of the bovine heart pyruvate dehydrogenase complex – effects on dihydrolipoamide dehydrogenase (E3) affinity and enzymatic properties of the complex. Biochem J 278:423–427

Neupert W (1997) Protein import into mitochondria. Annu Rev Biochem 66:863–917. doi:10.1146/annurev.biochem.66.1.863

Ng F, Tang BL (2014) Pyruvate dehydrogenase complex (PDC) export from the mitochondrial matrix. Mol Membr Biol 31(7–8):207–210. doi:10.3109/09687688.2014.987183

Pacini G, Carotenuto A, Rentier C, Nuti F, Real-Fernandez F, Brancaccio D, Sabatino G, Larregola M, Peroni E, Migliorini P, Novellino E, Battezzati PM, Selmi C, Papini AM, Rovero P (2015) Role of lipoylation of the immunodominant epitope of pyruvate dehydrogenase complex: toward a peptide-based diagnostic assay for primary biliary cirrhosis. J Med Chem 58(16):6619–6629. doi:10.1021/acs.jmedchem.5b00783

Patel MS, Korotchkina LG (2001) Regulation of mammalian pyruvate dehydrogenase complex by phosphorylation: complexity of multiple phosphorylation sites and kinases. Exp Mol Med 33(4):191–197

Patel KP, O'Brien TW, Subramony SH, Shuster J, Stacpoole PW (2012) The spectrum of pyruvate dehydrogenase complex deficiency: clinical, biochemical and genetic features in 371 patients. Mol Genet Metab 106(3):385–394

Perham RN (2000) Swinging arms and swinging domains in multifunctional enzymes: catalytic machines for multistep reactions. Annu Rev Biochem 69:961–1004. doi:10.1146/annurev.biochem.69.1.961

Perham RN, Reche PA (1998) Swinging arms in multifunctional enzymes and the specificity of post-translational modification. Biochem Soc Trans 26(3):299–303

Perham RN, Packman LC, Radford SE (1987) 2-oxo acid dehydrogenase multi-enzyme complexes: in the beginning and halfway there. Biochem Soc Symp 54:67–81

Popov KM, Hawes JW, Harris RA (1997) Mitochondrial alpha-ketoacid dehydrogenase kinases – a new family of protein kinases. In: Corbin JD, Francis SH (eds) Signal transduction in health and disease, Advances in second messenger and phosphoprotein research, vol 31. Lippincott-Raven, Philadelphia, pp 105–111

Quinlan CL, Perevoschikova IV, Goncalves RLS, Hey-Mogensen M, Brand MD (2013) The determination and analysis of site-specific rates of mitochondrial reactive oxygen species production. In: Cadenas E, Packer L (eds) Hydrogen peroxide and cell signaling, Pt A, Methods in Enzymology, vol 526. Academic, Amsterdam, pp 189–217. doi:10.1016/b978-0-12-405883-5.00012-0

Quinlan CL, Goncalves RLS, Hey-Mogensen M, Yadava N, Bunik VI, Brand MD (2014) The 2-oxoacid dehydrogenase complexes in mitochondria can produce superoxide/hydrogen peroxide at much higher rates than complex I. J Biol Chem 289(12):8312–8325. doi:10.1074/jbc.M113.545301

Rahmatullah M, Gopalakrishnan S, Andrews PC, Chang CL, Radke GA, Roche TE (1989) Subunit associations in the mammalian pyruvate-dehydrogenase complex – structure and role of protein X and the pyruvate dehydrogenase component binding domain of the dihydrolipoyl transacetylase component. J Biol Chem 264(4):2221–2227

Reed LJ, Oliver RM (1982) Structure-function-relationships in pyruvate and alpha-ketoglutarate dehydrogenase complexes. Adv Exp Med Biol 148:231–241

Reed LJ, Lawson JE, Niu XD, Yazdi MA, Fussey SP (1992) Biochemical and molecular genetic aspects of eukaryotic pyruvate dehydrogenase multienzyme complexes. J Nutr Sci Vitaminol (Tokyo) Spec No:46–51

Ricaud PM, Howard MJ, Roberts EL, Broadhurst RW, Perham RN (1996) Three-dimensional structure of the lipoyl domain from the dihydrolipoyl succinyltransferase component of the 2-oxoglutarate dehydrogenase multienzyme complex of *Escherichia coli*. J Mol Biol 264(1):179–190. doi:10.1006/jmbi.1996.0632

Roche TE, Hiromasa Y, Turkan A, Gong X, Peng T, Yan X, Kasten SA, Bao H, Dong J (2003) Essential roles of lipoyl domains in the activated function and control of pyruvate dehydrogenase kinases and phosphatase isoform 1. Eur J Biochem 270(6):1050–1056. doi:10.1046/j.1432-1033.2003.03468.x

Sanderson SJ, Miller C, Lindsay JG (1996) Stoichiometry, organisation and catalytic function of protein X of the pyruvate dehydrogenase complex from bovine heart. Eur J Biochem 236:68–77

Saunier E, Benelli C, Bortoli S (2016) The pyruvate dehydrogenase complex in cancer: an old metabolic gatekeeper regulated by new pathways and pharmacological agents. Int J Cancer 138(4):809–817. doi:10.1002/ijc.29564

Schutte KM, Fisher DJ, Burdick MD, Mehrad B, Mathers AJ, Mann BJ, Nakamoto RK, Hughes MA (2016) *Escherichia coli* pyruvate dehydrogenase complex is an important component of CXCL10-mediated antimicrobial activity. Infect Immun 84(1):320–328. doi:10.1128/iai.00552-15

Selmi C, Mackay IR, Gershwin ME (2011) The autoimmunity of primary biliary cirrhosis and the clonal selection theory. Immunol Cell Biol 89(1):70–80. doi:10.1038/icb.2010.126

Sergienko EA, Jordan F (2002) New model for activation of yeast pyruvate decarboxylase by substrate consistent with the alternating sites mechanism: demonstration of the existence of two active forms of the enzyme. Biochemistry 41(12):3952–3967. doi:10.1021/bi011860a

Sergienko EA, Wang J, Polovnikova L, Hasson MS, McLeish MJ, Kenyon GL, Jordan F (2000) Spectroscopic detection of transient thiamin diphosphate-bound intermediates on benzoylformate decarboxylase. Biochemistry 39(45):13862–13869. doi:10.1021/bi001214w

Singh G (2008) Analysis of genetic mutations using a recombinant model of the mammalian pyruvate dehydrogenase complex. PhD, University of Glasgow, Glasgow

Smolle M, Prior AE, Brown AE, Cooper A, Byron O, Lindsay JG (2006) A new level of architectural complexity in the human pyruvate dehydrogenase complex. J Biol Chem 281(28):19772–19780

Soubannier V, McLelland G-L, Zunino R, Braschi E, Rippstein P, Fon EA, McBride HM (2012) A vesicular transport pathway shuttles cargo from mitochondria to lysosomes. Curr Biol 22(2):135–141. doi:10.1016/j.cub.2011.11.057

Stoops JK, Baker TS, Schroeter JP, Kolodziej SJ, Niu XD, Reed LJ (1992) 3-dimensional structure of the truncated core of the *Saccharomyces cerevisiae* pyruvate dehydrogenase complex determined from negative stain and cryoelectron microscopy images. J Biol Chem 267(34):24769–24775

Stoops JK, Cheng RH, Yazdi MA, Maeng CY, Schroeter JP, Klueppelberg U, Kolodziej SJ, Baker TS, Reed LJ (1997) On the unique structural organization of the *Saccharomyces cerevisiae* pyruvate dehydrogenase complex. J Biol Chem 272:5757–5764

Stott KM, Yusof AM, Perham RN, Jones DD (2009) A surface loop directs conformational switching of a lipoyl domain between a folded and a novel misfolded structure. Structure 17(8):1117–1127. doi:10.1016/j.str.2009.07.001

Sugden MC, Holness MJ (2003) Recent advances in mechanisms regulating glucose oxidation at the level of the pyruvate dehydrogenase complex by PDKs. Am J Physiol Endocrinol Metab 284(5):E855–E862. doi:10.1152/ajpendo.00526.2002

Sugiura A, McLelland G-L, Fon EA, McBride HM (2014) A new pathway for mitochondrial quality control: mitochondrial-derived vesicles. EMBO J 33(19):2142–2156. doi:10.15252/embj.201488104

Sun W, Liu Q, Leng J, Zheng Y, Li J (2015) The role of pyruvate dehydrogenase complex in cardiovascular diseases. Life Sci 121:97–103. doi:10.1016/j.lfs.2014.11.030

Sutendra G, Kinnaird A, Dromparis P, Paulin R, Stenson TH, Haromy A, Hashimoto K, Zhang N, Flaim E, Michelakis ED (2014) A nuclear pyruvate dehydrogenase complex is important for the generation of acetyl-CoA and histone acetylation. Cell 158(1):84–97. doi:10.1016/j.cell.2014.04.046

Toyoda T, Suzuki K, Sekiguchi T, Reed LJ, Takenaka A (1998) Crystal structure of eukaryotic E3, lipoamide dehydrogenase from yeast. J Biochem (Tokyo) 123(4):668–674

Tretter L, Adam-Vizi V (2005) Alpha-ketoglutarate dehydrogenase: a target and generator of oxidative stress. Philos Trans Roy Soc B-Biol Sci 360(1464):2335–2345. doi:10.1098/rstb.2005.1764

Vassylyev DG, Symersky J (2007) Crystal structure of pyruvate dehydrogenase phosphatase 1 and its functional implications. J Mol Biol 370(3):417–426. doi:10.1016/j.jmb.2007.05.002

Vijayakrishnan S, Kelly SM, Gilbert RJC, Callow P, Bhella D, Forsyth T, Lindsay JG, Byron O (2010) Solution structure and characterisation of the human pyruvate dehydrogenase complex core assembly. J Mol Biol 399:71–93. doi:10.1016/j.jmb.2010.03.043

Vijayakrishnan S, Callow P, Nutley MA, McGow D, Gilbert D, Kropholler P, Cooper A, Byron O, Lindsay JG (2011) Variation in the organisation and subunit composition of the mammalian pyruvate dehydrogenase complex E2/E3BP core assembly. Biochem J 437(3):565–574

Wagenknecht T, Grassucci R, Schaak D (1990) Cryoelectron microscopy of frozen-hydrated alpha ketoacid dehydrogenase complexes from *Escherichia coli*. J Biol Chem 265(36):22402–22408

Wang L, Wang F-S, Chang C, Gershwin ME (2014) Breach of tolerance: primary biliary cirrhosis. Semin Liver Dis 34(3):297–317. doi:10.1055/s-0034-1383729

Wang J, Kumaran S, Zhou J, Nerneria NS, Tao H, Kakalis L, Park Y-H, Birkaya B, Patel MS, Jordan F (2015) Elucidation of the interaction loci of the human pyruvate dehydrogenase complex E2.E3BP core with pyruvate dehydrogenase kinase 1 and kinase 2 by H/D exchange mass spectrometry and nuclear magnetic resonance. Biochemistry 54(1):69–82. doi:10.1021/bi5013113

Wynn RM, Kato M, Chuang JL, Tso S-C, Li J, Chuang DT (2008) Pyruvate dehydrogenase kinase-4 structures reveal a metastable open conformation fostering robust core-free basal activity. J Biol Chem 283(37):25305–25315. doi:10.1074/jbc.M802249200

Yamagiwa S, Kamimura H, Takamura M, Aoyagi Y (2014) Autoantibodies in primary biliary cirrhosis: recent progress in research on the pathogenetic and clinical significance. World J Gastroenterol 20(10):2606–2612. doi:10.3748/wjg.v20.i10.2606

Yi JZ, Nemeria N, McNally A, Jordan F, Machado RS, Guest JR (1996) Effect of substitutions in the thiamin diphosphate-magnesium fold on the activation of the pyruvate dehydrogenase complex from *Escherichia coli* by cofactors and substrate. J Biol Chem 271(52):33192–33200

Yu X, Hiromasa Y, Tsen H, Stoops JK, Roche TE, Zhou ZH (2008) Structures of the human pyruvate dehydrogenase complex cores: a highly conserved catalytic center with flexible N-terminal domains. Structure 16(1):104–114. doi:10.1016/j.str.2007.10.024

Zhou ZH, Liao WC, Cheng RH, Lawson JE, McCarthy DB, Reed LJ, Stoops JK (2001a) Direct evidence for the size and conformational variability of the pyruvate dehydrogenase complex revealed by three-dimensional electron microscopy – the "breathing" core and its functional relationship to protein dynamics. J Biol Chem 276(24):21704–21713. doi:10.1074/jbc.M101765200

Zhou ZH, McCarthy DB, O'Connor CM, Reed LJ, Stoops JK (2001b) The remarkable structural and functional organization of the eukaryotic pyruvate dehydrogenase complexes. Proc Natl Acad Sci U S A 98(26):14802–14807. doi:10.1073/pnas.011597698

Chapter 20
Structure and Assembly of Clathrin Cages

Mary Halebian, Kyle Morris, and Corinne Smith

Abstract The unusual structure of clathrin, combined with its ability to assemble and disassemble rapidly in cells provides a model system for us to learn about the ways in which proteins can contribute mechanically to a functioning cell. In this article, we discuss the structural properties of clathrin cages and the triskelions which assemble to form them. The function of clathrin depends on the structure of these triskelions and the interactions they make both with each other during assembly and with the adaptor protein network that drives coated vesicle formation. The atomic resolution structure of clathrin domains has been revealed by X-ray crystallography while scattering studies have enabled the shape of a triskelion in solution to be deduced. Cryo-electron microscopy maps have shown the secondary structure of entire cages, how individual triskelion legs are arranged to form a cage and enabled some bound adaptor proteins to be located. Cage formation itself is energetically finely balanced and requires specific interactions between triskelion legs to be productive, as biochemical studies and *in silico* modeling have shown. Theoretical, structural and cell biological investigations over many years have contributed to our knowledge of clathrin structure and assembly. It now remains to determine the precise nature of the interactions which occur between clathrin triskelions, light chain and heavy chain and the adaptor protein network.

Keywords 3D protein structure • Macromolecular assemblies • Cryo-electron microscopy • X-ray crystallography • Atomic force microscopy • AFM • Coarse grain modeling • Endocytosis • Membrane trafficking • Coated vesicle

---

M. Halebian • K. Morris • C. Smith (✉)
School of Life Sciences, University of Warwick, Coventry CV4 7AL, UK
e-mail: M.halebian@warwick.ac.uk; Kyle.morris@warwick.ac.uk; Corinne.smith@warwick.ac.uk

J.R. Harris, J. Marles-Wright (eds.), *Macromolecular Protein Complexes*, Subcellular Biochemistry 83, DOI 10.1007/978-3-319-46503-6_20

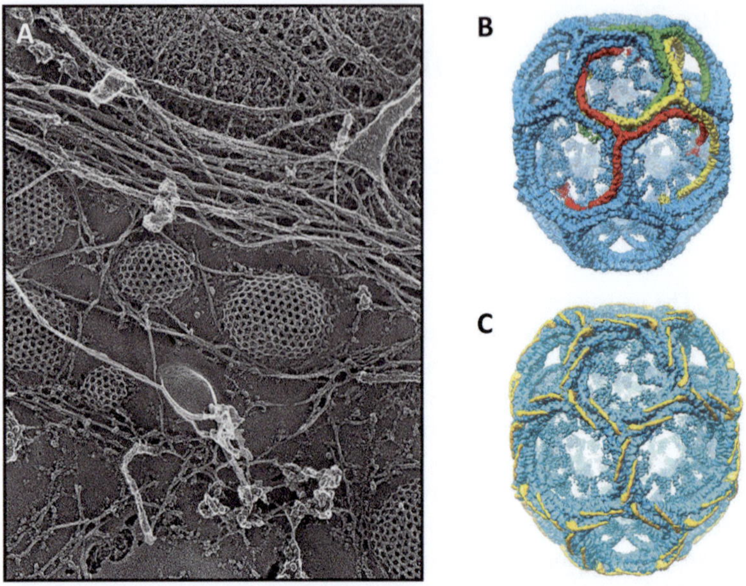

Fig. 20.1 (**a**). Freeze fracture, deep-etch rotary shadowed scanning electron micrograph of clathrin-coated vesicles showing the striking polyhedral lattice formed by clathrin around the vesicles (Image published with permission (pending). Copyright Heuser Lab). (**b**) Cryo-EM map of a clathrin cage with individual triskelions highlighted in *red, yellow* and *green*. (**c**) Cryo-EM map of clathrin cage with light chain subunits highlighted in *yellow* (**b** and **c** are reprinted by permission from Macmillan Publishers Ltd.: Nature (Fotin et al. 2004), copyright 2004)

20.1 Introduction

Clathrate cages are seen in many contexts in the natural world. In chemistry, clathrate compounds form geometric structures which often trap other molecules inside. Water forms clathrate hydrates around gas molecules such as methane, trapping it in ice crystals, and carbon was shown to adopt an icosahedral clathrate structure named Buckminsterfullerene (Kroto et al. 1985). In biology, pollen grains form clathrate lattices and, some orders of magnitude larger in size, fungi such as *Clathrus rubus* and *Phallus indusiata* grow into intricate clathrate structures, creating dramatic polyhedral cages. In cell biology, some of the transport vesicles responsible for moving molecules within cells were, as early as the 1960s, shown to be covered in a latticed coat (Kanaseki and Kadota 1969; Roth and Porter 1964). It was the study of the first of these types of coated vesicles to be identified that resulted in the discovery of a protein called clathrin, which was named for its ability to surround this class of vesicles with a lattice cage structure (see Fig. 20.1) (Crowther et al. 1976; Pearse 1975) . It is the remarkable ability of clathrin to form cage structures, both in cells as part of its biological function and *in vitro,* that has stimulated research into its precise structural composition and how this determines its assembly into cages.

In this article we draw together experimental and theoretical work from more than three decades, which addresses this intriguing question.

One of the best characterized functions of clathrin in cells is that of clathrin-mediated endocytosis. Endocytosis is a process through which cells absorb specific molecules and deliver them to destinations inside the cell via carrier vesicles composed of portions of the plasma membrane. Mechanisms of endocytosis have been classified into clathrin-dependent, clathrin-independent and caveolae-dependent and are covered in detail elsewhere (Grant and Donaldson 2009; McMahon and Boucrot 2011; Doherty and McMahon 2009; Mayor et al. 2014). Clathrin-mediated endocytosis was one of the earliest endocytic pathways to be discovered and characterized (Pearse and Bretscher 1981). In this pathway, the adaptor protein complex, AP2, binds to cytoplasmic motifs on receptors bound to ligands, such as transferrin or low density lipoprotein (LDL), enabling their specific incorporation into the coated vesicle. Additional adaptor proteins assemble around the newly forming coated vesicle, and some of these, including AP2, recruit clathrin to the outer layer of the coat, enabling the characteristic lattice to form (Traub 2009; Ungewickell and Hinrichsen 2007). In addition to forming coats around coated vesicles during endocytosis, clathrin also forms coats around intracellular coated vesicles cycling between the trans-Golgi network and endosomes (Robinson 1987). In addition it has been shown to be associated with multi-vesicular bodies (Prekeris et al. 1998, 1999). Clathrin plays a role in a variety of distinct cellular transport processes. However, this is not the whole story. In 2005, clathrin was shown to have a novel function in mitosis, unrelated to membrane trafficking, in stabilizing the mitotic spindle. It achieves this by crosslinking the microtubules which form the kinetochore fibres in the mitotic spindle through its interaction with two complexes; transforming acidic coiled-coil protein 3 (TACC3) and colonic hepatic tumour overexpressed gene (ch-TOG), which is alternatively named cytoskeleton-associated protein 5 or CKAP5 (Royle 2012; Royle et al. 2005).

In vertebrates, there are two genes located on chromosomes 17 and 22 which give rise to two different clathrin heavy chain isoforms, named CHC17 and CHC22. CHC22 is expressed most abundantly in skeletal muscle and has been shown to play a role in trafficking of the insulin-responsive glucose transporter GLUT4 (Vassilopoulos et al. 2009). CHC17 is the more abundant isoform of clathrin and is the most exhaustively studied of the two forms. We have focused on this isoform in this chapter since the biochemical properties of CHC22 remain to be fully elucidated (Brodsky 2012).

Thus, clathrin is a versatile molecule that performs a variety of different functions in cells. One key aspect of these functions is its ability to form lattice structures and to engage with multiple binding partners. The following sections discuss how the structure and biochemistry of clathrin contributes to these functions and influences its assembly properties.

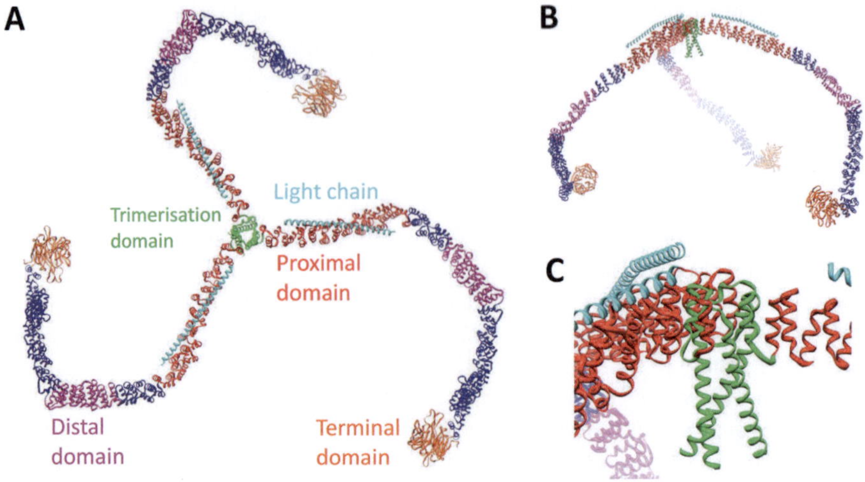

Fig. 20.2 Illustration of the structural features of a clathrin triskelion extracted from the alpha carbon model of Fotin et al. 2004. Image constructed using 3iyv.pdb in UCSF Chimera (Pettersen et al. 2004). (**a**) View from above with domains labelled as follows. Terminal domain, *orange*, residues 1–1130; distal domain, *pink*, residues 839–1073; proximal domain, *red*, residues 1198–1575, trimerisation domain, *green*, residues 1576–1675 and light chain, cyan. (**b**) Side view of a clathrin triskelion. (**c**) Close up view of the trimerisation domain using the same colour scheme. The helical tripod can be seen extending away from the junction of the three heavy chains. In a later model from Wilbur et al. (2010b) based on a crystallographic structure, two additional light chain helices per subunit were resolved. This structure is shown in Fig. 20.3

20.2 The Clathrin Triskelion

Barbara Pearse first purified clathrin from clathrin-coated vesicles in 1976, naming it clathrin for its ability to form lattice structures (Pearse 1976). Now that the primary component of these coated vesicles had been purified, the precise nature of the unit, which assembled to form these cage structures, became of interest. Disassembled clathrin was imaged in 1981 by Ungewickell and Branton using rotary-shadowing methods (Ungewickell and Branton 1981). These studies revealed a three-legged structure as the basic building block for clathrin assembly. Clathrin is composed of three ~190 kDa heavy chain subunits, each of which associates with a ~25 kDa light chain subunit. The heavy chains are long and thin and tightly associate at their C-termini, giving the resulting complex the appearance of an 'Isle of Man' three-legged structure or 'triskelion' (see Fig. 20.2).

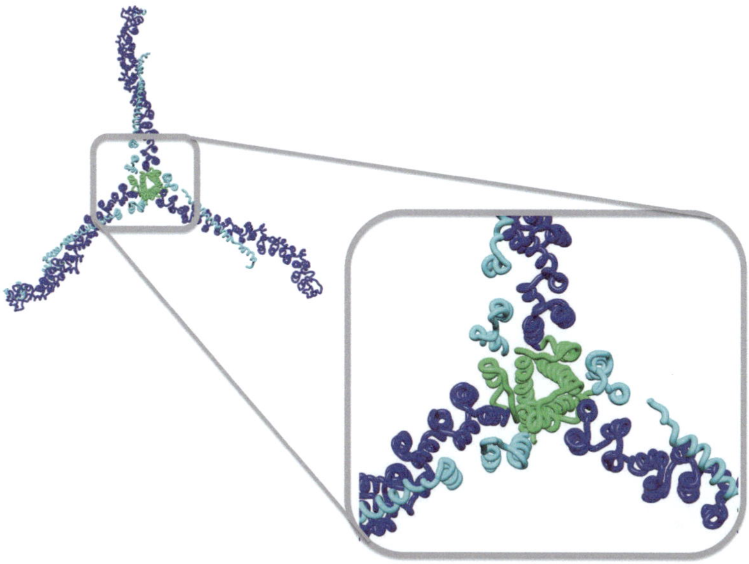

Fig. 20.3 View of the truncated 'hub' form of a clathrin triskelion (shown in *blue*) bound to light chains (*cyan*). The structural model used in this diagram, 3LVG.pdb, was obtained by Wilbur et al. (2010a, b) using X-ray crystallography. The structure revealed additional pairs of light chain helices (*cyan*) lying close to the trimerisation domain (*green*). These are shown in detail in the expanded window. Figure made using UCSF Chimera (Pettersen et al. 2004)

20.2.1 Trimerisation Domain

The question now arose as to how the primary structure of the protein determined this unusual triskelion shape and interest focused on the nature of the very stable linkage between the three heavy chains, termed the trimerisation domain (Fig. 20.2). Lemmon et al. (1991) showed that the C-terminal end of the yeast heavy chain, corresponding to residues 1590–1675 in rat, was not responsible for trimerisation. Näthke et al. (1992) using proteolytic digestion and monoclonal antibody binding, mapped the residues required for trimerisation to 1488–1587 for bovine clathrin, confirming the results in yeast. Liu et al. expressed recombinant fragments of clathrin heavy chain, which trimerised to form hub structures (Liu et al. 1995). Using deletion analysis, they established that the C-terminal limit of the trimerisation domain was around residues 1608–1615 and that, for hub fragments expressed as fusion proteins with maltose binding protein at the N-terminus, the shortest fragment to form a trimer consisted of residues 1550–1675. Thus, the trimerisation domain was narrowed down to residues 1550–1615. The nature of the interaction that knits the trimersation together is intriguing because of its high stability. Using the expressed hub domains as a basis for their study, Ybe et al. investigated the role of cysteine in the stabilization of the trimerisation domain (Ybe et al. 2003). They identified three cysteines which were clustered together at the trimerisation domain at positions 1565, 1569 and 1573, which were conserved in a variety of higher mammals. Mutation of these residues, individually and in pairs, revealed that the cysteine at position 1573 played a significant role in the stabilization of the trimerisation domain (Ybe et al. 2007). In 2004, the first structural view of the trimerisation domain was seen in the cryoEM map of the clathrin cage by Fotin et al., which revealed a remarkable helical tripod structure at the triskelion hub, modelled to encompass residues 1598–1630 (Fotin et al. 2004). It is tempting to speculate that the close proximity of three cysteine residues from each of the three heavy chain subunits leads to formation of disulphide bonds, which could explain the strength of this three way connection between subunits. Intramolecular disulphide bonds in cytoplasmic proteins do not usually form due to the redox conditions inside cells, but it is possible that a favorable local chemical environment could drive their formation.

20.2.2 Light Chains

The clathrin triskelion is not simply a trimer of heavy chains (Fig. 20.2). To each heavy chain is bound a smaller light chain, of ~25 kDa, which is tightly associated, requiring strong denaturants, such as sodium isothiocyanate, to achieve dissociation (Winkler and Stanley 1983). There are two vertebrate genes encoding light chains, giving rise to light chain types a and b, and in neurones splice variants have been identified for each type. The biological role of the light chains (reviewed in (Brodsky 2012)) has not yet been fully established but the consensus region shared by both light chains has been shown to bind to the Hip/Sla2p protein family, members of which interact with actin and actin-associated proteins (Chen and Brodsky 2005) and light chain binding to Hip1 and Hip1R reduces their affinity for actin (Wilbur et al. 2008, 2010a). They have also been shown to be necessary for the uptake of some G-protein coupled receptors (Ferreira et al. 2012).

Biochemical and biophysical studies showed that light chains are able to negatively regulate clathrin assembly *in vitro* and play a significant role in stabilizing the trimerisation domain. They may also regulate the degree of bend that the triskelion legs adopts (Wilbur et al. 2010b). Interestingly, clathrin light chains have been shown to influence the stiffness of flat clathrin lattices (Dannhauser et al. 2015). Atomic force microscopy (AFM) experiments by Dannhauser et al. (2015) showed that the presence of light chains allowed flat clathrin lattices to be reversibly compressed from a height of 12 to 6 nm above the surface by repeated application of orthogonal forces, suggesting that the light chains promote a flexible, expanded lattice conformation.

Liu et al. investigated the effect of co-expression of light chain b upon assembly of clathrin hub domains (Liu et al. 1995). They noted that the light chain had an inhibitory effect on hub assembly, preventing assembly at pH 7. These studies with clathrin hubs enabled the binding site for the light chain on the heavy chain to be tracked. Residues 1213–1522, when expressed, formed the smallest fragments able still to bind light chain. Interestingly, the addition of 3 mM calcium to hubs with light chain enhanced assembly. Light chain b has been shown to bind calcium with an affinity of 20–25 μM (Liu et al. 1995). It is intriguing that calcium has this effect, but the biological relevance of this is not clear as the intracellular calcium concentration is thought to be much lower, at 0.1–1uM.

Ybe et al. (1998) mapped the light chain residues responsible for controlling assembly to residues 23–25, with sequence GEED on light chain a and AEED on light chain b. This sequence is highly conserved across species. Removal of the GEED sequence in light chain b relieved the inhibitory effect of the light chain on assembly (Ybe et al. 1998). Therefore, this highly conserved motif appears to be responsible for the negative regulation of clathrin assembly.

Chen et al. used yeast two-hybrid approaches to investigate the critical residues for light chain-heavy chain interactions by looking for:

1. Mutations in the light chain that prevented binding to heavy chain
2. Mutations in the heavy chain that caused regain of light chain binding in those light chain mutants (Chen et al. 2002)

They showed that residues 1267–1522 encompassed the minimal domain required for light chain b binding. On the light chain, residues 90–157 were the shortest portion of the sequence able to bind well to the heavy chain. Neither residues 82–93 on light chain b, which contain the calcium binding domain, nor residues 155–172, an inserted sequence found in neurons, were critical for heavy chain binding. Yeast-2-hybrid studies revealed that residue 127 (though observation of the W127R mutant) on light chain b was important for binding. This was rescued by the K1415E mutation on the heavy chain. This was also the case for light chain a. In addition, double mutations W105R and W138R disrupted light chain binding. Some binding remained, suggesting that these residues contributed less to binding than W127R. Their effect was suppressed by K1326E.

These studies have provided functional evidence for the locations of residues critical for light chain / heavy chain interaction. Despite years of effort, however, there is not yet an atomic level structure describing light chain binding to the clathrin heavy chain. A significant portion of the clathrin light chain has been visualized in the cryo-EM structure of a clathrin cage by Fotin et al. and was shown to be bound as a single alpha helix to the outer face of the proximal domain (see Fig. 20.1c) (Fotin et al. 2004). In this map, structures with and without light chains were obtained and a difference map was used to reveal the location of the light chain density. Later, in the crystallographic study of the clathrin hub domain by Wilbur et al., new density at the trimerisation domain was identified and attributed to two short, paired helices at the C-terminus of the light chain (Wilbur et al. 2010b). On inspection of the previously published cryo-EM structure of Fotin et al., density corresponding to these helices could be observed, consistent with the new model. This illuminates the findings of Ybe et al. (2007) who found that residues 190–228 (encompassing the entire C-terminus of light chain) were required to stabilize clathrin hub trimers. The discovery of these two helices, which are very closely embedded in the structure of the trimerisation domain, may well explain the contribution of the light chain to the stability of this domain.

20.3 Formation of Clathrin Cages

The assembly of clathrin triskelions into cages is central to the cellular function of clathrin-coated vesicles. Clathrin triskelions purified from native source material, such as brain or liver, assemble spontaneously into cage structures at low pH and in the presence of magnesium or calcium ions. Therefore, the information required to determine cage formation is present in the clathrin molecule. Greene et al. explored the sequence requirements for cage formation and found that clathrin hubs, which contain only the trimerisation and proximal domains of clathrin heavy chain, could be assembled into cage structures with expressed fragments corresponding to residues 1–1074 (the terminal through to distal domains) (Greene et al. 2000). Thus, the proximal and distal sections of the triskelion leg did not have to be joined together for cages to form and the addition of distal fragments facilitated the formation of curved lattices. It appears that (Greene et al. 2000) distal-proximal leg interactions provide the curvature and directionality required for cage formation.

Wakeham et al., investigated the affinity of the proximal and distal legs for one another (Wakeham et al. 2003). They found that when expressed independently as monomers, it was not possible to measure an affinity between proximal and distal leg sections. However, when induced to form trimers through the use of an artificial trimerising sequence, proximal trimers and distal trimers interacted more strongly. Thus Wakeham et al. demonstrated the importance of trimerisation for increasing the affinity of the interactions between the triskelion legs required for cage assembly.

Clathrin cage assembly is promoted by below neutral pH, suggesting the possible involvement of histidine, which has a p*K*a close to six, depending on its local environment. Sequence comparisons between organisms revealed three histidine residues (His1279, 1313 and 1335) that were conserved in higher eukaryotes (Ybe et al. 1998). Ybe et al. proposed that these histidines could aid cage assembly through formation of salt bridges with glutamate residues (Ybe et al. 1998). Böcking et al. (2014) altered these histidine residues in pairs as follows: 867/876 (H8), 1275/1279 (H12) and 1432/1458 (H14), and monitored their contribution to cage stability. Mutations of H8, H12 and H14 histidine pairs resulted in clathrin that is only assembly competent at elevated concentrations, supporting a key role for these residues in pH dependent clathrin assembly. Mutating H12 and H14 to glutamines in a double mutant resulted in a transition from ~80 nm (barrel sized) to ~60 nm cages (mini coat sized) suggesting a role for these interactions in stabilizing larger cage types (Boecking et al. 2014).

20.4 Understanding the Structure of Clathrin Cages

Clathrin-coated vesicles in cells can vary in size within a range of 50 to 80 nm and, *in vitro*, clathrin triskelia assemble to form cages with a wide variety of different structures. Some of these have symmetry. One symmetrical form, the D6 hexagonal barrel cage type, has proved useful for structure determination by cryo-electron microscopy and single particle analysis. The variability of the cage structures which form spontaneously has made X-ray crystallographic approaches unfeasible for entire cages but vital information has been gained through crystallographic analysis of fragments of clathrin heavy chain. Combination of the information derived from X-ray and cryo-EM data has given us our current picture of the structure of the clathrin cage. In 1986 the first cryoEM map of a clathrin cage was published by Vigers, Crowther and Pearse (Vigers et al. 1986b). This was a very early example of single particle analysis and was achieved through averaging structures of clathrin hexagonal barrel cage types determined through tomography. A second paper from these authors revealed clathrin assembled in the presence of adaptor proteins, giving the first indication of the structure of a clathrin coat (Vigers et al. 1986a). The maps were at a nominal resolution of 50 Å, so compared to what is now achievable the resolution was very low, but this work represented a substantial technical advance in cryoEM, since tomography and single particle analysis methods were by no means established at the time. Later, a paper by Smith et al. described the cryo-EM structure at 21 Å of a clathrin hexagonal barrel where the arrangement of triskelions in the cage was first revealed (Smith et al. 1998).

In order to better understand the architecture of clathrin cages, the sequence of the clathrin heavy chains, which form the legs of the triskelion, has been categorized as consisting of 4 domains: terminal domain (1–330), distal domain (839–1073), proximal domain (1198–1575) and trimerisation domain (1576–1675), as illustrated in Fig. 20.2. The terminal domain is found at the end of the triskelion legs, at the heavy chain N-terminus. It forms a seven-bladed beta-propeller structure (ter Haar et al. 2000) which has been shown to bind to adaptor protein 'clathrin box' motifs. These are short linear amino acid sequences found in many clathrin binding proteins. The motifs bind in an extended form by slotting between two blades of the beta-propeller (ter Haar et al. 2000). A linker section of paired alpha helices continues the structure beyond the beta-propeller domain, and was the first indication of how the long flexible leg of a triskelion was formed. Further up the leg, closer to the trimerisation domain, the proximal domain was shown by crystallography (Ybe et al. 1999) to adopt tandem alpha-helical repeats up until the trimerisation domains at the C-terminus (Wilbur et al. 2010b). The proximal domain structure allowed structural homology searches to be carried out and resulted in identification of a series of repeating units within the clathrin heavy chain sequence and the prediction that the whole structure between proximal and terminal domains adopted a similar helical structure to those already seen in the crystal structures. This structural data was then combined with a sub-8 Angstrom cryoEM map (Fotin et al. 2004) of one clathrin cage architecture, the D6 hexagonal barrel, to produce a full alpha-carbon model of all the triskelions within the cage. The cryoEM map revealed detail of the trimerisation domain secondary structure, particularly a helical tripod which emerged below the clathrin vertex to form contacts with other triskelion legs and showed key points of interaction between the legs supporting the cage structure. Fotin et al. also revealed the location of the clathrin light chains, via a difference map with light chain free clathrin. Through this, a section of the light chain could be observed that was bound to the outer edge of the proximal domain as a single helix running parallel to the direction of the leg.

In addition, a number of additional cryoEM structures have been obtained bound to regulatory adaptors Hsc70 and auxilin, which cause clathrin disassembly (Fotin et al. 2004; Young et al. 2013; Xing et al. 2010). The highest resolution structures revealed that Hsc70 and the auxilin J-domain bound beneath the clathrin vertex, well placed to achieve disassembly through disruption of multiple leg contacts at that point. However, despite these advances, much remains to be learnt about the mode of interaction of the clathrin cage with its adaptor proteins.

20.5 Investigations of Clathrin Coated Vesicle Structure

The heterogeneous populations of clathrin-coated vesicles isolated from cells have prompted scientists to seek other structural techniques, such as cryo-electron tomography (cryo-ET), to explore their structure. In early studies, the lattice organization of individual clathrin-coated vesicles and the disposition of the captured vesicle with respect to the surrounding coat were studied (Cheng et al. 2007). *In vitro* assembled clathrin lattices can have more uniform and symmetrical arrangements (e.g. lattices with icosahedral, D6 barrel or tetrahedral symmetry), particularly if assembled with the endocytic adaptor protein AP2. However, in further tomography studies, the clathrin lattice was shown to have variability in its lattice pattern, with pentagonal, hexagonal and occasionally heptagonal facets enabling the coat to engulf a membrane vesicle of about 350 Å diameter (Cheng et al. 2007). Additionally, the eccentric position of the vesicle with respect to the surrounding clathrin coat may correspond to preferential locations of adaptors and other proteins possibly reflecting the region within a coat at which assembly began. This highlights the polarity of initiation of coat formation until membrane pinching and completion of lattice assembly even in the smallest of the many lattices that can enclose a vesicle (Cheng et al. 2007).

Cryo-ET was used to investigate clathrin coat polymorphism further by Heymann et al. (2013). Measuring the mass of the coat components, vesicles enclosed in the coat were observed to form closer contacts with the clathrin N-termini on one side with adaptor proteins occupying the other side. The number of adaptor proteins was found to correlate with the number of triskelions in a coated vesicle, giving a ratio of 0.92 adaptor proteins/triskelion for coated vesicles. In summary, the clathrin coat, adaptor proteins, and vesicle membrane contribute almost all of the mass of a clathrin vesicle with most cargos accounting for only a few percent.

20.6 Theoretical and Energetic Studies of Clathrin Cage Structure

The images produced by Heuser et al. (1987) of curved clathrin coated vesicles emerging from flat lattices stimulated questions as how the lattice could rearrange to generate a curved coated vesicle. Jin and Nossal (1993) considered this by investigating the topology of clathrin cage formation. In order to introduce the curvature necessary, the triskelions must rearrange to form a number of pentagons. The precise way in which this could happen is more complex than it might at first appear because when a pentagon is formed from a series of hexagons, free triskelion legs remain which must find a place to bind. Jin and Nossal (1993) considered the topological implications of this and concluded that the simplest way to introduce a pentagon into a flat hexagonal clathrin lattice was to introduce a pair of new triskelions to the lattice, which results in formation of two pentagons and two heptagons and the loss of three hexagons. Thus, if correct, this theory would suggest that some heptagons should be observed as well as pentagons in the lattice surrounding a coated vesicle.

How the pentagon itself is actually formed was considered by Shraiman (1997) who proposed a local thermal ratchet mechanism of assembly based on analysis of data from Zaremba and Keen (1983). In this mechanism, formation of a pentagon requires local thermal fluctuations of the membrane to release free triskelion legs for binding. Once a pentagon has formed, Shraiman proposed that the fluctuations become restrained as the structure is stabilized.

Nossal (2001) built on the work of Jin and Nossal (1993) and Shraiman (1997) and proposed a model for the free energy of cage formation for a given number of triskelia. This model is based on the sum of the energy required to distort a triskelion and the energy gained as a result of stable interactions between two triskelion legs. From this model a number of predictions can be made about the factors affecting the stability of a cage. First the size of a cage must decrease if the stiffness of triskelia increases. Second the stronger the leg-leg interactions are, the greater the size of the cage should be. Based on this, Nossal calculated a table of estimated basket energies for a given number of triskelia. The model proposed was found to be consistent with the data of Zaremba and Keen (1983). In all, Nossal (2001), concludes that triskelion stiffness and the strength of leg-leg interactions are key factors in determining the size and stability of a clathrin-coated vesicle.

In an alternative model, Mashl and Bruinsma (1998) propose that it is curvature of the membrane itself that can aid clathrin coat formation by causing 'dislocation unbinding' of triskelions and subsequent remodeling of the clathrin coat. This idea is supported by recent cryo-electron tomography and correlative light and electron microscopy data from Avinoam et al. (2015) which indicate that the clathrin coat forms as the vesicle curvature changes.

The mechanical properties of clathrin may also affect how it assembles and the degree to which it could bend membrane. There has been some debate over whether clathrin plays any independent role in driving curvature of the membrane. It seems likely that the clathrin lattice could stabilize a forming vesicle via the extensive interactions the triskelions form around its surface. To investigate this, Jin and Nossal (2000) carried out an analysis of the rigidity of clathrin triskelions using rotary-shadowed electron microscopy images of triskelions. By assuming that the legs were a thin wire-like body, they estimated the bending energy per unit length of the triskelion legs. One assumption in this study is that leg conformations are not perturbed by triskelion-mica interactions. Based on their calculations Jin and Nossal estimate that the rigidity of a triskelion leg is approximately 35 k_BT.nm and that the rigidity of a cage edge is approximately 560 k_BT.nm. This is in the same range for membrane rigidity (320–906 k_BT.nm) that the authors had previously estimated suggesting that any influence on membrane bending by clathrin would be very finely balanced energetically.

In addition to rigidity, the shape that a triskelion adopts in solution could also be important in explaining cage assembly and disassembly properties. To investigate this, Ferguson et al. (2006) used static and dynamic light scattering to determine the radius of gyration of a clathrin triskelion (Ferguson et al. 2006). Since pH is a key factor in determining clathrin cage assembly, they determined values for Rh (hydrodynamic radius) and Rg (radius of gyration) for a range of pH values. The radii did not alter significantly with pH, suggesting that assembly is not driven by alteration of the leg structure itself with pH. Using this method, Ferguson et al. obtained values for Rh of approximately 16 nm and for Rg of approximately 22 nm and went on further to calculate which pucker angles, PHI and PSI, (see Fig. 20.4c) would be consistent with the experimentally determined Rg and Rh values. The resulting models suggested that the conformation of a triskelion in solution was similar to that found in the D6 barrel cryo-electron microscopy structure. Overall, this work suggests that, despite the flexibility of the triskelion, the range of movement of the legs is sufficiently limited that its final shape is not overly constrained as a result of incorporation into a cage structure.

In further studies, Ferguson and colleagues (Ferguson et al. 2008) combined dynamic light scattering with neutron scattering to calculate the shape of a triskelion in solution. They fitted the combined scattering data with bead models of triskelia defined by a spring constant and persistence length. The parameters that fitted the scattering data most closely gave estimations of 100 Da ps^{-2} for the spring constant and 240 Å for the persistence length. The data also permitted fitting of a range of pucker angles, allowing the range of possible conformations available to the triskelion in solution to be estimated. On this basis it was possible to rule out the extremes of completely random leg orientations and a planar triskelion conformation and to show that some degree of pucker (such that the legs come together to form a tripod structure) was required for a good fit to the scattering data.

In 2010, Kotova et al. (2010) directly visualized clathrin triskelions in solution using atomic force microscopy (AFM). Changes in triskelion shape were monitored with this technique at 2 min intervals between 0 and 14 min (see Fig. 20.4a, b). Images of dimers and larger assemblies suggested that the variation in shape was uniform across the whole leg structure. This supports assumptions used in previous calculations (Jin and Nossal 2000) of leg rigidity and corresponds with structural studies which indicate that the triskelion leg between trimerisation domain and terminal

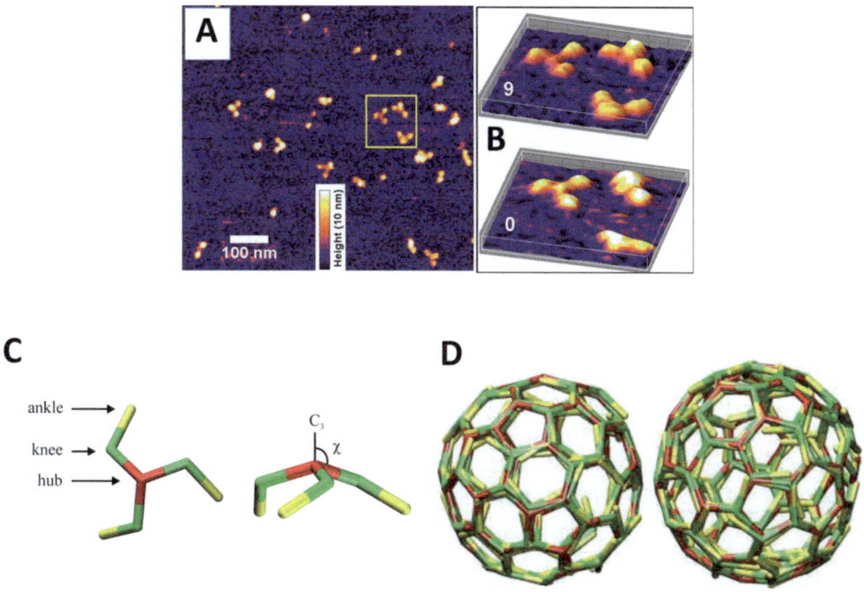

Fig. 20.4 (**a**) AFM images of clathrin triskelia with three triskelia with height indicated by *colour* scheme. (**b**) Three triskelia from (**a**) represented in 3D. Two sequential frames are shown, 9 min apart (**a** and **b** are reprinted from Kotova et al. 2010, Elsevier with permission). (**c**) The triskelion simulation model used by den Otter et al. 2010. Top view (*left*) and side view (*right*) are shown. The central hub is shown in *red*. The first bend of the triskelion, termed the knee is shown in *green* with the final 'ankle' section shown in *yellow*. The pucker angle χ is defined in the diagram. (**d**) This illustrates two self-assembled cages which formed at ε=8 k_BT using the patchy form of the simulation model (**c** and **d** are reprinted from den Otter et al. 2010, © 2010, Elsevier with permission)

domain has a uniform structure of paired alpha helices running perpendicular to the long axis of the leg.

In order to test theoretical models for clathrin assembly, coarse grain modeling approaches were adopted by (den Otter et al. 2010). Den Otter et al. used a simplified 'stick' model of a clathrin triskelion with legs kinked in one place to model the effects of four interaction potentials (Fig. 20.4c). These related to the effect of (1) salt bridges, (2) paired histidines, (3) weak hydrophobic interactions and (4) ankle brace interactions. The ankle brace was proposed by Fotin et al. (2004) to stabilise the cage. For the coarse grain modelling, ε, the segmental binding energy per pair of well-aligned proximal or distal legs, was varied and the effect on cage formation *in silico* noted. With only non-specific interaction energies in place, aggregates requried ε=7 k_BT to form and at energies of ε below 5 k_BT cages were unstable and fell apart. At 6 k_BT, pre-existing hemispherical cages grew slowly but did not necessarily form complete cages and if complete cages were preformed, they remained stable at 6 k_BT over long simulations. Therefore additional factors were needed in order to increase the number and stability of complete cages in the simulations. To address this, Den Otter et al., introduced patches at defined points on the surface of the triskelion legs that formed favourable interactions with each other. This modification to the simulation resulted in seeded cages growing to completion at $\varepsilon=6\ k_BT$ and when ε was increased to $8\ k_BT$, spontaneous formation of complete cages occurred. Thus den Otter et al., were able to demonstrate that asymmetric favourable interactions were critical for driving cage assembly and for formation of complete cages.

20.7 Conclusions

Clathrin's unusual structure, combined with its ability to assemble and disassemble rapidly in cells provides a model system for us to learn about the ways in which protein can contribute mechanically to a functioning cell. The properties of clathrin depend on triskelions and the interactions they make both with each other during assembly and with the adaptor protein network that drives coated vesicle formation. Scattering studies have revealed the triskelion to adopt a relatively stiff structure that does not require extensive alteration to become integrated into a cage. Cage formation however is complex and requires specific interactions between triskelion legs to be productive as biochemical studies and *in silico* modeling have shown. Theoretical studies predicted that heptagons should be part of a growing cage if triskelions are added to a flat lattice to form pentagons and in tomograms of coated vesicles in cells, heptagons are indeed seen. In a more recent study, direct evidence that triskelions are added to a growing coated vesicle was obtained through tomography and correlative light and electron microscopy (Avinoam et al. 2015).

Theoretical, structural and cell biological investigations have contributed to what we know of the mechanism by which clathrin adapts to the growing shape of a coated vesicle. The next stage is to map the precise molecular structure of interactions between clathrin triskelions, light chain and heavy chain and the adaptor protein network. Once we have this information, our understanding of clathrin-mediated endocytosis will be at a level where control of specific processes could become a reality, providing tools for dissection of multiple endocytic pathways and opening up new avenues for medical research.

Acknowledgements C.J.S and K.M gratefully acknowledge BBSRC grant (BB/K003461/1). M.H thanks the Medical Research Council for support. Molecular graphics figures were performed using the UCSF Chimera package. Chimera is developed by the Resource for Biocomputing, Visualization, and Informatics at the University of California, San Francisco (supported by NIGMS P41-GM103311).

References

Avinoam O, Schorb M, Beese CJ, Briggs JAG, Kaksonen M (2015) Endocytic sites mature by continuous bending and remodeling of the clathrin coat. Science 348(6241):1369–1372. doi:10.1126/science.aaa9555

Boecking T, Aguet F, Rapoport I, Banzhaf M, Yu A, Zeeh JC, Kirchhausen T (2014) Key interactions for clathrin coat stability. Structure 22(6):819–829. doi:10.1016/j.str.2014.04.002

Brodsky FM (2012) Diversity of clathrin function: new tricks for an old protein. Annu Rev Cell Dev Biol 28:309–336. doi:10.1146/annurev-cellbio-101011-155716

Chen CY, Brodsky FM (2005) Huntingtin-interacting protein 1 (Hip1) and Hip1-related protein (Hip1R) bind the conserved sequence of clathrin light chains and thereby influence clathrin assembly in vitro and actin distribution *in vivo*. J Biol Chem 280(7):6109–6117. doi:10.1074/jbc.M408454200

Chen CY, Reese ML, Hwang PK, Ota N, Agard D, Brodsky FM (2002) Clathrin light and heavy chain interface: alpha-helix binding superhelix loops via critical tryptophans. EMBO J 21(22):6072–6082. doi:10.1093/emboj/cdf594

Cheng Y, Boll W, Kirchhausen T, Harrison SC, Walz T (2007) Cryo-electron tomography of clathrin-coated vesicles: structural implications for coat assembly. J Mol Biol 365(3):892–899. doi:10.1016/j.jmb.2006.10.036

Crowther RA, Finch JT, Pearse BM (1976) On the structure of coated vesicles. J Mol Biol 103(4):785–798

Dannhauser PN, Platen M, Böning H, Ungewickell H, Schaap IA, Ungewickell EJ. (2015) Effect of clathrin light chains on the stiffness of clathrin lattices and membrane budding. Traffic 16(5):519–533. doi: 10.1111/tra.12263

den Otter WK, Renes MR, Briels WJ (2010) Asymmetry as the key to clathrin cage assembly. Biophys J 99(4):1231–1238. doi:10.1016/j.bpj.2010.06.011

Doherty GJ, McMahon HT (2009) Mechanisms of endocytosis. Annu Rev Biochem 78:857–902. doi:10.1146/annurev.biochem.78.081307.110540

Ferguson ML, Prasad K, Sackett DL, Boukari H, Lafer EM, Nossal R (2006) Conformation of a clathrin triskelion in solution. Biochemistry 45(18):5916–5922. doi:10.1021/bi052568w

Ferguson ML, Prasad K, Boukari H, Sackett DL, Krueger S, Lafer EM, Nossal R (2008) Clathrin triskelia show evidence of molecular flexibility. Biophys J 95(4):1945–1955. doi:10.1529/biophysj.107.126342

Ferreira F, Foley M, Cooke A, Cunningham M, Smith G, Woolley R, Henderson G, Kelly E, Mundell S, Smythe E (2012) Endocytosis of G protein-coupled receptors is regulated by clathrin light chain phosphorylation. Curr Biol 22(15):1361–1370. doi:10.1016/j.cub.2012.05.034

Fotin A, Cheng Y, Sliz P, Grigorieff N, Harrison SC, Kirchhausen T, Walz T (2004) Molecular model for a complete clathrin lattice from electron cryomicroscopy. Nature 432(7017):573–579. doi:10.1038/nature03079

Grant BD, Donaldson JG (2009) Pathways and mechanisms of endocytic recycling. Nat Rev Mol Cell Biol 10(9):597–608. doi:10.1038/nrm2755

Greene B, Liu SH, Wilde A, Brodsky FM (2000) Complete reconstitution of clathrin basket formation with recombinant protein fragments: adaptor control of clathrin self-assembly. Traffic 1(1):69–75. doi:10.1034/j.1600-0854.2000.010110.x

Heuser JE, Keen JH, Amende LM, Lippoldt RE, Prasad K (1987) Deep-etch visualization of 27s-clathrin – a tetrahedral tetramer. J Cell Biol 105(5):1999–2009. doi:10.1083/jcb.105.5.1999

Heymann JB, Winkler DC, Yim YI, Eisenberg E, Greene LE, Steven AC (2013) Clathrin-coated vesicles from brain have small payloads: a cryo-electron tomographic study. J Struct Biol 184(1):43–51. doi:10.1016/j.jsb.2013.05.006

Jin AJ, Nossal R (1993) Topological mechanisms involved in the formation of clathrin-coated vesicles. Biophys J 65(4):1523–1537

Jin AJ, Nossal R (2000) Rigidity of triskelion arms and clathrin nets. Biophys J 78(3):1183–1194

Kanaseki T, Kadota K (1969) The "vesicle in a basket". A morphological study of the coated vesicle isolated from the nerve endings of the guinea pig brain, with special reference to the mechanism of membrane movements. J Cell Biol 42(1):202–220

Kotova S, Prasad K, Smith PD, Lafer EM, Nossal R, Jin AJ (2010) AFM visualization of clathrin triskelia under fluid and in air. FEBS Lett 584(1):44–48. doi:10.1016/j.febslet.2009.11.039

Kroto HW, Heath JR, O'Brien SC, Curl RF, Smalley RE (1985) C60: buckminsterfullerene. Nature 318(6042):162–163

Lemmon SK, Pellicenapalle A, Conley K, Freund CL (1991) Sequence of the clathrin heavy-chain from saccharomyces-cerevisiae and requirement of the cooh terminus for clathrin function. J Cell Biol 112(1):65–80. doi:10.1083/jcb.112.1.65

Liu SH, Wong ML, Craik CS, Brodsky FM (1995) Regulation of clathrin assembly and trimerization defined using recombinant triskelion hubs. Cell 83(2):257–267

Mashl RJ, Bruinsma RF (1998) Spontaneous-curvature theory of clathrin-coated membranes. Biophys J 74(6):2862–2875. doi:10.1016/S0006-3495(98)77993-7

Mayor S, Parton RG, Donaldson JG (2014) Clathrin-independent pathways of endocytosis. Cold Spring Harb Perspect Biol 6(6). doi:10.1101/cshperspect.a016758

McMahon HT, Boucrot E (2011) Molecular mechanism and physiological functions of clathrin-mediated endocytosis. Nat Rev Mol Cell Biol 12(8):517–533. doi:10.1038/nrm3151

Nathke IS, Heuser J, Lupas A, Stock J, Turck CW, Brodsky FM (1992) Folding and trimerization of clathrin subunits at the triskelion hub. Cell 68(5):899–910

Nossal R (2001) Energetics of clathrin basket assembly. Traffic 2(2):138–147. doi:10.1034/j.1600-0854.2001.020208.x

Pearse BM (1975) Coated vesicles from pig brain: purification and biochemical characterization. J Mol Biol 97(1):93–98

Pearse BMF (1976) Clathrin – unique protein associated with intracellular transfer of membrane by coated vesicles. Proc Natl Acad Sci U S A 73(4):1255–1259. doi:10.1073/pnas.73.4.1255

Pearse BM, Bretscher MS (1981) Membrane recycling by coated vesicles. Annu Rev Biochem 50:85–101. doi:10.1146/annurev.bi.50.070181.000505

Pettersen EF, Goddard TD, Huang CC, Couch GS, Greenblatt DM, Meng EC, Ferrin TE (2004) UCSF Chimera – a visualization system for exploratory research and analysis. J Comput Chem 25(13):1605–1612

Prekeris R, Klumperman J, Chen YA, Scheller RH (1998) Syntaxin 13 mediates cycling of plasma membrane proteins via tubulovesicular recycling endosomes. J Cell Biol 143(4):957–971. doi:10.1083/jcb.143.4.957

Prekeris R, Yang B, Oorschot V, Klumperman J, Scheller RH (1999) Differential roles of syntaxin 7 and syntaxin 8 in endosomal trafficking. Mol Biol Cell 10(11):3891–3908

Robinson M (1987) 100-kD coated vesicle proteins: molecular heterogeneity and intracellular distribution studied with monoclonal antibodies. J Cell Biol 104(4):887–895. doi:10.1083/jcb.104.4.887

Roth TF, Porter KR (1964) Yolk protein uptake in the oocyte of the mosquito *Aedes Aegypti* L. J Cell Biol 20:313–332

Royle SJ (2012) The role of clathrin in mitotic spindle organisation. J Cell Sci 125(1):19–28. doi:10.1242/jcs.094607

Royle SJ, Bright NA, Lagnado L (2005) Clathrin is required for the function of the mitotic spindle. Nature 434(7037):1152–1157. doi:10.1038/nature03502

Shraiman BI (1997) On the role of assembly kinetics in determining the structure of clathrin cages. Biophys J 72(2 Pt 1):953–957

Smith CJ, Grigorieff N, Pearse BMF (1998) Clathrin coats at 21 angstrom resolution: a cellular assembly designed to recycle multiple membrane receptors. EMBO J 17(17):4943–4953. doi:10.1093/emboj/17.17.4943

ter Haar E, Musacchio A, Harrison SC, Kirchhausen T (1998) Atomic structure of clathrin: a beta propeller terminal domain joins an alpha zigzag linker. Cell 95(4):563–573. doi:10.1016/s0092-8674(00)81623-2 ter Haar E, Harrison SC, Kirchhausen T (2000) Peptide-in-groove interactions link target proteins to the beta-propeller of clathrin. Proc Natl Acad Sci U S A 97(3):1096–1100. doi:10.1073/pnas.97.3.1096

Traub LM (2009) Tickets to ride: selecting cargo for clathrin-regulated internalization. Nat Rev Mol Cell Biol 10(9):583–596. doi:10.1038/nrm2751

Ungewickell E, Branton D (1981) Assembly units of clathrin coats. Nature 289(5796):420–422. doi:10.1038/289420a0

Ungewickell EJ, Hinrichsen L (2007) Endocytosis: clathrin-mediated membrane budding. Curr Opin Cell Biol 19(4):417–425. doi:10.1016/j.ceb.2007.05.003

Vassilopoulos S, Esk C, Hoshino S, Funke BH, Chen C-Y, Plocik AM, Wright WE, Kucherlapati R, Brodsky FM (2009) A role for the CHC22 clathrin heavy-chain isoform in human glucose metabolism. Science 324(5931):1192–1196. doi:10.1126/science.1171529

Vigers GP, Crowther RA, Pearse BM (1986a) Location of the 100 kd-50 kd accessory proteins in clathrin coats. EMBO J 5(9):2079–2085

Vigers GPA, Crowther RA, Pearse BMF (1986b) 3-dimensional structure of clathrin cages in ice. EMBO J 5(3):529–534

Wakeham DE, Chen CY, Greene B, Hwang PK, Brodsky FM (2003) Clathrin self-assembly involves coordinated weak interactions favorable for cellular regulation. EMBO J 22(19):4980–4990. doi:10.1093/emboj/cdg511

Wilbur JD, Chen C-Y, Manalo V, Hwang PK, Fletterick RJ, Brodsky FM (2008) Actin binding by Hip1 (huntingtin-interacting protein 1) and Hip1R (Hip1-related protein) is regulated by clathrin light chain. J Biol Chem 283(47):32870–32879. doi:10.1074/jbc.M802863200

Wilbur JD, Hwang PK, Frances MB, Fletterick RJ (2010a) Accommodation of structural rearrangements in the huntingtin-interacting protein 1 coiled-coil domain. Acta Crystallogr Sect D-Biol Crystallogr 66:314–318. doi:10.1107/s0907444909054535

Wilbur JD, Hwang PK, Ybe JA, Lane M, Sellers BD, Jacobson MP, Fletterick RJ, Brodsky FM (2010b) Conformation switching of clathrin light chain regulates clathrin lattice assembly. Dev Cell 18(5):841–848. doi:10.1016/j.devcel.2010.04.007

Winkler FK, Stanley KK (1983) Clathrin heavy chain, light chain interactions. EMBO J 2(8):1393–1400

Xing Y, Bocking T, Wolf M, Grigorieff N, Kirchhausen T, Harrison SC (2010) Structure of clathrin coat with bound Hsc70 and auxilin: mechanism of Hsc70-facilitated disassembly. EMBO J 29(3):655–665. doi:10.1038/emboj.2009.383

Ybe JA, Greene B, Liu S-H, Pley U, Parham P, Brodsky FM (1998) Clathrin self-assembly is regulated by three light-chain residues controlling the formation of critical salt bridges. EMBO J 17(5):1297–1303

Ybe JA, Brodsky FM, Hofmann K, Lin K, Liu SH, Chen L, Earnest TN, Fletterick RJ, Hwang PK (1999) Clathrin self-assembly is mediated by a tandemly repeated superhelix. Nature 399(6734):371–375. doi:10.1038/20708

Ybe JA, Ruppel N, Mishra S, VanHaaften E (2003) Contribution of cysteines to clathrin trimerization domain stability and mapping of light chain binding. Traffic 4(12):850–856. doi:10.1046/j.1600-0854.2003.00139.x

Ybe JA, Perez-Miller S, Niu Q, Coates DA, Drazer MW, Clegg ME (2007) Light chain C-terminal region reinforces the stability of clathrin heavy chain trimers. Traffic 8(8):1101–1110. doi:10.1111/j.1600-0854.2007.00597.x

Young A, Stoilova-McPhie S, Rothnie A, Vallis Y, Harvey-Smith P, Ranson N, Kent H, Brodsky FM, Pearse BMF, Roseman A, Smith CJ (2013) Hsc70-induced changes in clathrin-auxilin cage structure suggest a role for clathrin light chains in cage disassembly. Traffic :987–996. doi:10.1111/tra.12085

Zaremba S, Keen JH (1983) Assembly polypeptides from coated vesicles mediate reassembly of unique clathrin coats. J Cell Biol 97(5 Pt 1):1339–1347

Index

J.R. Harris, J. Marles-Wright (eds.), *Macromolecular Protein Complexes*, Subcellular Biochemistry 83, DOI 10.1007/978-3-319-46503-6

V

W

X

Y